Environmental Science 16e

G. Tyler Miller

Scott E. Spoolman

Australia • Brazil • Mexico • Singapore • United Kingdom • United States

About the Cover

A major new theme for this edition of *Environmental Science* is *biomimicry* or *learning from nature*. In recent years, scientists have been studying nature in an effort to learn how a variety of life has existed on the earth for 3.8 billion years despite several catastrophic changes in the planet's environmental conditions. They include strikes by huge meteorites, long warming periods and ice ages, and five mass extinctions—each wiping out 30% to 90% of the world's species.

Examples of how life on the earth has sustained itself for 3.8 billion years are being used to help us develop technologies and solutions to the environmental problems we face and to learn how to live more sustainably. For example, the front cover of this book shows a humpback whale. These whales, which can be 12 to 15 meters (40 to 50 feet) long and weigh as much as 36,400 kilograms (80,000 pounds), can move swiftly and turn quickly as they swim in ocean water. Research, including wind-tunnel tests, shows that this dexterity is due to bumps called tubercles along the edges of their flippers (see photo at left), which somehow help whales move efficiently. A company called WhalePower is using this lesson from nature in designing the blades of wind turbines (see photo below) to be more efficient in producing electricity. Wind power is the world's second-fastest growing source of electricity, as discussed in detail in Chapter 13. Throughout this book, we provide a number of other examples of biomimicry, or learning from the earth.

Environmental Science, **Sixteenth Edition**
G. Tyler Miller, Scott E. Spoolman

Product Director: Dawn Giovanniello

Product Manager: April Cognato

Content Developer: Oden Connolly

Product Assistant: Marina Starkey

Executive Marketing Manager: Tom Ziolkowski

Content Project Manager: Hal Humphrey

Senior Designer: Michael Cook

Manufacturing Planner: Karen Hunt

Program Manager: Valarmathy Munuswamy, Lumina Datamatics

Intellectual Property Project Manager: Erika Mugavin

Intellectual Property Analyst: Christine Myaskovsky

Photo Researcher: Venkat Narayanan, Lumina Datamatics

Text Researcher: Ramesh Muthuramalingam, Lumina Datamatics

Copy Editor: Editorial Services, Lumina Datamatics

Illustrator: Patrick Lane, ScEYEnce Studios

Text Designer: Jeanne Calabrese

Cover Designer: Michael Cook

Cover Image: The Sweets/Moment/Getty Images

Compositor: Lumina Datamatics

For product information and technology assistance, contact us at **Cengage Customer & Sales Support, 1-800-354-9706 or support.cengage.com.**

For permission to use material from this text or product, submit all requests online at **www.cengage.com/permissions.**

Library of Congress Control Number: 2017942675

Paperback Student Edition:
ISBN: 978-1-337-56961-3

Loose-leaf Student Edition:
ISBN: 978-1-337-61275-3

Cengage
20 Channel Street
Boston, MA 02210
USA

Cengage is a leading provider of customized learning solutions with employees residing in nearly 40 different countries and sales in more than 125 countries around the world. Find your local representative at: **www.cengage.com.**

Cengage products are represented in Canada by Nelson Education, Ltd.

To learn more about Cengage platforms and services, register or access your online learning solution, or purchase materials for your course, visit **www.cengage.com.**

Printed at CLDPC, USA, 03-20

Brief Contents

Preface   xv
About the Authors   xxii
From the Authors   xxiii
Learning Skills   xxiv

Humans and Sustainability: An Overview

1   The Environment and Sustainability   2

Ecology and Sustainability

2   Science, Matter, Energy, and Systems   26
3   Ecosystems: What Are They and How Do They Work?   44
4   Biodiversity and Evolution   68
5   Species Interactions, Ecological Succession, and Population Control   88
6   The Human Population and Urbanization   106
7   Climate and Biodiversity   134

Sustaining Biodiversity

8   Sustaining Biodiversity: Saving Species   166
9   Sustaining Biodiversity: Saving Ecosystems   192

Sustaining Resources and Environmental Quality

10   Food Production and The Environment   224
11   Water Resources and Water Pollution   260
12   Geology and Nonrenewable Mineral Resources   300
13   Energy Resources   326
14   Environmental Hazards and Human Health   376
15   Air Pollution, Climate Change, and Ozone Depletion   404
16   Solid and Hazardous Waste   448

Sustaining Human Societies

17   Environmental Economics, Politics, and Worldviews   472
Glossary   G1
Index   I1

Contents

Preface **xv**

About the Authors **xxii**

From the Authors **xxiii**

Learning Skills **xxiv**

Humans and Sustainability: An Overview

1 The Environment and Sustainability **2**

Key Questions **3**

CORE CASE STUDY Learning from the Earth **4**

1.1 What Are Some Key Principles of Sustainability? **5**

Individuals Matter 1.1 Janine Benyus: Using Nature to Inspire Sustainable Design and Living **9**

1.2 How Are We Affecting the Earth? **10**

1.3 What Causes Environmental Problems? **14**

SCIENCE FOCUS 1.1 Some Biomimicry Principles **19**

1.4 What Is an Environmentally Sustainable Society? **21**

TYING IT ALL TOGETHER Learning from the Earth and Sustainability **22**

Chapter Review **22**

Critical Thinking **23**

Doing Environmental Science **24**

Global Environment Watch Exercise **24**

Ecological Footprint Analysis **24**

Ecology and Sustainability

2 Science, Matter, Energy, and Systems **26**

Key Questions **27**

CORE CASE STUDY How Do Scientists Learn about Nature? Experimenting with a Forest **28**

2.1 What Do Scientists Do? **29**

Individuals Matter 2.1 Jane Goodall: Chimpanzee Researcher and Protector **30**

2.2 What Is Matter and What Happens When It Undergoes Change? **31**

2.3 What Is Energy and What Happens When It Undergoes Change? **35**

2.4 What Are Systems and How Do They Respond to Change? **38**

TYING IT ALL TOGETHER The Hubbard Brook Forest Experiment and Sustainability **40**

Chapter Review **40**

Critical Thinking **41**

Doing Environmental Science **42**

Global Environmental Watch Exercise **42**

Data Analysis **43**

3 Ecosystems: What Are They and How Do They Work? **44**

Key Questions **45**

CORE CASE STUDY Tropical Rain Forests Are Disappearing **46**

3.1 How Does The Earth's Life-Support System Work? **47**

3.2 What Are The Major Components of an Ecosystem? **48**

SCIENCE FOCUS 3.1 Many of the World's Most Important Organisms Are Invisible to Us **51**

3.3 What Happens to Energy in an Ecosystem? **53**

3.4 What Happens to Matter in an Ecosystem? **56**

SCIENCE FOCUS 3.2 Water's Unique Properties **57**

Individuals Matter 3.1 Thomas E. Lovejoy—Forest Researcher and Biodiversity Educator **62**

3.5 How Do Scientists Study Ecosystems? **62**

SCIENCE FOCUS 3.3 Planetary Boundaries **63**

TYING IT ALL TOGETHER Tropical Rain Forests and Sustainability **64**

Chapter Review **64**

Critical Thinking **65**

Doing Environmental Science **66**

Global Environment Watch Exercise **66**

Data Analysis **67**

4 Biodiversity and Evolution **68**

Key Questions **69**

CORE CASE STUDY Why Are Amphibians Vanishing? **70**

4.1 What Is Biodiversity and Why Is It Important? **71**

SCIENCE FOCUS 4.1 Insects Play a Vital Role in Our World **72**

Individuals Matter 4.1 Edward O. Wilson: A Champion of Biodiversity **73**

4.2 What Roles Do Species Play in Ecosystems? **74**

CASE STUDY The American Alligator—A Keystone Species That Almost Went Extinct **75**

CASE STUDY Sharks as Keystone Species **76**

4.3 How Does the Earth's Life Change Over Time? **77**

SCIENCE FOCUS 4.2 Causes of Amphibian Declines **78**

4.4 What Factors Affect Biodiversity? **80**

SCIENCE FOCUS 4.3 Geological Processes Affect Biodiversity **81**

CASE STUDY The Threatened Monarch Butterfly **83**

TYING IT ALL TOGETHER Amphibians and Sustainability **85**

Chapter Review **85**

Critical Thinking **86**

Doing Environmental Science **86**

Global Environment Watch Exercise **87**

Data Analysis **87**

5 **Species Interactions, Ecological Succession, and Population Control** **88**

Key Questions **89**

CORE CASE STUDY The Southern Sea Otter: A Species in Recovery **90**

5.1 How Do Species Interact? **91**

SCIENCE FOCUS 5.1 Threats to Kelp Forests **92**

5.2 How Do Communities and Ecosystems Respond to Changing Environmental Conditions? **95**

5.3 What Limits the Growth of Populations? **97**

SCIENCE FOCUS 5.2 The Future of California's Southern Sea Otters **101**

TYING IT ALL TOGETHER Southern Sea Otters and Sustainability **102**

Chapter Review **102**

Critical Thinking **103**

Doing Environmental Science **104**

Global Environment Watch Exercise **104**

Data Analysis **105**

6 **The Human Population and Urbanization** **106**
Key Questions **107**
CORE CASE STUDY Planet Earth: Population
7.4 Billion **108**

6.1 How Many People Can the Earth Support? **109**

6.2 What Factors Influence the Size of the Human
Population? **109**

SCIENCE FOCUS 6.1 How Long Can the Human
Population Keep Growing? **110**

CASE STUDY The U.S. Population—Third Largest
and Growing **110**

6.3 How Does a Population's Age Structure Affect
Its Growth or Decline? **113**

CASE STUDY The American Baby Boom **114**

6.4 How Can We Slow Human Population Growth? **115**

CASE STUDY Population Growth in India **116**

CASE STUDY Slowing Population Growth in China **118**

6.5 What Are the Major Urban Resource and
Environmental Problems? **119**

CASE STUDY Urbanization in the United States **119**

CASE STUDY Mexico City **124**

6.6 How Does Transportation Affect Urban
Environmental Impacts? **124**

6.7 How Can We Make Cities More Sustainable and
Livable? **126**

CASE STUDY The Eco-City Concept in Curitiba, Brazil **128**

TYING IT ALL TOGETHER Population Growth,
Urbanization, and Sustainability **130**

Chapter Review **130**

Critical Thinking **131**

Doing Environmental Science **132**

Global Environment Watch Exercise **132**

Data Analysis **133**

7 Climate and Biodiversity 134

Key Questions 135

CORE CASE STUDY African Savanna 136

7.1 What Factors Influence Climate? 137

SCIENCE FOCUS 7.1 Greenhouse Gases and Climate 138

7.2 What Are the Major Terrestrial Ecosystems and How Are Human Activities Affecting Them? 140

SCIENCE FOCUS 7.2 Staying Alive in the Desert 144

SCIENCE FOCUS 7.3 Revisiting the Savanna: Elephants as a Keystone Species 146

Individuals Matter 7.1 Tuy Sereivathana: Elephant Protector 147

7.3 What Are the Major Types of Marine Aquatic Systems and How Are Human Activities Affecting Them? 152

SCIENCE FOCUS 7.4 Coral Reefs 156

7.4 What Are the Major Types of Freshwater Systems and How Are Human Activities Affecting Them? 157

Individuals Matter 7.2 Alexandra Cousteau: Environmental Storyteller and National Geographic Explorer 162

TYING IT ALL TOGETHER Tropical African Savanna and Sustainability 163

Chapter Review 163

Critical Thinking 164

Doing Environmental Science 165

Global Environment Watch Exercise 165

Data Analysis 165

Sustaining Biodiversity

8 Sustaining Biodiversity: Saving Species 166

Key Questions 167

CORE CASE STUDY Where Have All the European Honeybees Gone? 168

8.1 What Role Do Humans Play in the Loss of Species and Ecosystem Services? 169

SCIENCE FOCUS 8.1 Estimating Extinction Rates 170

8.2 Why Should We Try to Sustain Wild Species and the Ecosystem Services They Provide? 170

8.3 How Do Humans Accelerate Species Extinction and Degradation of Ecosystem Services? 173

SCIENCE FOCUS 8.2 Honeybee Losses: A Search for Causes 180

Individuals Matter 8.1 Juliana Machado Ferreira: Conservation Biologist and National Geographic Explorer 182

CASE STUDY A Disturbing Message from the Birds 183

Individuals Matter 8.2 Çağan Hakkı Sekercioğlu: Protector of Birds and National Geographic Emerging Explorer 184

8.4 How Can We Sustain Wild Species and the Ecosystem Services They Provide? 184

CASE STUDY The U.S. Endangered Species Act 185

TYING IT ALL TOGETHER Honeybees and Sustainability 189

Chapter Review 189

Critical Thinking 190

Doing Environmental Science 191

Global Environment Watch Exercise 191

Data Analysis 191

9 Sustaining Biodiversity: Saving Ecosystems 192

Key Questions 193

CORE CASE STUDY Costa Rica—A Global Conservation Leader 194

9.1 What Are the Major Threats to Forest Ecosystems? 195

SCIENCE FOCUS 9.1 Putting a Price Tag on Nature's Ecosystem Services 196

CASE STUDY Many Cleared Forests in the United States Have Grown Back 198

9.2 How Can We Manage and Sustain Forests? 201

9.3 How Can We Manage and Sustain Grasslands? 204

9.4 How Can We Sustain Terrestrial Biodiversity? 205

CASE STUDY Stresses on U.S. Public Parks 207

CASE STUDY Identifying and Protecting Biodiversity in Costa Rica 208

SCIENCE FOCUS 9.2 Reintroducing the Gray Wolf to Yellowstone National Park **209**

CASE STUDY Ecological Restoration of a Tropical Dry Forest in Costa Rica **212**

9.5 How Can We Sustain Aquatic Biodiversity? **213**

SCIENCE FOCUS 9.3 Ocean Acidification: The Other CO_2 Problem **214**

CASE STUDY Upsetting Marine Ecosystems: Jellyfish Invasions **215**

Individuals Matter 9.1 Sylvia Earle—Advocate for the Oceans **218**

TYING IT ALL TOGETHER Sustaining Costa Rica's Biodiversity **220**

Chapter Review **220**

Critical Thinking **221**

Doing Environmental Science **222**

Global Environment Watch Exercise **222**

Ecological Footprint Analysis **223**

Sustaining Resources and Environmental Quality

10 Food Production and the Environment **224**
Key Questions **225**

CORE CASE STUDY Growing Power—An Urban Food Oasis **226**

10.1 What Is Food Security and Why Is It Difficult to Attain? **227**

10.2 How Is Food Produced? **228**

CASE STUDY Industrialized Food Production in the United States **232**

10.3 What Are the Environmental Effects of Industrialized Food Production? **233**

10.4 How Can We Protect Crops from Pests More Sustainably? **242**

CASE STUDY Ecological Surprises: The Law of Unintended Consequences **244**

10.5 How Can We Produce Food More Sustainably? **247**

Individuals Matter 10.1 David Tilman—Polyculture Researcher **252**

10.6 How Can We Improve Food Security? **252**

SCIENCE FOCUS 10.1 Perennial Polyculture and the Land Institute **253**

Individuals Matter 10.2 Jennifer Burney: Environmental Scientist and National Geographic Explorer **254**

TYING IT ALL TOGETHER Growing Power and Sustainability **256**

Chapter Review **256**

Critical Thinking **258**

Doing Environmental Science **258**

Global Environment Watch Exercise **258**

Ecological Footprint Analysis **259**

11 Water Resources and Water Pollution **260**
Key Questions **261**

CORE CASE STUDY The Gulf of Mexico's Annual Dead Zone **262**

11.1 Will We Have Enough Usable Water? **263**

CASE STUDY Freshwater Resources in the United States **265**

CASE STUDY The Colorado River **268**

11.2 How Can We Increase Freshwater Supplies? **269**

CASE STUDY Overpumping the Ogallala Aquifer **269**

CASE STUDY How Dams Can Kill a Delta **272**

CASE STUDY The Aral Sea Disaster: An Example of Unintended Effects **275**

11.3 How Can We Use Freshwater More Sustainably? **276**

Individuals Matter 11.1 Sandra Postel: National Geographic Explorer and Freshwater Conservationist **280**

11.4 How Can We Reduce Water Pollution? **280**

CASE STUDY Is Bottled Water a Good Option? **287**

CASE STUDY Lead in Drinking Water **288**

SCIENCE FOCUS 11.1 Treating Sewage by Learning from Nature **295**

TYING IT ALL TOGETHER Dead Zones and Sustainability **296**

Chapter Review **296**

Critical Thinking **297**

Doing Environmental Science **298**

Global Environment Watch Exercise **298**

Data Analysis **299**

**12   Geology and Nonrenewable
       Mineral Resources 300**

Key Questions **301**

CORE CASE STUDY The Real Cost of Gold **302**

12.1 What Are the Earth's Major Geological Processes
      and What Are Mineral Resources? **303**

12.2 How Long Might Supplies of Nonrenewable Mineral
      Resources Last? **305**

CASE STUDY The Crucial Importance of Rare
Earth Metals **306**

12.3 What Are The Environmental Effects of Using
      Nonrenewable Mineral Resources? **309**

12.4 How Can We Use Mineral Resources
      More Sustainably? **312**

Individuals Matter 12.1 Maria Gunnoe: Fighting to Save
Mountains **313**

12.5 What Are the Earth's Major Geological
      Hazards? **314**

SCIENCE FOCUS 12.1 The Nanotechnology Revolution **315**

Individuals Matter 12.2 Yu-Guo Guo: Designer of
Nanotechnology Batteries and National Geographic
Explorer 316

Individuals Matter 12.3 Robert Ballard, Ocean Explorer 318

TYING IT ALL TOGETHER The Real Cost of Gold and
Sustainability **322**

Chapter Review **322**

Critical Thinking **323**

Doing Environmental Science **324**

Global Environment Watch Exercise **324**

Data Analysis **325**

13 Energy Resources 326

Key Questions 327

CORE CASE STUDY Using Hydrofracking to Produce Oil and Natural Gas 328

13.1 What Is Net Energy and Why Is It Important? 329

13.2 What Are the Advantages and Disadvantages of Using Fossil Fuels? 330

CASE STUDY Oil Production and Consumption in the United States 332

SCIENCE FOCUS 13.1 Environmental Effects of Natural Gas Production and Fracking in the United States 336

13.3 What Are the Advantages and Disadvantages of Using Nuclear Power? 339

13.4 Why Is Energy Efficiency an Important Energy Resource? 345

SCIENCE FOCUS 13.2 The Search for Better Batteries 349

13.5 What Are the Advantages and Disadvantages of Using Renewable Energy Resources? 352

SCIENCE FOCUS 13.3 Making Wind Turbines Safer for Birds and Bats 363

Individuals Matter 13.1 Andrés Ruzo—Geothermal Energy Sleuth and National Geographic Explorer 366

13.6 How Can We Make the Transition to a More Sustainable Energy Future? 367

CASE STUDY Germany Is a Renewable Energy Superpower 369

TYING IT ALL TOGETHER Energy Resources and Sustainability 371

Chapter Review 371

Critical Thinking 373

Doing Environmental Science 374

Global Environment Watch Exercise 374

Ecological Footprint Analysis 374

14 Environmental Hazards and Human Health 376

Key Questions 377

CORE CASE STUDY Mercury's Toxic Effects 378

14.1 What Major Health Hazards Do We Face? 379

14.2 How Do Biological Hazards Threaten Human Health? **379**

CASE STUDY The Global Threat from Tuberculosis **380**

SCIENCE FOCUS 14.1 Genetic Resistance to Antibiotics **381**

Individuals Matter 14.1 Hayat Sindi: Health Science Entrepreneur **382**

CASE STUDY The Global HIV/AIDS Epidemic **383**

CASE STUDY Malaria—The Spread of a Deadly Parasite **384**

14.3 How Do Chemical Hazards Threaten Human Health? **385**

SCIENCE FOCUS 14.2 The Controversy over BPA **389**

14.4 How Can We Evaluate Risks from Chemical Hazards? **390**

CASE STUDY Pollution Prevention Pays: The 3M Company **394**

14.5 How Do We Perceive and Avoid Risks? **395**

CASE STUDY Cigarettes and E-Cigarettes **396**

TYING IT ALL TOGETHER Mercury's Toxic Effects and Sustainability **400**

Chapter Review **400**

Critical Thinking **401**

Doing Environmental Science **402**

Global Environment Watch Exercise **402**

Data Analysis **403**

15 **Air Pollution, Climate Change, and Ozone Depletion** **404**

Key Questions **405**

CORE CASE STUDY Melting Ice in Greenland **406**

15.1 What Is the Nature of the Atmosphere? **407**

15.2 What Are the Major Air Pollution Problems? **407**

15.3 How Should We Deal With Air Pollution? **417**

15.4 How and Why Is the Earth's Climate Changing? **420**

SCIENCE FOCUS 15.1 Using Models to Project Future Changes in Atmospheric Temperatures **426**

15.5 What Are the Likely Effects of Climate Change? **427**

Individuals Matter 15.1 James Balog: Watching Glaciers Melt **430**

CASE STUDY Alaska: A Preview of the Effects of Climate Change **432**

15.6 How Can We Slow Climate Change? **433**

15.7 How Have We Depleted Ozone in the Stratosphere and What Can We Do About It? **441**

Individuals Matter 15.2 Sherwood Rowland and Mario Molina—A Scientific Story of Expertise, Courage, and Persistence **443**

TYING IT ALL TOGETHER Melting Ice in Greenland and Sustainability **444**

Chapter Review **444**

Critical Thinking **446**

Doing Environmental Science **446**

Global Environment Watch Exercise **447**

Data Analysis **447**

16 **Solid and Hazardous Waste** **448**

Key Questions **449**

CORE CASE STUDY Cradle-to-Cradle Design **450**

16.1 What Environmental Problems Are Related to Solid and Hazardous Wastes? **451**

CASE STUDY Solid Waste in the United States **451**

CASE STUDY Ocean Garbage Patches: There Is No Away **452**

CASE STUDY E-Waste—A Serious Hazardous Waste Problem **453**

16.2 How Should We Deal with Solid Waste? **453**

16.3 Why Are Refusing, Reducing, Reusing, and Recycling So Important? **455**

Individuals Matter 16.1 William McDonough **456**

SCIENCE FOCUS 16.1 Bioplastics **458**

16.4 What Are the Advantages and Disadvantages of Burning or Burying Solid Waste? **459**

16.5 How Should We Deal with Hazardous Waste? **461**

CASE STUDY Recycling E-Waste **462**

CASE STUDY Hazardous Waste Regulation in the United States **464**

16.6 How Can We Shift to A Low-Waste Economy? **465**

CASE STUDY Biomimicry and Industrial Ecosystems: Copying Nature **466**

TYING IT ALL TOGETHER The Cradle-to-Cradle Approach and Sustainability **469**

Chapter Review **469**

Critical Thinking **470**

Doing Environmental Science **471**

Global Environment Watch Exercise **471**

Ecological Footprint Analysis **471**

Sustaining Human Societies

17 **Environmental Economics, Politics, and Worldviews** **472**

Key Questions **473**

CORE CASE STUDY The Greening of American Campuses **474**

17.1 How Are Economic Systems Related to the Biosphere? **475**

17.2 How Can We Use Economic Tools to Deal with Environmental Problems? **478**

CASE STUDY Microlending **483**

17.3 How Can We Implement More Sustainable and Just Environmental Policies? **484**

CASE STUDY U.S. Environmental Laws **486**

CASE STUDY Managing Public Lands in the United States—Politics in Action **487**

Individuals Matter 17.1 Xiuhtezcatl Roske-Martinez **492**

17.4 What Are Some Major Environmental Worldviews? **494**

SCIENCE FOCUS 17.1 Biosphere 2: A Lesson in Humility **495**

17.5 How Can We Live More Sustainably? **496**

Individuals Matter 17.2 Juan Martinez—Reconnecting People with Nature **498**

TYING IT ALL TOGETHER Greening College Campuses and Sustainability **501**

Chapter Review **501**

Critical Thinking **503**

Doing Environmental Science **504**

Global Environment Watch Exercise **504**

Ecological Footprint Analysis **505**

Glossary **G1**

Index **I1**

Preface

We wrote this book to help you achieve three important goals: *first*, to explain to your students the basics of environmental science; *second*, to help your students in using this scientific foundation to understand the environmental problems that we face and to evaluate possible solutions to them; and *third*, to inspire them to make a difference in how we treat the earth on which our lives and economies depend, and thus in how we treat ourselves and our descendants.

We view environmental problems and possible solutions to them through the lens of *sustainability*—the integrating theme of this book. We believe that most people can live comfortable and fulfilling lives and that societies will be more prosperous when sustainability becomes one of the chief measures by which personal choices and public policies are made. Our belief in a sustainable future is foundational to this textbook, and we consistently challenge students to work toward attaining it.

For this reason, we are happy to be continuing our partnership with *National Geographic Learning*. One result has been the addition of many stunning and informative photographs, numerous maps, and several stories of National Geographic Explorers—people who are making a positive difference in the world. With these tools, we continue to tell of the good news from various fields of environmental science, hoping to inspire young people to commit themselves to making our world a more sustainable place to live for their own and future generations.

What's New in This Edition?

- *An emphasis on learning from nature*: We establish this in the Core Case Study for Chapter 1, *Learning from the Earth*, which introduces the principles of biomimicry. We further explore the principles and applications of biomimicry in a Science Focus box and a feature article on biomimicry pioneer Janine Benyus later in the chapter. In our research, we have found that biomimicry presents a growing number of opportunities for using nature's genius, as Benyus puts it, to make our own economies and lifestyles more sustainable.

- A new feature called *Learning from Nature*—a set of brief summaries of specific applications of biomimicry in various industries and fields of research—appearing in most chapters.

- An *attractive and efficient new design* with visual elements inspired by National Geographic Learning to capture and hold students' attention.

- *New Core Case Studies* for 8 of the book's 17 chapters bring important real-world stories to the forefront for use in applying those chapters' concepts and principles.

- A *heavier emphasis on data analysis*, with new questions added to the captions of all figures that involve data graphs, designed to get students to analyze the data represented in the figure. These complement the exercises we provide at the ends of chapters.

- A new feature called *Econumbers*, which highlight key statistics that will be helpful for students to remember.

- *New treatment of the history* of environmental conservation and protection in the United States.

Sustainability Is the Integrating Theme of This Book

Sustainability, a watchword of the 21st century for those concerned about the environment, is the overarching theme of this textbook. You can see the sustainability emphasis by looking at the Brief Contents (p. v).

Six principles of sustainability play a major role in carrying out this book's sustainability theme. These principles are introduced in Chapter 1. They are depicted in Figure 1.2 (p. 6), Figure 1.7 (p. 9), and on the inside back cover of the book and are used throughout the book, with each reference marked in the margin by (see pp. 47 and 314).

We use the following five major subthemes to integrate material throughout this book:

- *Natural Capital.* Sustainability depends on the natural resources and ecosystem services that support all life and economies. See Figures 1.3, p. 7, and 7.16, p. 152.

- *Natural Capital Degradation.* We describe how human activities can degrade natural capital. See Figures 6.3, p. 111, and 10.11, p. 236.

- *Solutions.* We present existing and proposed solutions to environmental problems in a balanced manner and challenge students to use critical thinking to evaluate them. See Figures 9.12, p. 202, and 13.23, p. 346.

- *Trade-Offs.* The search for solutions involves trade-offs, because any solution requires weighing advantages against disadvantages. Our Trade-Offs diagrams located in several chapters present the benefits and drawbacks of various environmental technologies and solutions to environmental problems. See Figures 10.18, p. 242, and 16.10, p. 458.

- *Individuals Matter.* Throughout the book, Individuals Matter boxes and some of the Case Studies describe what various scientists and concerned citizens (including several National Geographic Explorers) have done to help us work toward sustainability (see pp. IM 1.1, p. 9, IM 7.1, p. 147, and IM 15.1, p. 430). Also, a number of What Can You Do? diagrams describe how readers can deal with the problems we face (see Figures 8.11, p. 178, and 11.20, p. 279). Eight especially important ways in which individuals can live more lightly on the earth are summarized in Figure 17.24 (p. 499).

Other Successful Features of This Textbook

- **Up-to-Date Coverage.** Our textbooks have been widely praised for keeping users up to date in the rapidly changing field of environmental science. Since the last edition, we have updated the information and concepts in this book using thousands of articles and reports published between 2013 and 2017. Major new or updated topics include biomimicry, fracking, the growing problem of lead poisoning in public water supplies, ocean acidification, and developments in battery technology. Other such topics include synthetic biology; threats to the Monarch butterfly; Chinese, Indian, and U.S. population trends; African Savanna; elephants as keystone species; climate change and species extinction; wildfires in the western United States; jellyfish populations explosion; marine protected areas and marine reserves; effects of overfertilization; aquaculture effects on mangroves; organic no-till farming; deep-sea mining; costs of producing heavy oil from tar sands; increased natural gas production in the United States; methane leaks from natural gas production; coal burning and air pollution in China; shared (community) solar power; C. diff superbug; Ebola virus; effects of smoking and e-cigarette use; deaths from air pollution in China and India; case study on climate change in Alaska; and the overall drop in coal use.

- **Concept-Centered Approach.** To help students focus on the main ideas, we built each major chapter section around a key question and one to three key concepts, which state the section's most important take-away messages. In each chapter, all key questions are listed at the front of the chapter, and each chapter section begins with its key question and concepts (see pp. 3 and 89). Also, the concept applications are highlighted and referenced throughout each chapter.

- **Science-Based.** Chapters 2–7 cover scientific principles important to the course and discuss how scientists work (see Brief Contents, p. v). Important environmental science topics are explored in depth in Science Focus boxes distributed among the chapters throughout the book (see pp. 19 and 76) and integrated throughout the book in various Case Studies (see pp. 76 and 83) and in numerous figures.

- **Global Coverage.** This book also provides a global perspective, first on the ecological level, revealing how all the world's life is connected and sustained within the biosphere, and second, through the use of information and images from around the world. This includes more than 30 maps in the basic text and available on the Learning Path. At the end of each chapter is a Global Environment Watch exercise that applies this global perspective.

- **Core Case Studies.** Each chapter opens with a Core Case Study (see pp. 28 and 90), which is applied throughout the chapter. These applications are indicated by the notation (**Core Case Study**) wherever they occur (see pp. 9 and 74). Each chapter ends with a *Tying It All Together* box (see pp. 64 and 163), which connects the Core Case Study and other material in the chapter to some or all of the principles of sustainability.

- **Case Studies.** In addition to the 17 Core Case Studies, more than 40 additional Case Studies (see pp. 76, 83, and 110) appear throughout the book (and are listed in the Detailed Contents, pp. vi – xiv). Each of these provides an in-depth look at specific environmental problems and their possible solutions.

- **Critical Thinking.** The Learning Skills section (p. xxiv) describes critical thinking skills, and specific critical thinking exercises are used throughout the book in several ways:

  - In dozens of *Thinking About* exercises that ask students to analyze material immediately after it is presented (see pp. 31 and 121).

  - In all *Science Focus* boxes.

  - In dozens of *Connections* boxes that stimulate critical thinking by exploring often surprising connections related to environmental problems (see pp. 53 and 122).

  - In the captions of many of the book's figures (see Figures 1.11, p. 14, and 3.10, p. 53).

  - In end-of-chapter *Critical Thinking* questions (see pp. 41 and 164).

- **Visual Learning.** With a new design heavily influenced by material from National Geographic and new photographs, many of them from the archives of National Geographic, this is the most visually interesting environmental science textbook available (see Figure 1.6, p. 8; chapter-opening photo, pp. 26-27; and Figure 5.10, p. 98). Add in the more than 130 diagrams, each designed to present complex ideas in understandable ways relating to the real world (see Figures 3.12, p. 54, and 7.8, p. 141), and you also have one of the most visually informative textbooks available.

- **Flexibility.** To meet these diverse needs of hundreds of widely varying environmental science courses, we have designed a highly flexible book that allows instructors to vary the order of chapters and sections within chapters without exposing students to terms and concepts that could confuse them. We recommend that instructors start with Chapter 1, which defines basic terms and gives an overview of sustainability, population, pollution, resources, and economic development issues that are discussed throughout the book. This provides a springboard for instructors to use other chapters in almost any order. One often-used strategy is to follow Chapter 1 with Chapters 2–7, which introduce basic science and ecological concepts. Instructors can then use the remaining chapters in any order desired. Some instructors follow Chapter 1 with Chapter 17 on environmental economics, politics, and worldviews, before proceeding to the chapters on basic science and ecological concepts. Instructors whose students have access to MindTap have a second level of flexibility in the supplemental information, maps, and graphs provided there. Examples include basic chemistry (Supplement 3), maps and map analysis (Supplement 4), and environmental data and data analysis (Supplement 5).

- **In-Text Study Aids.** Each chapter begins with a list of *Key Questions* showing how the chapter is organized (see p. 107). Wherever a new *key term* is introduced and defined, it appears in boldface type and all such terms are summarized in the glossary at the end of the book. In most chapters, *Thinking About* exercises reinforce learning by asking students to think critically about the implications of various environmental issues and solutions immediately after they are discussed in the text (see pp. 13 and 121). The captions of many figures contain similar questions that get students to think about the figure content (see pp. 14 and 53). In their reading, students also encounter *Connections* boxes, which briefly describe connections between human activities and environmental consequences, environmental and social issues, and environmental issues and solutions (see pp. 53 and 122). New to this edition is a set of *Learning from Nature* boxes that give quick summaries of

biomimicry applications (see pp. 53 and 77). The text of each chapter concludes with three *Big Ideas* (see pp. 39 and 129), which summarize and reinforce three of the major take-away messages from each chapter. Finally, a *Tying It All Together* section relates the Core Case Study and other chapter content to the principles of sustainability (see pp. 22 and 85). These concluding features reinforce the main messages of the chapter along with the themes of sustainability to give students a stronger understanding of how they all tie together.

Each chapter ends with a *Chapter Review* section containing a detailed set of review questions that include all the chapter's key terms in bold type; *Critical Thinking* questions that encourage students to think about and apply what they have learned to their lives; *Doing Environmental Science*—an exercise that will help students experience the work of various environmental scientists; a *Global Environment Watch* exercise taking student to Cengage's GREENR site where they can use this tool for interesting research related to chapter content; and a *Data Analysis* or *Ecological Footprint Analysis* problem built around ecological footprint data or some other environmental data set (see pp. 102–105 and 256–259).

Supplements for Instructors

- **MindTap.** MindTap is a new approach to highly person-alized online learning. Beyond an eBook, homework solution, digital supplement, or premium website, MindTap is a digital learning platform that works alongside your campus Learning Management System (LMS) to deliver course curriculum across the range of electronic devices in your life. MindTap is built on an "app" model allowing enhanced digital collaboration and delivery of engaging content across a spectrum of Cengage and non-Cengage resources. Visit the Instructor's Companion Site for tips on maximizing your MindTap course.

- **Instructor's Companion Site.** Everything you need for your course in one place! This collection of book-specific lecture and class tools is available online via www.cengage.com/login. Access and download PowerPoint presentations, images, instructor's manual, videos, and more.

- **Cognero Test Bank.** Available to adopters. Cengage Learning Testing Powered by Cognero is a flexible, online system that allows you to:

  - author, edit, and manage test bank content from multiple Cengage Learning solutions;
  - create multiple test versions in an instant; and deliver tests from your LMS, your classroom, or wherever you want.

Help Us Improve This Book or Its Supplements

Let us know how you think this book can be improved. If you find any errors, bias, or confusing explanations, please e-mail us about them at:

mtg89@hotmail.com
spoolman@tds.net

Most errors can be corrected in subsequent printings of this edition, as well as in future editions.

Acknowledgments

We wish to thank the many students and teachers who have responded so favorably to the 15 previous editions of *Environmental Science*, the 19 editions of *Living in the Environment*, the 11 editions of *Sustaining the Earth*, and the 8 editions of *Essentials of Ecology*, and who have corrected errors and offered many helpful suggestions for improvement. We are also deeply indebted to the more than 300 reviewers, who pointed out errors and suggested many important improvements in the various editions of these three books.

It takes a village to produce a textbook, and the members of the talented production team, listed on the copyright page, have made vital contributions. Our special thanks go to content developer Oden Connolly; production managers Hal Humphrey and Valarmathy Munuswamy; the copy editors of Editorial Services, Lumina Datamatics; compositor Lumina Datamatics; photo researcher Venkat Narayanan; artist Patrick Lane; development manager Lauren Oliveira; and Cengage Learning's hard-working sales staff. Finally, we are very fortunate to have the guidance, inspiration, and unfailing support of our Project Manager April Cognato and her dedicated team of highly talented people who have made this and other book projects such a pleasure to work on.

G. Tyler Miller

Scott E. Spoolman

Pedagogy Contributors

Dr. Dean Goodwin and his colleagues, Berry Cobb, Deborah Stevens, Jeannette Adkins, Jim Lehner, Judy Treharne, Lonnie Miller, and Tom Mowbray provided excellent contributions to the Data Analysis and Ecological Footprint Analysis exercises. Mary Jo Burchart of Oakland Community College wrote the in-text Global Environment Watch exercises.

Cumulative List of Reviewers

Barbara J. Abraham, Hampton College; Donald D. Adams, State University of New York at Plattsburgh; Larry G. Allen, California State University, Northridge; Susan Allen-Gil, Ithaca College; James R. Anderson, U.S. Geological Survey; Mark W. Anderson, University of Maine; Kenneth B. Armitage, University of Kansas; Samuel Arthur, Bowling Green State University; Gary J. Atchison, Iowa State University; Thomas W. H. Backman, Lewis-Clark State College; Marvin W. Baker, Jr., University of Oklahoma; Virgil R. Baker, Arizona State University; Stephen W. Banks, Louisiana State University in Shreveport; Ian G. Barbour, Carleton College; Albert J. Beck, California State University, Chico; Marilynn Bartels, Black Hawk College; Eugene C. Beckham, Northwood University; Diane B. Beechinor, Northeast Lakeview College; W. Behan, Northern Arizona University; David Belt, Johnson County Community College; Keith L. Bildstein, Winthrop College; Andrea Bixler, Clarke College; Jeff Bland, University of Puget Sound; Roger G. Bland, Central Michigan University; Grady Blount II, Texas A&M University, Corpus Christi; Barbara I. Bonder, Flagler College; Lisa K. Bonneau, University of Missouri–Kansas City; Georg Borgstrom, Michigan State University; Arthur C. Borror, University of New Hampshire; John H. Bounds, Sam Houston State University; Leon F. Bouvier, Population Reference Bureau; Daniel J. Bovin, Université Laval; Jan Boyle, University of Great Falls; James A. Brenneman, University of Evansville; Michael F. Brewer, Resources for the Future, Inc.; Mark M. Brinson, East Carolina University; Dale Brown, University of Hartford; Patrick E. Brunelle, Contra Costa College; Terrence J. Burgess, Saddleback College North; David Byman, Pennsylvania State University Worthington Scranton; Michael L. Cain, Bowdoin College; Lynton K. Caldwell, Indiana University; Faith Thompson Campbell, Natural Resources Defense Council, Inc.; John S. Campbell, Northwest College; Ray Canterbery, Florida State University; Deborah L. Carr, Texas Tech University; Ted J. Case, University of San Diego; Ann Causey, Auburn University; Richard A. Cellarius, Evergreen State University; William U. Chandler, Worldwatch Institute; F. Christman, University of North Carolina, Chapel Hill; Peter Chen, College of DuPage; Lu Anne Clark, Lansing Community College; Preston Cloud, University of California, Santa Barbara; Bernard C. Cohen, University of Pittsburgh; Richard A. Cooley, University of California, Santa Cruz; Dennis J. Corrigan; George Cox, San Diego State University; John D. Cunningham, Keene State College; Herman E. Daly, University of Maryland; Raymond F. Dasmann, University of California, Santa Cruz; Kingsley Davis, Hoover Institution; Edward E. DeMartini, University of California, Santa Barbara; James Demastes, University of Northern Iowa; Robert L. Dennison, Heartland Community College; Charles E. DePoe, Northeast Louisiana University; Thomas

R. Detwyler, University of Wisconsin; Bruce DeVantier, Southern Illinois University at Carbondale; Peter H. Diage, University of California, Riverside; Stephanie Dockstader, Monroe Community College; Lon D. Drake, University of Iowa; Michael Draney, University of Wisconsin–Green Bay; David DuBose, Shasta College; Dietrich Earnhart, University of Kansas; Robert East, Washington & Jefferson College; T. Edmonson, University of Washington; Thomas Eisner, Cornell University; Michael Esler, Southern Illinois University; David E. Fairbrothers, Rutgers University; Paul P. Feeny, Cornell University; Richard S. Feldman, Marist College; Vicki Fella-Pleier, La Salle University; Nancy Field, Bellevue Community College; Allan Fitzsimmons, University of Kentucky; Andrew J. Friedland, Dartmouth College; Kenneth O. Fulgham, Humboldt State University; Lowell L. Getz, University of Illinois at Urbana-Champaign; Frederick F. Gilbert, Washington State University; Jay Glassman, Los Angeles Valley College; Harold Goetz, North Dakota State University; Srikanth Gogineni, Axia College of University of Phoenix; Jeffery J. Gordon, Bowling Green State University; Eville Gorham, University of Minnesota; Michael Gough, Resources for the Future; Ernest M. Gould, Jr., Harvard University; Peter Green, Golden West College; Katharine B. Gregg, West Virginia Wesleyan College; Stelian Grigoras, Northwood University; Paul K. Grogger, University of Colorado at Colorado Springs; L. Guernsey, Indiana State University; Ralph Guzman, University of California, Santa Cruz; Raymond Hames, University of Nebraska, Lincoln; Robert Hamilton IV, Kent State University, Stark Campus; Raymond E. Hampton, Central Michigan University; Ted L. Hanes, California State University, Fullerton; William S. Hardenbergh, Southern Illinois University at Carbondale; John P. Harley, Eastern Kentucky University; Cindy Harmon, State Fair Community College; Neil A. Harriman, University of Wisconsin, Oshkosh; Grant A. Harris, Washington State University; Harry S. Hass, San Jose City College; Arthur N. Haupt, Population Reference Bureau; Denis A. Hayes, environmental consultant; Stephen Heard, University of Iowa; Gene Heinze-Fry, Department of Utilities, Commonwealth of Massachusetts; Jane Heinze-Fry, environmental educator; Keith R. Hench, Kirkwood Community College; John G. Hewston, Humboldt State University; David L. Hicks, Whitworth College; Kenneth M. Hinkel, University of Cincinnati; Eric Hirst, Oak Ridge National Laboratory; Doug Hix, University of Hartford; Kelley Hodges, Gulf Coast State College; S. Holling, University of British Columbia; Sue Holt, Cabrillo College; Donald Holtgrieve, California State University, Hayward; Michelle Homan, Gannon University; Michael H. Horn, California State University, Fullerton; Mark A. Hornberger, Bloomsberg University; Marilyn Houck, Pennsylvania State University; Richard D. Houk, Winthrop College; Robert J. Huggett, College of William and Mary; Donald Huisingh, North Carolina State University; Catherine Hurlbut, Florida Community College at Jacksonville; Marlene K. Hutt, IBM; David R. Inglis, University of Massachusetts; Robert Janiskee, University of South Carolina; Hugo H. John, University of Connecticut; Brian A. Johnson, University of Pennsylvania, Bloomsburg; David I. Johnson, Michigan State University; Mark Jonasson, Crafton Hills College; Zoghlul Kabir, Rutgers, New Bruns-wick; Agnes Kadar, Nassau Community College; Thomas L. Keefe, Eastern Kentucky University; David Kelley, University of St. Thomas; William E. Kelso, Louisiana State University; Nathan Keyfitz, Harvard University; David Kidd, University of New Mexico; Pamela S. Kimbrough; Jesse Klingebiel, Kent School; Edward J. Kormondy, University of Hawaii–Hilo/West Oahu College; John V. Krutilla, Resources for the Future, Inc.; Judith Kunofsky, Sierra Club; E. Kurtz; Theodore Kury, State University of New York at Buffalo; Troy A. Ladine, East Texas Baptist University; Steve Ladochy, University of Winnipeg; Anna J. Lang, Weber State University; Mark B. Lapping, Kansas State University; Michael L. Larsen, Campbell University; Linda Lee, University of Connecticut; Tom Leege, Idaho Department of Fish and Game; Maureen Leupold, Genesee Community College; William S. Lindsay, Monterey Peninsula College; E. S. Lindstrom, Pennsylvania State University; M. Lippiman, New York University Medical Center; Valerie A. Liston, University of Minnesota; Dennis Livingston, Rensselaer Polytechnic Institute; James P. Lodge, air pollution consultant; Raymond C. Loehr, University of Texas at Austin; Ruth Logan, Santa Monica City College; Robert D. Loring, DePauw University; Paul F. Love, Angelo State University; Thomas Lovering, University of California, Santa Barbara; Amory B. Lovins, Rocky Mountain Institute; Hunter Lovins, Rocky Mountain Institute; Gene A. Lucas, Drake University; Claudia Luke, University of California, Berkeley; David Lynn; Timothy F. Lyon, Ball State University; Stephen Malcolm, Western Michigan University; Melvin G. Marcus, Arizona State University; Gordon E. Matzke, Oregon State University; Parker Mauldin, Rockefeller Foundation; Marie McClune, The Agnes Irwin School (Rosemont, Pennsylvania); Theodore R. McDowell, California State University;

Vincent E. McKelvey, U.S. Geological Survey; Robert T. McMaster, Smith College; John G. Merriam, Bowling Green State University; A. Steven Messenger, Northern Illinois University; John Meyers, Middlesex Community College; Raymond W. Miller, Utah State University; Arthur B. Millman, University of Massachusetts, Boston; Sheila Miracle, Southeast Kentucky Community & Technical College; Fred Montague, University of Utah; Rolf Monteen, California Polytechnic State University; Debbie Moore, Troy University Dothan Campus; Michael K. Moore, Mercer University; Ralph Morris, Brock University, St. Catherine's, Ontario, Canada; Angela Morrow, Auburn University; William W. Murdoch, University of California, Santa Barbara; Norman Myers, environmental consultant; Brian C. Myres, Cypress College; A. Neale, Illinois State University; Duane Nellis, Kansas State University; Jan Newhouse, University of Hawaii, Manoa; Jim Norwine, Texas A&M University, Kingsville; John E. Oliver, Indiana State University; Mark Olsen, University of Notre Dame; Bruce Olszewski, San Jose State University; Carol Page, copy editor; Bill Paletski, Penn State University; Eric Pallant, Allegheny College; Charles F. Park, Stanford University; Richard J. Pedersen, U.S. Department of Agricul-ture, Forest Service; David Pelliam, Bureau of Land Management, U.S. Department of the Interior; Barry Perlmutter, College of Southern Nevada; Murray Paton Pendarvis, Southeastern Louisiana University; Dave Perault, Lynchburg College; Carolyn J. Peters, Spoon River College; Rodney Peterson, Colorado State University; Julie Phillips, De Anza College; John Pichtel, Ball State University; William S. Pierce, Case Western Reserve University; David Pimentel, Cornell University; Peter Pizor, Northwest Community College; Mark D. Plunkett, Bellevue Community College; Grace L. Powell, University of Akron; James H. Price, Oklahoma College; Alan D. Redmond, East Tennessee State University; Marian E. Reeve, Merritt College; Carl H. Reidel, University of Vermont; Charles C. Reith, Tulane University; Erin C. Rempala, San Diego City College; Roger Revelle, California State University, San Diego; L. Reynolds, University of Central Arkansas; Ronald R. Rhein, Kutztown University of Pennsylvania; Charles Rhyne, Jackson State University; Robert A. Richardson, University of Wisconsin; Benjamin F. Richason III, St. Cloud State University; Jennifer Rivers, Northeastern University; Ronald Robberecht, University of Idaho; William Van B. Robertson, School of Medicine, Stanford University; C. Lee Rockett, Bowling Green State University; Terry D. Roelofs, Humboldt State University; Daniel Ropek, Columbia George Community College; Christopher Rose, California Polytechnic State University; Richard G. Rose, West Valley College; Stephen T. Ross, University of Southern Mississippi; Robert E. Roth, Ohio State University; Dorna Sakurai, Santa Monica College; Arthur N. Samel, Bowling Green State University; Shamili Sandiford, College of DuPage; Floyd Sanford, Coe College; David Satterthwaite, I.E.E.D., London; Stephen W. Sawyer, University of Maryland; Arnold Schecter, State University of New York; Frank Schiavo, San Jose State University; William H. Schlesinger, Ecological Society of America; Stephen H. Schneider, National Center for Atmospheric Research; Clarence A. Schoenfeld, University of Wisconsin, Madison; Madeline Schreiber, Virginia Polytechnic Institute; Henry A. Schroeder, Dartmouth Medical School; Lauren A. Schroeder, Youngstown State University; Norman B. Schwartz, University of Delaware; George Sessions, Sierra College; David J. Severn, Clement Associates; Don Sheets, Gardner-Webb University; Paul Shepard, Pitzer College and Claremont Graduate School; Michael P. Shields, Southern Illinois University at Carbondale; Kenneth Shiovitz; F. Siewert, Ball State University; E. K. Silbergold, Environmental Defense Fund; Joseph L. Simon, University of South Florida; William E. Sloey, University of Wisconsin, Oshkosh; Michelle Smith, Windward Community College; Robert L. Smith, West Virginia University; Val Smith, University of Kansas; Howard M. Smolkin, U.S. Environmental Protection Agency; Patricia M. Sparks, Glassboro State College; John E. Stanley, University of Virginia; Mel Stanley, California State Polytechnic University, Pomona; Richard Stevens, Monroe Commu-nity College; Norman R. Stewart, University of Wisconsin, Milwaukee; Frank E. Studnicka, University of Wisconsin, Platteville; Chris Tarp, Contra Costa College; Roger E. Thibault, Bowling Green State University; Nathan E. Thomas, University of South Dakota; William L. Thomas, California State University, Hayward; Jamey Thompson, Hudson Valley Community College; Kip R. Thompson, Ozarks Technical Community College; Shari Turney, copy editor; John D. Usis, Youngstown State University; Tinco E. A. van Hylckama, Texas Tech University; Robert R. Van Kirk, Humboldt State University; Donald E. Van Meter, Ball State University; Rick Van Schoik, San Diego State University; Gary Varner, Texas A&M University; John D. Vitek, Oklahoma State University; Harry A. Wagner, Victoria College; Lee B. Waian, Saddleback College; War-ren C. Walker, Stephen F. Austin State University; Thomas D. Warner, South Dakota State University; Kenneth E. F. Watt, University of California, Davis; Alvin M. Weinberg, Institute of Energy Analysis, Oak Ridge Associated Univer-sities; John F. Weishampel, University of Central Florida; Brian Weiss; Margery Weitkamp, James Monroe High

School (Granada Hills, California); Anthony Weston, State University of New York at Stony Brook; Raymond White, San Francisco City College; Douglas Wickum, University of Wisconsin, Stout; Charles G. Wilber, Colorado State University; Nancy Lee Wilkinson, San Francisco State Univer-sity; John C. Williams, College of San Mateo; Ray Williams, Rio Hondo College; Roberta Williams, University of Nevada, Las Vegas; Samuel J. Williamson, New York University; Dwina Willis, Freed-Hardeman University; Ted L. Willrich, Oregon State University; James Winsor, Pennsylvania State University; Fred Witzig, University of Minnesota at Duluth; Martha Wolfe, Elizabethtown Community and Technical College; George M. Woodwell, Woods Hole Research Center; Peggy J. Wright, Columbia College; Todd Yetter, University of the Cumberlands; Robert Yoerg, Belmont Hills Hospital; Hideo Yonenaka, San Francisco State University; Brenda Young, Daemen College; Anita Závodská, Barry University; Malcolm J. Zwolinski, University of Arizona.

G. TYLER MILLER

G. Tyler Miller has written 64 textbooks for introductory courses in environmental science, basic ecology, energy, and environmental chemistry. Since 1975, Miller's books have been the most widely used textbooks for environmental science in the United States and throughout the world. They have been used by almost 3 million students and have been translated into eight languages.

Miller has a professional background in chemistry, physics, and ecology. He has a PhD from the University of Virginia and has received two honorary doctoral degrees for his contributions to environmental education. He taught college for 20 years, developed one of the nation's first environmental studies programs, and developed an innovative interdisciplinary undergraduate science program before deciding to write environmental science textbooks full time in 1975. Currently, he is the president of Earth Education and Research, devoted to improving environmental education.

He describes his hopes for the future as follows:

If I had to pick a time to be alive, it would be the next 75 years. Why? First, there is overwhelming scientific evidence that we are in the process of seriously degrading our own life-support system. In other words, we are living unsustainably. Second, within your lifetime we have the opportunity to learn how to live more sustainably by working with the rest of nature, as described in this book.

I am fortunate to have three smart, talented, and wonderful sons—Greg, David, and Bill. I am especially privileged to have Kathleen as my wife, best friend, and research associate. It is inspiring to have a brilliant, beautiful (inside and out), and strong woman who cares deeply about nature as a lifemate. She is my hero. I dedicate this book to her and to the earth.

SCOTT E. SPOOLMAN

Scott Spoolman is a writer with more than 30 years of experience in educational publishing. He has worked with Tyler Miller since 2003 as a contributing editor and lately as coauthor of *Living in the Environment, Environmental Science,* and *Sustaining the Earth.* With Norman Myers, he coauthored *Environmental Issues and Solutions: A Modular Approach.*

Spoolman holds a master's degree in science journalism from the University of Minnesota. He has authored numerous articles in the fields of science, environmental engineering, politics, and business. He has also worked as a consulting editor in the development of over 70 college and high school textbooks in the fields of the natural and social sciences.

In his free time, he enjoys exploring the forests and waters of his native Wisconsin along with his family—his wife, environmental educator Gail Martinelli, and his children, Will and Katie.

Spoolman has the following to say about his collaboration with Tyler Miller:

I am honored to be working with Tyler Miller as a coauthor to continue the Miller tradition of thorough, clear, and engaging writing about the vast and complex field of environmental science. I share Tyler Miller's passion for ensuring that these textbooks and their multimedia supplements will be valuable tools for students and instructors. To that end, we strive to introduce this interdisciplinary field in ways that will be not only informative and sobering but also tantalizing and motivational.

If the flip side of any problem is an opportunity, then this truly is one of the most exciting times in history for students to start an environmental career. Environmental problems are numerous, serious, and daunting, but their possible solutions generate exciting new career opportunities. We place high priorities on inspiring students with these possibilities, challenging them to maintain a scientific focus, pointing them toward rewarding and fulfilling careers, and in doing so, working to help sustain life on Earth.

My Environmental Journey— *G. Tyler Miller*

My environmental journey began in 1966 when I heard a lecture on population and pollution problems by Dean Cowie, a biophysicist with the U.S. Geological Survey. It changed my life. I told him that if even half of what he said was valid, I would feel ethically obligated to spend the rest of my career teaching and writing to help students learn about the basics of environmental science. After spending six months studying the environmental literature, I concluded that he had greatly underestimated the seriousness of these problems.

I developed an undergraduate environmental studies program and in 1971 published my first introductory environmental science book, an interdisciplinary study of the connections between energy laws (thermodynamics), chemistry, and ecology. In 1975, I published the first edition of *Living in the Environment*. Since then, I have completed multiple editions of this textbook, and of three, others derived from it, along with other books.

Beginning in 1985, I spent 10 years in the deep woods living in an adapted school bus that I used as an environmental science laboratory and writing environmental science textbooks. I evaluated the use of passive solar energy design to heat the structure; buried earth tubes to bring in air cooled by the earth (geothermal cooling) at a cost of about $1 per summer; set up active and passive systems to provide hot water; installed an energy-efficient instant hot water heater powered by LPG; installed energy-efficient windows and appliances and a composting (waterless) toilet; employed biological pest control; composted food wastes; used natural planting (no grass or lawnmowers); gardened organically; and experimented with a host of other potential solutions to major environmental problems that we face.

I also used this time to learn and think about how nature works by studying the plants and animals around me. My experience from living in nature is reflected in much of the material in this book. It also helped me develop the six simple principles of sustainability that serve as the integrating theme for this textbook and to apply these principles to living my life more sustainably.

I came out of the woods in 1995 to learn about how to live more sustainably in an urban setting where most people live. Since then, I have lived in two urban villages, one in a small town and one within a large metropolitan area.

Since 1970, my goal has been to use a car as little as possible. Since I work at home, I have a "low-pollute commute" from my bedroom to a chair and a laptop computer. I usually take one or two airplane trips a year to visit my sister and my publisher.

As you will learn in this book, life involves a series of environmental trade-offs. Like most people, I still have a large environmental impact, but I continue to struggle to reduce it. I hope you will join me in striving to live more sustainably and sharing what you learn with others. It is not always easy, but it sure is fun.

Cengage Learning's Commitment to Sustainable Practices

We the authors of this textbook and Cengage Learning, the publisher, are committed to making the publishing process as sustainable as possible. This involves four basic strategies: ■ *Using sustainably produced paper.* The book publishing industry is committed to increasing the use of recycled fibers, and Cengage Learning is always looking for ways to increase this content. Cengage Learning works with paper suppliers to maximize the use of paper that contains only wood fibers that are certified as sustainably produced, from the growing and cutting of trees all the way through paper production. ■ *Reducing resources used per book.* The publisher has an ongoing program to reduce the amount of wood pulp, virgin fibers, and other materials that go into each sheet of paper used. New, specially designed printing presses also reduce the amount of scrap paper produced per book. ■ *Recycling.* Printers recycle the scrap paper that is produced as part of the printing process. Cengage Learning also recycles waste cardboard from shipping cartons, along with other materials used in the publishing process. ■ *Process improvements.* In years past, publishing has involved using a great deal of paper and ink for the writing and editing of manuscripts, copyediting, reviewing page proofs, and creating illustrations. Almost all of these materials are now saved through use of electronic files. Very little paper and ink were used in the preparation of this textbook.

Learning Skills

Students who can begin early in their lives to think of things as connected, even if they revise their views every year, have begun the life of learning.

Mark Van Doren

Why Is It Important to Study Environmental Science?

Welcome to **environmental science**—an *interdisciplinary* study of how the earth works, how we interact with the earth, and how we can deal with the environmental problems we face. Because environmental issues affect every part of your life, the concepts, information, and issues discussed in this book and the course you are taking will be useful to you now and throughout your life.

Understandably, we are biased, but *we strongly believe that environmental science is the single most important course that you could take.* What could be more important than learning about the earth's life-support system, how our choices and activities affect it, and how we can reduce our growing environmental impact? Evidence indicates strongly that we will have to learn to live more sustainably by reducing our degradation of the planet's life-support system. We hope this book will inspire you to become involved in this change in the way we view and treat the earth, which sustains us, our economies, and all other living things.

You Can Improve Your Study and Learning Skills

Maximizing your ability to learn involves trying to *improve your study and learning skills*. Here are some suggestions for doing so:

Develop a passion for learning. This is a key to success.

Get organized. Planning is a key life skill.

Make daily to-do lists in writing. Put items in order of importance, focus on the most important tasks, and assign a time to work on these items. Shift your schedule as needed to accomplish the most important items.

Set up a study routine in a distraction-free environment. Develop a written daily study schedule and stick to it. Study in a quiet, well-lit space. Take breaks every hour or so. During each break, take several deep breaths and move around; this will help you stay more alert and focused.

Avoid procrastination. Do not fall behind on your reading and other assignments. Set aside a particular time for studying each day and make it a part of your daily routine.

Make hills out of mountains. It can be difficult to read an entire chapter or book, write a paper, or cram for a test within a short period of time. Instead, break these large tasks (mountains) down into a series of small tasks (hills). Each day, read a few pages of the assigned book or chapter, write a few paragraphs of the paper, and review what you have studied and learned.

Ask and answer questions as you read. For example, "What is the main point of a particular subsection or paragraph?" Relate your own questions to the key questions and key concepts addressed in each major chapter section and listed in the review section at the end of each chapter.

Focus on key terms. Use the glossary in your textbook to look up the meaning of terms or words you do not understand. This book shows all key terms in **bold** type and lesser, but still important, terms in *italicized* type. The *Chapter Review* questions at the end of each chapter also include the chapter's key terms in bold. Flash cards for testing your mastery of key terms for each chapter are available on the website for this book, or you can make your own.

Interact with what you read. You could mark key sentences and paragraphs with a highlighter or pen or with asterisks and notes in the margin. You might also mark important pages that you want to return to by adding notes or highlighting material or by folding down page corners.

Review to reinforce learning. Before each class session, review the material you learned in the previous session and read the assigned material.

Become a good note taker. Learn to write down the main points and key information from any lecture. Review, fill in, and organize your notes as soon as possible after each class.

Check what you have learned. At the end of each chapter, you will find review questions that cover all of the key material in each chapter section. We suggest that you try to answer each of these questions after studying each chapter section. Waiting to do this for the entire chapter after you complete it can be overwhelming.

Write out answers to questions to focus and reinforce learning. Write down your answers to the critical thinking questions found in the *Thinking About* boxes throughout the chapters, in many figure captions, and at the end of each chapter. These questions are designed to inspire you to think critically about key ideas and connect them to other ideas and to your own life. Also, write down your answers to all chapter-ending review questions. The website for each chapter has an additional detailed list of review questions for that chapter. Save your answers for review and test preparation.

Use the buddy system. Study with a friend or become a member of a study group to compare notes, review material, and prepare for tests. Explaining something to someone else is a great way to focus your thoughts and reinforce your learning. Attend any review sessions offered by instructors or teaching assistants.

Learn your instructor's test style. Does your instructor emphasize multiple-choice, fill-in-the-blank, true-or-false, factual, or essay questions? How much of the test will come from the textbook and how much from lecture material? Adapt your learning and studying methods to this style.

Become a good test taker. Avoid cramming. Eat well and get plenty of sleep before a test. Arrive on time or early. Calm yourself and increase your oxygen intake by taking several deep breaths. (Do this also about every 10–15 minutes while taking the test.) Look over the test and answer the questions you know well first. Then work on the harder ones. Use the process of elimination to narrow down the choices for multiple-choice questions. For essay questions, organize your thoughts before you start writing. If you don't understand what a question means, make an educated guess. You might earn some partial credit and avoid getting a zero. Another strategy for getting some credit is to show your knowledge and reasoning by writing something like this: "If this question means so and so, then my answer is _____."

Develop an optimistic but realistic outlook. Try to be a "glass is half-full" rather than a "glass is half-empty" person. Pessimism, fear, anxiety, and excessive worrying (especially over things you cannot control) are destructive and lead to inaction.

Take time to enjoy life. Every day, take time to laugh and enjoy nature, beauty, and friendship.

You Can Improve Your Critical Thinking Skills

Critical thinking involves developing skills to analyze information and ideas, judge their validity, and make decisions. Critical thinking helps you distinguish between facts and opinions, evaluate evidence and arguments, and take and defend informed positions on issues. It also helps you integrate information, see relationships, and apply your knowledge to dealing with various problems and decisions. Here are some basic skills for learning how to think more critically.

Question everything and everybody. Be skeptical, as any good scientist is. Do not believe everything you hear and read, including the content of this textbook, without evaluating the information you receive. Seek other sources and opinions.

Identify and evaluate your personal biases and beliefs. Each of us has biases and beliefs taught to us by our parents, teachers, friends, role models, and our own experience. What are your basic beliefs, values, and biases? Where did they come from? What assumptions are they based on? How sure are you that your beliefs, values, and assumptions are right and why? According to the American psychologist and philosopher William James, "A great many people think they are thinking when they are merely rearranging their prejudices."

Be open-minded and flexible. Be open to considering different points of view. Suspend judgment until you gather more evidence, and be willing to change your mind. Recognize that there may be a number of useful and acceptable solutions to a problem, and that very few issues are either black or white. Try to take the viewpoints of those you disagree with to better understand their thinking. There are trade-offs involved in dealing with any environmental issue, as you will learn in this book.

Be humble about what you know. Some people are so confident in what they know that they stop thinking and questioning. To paraphrase American writer Mark Twain, "It's what we know is true, but just ain't so, that hurts us."

Find out how the information related to an issue was obtained. Are the statements you heard or read based on firsthand knowledge and research or on hearsay? Are unnamed sources used? Is the information based on reproducible and widely accepted scientific studies or on preliminary scientific results that may be valid but need further testing? Is the information based on a few isolated stories or experiences or on carefully controlled studies that have been reviewed by experts in the field involved? Is it based on unsubstantiated and dubious scientific information or beliefs?

Question the evidence and conclusions presented. What are the conclusions or claims based on the information you're considering? What evidence is presented to support them? Does the evidence support them? Is there a need to gather more evidence to test the conclusions? Are there other, more reasonable conclusions?

Try to uncover differences in basic beliefs and assumptions. On the surface, most arguments or disagreements involve differences of opinion about the validity or meaning of certain facts or conclusions. Scratch a little deeper and you will find that many disagreements are based on different (and often hidden) basic assumptions concerning how we look at and interpret the world around us. Uncovering these basic differences can allow the parties involved to understand one another's viewpoints and to agree or disagree about their basic assumptions, beliefs, or principles.

Try to identify and assess any motives on the part of those presenting evidence and drawing conclusions. What is their expertise in this area? Do they have any unstated assumptions, beliefs, biases, or values? Do they have a personal agenda? Can they benefit financially or politically from acceptance of their evidence and conclusions? Would investigators with different basic assumptions or beliefs take the same data and come to different conclusions?

Expect and tolerate uncertainty. Recognize that scientists cannot establish absolute proof or certainty about anything. However, the goal of science is to provide a high degree of certainty (at least 90%) about its data and the scientific theories used to explain the data.

Check the arguments you hear and read for logical fallacies and debating tricks. Here are six of many examples of such debating tricks. *First*, attack the presenter of an argument rather than the argument itself. *Second*, appeal to emotion rather than facts and logic. *Third*, claim that if one piece of evidence or one conclusion is false, then all other related pieces of evidence and conclusions are false. *Fourth*, say that a conclusion is false because it has not been scientifically proven. Scientists never prove anything absolutely, but they strive to establish a high degree of certainty (at least 90%) about their results and theories. *Fifth*, inject irrelevant or misleading information to divert attention from important points. *Sixth*, present only either/or alternatives when there may be a number of options.

Do not believe everything you read on the Internet. The Internet is a wonderful and easily accessible source of information that includes alternative explanations and opinions on almost any subject or issue—much of it not available in the mainstream media and scholarly articles. Blogs of all sorts have become a major source of information, more important than standard news media for some people. However, because the Internet is so open, anyone can post anything they want to some blogs and other websites with no editorial control or review by experts. As a result, evaluating information on the Internet is one of the best ways to put into practice the principles of critical thinking discussed here. Use and enjoy the Internet, but think critically and proceed with caution.

Develop principles or rules for evaluating evidence. Develop a written list of principles to serve as guidelines for evaluating evidence and claims. Continually evaluate and modify this list on the basis of your experience.

Become a seeker of wisdom, not a vessel of information. Many people believe that the main goal of their education is to learn as much as they can by gathering more and more information. We believe that the primary goal is to learn how to sift through mountains of facts and ideas to find the few *nuggets of wisdom* that are the most useful for understanding the world and for making decisions. This book is full of facts and numbers, but they are useful only to the extent that they lead to an understanding of key ideas, concepts, connections, and scientific laws and theories. The major goals of the study of environmental science are to find out how nature works and sustains itself (*environmental wisdom*) and to use *principles of environmental wisdom* to help make human societies and economies more sustainable, more just, and more beneficial and enjoyable for all. As writer Sandra Carey observed, "Never mistake knowledge for wisdom. One helps you make a living; the other helps you make a life."

To help you practice critical thinking, we have supplied questions throughout this book, found within each chapter in brief boxes labeled *Thinking About*, in the captions of many figures, and at the end of each chapter. There are no right or wrong answers to many of these questions. A good way to improve your critical thinking skills is to compare your answers with those of your classmates and to discuss how you arrived at your answers.

Use the Learning Tools We Offer in This Book

We have included a number of tools throughout this textbook that are intended to help you improve your learning skills and apply them. First, consider the *Key Concepts* list at the beginning of each chapter section. You can use these to preview a chapter and to review the material after you've read it.

Next, note that we use three different special notations throughout the text. Each chapter opens with a **Core Case Study**, and each time we tie material within the chapter back to this core case, we note it in bold, colored type as we did in this sentence. You will also see two icons appearing regularly in the text margins. When you see the *sustainability* icon, you will know that you have just read something that relates directly to the overarching theme of this text, summarized by our six **principles of sustainability**, which are introduced in Figures 1.2, p. 6, and 1.7, p. 9, and summarized on the inside back cover of this book. The *Good News* icon appears near each of many examples of successes that people have had in dealing with the environmental challenges we face.

We also include several brief *Connections* boxes to show you some of the often surprising connections between environmental problems or processes and some of the products and services we use every day or some of the activities we partake in. These, along with the *Thinking About* boxes scattered throughout the text (both designated by the *Consider This* heading), are intended to get you to think carefully about the activities and choices we take for granted and about how they might affect the environment.

New to this edition is a third Consider This feature called *Learning from Nature*. Most chapters contain one or more of these, each of which gives an example of how scientists and engineers are applying nature's lessons through biomimicry (a major new theme of this edition) to solve a problem or to improve a technology.

At the end of the chapter, we list what we consider to be the *three big ideas* that you should take away from each chapter. Following that list in each chapter is a *Tying It All Together* box. This feature quickly reviews the Core Case Study and how key chapter material relates to it, and it explains how the principles of sustainability can be applied to deal with challenges discussed in the core case study and throughout the chapter.

Finally, we have included a *Chapter Review* section at the end of each chapter, with questions listed for each chapter section. These questions cover all of the key material and key terms in each chapter. In each chapter, they are followed by *Critical Thinking* questions that help you apply chapter material to the real world and to your own life; a *Doing Environmental Science* exercise to help you experience the work of scientists; a *Global Environment Watch* exercise, in which you can use the GREENR online global environmental database; and a *Data Analysis* or *Ecological Footprint Analysis* exercise to help you learn how to interpret and use scientific research data.

Know Your Own Learning Style

People have different ways of learning and it can be helpful to know your own learning style. *Visual learners* learn best from reading and viewing illustrations and diagrams. *Auditory learners* learn best by listening and discussing. They might benefit from reading aloud while studying and using a tape recorder in lectures for study and review. *Logical learners* learn best by using concepts and logic to uncover and understand a subject rather than relying mostly on memory.

This book and its supporting website material contain plenty of tools for all types of learners. Visual learners can benefit from using flash cards (available on the website) to memorize key terms and ideas. This is a highly visual book with many photographs and diagrams carefully selected to illustrate important ideas, concepts, and processes. Auditory learners can make use of our *ReadSpeaker app* in MindTap, which can read the chapter aloud in various voices and speeds. For logical learners, the book is organized by key concepts that are revisited throughout any chapter and related carefully to other concepts, major principles, and case studies and other examples. We urge you to become aware of your own learning style and make the most of these various tools.

This Book Presents a Positive, Realistic Environmental Vision of the Future

Our goal is to present a positive vision of our environmental future based on realistic optimism. To do so, we strive not only to present the facts about environmental issues but also to give a balanced presentation of different viewpoints. We consider the advantages and disadvantages of various technologies and proposed solutions to environmental problems. We argue that environmental solutions usually require *trade-offs* among opposing parties, and that the best solutions are *win-win* solutions where everyone benefits. We also present the good news as well as the bad news about efforts to deal with environmental problems.

One cannot study a subject as important and complex as environmental science without forming conclusions, opinions, and beliefs. However, we argue that any such results should be based on use of critical thinking to evaluate conflicting positions and to understand the trade-offs involved in most environmental solutions. To that end, we emphasize critical thinking throughout this textbook, and we encourage you to develop a practice of thinking critically about everything you read and hear, both in school and throughout your life.

Help Us Improve This Book

Researching and writing a book that covers and connects the numerous major concepts from the wide variety of environmental science disciplines is a challenging and exciting task. Almost every day, we learn about some new connection in nature. However, in a book this complex, there are bound to be some errors—some typographical mistakes that slip through and some statements that you might question, based on your knowledge and research. We invite you to contact us to correct any errors you find, point out any bias you see, and suggest ways to improve this book. Please e-mail your suggestions to Tyler Miller at mtg89@hotmail.com or Scott Spoolman at spoolman@tds.net.

Now start your journey into this fascinating and important study of how the earth's life-support system works and how we can leave the planet in a condition at least as good as what we now enjoy. Have fun.

Supplements for Students

You have a large variety of electronic and other supplemental materials available to you to help you take your learning experience beyond this textbook:

- **Environmental Science MindTap.** MindTap provides you with the tools you need to better manage your limited time. You can complete assignments whenever and wherever you are ready to learn with course material specifically customized for you by your instructor and streamlined in one proven, easy-to-use interface. MindTap includes an online homework solution that helps you learn and understand key concepts through focused assignments, exceptional text-art integration, and immediate feedback. With these resources and an array of tools and apps—from note taking to flashcards—you'll get a true understanding of course concepts, helping you achieve better grades and setting the groundwork for your future courses.

- **Global Environment Watch.** Integrated within MindTap and updated several times a day, the Global Environment Watch is a focused portal into GREENR—the Global Reference on the Environment, Energy, and Natural Resources—an ideal one-stop site for classroom discussion and research projects. This resource center keeps courses up to date with the most current news on the environment. Users get access to information from trusted academic journals, news outlets, and magazines, as well as statistics, an interactive world map, videos, primary sources, case studies, podcasts, and much more.

Other student learning tools include:

- **Essential Study Skills for Science Students** by Daniel D. Chiras. This book includes chapters on developing good study habits; sharpening memory; getting the most out of lectures, labs, and reading assignments; improving test-taking abilities; and becoming a critical thinker. Available for students on instructor's request.

- **Lab Manual.** Edited by Edward Wells, this lab manual includes both hands-on and data analysis labs to help your students develop a range of skills. Create a custom version of this Lab Manual by adding labs you have written or ones from our collection with Cengage Custom Publishing. An Instructor's Manual for the labs will be available to adopters.

- **What Can You Do?** This guide presents students with a variety of ways that they can affect the environment and shows them how to track the effect their actions have on their carbon footprint. Available for students on instructor's request.

Environmental Science 16e

CHAPTER 1

The Environment and Sustainability

No civilization has survived the ongoing destruction of its natural support system. Nor will ours.

LESTER R. BROWN

Key Questions

1.1 What are some key principles of sustainability?

1.2 How are we affecting the earth?

1.3 Why do we have environmental problems?

1.4 What is an environmentally sustainable society?

Forests such as this one in California's Sequoia National Park help to sustain all life and economies.

robertharding/Alamy Stock Photo

3

Learning from the Earth

Sustainability is the capacity of the earth's natural systems that support life and human economic systems to survive or adapt to changing environmental conditions indefinitely. Sustainability is the big idea and the integrating theme of this book.

The earth is a remarkable example of a sustainable system. Life has existed on the earth for about 3.8 billion years. During this time, the planet has experienced several catastrophic environmental changes. They include gigantic meteorite impacts, ice ages lasting millions of years, long warming periods that melted land-based ice and raised sea levels by hundreds of feet, and five mass extinctions—each wiping out more than half of the world's species. Despite these dramatic environmental changes, an astonishing variety of life has survived.

How has life survived such challenges? Long before humans arrived, organisms had developed abilities to use sunlight to make their food and to recycle all of the nutrients they needed

for survival. Organisms also developed a variety of abilities to find food and survive. For example, spiders create webs that are strong enough to capture fast-moving flying insects. Bats have a radar system for finding prey and avoiding collisions. These and many other abilities and materials were developed without the use of the high-temperature or high-pressure processes or the harmful chemicals that we employ in manufacturing.

This explains why many scientists urge us to focus on learning from the earth about how to live more sustainably. Biologist Janine Benyus is a pioneer in this area. In 1997, she coined the term **biomimicry** to describe the rapidly growing scientific effort to understand, mimic, and catalog the ingenious ways in which nature has sustained life on the earth for 3.8 billion years. She views the earth's life-support system as the world's longest and most successful research and development laboratory.

How do geckos (Figure 1.1, left) cling to and walk on windows, walls, and ceilings? Scientists have learned that these little lizards have many thousands of tiny hairs growing in ridges on the toes of their feet and that each hair is divided into a number of segments that they use to grasp the tiniest ridges and cracks on a surface (Figure 1.1, right). They release their iron grip by tipping their foot until the hairs let go.

This discovery led to the development of a sticky, toxin-free "gecko tape" that could replace toxin-containing glues and tapes. It is an excellent example of biomimicry, or earth wisdom, and you will see many more of such examples throughout this book.

Nature can teach us how to live more sustainably on the amazing planet that is our only home. As Benyus puts it, after billions of years of trial-and-error research and development: "Nature knows what works, what is appropriate, and what lasts here on Earth." ●

FIGURE 1.1 The gecko (left) has an amazing ability to cling to surfaces because of projections from many thousands of tiny hairs on its toes (right).

1.1 WHAT ARE SOME KEY PRINCIPLES OF SUSTAINABILITY?

CONCEPT 1.1A Life on the earth has been sustained for billions of years by solar energy, biodiversity, and chemical cycling.

CONCEPT 1.1B Our lives and economies depend on energy from the sun and on natural resources and ecosystem services (*natural capital*) provided by the earth.

CONCEPT 1.1C We can live more sustainably by following six principles of sustainability.

Environmental Science Is a Study of Connections in Nature

The **environment** is everything around you. It includes energy from the sun and all the living things (such as plants, animals, and bacteria) and the nonliving things (such as air, water, and sunlight) with which you interact. Despite humankind's many scientific and technological advances, our lives depend on sunlight and the earth for clean air and water, food, shelter, energy, fertile soil, a livable climate, and other components of the planet's *life-support system*.

Environmental science is a study of connections in the natural environment nature. It is an interdisciplinary study of **(1)** how the earth (nature) works and has survived and thrived, **(2)** how humans interact with the environment, and **(3)** how humans can live more sustainably. It strives to answer several questions: What environmental problems do we face? How serious are they? How do they interact? What are their causes? How has nature solved such problems? How can we solve such problems? To answer such questions, environmental science integrates information and ideas from fields such as biology, chemistry, geology, engineering, geography, economics, political science, and ethics.

A key component of environmental science is **ecology**, the branch of biology that focuses on how living organisms interact with the living and nonliving parts of their environment. Each organism, or living thing, belongs to a **species**—a group of organisms having a unique set of characteristics that set it apart from other groups.

A major focus of ecology is the study of ecosystems. An **ecosystem** is a biological community of organisms within a defined area of land or volume of water that interact with one another and with the nonliving chemical and physical factors in their environment. For example, a forest ecosystem consists of plants, animals, and organisms that decompose organic materials, all interacting with one another and the chemicals in the forest's air, water, and soil.

Environmental science and ecology should not be confused with **environmentalism**, or **environmental activism**, which is a social movement dedicated to protecting the earth's life-support system for humans and other species.

Learning from the Earth: Three Scientific Principles of Sustainability

Modern humans have been around for about 200,000 years—less than the blink of an eye, relative to the 3.8 billion years that life has existed on the earth. During their short time on the earth, and especially since 1900, humans have expanded into and dominated almost all of the earth's ecosystems.

This large and growing human impact threatens the existence of many species and biological centers of life such as tropical rainforests and coral reefs. It also adds pollutants to the earth's air, water, and soil. Many environmental scientists warn that humans are degrading the earth's life-support system that supports all life and human economies.

Scientific studies of how the earth works reveal that three natural factors play key roles in the long-term sustainability of the planet's life, as summarized below and in Figure 1.2 (**Concept 1.1A**). Understanding these three **scientific principles of sustainability**, or major *lessons from nature*, can help us move toward a more sustainable future.

- **Solar energy:** The sun's energy warms the planet and provides energy that plants use to produce **nutrients**, the chemicals that plants and animals need to survive.

- **Biodiversity:** The variety of genes, species, ecosystems, and ecosystem processes are referred to as **biodiversity** (short for *biological diversity*). Interactions among species provide vital ecosystem services and keep any population from growing too large. Biodiversity also provides ways for species to adapt to changing environmental conditions and for new species to arise and replace those wiped out by catastrophic environmental changes.

- **Chemical cycling:** The circulation of nutrients from the environment (mostly from soil and water) through various organisms and back to the environment is called **chemical cycling**, or **nutrient cycling**. The earth receives a continuous supply of energy from the sun but it receives no new supplies of life-supporting chemicals. Through billions of years of interactions with their living and nonliving environment, organisms have developed ways to recycle the chemicals they need to survive. This means that the wastes and decayed bodies of organisms become nutrients or raw materials for other organisms. In nature, **waste = useful resources.**

Key Components of Sustainability

Sustainability, the integrating theme of this book, has several key components that we use as subthemes. One is **natural capital**—natural resources and ecosystem

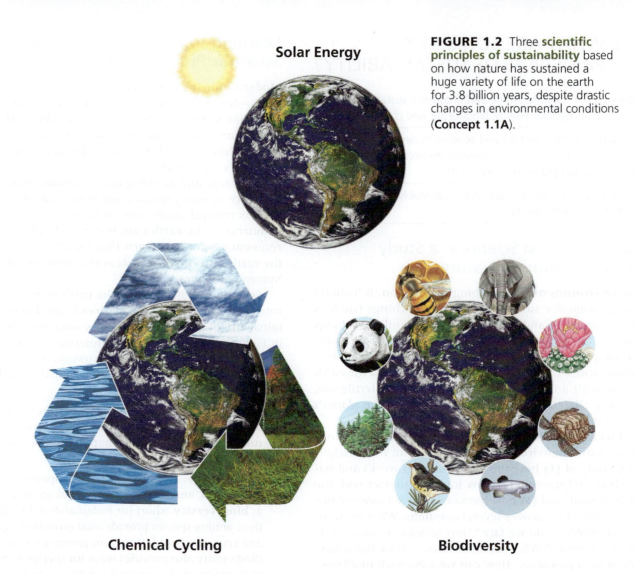

Solar Energy

Chemical Cycling

Biodiversity

FIGURE 1.2 Three **scientific principles of sustainability** based on how nature has sustained a huge variety of life on the earth for 3.8 billion years, despite drastic changes in environmental conditions (**Concept 1.1A**).

services that keep humans and other species alive and that support human economies (Figure 1.3).

Natural resources are materials and energy provided by nature that are essential or useful to humans. They fall into three categories: *inexhaustible resources, renewable resources,* and *nonrenewable (exhaustible) resources* (Figure 1.4). An **inexhaustible resource** is one that is expected to last forever on a human timescale. **Solar energy** is such a resource, expected to last for at least 5 billion years until the death of the star we call the sun. A **renewable resource** is a resource that can be used repeatedly because it is replenished through natural processes as long as it is not used up faster than nature can renew it. Examples are forests, grasslands, fertile topsoil, fishes, clean air, and fresh water. The highest rate at which people can use a renewable resource indefinitely without reducing its available supply is called its **sustainable yield**.

Nonrenewable or **exhaustible resources** are those that exist in a fixed amount, or *stock*, in the earth's crust. They take millions to billions of years to form

through geological processes. On the much shorter human timescale, we can use these resources faster than nature can replace them. Examples of nonrenewable resources are oil, natural gas, and coal (Figure 1.5), and metallic mineral resources such as copper and aluminum.

Ecosystem services are the natural services provided by healthy ecosystems that support life and human economies at no monetary cost (Figure 1.3). For example, forests help purify air and water, reduce soil erosion, regulate climate, and recycle nutrients. Thus, our lives and economies are sustained by energy from the sun and by natural resources and ecosystem services (natural capital) provided by the earth (**Concept 1.1B**).

Key ecosystem services include purification of air and water, renewal of topsoil, pollination, and pest control. Another important example is nutrient cycling, which is a **scientific principle of sustainability**. Without nutrient cycling in topsoil, there would be no land plants, no pollinators, and no food for us and other animals.

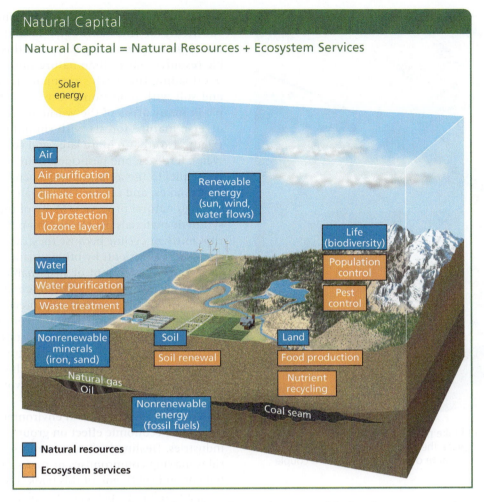

FIGURE 1.3 Natural Capital consists of natural resources (blue) and ecosystem services (orange) that support and sustain the earth's life and human economies (**Concept 1.1B**).

Inexhaustible
Solar energy
Wind energy
Geothermal energy

Renewable
Trees
Topsoil
Freshwater

Nonrenewable (Exhaustible)
Fossil fuels (oil, natural gas, coal)
Iron and copper

FIGURE 1.4 We depend on a combination of inexhaustible, renewable, and nonrenewable (exhaustible) natural resources.

FIGURE 1.5 It would take more than a million years for natural processes to replace the coal that was removed from this strip mine in the U.S. state of Wyoming within a couple of decades.

A second component of sustainability—and another subtheme of this textbook—is that human activities can *degrade natural capital*. We do this by using renewable resources faster than nature can restore them and by overloading the earth's normally renewable air, water, and soil with pollution and wastes. For example, people in many areas of the world are replacing biologically diverse mature forests with simplified crop plantations (Figure 1.6) that require large and costly inputs of energy, water, fertilizer, and pesticides. Many human activities add pollutants to the air and dump chemicals and wastes into rivers, lakes, and oceans faster than they can be cleansed through natural processes. Many of the plastics and other synthetic materials people use can poison wildlife and disrupt nutrient cycling because they cannot be broken down and used as nutrients by other organisms.

A third component of sustainability involves people finding *solutions* to the environmental problems we face. People can work together to protect the earth's natural capital and to use it sustainably. For example, a solution to the loss of forests is to stop burning or cutting down mature forests faster than they can grow back (Figure 1.6). This requires that citizens become educated about the ecosystem services forests provide and work to see that forests are used sustainably.

Conflicts can arise when environmental protection has a negative economic effect on groups of people or certain industries. Dealing with such conflicts often involves both sides making compromises or *trade-offs*—the fourth component and subtheme of this book. For example, a timber company might be persuaded to plant and harvest trees in an area that it had already cleared or degraded instead of clearing an undisturbed area of a mature forest. In return, the government may subsidize (pay part of the cost of) planting new trees.

Each individual—including you—plays an important role in learning how to live more sustainably. Thus, *individuals matter*—the fifth component of sustainability and subtheme of this book.

FIGURE 1.6 Small remaining area of once diverse Amazon rain forest surrounded by vast soybean fields in the Brazilian state of Mato Grosso.

Janine Benyus: Using Nature to Inspire Sustainable Design and Living

Janine Benyus has had a lifelong interest in learning how nature works and how to live more sustainably. She realized that 99% of the species that have lived on the earth became extinct because they could not adapt to changing environmental conditions. She views the surviving species as examples of *natural genius* that we can learn from.

Benyus says that when we need to solve a problem or design a product, we should ask: Has nature done this and how did it do it? We should also think about what nature does not do as a clue to what we should not do, she argues. For example, nature does not produce waste materials or chemicals that cannot be broken down and recycled.

Benyus has set up the nonprofit Biomimicry Institute that has developed a curriculum for K–12 and university students and a two-year program to train biomimicry professionals. She has also established a network called Biomimicry 3.8, named for the 3.8 billion years during which organisms have developed what Benyus calls their *genius for surviving*. It is a network of scientists, engineers, architects, and designers who share examples of biomimicry.

ecowatch.com

Three Additional Principles of Sustainability

Our research in economics, politics, and ethics has provided us with three additional **principles of sustainability** (Figure 1.7):

- **Full-cost pricing** (from economics): Some economists urge us to find ways to include in market prices the harmful environmental and health costs of producing and using goods and services. This practice, called **full-cost pricing**, would give consumers information about the harmful environmental impacts of the goods and services that they use.

- **Win–win solutions** (from political science): Political scientists urge us to look for *win–win solutions* to environmental problems, based on cooperation and compromise, that will benefit the largest number of people as well as the environment:

- **Responsibility to future generations** (from ethics): Ethics is a branch of philosophy devoted to studying ideas about what is right and what is wrong. According to environmental ethicists, we have a responsibility to leave the planet's life-support systems in a condition as good as or better than what we inherited for the benefit of future generations and for other species.

These six **principles of sustainability** (see inside back cover of this book) can serve as guidelines to help us live more sustainably. This includes using biomimicry as a tool for learning from the earth about how to live more sustainably (**Core Case Study** and Individuals Matter 1.1).

FIGURE 1.7 Three **principles of sustainability based on economics, political science, and ethics** can help us make a transition to a more environmentally and economically sustainable future.

Left: Minerva Studio/Shutterstock.com. Center: mikeledray/Shutterstock.com. Right: iStock.com/Kali Nine LLC

Countries Differ in Their Resource Use and Environmental Impact

The United Nations (UN) classifies the world's countries as economically more developed or less developed, based primarily on their average income per person. **More-developed countries** are industrialized nations with high average income per person. They include the United States, Japan, Canada, Australia, Germany, and most other European countries. These countries, with 17% of the world's population, use about 70% of the earth's natural resources. The United States, with only 4.4% of the world's population, uses about 30% of the world's resources.

All other nations are classified as **less-developed countries**, most of them in Africa, Asia, and Latin America. Some are *middle-income, moderately developed countries* such as China, India, Brazil, Thailand, and Mexico. Others are *low-income, least-developed countries* including Nigeria, Bangladesh, Congo, and Haiti. The less-developed countries, with 83% of the world's population, use about 30% of the world's natural resources.

1.2 HOW ARE WE AFFECTING THE EARTH?

CONCEPT 1.2A Humans dominate the earth with the power to sustain, add to, or degrade the natural capital that supports all life and human economies.

CONCEPT 1.2B As our ecological footprints grow, we deplete and degrade more of the earth's natural capital that sustains us.

Good News: Many People Have a Better Quality of Life

As the world's dominant animal, humans have an awesome power to degrade or sustain the earth's life-support system. For example, humans decide whether forests are preserved or cut down. Human activities affect the temperature of the atmosphere, the temperature and acidity of ocean waters, and which species survive or become extinct. At the same time, creative thinking, scientific research, political pressure by citizens, and regulatory laws have improved the quality of life for many of the earth's people, especially in the more-developed countries.

Humans have developed an amazing array of useful materials and products. We have learned how to use wood, fossil fuels, the sun, wind, flowing water, the nuclei of certain atoms, and the earth's heat (geothermal energy) to supply us with enormous amounts of energy. Most people live and work in artificial environments within buildings and cities. We have invented computers to extend our brainpower, robots to perform repetitive tasks with great precision, and electronic networks to enable instantaneous global communication.

Globally, life spans are increasing, infant mortality is decreasing, education is on the rise, some diseases are being conquered, and the population growth rate has slowed. While one out of seven people live in extreme poverty, we have witnessed the greatest reduction in poverty in human history. The food supply is generally more abundant and safer, air and water are getting cleaner in many parts of the world, and exposure to toxic chemicals is more avoidable. People have protected some endangered species and ecosystems and restored some grasslands and wetlands, and forests are growing back in some areas.

Scientific research and technological advances financed by affluence helped achieve these improvements in life and environmental quality. Education also spurred many citizens to insist that businesses and governments work toward improving environmental quality. We are a globally connected species with growing access to information that could help us to shift to a more sustainable path.

Bad News: On the Whole, We Are Living Unsustainably

According to a large body of scientific evidence, humans are living unsustainably. People continually waste, deplete, and degrade much of the earth's life-sustaining natural capital—a process known as **environmental degradation**, or **natural capital degradation** (Figure 1.8).

According to research by the Wildlife Conservation Society and the Columbia University Center for International Earth Science Information Network, human activities directly affect about 83% of the earth's land surface (excluding Antarctica) as human ecological footprints have impacted the earth (Figure 1.9). This land is used for important purposes such as urban development, growing crops, grazing livestock, mining, timber cutting, and energy production.

In many parts of the world, however, renewable forests are shrinking (Figure 1.6), deserts are expanding, and topsoil is eroding. The lower atmosphere is warming, floating ice and many glaciers are melting at unexpected rates, sea levels are rising, and ocean acidity is increasing. There are more intense floods, droughts, severe weather, and forest fires in many areas. In a number of regions, rivers are running dry, 20% of the world's species-rich coral reefs are gone, and others are threatened. Species are becoming extinct at least 100 times faster than in prehuman times. And extinction rates are projected to increase at least another 100-fold during this century, creating a 6th mass extinction caused by human activities.

In 2005, the United Nations released its *Millennium Ecosystem Assessment*, a four-year study by 1,360 experts from 95 countries. According to this study, human activities have overused about 60% of the earth's ecosystem services

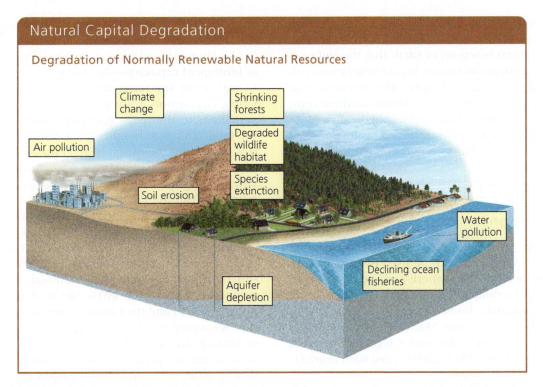

Degradation of Normally Renewable Natural Resources

Climate change

Shrinking forests

Degraded wildlife habitat

Air pollution

Species extinction

Soil erosion

Water pollution

Declining ocean fisheries

Aquifer depletion

FIGURE 1.8 Natural Capital Degradation: Degradation of normally renewable natural resources and ecosystem services (Figure 1.3), caused by growing human ecological footprints mostly as a result of population growth and rising rates of resource use per person.

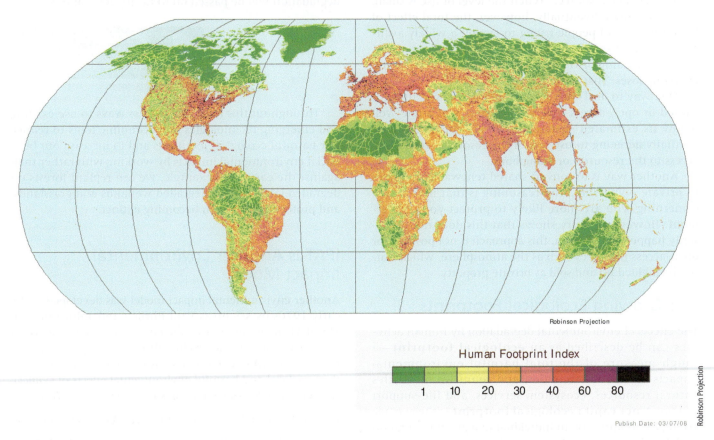

Robinson Projection

Human Footprint Index

1   10   20   30   40   60   80

Publish Date: 03/07/08

Robinson Projection

FIGURE 1.9 Natural Capital Use and Degradation: The human ecological footprint has an impact on about 83% of the earth's total land surface.

(see orange boxes in Figure 1.3), mostly since 1950. According to these researchers, "human activity is putting such a strain on the natural functions of Earth that the ability of the planet's ecosystems to sustain future generations can no longer be taken for granted." They also concluded that scientific, economic, and political solutions to these complex problems could be implemented within decades. Since that study, the harmful environmental and health impact of human activities on the planet's ecosystems has increased.

Degrading Commonly Shared Renewable Resources: The Tragedy of the Commons

Some renewable resources, called *open-access resources*, are not owned by anyone and can be used by almost anyone. Examples are the atmosphere and the open ocean and its fish. Other examples of less open, but often *shared resources*, are grasslands, forests, streams, and *aquifers*, or underground bodies of water. Many of these renewable resources have been environmentally degraded. In 1968, biologist Garrett Hardin (1915–2003) called such degradation the *tragedy of the commons*.

Degradation of such shared or open-access renewable resources occurs because each user reasons, "The little bit that I use or pollute is not enough to matter, and anyway, it's a renewable resource." When the level of use is small, this logic works. Eventually, however, the total effect of large numbers of people trying to exploit a widely available or shared renewable resource can degrade it, eventually exhausting or ruining it. Then no one benefits and everyone loses. That is the tragedy.

One way to deal with this difficult problem is to use a shared or open-access renewable resource at a rate well below its estimated sustainable yield. This is done by mutually agreeing to use less of the resource, regulating access to the resource, or doing both.

Another way is to convert shared renewable resources to private ownership. The reasoning is that if you own something, you are more likely to protect your investment. However, history shows that this does not necessarily happen. In addition, this approach is not possible for open-access resources such as the atmosphere, which cannot be divided up and sold as private property.

Our Growing Ecological Footprints

The effects of environmental degradation by human activities can be described as an **ecological footprint**—a rough measure of the total harmful environmental impacts of individuals, cities, and countries on Earth's natural resources, ecosystem services, and life-support system. A **per capita ecological footprint** is the average ecological footprint of an individual in a given population or defined area. Figure 1.9 shows that the human ecological footprint has impacted 83% of the earth's land surface,

and Figure 1.10 shows the human ecological footprint in North America.

An important measure of sustainability is **biocapacity**, or **biological capacity**—the ability of an area's ecosystems to regenerate the renewable resources used by a population, city, region, country, or the world in a given time period and to absorb the resulting wastes and pollution. If the total ecological footprint in a defined area (such as a city, country, or the world) is larger than its biocapacity, the area is said to have an *ecological deficit*. Such a deficit occurs when people are living unsustainably by depleting natural capital instead of living off the renewable resources and ecosystem services provided by such capital. Figure 1.11 is a map of ecological debtor and creditor countries.

Ecological footprint data and models have been in use since the 1990s. Though imperfect, they provide useful rough estimates of individual, national, and global environmental impacts. In 2016, the World Wide Fund for Nature (WWF) and the Global Footprint Network estimate that we would need the equivalent of 1.6 planet Earths to sustain the world's average 2014 rate of renewable resource use per person far into the future. They estimated that by 2030, we would need the equivalent of two planet Earths and, by 2050, three planet Earths. The current and projected future overdraft of the earth's natural resources and ecosystem services and the resulting environmental degradation will be passed on to future generations.

1.6 Number of earths needed to sustain the 2014 global rate of renewable resource per person use indefinitely

Throughout this book, we discuss ways to use existing and emerging technologies and economic tools to reduce our harmful ecological footprints and to increase our beneficial environmental impacts by working with rather than against the earth. For example, we can replant forests on degraded land, restore degraded wetlands and grasslands, and protect species from becoming extinct.

IPAT Is Another Environmental Impact Model

Another environmental impact model was developed in the early 1970s by scientists Paul Ehrlich and John Holdren. This IPAT model shows that the environmental impact (**I**) of human activities is the product three factors: *population size* (**P**), *affluence* (**A**) or resource consumption per person, and the beneficial and harmful environmental effects of *technologies* (**T**). The following equation summarizes the IPAT model:

Impact (**I**) = Population (**P**) × Affluence (**A**) × Technology (**T**)

The **T** factor can be either harmful or beneficial. Some forms of technology such as polluting factories, gas-guzzling

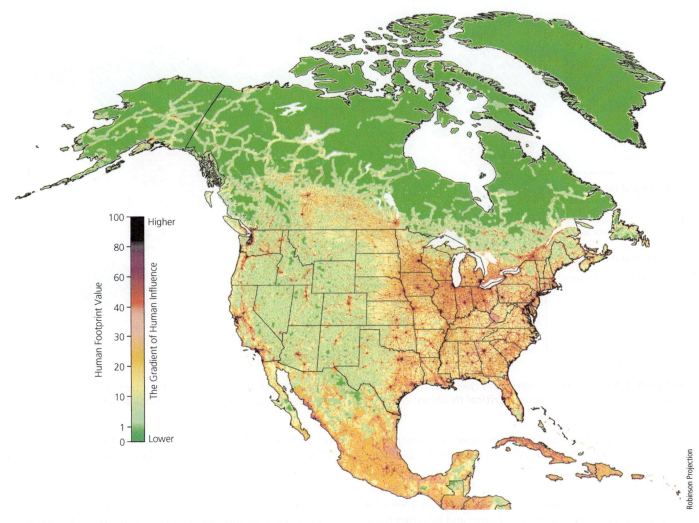

FIGURE 1.10 Natural Capital Use and Degradation: The human ecological footprint in North America. Colors represent the percentage of each area influenced by human activities.

Compiled by the authors using data from Wildlife Conservation Society and Center for International Earth Science Information Network at Columbia University.

motor vehicles, and coal-burning power plants increase our harmful environmental impact by raising the T factor. Other technologies reduce our harmful environmental impact by decreasing the T factor. Examples are pollution control and prevention technologies, fuel-efficient cars, and wind turbines and solar cells that generate electricity with a low environmental impact. By developing technologies that mimic natural processes (**Core Case Study**), scientists and engineers are finding ways to have positive environmental impacts, and we introduce such developments in biomimicry throughout this book.

In a moderately developed country such as India, population size is a more important factor than affluence, or resource use per person, in determining the country's environmental impact. In a highly developed country such as the United States with a much smaller population, resource use per person and the ability to develop environmentally beneficial technologies play key roles in the country's environmental impact.

Cultural Changes Can Increase or Shrink Our Ecological Footprints

Until about 10,000 to 12,000 years ago, humans were mostly *hunter–gatherers* who obtained food by hunting wild animals or scavenging their remains, and gathering wild plants. Our hunter–gatherer ancestors lived in small groups, consumed few resources, had few possessions, and moved as needed to find enough food to survive.

Since then, three major cultural changes have occurred. *First* was the *agricultural revolution*. It began around 10,000 years ago when humans learned how to grow and breed plants and animals for food, clothing, and other purposes and began living in villages instead of frequently moving to find food. They had a more reliable source of food, lived longer, and produced more children who survived to adulthood.

Second was the *industrial–medical revolution*, beginning about 300 years ago when people invented machines for

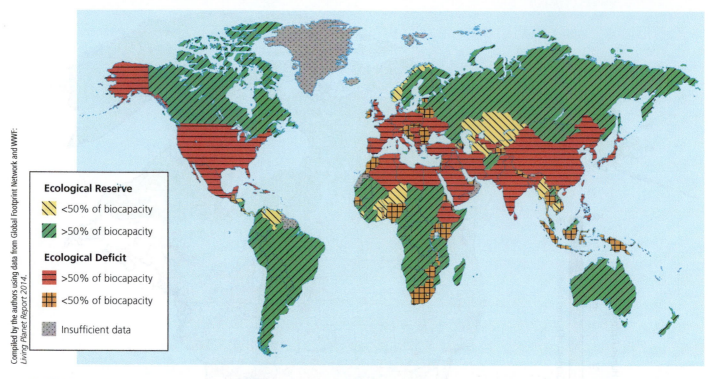

Compiled by the authors using data from Global Footprint Network and WWF:
Living Planet Report 2014.

Ecological Reserve

<50% of biocapacity

>50% of biocapacity

Ecological Deficit

>50% of biocapacity

<50% of biocapacity

Insufficient data

FIGURE 1.11 *Ecological debtors and creditors.* The ecological footprints of some countries exceed their biocapacity, while other countries have ecological reserves. *Critical thinking:* Why do you think that the United States is an ecological debtor country?

the large-scale production of goods in factories. Many people moved from rural villages to cities to work in the factories. This shift involved learning how to get energy from fossil fuels (such as coal and oil) and how to grow large quantities of food. It also included medical advances that allowed a growing number of people to have longer and healthier lives.

Third, about 50 years ago, the *information–globalization revolution* began when we developed new technologies for gaining rapid access to all kinds of information and resources on a global scale.

Each of these three cultural changes gave us more energy and new technologies with which to alter and control more of the planet's resources to meet our basic needs and increasing wants. They also allowed expansion of the human population, mostly because of larger food supplies and longer life spans. In addition, these cultural changes resulted in greater resource use, pollution, and environmental degradation and expanding ecological footprints (Figures 1.9 and 1.10).

On the other hand, some technological leaps have enabled us to shrink our ecological footprints by reducing our use of energy and matter resources and our production of wastes and pollution. For example, the use of energy-efficient LED light bulbs, energy-efficient cars and buildings, recycling, sustainable farming, and solar energy and wind energy to produce electricity is on the rise.

Many environmental scientists and other analysts see such developments as evidence of an emerging fourth major cultural change: a **sustainability revolution**, in which we

could learn to live more sustainably during this century. This would involve avoiding degradation and depletion of the natural capital that supports all life and our economies and restoring natural capital that we have degraded (Figure 1.3). Making this shift would involve learning how nature has sustained life for over 3.8 billion years and using these lessons from nature to shrink our ecological footprints and increase our beneficial environmental impacts. GOOD NEWS

1.3 WHAT CAUSES ENVIRONMENTAL PROBLEMS?

CONCEPT 1.3A Basic causes of environmental problems are population growth, wasteful and unsustainable resource use, poverty, avoidance of full-cost pricing, and increasing isolation from nature.

CONCEPT 1.3B Our environmental worldviews play a key role in determining whether we live unsustainably or more sustainably.

The Human Population Is Growing at a Rapid Rate

Exponential growth occurs when a quantity increases at a fixed percentage per unit of time, such as 0.5% or 2% per year. Exponential growth starts slowly, but after a few doublings, it grows to enormous numbers because each doubling is twice the total of all earlier growth. When we plot the data for an exponentially growing quantity, we get a curve that looks like the letter J.

For an example of the awesome power of exponential growth, consider a simple form of bacterial reproduction in which one bacterium splits into two every 20 minutes. Starting with 1 bacterium, after 20 minutes, there would be 2; after an hour, there would be 8; 10 hours later, there would be more than 1,000; and after just 36 hours (assuming that nothing interfered with their reproduction), there would be enough bacteria to form a layer of 0.3 meters (1 foot) deep over the entire earth's surface.

The human population has grown exponentially (Figure 1.12) to the current population of 7.4 billion people. In 2016, the rate of growth was 1.21%. Although this rate of growth seems small, it added 89.7 million people to the world's 7.4 billion people. By 2050, the population could reach 9.9 billion—an addition of 2.5 billion people within your lifetime.

No one knows how many people the earth can support indefinitely. No one knows how much average resource consumption per person will seriously degrade the planet's natural capital. However, humanity's large and expanding ecological footprints and the resulting widespread natural capital degradation are disturbing warning signs (**Concept 1.3A**).

Some analysts call for us to reduce severe environmental degradation by slowing population growth with the goal of leveling it off at around 8 billion by 2050 instead of 9.9 billion. Some ways to do this include reducing poverty through economic development, promoting family planning, and elevating the status of women, as discussed in Chapter 6.

Affluence and Unsustainable Resource Use

The lifestyles of many of the world's expanding population of consumers are built on growing affluence, or resource consumption per person, as more people earn higher incomes. As total resource consumption and average resource consumption per person increase, so

CONSIDER THIS ...

CONNECTIONS Exponential Growth and Doubling Time: The Rule of 70

The approximate doubling time of the human population can be calculated by using the rule of 70. (You can apply this rule to any quantity that is growing exponentially.)

doubling time (years) = 70 / annual growth rate (%).

The world's population is growing at about 1.21% per year. At this rate, about how long will it take the human population to double?

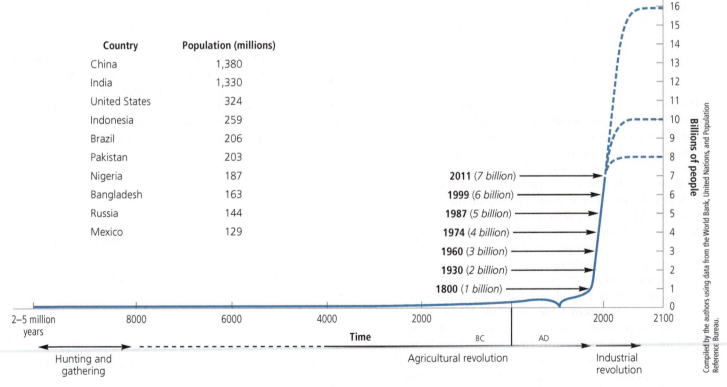

Country	Population (millions)
China	1,380
India	1,330
United States	324
Indonesia	259
Brazil	206
Pakistan	203
Nigeria	187
Bangladesh	163
Russia	144
Mexico	129

2011 (*7 billion*)
1999 (*6 billion*)
1987 (*5 billion*)
1974 (*4 billion*)
1960 (*3 billion*)
1930 (*2 billion*)
1800 (*1 billion*)

2–5 million years
8000
6000
4000
2000
Time
BC
AD
2000
2100

Hunting and gathering
Agricultural revolution
Industrial revolution

Billions of people

Compiled by the authors using data from the World Bank, United Nations, and Population Reference Bureau.

FIGURE 1.12 *Exponential growth:* The J-shaped curve represents past exponential world population growth, with projections to 2100 showing possible population stabilization as the J-shaped curve of growth changes to an S-shaped curve. The top 10 countries (left) represent nearly 60% of the world's total population. **Data analysis:** By what percentage did the world's population increase between 1960 and 2016? (This figure is not to scale.)

do environmental degradation, resource waste, and pollution, unless individuals can live more sustainably (**Concept 1.3A**).

The effects of affluence can be dramatic. The WWF and the Global Footprint Network estimate that the United States, with only 4.4% of the world's population, is responsible for about 23% of the global ecological footprint. The average American consumes about 30 times the amount of resources that the average Indian consumes and 100 times the amount consumed by the average person in the world's poorest countries. The WWF has projected that we would need the equivalent of five planet Earths to sustain the world's population indefinitely if everyone used renewable resources at the same rate as the average American did in 2014.

5 Number of Earths needed to sustain the world's population indefinitely at average per-person U.S. resource consumption rate

On the other hand, affluence can allow for widespread and better education, which can lead people to become more concerned about environmental quality. Affluence also makes more money available for developing technologies to reduce pollution, environmental degradation, and resource waste. It also provides other ways for humans to increase their beneficial environmental impacts.

Poverty Has Harmful Environmental and Health Effects

Poverty is a condition in which people lack enough money to fulfill their basic needs for food, water, shelter, health care, and education. According to the World Bank, about one of every three people, or 2.5 billion people, struggled to live on the equivalent of less than $3.10 a day in 2014. In addition, nearly 900 million people—almost three times the U.S. population—live in *extreme poverty* on the equivalent of less than $1.90 a day, according to the World Bank. This is less than what many people spend for a bottle of water or a cup of coffee. Could you do this? On the other hand, the percentage of the world's population living in extreme poverty decreased from 52% in 1981 to 14% in 2014.

Poverty causes a number of harmful environmental and health effects (**Concept 1.3A**). The daily lives of the world's poorest people center on getting enough food, water, and fuel for cooking and heating to survive. These individuals are too desperate for short-term survival to worry about long-term environmental quality or sustainability. Thus, collectively, they may degrade forests, topsoil, and grasslands, and deplete fisheries and wildlife populations to stay alive.

Poverty does not always lead to environmental degradation. Some of the poor increase their beneficial

environmental impact by planting and nurturing trees and conserving the soil that they depend on as a part of their short-term and long-term survival strategy.

CONSIDER THIS . . .

CONNECTIONS Poverty and Population Growth

To many poor people, having more children is a matter of survival. Their children help them gather firewood, haul water, and tend crops and livestock, and some have to work at jobs. The children also help take care of their aging parents, most of whom do not have social security, health care, and retirement funds. This daily struggle for survival is largely why populations in some of the poorest countries continue to grow at high rates.

Environmental degradation can have severe health effects on the poor. One problem is life-threatening *malnutrition*, a lack of protein and other nutrients needed for good health (Figure 1.13). Another effect is illness caused by limited access to adequate sanitation facilities and clean drinking water. More than one-third of world's people have no bathroom facilities and are forced to use backyards, alleys, ditches, and streams. As a result, one of every nine of the world's people gets water for drinking, washing, and cooking from sources polluted by human and animal feces. Another problem for many poor people is indoor air pollution, mostly from the smoke from open fires or poorly vented stoves (Figure 1.14) used for heating and cooking. This form of indoor air pollution kills about 4.3 million people a year in less-developed countries, according to the World Health Organization (WHO).

In 2010, the WHO estimated that these factors, mostly related to poverty, were killing about 7 million children under age 5 each year—an average of 19,000 young children per day. This is equivalent to *95 fully loaded 200-passenger airliners crashing every day with no survivors*. The news media rarely cover this ongoing human tragedy.

CONSIDER THIS . . .

THINKING ABOUT The Poor, the Affluent, and Environmental Harm

Some see the rapid population growth in less-developed countries as the primary cause of our environmental problems. Others say that the high rate of resource use per person in more-developed countries is a more important factor. Which factor do you think is more important? Why?

Prices of Goods and Services Rarely Include Their Harmful Environmental and Health Costs

Another basic cause of environmental problems has to do with how the marketplace prices goods and services (**Concept 1.3A**). Companies using resources to provide

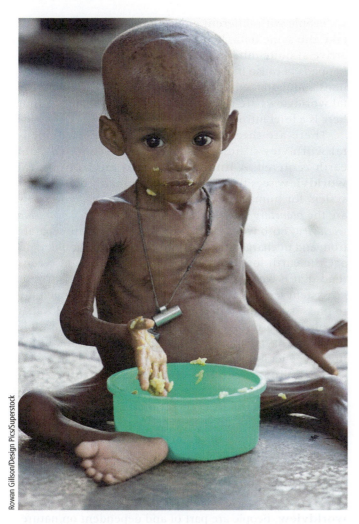

Rowan Gillson/Design Pics/Superstock

FIGURE 1.13 One of every three children younger than age 5 in less-developed countries, such as this starving child in Bangladesh, suffers from severe malnutrition caused by a lack of calories and protein.

Ted Spiegel/National Geographic Creative

FIGURE 1.14 Indoor air pollution from open fires and poorly vented stoves is a major health threat to many poor people in less-developed countries.

goods for consumers generally are not required to pay for most of the harmful environmental and health costs of supplying such goods. For example, timber companies pay the cost of clear-cutting forests but do not pay for the resulting environmental degradation and loss of wildlife habitat.

The primary goal of a company is to maximize profits for its owners or stockholders, so it is not inclined to add these costs to its prices voluntarily. Because the prices of goods and services do not include most of their harmful environmental and health costs, consumers and decision makers have no effective way to evaluate these harmful effects.

Another problem can arise when governments give companies *subsidies* such as tax breaks and payments to assist them with using resources to run their businesses. This helps to create jobs and stimulate economies, but environmentally harmful subsidies encourage the depletion and degradation of natural capital.

According to environmental economists, people could live more sustainably and increase their beneficial

environmental impact if the harmful environmental and health costs of the goods and services were included in market prices of the goods they buy and if we place a monetary value on the natural capital that supports all economies. Such full-cost pricing is a powerful economic tool and is one of the six **principles of sustainability**.

Economists propose two ways to implement full-cost pricing over the next two decades. One is to shift from environmentally harmful government subsidies to environmentally beneficial subsidies that sustain or restore natural capital. Examples of environmentally beneficial subsidies are those that reward sustainable forest management, replanting degraded lands, sustainable agriculture, and increased use of wind and solar power to produce electricity. A second way to implement full-cost pricing is to increase taxes on pollution and wastes and reduce taxes on income and wealth. We discuss such *subsidy shifts* and *tax shifts* in Chapter 17.

People Are Increasingly Isolated from Nature

Today, more than half of the world's people and three out of four people in more-developed countries live in urban areas, and this shift from rural to urban living is continuing at a rapid pace. Urban environments and the increasing use of cell phones, computers, and other electronic devices are isolating people, especially children, from the natural world.

Some argue that this has led to a phenomenon called *nature deficit disorder*. People with this disorder may suffer from stress, anxiety, depression, and other problems. Research indicates that experiencing nature (see chapter-opening photo) can reduce stress, improve mental abilities, activate one's imagination and creativity, and lead to better health. The research also shows that when people are isolated from nature, they are less likely to act in ways that will lessen their harmful environmental impacts (**Concept 1.3A**), because they are not aware of their impacts.

Differing Environmental Worldviews

Another reason why environmental problems persist is that people differ over the nature and seriousness of the world's environmental problems, as well as how to solve them (**Concept 1.3B**). These differences arise mostly because of differing environmental worldviews. Your **environmental worldview** is your set of assumptions and values about how the natural world works and how you think you should interact with the environment.

Your environmental worldview is determined partly by your **environmental ethics**—what you believe about what is right and what is wrong in your behavior toward the environment. Here are some important *ethical questions* relating to the environment:

- Why should we care about the environment?

- Are humans the most important species on the planet or are they just another one of the earth's millions of life forms?

- Do people have an obligation to see that their activities do not cause the extinction of other species? If so, should people try to protect all species or only some? How does society decide which ones to protect?

- Does the current human generation have an ethical obligation to pass the natural world on to future generations in a condition as good as or better than what they inherited?

- Should every person be entitled to equal protection from environmental hazards regardless of race, gender, age, national origin, income, social class, or any other factor?

- Should individuals and society as a whole seek to live more sustainably, and, if so, how?

People with different environmental worldviews can take the same data, be logically consistent with it, and arrive at quite different answers to such questions. This happens because individuals start with different assumptions and moral, ethical, or religious beliefs. Environmental worldviews are discussed in detail in Chapter 17, but here is a brief introduction.

There are three major categories of environmental worldviews: human-centered, life-centered, and earth-centered. A **human-centered environmental worldview** sees the natural world primarily as a support system for human life. Two variations in this worldview are the *planetary management worldview* and the *stewardship worldview*. Both worldviews hold that humans are separate from and in charge of nature and that society should manage the earth for the benefit of humans. They also contend that if we degrade or deplete a natural resource or ecosystem service, we can use our technological ingenuity to find a substitute. The stewardship worldview adds that people have a responsibility to be caring and responsible managers, or *stewards*, of the planet for current and future human generations.

According to the **life-centered environmental worldview**, all species have value in fulfilling their ecological roles, regardless of their potential or actual use to society. Eventually, all species become extinct. However, most people with a life-centered worldview believe that we ought to avoid hastening the extinction of species through human activities because each species is a unique part of the biosphere that sustains all life.

According to the **earth-centered environmental worldview**, people are part of and dependent on nature, and the earth's natural capital exists for all species, not just for humans. According to this view, our economic success and the long-term survival of our cultures, our species, and many other species depend on learning how life on the earth has sustained itself for billions of years (Figure 1.2) and integrating such lessons from nature (**Core Case Study** and Science Focus 1.1) into the ways people think and act.

The Rise of Environmental Conservation and Protection in the United States

When European colonists arrived in North America in the early 1600s, Native American tribes had been living sustainably on the continent for thousands of years. The colonists viewed North America as a land with inexhaustible resources and a wilderness to be conquered and managed for human use. As settlers spread across the continent, they cleared forests to build settlements, plowed up grasslands to plant crops, and mined for gold, lead, and other minerals.

In 1864, George Perkins Marsh, a scientist and member of Congress from Vermont, questioned the idea that America's resources were inexhaustible. He used scientific

Some Biomimicry Principles

According to Janine Benyus (Individuals Matter 1.1), "The study of biomimicry reveals that life creates conditions conducive to life." She calls for us to evaluate each of the goods and services we produce and use by asking: Is it something nature would do? Does it help sustain life? Will it last?

Benyus recognizes three levels of biomimicry. The first involves mimicking the characteristics of species, such as bumps on a whale's fins or the wing and feather designs of birds, which are believed to have enhanced the long-term survival of such species. The second and deeper level involves mimicking the processes that species use to make shells, feathers, and other parts that benefit their long-term survival without using or producing toxins and without using the high-temperature or high-pressure processes we use in manufacturing. The third and deepest level involves mimicking the long-term survival strategies and beneficial environmental effects of natural ecosystems such as forests and coral reefs.

Since 1997, scientists, engineers, and others working in the field of biomimicry have identified several principles that have sustained life on the earth for billions of years. They have found that life

- runs on sunlight;
- does not waste energy;
- adapts to changing environmental conditions;
- depends on biodiversity for population control and adaptation;
- creates no waste because the matter outputs of one organism are resources for other organisms;
- does not pollute its own environment; and
- does not produce chemicals that cannot be recycled by the earth's chemical cycles.

By learning from nature and using such principles, innovative scientists, engineers, and business people are leading a *biomimicry revolution* by creating life-friendly goods and services and profitable businesses that could enrich and sustain humanity and its economies far into the future.

CRITICAL THINKING

Which, if any, of these principles of biomimicry do you follow in your life? How might your lifestyle change if you followed all of these principles? Would you resist or embrace doing this, and why?

---

studies and case studies to show how the rise and fall of past civilizations were linked to the use and misuse of their soils, water supplies, and other resources. Marsh was one of the founders of the conservation movement in the United States.

Early in the 20th century, this movement split into two factions that differed over how to use U.S. public lands owned jointly by all American citizens. The *preservationist view*, led by naturalist John Muir (Figure 1.15), wanted wilderness areas on some public lands to be left untouched so they would be preserved indefinitely. The *conservationist view*, promoted by President Theodore "Teddy" Roosevelt (Figure 1.16) and Gifford Pinchot, the first chief of the U.S. Forest Service, held that all public lands should be managed wisely and scientifically, primarily to provide resources for people.

Aldo Leopold (Figure 1.17)—wildlife manager, professor, writer, and conservationist—was trained in the conservationist view but shifted toward the preservationist view. He became a pioneer in forestry, soil conservation, wildlife ecology, and wilderness preservation. In 1935, he helped found the U.S. Wilderness Society. Through his writings, especially his 1949 book *A Sand County Almanac,* he laid the groundwork for the field of environmental ethics. He argued that the role of the human species should be to protect nature, not conquer it.

FIGURE 1.15 As leader of the preservationist movement, John Muir (1838–1914) called for setting aside some of the country's public lands as protected wilderness, an idea that was not enacted into law until 1964. Muir was also largely responsible for establishing Yosemite National Park in 1890, and in 1892, he founded the Sierra Club, which is to this day a political force working on behalf of the environment.

FIGURE 1.16 Effective protection of forests and wildlife on federal lands did not begin until Theodore "Teddy" Roosevelt (1858–1919) became president. His term of office, 1901–1909, has been called the country's *Golden Age of Conservation*. He established 36 national wildlife reserves and more than tripled the size of the national forest reserves.

U.S. Fish and Wildlife Service

FIGURE 1.18 Rachel Carson (1907–1964) alerted us to the harmful effects of the widespread use of pesticides. Many environmental historians mark Carson's wake-up call as the beginning of the modern environmental movement in the United States.

Photo courtesy of the Aldo Leopold Foundation, www.aldoleopold.org

FIGURE 1.17 Aldo Leopold (1887–1948) became a leading conservationist and his book, *A Sand County Almanac*, is considered an environmental classic that helped to inspire the modern conservation and environmental movements.

Later in the 20th century, the concept of resource conservation was broadened to include preservation of the *quality* of the planet's air, water, soil, and wildlife.

A prominent pioneer in that effort was biologist Rachel Carson (Figure 1.18), whose book *Silent Spring* was published in 1962. Carson's book documented the pollution of air, water, and wildlife from the widespread use of pesticides such as DDT. This influential book heightened public awareness of pollution problems and led to the regulation of several dangerous pesticides.

Between 1940 and 1970, the United States underwent rapid economic growth and industrialization. The by-products of industrialization were increased air and water pollution and large quantities of solid and hazardous wastes. Air pollution was so bad in many cities that drivers had to use their car headlights during the daytime. Thousands died each year from the harmful effects of air pollution. A stretch of the Cuyahoga River, running through Cleveland, Ohio, was so polluted with oil and other flammable pollutants that it caught fire several times. A devastating oil spill off the California coast occurred in 1969. Well-known wildlife species such as the American bald eagle, the grizzly bear, the whooping crane, and the peregrine falcon became endangered.

Growing publicity over these problems led the American public to demand government action. When the first Earth Day was held on April 20, 1970, some 20 million people in more than 2,000 U.S. communities and college and university campuses attended rallies to demand improvements in environmental quality. Earth Day and the resulting bottom-up political pressure it created led the U.S. government to establish the Environmental Protection Agency (EPA) in 1970 and to pass most of the U.S. environmental laws now in place during the 1970s, which became known as the *decade of the environment.*

Since 1970, many grassroots environmental organizations have sprung up to help deal with environmental threats. Interest in environmental issues has grown on many college and university campuses, resulting in the expansion of environmental science and environmental studies courses and programs. In addition, awareness of critical, complex, and largely invisible environmental issues has increased. These issues include threats to biodiversity, depletion of underground water supplies (aquifers), ocean warming, ocean acidification, atmospheric warming, and climate change.

In the 1980s, there was a backlash against environmental laws and regulations led by some corporate leaders, landowners, and state and local government officials who resented having to implement environmental laws and regulations with little or no federal funding. They contended that environmental laws were hindering economic growth and threatening private property rights and jobs. Since 1980, they have pushed to weaken or eliminate many environmental laws passed during the 1970s and to eliminate the EPA. Since the 1980s, environmental leaders and their supporters have had to spend much of their time and resources fighting to keep key environmental laws from being weakened or repealed.

Imagine that you win $1 million in a lottery. Suppose you invest this money (your capital) and earn 10% interest per year. If you live on just the interest income made by your capital, you will have a sustainable annual income of $100,000. You can spend $100,000 each year indefinitely and not deplete your capital. However, if you consistently spend more than your income, you will deplete your capital. Even if you spend just $10,000 more per year while still allowing the interest to accumulate, your money will be gone within 18 years.

This lesson is an old one: *Protect your capital and live on the income it provides.* Deplete or waste your capital and you will move from a sustainable to an unsustainable lifestyle.

The same lesson applies to using the earth's natural capital (Figure 1.3). This natural capital is a global trust fund of natural resources and ecosystem services that are available to people now and in the future and to all of the earth's other species. *Living sustainably* means living on **natural income**, which is the renewable resources such as plants, animals, soil, clean air, and clean water, provided by the earth's natural capital. By preserving and replenishing the earth's natural capital that supplies this natural income, people can reduce their ecological footprints and expand their beneficial environmental impact (**Concept 1.4**).

One of our goals in writing this book has been to provide a realistic vision of how we can live more sustainably. We base this vision not on immobilizing fear, gloom, and doom, but on providing education about how the earth sustains life and human economies and on energizing and realistic hope.

1.4 WHAT IS AN ENVIRONMENTALLY SUSTAINABLE SOCIETY?

CONCEPT 1.4 Living sustainably means living on the earth's natural income without depleting or degrading the natural capital that supplies it.

Protecting Natural Capital and Living on Its Income

An **environmentally sustainable society** protects natural capital and lives on its income. Such a society would meet the current and future basic resource needs of its people without compromising the ability of future generations to meet their basic resource needs. This is in keeping with the ethical **principle of sustainability**.

BIG IDEAS

- We can ensure a more sustainable future by relying more on energy from the sun and other renewable energy sources, protecting biodiversity through the preservation of natural capital, and avoiding the disruption of the earth's vital chemical cycles.

- A major goal for achieving a more sustainable future is full-cost pricing—the inclusion of harmful environmental and health costs in the market prices of goods and services.

- We will benefit ourselves and future generations if we commit ourselves to finding win–win solutions to environmental problems and to leaving the planet's life-support system in a condition as good as or better than what we inherited.

Learning from the Earth and Sustainability

Pecold/Shutterstock.com

We opened this chapter with a Core Case Study about learning from nature by understanding how the earth—the only truly sustainable system—has sustained an incredible diversity of life for 3.8 billion years despite drastic and long-lasting changes in the planet's environmental conditions. Part of the answer involves learning how to apply the six **principles of sustainability** (see Figures 1.2 and 1.7 and inside the back cover of this book) to the design and management of our economic and social systems, and to our individual lifestyles.

We can use such strategies to slow the rapidly expanding losses of biodiversity, to sharply reduce our production of wastes and pollution, to switch to more sustainable sources of energy, and to promote more sustainable forms of agriculture and uses of land and water. We can also use these principles to sharply reduce poverty and slow human population growth.

You are a member of the 21st century's *transition generation* that will play a major role in deciding whether humanity creates a more sustainable future or continues on an unsustainable path toward further environmental degradation and disruption. It is an incredibly exciting and challenging time to be alive as we struggle to develop a more sustainable relationship with the earth that keeps us alive and supports our economies.

Chapter Review

Core Case Study

1. What is **sustainability**? What is **biomimicry**? Explain why learning from the earth is a key to learning how to live more sustainably.

Section 1.1

2. What are the three key concepts for this section? Define **environment**. Distinguish among **environmental science, ecology**, and **environmentalism (environmental activism)**. Define species. What is an **ecosystem**? Define **solar energy, biodiversity, nutrients**, and **chemical cycling (nutrient cycling)**; and explain why they are important to life on the earth.

3. Define **natural capital**. Define **natural resources** and distinguish among **inexhaustible, renewable**, and **nonrenewable (exhaustible) resources**. What is a **sustainable yield**? Define **ecosystem services** and give two examples. Give three examples of how we are degrading natural capital. Explain

how finding solutions to environmental problems involves making trade-offs. Explain why individuals matter in dealing with the environmental problems we face. What are three economic, political, and ethical **principles of sustainability**? What is **full-cost pricing** and why is it important? Describe the role of Janine Benyus in promoting the important and growing field of biomimicry.

4. Define and distinguish between **more-developed countries** and **less-developed countries**; and give one example each of a high-income, middle-income, and low-income country.

Section 1.2

5. What are the two key concepts for this section? How have humans improved the quality of life for many people? How are humans living unsustainably? Define and give three examples of **environmental degradation (natural capital degradation)**. About

what percentage of the earth's natural or ecosystem services have been degraded by human activities? What is the tragedy of the commons? What are two ways to deal with this effect?

6. What is an **ecological footprint**? What is a **per capita ecological footprint**? What is **biocapacity**, or **biological capacity**, and what is an ecological deficit? Use the ecological footprint concept to explain how we are living unsustainably. What is the IPAT model for estimating our environmental impact? Explain how three major cultural changes taking place over the last 10,000 years have increased our overall environmental impact. What would a **sustainability revolution** involve?

Section 1.3

7. What are the two key concepts for this section? Identify five basic causes of the environmental problems that we face. What is **exponential growth**? What is the rule of 70? What is the current size of the human population? About how many people are added each year? How big is the world's population projected to be in 2050? Summarize the potentially harmful and beneficial environmental effects of affluence.

8. What is **poverty** and what are three of its harmful environmental and health effects? About what percentage of the world's people struggle to live on the equivalent of $1.90 a day? About what percentage have to live on $3.10 a day? How are poverty and population growth connected? List three major health problems faced by many of the poor.

Critical Thinking

1. Why is biomimicry so important? Find an example of something in nature that you think could be mimicked for some beneficial purpose. Explain that purpose and how biomimicry could apply.

2. What do you think are the three most environmentally unsustainable components of your lifestyle? List two ways in which you could apply each of the six **principles of sustainability** (Figures 1.2 and 1.7 and inside back cover of book) to making your lifestyle more environmentally sustainable.

3. For each of the following actions, state one or more of the three **scientific principles of sustainability** that are involved: **(a)** recycling aluminum cans; **(b)** using a rake instead of a leaf blower; **(c)** walking or bicycling to class instead of driving; **(d)** taking your own reusable bags to a store to carry your purchases home; and **(e)** volunteering to help restore a prairie or other degraded ecosystem.

9. Explain how excluding the harmful environmental and health costs of production from the prices of goods and services affects the environmental and health problems we face. What is the connection between government subsidies, resource use, and environmental degradation? What are two ways to include the harmful environmental and health costs of the goods and services in their market prices? Explain how a lack of knowledge about nature and the importance of natural capital, along with our increasing isolation from nature, can intensify the environmental problems we face. What is an **environmental worldview**? What is **environmental ethics**? What are five important ethical questions relating to the environment? Distinguish among the **human-centered**, **life-centered**, and **earth-centered environmental worldviews**. What are three levels of biomimicry? List seven key biomimicry principles. Summarize the rise of environmental conservation and protection in the United States.

Section 1.4

10. What is the key concept for this section? What is an **environmentally sustainable society**? What is **natural income** and how is it related to sustainability? What are this chapter's three big ideas?

_____

Note: Key terms are in bold type. Knowing the meanings of these terms will help you in the course you are taking.

4. Explain why you agree or disagree with the following propositions:
   a. Stabilizing population is not desirable because, without more consumers, economic growth would stop.
   b. The world will never run out of resources because we can use technology to find substitutes and to help us reduce resource waste.
   c. We can shrink our ecological footprints while creating beneficial environmental impacts.

5. Should nations with large ecological footprints reduce their footprints to decrease their harmful environmental impact and leave more resources for nations with smaller footprints and for future generations? Explain.

6. When you read that at least 19,000 children of ages 5 and younger die each day (13 per minute) from preventable malnutrition and infectious disease, what is your response? How would you deal with this problem?

7. Explain why you agree or disagree with each of the following statements: **(a)** humans are superior to other forms of life; **(b)** humans are in charge of the earth; **(c)** the value of other forms of life depends only on whether they are useful to humans; **(d)** all forms of life have a right to exist; **(e)** all economic growth is good; **(f)** nature has an almost unlimited storehouse of resources for human use; **(g)** technology can solve our environmental problems; **(h)** I don't have any obligation to future generations; and **(i)** I don't have any obligation to other forms of life.

8. What are the basic beliefs of your environmental worldview? Record your answer. At the end of this course, return to your answer to see if your environmental worldview has changed. Are the beliefs included in your environmental worldview consistent with the answers you gave for Question 7? Are your actions that affect the environment consistent with your environmental worldview? Explain.

Doing Environmental Science

Estimate your own ecological footprint by using one of the many estimator tools available on the Internet. Is your ecological footprint larger or smaller than you thought it would be, according to this estimate? Why do you think this is so? List three ways in which you could reduce your ecological footprint. Try one of them for a week, and write a report on this change. List three ways you could increase your beneficial environmental impact.

Global Environment Watch Exercise

Go to your MindTap course to access the GREENR database. Use the information on pages 9 and 10 in this chapter to choose one more-developed country and one less-developed country to compare their ecological footprints. Use the "World Map" link at the top of the page to access information about the countries you have chosen to research. Once on the country page, view the "Quick Facts" panel at the right. Click on the ecological footprint number to view a graph of both the ecological footprint and biocapacity of each country. Using those graphs, determine whether these countries are living sustainably or not. What would be some reasons for these trends?

Ecological Footprint Analysis

If the *ecological footprint per person* of a country or the world is larger than its *biological capacity per person* to replenish its renewable resources and absorb the resulting waste products and pollution, the country or the world is said to have an *ecological deficit*. If the reverse is true, the country or the world has an *ecological credit* or *reserve*. See Figure 1.11 for a map of the world's ecological debtor and creditor countries. Use the data to the right to calculate the ecological deficit or credit for the countries listed. (As an example, this value has been calculated and filled in for the world.)

1. Which three countries have the largest ecological deficits? For each of these countries, why do you think it has a deficit?

2. Rank the countries with ecological credits in order from highest to lowest credit. For each country, why do you think it has an ecological credit?

3. Rank all of the countries in order from the largest to the smallest per capita ecological footprint.

Place	Per Capita Ecological Footprint (hectares per person)	Per Capita Biological Capacity (hectares per person)	Ecological Credit (+) or Deficit (−) (hectares per person)
World	2.6	1.8	−0.8
United States	6.8	3.8	
Canada	7.0	13	
Mexico	2.4	1.3	
Brazil	2.5	9	
South Africa	2.5	1.2	
United Arab Emirates	8.0	0.7	
Israel	4.6	0.3	
Germany	4.3	1.9	
Russian Federation	4.4	6.6	
India	0.9	0.4	
China	0.5	0.8	
Australia	7.5	15	
Bangladesh	0.65	0.35	
Denmark	4.0	4.0	
Japan	3.7	0.7	
United Kingdom	4.0	1.1	

Compiled by the authors using data from World Wide Fund for Nature Living Planet Report 2014.

CHAPTER 2

Science, Matter, Energy,
and Systems

> Science is built up of facts, as a house is built of stones; but an accumulation of facts is no more a science than a heap of stones is a house.
>
> HENRI POINCARÉ

Key Questions

2.1 What do scientists do?

2.2 What is matter and what happens when it undergoes change?

2.3 What is energy and what happens when it undergoes change?

2.4 What are systems and how do they respond to change?

Researchers measuring a 3,200-year-old giant sequoia in California's Sequoia National Park.

Michael Nichols/National Geographic Creative

How Do Scientists Learn about Nature? Experimenting with a Forest

Suppose a logging company plans to cut down all of the trees on land behind your house. You are concerned and want to know about the possible harmful environmental effects.

One way to learn about such effects is to conduct a *controlled experiment*. To conduct such an experiment, scientists begin by identifying key *variables*, such as water loss and soil nutrient content that might change after the trees are cut down. Then they set up two groups. One is the *experimental group*, in which a chosen variable is changed in a known way. The other is the *control group*, in which the chosen variable is not changed. Then they compare the results from the two groups.

In 1963, botanist F. Herbert Bormann and forest ecologist Gene Likens began carrying out such a controlled experiment. Their goal was to compare the loss of water and soil nutrients from an area of uncut forest (the *control site*) with one

that had been stripped of its trees (the *experimental site*).

The researchers built V-shaped concrete dams across the creeks at the bottoms of several forested valleys in the Hubbard Brook Experimental Forest in New Hampshire (Figure 2.1). They designed the dams so that all surface water leaving each forested valley had to flow across a dam, where they could measure its volume and dissolved nutrient content.

First, the researchers measured the amounts of water and dissolved soil nutrients flowing from an undisturbed forested area in one of the valleys (the control site) (Figure 2.1, left). These measurements showed that an undisturbed mature forest is very efficient at storing water and retaining chemical nutrients in its soils.

Next, they set up an experimental forest area in a nearby valley (Figure 2.1, right). One winter, they cut down all the trees and shrubs in that valley, left them where they fell, and sprayed the area with

herbicides to prevent regrowth of vegetation. Then, for 3 years, the researchers compared the outflow of water and nutrients in this experimental site with those from the control site.

The scientists found that, with no plants to help absorb and retain water, the amount of water flowing out of the deforested valley increased by 30–40%. As this excess water ran over the ground rapidly, it eroded soil and removed dissolved nutrients, such as nitrates out of the topsoil. Overall, the loss of key soil nutrients from the experimental forest was six to eight times that in the nearby uncut control forest.

In this chapter, you will learn more about how scientists study nature and about the matter and energy that make up the world. You will also learn about the important difference between a scientific hypothesis and a scientific theory. And you will learn about three scientific laws that govern the changes in matter and energy.

FIGURE 2.1 This controlled field experiment measured the loss of water and soil nutrients from a forest due to deforestation. The forested valley (left) was the control site; the cutover valley (right) was the experimental site.

2.1 WHAT DO SCIENTISTS DO?

CONCEPT 2.1 Scientists collect data and develop hypotheses, theories, and laws about how nature works.

Scientists Collect Evidence to Learn How Nature Works

Science is a field of study focused on discovering how nature works and using that knowledge to describe what is likely to happen in nature. Science is based on the assumption that events in the natural world follow orderly cause-and-effect patterns. These patterns can be understood through careful *observations* (through the use of our senses and instruments that expand our senses), *measurements*, and *experiments* such as the one described in the Core Case Study.

Scientists use a variety of **scientific methods**, or practices, to advance knowledge and understanding of how the natural world works. Figure 2.2 summarizes these practices. In carrying out such research, scientists identify a problem for study, gather relevant data, propose a hypothesis that explains the data, gather data to test the hypothesis, and modify the hypothesis as needed. Within this process, scientists use many different methods to learn more about how nature works (**Concept 2.1**).

There is nothing mysterious about the scientific process. You use it all the time in making decisions. As the famous physicist Albert Einstein put it, "The whole of science is nothing more than a refinement of everyday thinking."

In this chapter's Core Case Study, Bormann and Likens used the scientific process to find out how clearing forested land can affect its ability to store water and retain soil nutrients. They designed an experiment to collect **data**, or information, to answer their question (Figure 2.1). They then proposed a **scientific hypothesis**—a testable explanation of the data they collected. Hypotheses can be written as "*if, then*" statements. Bormann and Likens came up with the following hypothesis to explain their data: *If* land is cleared of vegetation and exposed to rain and melting snow, *then* the land retains less water and loses soil nutrients. They tested this hypothesis for the soil nutrient nitrogen and then repeated their controlled experiment for phosphorus.

The experimenters wrote scientific articles describing their research and other scientists in their fields evaluated them. These reviews and further research by other scientists supported their results and hypothesis.

Another way to study nature is to develop a **model**, an approximate physical or mathematical representation that is used to understand or explain the behavior of complex natural systems. Data from the research carried out by Bormann, Likens, and others were fed into such models, which also supported their hypothesis.

A well-tested and widely accepted scientific hypothesis or a group of related hypotheses is called a **scientific theory**. It is one of the most important and certain results of science and based on a large body of evidence. The

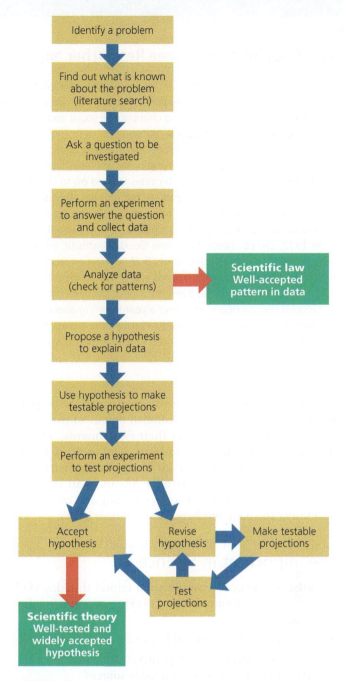

FIGURE 2.2 The general process that scientists use for discovering and testing ideas about how the natural world works.

research conducted by Bormann and Likens and other scientists led to the scientific theory that trees and other plants hold soil in place and help the soil retain water and nutrients that support the plants.

Scientists Are Curious and Skeptical

Good scientists are curious about how nature works (see Individuals Matter 2.1) but also skeptical about new data and hypotheses. They say, "Show me your evidence. Explain the reasoning behind the scientific ideas or hypotheses that you propose to explain your data."

Jane Goodall: Chimpanzee Researcher and Protector

JENS SCHLUETER/Getty Images

Jane Goodall is a scientist who studies animal behavior. She has a PhD from England's Cambridge University and is a National Geographic Explorer. At age 26, she began a decades-long career of studying chimpanzee social and family life in the Gombe Stream Game Reserve in Tanzania, Africa.

One of her major scientific discoveries was that chimpanzees make and use tools. She watched chimpanzees modifying twigs or blades of grass and then poking them into termite mounds. When the termites latched on to these primitive tools, the chimpanzees pulled them out and ate the termites. Goodall and several other scientists have also observed that chimpanzees, including captive chimpanzees, can learn simple sign language, do simple arithmetic, play computer games, develop relationships, and worry about and protect one another.

In 1977, she established the Jane Goodall Institute, an organization that works to preserve great ape populations and their habitats. In 1991, Goodall started *Roots & Shoots*, an environmental education program for youth with chapters in more than 130 countries. She has received many awards and prizes for her scientific contributions and conservation efforts. She has written 27 books for adults and children and has been involved with more than a dozen films about the lives and importance of chimpanzees.

Goodall spends nearly 300 days a year traveling and educating people throughout the world about chimpanzees, which are an endangered species, and the need to protect the environment. She says, "I can't slow down … If we're not raising new generations to be better stewards of the environment, what's the point?"

An important part of the scientific process is **peer review**, in which scientists publish details of the methods they used, the results of their experiments, and the reasoning behind their hypotheses, and other scientists working in the same field (their *peers*) evaluate what they published. Scientific knowledge advances in this self-correcting way, with scientists questioning and confirming or revising the data and hypotheses of their peers.

Critical Thinking and Creativity Are Important in Science

Scientists use logical reasoning and critical thinking skills (p. xxv) to learn about nature. Thinking critically involves three steps:

- Be skeptical about everything you read or hear.
- Evaluate evidence and hypotheses using inputs and opinions from a variety of reliable sources.
- Identify and evaluate your personal assumptions, biases, and beliefs and distinguish between facts and opinions before coming to a conclusion.

Logic and critical thinking are important tools in science, but imagination, creativity, and intuition are also vital. According to Albert Einstein, "There is no completely logical way to a new scientific idea."

Theories and Laws Are the Most Important and Certain Results of Science

We should never take a scientific theory lightly. It has been tested widely, is supported by extensive evidence, and is accepted by most scientists in a field or related fields of study as being a useful explanation of some phenomenon. So when you hear someone say, "Oh, that's just a theory," you will know that he or she does not have a clear understanding of what a scientific theory is and how it is one of key outcomes of science.

Another important and reliable outcome of science is a **scientific law**, or **law of nature**—a well-tested and widely accepted description of observations of what we find always happening in the same way in nature. An example is the *law of gravity*. After making many thousands of observations and measurements of objects falling from different heights, scientists developed the following scientific law: All objects fall to the earth's surface at predictable speeds. You can break a society's law, for example, by driving faster than the speed limit. However, you cannot break a scientific law such as the law of gravity.

Science Can Be Reliable, Unreliable, and Tentative

Reliable science consists of data, hypotheses, models, theories, and laws that are accepted by most of the scientists who are considered experts in the field under study. Scientific results and hypotheses that are presented as reliable without having undergone peer review, or are discarded as a result of peer review or additional research, are considered to be **unreliable science**.

Preliminary scientific results that have not undergone adequate testing and peer review are viewed as **tentative science**. Some of these results and hypotheses will be validated and classified as reliable. Others may be discredited and classified as unreliable. This is how scientific knowledge advances.

Science Has Limitations

Environmental science and science in general have several limitations. *First*, scientific research cannot prove that any scientific theory is absolutely true. This is because there is always some degree of uncertainty in measurements, observations, models, and the resulting hypotheses and theories. Instead, scientists try to establish that a particular scientific theory has a very high *probability* or *certainty* (typically 90–95%) of being useful for understanding some aspect of the natural world.

Many scientists do not use the word *proof* because it can falsely imply "absolute proof." For example, most scientists would not say: "Science has proven that cigarettes cause lung cancer." Instead, they might say: "Overwhelming evidence from thousands of studies indicates that people who smoke regularly for many years have a greatly increased chance of developing lung cancer."

CONSIDER THIS . . .

THINKING ABOUT Scientific Proof

Does the fact that science can never prove anything absolutely mean that its results are not valid or useful? Explain.

A *second* limitation of science is that scientists are not always free of bias about their own results and hypotheses. However, the high standards for evidence and peer review uncover or greatly reduce personal bias and any tendency to falsify scientific results.

A *third* limitation is that many systems in the natural world involve a huge number of variables with complex interactions. This makes it too difficult, costly, and time consuming to test one variable at a time in controlled experiments such as the one described in this chapter's **Core Case Study**. To deal with this, scientists develop *mathematical models* that can take into account the interactions of many variables, and they run the models on high-speed computers.

A *fourth* limitation of science involves the use of statistical tools. For example, there is no way to measure accurately the number of metric tons of soil eroded annually worldwide. Instead, scientists use statistical sampling and mathematical methods to estimate such numbers.

Despite these limitations, science is the most useful way of learning about how nature works and projecting how it might behave in the future.

2.2 WHAT IS MATTER AND WHAT HAPPENS WHEN IT UNDERGOES CHANGE?

CONCEPT 2.2A Matter consists of elements and compounds, which, in turn, are made up of atoms, ions, or molecules.

CONCEPT 2.2B Whenever matter undergoes a physical or chemical change, no atoms are created or destroyed (*law of conservation of matter*).

Matter Consists of Elements and Compounds

Matter is anything that has mass and takes up space. Matter can exist in one of three *physical states*—solid, liquid, and gas—at a given pressure and temperature and in two *chemical forms*—elements and compounds (**Concept 2.2A**).

An **element** such as gold or mercury (Figure 2.3) is a fundamental type of matter with a unique set of properties and that cannot be broken down into simpler substances by chemical means. Chemists refer to each element with a unique one- or two-letter symbol such as C for carbon and Au for gold. Scientists have arranged the known elements based on their chemical behavior on a chart is called the **periodic table of elements**. Table 2.1 lists the elements and their symbols that you need to know to understand the material in this book.

Some matter is composed of one element, such as carbon (C) and oxygen gas (O_2). However, most matter consists of **compounds**, which are combinations of two or more different elements held together in fixed proportions.

FIGURE 2.3 Mercury (left) and gold (right) are chemical elements. Each has a unique set of properties and cannot be broken down into simpler substances.

TABLE 2.1 Chemical Elements Used in This Book

Element	Symbol	Element	Symbol
Arsenic	As	Lead	Pb
Bromine	Br	Lithium	Li
Calcium	Ca	Mercury	Hg
Carbon	C	Nitrogen	N
Copper	Cu	Phosphorus	P
Chlorine	Cl	Sodium	Na
Fluorine	F	Sulfur	S
Gold	Au	Uranium	U

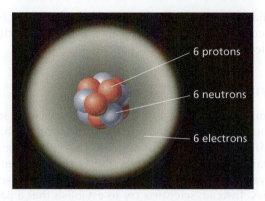

FIGURE 2.4 Simplified model of a carbon-12 atom. It consists of a nucleus containing six protons, each with a positive electrical charge, and six neutrons with no electrical charge. Six negatively charged electrons are found outside its nucleus.

For example, water (H_2O) is a compound containing the elements hydrogen and oxygen, and sodium chloride (NaCl) contains the elements sodium and chlorine.

Elements and Compounds Are Made of Atoms, Molecules, and Ions

The basic building block of matter is an **atom**—the smallest unit of matter into which an element can be divided and still have its distinctive chemical properties. The idea that all elements are made up of atoms is called the **atomic theory** and is the most widely accepted scientific theory in chemistry.

Atoms are incredibly small. More than 3 million hydrogen atoms could sit side by side on the period at the end of this sentence. If you could view atoms with a super microscope, you would find that each different type of atom contains a certain number of three types of *subatomic particles:* **neutrons**, with no electrical charge; **protons**, each with a positive electrical charge (+); and **electrons**, each with a negative electrical charge (−).

Each atom has an extremely small center called the **nucleus**, which contains one or more protons and, in most cases, one or more neutrons. Outside of the nucleus, we find one or more electrons in rapid motion (Figure 2.4).

Each element has a unique **atomic number** equal to the number of protons in the nucleus of its atom. Carbon (C), with 6 protons in its nucleus, has an atomic number of 6, whereas uranium (U) has 92 protons in its nucleus and thus has an atomic number of 92.

Because electrons have so little mass compared to protons and neutrons, most of an atom's mass is concentrated in its nucleus. The mass of an atom is described by its **mass number**, the total number of neutrons and protons in its nucleus. For example, a carbon atom with 6 protons and 6 neutrons in its nucleus (Figure 2.4) has a mass number of 12 (6 + 6 =12) and a uranium atom with 92 protons and 143 neutrons in its nucleus has a mass number of 235 (92 + 143 = 235).

Each atom of a particular element has the same number of protons in its nucleus. However, the nuclei of atoms of a particular element can vary in the number of neutrons they contain and, therefore, in their mass numbers. The forms of an element having the same atomic number but different mass numbers are called **isotopes** of that element. Scientists identify isotopes by attaching their mass numbers to the name or symbol of the element. For example, the three most common isotopes of carbon are carbon-12 (with six protons and six neutrons, Figure 2.4), carbon-13 (with six protons and seven neutrons), and carbon-14 (with six protons and eight neutrons).

A second building block of matter is a **molecule**, a combination of two or more atoms of the same or different elements held together by forces known as *chemical bonds*. Molecules are the basic building blocks of many compounds. Examples are water (H_2O) and hydrogen gas (H_2).

A third building block of some types of matter is an **ion**. It is an atom or a group of atoms with one or more net positive (+) or negative (−) electrical charges resulting from the loss or gain of negatively charged electrons. Chemists use a superscript after the symbol of an ion to indicate the number of positive or negative electrical charges. The hydrogen ion (H^+) and sodium ion (Na^+) are examples of positive ions. Examples of negative ions are the hydroxide ion (OH^-), the chloride ion (Cl^-), and the nitrate ion (NO_3^-), a nutrient that is essential for plant growth. In this chapter's **Core Case Study**, Bormann and Likens measured the loss of nitrate ions (Figure 2.5) from the deforested area (Figure 2.1, right) in their controlled experiment. Table 2.2 lists the chemical ions used in this book.

Ions are important for measuring a substance's **acidity** in a water solution. Acidity is measure of the comparative amounts of hydrogen ions (H^+) and hydroxide ions (OH^-) in a particular volume of a water solution. Scientists use **pH** as a measure of acidity. Pure water (not tap water or rainwater) has an equal number of H^+ and OH^- ions. It is called a *neutral solution* and has a pH of 7. A solution

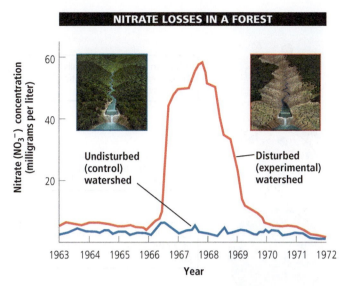

NITRATE LOSSES IN A FOREST

Undisturbed (control) watershed

Disturbed (experimental) watershed

FIGURE 2.5 Loss of nitrate ions (NO_3^-) from a deforested watershed in the Hubbard Brook Experimental Forest (**Core Case Study**, Figure 2.1, right). *Data analysis*: By what percent did the nitrate concentration increase between 1965 and the time of peak concentration between 1967 and 1968.

Compiled by the authors using data from F. H. Bormann and Gene Likens

TABLE 2.2 Chemical Ions Used in This Book

Positive Ion	Symbol	Negative Ion	Symbol
Hydrogen ion	H^+	Chloride ion	Cl^-
Sodium ion	Na^+	Hydroxide ion	OH^-
Calcium ion	Ca^{2+}	Nitrate ion	NO_3^-
Aluminum ion	Al^{3+}	Carbonate ion	CO_3^{2-}
Ammonium ion	NH_4^+	Sulfate ion	SO_4^{2-}
		Phosphate ion	PO_4^{3-}

TABLE 2.3 Compounds Used in This Book

Compound	Formula	Compound	Formula
Sodium chloride	NaCl	Methane	CH_4
Sodium hydroxide	NaOH	Glucose	$C_6H_{12}O_6$
Carbon monoxide	CO	Water	H_2O
Carbon dioxide	CO_2	Hydrogen sulfide	H_2S
Nitric oxide	NO	Sulfur dioxide	SO_2
Nitrogen dioxide	NO_2	Sulfuric acid	H_2SO_4
Nitrous oxide	N_2O	Ammonia	NH_3
Nitric acid	HNO_3	Calcium carbonate	$CaCO_3$

that has more hydrogen ions than hydroxide ions is an *acidic solution* and has a pH less than 7. A *basic solution* has more hydroxide ions than hydrogen ions and has a pH greater than 7. Each single unit change on the pH scale represents a tenfold increase or decrease in the concentration of hydrogen ions in a liter of solution. For example, an acidic solution with a pH of 3 is 10 times more acidic than a solution with a pH of 4.

Chemists use a **chemical formula** to show the number of each type of atom or ion in a compound. The formula contains the symbol for each element present and uses subscripts to show the number of atoms or ions of each element in the compound's basic structural unit. Examples of compounds and their formulas encountered in this book are sodium chloride (NaCl) and water (H_2O, read as "H-two-O"). These and other compounds important to the study of environmental science in this textbook are listed in Table 2.3.

Organic Compounds Are the Chemicals of Life

Plastics, table sugar, vitamins, aspirin, penicillin, and most of the chemicals in your body are called **organic compounds**, which contain at least two carbon atoms combined with atoms of one or more other elements. The exception is methane (CH_4), with only one carbon atom.

The millions of known organic (carbon-based) compounds include *hydrocarbons*—compounds of carbon and hydrogen atoms—such as methane (CH_4), the main component of natural gas. They also include *simple carbohydrates (simple sugars)* that contain carbon, hydrogen, and oxygen atoms. An example is glucose ($C_6H_{12}O_6$), which most plants and animals break down in their cells to obtain energy.

Several types of larger and more complex organic compounds essential to life are called *polymers*. They form when a number of simple organic molecules *(monomers)* are linked together by chemical bonds, somewhat like rail cars in a freight train. Three major types of organic polymers are *carbohydrates* such as glucose, *proteins*, which play many vital roles in the body, and *nucleic acids* such as RNA and DNA (Figure 2.6), formed by monomers called *nucleotides* and critical for reproduction.

Matter Comes to Life through Cells, Genes, and Chromosomes

All organisms are composed of one or more **cells**—the fundamental structural and functional units of life. The idea that all living things are composed of cells is called the **cell theory**. It is the most widely accepted scientific theory in biology.

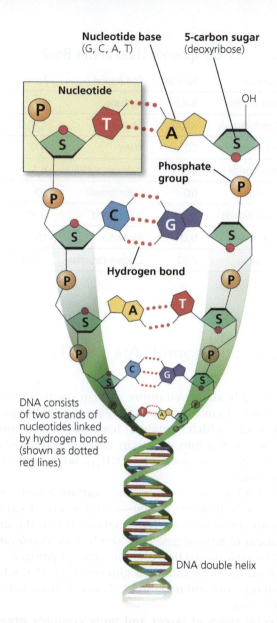

Nucleotide base
(G, C, A, T)

5-carbon sugar
(deoxyribose)

Nucleotide

Phosphate group

Hydrogen bond

DNA consists of two strands of nucleotides linked by hydrogen bonds (shown as dotted red lines)

DNA double helix

FIGURE 2.6 Portion of a DNA molecule, which is composed of spiral (helical) strands of nucleotides. Each nucleotide contains three units: phosphate (P), a sugar (S), which is deoxyribose), and one of four different nucleotide bases represented by the letters A, G, C, and T.

Within some DNA molecules (Figure 2.6) are certain sequences of nucleotides called **genes**. Each of these segments of DNA contains instructions, or codes, called *genetic information*, for making specific proteins. *Each coded unit of information leads to* a **trait**, or characteristic, that passes from parents to offspring during reproduction in an animal or plant.

Thousands of genes make up a single **chromosome**, a double helix DNA molecule wrapped around one or more proteins. Genetic information coded in your chromosomal DNA is what makes you different from an oak leaf, a mosquito, and your parents. Figure 2.7 shows the relationships of genetic material to cells.

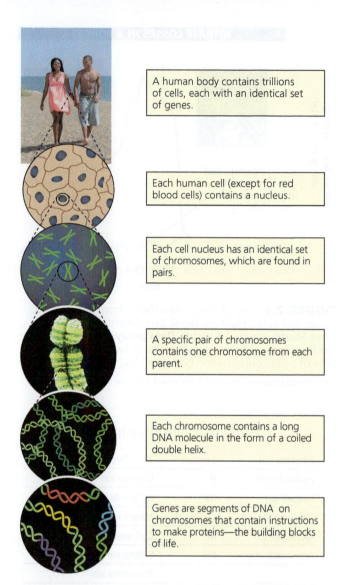

A human body contains trillions of cells, each with an identical set of genes.

Each human cell (except for red blood cells) contains a nucleus.

Each cell nucleus has an identical set of chromosomes, which are found in pairs.

A specific pair of chromosomes contains one chromosome from each parent.

Each chromosome contains a long DNA molecule in the form of a coiled double helix.

Genes are segments of DNA on chromosomes that contain instructions to make proteins—the building blocks of life.

FIGURE 2.7 The relationships among cells, nuclei, chromosomes, DNA, and genes.

Flashon Studio/Shutterstock.com

Physical and Chemical Changes

Matter can undergo physical and chemical changes. When matter undergoes a **physical change**, there is no change in its chemical composition. A piece of aluminum foil cut into small pieces is still aluminum foil. When solid water (ice) melts and when liquid water boils, the resulting liquid water and water vapor remain as H_2O molecules.

When a **chemical change**, or **chemical reaction**, takes place, there is a change in the chemical composition of the substances involved. Chemists use a *chemical equation* to show how chemicals are rearranged in a chemical reaction. For example, coal is made up almost entirely of the element carbon (C). When coal is burned completely in a power plant, the solid carbon in the coal combines with oxygen gas (O_2) from the atmosphere to form the gaseous compound carbon dioxide (CO_2). Chemists use the following shorthand chemical equation to represent this chemical reaction:

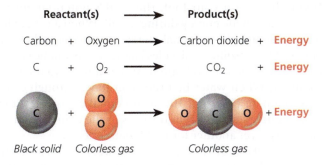

Reactant(s) ——→ Product(s)

Carbon + Oxygen ——→ Carbon dioxide + **Energy**

C + O_2 ——→ CO_2 + **Energy**

Black solid Colorless gas Colorless gas + **Energy**

Law of Conservation of Matter

Elements and compounds can change from one physical or chemical form to another. However, atoms are never created or destroyed in any physical or chemical change. Instead, atoms, ions, or molecules can only be rearranged into different spatial patterns (physical changes) or chemical combinations (chemical changes). This finding, based on many thousands of measurements, describes an unbreakable scientific law known as the **law of conservation of matter**: Whenever matter undergoes a physical or chemical change, no atoms are created or destroyed (**Concept 2.2B**).

Chemists obey this scientific law by balancing the equation for a chemical reaction to account for the fact that no atoms are created or destroyed. Passing electricity through water (H_2O) can break it down into hydrogen (H_2) and oxygen (O_2), as represented by the following equation:

$$H_2O \rightarrow H_2 + O_2$$

2 H atoms   2 H atoms   2 O atoms

1 O atom

This equation is unbalanced because one atom of oxygen is on the left side of the equation but two oxygen atoms are on the right side. We cannot change the subscripts of any of the formulas to balance this equation because that would change the arrangements of the atoms, leading to different substances. Instead, we must use different numbers of the molecules involved to balance the equation. For example, we could use two water molecules:

$$2 H_2O \rightarrow H_2 + O_2$$

4 H atoms   2 H atoms   2 O atoms

2 O atoms

This equation is still unbalanced. Although the numbers of oxygen atoms on both sides of the equation are now equal, the numbers of hydrogen atoms are not. We can correct this problem by recognizing that the reaction must produce two hydrogen molecules:

$$2 H_2O \rightarrow 2 H_2 + O_2$$

4 H atoms   4 H atoms   2 O atoms

2 O atoms

Now the equation is balanced, and the law of conservation of matter has been observed.

2.3 WHAT IS ENERGY AND WHAT HAPPENS WHEN IT UNDERGOES CHANGE?

CONCEPT 2.3A Whenever energy is converted from one form to another in a physical or chemical change, no energy is created or destroyed (*first law of thermodynamics*).

CONCEPT 2.3B Whenever energy is converted from one form to another in a physical or chemical change, we end up with lower-quality or less-usable energy than we started with (*second law of thermodynamics*).

Energy Comes in Many Forms

Scientists define **energy** as the capacity to do work or to transfer heat. Suppose you pick this book up off the floor and put it on your desktop. In doing this, you have to do *work*, or use a certain amount of muscular force to move the book from one place to another. In scientific terms, work is done when any object is moved a certain distance (work = force × distance). When you touch a hot object such as a stove, *heat* (or thermal energy) flows from the stove to your finger.

There are two major types of energy: *moving energy* (called kinetic energy) and *stored energy* (called potential energy). Matter in motion has **kinetic energy**, or energy associated with motion. Examples are flowing water, a car speeding down the highway, electricity (electrons flowing through a wire or other conducting material), and wind (a mass of moving air that we can use to produce electricity, as shown in Figure 2.8).

Another form of kinetic energy is **heat**, or **thermal energy**. It is the total kinetic energy of all moving atoms, ions, or molecules in an object, a body of water, or a volume of gas such as the atmosphere. The hotter an object is, the faster the motion of the atoms, ions, or molecules inside that object. **Temperature** is a measure of the average heat or thermal energy of the atoms, ions, or molecules in a sample of matter. When two objects at different temperatures make contact, heat flows from the warmer object to the cooler object. You learned this the first time you touched a hot stove.

In another form of kinetic energy called **electromagnetic radiation**, energy travels from one place to another in the form of *waves* formed from changes in electrical and magnetic fields. There are many different forms of electromagnetic radiation (Figure 2.9). Each form has a different *wavelength*—the distance between successive peaks or troughs in the wave—and *energy content*. Those with short wavelengths have more energy than do those with longer wavelengths.

The other major type of energy is **potential energy**, which is stored and potentially available for use. Examples of this type of energy include a rock held in your hand, the water in a reservoir behind a dam, and the chemical energy stored in the carbon atoms of coal or in the molecules of any food you eat.

FIGURE 2.8 Kinetic energy, created by the gaseous molecules in a mass of moving air, turns the blades of these wind turbines. The turbines then convert this kinetic energy to electrical energy, which is another form of kinetic energy.

You can change potential energy to kinetic energy. If you hold this book in your hand, it has potential energy. If you drop it on your foot, the book's potential energy changes to kinetic energy during its fall. When a car engine burns gasoline, the potential energy stored in the chemical bonds of the gasoline molecules changes into kinetic energy that propels the car, and into heat that flows into the environment. When water in a reservoir flows through channels in a dam (Figure 2.10), its potential energy becomes kinetic energy used to spin turbines in the dam to produce electricity—yet another form of kinetic energy.

About 99% of the energy that keeps us warm and supports the plants that we and other organisms eat is electromagnetic radiation that comes from the sun at no cost to us, in keeping with the solar energy **principle of sustainability** (see inside back cover). Without this essentially inexhaustible solar energy, the earth would be frozen and life as we know it would not exist.

Commercial energy—energy that is sold in the marketplace—makes up the remaining 1% of the energy we use to supplement the earth's direct input of solar energy. About 90% of the commercial energy used in the world and 90% of that used in the United States comes from the burning of nonrenewable *fossil fuels*—oil, coal, and natural gas. They are called fossil fuels because they were formed over hundreds of thousands to millions of years as layers of the decaying remains of ancient plants and animals were exposed to intense heat and pressure within the earth's crust.

99% Percentage of all energy used by all life that comes directly from the sun

Energy Varies in Its Quality

Some types of energy are more useful than others. **Energy quality** is a measure of the capacity of energy to do useful work. **High-quality energy** is concentrated energy that has a high capacity to do useful work. Examples are

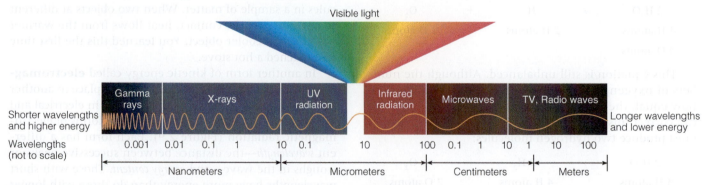

FIGURE 2.9 The *electromagnetic spectrum* consists of a range of electromagnetic waves, which differ in wavelength (the distance between successive peaks or troughs) and energy content.

FIGURE 2.10 The water stored in this reservoir behind a dam has potential energy, which becomes kinetic energy when the water flows through channels built into the dam where it spins a turbine and produces electricity—another form of kinetic energy.

high-temperature heat, concentrated sunlight, high-speed wind, and the energy released when we burn wood, gasoline, natural gas, or coal.

By contrast, **low-quality energy** is so dispersed that it has little capacity to do useful work. For example, the enormous number of moving molecules in the atmosphere or in an ocean together have a huge amount of energy. However, it is low-quality energy because it is greatly dispersed and has such a low temperature, that we cannot use it to do work.

Energy Changes Are Governed by Two Scientific Laws

From millions of observations and measurements of energy changing from one form to another in physical and chemical changes, scientists have summarized their results

in the **first law of thermodynamics**, also known as the **law of conservation of energy**. According to this scientific law, whenever energy is converted from one form to another in a physical or chemical change, no energy is created or destroyed (**Concept 2.3A**).

No matter how hard we try or how clever we are, we cannot get more energy out of a physical or chemical change than we put in. This scientific law is one of nature's basic rules that we cannot violate.

Because energy cannot be created or destroyed, only converted from one form to another, you might think we will never run out of energy. Think again. If you fill a car's tank with gasoline and drive around all day or run your cell phone battery down, something has been lost. What is it? The answer is *energy quality*, the amount of energy available for performing useful work.

Thousands of experiments have shown that whenever energy is converted from one form to another in a physical or chemical change, we end up with lower-quality or less-usable energy than we started with (**Concept 2.3B**). This is a statement of the **second law of thermodynamics**. The low-quality energy usually takes the form of heat that flows into the environment. The random motion of air or water molecules further disperses this heat, decreasing its temperature to the point where its energy quality is too low to do much useful work.

In other words, *when energy is changed from one form to another, it always goes from a more useful to a less useful form.* This means we cannot recycle or reuse high-quality energy to perform useful work. Once the high-quality energy in a serving of food, a tank of gasoline, or a chunk of coal is released, it is degraded to low-quality heat and dispersed into the environment. The second law of thermodynamics is another basic rule of nature that we cannot violate.

information is fed back into a system as input and leads to changes in that system (**Concept 2.4**). A **positive feedback loop** causes a system to change in the same direction. For example, when researchers removed the vegetation from a stream valley in the Hubbard Brook Experimental Forest (**Core Case Study**), they found that flowing water from precipitation caused erosion and losses of nutrients, which caused more vegetation to die (Figure 2.12). With even less vegetation to hold soil

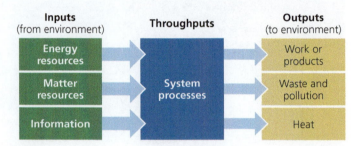

Inputs (from environment)	Throughputs	Outputs (to environment)
Energy resources		Work or products
Matter resources	System processes	Waste and pollution
Information		Heat

FIGURE 2.11 Simplified model of a system.

2.4 WHAT ARE SYSTEMS AND HOW DO THEY RESPOND TO CHANGE?

CONCEPT 2.4 Systems have inputs, flows, and outputs of matter and energy, and feedback can affect their behavior.

Systems and Feedback Loops

A **system** is any set of components that function and interact in some regular way. Examples are a cell, the human body, a forest, an economy, a car, and the earth.

Most systems have three key components: **inputs** of matter, energy, and information from the environment, **flows** or **throughputs** of matter, energy, and information within the system, and **outputs** of products, wastes, and degraded energy (usually heat) to the environment (Figure 2.11) (**Concept 2.4**). A system can become unsustainable if the throughputs are greater than the ability of the environment to provide the required inputs and to absorb or dilute the system's outputs of matter and energy.

Most systems are affected by **feedback**, any process that increases (positive feedback) or decreases (negative feedback) a change in a system. Such a process, called a **feedback loop**, occurs when an output of matter, energy, or

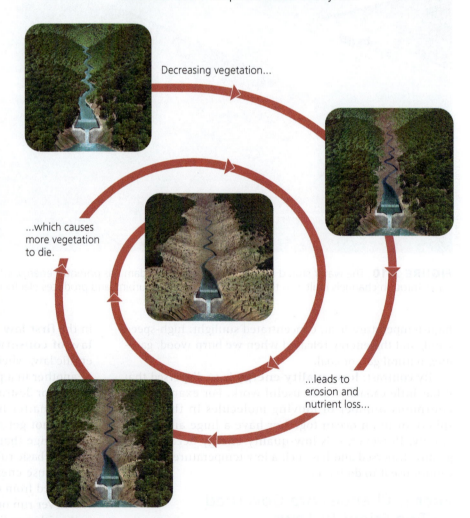

FIGURE 2.12 A *positive feedback loop.* Decreasing vegetation in a valley causes increasing erosion and nutrient losses that in turn cause more vegetation to die, resulting in more erosion and nutrient losses. ***Question***: Can you think of another positive feedback loop in nature?

Decreasing vegetation...

...leads to erosion and nutrient loss...

...which causes more vegetation to die.

in place, flowing water caused even more erosion and nutrient loss, which caused even more plants to die.

When a natural system becomes locked into a positive feedback loop, it can reach an **ecological tipping point**, beyond which, the system can change so drastically that it can suffer severe degradation or collapse. Reaching and exceeding a tipping point is somewhat like stretching a rubber band. We can stretch it to several times its original length, but at some point, we reach an irreversible tipping point where the rubber band breaks. Many types of ecological tipping points will be discussed throughout this book.

A **negative**, or **corrective, feedback loop** causes a system to change in the opposite direction. An example of a negative feedback loop is a thermostat, a device that measures the temperature of a house and uses this information to turn its heating or cooling system on or off to achieve a desired temperature (Figure 2.13).

Another example of a negative feedback loop is the recycling of aluminum. An aluminum can is an output of mining and manufacturing systems that require large inputs of energy and matter and produce pollution and solid waste. When we recycle the output (the used can), it becomes an input that reduces the need for mining aluminum and manufacturing the can. This reduces the energy and matter inputs and the harmful environmental effects. This is an application of the chemical cycling **principle of sustainability**.

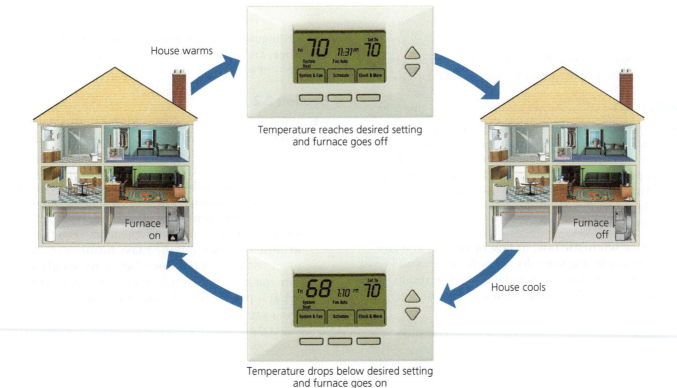

House warms

Temperature reaches desired setting and furnace goes off

House cools

Temperature drops below desired setting and furnace goes on

Furnace on

Furnace off

FIGURE 2.13 A *negative feedback loop*. When a house being heated by a furnace gets to a certain temperature, its thermostat is set to turn off the furnace, and the house begins to cool instead of continuing to get warmer. When the house temperature drops below the set point, this information is fed back to turn the furnace on until the desired temperature is reached again.

The Hubbard Brook Forest Experiment and Sustainability

In the controlled experiment discussed in this chapter's **Core Case Study**, the clearing of a mature forest degraded some of its natural capital (see Figure 1.3, p. 7, and photo at right). Specifically, the loss of trees and vegetation altered the ability of the forest to retain and recycle water and other critical plant nutrients—a crucial ecological function based on the chemical cycling **principle of sustainability**.

This clearing of vegetation also violated the solar energy and biodiversity **principles of sustainability**. For example, the cleared forest lost most of its plants that had used solar energy to produce food for the forest's animals, which supplied nutrients to the soil when they died. Thus, the forest lost many of its key nutrients that would normally have been recycled. It also lost much of its life-sustaining biodiversity.

Many of the results of environmental science are based on this sort of experimentation. Throughout this textbook, we explore other examples of how scientists learn about nature. We will see how we can use these results to help us understand how the earth works, how our actions affect the environment, and how we can solve some of our environmental problems.

steve estvanik/Shutterstock.com

Chapter Review

Core Case Study

1. Describe the controlled scientific experiment carried out in the Hubbard Brook Experimental Forest.

Section 2.1

2. What is the key concept for this section? What is **science**? List the steps involved in a scientific method. What are **data**? Define **scientific hypothesis**. What is a **model**? What is a **scientific theory**? What is **peer review** and why is it important? Summarize scientist Jane Goodall's achievements.

3. Define **scientific law**, or **law of nature**, and give an example. Explain why scientific theories and laws are the most important and most certain results of science and why people often use the term *theory* incorrectly.

4. Distinguish among **reliable science, unreliable science**, and **tentative science**. What are four limitations of science?

Section 2.2

5. What are the two key concepts for this section? What is **matter**, and what are its three physical states? Distinguish between an **element** and a **compound** and give an example of each. What is the **periodic table of elements**? Define **atom** and state the **atomic theory**. Distinguish among **protons, neutrons**, and **electrons**. What is the **nucleus** of an atom? Distinguish between the **atomic number** and the **mass number** of an element. What is an **isotope**? Define **molecule** and **ion** and give an example of each. What is **acidity**? What is **pH**? Define **chemical formula** and give two examples.

6. Define and give two examples of an **organic compound**. What are three types of organic polymers that are important to life? Define **cell** and state the **cell theory**. What is **DNA**? Define **gene, trait**, and **chromosome**. Define and distinguish between a **physical change** and a **chemical change (chemical reaction)** in matter and give an example of each. What is the **law of conservation of matter**?

Section 2.3

7. What are the two key concepts for this section? What is **energy**? Define and give two examples of **kinetic energy**. What is **heat (thermal energy)**? Define **temperature**. Define and give two examples of **electromagnetic radiation.** Define and give two examples of **potential energy.** What is commercial energy and what percentage of it is provided by fossil fuels? What percentage of all energy comes from the sun?

8. What is **energy quality**? Distinguish between **high-quality energy** and **low-quality energy** and give an example of each. What is the **first**

law of thermodynamics (law of conservation of energy)** and why is it important? What is the **second law of thermodynamics** and why is it important? Explain why the second law means that we can never recycle or reuse high-quality energy.

Section 2.4

9. What is the key concept for this section? Define and give an example of a **system.** Distinguish among the **inputs, flows (throughputs)**, and **outputs** of a system. What is **feedback** and what is a **feedback loop**? Distinguish between a **positive feedback loop** and a **negative (corrective) feedback loop**, and give an example of each. What is an **ecological tipping point**?

10. What are this chapter's *three big ideas*? Explain how the Hubbard Brook Experimental Forest controlled experiments illustrated the three **scientific principles of sustainability.**

Note: Key terms are in bold type.

Critical Thinking

1. What ecological lesson can we learn from the controlled experiment described in the **Core Case Study** that opened this chapter?

2. Suppose you observe that all of the fish in a pond have disappeared. How might you use the scientific process described in the **Core Case Study** and in Figure 2.2 to determine the cause of this fish kill?

3. Respond to the following statements:
   a. Scientists have not absolutely proven that anyone has ever died from smoking cigarettes.
   b. The *natural greenhouse effect*—the warming effect of certain gases such as water vapor and carbon dioxide in the lower atmosphere—is not a reliable idea because it is just a scientific theory.

4. A tree grows and increases its mass. Explain why this is not a violation of the law of conservation of matter.

5. If there is no "away" where organisms can get rid of their wastes due to the law of conservation of matter, why is the world not filled with waste matter?

6. Suppose someone wants you to invest money in an automobile engine, claiming that it will produce more energy than is found in the fuel used to run it. What would be your response? Explain.

7. Use the second law of thermodynamics to explain why we can use oil only once as a fuel, or in other words, why we cannot recycle its high-quality energy.

8. Imagine that for one day **(a)** you have the power to revoke the law of conservation of matter, and **(b)** you have the power to violate the first law of thermodynamics. For each of these scenarios, list three ways in which you would use your new power. Explain your choices.

Doing Environmental Science

Find a newspaper or magazine article or a report on the Web that attempts to discredit a scientific hypothesis because it has not been proven, or a report of a new scientific hypothesis that has the potential to be controversial. Analyze the piece by doing the following: **(1)** determine its source (author or organization); **(2)** detect an alternative hypothesis, if any, that is offered by the author; **(3)** determine the primary objective of the author (for example, to debunk the original hypothesis, to state an alternative hypothesis, or to raise new questions); **(4)** summarize the evidence given by the authors for their position; and **(5)** compare the authors' evidence with the evidence for the original hypothesis. Write a report summarizing your analysis and compare it with those of your classmates.

Global Environmental Watch Exercise

Go to your MindTap course to access the GREENR database. Starting on the home page, under "Browse Issues and Topics", click on *Resource Management*, then select *Forests and Deforestation*. Browse the articles listed there and find one that involves a controlled experiment or some other form of scientific research in a forest.

Determine what the hypothesis was that the researchers were testing. Summarize their research methods and any conclusions that were reached. Was the research similar in any way to that described in the **Core Case Study**? Explain.

Data Analysis

Consider the accompanying graph, which compares the losses of calcium from the experimental and control sites in the Hubbard Brook Experimental Forest (**Core Case Study**). Note that this figure is very similar to Figure 2.5, which compares loss of nitrates from the two sites. After studying this graph, answer the following questions.

1. In what year did the calcium loss from the experimental site begin a sharp increase? In what year did it peak? In what year did it again level off?

2. In what year were the calcium losses from the two sites closest together? In the span of time between 1963 and 1972, did they ever get that close again?

3. Does this graph support the hypothesis that cutting the trees from a forested area causes the area to lose nutrients more quickly than leaving the trees in place? Explain.

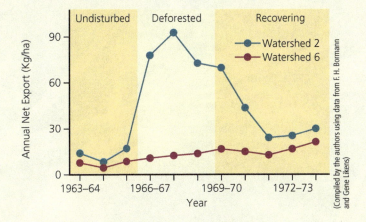

CENGAGE brain To access course materials, including Aplia homework, please visit www.cengagebrain.com.

WWW.CENGAGEBRAIN.COM   43

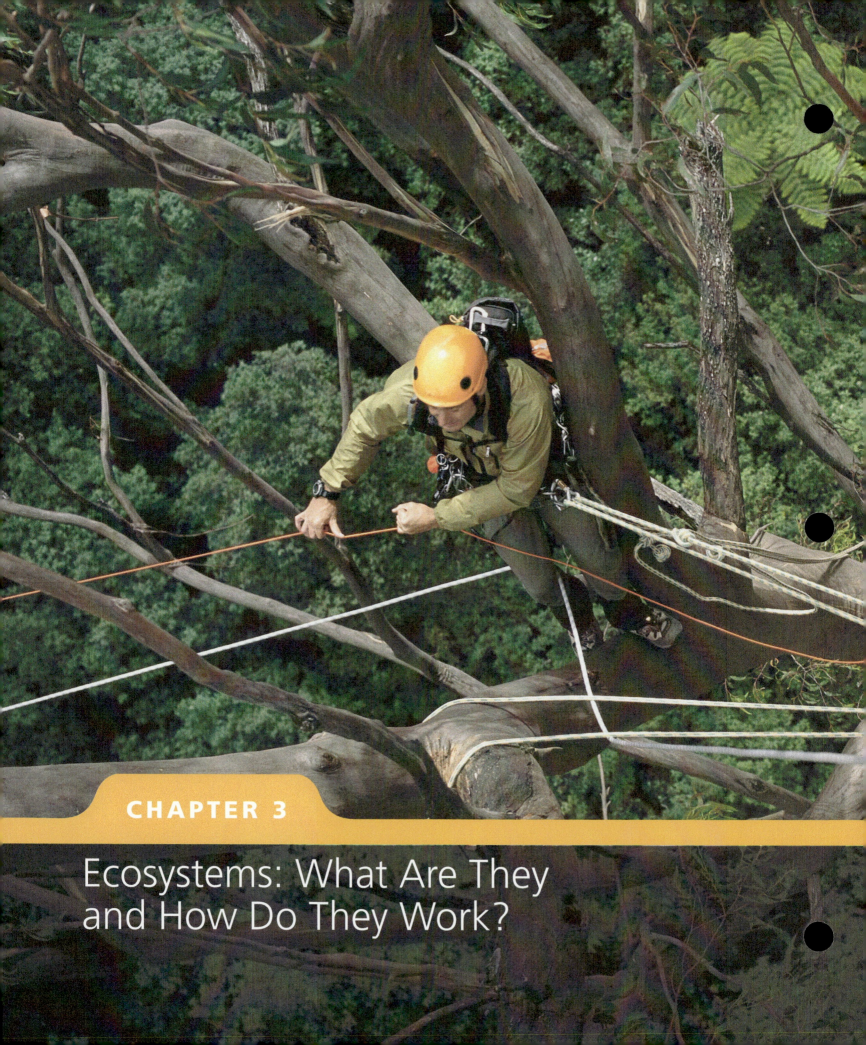

Ecosystems: What Are They and How Do They Work?

Key Questions

3.1 How does the earth's life-support system work?

3.2 What are the major components of an ecosystem?

3.3 What happens to energy in an ecosystem?

3.4 What happens to matter in an ecosystem?

3.5 How do scientists study ecosystems?

Scientists studying life in the treetops of a tropical forest.

Bill Hatcher/National Geographic Creative

Tropical Rain Forests Are Disappearing

Tropical rain forests are found near the earth's equator and contain an incredible variety of life. Rain forests cover only 7% of the earth's dry land surface but contain up to half of the world's known plant and animal species found on land. These lush forests are warm and humid year round because of their almost daily rainfall and nearness to the equator. The biodiversity of tropical rain forests makes them an excellent natural laboratory for the study of ecosystems (see chapter-opening photo).

To date, human activities have destroyed or degraded more than half of the earth's tropical rain forests. People continue clearing the forests to grow crops, graze cattle, and build settlements (Figure 3.1). Ecologists warn that without protection, most of these forests will be gone or severely degraded by the end of this century.

Why should we care that tropical rain forests are disappearing? Scientists give three reasons. *First,* clearing these forests causes the extinction of many of their plant and animal species by destroying the habitats where they live. The loss of key species in these forests can have a ripple effect that leads to the extinction of other species that they help support.

Second, destroying these forests warms the atmosphere and speeds up climate change. How does this occur? Eliminating large areas of trees faster than they can grow back means that there are fewer plants using photosynthesis to remove some of the human-generated emissions of carbon dioxide (CO_2), which are caused mostly by the burning of huge quantities of fossil fuels. The resulting increased levels of CO_2 in the atmosphere contributes to atmospheric warming and climate change, which you will learn more about in Chapter 15.

Third, large-scale losses of tropical rain forests can change regional weather patterns in ways that can prevent the forest from returning in cleared or degraded areas. When this irreversible *ecological tipping point* is reached, the tropical rain forests in such areas become drier and less-diverse tropical grasslands.

In this chapter, you will learn about the living and nonliving components of tropical rain forests and other ecosystems and how they work, how human activities are affecting them, and how we can help sustain them. ●

17 Jun 1975

6 May 2003

Left: United Nations Environment Programme ; Right: United Nations Environment Programme

FIGURE 3.1 Natural Capital Degradation: Satellite image of the loss of tropical rain forest, cleared for farming, cattle grazing, and settlements, near the Bolivian city of Santa Cruz between June 1975 (left) and May 2003 (right). This is the latest available view of the area, but forest degradation has continued since 2003.

3.1 HOW DOES THE EARTH'S LIFE-SUPPORT SYSTEM WORK?

CONCEPT 3.1A The four major components of the earth's life-support system are the atmosphere (air), the hydrosphere (water), the geosphere (rock, soil, and sediment), and the biosphere (living things).

CONCEPT 3.1B Life is sustained by the flow of energy from the sun through the biosphere, the cycling of nutrients within the biosphere, and gravity.

Earth's Life-Support System Has Four Major Components

The earth's life-support system consists of four main spherical systems (Figure 3.2) that interact with one another. They are the atmosphere (air), the hydrosphere (water), the geosphere (rock, soil, and sediment), and the biosphere (living things) (**Concept 3.1A**).

The **atmosphere** is a spherical mass of air surrounding the earth's surface that is held to the earth by gravity. Its innermost layer, the **troposphere**, extends about 19 kilometers (12 miles) above sea level at the equator and about 6 kilometers (4 miles) above the earth's North and

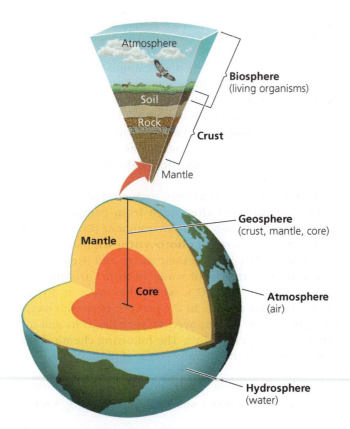

FIGURE 3.2 Natural Capital: The earth consists of a land sphere (*geosphere*), an air sphere (*atmosphere*), a water sphere (*hydrosphere*), and a life sphere (*biosphere*) (**Concept 3.1A**).

South Poles. The troposphere contains the air we breathe. It is 78% nitrogen (N_2) and 21% oxygen (O_2). The remaining 1% of air is mostly water vapor, carbon dioxide, and methane. The troposphere is the layer in which the earth's weather occurs and where life can survive.

The **stratosphere** is the atmospheric layer above the troposphere. It reaches 17 to 50 kilometers (11 to 31 miles) above the earth's surface. The lower stratosphere, called the *ozone layer*, contains enough ozone (O_3) gas to filter out about 95% of the sun's harmful *ultraviolet (UV) radiation*. It acts as a global sunscreen that allows life to exist on the earth's surface.

The **hydrosphere** contains all of the water on or near the earth's surface. It is found as *water vapor* in the atmosphere, as *liquid water* on the surface and underground, and as *ice*—polar ice, icebergs, glaciers, and ice in frozen soil-layers called *permafrost*. Salty oceans that cover that about 71% of the earth's surface contain 97% of the planet's water and support almost half of the world's species. About 2.5% of the earth's water is fresh water and three-fourths of that is ice.

The **geosphere** contains the earth's rocks, minerals, and soil. It consists of an intensely hot *core*, a thick *mantle* of very hot rock, and a thin outer *crust* of rock and soil. The crust's upper portion contains soil chemicals or nutrients that organisms need to live, grow, and reproduce. It also contains nonrenewable *fossil fuels*—coal, oil, and natural gas—and minerals that we extract and use.

The **biosphere** consists of the parts of the atmosphere, hydrosphere, and geosphere where life is found. If the earth were the size of an apple, the biosphere would be no thicker than the apple's skin.

Three Factors Sustain the Earth's Life

Life on the earth depends on three interconnected factors (**Concept 3.1B**):

- *One-way flow of high-quality energy* from the sun. The sun's energy supports plant growth, which provides energy for plants and animals, in keeping with the solar energy **principle of sustainability**. As solar energy interacts with carbon dioxide (CO_2), water vapor, and several other gases in the troposphere, it warms the troposphere—a process known as the **greenhouse effect** (Figure 3.3). Without this natural process, the earth would be too cold to support humans and most other forms of life found on the earth today.

- *Cycling of nutrients through parts of the biosphere.* **Nutrients** are chemicals that organisms need to survive. Because the earth does not get significant inputs of matter from space, its fixed supply of nutrients must be recycled to support life. This is in keeping with the chemical cycling **principle of sustainability**.

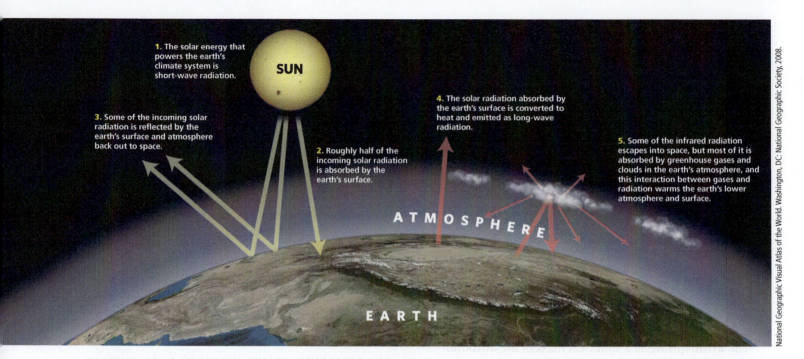

1. The solar energy that powers the earth's climate system is short-wave radiation.

3. Some of the incoming solar radiation is reflected by the earth's surface and atmosphere back out to space.

2. Roughly half of the incoming solar radiation is absorbed by the earth's surface.

4. The solar radiation absorbed by the earth's surface is converted to heat and emitted as long-wave radiation.

5. Some of the infrared radiation escapes into space, but most of it is absorbed by greenhouse gases and clouds in the earth's atmosphere, and this interaction between gases and radiation warms the earth's lower atmosphere and surface.

SUN

ATMOSPHERE

EARTH

FIGURE 3.3 *Greenhouse Earth.* High-quality solar energy flows from the sun to the earth. It is degraded to lower-quality energy (mostly heat) as it interacts with the earth's air, water, soil, and life forms, and eventually some of it returns to space. Certain gases in the earth's atmosphere retain enough of the sun's incoming energy as heat to warm the planet in what is known as the *greenhouse effect.*

- *Gravity* allows the planet to hold on to its atmosphere and enables the movement and cycling of chemicals through air, water, soil, and organisms.

3.2 WHAT ARE THE MAJOR COMPONENTS OF AN ECOSYSTEM?

CONCEPT 3.2A Some organisms produce the nutrients they need, others get the nutrients they need by consuming other organisms, and some recycle nutrients back to producers by decomposing the wastes and remains of other organisms.

CONCEPT 3.3B Soil is a renewable resource that provides nutrients that support terrestrial plants and helps purify water and control the earth's climate.

Ecosystems Have Several Important Components

Ecology is the science that focuses on how organisms interact with one another and with their nonliving physical environment of matter and energy. Scientists classify matter into levels of organization ranging from atoms to galaxies. Ecologists study five levels of matter: the **biosphere**, **ecosystems**, **communities**, **populations**, and **organisms**, all shown and defined in Figure 3.4.

The biosphere and its ecosystems are made up of living (*biotic*) and nonliving (*abiotic*) components. Examples of living components include plants, animals, and microbes. Nonliving components include water, air, nutrients, rocks, heat, and solar energy.

Ecologists assign each organism in an ecosystem to a *feeding level*, or **trophic level**, depending on its source of nutrients. Organisms are classified as producers and consumers based on whether they make (produce) or find (consume) food.

Producers are organisms, such as green plants, that make the nutrients they need from compounds and energy obtained from their environment (**Concept 3.3A**). In the process known as **photosynthesis**, plants capture solar energy that falls on their leaves and use it to combine carbon dioxide and water to form carbohydrates, such as glucose ($C_6H_{12}O_6$), which they store as a source of the chemical energy. In the process, they emit oxygen (O_2) gas into the atmosphere. Oxygen keeps us and most other animal species alive. The following chemical reaction summarizes the overall process of photosynthesis.

carbon dioxide + water + **solar energy** → glucose + oxygen

$$6\ CO_2 + 6\ H_2O + \textbf{solar energy} \rightarrow C_6H_{12}O_6 + 6\ O_2$$

About 2.8 billion years ago, producer organisms called *cyanobacteria* started carrying out photosynthesis and adding oxygen to the atmosphere. After several hundred million years, oxygen levels reached about 21%—high

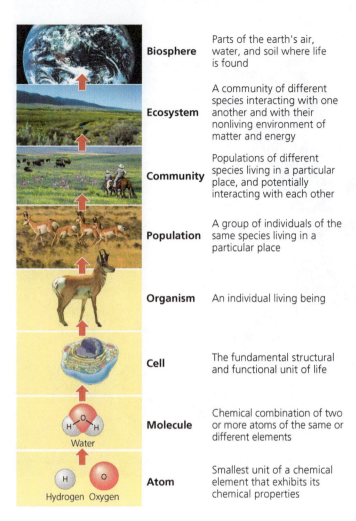

Biosphere	Parts of the earth's air, water, and soil where life is found
Ecosystem	A community of different species interacting with one another and with their nonliving environment of matter and energy
Community	Populations of different species living in a particular place, and potentially interacting with each other
Population	A group of individuals of the same species living in a particular place
Organism	An individual living being
Cell	The fundamental structural and functional unit of life
Molecule	Chemical combination of two or more atoms of the same or different elements
Atom	Smallest unit of a chemical element that exhibits its chemical properties

FIGURE 3.4 Ecology focuses on the top five of these levels of the organization of matter in nature.

enough to keep humans and other oxygen-breathing animals alive.

On land, most producers are green plants such as trees and grasses. In freshwater and ocean ecosystems, algae and aquatic plants growing near shorelines are the major producers. In open water, the dominant producers are *phytoplankton*—mostly microscopic organisms that float or drift in the water.

The other organisms in an ecosystem are **consumers** that cannot produce their own food (**Concept 3.2**). They get the nutrients they need by feeding on other producers or other consumers, or on the wastes and remains of producers and consumers.

There are several types of consumers. **Primary consumers**, or **herbivores** (plant eaters), are animals that eat mostly green plants or algae. Examples are caterpillars, giraffes, and zooplankton (tiny sea animals that feed on phytoplankton). **Carnivores** (meat eaters) are animals that feed on the flesh of other animals. Some carnivores such as spiders, lions (Figure 3.5), and most small fishes

are **secondary consumers** that feed on the flesh of herbivores. Other carnivores such as tigers, hawks, and killer whales (orcas) are **tertiary** (or higher-level) **consumers** that feed on the flesh of herbivores and other carnivores. Some of these relationships are shown in Figure 3.6. **Omnivores** such as pigs, rats, and humans eat both plants and animals.

CONSIDER THIS ...

THINKING ABOUT What You Eat

When you ate your most recent meal, were you an herbivore, a carnivore, or an omnivore?

Decomposers are consumers that get their nutrients by breaking down (decomposing) the wastes or remains of plants and animals. The process of decomposition returns these nutrients to the soil, water, and air for reuse by producers (**Concept 3.2A**). Most decomposers are bacteria and fungi. Other consumers, called **detritus feeders**, or **detritivores**, get their nutrients by feeding on the wastes or dead bodies (detritus) of other organisms. Examples are earthworms, some soil insects, hyenas, and vultures.

Detritivores and decomposers can transform a fallen tree trunk into simple inorganic molecules that plants can absorb as nutrients (Figure 3.7). In natural ecosystems, the wastes and dead bodies of organisms are resources for other organisms in keeping with the chemical cycling **principle of sustainability**. Without decomposers and detritivores, many of which are microscopic organisms (Science Focus 3.1), the planet would be buried in plant litter, animal wastes, dead animal bodies, dead and fallen trees, and garbage.

Producers, consumers, and decomposers use the chemical energy stored in glucose and other organic compounds to fuel their life processes through cellular respiration. In most cells, this energy is released by **aerobic respiration**, which uses oxygen to convert glucose and other organic compounds back into carbon dioxide and water, as shown below.

glucose + oxygen → carbon dioxide + water + **energy**

$C_6H_{12}O_6 + 6 O_2 \rightarrow \quad 6 CO_2 \quad + 6 H_2O + $ **energy**

Some decomposers, such as yeast and some bacteria get the energy they need by breaking down glucose (or other organic compounds) in the *absence* of oxygen. This form of cellular respiration is called **anaerobic respiration**, or **fermentation**. Instead of carbon dioxide and water, the products of this process are compounds such as methane gas (CH_4), ethyl alcohol (C_2H_6O), acetic acid ($C_2H_4O_2$, the key component of vinegar), and hydrogen sulfide (H_2S, a highly poisonous gas that smells like rotten eggs). Note that all organisms get their energy from

FIGURE 3.5 Lions feeding on prey.

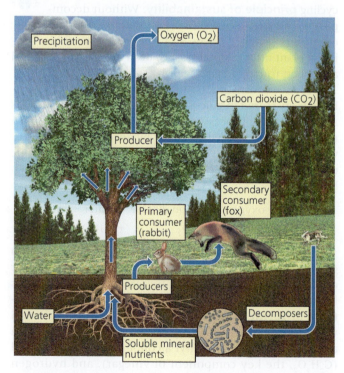

FIGURE 3.6 Key living (biotic) and nonliving (abiotic) components of an ecosystem in a field.

aerobic or anaerobic respiration but only plants carry out photosynthesis.

To summarize, ecosystems and the biosphere are sustained by the *one-way energy flow* from the sun and the *nutrient cycling* of key materials (**Concept 3.1B**)—in keeping with two of the **scientific principles of sustainability** (Figure 3.8).

Soil is the Foundation of Life on Land

Soil is a complex mixture of rock pieces and particles, mineral nutrients, decaying organic matter, water, air, and living organisms that support plant life, which, in turn, supports animal life (**Concept 3.2B**). Soil is one of the most important components of the earth's natural capital. It purifies water and supplies most of the nutrients needed for plant growth. Through aerobic respiration, organisms living in soil remove some of the carbon dioxide from the atmosphere and store it as organic carbon compounds, thereby helping to control the earth's climate.

Most *mature soils* contain several horizontal layers or *horizons*. A cross-sectional view of the horizons of a soil is called a **soil profile** (Figure 3.9, right). The major horizons in a mature soil are **O** (leaf litter), **A** (topsoil),

Many of the World's Most Important Organisms Are Invisible to Us

They are everywhere. Trillions can be found inside your body, on your body, in a handful of soil, and in a cup of ocean water.

These mostly invisible rulers of the earth are *microbes,* or *microorganisms,* catchall terms for many thousands of species of bacteria, protozoa, fungi, and floating phytoplankton. They play key roles in the earth's life-support system.

Bacteria in our intestinal tracts break down the food we eat, and microbes in our noses help prevent harmful bacteria from reaching our lungs. Other microbes

help purify the water we drink by breaking down plant and animal wastes in the water. Bacteria and fungi in the soil decompose organic wastes into nutrients that can be taken up by plants that are then eaten by humans and other plant eaters. Without these tiny creatures, we would go hungry and be up to our necks in waste matter.

Some microorganisms, particularly phytoplankton in the ocean, provide much of the planet's oxygen. They also help regulate the atmosphere's average temperature by removing some of the

carbon dioxide produced when we burn coal, natural gas, and gasoline. Other microbes control diseases that harm plants and limit populations of insects that attack our food crops. In short, microbes are a vital part of the earth's natural capital.

CRITICAL THINKING

What are two advantages that microbes have over humans for thriving in the world?

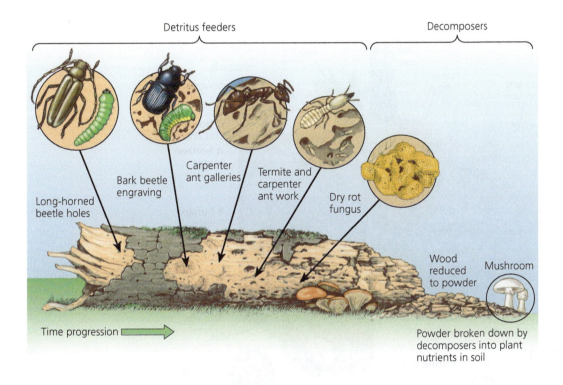

FIGURE 3.7 Various detritivores and decomposers (mostly fungi and bacteria) can "feed on" or digest parts of a log and eventually convert its complex organic chemicals into simpler inorganic nutrients that can be taken up by producers.

Detritus feeders

Decomposers

Long-horned beetle holes

Bark beetle engraving

Carpenter ant galleries

Termite and carpenter ant work

Dry rot fungus

Wood reduced to powder

Mushroom

Time progression

Powder broken down by decomposers into plant nutrients in soil

B (subsoil), and **C** (weathered parent material), which build up over the parent material.

The roots of most plants and the majority of a soil's organic matter are found in the soil's two upper layers: the O horizon of leaf litter and the A horizon of topsoil. In a fertile soil, these two layers teem with bacteria, fungi, earthworms, and numerous small insects, all interacting by feeding on and decomposing one another.

Every handful of topsoil contains billions of bacteria and other decomposer organisms. They break down some of the soil's complex organic compounds into a mixture of the partially decomposed plant and animal remains, called *humus.* A fertile soil that produces high crop yields has a thick topsoil layer with a lot of humus mixed with mineral particles from weathered plant material.

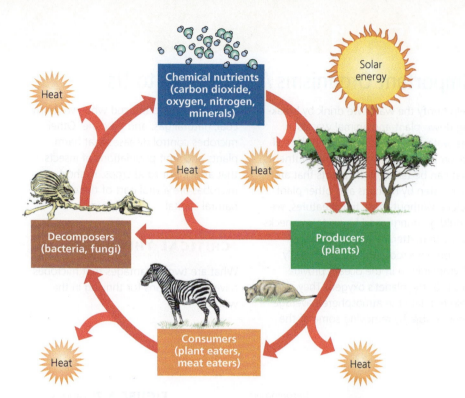

FIGURE 3.8 Natural Capital: The main components of an ecosystem are energy, chemicals, and organisms. Nutrient cycling and the flow of energy—first from the sun, then through organisms, and finally into the environment as low-quality heat—link these components.

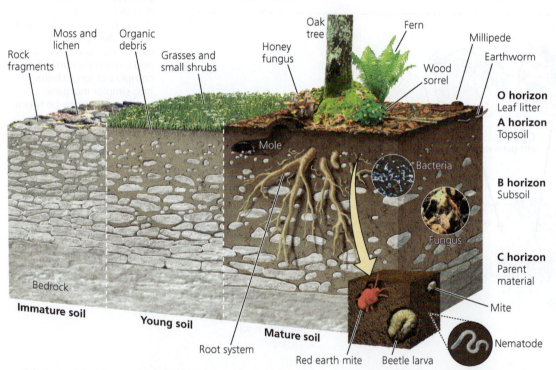

FIGURE 3.9 Natural Capital: Generalized soil formation and soil profile. *Critical thinking:* What role do you think the tree in this figure plays in soil formation? How might the soil-formation process change if the tree were removed?

Soil is a renewable resource but it is renewed very slowly and becomes a nonrenewable resource if we deplete it faster than natural processes can renew it. The formation of just 2.5 centimeters (1 inch) of topsoil can take hundreds to thousands of years. Removing plant cover from soil exposes its topsoil to erosion by water and wind. This explains why protecting and renewing topsoil is a key to sustainability. You will learn more about soil erosion and soil conservation in Chapter 10.

3.3 WHAT HAPPENS TO ENERGY IN AN ECOSYSTEM?

CONCEPT 3.3 As energy flows through ecosystems in food chains and food webs, the amount of high-quality chemical energy available to organisms decreases at each successive feeding level.

Energy Flows through Ecosystems in Food Chains and Food Webs

Chemical energy stored as nutrients in the bodies and wastes of organisms flows through ecosystems from one trophic (feeding) level to another in food chains and food webs. A sequence of organisms with each serving as a source of nutrients or energy for the next level of organisms is called a **food chain** (Figure 3.10). Every use and transfer of energy by organisms involves a loss of some high-quality energy to the environment as low-quality energy in the form of heat, as required by the second law of thermodynamics. A graphic display of the energy loss at each trophic level is called a **pyramid of energy flow.** Figure 3.11 illustrates this energy loss for a food chain, assuming a 90% energy loss for each level of the chain.

CONSIDER THIS ...

CONNECTIONS Energy Flow and Feeding People

Energy flow pyramids explain why the earth could support more people if they all ate at a low trophic level by consuming grains, vegetables, and fruits directly rather than passing such crops through another trophic level and eating the flesh of herbivores such as cattle, pigs, sheep, and chickens. About two-thirds of the world's people survive primarily by eating wheat, rice, and corn at the first trophic level mostly because they cannot afford to eat much meat.

In natural ecosystems, most consumers feed on more than one type of organism, and most organisms are eaten or decomposed by more than one type of consumer. Because of this, organisms in most ecosystems form a complex network of interconnected food chains called a **food web**. Food chains and food webs show how producers, consumers, and decomposers are connected to one another as energy flows through trophic levels in an ecosystem. Figure 3.12 shows an aquatic food web and Figure 3.13 shows a terrestrial food web.

CONSIDER THIS ...

LEARNING FROM NATURE

There is no waste in nature because the wastes and remains of one organism become food for other organisms. Scientists and engineers study food webs to learn how to reduce or eliminate food waste and other forms of waste produced by humans.

Some Ecosystems Produce Plant Matter Faster than Others Do

Scientists measure the rates at which ecosystems produce chemical energy to compare ecosystems and understand how they interact. **Gross primary productivity (GPP)** is the *rate* at which an ecosystem's producers (such as plants and phytoplankton) convert solar energy into chemical energy, which they store as compounds in their bodies. To stay alive, grow, and reproduce, producers must use some of their stored chemical energy for their own aerobic respiration.

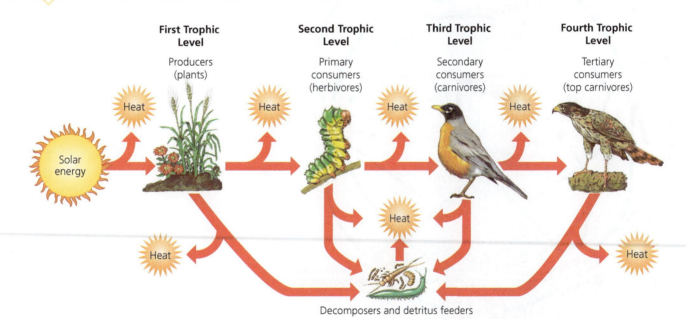

FIGURE 3.10 In a food chain, chemical energy in nutrients flows through various trophic levels. *Critical thinking:* Think about what you ate for breakfast. At what level or levels on a food chain were you eating?

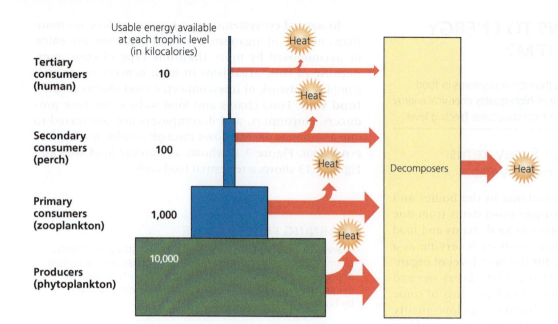

Usable energy available
at each trophic level
(in kilocalories)

Tertiary consumers (human) 10

Secondary consumers (perch) 100

Primary consumers (zooplankton) 1,000

Producers (phytoplankton) 10,000

Heat

Heat

Heat

Heat

Heat

Decomposers

Heat

FIGURE 3.11 Generalized *pyramid of energy flow* showing the decrease in usable chemical energy available at each succeeding trophic level in a food chain or food web. This model assumes that with each transfer from one trophic level to another, there is a 90% loss of usable energy to the environment in the form of low-quality heat. Calories and joules are used to measure energy. 1 kilocalorie = 1,000 calories = 4,184 joules. **Critical thinking:** Why is a vegetarian diet more energy efficient than a meat-based diet?

FIGURE 3.12 This is a greatly simplified aquatic food web found in the southern hemisphere. The shaded middle area shows a simple food chain that is part of these complex interacting feeding relationships. Many more participants in the web, including an array of decomposer and detritus feeder organisms, are not shown here. **Critical thinking:** Can you imagine a food web of which you are a part? Try drawing a simple diagram of it.

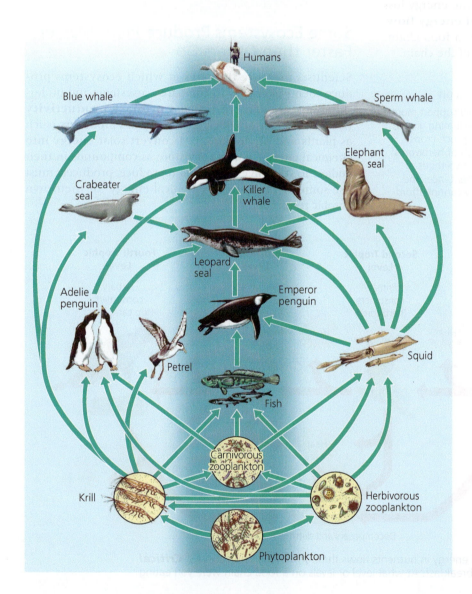

Humans

Blue whale

Sperm whale

Crabeater seal

Killer whale

Elephant seal

Leopard seal

Adelie penguin

Emperor penguin

Petrel

Squid

Fish

Carnivorous zooplankton

Krill

Herbivorous zooplankton

Phytoplankton

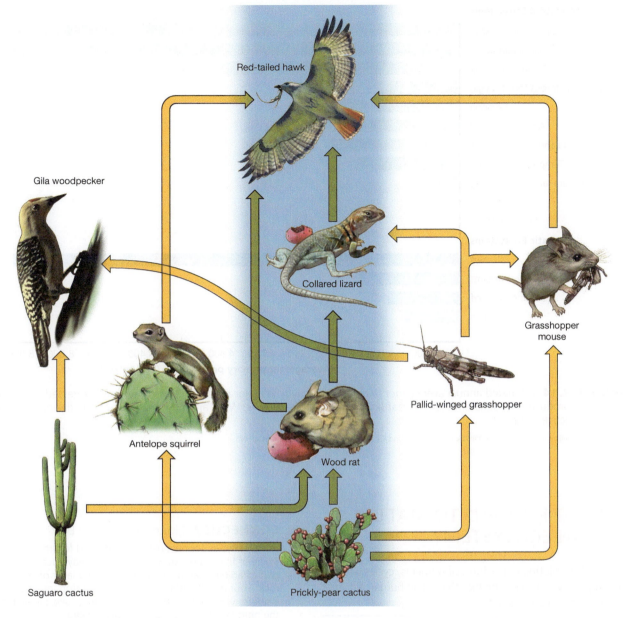

FIGURE 3.13 Greatly simplified terrestrial food web in a desert ecosystem. The shaded middle area shows a simple food chain that is part of these complex interacting feeding relationships. Many more participants in the web, including an array of decomposer and detritus feeder organisms, are not shown here.

Net primary productivity (NPP) is the *rate* at which producers use photosynthesis to produce and store chemical energy *minus* the *rate* at which they use some of this stored chemical energy through aerobic respiration. NPP measures how fast producers can make the chemical energy that is potentially available to the consumers in an ecosystem.

Terrestrial ecosystems and aquatic life zones differ in their NPP as illustrated in Figure 3.14. Despite its low NPP, the open ocean produces more of the earth's biomass per year than any other ecosystem or life zone. This happens because oceans cover 71% of the earth's surface and contain huge numbers of phytoplankton and other producers.

Tropical rain forests have a very high NPP because they have an abundance and variety of producer trees and other plants to support a large number of consumers. When such forests are cleared (**Core Case Study**) or burned to make way for crops or for grazing cattle, they suffer a sharp drop in NPP and lose many of their plant and animal species.

Only the plant matter represented by NPP is available as nutrients for consumers. Thus, *the planet's NPP ultimately limits the number of consumers (including humans) that can survive on the earth*. This is one of nature's important lessons.

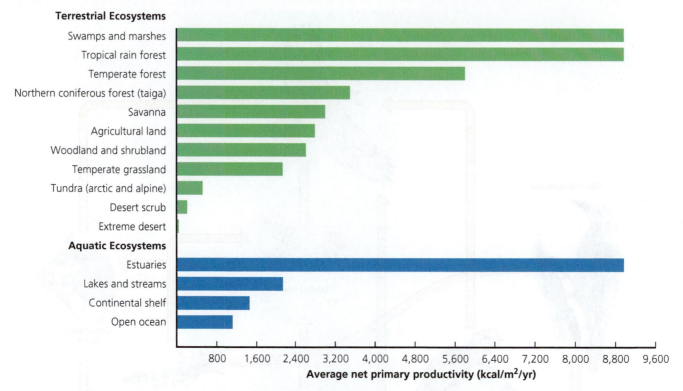

FIGURE 3.14 Estimated annual average *net primary productivity* in major life zones and ecosystems expressed as kilocalories of energy produced per square meter per year (kcal/m²/yr). **Data analysis:** What are the three most productive and the three least productive systems?

Compiled by the authors using data from R. H. Whittaker, *Communities and Ecosystems,* 2nd ed., New York: Macmillan, 1975.

3.4 WHAT HAPPENS TO MATTER IN AN ECOSYSTEM?

CONCEPT 3.4 Matter, in the form of nutrients, cycles within and among ecosystems and the biosphere, and human activities are altering these chemical cycles.

Nutrients Cycle within and among Ecosystems

The elements and compounds that make up nutrients move continually through air, water, soil, rock, and living organisms within ecosystems, in cycles called **nutrient cycles**, or *biogeochemical cycles*. They represent the chemical cycling **principle of sustainability** in action. These cycles are driven directly or indirectly by energy from the sun and by the earth's gravity. They include the hydrologic (water), carbon, nitrogen, and phosphorus cycles. Human activities are altering these important components of the earth's natural capital (see Figure 1.3, p. 7). (**Concept 3.4**)

The Water Cycle

Water (H_2O) is an amazing substance (Science Focus 3.2) that is essential for life on the earth. The **hydrologic cycle**, or **water cycle**, collects, purifies, and distributes the earth's fixed supply of water, as shown in Figure 3.15.

The sun provides the energy needed to power the water cycle. Incoming solar energy causes *evaporation*—the conversion of some of the liquid water in the earth's oceans, lakes, rivers, soil, and plants to vapor. Most water vapor rises into the atmosphere, where it condenses into droplets in clouds. Gravity then draws the water back to the earth's surface as *precipitation*, such as rain, snow, or sleet.

Water's Unique Properties

Without water, the earth would be a lifeless planet. Water is a remarkable compound with a unique combination of properties:

- *Water exists as a liquid over a wide temperature range because of forces of attraction between its molecules.* If liquid water had a much narrower range of temperatures between freezing and boiling, the ocean would probably have frozen solid or boiled away long ago.

- *Liquid water changes temperature slowly because it can store a large amount of heat without a large change in its own temperature.* This property helps protect living organisms from temperature changes, moderates the earth's climate, and makes water an excellent coolant for car engines and power plants.

- *It takes a large amount of energy to evaporate water because of the attractive forces between its molecules.*

Water absorbs large amounts of heat as it changes into water vapor and releases this heat as the vapor condenses back to liquid water. This helps to distribute heat throughout the world and to determine regional and local climates. It also makes evaporation a cooling process—explaining why you feel cooler when perspiration evaporates from your skin.

- *Liquid water can dissolve more compounds than other liquids.* Water carries dissolved nutrients into the tissues of living organisms, flushes waste products out of those tissues, serves as an all-purpose cleanser, and helps to remove and dilute the water-soluble wastes of civilization. This property also means that water-soluble wastes can easily pollute water.

- *Water filters out wavelengths of the sun's ultraviolet radiation that would harm some aquatic organisms*

(see Figure 2.9, p. 36). This allows life to exist in the upper layer of aquatic systems.

- *Unlike most liquids, water expands when it freezes.* This means that ice floats on water because it has a lower density (mass per unit of volume) than liquid water has. Otherwise, lakes and streams in cold climates would freeze solid from the bottom up and lose most of their aquatic life. Because water expands upon freezing, it can break pipes, crack a car's engine block (if it doesn't contain antifreeze), break up pavement, and fracture rocks (which helps form soil, see Figure 3.9).

CRITICAL THINKING

Pick two of the special properties listed above and, for each property, explain how life on the earth would be different if it did not exist.

Most precipitation falling on terrestrial ecosystems becomes **surface runoff**. This water flows over land surfaces into streams, rivers, lakes, wetlands, and the ocean from which some of the water evaporates. Some of the water seeps into the upper layers of soils and is used by plants, and some evaporates from the soils back into the atmosphere.

Some precipitation seeps into the soil. Water that seeps deeper into the soil is known as **groundwater**. Groundwater collects in **aquifers**, which are underground layers of sand, and water-bearing rock. Some precipitation is converted to ice that is stored in *glaciers*.

Only about 0.024% of the earth's vast water supply is available to humans and other species as liquid freshwater in accessible groundwater deposits and in lakes, rivers, and streams. The rest of the planet's water is too salty, is too deep underground to extract at affordable prices, or is stored as ice in glaciers.

Human activities alter the water cycle in three major ways (see the red arrows and boxes in Figure 3.15). *First,* people sometimes withdraw freshwater from rivers, lakes, and aquifers, at rates faster than natural processes can replace it. As a result, some aquifers are being depleted and some rivers no longer flow to the ocean.

Second, people clear vegetation from land for agriculture, mining, road building, and other activities, and cover much of the land with buildings, concrete, and asphalt. This increases water runoff and reduces infiltration that would normally recharge groundwater supplies.

Third, people drain and fill wetlands for farming and urban development. Left undisturbed, wetlands provide the ecosystem service of flood control. Wetlands act like sponges to absorb and hold overflows of water from drenching rains and rapidly melting snow.

The Carbon Cycle

Carbon is the basic building block of the carbohydrates, fats, proteins, DNA, and other organic compounds required for life. Various compounds of carbon circulate through the biosphere, the atmosphere, and parts of the

0.024% Percentage of the earth's freshwater supply that is available to humans and other species

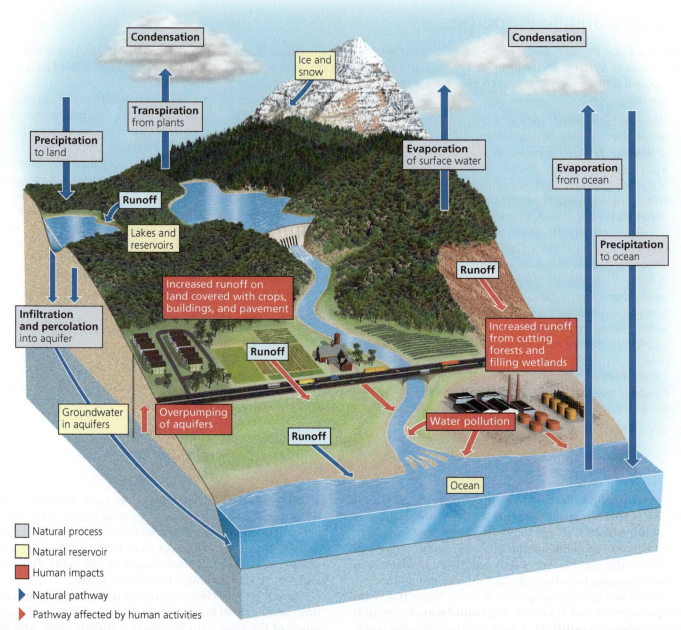

Condensation

Ice and snow

Condensation

Transpiration from plants

Precipitation to land

Evaporation of surface water

Evaporation from ocean

Runoff

Lakes and reservoirs

Precipitation to ocean

Increased runoff on land covered with crops, buildings, and pavement

Runoff

Increased runoff from cutting forests and filling wetlands

Infiltration and percolation into aquifer

Runoff

Groundwater in aquifers

Overpumping of aquifers

Water pollution

Runoff

Ocean

☐ Natural process

☐ Natural reservoir

☐ Human impacts

▶ Natural pathway

▶ Pathway affected by human activities

FIGURE 3.15 Natural Capital: Simplified model of the *water cycle*, or *hydrologic cycle*, in which water circulates in various physical forms within the biosphere. The red arrows and boxes identify major effects of human activities on this cycle. ***Critical thinking:*** What are three ways in which your lifestyle directly or indirectly affects the hydrologic cycle?

hydrosphere and geosphere in the **carbon cycle** shown in Figure 3.16.

A key component of the carbon cycle is carbon dioxide (CO_2) gas. It makes up about 0.040% of the volume of the troposphere. The amount of carbon dioxide (along with water vapor in the water cycle) has a large effect on the temperature of the earth's atmosphere (the greenhouse effect, see Figure 3.3) and thus plays a major role in determining the earth's climate.

Carbon is cycled through the biosphere by a combination of *photosynthesis* by producers that remove CO_2 from the air and water, and *aerobic respiration* by producers, consumers, and decomposers that add CO_2 to the atmosphere. Typically, CO_2 remains in the atmosphere for 100 years or longer. Some of the CO_2 in the atmosphere dissolves in ocean waters. In the ocean, decomposers release carbon that is stored as insoluble carbonate minerals and rocks in bottom sediment for long periods.

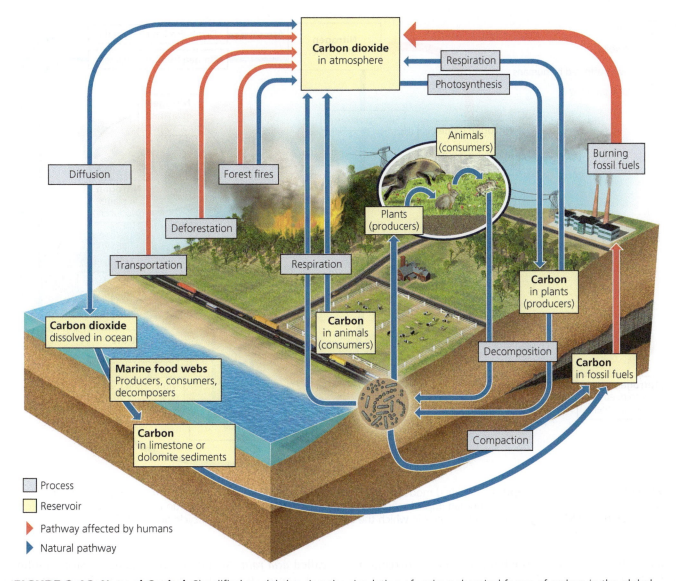

FIGURE 3.16 Natural Capital: Simplified model showing the circulation of various chemical forms of carbon in the global *carbon cycle*. Red arrows show major harmful impacts of human activities. (Yellow box sizes do not show relative reservoir sizes.) **Critical thinking:** What are three ways in which you directly or indirectly affect the carbon cycle?

Over millions of years, some of the carbon in deeply buried deposits of dead plant matter and algae has been converted into carbon-containing *fossil fuels* such as coal, oil, and natural gas (Figure 3.16). Within a few hundred years, we have extracted and burned huge quantities of fossil fuels that took millions of years to form. This has added large quantities of CO_2 to the atmosphere (see the red arrows in Figure 3.16) faster than the carbon cycle can recycle it. There is considerable scientific evidence that this disruption of the carbon cycle is helping to warm the atmosphere and change the earth's climate. The oceans remove some of this CO_2 but as a result, the acidity of ocean waters is rising. This is bad news for organisms that are adapted to less-acidic ocean waters.

Another way in which we alter the cycle is by clearing carbon-absorbing vegetation from many forests, especially tropical forests (Figure 3.1), faster than it can grow back (**Core Case Study**). This reduces the ability of the carbon cycle to remove excess CO_2 from the atmosphere and it contributes to climate change. We discuss the major environmental problems of *ocean acidification* in Chapter 9 and *climate change* in Chapter 15.

The Nitrogen Cycle

Nitrogen gas (N_2) makes up 78% of the volume of the atmosphere and is a crucial component of proteins, many vitamins, and DNA. However, N_2 in the atmosphere cannot be absorbed and used directly as a nutrient by plants or other organisms. It becomes a plant nutrient only as a component of nitrogen-containing ammonia (NH_3), ammonium ions (NH_4^+), and nitrate ions (NO_3^-), which are

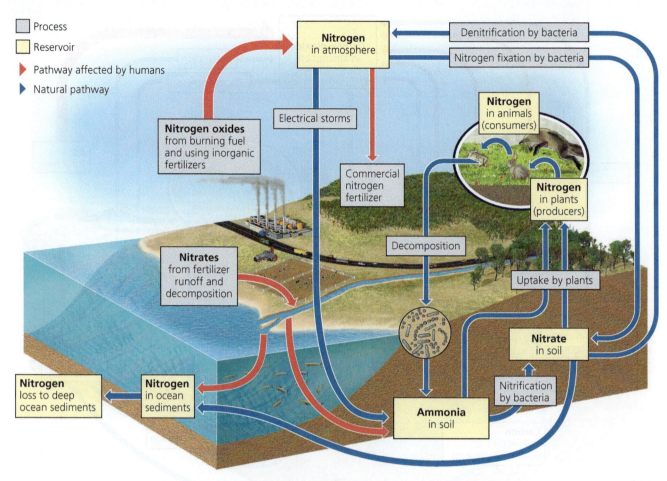

Nitrogen in atmosphere

Denitrification by bacteria

Nitrogen fixation by bacteria

Electrical storms

Nitrogen in animals (consumers)

Nitrogen oxides from burning fuel and using inorganic fertilizers

Commercial nitrogen fertilizer

Nitrogen in plants (producers)

Decomposition

Nitrates from fertilizer runoff and decomposition

Uptake by plants

Nitrate in soil

Nitrogen loss to deep ocean sediments

Nitrogen in ocean sediments

Nitrification by bacteria

Ammonia in soil

FIGURE 3.17 Natural Capital: Simplified model showing the circulation of various chemical forms of nitrogen in the *nitrogen cycle*, with major harmful human impacts shown by the red arrows. (Yellow box sizes do not show relative reservoir sizes.) ***Critical thinking:*** What are two ways in which the carbon cycle and the nitrogen cycle are linked?

circulated through parts of the biosphere in the **nitrogen cycle** (Figure 3.17).

These chemical forms of nitrogen are created by lightning which converts N_2 to NH_3 and by specialized bacteria in topsoil. Other bacteria in topsoil and bottom sediments convert NH_3 to NH_4^+ and nitrate ions (NO_3^-) that are taken up by the roots of plants. Plants then use these forms of nitrogen to produce various proteins, nucleic acids, and vitamins necessary for their own survival and that of other organisms. Animals that eat plants consume these nitrogen-containing compounds, as do detritus feeders and decomposers. Bacteria in waterlogged soil and bottom sediments of lakes, oceans, swamps, and bogs convert nitrogen compounds into nitrogen gas (N_2), which is released to the atmosphere to begin the nitrogen cycle again.

Human activities interfere with the nitrogen cycle in several ways (see the red arrows in Figure 3.17). When we burn gasoline and other fuels, the resulting high temperatures convert some of the N_2 and O_2 in air to nitric oxide (NO). In the atmosphere, NO can be converted to nitrogen dioxide gas (NO_2) and nitric acid vapor (HNO_3), which can return to the earth's surface as *acid deposition*, commonly

called *acid rain*. Acid deposition damages stone buildings and statues. It can also kill forests and other plant ecosystems, and wipe out life in ponds and lakes.

We also remove large amounts of N_2 from the atmosphere to make ammonia (NH_3) and ammonium ions (NH_4^+), used to make fertilizers. In addition, we add the greenhouse gas nitrous oxide (N_2O) to the atmosphere through the action of anaerobic bacteria on nitrogen-containing fertilizer or organic animal manure applied to the soil.

People also alter the nitrogen cycle in aquatic ecosystems by adding excess nitrates (NO_3^-). These nitrates contaminate bodies of water through agricultural runoff of fertilizers, animal manure, and discharges from municipal sewage treatment systems. This plant nutrient can cause excessive growths of algae that can disrupt aquatic systems.

The Phosphorus Cycle

Phosphorus (P) is an element that is essential for living things. It is necessary for the production of DNA and cell membranes, and is important for the formation of bones and teeth.

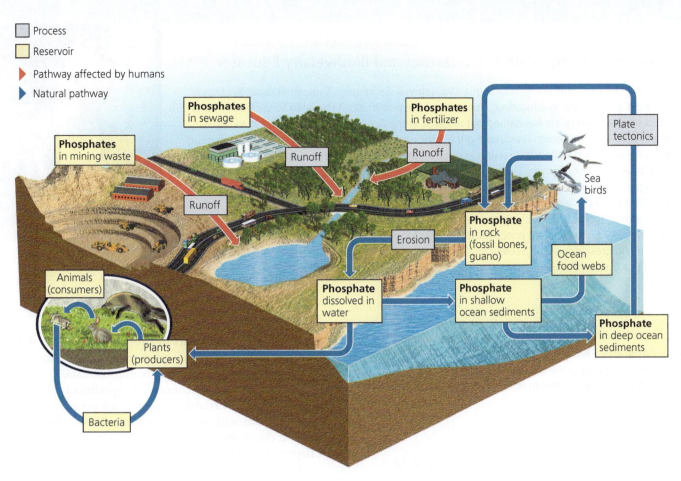

Phosphates in mining waste

Phosphates in sewage

Phosphates in fertilizer

Plate tectonics

Runoff

Runoff

Runoff

Sea birds

Erosion

Phosphate in rock (fossil bones, guano)

Ocean food webs

Animals (consumers)

Phosphate dissolved in water

Phosphate in shallow ocean sediments

Phosphate in deep ocean sediments

Plants (producers)

Bacteria

FIGURE 3.18 Natural Capital: Simplified model showing the circulation of various chemical forms of phosphorus (mostly phosphates) in the *phosphorus cycle*, with major harmful human impacts shown by the red arrows. (Yellow box sizes do not show relative reservoir sizes.) ***Critical thinking:*** What are two ways in which the phosphorus cycle and the nitrogen cycle are linked? What are two ways in which the phosphorus cycle and the carbon cycle are linked?

The cyclic movement of phosphorus (P) through water, the earth's crust, and living organisms is called the **phosphorus cycle** (Figure 3.18). Most of phosphorus compounds in this cycle contain *phosphate* ions (PO_4^{3-}), which are an important plant nutrient. Phosphorus does not cycle through the atmosphere because few of its compounds exist as a gas. Phosphorous also cycles slower than water, carbon, and nitrogen.

As water runs over exposed rocks, it slowly erodes inorganic compounds that contain phosphate ions. Water carries these ions into the soil, where they are absorbed by the roots of plants and by other producers. Phosphate compounds are then transferred by food webs from producers to consumers and eventually to detritus feeders and decomposers.

Much of the phosphate that erodes from rocks is carried into rivers and streams and into the ocean. In the ocean, phosphates can be deposited as marine sediments and remain trapped for millions of years. Over time, geological processes uplift and expose some of these seafloor deposits. The exposed rocks are then eroded, freeing up the phosphorus to re-enter the phosphorus cycle.

Most soils contain little phosphate, which often limits plant growth on land. For this reason, people often fertilize soil by adding phosphorous as phosphates mined from the ground. Lack of phosphorus also limits the growth of producer populations in many freshwater streams and lakes. This is because phosphate compounds are only slightly soluble in water and do not release many phosphate ions to producers in aquatic systems.

Human activities, including the removal of large amounts of phosphate from the earth to make fertilizer, disrupt the phosphorus cycle (see red arrows in Figure 3.18). By clearing tropical forests (**Core Case Study**), we expose the topsoil to increased erosion, which reduces phosphate levels in the tropical soils.

Eroded topsoil and fertilizer washed from fertilized crop fields, lawns, and golf courses carry large quantities of phosphate ions into streams, lakes, and oceans. There they stimulate the growth of producers such as algae and various aquatic plants, which can upset chemical cycling and other processes in bodies of water.

Thomas E. Lovejoy—Forest Researcher and Biodiversity Educator

For several decades, conservation biologist and National Geographic Explorer Thomas E. Lovejoy has played a major role in educating scientists and the public about the need to understand and protect tropical forests. He has carried out research in the Amazon forests of Brazil since 1965. A major goal of this research is estimating the minimum area necessary for sustaining biodiversity in national parks and biological reserves in tropical forests. In 1980, he coined the term *biodiversity*, or biological diversity.

Lovejoy served as the principal adviser for the popular and widely acclaimed public television series *Nature*. He has also written numerous articles and books on issues related to conserving biodiversity. In addition to teaching environmental science and policy at George Mason University, he has held several prominent posts, including director of the World Wildlife Fund's conservation program and president of the Society for Conservation Biology. In 2012, he was awarded the Blue Planet Prize for his efforts to understand and sustain the earth's biodiversity.

CONSIDER THIS ...

LEARNING FROM NATURE

Scientists study the water, carbon, nitrogen, and phosphorus cycles to help us learn how to reuse and recycle the wastes we create.

3.5 HOW DO SCIENTISTS STUDY ECOSYSTEMS?

CONCEPT 3.5 Scientists use field research and laboratory research, and mathematical and other type of models, to learn about ecosystems and how much stress they can take.

Studying Ecosystems Directly

Ecologists and other scientists use several approaches to increase their scientific understanding of ecosystems. These approaches include field and laboratory research and mathematical and other types of models (**Concept 3.5**).

Field research involves going into forests and other natural settings to study ecosystems. Ecologists use a variety of methods for field research. They include collecting water and soil samples, identifying and studying the species in an area, observing feeding behaviors, and using global positioning system (GPS) to track the movements of animals. Most of what we know about ecosystems has come from such research (Individuals Matter 3.1). **GREEN CAREER: Ecologist**

Scientists also use a variety of methods to study tropical rain forests (**Core Case Study**). Some erect construction cranes to reach the canopies. Others climb the trees and install ropes and pulleys (see chapter-opening photo) and temporary platforms in the treetops. These devices help scientists to identify and observe the diversity of species living or feeding in these treetop habitats.

Ecologists carry out controlled experiments by isolating and changing a variable in part of an area and comparing the results with nearby unchanged areas. You learned about a classic example of this in the **Core Case Study** of Chapter 2 (p. 28).

Ecologists also use aircraft and satellites equipped with sophisticated cameras and other *remote sensing* devices to scan and collect data about the earth's surface. In addition, they use *geographic information system (GIS)* software to capture, store, analyze, and display such data. For example, a GIS can convert digital satellite images into global, regional, and local maps. These maps show variations in vegetation, GPP, deforestation, soil erosion, air pollution, drought, flooding, and other variables.

Some researchers attach tiny radio transmitters to animals and use GPS to learn about the animals by tracking where and how far animals go. Scientists also study nature by mounting time-lapse cameras or video cameras on small drones and on stationary objects such as trees to capture images of wildlife. **GREEN CAREERS: GIS analyst; remote sensing analyst**

Laboratory Research and Models

Ecologists supplement their field research by conducting *laboratory research*. In laboratories, scientists create simplified systems in containers such as culture tubes, bottles, aquariums, and greenhouses, and in indoor and outdoor chambers. In these structures, they control temperature, light, CO_2, humidity, and other variables.

Planetary Boundaries

For most of the past 10,000–12,000 years, humans have been living in an era called the *Holocene*—a period of relatively stable climate and other environmental conditions. This general stability has allowed the human population to grow, develop agriculture, and take over a large share of the earth's land and other resources (Figure 1.9, p. 11).

Most geologists contend that we are still living in the Holocene era but some scientists disagree. According to them, when the Industrial Revolution began (around 1750), we entered an era called the *Anthropocene* (the era of man). In this new era, our ecological footprints have expanded significantly and are changing and stressing the earth's life-support system, especially since 1950.

In 2015, an international team of 18 highly regarded researchers, led by Will Steffen and Johan Rockstrom of the Stockholm Resilience Centre, published a paper estimating how close we are to exceeding 9 major *planetary boundaries* or *ecological tipping points* as a result of certain human activities. They warn that exceeding them could trigger abrupt and long-lasting or irreversible environmental changes that could seriously degrade the earth's life-support system and our economies.

In 2015, the researchers estimated that we have exceeded four of these planetary boundaries. They are **(1)** *disruption of the nitrogen and phosphorous cycles* mostly due to greatly increased use of fertilizers to produce food, **(2)** *biodiversity loss* caused by replacement of biologically diverse forests and grasslands with *monocultures*, or fields of single crops, **(3)** *land system change* due to agriculture and urban development, and **(4)** *climate change* caused by our disruption of the carbon cycle, mostly through excessive emissions of carbon dioxide from the burning of fossil fuels.

These researchers warn that we need to act to reverse or reduce these impacts and to avoid exceeding the following additional boundaries: **(1)** freshwater use, **(2)** ocean acidification, **(3)** ozone depletion in the stratosphere, **(4)** fine-particle air pollution, and **(5)** pollution from chemicals such as toxic heavy metals and chemicals that can disrupt the human endocrine system.

There is an urgent need for more research to fill gaps in the data on planetary boundaries. Such information could help us avoid exceeding such boundaries by shrinking our ecological footprints while expanding our beneficial environmental impacts.

CRITICAL THINKING

Which two of these boundaries do you think are the most important?

These systems make it easier for scientists to carry out controlled experiments. Laboratory experiments are often faster and less costly than similar experiments in the field. However, scientists must consider how well their scientific observations and measurements in simplified, controlled systems in laboratory conditions reflect what takes place under the more complex and often-changing conditions found in nature.

Since the late 1960s, ecologists have developed mathematical models that simulate ecosystems and they run these models on high-speed supercomputers. The models help them understand large and complex systems such as lakes, oceans, forests, and the earth's climate, which cannot be adequately studied in field or laboratory research.
GREEN CAREER: Ecosystem modeler

Ecologists call for greatly increased research on the condition of the world's ecosystems to see how they are changing. This would help scientists develop strategies for preventing or slowing natural capital degradation. It would also help us to avoid going beyond ecological tipping points, which could cause severe degradation or collapse of ecosystems (Science Focus 3.3).

BIG IDEAS

- Life is sustained by the flow of energy from the sun through the biosphere, the cycling of nutrients within the biosphere, and gravity.

- Some organisms produce the nutrients they need, others survive by consuming other organisms, and still others live on the wastes and remains of organisms while recycling nutrients that are used again by producer organisms.

- Human activities are altering the chemical cycling of nutrients and the flow of energy through food chains and webs in ecosystems.

Tropical Rain Forests and Sustainability

This chapter began with a discussion of the importance of the world's incredibly diverse tropical rain forests (**Core Case Study**). These ecosystems showcase the functioning of the three **scientific principles of sustainability**, which apply as well to the world's other ecosystems.

First, producers within rain forests rely on *solar energy* to produce a vast amount of biomass through photosynthesis. *Second*, species living in the forests take part in and depend on the *cycling of nutrients* and the flow of energy within the forests and throughout the biosphere. *Third*, tropical rain forests contain a huge and vital part of the earth's *biodiversity*, and interactions among species living in these forests help to sustain these complex ecosystems.

We also reported recent research on the possible long-lasting, harmful effects of our exceeding any key planetary boundaries. In many of the chapters to follow, we will further examine such risks, and we will consider ways in which we can apply the six **principles of sustainability** (see inside back cover of this book) to try to stay within the key planetary boundaries, to live more

Anneka/Shutterstock.com

sustainably, and to create and expand beneficial environmental impacts.

Chapter Review

Core Case Study

1. What are three harmful effects of the clearing and degradation of tropical rain forests?

Section 3.1

2. What are the two key concepts for this section? Define and distinguish among the **atmosphere, troposphere, stratosphere, hydrosphere, geosphere,** and **biosphere**. What three interconnected factors sustain life on the earth? Describe the flow of energy to and from the earth. What is the **greenhouse effect** and why is it important?

Section 3.2

3. What are the two key concepts for this section? Define **ecology**. Define **organism, population, community**, and **ecosystem**, and give an example of each.

4. Distinguish between the living and nonliving components in ecosystems and give two examples of each.

5. What is a **trophic level**? Distinguish among **producers, consumers, decomposers,** and **detritus feeders (detritivores)**, and give an example of each. Summarize the processes of **photosynthesis**. Distinguish among **primary consumers (herbivores), carnivores, secondary consumers, tertiary consumers,** and **omnivores**, and give an example of each.

6. Explain the importance of microbes. What is **aerobic respiration (fermentation)**? What two processes sustain ecosystems and the biosphere and how are they linked? Define **soil** and **soil profile**. What are soil horizons? Name four major horizons. What is humus and how does it relate to fertile soil?

Section 3.3

7. What is the key concept for this section? Define and distinguish between a **food chain** and a **food web**. Explain what happens to energy as it flows through food chains and food webs. What is a **pyramid of energy flow**?

8. Distinguish between **GPP** and **NPP**, and explain their importance. What are the two most productive land ecosystems and the two most productive aquatic ecosystems?

Section 3.4

9. What is the key concept for this section? What happens to matter in an ecosystem? What is a **nutrient cycle**? Explain how nutrient cycles connect past, present, and future life. Describe the **hydrologic cycle**, or **water cycle**. What is **surface runoff**? Define **groundwater**. What is an **aquifer**? What percentage of the earth's water supply is available to humans and other species as liquid freshwater? Summarize the unique properties of water. Explain how human activities are affecting the water cycle. Describe the **carbon**, **nitrogen**, and **phosphorus cycles**, and explain how human activities are affecting each cycle.

Section 3.5

10. What is the key concept for this section? Describe three ways in which scientists study ecosystems. Explain why we need much more basic data about the structure and condition of the world's ecosystems. Distinguish between the Holocene and Anthropocene eras. List four planetary boundaries that we have exceeded, according to some scientists. What are this chapter's *three big ideas*? How are the three **scientific principles of sustainability** showcased in tropical rain forests?

_____

Note: Key terms are in bold type.

Critical Thinking

1. How would you explain the importance of tropical rain forests (**Core Case Study**) to people who think that such forests have no connection to their lives?

2. Explain **(a)** why the flow of energy through the biosphere depends on the cycling of nutrients, and **(b)** why the cycling of nutrients depends on gravity.

3. Explain why microbes are so important. What are two ways in which they benefit your health or lifestyle? Write a brief description of what you think would happen to you if microbes were eliminated from the earth.

4. Make a list of the foods you ate for lunch or dinner today. Trace each type of food back to a particular producer species. Describe the sequence of feeding levels that led to your feeding.

5. Use the second law of thermodynamics (see Chapter 2, p. 38) to explain why many poor people in less-developed countries live on a mostly vegetarian diet.

6. How might your life and the lives of any children or grandchildren you might have be affected if human activities as a whole continue to intensify the water cycle?

7. What would happen to an ecosystem if **(a)** all of its decomposers and detritus feeders were eliminated, **(b)** all of its producers were eliminated, and **(c)** all of

its insects were eliminated? Could an ecosystem function with only producers and decomposers and no consumers? Explain.

8. Describe how exceeding each of the planetary boundaries—*disruption of the nitrogen and phosphorous cycles, biodiversity loss, land system change,* and *climate change*—might affect **(a)** you, **(b)** any child you might have, and **(c)** any grandchild you might have.

Doing Environmental Science

Visit a nearby terrestrial ecosystem or aquatic life zone and try to identify major producers, primary and secondary consumers, detritus feeders, and decomposers. Take notes and describe at least one example of each of these types of organisms. Make a simple sketch showing how these organisms might be related to each other or to other organisms in a food chain or food web. Think of two ways in which this food chain or web could be disrupted. Write a report summarizing your research and conclusions.

Global Environment Watch Exercise

Go to your MindTap course to access the GREENR database. Using the "Basic Search" box at the top of the page, search for *Nitrogen Cycle* and look for information on how humans are affecting the nitrogen cycle. Specifically look for impacts on the atmosphere and on human health from emissions of nitrogen oxides, and look for the harmful ecological effects of the runoff of nitrate fertilizers into rivers and lakes. Make a list of these impacts and use this information to review your daily activities. Find three things that you do regularly that contribute to these impacts.

Data Analysis

Recall that NPP is the *rate* at which producers can make the chemical energy that is stored in their tissues and that is potentially available to other organisms (consumers) in an ecosystem. In Figure 3.14, it is expressed as units of energy (kilocalories, or *kcal*) produced in a given area (square meters, or m^2) over a period of time (a year). Look again at Figure 3.14 and consider the differences in NPP among various ecosystems. Then answer the following questions:

1. What is the approximate NPP of a tropical rain forest in kcal/m^2/yr? Which terrestrial ecosystem produces at about one-third of that rate? Which aquatic ecosystem has about the same NPP as a tropical rain forest?

2. Early in the 20th century, large areas of temperate forestland in the United States were cleared to make way for agricultural land. For each unit of this forest area that was cleared and replaced by farmland, by about how much was NPP reduced?

3. Why do you think deserts and grasslands have dramatically lower NPP than swamps and marshes?

4. About how many times higher is NPP in estuaries than in lakes and streams? Why do you think this is so?

CENGAGE**brain**.com For access to MindTap and additional study materials visit **www.cengagebrain.com**.

WWW.CENGAGEBRAIN.COM • **67**

CHAPTER 4

Biodiversity and Evolution

Nothing in biology makes sense except in the light of evolution.
THEODOSIUS DOBZHANSKY

Key Questions

4.1 What is biodiversity and why is it important?

4.2 What roles do species play in ecosystems?

4.3 How does the earth's life change over time?

4.4 What factors affect biodiversity?

Endangered San Lucas marsupial frog in Ecuador.

Pete Oxford/Minden Pictures

Why Are Amphibians Vanishing?

Amphibians are a class of animals that includes frogs, toads, and salamanders. Amphibians were among the first vertebrates (animals with backbones) to leave the earth's waters and live on land. They have adjusted to and survived environmental changes more effectively than many other species but their world is changing rapidly.

An amphibian lives part of its life in water and part on land. Human activities such as the use of pesticides and other chemicals can pollute the land and water habitats of amphibians. Many of the more than 7,500 known amphibian species have problems adapting to these changes.

Since 1980, populations of hundreds of amphibian species have declined or vanished (Figure 4.1). According to the International Union for Conservation of Nature (IUCN), about 33% of known amphibian species face *extinction—the end of their existence*. Studies indicate that frogs are going extinct 10,000 times faster than their historical rates.

No single cause can account for the decline of many amphibian species, but scientists have identified a number of factors that affect amphibians at various points in their life cycles. For example, frog eggs lack shells to protect embryos inside from water pollutants, and adult frogs ingest insecticides contained in many of the insects they eat. We explore these and other factors later in this chapter.

Why should we care if some amphibian species become extinct? Scientists give three reasons. *First*, amphibians are sensitive *biological indicators* of changes in environmental conditions. These changes include habitat loss, air and water pollution, ultraviolet (UV) radiation, and a warming climate. The growing threats to the survival of an increasing number of amphibian species indicate that environmental conditions for amphibians and

Joel Sartore/National Geographic Creative

FIGURE 4.1 Specimens of some of the nearly 200 amphibian species that have gone extinct since the 1970s.

many other species are deteriorating in many parts of the world.

Second, adult amphibians play important roles in biological communities. They eat more insects (including mosquitoes) than do many species of birds. In some habitats, the extinction of certain amphibian species could lead to population declines or extinction of animals that eat amphibians or their larvae, such as aquatic insects, reptiles, birds, fish, mammals, and other amphibians.

Third, amphibians play a role in human health. A number of pharmaceutical products come from compounds found in secretions from the skin of certain amphibians. Many of these compounds have been isolated and used as painkillers and antibiotics and in treatments for burns and heart disease. If amphibians vanish, these potential medical benefits and others that scientists have not yet discovered would vanish with them.

The threat to amphibians is part of a greater threat to the earth's biodiversity. In this chapter, we will learn about biodiversity, how it arose on the earth, why it is important, and how it is threatened. We will also consider possible solutions to these threats. ●

4.1 WHAT IS BIODIVERSITY AND WHY IS IT IMPORTANT?

CONCEPT 4.1 The biodiversity found in genes, species, ecosystems, and ecosystem processes is vital to sustaining life on the earth.

Biodiversity Is the Variety of Life

Biodiversity, or **biological diversity**, is the variety of life on the earth. It has four components, as shown in Figure 4.2. One is **species diversity**, the number and abundance of the different kinds of species living in an ecosystem. Estimates of the number of species on the earth range from 7 million to 100 million, with a best guess of 7–10 million species. So far, biologists have identified about 2 million species—most of being insects (Science Focus 4.1).

The second component of biodiversity is **genetic diversity**, which is the variety of genes found in a population or in a species (Figure 4.3). Genes contain information that gives rise to traits that can be passed on to offspring during reproduction. Species whose populations have greater genetic diversity have a better chance of surviving and adapting to environmental changes.

The third component, **ecosystem diversity**, refers to the earth's diversity of biological communities such

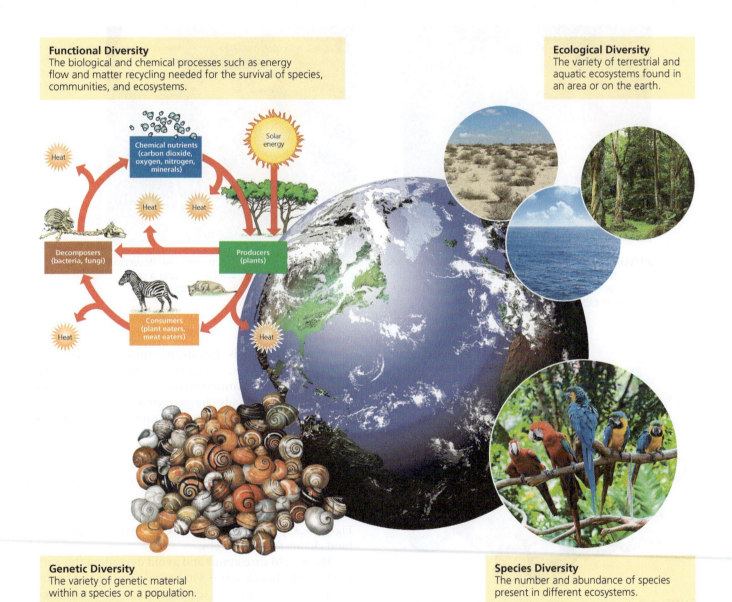

Functional Diversity
The biological and chemical processes such as energy flow and matter recycling needed for the survival of species, communities, and ecosystems.

Ecological Diversity
The variety of terrestrial and aquatic ecosystems found in an area or on the earth.

Solar energy

Chemical nutrients (carbon dioxide, oxygen, nitrogen, minerals)

Heat

Heat   Heat

Decomposers (bacteria, fungi)

Producers (plants)

Consumers (plant eaters, meat eaters)

Heat

Heat

Genetic Diversity
The variety of genetic material within a species or a population.

Species Diversity
The number and abundance of species present in different ecosystems.

FIGURE 4.2 Natural Capital: The major components of the earth's *biodiversity*—one of the planet's most important renewable resources and a key component of its natural capital (see Figure 1.3, p. 7).

Insects Play a Vital Role in Our World

Almost half of the world's identified species are insects. Many people view insect species as *pests* because they compete with us for food, bite or sting us, infect us with diseases such as malaria, and invade our lawns, gardens, and houses. This fear of insects fails to recognize the vital roles that insects play in sustaining the earth's life.

For example, *pollination* is a vital ecosystem service that allows flowering plants to reproduce. Many flowering plant species depend on insects to pollinate their flowers (Figure 4.A, left) so that reproduction occurs. In addition, insects that are decomposers or detritivores help return organic matter and nutrients to the soil (Figure 3.9). And insects that eat other insects—such as the praying mantis (Figure 4.A, right)—help control the populations of at least half the insect species we call pests. This free pest control is another vital ecosystem service.

CRITICAL THINKING

Can you think of three insect species not discussed above that benefit your life?

DARLYNE A. MURAWSKI/National Geographic Creative

Dr Morley Read/Shutterstock.com

FIGURE 4.A *Importance of insects:* Bees (left) and numerous other species of insects pollinate flowering plants that serve as food for plant eaters, including humans. This praying mantis, which is eating a moth (right), and many other insect species help to control the populations of most of the insect species we classify as pests.

FIGURE 4.3 *Genetic diversity* in this population of a species of Caribbean snail is reflected in the variations in shell color and banding patterns. Genetic diversity can also include other variations such as slight differences in chemical makeup, sensitivity to various chemicals, and behavior.

as deserts, grasslands, forests, mountains, oceans, lakes, rivers, and wetlands. Biologists classify terrestrial (land) ecosystems into **biomes**—large regions such as forests, deserts, and grasslands characterized by distinct climates and certain prominent species (especially vegetation). Figure 4.4 shows the major biomes lying across the mid-section of the United States. We discuss biomes in detail in Chapter 7.

The fourth component of biodiversity is **functional diversity**—the variety of processes such as energy flow and matter cycling that occur within ecosystems (see Figure 3.8, p. 52) as species interact with one another in food chains and food webs.

We should care about and avoid degrading the earth's biodiversity because it is vital to sustaining and increasing the natural capital (see Figure 1.3, p. 7) that keeps us alive and supports our economies. Humans use biodiversity as a source of food, medicine, building materials, and fuel. Biodiversity also provides natural ecosystem services such as air and water purification, renewal of

Edward O. Wilson: A Champion of Biodiversity

As a boy growing up in the southeastern United States, Edward O. Wilson became interested in insects at age 9. He has said, "Every kid has a bug period. I never grew out of mine."

Before entering college, Wilson had decided he would specialize in the study of ants. He became one of the world's experts on ants, unlocking secrets to their methods of communication and social behaviors.

Over time, Wilson widened his focus to include the entire biosphere. One of Wilson's landmark works is *The Diversity of Life*, published in 1992. In that book, he presented the principles and practical issues of biodiversity more completely than anyone had to that point. Today, he is recognized as one of the world's leading experts on biodiversity. He is now deeply involved in writing and lecturing about the need for global conservation efforts and working on Harvard University's *Encyclopedia of Life*, an online database of information on the world's known species.

Wilson has won more than 100 national and international awards and has written 33 books, two of which won the Pulitzer Prize for General Nonfiction. In 2013, he received the National Geographic Society's highest award, the Hubbard Medal. About the importance of biodiversity, he writes: "How can we save Earth's life forms from extinction if we don't even know what most of them are? … I like to call Earth a little known planet."

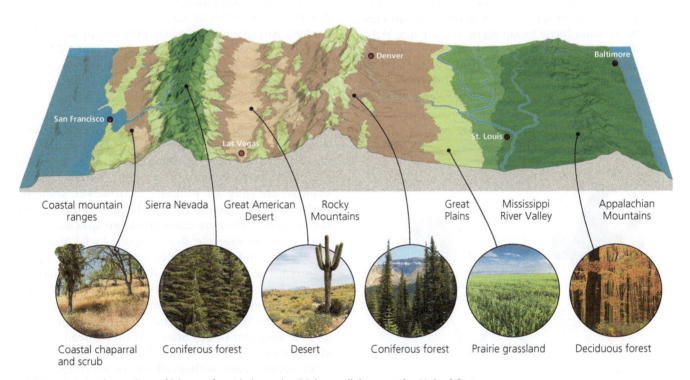

FIGURE 4.4 The variety of biomes found along the 39th parallel across the United States.

First: Zack Frank/Shutterstock.com; Second: Robert Crum/Shutterstock.com; Third: Joe Belanger/Shutterstock.com; Fourth: Protasov AN/Shutterstock.com; Fifth: Maya Kruchankova/Shutterstock.com; Sixth: Marc von Hacht/Shutterstock.com

topsoil, decomposition of wastes, and pollination. In addition, the earth's variety of genetic information, species, and ecosystems are needed for the evolution of new species and ecosystem services, as they respond to changing environmental conditions. Biodiversity is the earth's ecological insurance policy. We owe much of what we know about biodiversity to researchers such as Edward O. Wilson (Individuals Matter 4.1).

4.2 WHAT ROLES DO SPECIES PLAY IN ECOSYSTEMS?

CONCEPT 4.2A Each species plays a specific ecological role called its *niche*.

CONCEPT 4.2B Any given species may play one or more of four important roles—native, nonnative, indicator, or keystone—in a particular ecosystem.

Each Species Plays a Role

Each species plays a role within the ecosystem it inhabits (**Concept 4.2A**). Ecologists describe this role as a species' **ecological niche**, or simply its **niche**. It is a species' way of life in a community and includes everything that affects its survival and reproduction, such as how much water and sunlight it needs, how much space it requires, what it feeds on, what feeds on it, and the temperatures and other conditions it can tolerate. A species' niche should not be confused with its **habitat**, which is the place, or type of ecosystem, in which it lives and obtains what it needs to survive.

Ecologists use the niches of species to classify them mostly as *generalists* or *specialists*. A **generalist species** such as the raccoon has a broad niche (Figure 4.5, right curve). Generalists can live in many different places, eat a variety of foods, and often tolerate a wide range of environmental conditions. Other generalist species are cockroaches, rats, coyotes, and white-tailed deer.

In contrast, a **specialist species**, such as the giant panda, occupies a narrow niche (Figure 4.5, left curve).

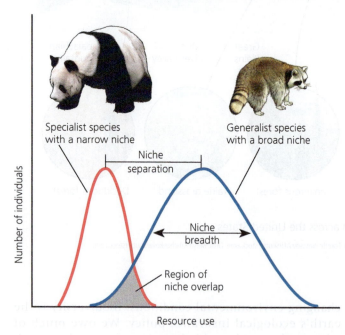

FIGURE 4.5 Specialist species such as the giant panda have a narrow niche (left curve) and generalist species such as the raccoon have a broad niche (right curve).

Such species may be able to live in only one type of habitat, eat only one or a few types of food, or tolerate a narrow range of environmental conditions. For example, some shorebirds are specialized to feed on crustaceans, insects, or other organisms found on sandy beaches and their adjoining coastal wetlands (Figure 4.6).

Because of their narrow niches, specialists are more likely to become endangered or extinct when environmental conditions change. For example, China's giant panda (Figure 4.5, left) is considered vulnerable because of a combination of habitat loss, low birth rate, and its specialized diet consisting mainly of bamboo.

Is it better to be a generalist or a specialist? It depends. When environmental conditions are constant, as in a tropical rain forest, specialists have an advantage because they have fewer competitors. Under rapidly changing environmental conditions, the more adaptable generalist usually is better off.

Species Play Four Major Ecosystem Roles

Niches can be classified further in terms of specific roles that certain species play within ecosystems. Ecologists describe these roles as *native, nonnative, indicator*, and *keystone*. Any given species may play one or more of these four roles in a particular ecosystem (**Concept 4.2B**).

Native species are those that normally live and thrive in a particular ecosystem. Other species that migrate into or are deliberately or accidentally introduced into a new ecosystem are called **nonnative species**.

Some people assume that nonnative species are harmful, which is often not the case. For example, most domesticated species, including certain food crops, flowers, chickens, cattle, and fishes, have benefited humans even where they are nonnative. However, some nonnative species compete with and reduce an ecosystem's native species. They are called **invasive species**.

In 1957, for example, Brazil imported wild African honeybees to help increase honey production. The opposite occurred. The more aggressive African bees displaced some of Brazil's native honeybee populations, which led to a reduced honey supply. Since then, African honeybees have spread across South and Central America and into Mexico and the southern United States. As they spread, they have killed thousands of domesticated animals and an estimated 1,000 people, many of them allergic to bee stings.

Indicator Species Can Sound the Alarm

Species that provide early warnings of environmental change in a community or an ecosystem are called **indicator species**. They are like biological smoke alarms.

For example, in this chapter's **Core Case Study**, you learned that some amphibians are classified as indicator species. One study found an apparent correlation between climate change caused by atmospheric warming and the

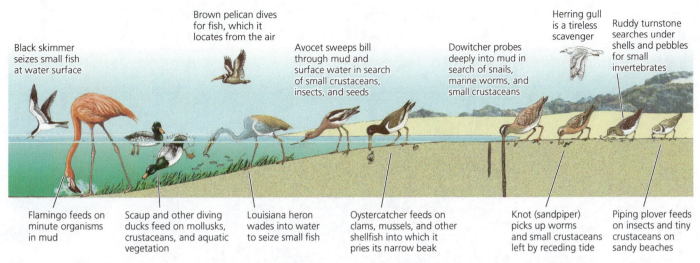

Black skimmer seizes small fish at water surface

Brown pelican dives for fish, which it locates from the air

Avocet sweeps bill through mud and surface water in search of small crustaceans, insects, and seeds

Dowitcher probes deeply into mud in search of snails, marine worms, and small crustaceans

Herring gull is a tireless scavenger

Ruddy turnstone searches under shells and pebbles for small invertebrates

Flamingo feeds on minute organisms in mud

Scaup and other diving ducks feed on mollusks, crustaceans, and aquatic vegetation

Louisiana heron wades into water to seize small fish

Oystercatcher feeds on clams, mussels, and other shellfish into which it pries its narrow beak

Knot (sandpiper) picks up worms and small crustaceans left by receding tide

Piping plover feeds on insects and tiny crustaceans on sandy beaches

FIGURE 4.6 Various bird species in a coastal wetland occupy specialized feeding niches. This specialization reduces competition and allows for sharing of limited resources.

extinction of about two-thirds of the known species of harlequin frogs in tropical forests of Central and South America. Other studies have found similar correlations. Scientists have been working hard to identify some of the possible causes of the declines in amphibian populations (Science Focus 4.2).

Keystone Species Play Critical Ecosystem Roles

A **keystone species** has a large effect on the types and abundance of other species in an ecosystem. Without the keystone species, the ecosystem would be dramatically different or might collapse.

Keystone species play several critical roles in helping to sustain ecosystems. One is the *pollination* of flowering plant species by butterflies, honeybees (Figure 4.A, left), hummingbirds, bats, and other species. In addition, top predator keystone species feed on and help regulate the populations of other species. Examples are wolves, leopards, lions, the American alligator (see the following Case Study), and some shark species (see second Case Study that follows).

The loss of a keystone species can lead to population crashes and extinctions of other species that depend on them for certain ecosystem services. This is why it is so important for scientists to identify keystone species and work to protect them.

The American Alligator—A Keystone Species That Almost Went Extinct

The American alligator (Figure 4.7) is a keystone species because it plays a number of important roles in the ecosystems where it is found in the southeastern United States.

American alligators dig deep depressions, or gator holes. These depressions hold freshwater during dry spells and serve as refuges for aquatic life. They also supply freshwater and food for fishes, insects, snakes, turtles, birds, and other animals.

The large nesting mounds that alligators build provide nesting and feeding sites for some herons and egrets, and red-bellied turtles lay their eggs in old gator nests. In addition, alligators eat large numbers of gar, a predatory fish, which helps to maintain populations of game fish that gar eat, such as bass and bream.

When alligators create gator holes and build nesting mounds, they help prevent vegetation from invading shorelines and open-water areas. Without this ecosystem service, freshwater ponds and coastal wetlands where alligators live would fill in with shrubs and trees, and dozens of species would disappear from these ecosystems.

In the 1930s, hunters began killing large numbers of these animals for their exotic meat and their soft belly skin, used to make expensive shoes, belts, and pocketbooks. Other people hunted alligators for sport or out of dislike for the large reptile. By the 1960s, hunters and poachers had wiped out 90% of the alligators in the state of Louisiana and driven the Florida Everglades population to near extinction.

In 1967, the U.S. government placed the American alligator on the endangered species list. By 1987, because it was protected, its populations had made a strong enough comeback to be removed from the list. Today, there are well over a million alligators in Florida. The state now allows property owners to kill alligators that stray onto their land.

To conservation biologists, the comeback of the American alligator is an important success story in wildlife conservation. Recently, however, large and

FIGURE 4.7 *Keystone species:* The American alligator plays an important ecological role in its marsh and swamp habitats in the southeastern United States by helping to support many other species.

Martha Marks/Shutterstock.com

rapidly reproducing Burmese and African pythons released deliberately or accidently by humans have invaded the Florida Everglades. These nonnative invaders feed on young alligators, threatening the long-term survival of American alligators in the Everglades.

Sharks as Keystone Species

The world's more than 480 known shark species vary widely in size. They range from the goldfish-sized dwarf dog shark to the threatened whale shark (Figure 4.8, left), which can grow to the length of a city bus and weigh as much as two full-grown African elephants.

Certain shark species are keystone species in their ecosystems. Sharks feeding at or near the tops of their food webs, such as the endangered scalloped hammerhead shark (Figure 4.8, right), remove injured and sick animals. Without this ecosystem service, the oceans would teem with dead and dying fish and marine mammals.

Media reports on shark attacks greatly exaggerate the dangers that sharks pose to humans. Every year, members of a few species, including the great white, bull, tiger, oceanic white tip, and hammerhead sharks, injure 60 to 80 people and typically kill 6 to 10 people worldwide. According to shark attack files at the Florida Museum of History, you are much more likely to be killed by a falling coconut or by falling out of bed than by a shark. In 2016,

FIGURE 4.8 The threatened whale shark (left), which feeds on plankton, is the ocean's largest fish and is quite friendly to humans. The scalloped hammerhead shark (right) is endangered.

more people in the world were killed taking selfies (127) than by shark attacks (4).

Each year human activities kill more than 200 million sharks. As many as 73 million sharks die each year after being caught for their valuable fins, a practice called *shark finning*. After capture, their fins are cut off and the sharks are thrown back into the ocean to bleed to death or drown because they can no longer swim. Sharks are also killed for their livers, meat, hides, and jaws, and because we fear them.

Harvested shark fins are used widely in Asia as an ingredient in expensive soup (up to $100 a bowl) and as a supposed pharmaceutical cure-all. According to the wildlife conservation group WildAid, there is no reliable evidence that the fins provide flavor or have any nutritional or medicinal value. The group also warns that consumption of shark fins and shark meat can threaten human health because these foods often contain high levels of mercury and other toxins.

According to a 2014 IUCN study, 25% of the world's open-ocean shark species are threatened with extinction, primarily due to overfishing. Because some sharks are keystone species, their extinction can threaten the ecosystems and the ecosystem services they provide. Sharks are especially vulnerable to population declines because they grow slowly, mature late, and have only a few offspring per generation. Today, they are among the earth's most vulnerable and least protected animals.

CONSIDER THIS ...

LEARNING FROM NATURE

Sharks glide easily through the water because tiny grooves in their skin form continuous channels for water flow. Scientists are studying this to design ship hulls that will save energy and money by moving through the water with less resistance.

4.3 HOW DOES THE EARTH'S LIFE CHANGE OVER TIME?

CONCEPT 4.3A The scientific theory of evolution through natural selection explains how life on the earth changes over time due to changes in the genes of populations.

CONCEPT 4.3B Populations evolve when genes mutate and give some individuals genetic traits that enhance their abilities to survive and to produce offspring with these traits (natural selection).

Evolution Explains How Life Changes over Time

How did the earth end up with such an amazing diversity of species? The scientific answer is **biological evolution** or simply **evolution**—the process by which species change genetically over time (**Concept 4.4A**).

Causes of Amphibian Declines

Herpetologists, or scientists who study amphibians, have identified natural and human-related factors that can cause the decline and disappearance of these indicator species.

One natural threat is *parasites* such as flatworms that feed on certain amphibian eggs. Research indicates that this has caused birth defects such as missing limbs or extra limbs in some amphibians.

Another natural threat comes from *viral and fungal diseases*. An example is the chytrid fungus that infects a frog's skin and causes it to thicken. This reduces the frog's ability to take in water through its skin and leads to death from dehydration. Such diseases can spread easily, because adults of many amphibian species congregate in large numbers to breed.

Habitat loss and fragmentation is another major threat to amphibians. It is mostly a human-caused problem resulting from the clearing of forests and the draining and filling of freshwater wetlands for farming and urban development.

Another human-related problem is *higher levels of UV radiation* from the sun. During the past few decades, ozone-depleting chemicals released into the atmosphere by human activities have

destroyed some the stratosphere's protective ozone. The resulting increase in UV radiation can kill embryos of amphibians in shallow ponds as well as adult amphibians basking in the sun for warmth.

Pollution from human activities also threatens amphibians. Frogs and other species are exposed to pesticides in ponds and in the bodies of insects that they eat. This can make them more vulnerable to bacterial, viral, and fungal diseases and to some parasites.

Climate change is also a concern. Amphibians are sensitive to even slight changes in temperature and moisture. Warmer temperatures may lead amphibians to breed too early. Extended dry periods also lead to a decline in amphibian populations by drying up ponds that frogs and other amphibians depend on for their survival (Figure 4.B).

Overhunting is another human-related threat, especially in areas of Asia and Europe, where frogs are hunted for their leg meat. Another threat is the invasion of amphibian habitats by *nonnative predators and competitors*, such as certain fish species. Some of this immigration is natural, but humans accidentally or deliberately transport many species to amphibian habitats.

Charles H. Smith/U.S. Fish and Wildlife Service

FIGURE 4.B This golden toad lived in Costa Rica's high-altitude Monteverde Cloud Forest Reserve. The species became extinct in 1989, apparently because its habitat dried up.

According to most amphibian experts, a combination of these factors, which vary from place to place, is responsible for most of the decline and extinctions of amphibian species.

CRITICAL THINKING

Of the factors listed above, which three do you think could be most effectively controlled by human efforts?

According to this scientific theory, species have evolved from earlier, ancestral species through **natural selection**—the process in which individuals with certain genetic traits are more likely to survive and reproduce under a specific set of environmental conditions. These individuals then pass these traits on to their offspring (**Concept 4.3B**).

A huge body of scientific evidence supports this idea. As a result, *biological evolution through natural selection* is the most widely accepted scientific theory that explains how the earth's life has changed over the past 3.8 billion years and why we have today's diversity of species.

Most of what we know about the history of life on the earth comes from **fossils**—the remains or traces of past organisms. Fossils include mineralized or petrified replicas

of skeletons, bones, teeth, shells, leaves, and seeds, or impressions of such items found in rocks. Scientists have discovered fossil evidence in successive layers of sedimentary rock such as limestone and sandstone. They have also studied evidence of ancient life contained in ice core samples drilled from glacial ice at the earth's poles and on mountaintops.

This total body of evidence is called the *fossil record*. It is uneven and incomplete because many past forms of life left no fossils and some fossils have decomposed. Scientists estimate that the fossils found so far represent probably only 1% of all species that have ever lived. There are still many unanswered scientific questions about the details of evolution by natural selection, and research continues in this area.

Evolution Depends on Genetic Variability and Natural Selection

The idea that organisms change over time and are descended from a single common ancestor has been discussed since the early Greek philosophers. There was no convincing explanation of how this could happen until 1858 when naturalists Charles Darwin (1809–1882) and Alfred Russel Wallace (1823–1913) independently proposed the concept of natural selection as a mechanism for biological evolution. Darwin gathered evidence for this idea and published it in his 1859 book, *On the Origin of Species by Means of Natural Selection*.

Biological evolution by natural selection involves changes in a population's genetic makeup through successive generations (**Concept 4.3A**). Populations—not individuals—evolve by becoming genetically different. The first step in this process is the development of **genetic variability**: a variety in the genetic makeup of individuals in a population. This occurs primarily through **mutations**, or changes in the coded genetic instructions in the DNA in a gene. During an organism's lifetime, the DNA in its cells is copied each time one of its cells divides and whenever it reproduces. In a lifetime, this happens millions of times and results in various mutations.

Most mutations result from random changes in the DNA's coded genetic instructions that occur in only a tiny fraction of these millions of divisions. Some mutations also occur from exposure to external agents such as radioactivity, UV radiation from the sun, and certain natural and human-made chemicals called *mutagens*.

Mutations can occur in any cell, but only those that take place in the genes of reproductive cells may be passed on to offspring. Sometimes a mutation can result in a new genetic trait, called a *heritable trait*, which can be passed from one generation to the next. In this way, populations develop genetic differences among their individuals.

The next step in biological evolution is natural selection, which explains how populations evolve in response to changes in environmental conditions by changing their genetic makeup. Through natural selection, environmental conditions favor increased survival and reproduction of certain individuals in a population. These favored individuals possess heritable traits that give them an advantage over other individuals in the population. Such a trait is called an **adaptation**, or **adaptive trait**. An adaptive trait improves the ability of an individual organism to survive and to reproduce at a higher rate than other individuals in a population can under current environmental conditions.

An example of natural selection at work is *genetic resistance*. It occurs when one or more organisms in a population have genes that can tolerate a chemical (like as a pesticide or antibiotic) that normally would be fatal. The resistant individuals survive and reproduce more rapidly than the members of the population that do not have such genetic traits. Genetic resistance can develop quickly in populations of organisms like bacteria and insects that can produce large numbers of offspring in a short time. For example, some disease-causing bacteria have developed genetic resistance to widely used antibacterial drugs, or *antibiotics* (Figure 4.9).

Through natural selection, humans have evolved traits that have enabled them to survive in many different environments and to reproduce successfully. If we think of the earth's 4.6 billion years of geological and biological history as one 24-hour day, the human species arrived about a tenth of one second before midnight. In that short time, we have dominated most of the earth's land and aquatic systems with a growing ecological footprint (see Figure 1.9, p. 11). Evolutionary biologists attribute our ability to dominate the earth to three major adaptations:

- *Strong opposable thumbs* allowed humans to grip and use tools better than the few other animals that have thumbs.

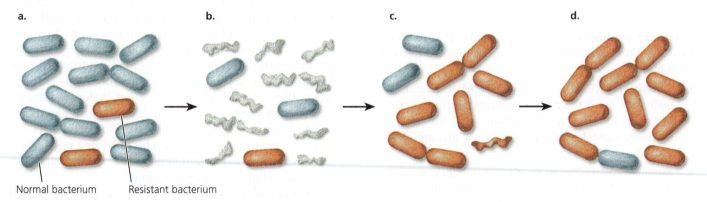

a.　　　　b.　　　　c.　　　　d.

Normal bacterium　　Resistant bacterium

FIGURE 4.9 *Evolution by natural selection:* **(a)** A population of bacteria is exposed to an antibiotic, which **(b)** kills all individuals except those possessing a trait that makes them resistant to the drug. **(c)** The resistant bacteria multiply and eventually **(d)** replace all or most of the nonresistant bacteria.

- *The ability to walk upright* gave humans agility and freed up their hands for many uses.

- *A complex brain* allowed humans to develop many skills, including the ability to communicate complex ideas.

To summarize the process of biological evolution by natural selection: genes mutate, certain individuals are selected, and populations that are better adapted to survive and reproduce under existing environmental conditions evolve (**Concept 4.3B**).

Limits to Adaptation through Natural Selection

In the not-too-distant future, will adaptations to new environmental conditions through natural selection protect us from harm? For example, will adaptations make the skin of our descendants more resistant to the harmful effects of UV radiation or enable their lungs to cope with air pollutants?

Scientists in this field say this is not likely because of two limitations on adaptation through natural selection. *First*, a change in environmental conditions leads to adaptation only for genetic traits already present in a population's gene pool, or if such traits arise from random mutations.

Second, even if a beneficial heritable trait is present in a population, the population's ability to adapt may be limited by its reproductive capacity. Populations of genetically diverse species that reproduce quickly often adapt to a change in environmental conditions in a short time (days to years). Examples are dandelions, mosquitoes, rats, bacteria, and cockroaches. By contrast, species that cannot produce large numbers of offspring rapidly—such as elephants, tigers, sharks, orangutans, and humans—take thousands or even millions of years to adapt through natural selection.

Myths about Evolution through Natural Selection

There are a number of misconceptions about biological evolution through natural selection. Here are five common myths:

- *Survival of the fittest means survival of the strongest.* To biologists, *fitness* is a measure of reproductive success, not strength. Thus, the fittest individuals are those that leave the most descendants, not those that are physically the strongest.

- *Evolution explains the origin of life.* It does not. However, it does explain how species evolved after life came into being around 3.8 billion years ago.

- *Humans evolved from apes or monkeys.* Fossil and other evidence shows that humans, apes, and monkeys evolved along different paths from a common ancestor that lived 5–8 million years ago.

- *Evolution by natural selection is part of a grand plan in nature in which species are to become more perfectly adapted.* There is no evidence of such a plan.

- *Evolution by natural selection is not important because it is just a theory.* This reveals a misunderstanding of the concept of a scientific theory, which is based on extensive evidence and accepted widely by the scientific experts in a particular field of study. Numerous polls show that evolution by natural selection is widely accepted by over 95% of biologists because it best explains the earth's biodiversity and how populations of different species have adapted to changes in the earth's environmental conditions over billions of years.

4.4 WHAT FACTORS AFFECT BIODIVERSITY?

CONCEPT 4.4A As environmental conditions change, the balance between the formation of new species and the extinction of existing species determines the earth's biodiversity.

CONCEPT 4.4B Human activities are decreasing biodiversity by causing the extinction of many species and by destroying or degrading habitats needed for the development of new species through natural selection.

How Do New Species Arise?

Under certain circumstances, natural selection can lead to an entirely new species. Through this process, called **speciation**, one species evolves into two or more different species.

Speciation, especially among sexually reproducing species, happens in two phases: first geographic isolation, and then reproductive isolation. **Geographic isolation** occurs when different groups of the same population of a species become physically isolated from one another for a long time. Part of a population may migrate in search of food and then begin living as a separate population in an area with different environmental conditions. Winds and flowing water may carry a few individuals far away where they establish a new population. A flooding stream, new road, a hurricane, earthquake, or volcanic eruption, and long-term geological processes (Science Focus 4.3) can also separate populations. The separated populations can develop quite different genetic characteristics because they are no longer exchanging genes.

In **reproductive isolation**, mutation and change by natural selection operate independently in the gene pools of geographically isolated populations. If this process continues for a long enough time, members of isolated populations of sexually reproducing species can become different in genetic makeup. Then they cannot produce

Geological Processes Affect Biodiversity

The earth's surface has changed dramatically over its long history. Scientists discovered that huge flows of molten rock within the earth's interior have broken its surface into a number of gigantic solid plates, called *tectonic plates*. For hundreds of millions of years, these plates have drifted slowly on the planet's mantle (Figure 4.C).

Rock and fossil evidence indicates that 200–250 million years ago, all of the earth's present-day continents were connected in a supercontinent called Pangaea (Figure 4.C, left). About 175 million years ago, Pangaea began splitting apart as the earth's tectonic plates moved. Eventually, tectonic movement resulted in the present-day locations of the continents (Figure 4.C, right).

The movement of tectonic plates has had two important effects on the evolution and distribution of life on the earth. *First*, the locations of continents and oceanic basins have greatly influenced the earth's climate, which plays a key role in where plants and animals can live. *Second*, the breakup, movement, and joining of continents have allowed species to move and adapt to new environments. This led to the formation of a large number of new species through speciation.

Along boundaries where they meet, tectonic plates may pull away from, collide with, or slide past each other. Tremendous forces produced by these interactions along plate boundaries can lead to earthquakes and volcanic eruptions. These geological activities can also affect biological evolution by causing fissures in the earth's crust, which can isolate populations of species on either side of the fissure. Over long periods, this can lead to the formation of new species as each isolated population changes genetically in response to new environmental conditions.

Volcanic eruptions that occur along the boundaries of tectonic plates can also affect extinction and speciation by destroying habitats and reducing, isolating, or wiping out populations of species. These geological processes are further discussed in Chapter 12.

CRITICAL THINKING

The earth's tectonic plates, including the one you are riding on, typically move at about the rate at which your fingernails grow. If they stopped moving, how might this affect the future biodiversity of the planet?

225 million years ago

Present

FIGURE 4.C Over millions of years, the earth's continents have moved very slowly on several gigantic tectonic plates. **Critical thinking:** How might an area of land splitting apart cause the extinction of a species?

live, fertile offspring if they are rejoined and attempt to interbreed. When that happens, speciation occurs and one species becomes two (Figure 4.10).

Artificial Selection, Genetic Engineering, and Synthetic Biology

For thousands of years, humans have used **artificial selection** to change the genetic characteristics of populations with similar genes. First, they select one or more desirable genetic traits that already exist in the population of a plant or animal. Then, they use *selective breeding*, or *crossbreeding*, to control which members of a population have the opportunity to reproduce to increase the numbers of individuals in a population with the desired traits.

Artificial selection is not a form of speciation. It is limited to crossbreeding between genetic varieties of the same species or between species that are genetically similar to one another. Most of the grains, fruits, and vegetables we eat are produced by artificial selection. Artificial selection has also given us food crops with higher yields, cows that give more milk, trees that grow faster, and many different varieties of dogs and cats. However, traditional crossbreeding is a slow process.

Scientists have learned how to speed this process of manipulating genes in order to select desirable traits or

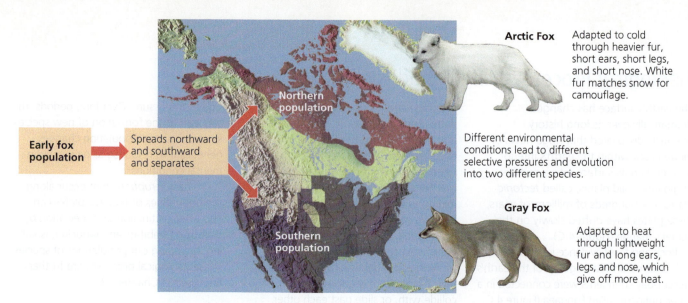

Arctic Fox Adapted to cold through heavier fur, short ears, short legs, and short nose. White fur matches snow for camouflage.

Early fox population

Spreads northward and southward and separates

Northern population

Southern population

Different environmental conditions lead to different selective pressures and evolution into two different species.

Gray Fox Adapted to heat through lightweight fur and long ears, legs, and nose, which give off more heat.

FIGURE 4.10 *Geographic isolation* can lead to reproductive isolation, divergence of gene pools, and speciation.

eliminate undesirable ones. They do this by transferring segments of DNA with a desired trait from one species to another through a process called **genetic engineering**. In this process, also known as *gene splicing*, scientists alter an organism's genetic material by adding, deleting, or changing segments of its DNA to produce desirable traits or to eliminate undesirable ones. Scientists have used genetic engineering to develop modified crop plants, new drugs, pest-resistant plants, and animals that grow rapidly.

The result is a **genetically modified organism (GMO)**—an organism with its genetic information modified in a way not found in natural organisms. Genetic engineering enables scientists to transfer genes between different species that would not interbreed in nature. For example, scientists can put genes from a cold-water fish species into a tomato plant to give it properties that help it resist cold weather. Genetic engineering has revolutionized agriculture and medicine. However, it is a controversial technology, as discussed in Chapter 10.

A new and rapidly growing form of genetic engineering is **synthetic biology**. It enables scientists to make new sequences of DNA and use such genetic information to design and create artificial cells, tissues, body parts, and organisms not found in nature.

Proponents of this new technology want to use it to create bacteria that can use sunlight to produce clean-burning hydrogen gas, which can be used to fuel motor vehicles. They also view it as a way to create new vaccines to prevent diseases and drugs to combat parasitic diseases such as malaria. Synthetic biology might also be used to create bacteria and algae that would break down oil, industrial wastes, toxic heavy metals, pesticides, and radioactive waste in contaminated soil and water. Scientists are a long way from achieving such goals but the potential is there.

The problem is that like any technology, synthetic biology can be used for good or bad. For example, it could also be used to create biological weapons such as deadly bacteria that spread new diseases, to destroy existing oil deposits, or to interfere with the chemical cycles that keep us alive. This is why many scientists call for increased monitoring and regulation of this new technology to help control its use.

Extinction Eliminates Species

Another factor affecting the number and types of species on the earth is **biological extinction**, or simply **extinction**, which occurs when an entire species ceases to exist. When environmental conditions change dramatically or rapidly, a population of a species faces three possible futures: *adapt* to the new conditions through natural selection, *migrate* (if possible) to another area with more favorable conditions, or *become extinct*.

Species found in only one area, called **endemic species**, are especially vulnerable to extinction. They exist on islands and in other isolated areas. These organisms are unlikely to be able to migrate or adapt to rapidly changing environmental conditions. Many of these endangered species are amphibians (**Core Case Study**), such as the now-extinct golden toad (Figure 4.A).

Fossils and other scientific evidence indicate that 99.9% of all the species that have existed on the earth are now extinct. Throughout most of the earth's long history, species have disappeared at a low rate, called the **background extinction rate**.

Evidence indicates that life on the earth has been sharply reduced by several periods of **mass extinction** during which there is a significant rise in extinction rates, well above the background rate. In such a catastrophic, widespread, and often global events, 50–95% of all species are wiped out primarily because of major, widespread environmental changes such as long-term climate change, massive flooding because of rising sea levels, and huge meteorites striking the earth's surface. Fossil and geological evidence indicate that there have been five mass extinctions (at intervals of 25–100 million years) during the past 500 million years (Figure 4.11).

A mass extinction provides an opportunity for the evolution of new species that can fill unoccupied ecological niches or newly created ones. Scientific evidence indicates that each mass extinction has been followed by an increase in species diversity as shown by the wedges in Figure 4.11.

As environmental conditions change, the balance between speciation and extinction determines the earth's biodiversity (**Concept 4.4A**). The existence of millions of species today means that speciation, on average, has kept ahead of extinction. However, evidence indicates that the global extinction rate is rising dramatically, as discussed more fully in Chapter 8. Many scientists argue we are experiencing the beginning of a new sixth mass extinction caused mostly by human activities (**Concept 4.4B**).

CASE STUDY

The Threatened Monarch Butterfly

The beautiful North American monarch butterfly (Figure 4.12) is in trouble. This species is known for its annual 3,200- to 4,800-kilometer (2,000- to 3,000-mile) migration from parts of the northern United States and Canada to a small number of tropical forest areas in central Mexico. They arrive on a predictable schedule and later return to their North American home. Another monarch population in the Midwestern United States makes a

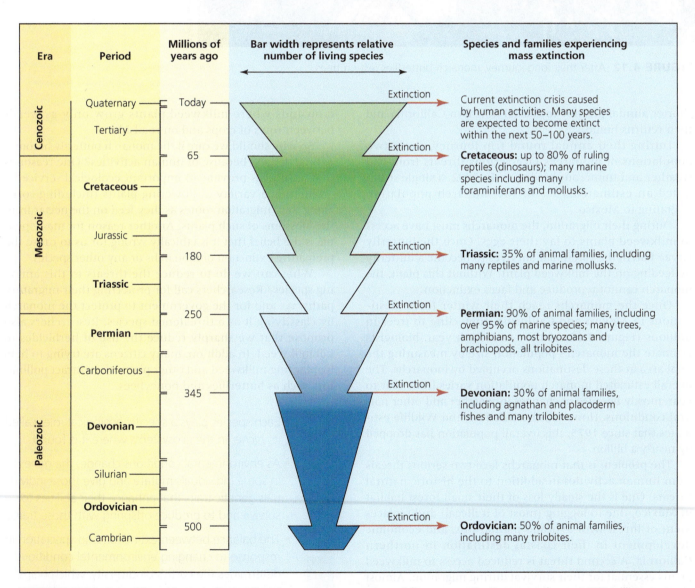

FIGURE 4.11 Scientific evidence indicates that the earth has experienced five mass extinctions over the past 500 million years and that human activities have initiated a new sixth mass extinction.

FIGURE 4.12 After their long journey, monarch butterflies rest on trees.

shorter annual journey to coastal northern California and then returns home.

During their annual round trip journey, these two populations of monarchs face serious threats from bad weather and numerous predators. In 2002, a single storm killed an estimated 75% of the monarch population migrating to Mexico.

During their migration, the monarchs must have access to milkweed plants to lay their eggs. Once the butterfly larvae hatch, the caterpillar survives to become a butterfly by feeding on the milkweed plant. Without this plant, the monarch cannot reproduce and faces extinction.

Once the monarchs reach their winter forest destinations in Mexico and California, they cling to trees in millions (Figure 4.12) as they rest. Each year, biologists estimate the monarch's population size by measuring the total area at these destinations occupied by monarchs. The overall estimated monarch population varies from year to year, mostly because of changes in weather and other natural conditions. However, the U.S. Fish and Wildlife estimates that since 1975, this overall population has dropped by nearly a billion.

The problem is that monarchs face two serious threats from human activities in addition to the historic natural threats. One is the steady loss of their small forest habitat in Mexico, due to logging (most of it illegal) and replacement of forest with avocado plantations, and economic development in their coastal destination in northern California. A second threat is reduced access to milkweed plants essential for their survival during migration. Almost all of the natural prairies in the United States, which were abundant with milkweed plants, have been replaced by croplands where milkweed plants grow only as weeds between rows of crops and on roadsides.

So why should we care if the monarch butterfly becomes extinct, largely because of human activities? One reason is that monarchs provide an important ecological service by pollinating a variety of flowering plants (including corn) along their migration routes as they feed on the nectar from the blossoms of such plants. Another reason for many people is the belief that it is ethically wrong for us to cause the premature extinction of monarchs or any other species.

What can we do to reduce the threats to this amazing species? Researchers call for protecting their migratory pathways and for the government to protect the monarch by classifying it as a threatened species. Researchers also propose that we sharply reduce the use of herbicides to kill milkweed. In addition, many citizens are trying to help by planting milkweed and other plants that attract pollinators such as butterflies and honeybees.

BIG IDEAS

- Each species plays a specific ecological role, called its *niche*, in the ecosystems where it is found.

- As environmental conditions change, the genes in some individuals mutate and give those individuals genetic traits that enhance their abilities to survive and to produce offspring with those traits.

- The balance between extinction and speciation in response to changing environmental conditions determines the earth's biodiversity, which helps to sustain the earth's life and our economies.

Tying It All Together

Amphibians and Sustainability

This chapter's Core Case Study describes the increasing losses of amphibian species and explains why these species are important, ecologically. In this chapter, we studied the importance of biodiversity—the numbers and varieties of species found in different parts of the world, along with genetic, ecosystem, and functional diversity.

We examined the variety of niches, or roles played by species in ecosystems. For example, we saw that some species, including many amphibians, are indicator species that warn us about threats to biodiversity, to ecosystems, and to the biosphere. Others such as the American alligator and some shark species are keystone species that play vital

roles in sustaining the ecosystems where they live.

We also studied the scientific theory of biological evolution through natural selection, which explains how life on the earth changes over time due to changes in the genes of populations and how new species can arise. We learned that the earth's biodiversity is the result of a balance between the formation of new species (speciation) and extinction of existing species in response to changing environmental conditions.

The ecosystems where amphibians and other species live are functioning examples of the three **scientific principles of sustainability** in action. They depend on solar energy,

Robert King/Shutterstock.com

the cycling of nutrients, and biodiversity. Disruptions of any of these forms of natural capital can result in degradation of these species' populations and their ecosystems.

Chapter Review

Core Case Study

1. Define **amphibian**. Describe the threats to many of the world's amphibian species. Explain why we should avoid hastening the extinction of amphibian species through our activities.

Section 4.1

2. What is the key concept for this section? Define **biodiversity (biological diversity)** and list and describe its four major components. Define and distinguish among **species diversity**, **genetic diversity**, **ecosystem diversity**, and **functional diversity**. Define and give three examples of **biomes**. Why is biodiversity important? Summarize the importance of insects. Summarize the scientific contributions of Edward O. Wilson.

Section 4.2

3. What are the two key concepts for this section? Define and distinguish between **ecological niche** and **habitat**. Distinguish between **generalist species** and **specialist species** and give an example of each.

4. Define and distinguish among **native**, **nonnative**, **indicator**, and **keystone species** and give

an example of each. What is an **invasive species**? List six factors that threaten many species of frogs and other amphibians with extinction. Describe the role of the American alligator as a keystone species. Explain why ecologists say we should protect sharks.

Section 4.3

5. What are the two key concepts for this section? Define **biological evolution (evolution)** and **natural selection** and explain how they are related. What is the scientific theory of biological evolution through natural selection? What are **fossils** and how do scientists use them to understand evolution?

6. What is **genetic variability**? What is a **mutation** and what role do mutations play in evolution through natural selection? What is an **adaptation**, or **adaptive trait**? Explain how harmful bacteria can become genetically resistant to antibiotics. What three genetic adaptations have helped humans to become such a powerful species?

7. What are two limitations on evolution through natural selection? What are five common myths about evolution through natural selection?

8. What are the two key concepts for this section? Define **speciation**. Distinguish between **geographic isolation** and **reproductive isolation**, and explain how they can lead to the formation of a new species. Explain how geological processes can affect biodiversity. What is **artificial selection**? Explain the process of **genetic engineering**. What is a **genetically modified organism (GMO)**? Define **synthetic biology**, explain how it differs from evolution by natural selection, and point out some of its potential benefits and dangers.

9. What is **biological extinction (extinction)**? What is an **endemic species** and why are such species vulnerable to extinction? Define and distinguish between the **background extinction rate** and a **mass extinction**. How many mass extinctions has the earth experienced? What is one of the leading causes of the rising rate of extinction? Explain why the monarch butterfly is threatened with extinction.

10. What are this chapter's *three big ideas*? How are ecosystems where amphibians and other species live functioning examples of the three **scientific principles of sustainability**?

_____

Note: Key terms are in bold type.

Critical Thinking

1. How might we and other species be affected if most or all amphibians (**Core Case Study**) were to go extinct?

2. Is the human species a keystone species? Explain. If humans were to become extinct, what are three species that might also become extinct and what are three species whose populations would probably grow?

3. Why should we care about saving the monarch butterfly from extinction caused by our actions?

4. If you were forced to choose between saving the giant panda from extinction and saving a shark species, which would you choose? Explain.

5. How would you respond to someone who tells you that:
   a. We should not believe in biological evolution because it is "just a theory"?

   b. We should not worry about air pollution because natural selection will enable humans to develop lungs that can detoxify pollutants?

6. How would you respond to someone who says that because extinction is a natural process, we should not worry about the loss of biodiversity when species become extinct largely because of our activities?

7. List three aspects of your lifestyle that could be contributing to some of the losses of the earth's biodiversity. For each of these, what are some ways to avoid making this contribution?

8. Congratulations! You are in charge of the future evolution of life on the earth. What are the three things that you would consider to be the most important to do?

Doing Environmental Science

Study an ecosystem of your choice, such as a meadow, a patch of forest, a garden, or an area of wetland. (If you cannot do this physically, do so virtually by reading about an ecosystem online or in a library.) Determine and list five major plant species and five major animal species in your ecosystem. Write hypotheses about **(a)** which of these species, if any, are indicator species and **(b)** which of them, if any, are keystone species. Explain how you arrived at your hypotheses. Then design an experiment to test each of your hypotheses, assuming you would have unlimited means to carry them out.

Global Environment Watch Exercise

Go to your MindTap course to access the GREENR database. Using the "Basic Search" box at the top of the page, search for *Amphibians* to find out more about the current state of these species with regard to threats to their existence (**Core Case Study**). What actions are being taken by various nations and organizations to protect amphibians? Write a short summary report on your research.

Data Analysis

The following table is a sample of a very large body of data reported by J. P. Collins, M. L. Crump, and T. E. Lovejoy III in their book *Extinction in Our Times– Global Amphibian Decline*. It compares various areas of the world in terms of the number of amphibian species found and the number of amphibian species that were endemic, or unique to each area. Scientists like to know these percentages because endemic species tend to be more vulnerable to extinction than do non-endemic species. Study the table below and then answer the following questions.

1. Fill in the fourth column to calculate the percentage of amphibian species that are endemic to each area.

2. Which two areas have the highest numbers of endemic species? Which two areas with the highest percentages of endemic species.

3. Which two areas have the lowest numbers of endemic species? Which two areas have the lowest percentages of endemic species?

4. Which two areas have the highest percentages of non-endemic species?

Area	Number of Species	Number of Endemic Species	Percentage Endemic
Pacific/Cascades/Sierra Nevada Mountains of North America	52	43	
Southern Appalachian Mountains of the United States	101	37	
Southern Coastal Plain of the United States	68	27	
Southern Sierra Madre of Mexico	118	74	
Highlands of Western Central America	126	70	
Highlands of Costa Rica and Western Panama	133	68	
Tropical Southern Andes Mountains of Bolivia and Peru	132	101	
Upper Amazon Basin of Southern Peru	102	22	

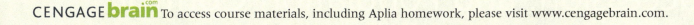

CENGAGE**brain**.com To access course materials, including Aplia homework, please visit www.cengagebrain.com.

CHAPTER 5

Species Interactions, Ecological Succession, and Population Control

In looking at nature, never forget that every single organic being around us may be said to be striving to increase its numbers.

CHARLES DARWIN, 1859

Key Questions

5.1 How do species interact?

5.2 How do communities and ecosystems respond to changing environmental conditions?

5.3 What limits the growth of populations?

A clownfish gains protection by living among sea anemones and helps protect the anemones from some of their predators.

Morrison/Dreamstime.com

The Southern Sea Otter: A Species in Recovery

Southern sea otters (Figure 5.1, left) live in giant kelp forests (Figure 5.1, right) in shallow waters along parts of the Pacific coast of North America. Most of the remaining members of this endangered species are found off the California coast of between the cities of Santa Cruz and Santa Barbara.

Southern sea otters are fast and agile swimmers that dive to the ocean bottom looking for shellfish and other prey. They swim on their backs on the ocean surface and use their bellies as a table to eat their prey (Figure 5.1, left). Each day, a sea otter consumes 20–35% of its weight in clams, mussels, crabs, sea urchins, abalone, and other species of bottom-dwelling organisms. Their thick, dense fur traps air bubbles and keeps the otters warm.

Estimates show that around 16,000 southern sea otters once lived in California's coastal waters. By the early 1900s, they had been hunted almost to extinction in this region by fur traders who killed them for their luxurious fur. Commercial fishers also killed otters, as they competed with them for valuable abalone and other shellfish.

The southern sea otter population grew from a low of 50 in 1938 to 1,850 by 1977 when the U.S. Fish and Wildlife Service listed the species as endangered. By 2016, the population had grown slowly to 3,511, which was above the size needed for it to be removed from the endangered species list, although it also must stay above that size for three consecutive years for such delisting.

Why should we care about the southern sea otters of California? One reason is ethical: Many people believe it is wrong to allow human activities to cause the extinction of a species. Another reason is that people love to look at these appealing and highly intelligent animals as they play in the water. As a result, otters help to generate millions of dollars a year in tourism revenues. A third reason—and a key reason in our study of environmental science—is that biologists classify the southern sea otter as a *keystone species* (see p. 75). Scientists hypothesize that in the absence of southern sea otters, sea urchins and other kelp-eating species would probably destroy the Pacific coast kelp forests and much of the rich biodiversity they support.

Biodiversity is an important part of the earth's natural capital and is the basis for one of the three **scientific principles of sustainability**. In this chapter, we look at how species interact and help control one another's population sizes. We also explore how communities, ecosystems, and populations of species respond to changes in environmental conditions. ●

FIGURE 5.1 An endangered southern sea otter in Monterey Bay, California, uses a stone to crack the shells of the clams that it feeds on (left). It lives in a bed of seaweed called giant kelp (right).

5.1 HOW DO SPECIES INTERACT?

CONCEPT 5.1 Five types of interactions among species—interspecific competition, predation, parasitism, mutualism, and commensalism—affect the resource use and population sizes of species.

Competition for Resources

Ecologists have identified five basic types of interactions among species as they share limited resources such as food, shelter, and space. These types of interactions are called *interspecific competition, predation, parasitism, mutualism*, and *commensalism*. They all affect the population sizes and the use of resources in an ecosystem (**Concept 5.1**).

Competition is the most common interaction among species. It occurs when members of one or more species interact to use the same limited resources such as food, water, light, and space. Competition between different species is called **interspecific competition**. It plays a larger role in most ecosystems than *intraspecific competition*—competition among members of the same species.

When two species compete with one another for the same resources, their niches *overlap* (see Figure 4.5, p. 74). The greater this overlap, the more they compete for key resources. If one species can take over the largest share of one or more key resources, each of the other competing species must move to another area (if possible), suffer a sharp population decline, or become extinct in that area.

Humans compete with many other species for space, food, and other resources. As our ecological footprints grow and spread (see Figure 1.9, p. 11), we take over or degrade the habitats of many of those species and deprive them of resources they need in order to survive.

If given enough time for natural selection to occur, populations can develop adaptations that enable them to reduce or avoid competition with other species. **Resource partitioning** occurs when different species competing for similar scarce resources evolve specialized traits that allow them to "share" the same resources. Sharing resources can mean using parts of the resources, or using the resources at different times or in different ways. Figure 5.2 shows resource partitioning by insect-eating bird species. Adaptations allow the birds to reduce competition by feeding in different portions of certain spruce trees and by feeding on different insect species.

Predation

In **predation**, a member of one species is the **predator** that feeds directly on all or part of a member of another species, the **prey**. A lion (the predator) and a zebra (the prey) are engaged in a **predator–prey relationship** (Figure 3.5, p. 50). This type of species interaction has a strong effect on population sizes and other factors in many ecosystems.

In a giant kelp forest ecosystem, sea urchins prey on kelp, a type of seaweed (Science Focus 5.1). As a keystone species, southern sea otters (**Core Case Study**) prey on the sea urchins and prevent them from destroying the kelp forests.

Predators have a variety of methods to help them capture prey. Herbivores can walk, swim, or fly to the plants they feed on. Many carnivores such as cheetahs use their speed to chase down and kill their prey. Eagles and hawks

Blackburnian Warbler   Black-throated Green Warbler   Cape May Warbler   Bay-breasted Warbler   Yellow-rumped Warbler

After R. H. MacArthur, "Population Ecology of Some Warblers in Northeastern Coniferous Forests," *Ecology* 36:533–535, 1958.

FIGURE 5.2 *Sharing the wealth:* Resource partitioning among five species of insect-eating warblers in the spruce forests of the U.S. state of Maine. Each species spends at least half of its feeding time in its associated yellow-highlighted areas of these spruce trees.

Threats to Kelp Forests

A kelp forest contains large concentrations of seaweed called *giant kelp*. Anchored to the ocean floor, its long blades grow toward the sunlit surface waters (Figure 5.1, right). Under good conditions, the blades can grow 0.6 meter (2 feet) in a day and the plant can grow as tall as a 10-story building. The blades are flexible and can survive all but the most violent storms and waves.

Kelp forests support many marine plants and animals and are one of the most biologically diverse marine ecosystems. They also reduce shore erosion by blunting the force of incoming waves and trapping some of the outgoing sand.

Sea urchins (Figure 5.A) prey on kelp plants. Large populations of these predators can rapidly devastate a kelp forest because they eat the bases of young kelp plants. Scientific studies by biologists, including James Estes of the University of California at Santa Cruz, indicate that the southern sea otter (**Core Case Study**) is

a keystone species that helps to sustain kelp forests by controlling populations of sea urchins.

Another threat to kelp forests is polluted water running off the land. The pollutants in this runoff include pesticides and herbicides that can kill kelp plants and other species and upset the food webs in these aquatic forests. Another runoff pollutant is fertilizer. Its plant nutrients (mostly nitrates) can cause excessive growth of algae and other aquatic plants. This growth blocks some of the sunlight needed to support the growth of giant kelp.

Some scientists warn that the current warming of the world's oceans is a growing threat to kelp forests, which require cool water. If coastal waters get warmer during this century, as projected by climate

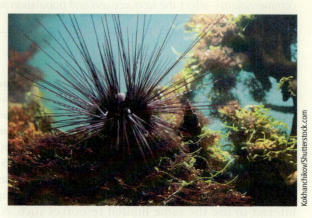

FIGURE 5.A The purple sea urchin inhabits the coastal waters of the U.S. state of California and feeds on kelp.

models, many or most of the California's coastal kelp forests could disappear.

CRITICAL THINKING

List three ways in which we could reduce the degradation of giant kelp forest ecosystems.

---

have keen enough eyesight to spot their prey from the air as they fly. Some predators such as female African lions work in groups to capture large or fast-running prey.

Other predators use *camouflage* to hide in plain sight and ambush their prey. For example, praying mantises (see Figure 4.A, right, p. 72) sit on flowers or plants of a color similar to their own and ambush visiting insects. White ermines (a type of weasel), snowy owls, and arctic foxes hunt their prey in snow-covered areas. People camouflage themselves to hunt wild game and use camouflaged traps to capture wild animals.

Some predators use *chemical warfare* to attack their prey. For example, some spiders and poisonous snakes use venom to paralyze their prey and to defend against their predators.

Prey species have evolved many ways to avoid predators. Some can run, swim, or fly fast; and some have highly developed senses of sight, sound, or smell that alert them to the presence of predators. Other adaptations include protective shells (abalone and turtles), thick bark (on giant sequoia trees), spines (porcupines and sea urchins), and thorns (cacti and rose bushes).

Other prey species use *camouflage* to blend into their surroundings. Some insect species resemble twigs (Figure 5.3a) or bird droppings on leaves. A leaf insect can

be almost invisible against its background (Figure 5.3b), as can an arctic hare in its white winter fur.

Prey species also use *chemical warfare*. Some discourage predators by containing or emitting chemicals that are poisonous (oleander plants), irritating (stinging nettles and bombardier beetles, Figure 5.3c), foul smelling (skunks and stinkbugs), or bad tasting (buttercups and monarch butterflies, Figure 5.3d). When attacked, some species of squid and octopus emit clouds of black ink, allowing them to escape by confusing their predators.

Many bad-tasting, bad-smelling, toxic, or stinging prey species have evolved *warning coloration* to warn others that eating them is risky. Examples are the brilliantly colored, foul-tasting monarch butterflies (Figure 5.3d) and poisonous frogs (Figure 5.3e). When a bird eats a monarch butterfly, it usually vomits and learns to avoid monarchs.

CONSIDER THIS ...

CONNECTIONS Coloration and Poison

Biologist Edward O. Wilson gives us two rules for evaluating the possible dangers posed by various brightly colored animal species. First, if they are small and strikingly beautiful, they are probably poisonous. Second, if they are strikingly beautiful and easy to catch, they are probably deadly.

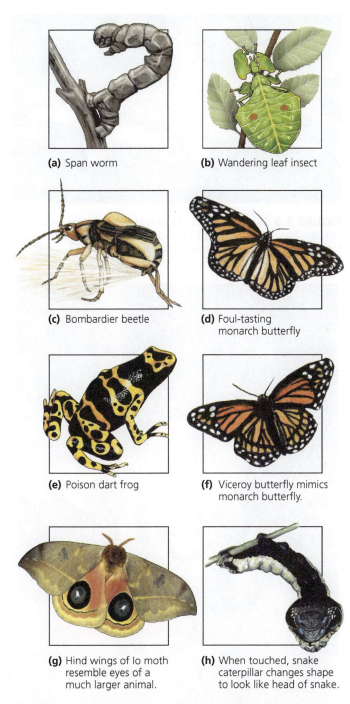

(a) Span worm

(b) Wandering leaf insect

(c) Bombardier beetle

(d) Foul-tasting monarch butterfly

(e) Poison dart frog

(f) Viceroy butterfly mimics monarch butterfly.

(g) Hind wings of Io moth resemble eyes of a much larger animal.

(h) When touched, snake caterpillar changes shape to look like head of snake.

FIGURE 5.3 These prey species have developed specialized ways to avoid their predators: (a, b) *camouflage*, (c, d, e) *chemical warfare*, (d, e, f) *warning coloration*, (f) *mimicry*, (g) *deceptive looks*, and (h) *deceptive behavior*.

Some butterfly species gain protection by looking and acting like other, more dangerous species, a protective device known as *mimicry*. For example, the nonpoisonous viceroy butterfly (Figure 5.3f) mimics the monarch butterfly. Other prey species use *behavioral strategies* to avoid predation. Some attempt to scare off predators by

puffing up (blowfish), spreading their wings (peacocks), or mimicking a predator (Figure 5.3h). Some moths have wings that look like the eyes of much larger animals (Figure 5.3g). Other prey species gain protection by living in large groups such as schools of fish and herds of antelope.

At the individual level, members of predator species benefit from their predation and members of the prey species are harmed. At the population level, predation plays a role in natural selection. Animal predators tend to kill the sick, weak, aged, and least fit members of a prey population because they are the easiest to catch. Individuals with better defenses against predation thus tend to survive longer and leave more offspring with adaptations that can help them avoid predation.

Coevolution

Over time, a prey species develops traits that make it more difficult to catch. Its predators then face selection pressures that favor traits that increase their ability to catch their prey. Then, the prey species must get better at eluding the more effective predators.

This back-and-forth adaptation is called **coevolution**, a natural selection process in which changes in the gene pool of one species lead to changes in the gene pool of another species. It can play an important role in controlling population growth of predator and prey species. When populations of two different species interact in such a way over a long period of time, changes in the gene pool of one population can lead to changes in the gene pool of the other. Such changes can help both competing species to become more competitive or to avoid or reduce competition.

For example, bats prey on certain species of moths (Figure 5.4) that they hunt at night using echolocation. They emit pulses of high-frequency sound that bounce off their prey and they capture the returning echoes that tell them where their prey is located. Over time, certain moth species have evolved ears that are sensitive to the sound frequencies that bats use to find them. When they hear these frequencies, they drop to the ground or fly evasively. Some bat species evolved ways to counter this defense by changing the frequency of their sound pulses. In turn, some moths evolved their own high-frequency clicks to jam the bats' echolocation systems. Some bat species then adapted by turning off their echolocation systems and using the moths' clicks to locate their prey. This is a classic example of coevolution.

Parasitism, Mutualism, and Commensalism

Parasitism occurs when one species (the *parasite*) lives in or on another organism (the *host*). The parasite benefits by extracting nutrients from the host. A parasite weakens

FIGURE 5.4 *Coevolution*: This bat is using ultrasound to hunt a moth. As the bats evolve traits to increase their chance of getting a meal, the moths evolve traits to help them avoid being eaten.

FIGURE 5.5 *Parasitism*: This blood-sucking, parasitic sea lamprey has attached itself to an adult lake trout from one of the Great Lakes of the United States and Canada.

its host but rarely kills it, since doing so eliminates the source of its benefits. Parasites can be plants, animals, or microorganisms.

Tapeworms are parasites that live part of their life cycle inside their hosts. Others such as mistletoe plants and blood-sucking sea lampreys (Figure 5.5) attach themselves to the outsides of their hosts and suck nutrients from them. Some parasites move from one host to another (fleas and ticks), while others (such as certain protozoa) spend their adult lives within a single host. Parasites harm their individual hosts and help keep the populations of their hosts in check.

In **mutualism**, two species behave in ways that benefit both by providing each with food, shelter, or some other resource. One example is pollination of flowering plants by species such as honeybees (Figure 4.A, left, p. 72), hummingbirds, and butterflies that feed on the nectar of flowers.

Figure 5.6 shows an example of a mutualistic relationship that combines *nutrition* and *protection*. It involves birds that ride on the backs or heads of large animals such as elephants, rhinoceroses, and impalas. The birds remove and eat parasites and pests (such as ticks and flies) from the animals' bodies and often make noises warning the larger animals when predators are approaching.

Another example of mutualism involves clownfish, which usually live within sea anemones (see chapter-opening photo), whose tentacles sting and

FIGURE 5.6 *Mutualism*: Oxpeckers feed on parasitic ticks that infest animals such as this impala and warn of approaching predators.

paralyze most fish that touch them. The clownfish, which are not harmed by the tentacles, gain protection from predators and feed on the waste matter left from the anemones' meals. The sea anemones benefit because the clownfish protect them from some of their predators and parasites.

Mutualism might appear to be a form of cooperation between species. However, each species is acting only for its own survival.

Commensalism is an interaction that benefits one species but has little, if any, beneficial or harmful effect on the other. For example, plants called *epiphytes* (air plants) attach themselves to the trunks or branches of trees (Figure 5.7) in tropical and subtropical forests. The epiphytes gain better access to sunlight, water from the humid air and rain, and nutrients falling from the trees' upper leaves and limbs. Their presence apparently does not harm the trees. Similarly, birds benefit by nesting in trees, generally without harming them.

FIGURE 5.7 *Commensalism*: this pitcher plant is attached to a branch of a tree without penetrating or harming the tree. This carnivorous plant feeds on insects that become trapped inside it.

5.2 HOW DO COMMUNITIES AND ECOSYSTEMS RESPOND TO CHANGING ENVIRONMENTAL CONDITIONS?

CONCEPT 5.2 The species composition of a community or ecosystem can change in response to changing environmental conditions through a process called *ecological succession*.

Communities and Ecosystems Change over Time: Ecological Succession

The types and numbers of species in biological communities and ecosystems change in response to changing environmental conditions caused by fires, volcanic eruptions, climate change, and the clearing of forests to plant crops. The normally gradual change in species composition in a given area is called **ecological succession** (**Concept 5.2**).

Ecologists recognize two major types of ecological succession, depending on the conditions present at the beginning of the process. **Primary ecological succession** involves the gradual establishment of communities of different species in lifeless areas where there is no soil in a terrestrial ecosystem or no bottom sediment in an aquatic ecosystem. Examples include bare rock exposed by a retreating glacier (Figure 5.8), newly cooled lava, an abandoned highway or parking lot, and a newly created shallow pond or reservoir. Primary succession can take hundreds to thousands of years because of the need to build up fertile soil or aquatic sediments to provide the nutrients needed to establish a community of producers.

Primary ecological succession can also take place in a lake basin gouged out by a glacier. When the glacier melts, the lake basin begins accumulating sediments and plant and animal life. Over hundreds to thousands of years, the lake can fill with sediments and become a terrestrial habitat.

The other, more common type of ecological succession is called **secondary ecological succession**, in which a community or ecosystem develops on the site of an existing community or system, replacing or adding to the existing set of resident species. This type of succession begins in an area where an ecosystem has been disturbed, removed, or destroyed, but where some soil or bottom sediment remains. Candidates for secondary ecological succession include abandoned farmland (Figure 5.9), burned or cut forests, heavily polluted streams, and flooded land. Because some soil or sediment is present, new vegetation can begin to grow, usually within a few weeks. Growth begins with the germination of seeds already in the soil and seeds imported by wind or in the droppings of birds and other animals.

Ecological succession is an important ecosystem service that tends to enrich the biodiversity of communities and ecosystems by increasing species diversity and

Exposed rocks

Lichens and mosses

Small herbs and shrubs

Heath mat

Jack pine, black spruce, and aspen

Balsam fir, paper birch, and white spruce forest community

Time

FIGURE 5.8 *Primary ecological succession*: Over almost a thousand years, these plant communities developed, starting on bare rock exposed by a retreating glacier on Isle Royal, Michigan, in western Lake Superior. The details of this process vary from one site to another.

interactions among species. Such interactions strengthen an ecosystem's sustainability by promoting population control and by increasing the complexity of food webs, which enhances energy flow and nutrient cycling. Primary and secondary ecological successions are examples of *natural ecological restoration*.

Living Systems Are Sustained through Constant Change

All living systems, from a cell to the biosphere, are constantly changing in response to changing environmental conditions. Living systems have processes that interact to provide some degree of stability, or sustainability. This *stability*, or capacity to withstand external stress and disturbance, is maintained by constant change in response to changing environmental conditions. In a mature tropical rain forest, some trees die and others take their places. However, unless the forest is cut, burned, or otherwise destroyed, you would still recognize it as a tropical rain forest 50 or 100 years from now.

Ecologists distinguish between two aspects of stability or sustainability in ecosystems. Ecological **inertia**,

or **persistence**, is the ability of an ecosystem to survive moderate disturbances. Ecological **resilience** is the ability of an ecosystem to be restored through secondary ecological succession after a severe disturbance.

Evidence suggests that some ecosystems have one of these properties but not the other. Tropical rain forests have high species diversity and high inertia. However, once a large tract of tropical rain forest is cleared or severely damaged, the resilience of the resulting degraded forest ecosystem may be so low that degradation reaches an ecological tipping point. Once it is exceeded, the forest might not be restored by secondary ecological succession. One reason is that most of the nutrients in a typical rain forest are stored in its vegetation, not in the topsoil. Once the nutrient-rich vegetation is gone, frequent rains can remove most of the remaining soil nutrients and thus prevent the return of a tropical rain forest on a large cleared area.

By contrast, grasslands are much less diverse than most forests. Thus, they have low inertia and can burn easily. Because most of their plant matter is stored in underground roots, these ecosystems have high resilience and can recover quickly after a fire because their root systems produce new

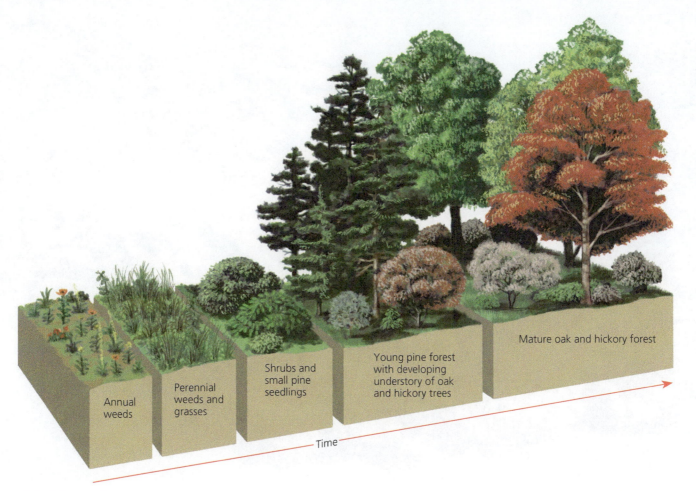

FIGURE 5.9 *Secondary ecological succession*: Natural restoration of disturbed land on an abandoned farm field in the U.S. state of North Carolina. It took 150–200 years after the farmland was abandoned for the area to become covered with a mature oak and hickory forest.

grasses. Grassland can be destroyed only if its roots are plowed up and something else is planted in its place, or if it is severely overgrazed by livestock or other herbivores.

5.3 WHAT LIMITS THE GROWTH OF POPULATIONS?

CONCEPT 5.3 No population can grow indefinitely because of limitations on resources and because of competition among species for those resources.

Populations Can Grow, Shrink, or Remain Stable

A **population** is a group of interbreeding individuals of the same species. Most populations live together in *clumps* or *groups* such as packs of wolves, schools of fish (Figure 5.10), and flocks of birds. Living in groups allows them to cluster where resources are available and provides some protection from predators.

Population size is the number of individual organisms in a population at a given time. Four variables—*births*, *deaths*, *immigration*, and *emigration*—govern changes in population size. A population increases through birth and immigration (the arrival of individuals from outside the population) and decreases through death and emigration (the departure of individuals from the population):

Population change = Individuals added − Individuals lost

Population change = (Births + Immigration) − (Deaths + Emigration)

A population's **age structure**—its distribution of individuals among various age groups—can have a strong effect on how rapidly it grows or declines. Age groups are usually described in terms of organisms not mature enough to reproduce (the *pre-reproductive stage*), those capable of reproduction (the *reproductive stage*), and those too old to reproduce (the *post-reproductive stage*).

The size of a population will likely increase if it is made up mostly of individuals in their reproductive stage, or

FIGURE 5.10 A population, or *school*, of Anthias fish on coral in Australia's Great Barrier Reef.

iStockphoto.com/Rich Carey

soon to enter this stage. In contrast, the size of a population dominated by individuals in their post-reproductive stage will tend to decrease over time.

Several Factors Can Limit Population Size

Each population in an ecosystem has a **range of tolerance**—a range of variations in its physical and chemical environmental conditions within which it is most likely to survive. For example, a trout population (Figure 5.11) will thrive within a narrow band of temperatures (*optimum level* or *range*), although a few individuals can survive above and below that band. If the water becomes too hot or too cold, none of the trout can survive.

Various physical or chemical factors can determine the number of organisms in a population and how fast a population grows or declines. These **limiting factors** are those that are more important than others in regulating population growth.

On land, precipitation often is a limiting factor. Low precipitation levels in desert ecosystems limit desert plant growth. Lack of key soil nutrients limits the growth of plants, which, in turn, limits populations of animals that eat plants, and animals that feed on such plant-eating animals.

Limiting physical factors for populations in *aquatic systems* include water temperature (Figure 5.11), depth, and clarity (allowing for more or less sunlight). Other important factors are nutrient availability, acidity, salinity, and the level of oxygen gas in the water (*dissolved oxygen content*).

Too much of a physical or chemical factor can also be limiting. For example, too much water or fertilizer can kill land plants. If acidity levels are too high in an aquatic environment, some of its organisms can be harmed.

An additional factor that can affect the sizes of some populations is **population density**, the number of individuals in a population found within a defined area or volume. *Density-dependent factors* become more important as a population's density increases. In a dense population, parasites and diseases can spread more easily, resulting in higher death rates. On the other hand, a higher population density can help sexually reproducing individuals to find mates more easily in order to produce offspring. Other factors such as drought and climate change are considered *density-independent*, because they can affect population sizes regardless of density.

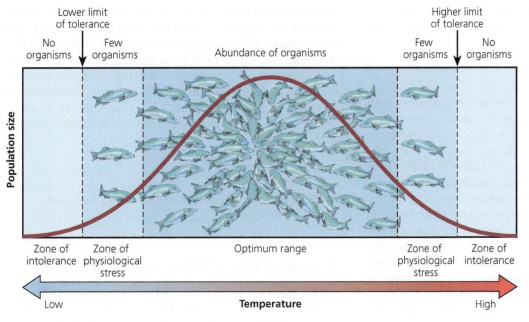

FIGURE 5.11 Range of tolerance for a population of trout to changes in water temperature.

Lower limit of tolerance

Higher limit of tolerance

No organisms | Few organisms | Abundance of organisms | Few organisms | No organisms

Population size

Zone of intolerance | Zone of physiological stress | Optimum range | Zone of physiological stress | Zone of intolerance

Low | **Temperature** | High

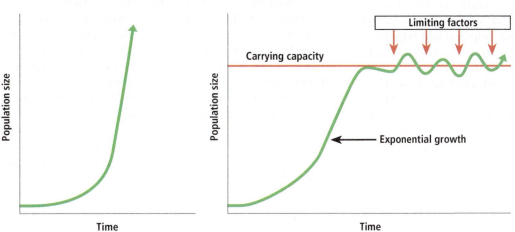

Limiting factors

Carrying capacity

Population size

Exponential growth

Time

Population size

Time

FIGURE 5.12 Populations of species can undergo *exponential growth* represented by a J-shaped curve (left) when resource supplies are plentiful. As resource supplies become limited, a population undergoes *logistic growth*, represented by an S-shaped curve (right), when the size of the population approaches the carrying capacity of its habitat.

No Population Can Grow Indefinitely: J-Curves and S-Curves

The populations of some species, such as bacteria and many insect species, have an incredible ability to increase their numbers exponentially. For example, with no controls on its population growth, a species of bacteria that can reproduce every 20 minutes would generate enough offspring to form a 0.3-meter-deep (1-foot-deep) layer over the surface of the entire earth in only 36 hours. Plotting such numbers against time yields a J-shaped curve of exponential growth (Figure 5.12, left). Members of such populations typically reproduce at an early age, have many offspring each time they reproduce, and reproduce many times, with short intervals between successive generations.

However, *there are always limits to population growth in nature* (**Concept 5.3**). Research reveals that a rapidly growing population of any species eventually reaches some size limit imposed by limiting factors. These factors can include

sunlight, water, temperature, space, nutrients, or exposure to predators or infectious diseases. **Environmental resistance** is the sum of all such factors in a habitat.

Limiting factors largely determine any area's **carrying capacity**, the maximum population of a given species that a particular habitat can sustain indefinitely. The carrying capacity for a population is not fixed and can rise or fall as environmental conditions change the factors that limit the population's growth. As a population approaches the carrying capacity of its habitat, the J-shaped curve of its exponential growth (Figure 5.12, left) is converted to an S-shaped curve of *logistic growth*, or growth that fluctuates around a certain level (Figure 5.12, right).

Some populations do not make a smooth transition from exponential growth to logistic growth. Instead, they use up their resource supplies and temporarily *overshoot*, or exceed, the carrying capacity of their environment. In such cases, the population suffers a sharp decline, called a *dieback*, or **population crash**, unless part of the population can switch to new resources or move to an area that has more resources. Such a crash occurred when reindeer were introduced onto a small island in the Bering Sea in the early 1900s (Figure 5.13).

Reproductive Patterns

Species vary in their reproductive patterns. Species with a capacity for a high rate of population increase (r) are called **r-selected species**. These species tend to have short life spans and produce many, usually small, offspring and to give them little or no parental care or protection. As a result, many of the offspring die at an early age. To overcome such losses, r-selected species produce large numbers of offspring so a few will likely survive and have many offspring to sustain the species. Examples of r-selected species include algae, bacteria, and most insects.

Such species tend to be *opportunists*. They reproduce and disperse rapidly when conditions are favorable or when a disturbance such as a fire or clear-cutting of a forest opens up a new habitat or niche for invasion. Once established, their populations may crash because of unfavorable changes in environmental conditions or invasion by more competitive species. This helps explain why most opportunist species go through irregular and unstable boom-and-bust cycles in their population sizes.

At the other extreme are **K-selected species**. They tend to reproduce later in life, have few offspring, and have long life spans. Typically, the offspring of K-selected mammal species develop inside their mothers where they are safe. After birth, they mature slowly and are cared for and protected by one or both parents. In some cases, they live in herds or groups, until they reach their reproductive stage.

The population size of K-selected species tends to be near the carrying capacity (K) of its environment (Figure 5.12, right). Examples of K-selected species include most large mammals such as elephants, whales,

and humans, birds of prey, and large and long-lived plants such as the saguaro cactus and most tropical rain forest trees. Many of these species—especially those with low reproductive rates, such as elephants, sharks, giant redwood trees, and California's southern sea otters (**Core Case Study** and Science Focus 5.2)—are vulnerable to extinction. Most organisms have reproductive patterns between the extremes of r-selected and K-selected species.

CONSIDER THIS ...

THINKING ABOUT *r*-Selected and *K*-Selected Species

If the earth experiences significant warming during this century as projected, is this likely to favor *r*-selected or *K*-selected species? Explain.

Species Vary in Their Typical Life Spans

Individuals of species with different reproductive strategies tend to have different *life expectancies*. This can be illustrated by a **survivorship curve**, which shows the percentages of the members of a population surviving at different ages. There are three generalized types of survivorship curves: late loss, early loss, and constant loss (Figure 5.14). A *late loss* population (such as elephants and rhinoceroses) typically has high survivorship to a certain age, then high mortality. A *constant loss* population (such as many songbirds) shows a roughly constant death rate at all ages. For an *early loss* population (such as annual plants and many bony fish species), survivorship is low early in life. These generalized survivorship curves only roughly model the realities of nature.

CONSIDER THIS ...

THINKING ABOUT Survivorship Curves

Which type of survivorship curve applies to the human species?

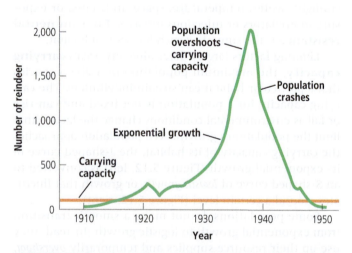

FIGURE 5.13 Exponential growth, overshoot, and population crash of a population of reindeer introduced onto the small Bering Sea island of St. Paul in 1910. *Data analysis*: By what percentage did the population of reindeer grow between 1923 and 1940?

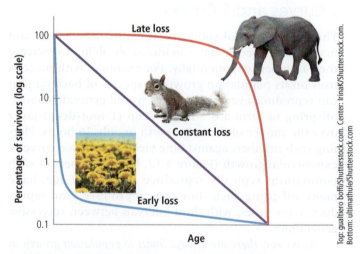

FIGURE 5.14 Survivorship curves for populations of different species, obtained by showing the percentages of the members of a population surviving at different ages.

The Future of California's Southern Sea Otters

The population size of southern sea otters (**Core Case Study**) has fluctuated in response to changes in environmental conditions (Figure 5.B). One change was a rise in populations of the orcas (killer whales) that feed on otters. Scientists hypothesize that orcas started feeding more on southern sea otters when populations of their normal prey, sea lions and seals, began declining. Also, between 2010 and 2016, the number of southern sea otters killed or injured by sharks increased.

Another factor affecting sea otters may be parasites that breed in the intestines of cats. Scientists hypothesize that some southern sea otters might be dying because coastal area cat owners flush feces-laden cat litter down their toilets or dump it into storm drains that empty into coastal waters. The feces contain parasites that then infect otters.

Toxic algae blooms also threaten otters. The algae thrive on urea, a key ingredient in fertilizer that washes into coastal waters. Other pollutants released by human activities include PCBs and other fat-soluble toxic chemicals. These chemicals can kill otters by accumulating to high levels in the tissues of the shellfish that otters eat. Because southern sea otters feed at high trophic levels and live close to the shore, they are vulnerable to these and other pollutants in coastal waters.

The factors listed here, mostly resulting from human activities, together with a low reproductive rate and a rising mortality rate, have hindered the ability of the endangered southern sea otter to rebuild its population (Figure 5.B). Since 2012, however, the sea otter population has increased, possibly because of an increase in the population of sea urchins, their preferred prey. In 2016, the sea otter population was 3,511, the highest it has been since 1985. The population must stay above 3,090 for 2 more consecutive years for the otter to be considered for removal from the federal endangered species list. If it is removed, the otters will still be protected under a California state law.

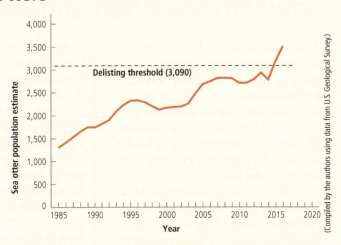

FIGURE 5.B Changes in the population size of southern sea otters off the coast of the U.S. state of California, 1985–2016.

(Compiled by the authors using data from U.S. Geological Survey.)

CRITICAL THINKING

How would you design a controlled experiment to test the hypothesis that cat litter flushed down toilets might be killing southern sea otters?

Humans Are Not Exempt from Nature's Population Controls

Humans are not exempt from population crashes. In 1845, Ireland experienced such a crash after a fungus destroyed its potato crop. About 1 million people died from hunger or diseases related to malnutrition, and millions more migrated to other countries, sharply reducing the Irish population.

During the 14th century, the *bubonic plague* spread through densely populated European cities and killed at least 25 million people—one-third of the European population. The bacterium causing this disease normally lives in rodents. It was transferred to humans by fleas that fed on infected rodents and then bit humans. The disease spread like wildfire through crowded cities, where sanitary conditions were poor and rats were abundant. Today several antibiotics can be used to treat bubonic plague.

So far, technological, social, and other cultural changes have expanded the earth's carrying capacity for the human species. We have used large amounts of energy and matter resources to occupy formerly uninhabitable areas. We have expanded agriculture and controlled the populations of other species that compete with us for resources. Some say we can keep expanding our ecological footprint in this way indefinitely because of our technological ingenuity. Others say that at some point, we will reach the limits that nature eventually imposes on any population that exceeds or degrades its resource base. We discuss these issues in Chapter 6.

BIG IDEAS

- Certain interactions among species affect their use of resources and their population sizes.

- The species composition and population sizes of a community or ecosystem can change in response to changing environmental conditions through a process called *ecological succession*.

- There are always limits to population growth in nature.

Southern Sea Otters and Sustainability

The sea otters of California are part of a complex ecosystem made up of underwater kelp forests, bottom-dwelling creatures, whales, and other species that depend on one another for survival. The sea otters act as a keystone species mostly by feeding on sea urchins and keeping them from destroying the kelp.

In this chapter, we focused on how biodiversity promotes sustainability, provides a variety of species to restore damaged ecosystems through ecological succession, and limits the sizes of populations. Populations of most plants and animals depend, directly or indirectly,

on solar energy, and all populations play roles in the cycling of nutrients in the ecosystems where they live. In addition, the biodiversity found in the variety of species in different terrestrial and aquatic ecosystems provides alternative paths for energy flow and nutrient cycling, better opportunities for natural selection as environmental conditions change, and natural population control mechanisms. When we disrupt these paths, we violate all

three **scientific principles of sustainability**.

fred goldstein/Shutterstock.com

Chapter Review

Core Case Study

1. How do southern sea otters act as a keystone species in their environment? Explain why we should care about protecting this species from extinction.

Section 5.1

2. What is the key concept for this section? Define and give an example of **interspecific competition**. How is it different from intraspecific competition? Define and give an example of **resource partitioning** and explain how it can increase species diversity. Define **predation**. Distinguish between a **predator** species and a **prey** species and give an example of each. What is a **predator–prey relationship** and why is it important?

3. Describe three threats to kelp forest and explain why kelp forests should be preserved. List three ways in which predators can increase their chances of feeding on their prey and three ways in which prey

species can avoid their predators. Define and give an example of **coevolution**.

4. Define **parasitism**, **mutualism**, and **commensalism** and give an example of each. Explain how each of these species interactions, along with predation, can affect the population sizes of species in ecosystems.

Section 5.2

5. What is the key concept for this section? What is **ecological succession**? Distinguish between **primary ecological succession** and **secondary ecological succession** and give an example of each.

6. Explain how living systems achieve some degree of sustainability by undergoing constant change in response to changing environmental conditions. In terms of the stability of ecosystems, distinguish between **inertia (persistence)** and **resilience** and give an example of each.

7. What is the key concept for this section? Define **population and population size**. Why do most populations live in clumps? List four variables that govern changes in population size. Write an equation showing how these variables interact. What is a population's **age structure** and what are the three major age groups called? Define **range of tolerance**. Define **limiting factor** and give an example. Define **population density** and explain how some limiting factors can become more important as a population's density increases.

8. Distinguish between the exponential and logistic growth of a population and describe the nature of their growth curves. Define **environmental resistance**. What is the **carrying capacity** of a habitat or ecosystem? Define and give an example of a **population crash**.

Critical Thinking

1. What difference would it make if the southern sea otter (**Core Case Study**) became extinct primarily because of human activities? What are three things we could do to help prevent the extinction of this species?

2. Use the second law of thermodynamics (Chapter 2, p. 38) and the concept of food chains and food webs to explain why predators are generally less abundant than their prey.

3. How would you reply to someone who argues that we should not worry about the effects that human activities have on natural systems because ecological succession will repair whatever damage we do?

4. How would you reply to someone who contends that efforts to preserve species and ecosystems are not worthwhile because nature is largely unpredictable?

9. Describe two different reproductive strategies for species. Distinguish between **r-selected species** and **K-selected species** and give an example of each. Define **survivorship curve** and describe three types of curves. Why has the recovery of the southern sea otters been slow and what factors are threatening this recovery? Explain why humans are not exempt from nature's population controls.

10. What are this chapter's *three big ideas*? Explain how the interactions among plant and animal species in any ecosystem are related to the **scientific principles of sustainability**.

Note: Key terms are in bold type.

5. Explain why most species with a high capacity for population growth (such as bacteria, flies, and cockroaches) tend to have small individuals, while those with a low capacity for population growth (such as humans, elephants, and whales) tend to have large individuals.

6. What is the reproductive strategy of most species of insect pests and harmful bacteria? Why does this make it difficult for us to control their populations?

7. List two factors that may limit human population growth in the future. Do you think that we are close to reaching those limits? Explain.

8. If the human species were to suffer a population crash, what are three species that might move in to occupy part of our ecological niche?

Doing Environmental Science

Visit a nearby land area, such as a partially cleared or burned forest, a grassland, or an abandoned crop field, and record signs of secondary ecological succession. Take notes on your observations and formulate a hypothesis about what sort of disturbance led to this succession. Include your thoughts about whether this disturbance was natural or caused by humans. Study the area carefully to see whether you can find patches that are at different stages of succession and record your thoughts about what sorts of disturbances have caused these differences. You might want to research the topic of ecological succession in such an area.

Global Environment Watch Exercise

Go to your MindTap course to access the GREENR database. Using the "Basic Search" function at the top of the page, search for *kelp forests* (also sometimes called *kelp beds*), and use the results to find sources of information about how a warmer ocean resulting from climate change might affect California's coastal kelp forests on which the southern sea otters depend (**Core Case Study**). Write a report on what you found. Try to include information on current effects of warmer water on the kelp beds as well as projections about future effects. Also, summarize any information you might find on possible ways to prevent harm to these kelp forests.

Data Analysis

The graph below shows changes in the size of an Emperor penguin population in terms of numbers of breeding pairs on the island of Terre Adelie in the Antarctic. Scientists used these data along with data on the penguins' shrinking ice habitat to project a general decline in the island's Emperor penguin population, to the point where they will be endangered in 2100. Use the graph to answer the questions on the right.

1. If the penguin population fluctuates around the carrying capacity, what was the approximate carrying capacity of the island for the penguin population

from 1960 to 1975? What was the approximate carrying capacity of the island for the penguin population from 1980 to 2010?

2. What was the overall percentage decline in the penguin population from 1975 to 2010?

3. What is the projected overall percentage decline in the penguin population between 2010 and 2100?

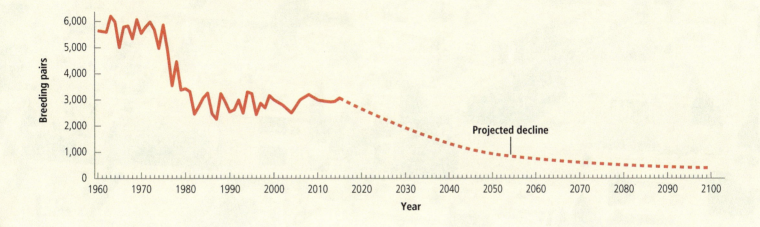

CENGAGE **brain** .com To access course materials, including Aplia homework, please visit www.cengagebrain.com.

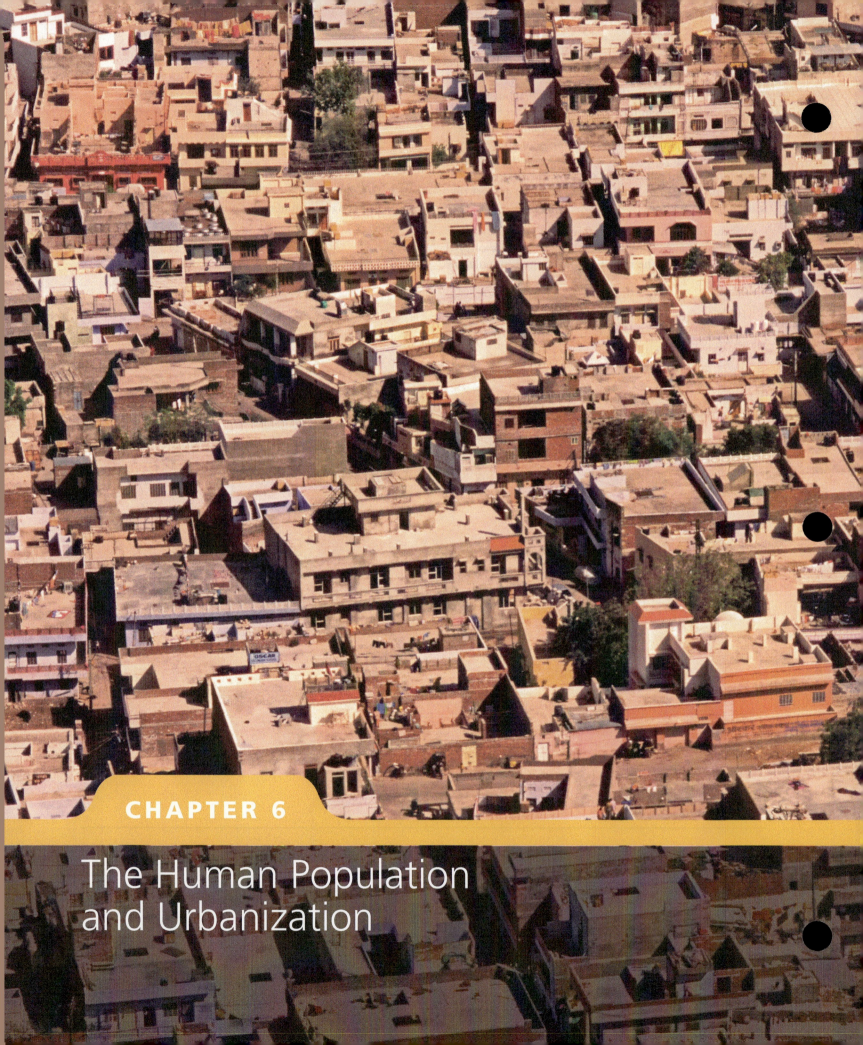

CHAPTER 6

The Human Population
and Urbanization

Either we limit our population growth or the natural world will do it for us.

SIR DAVID ATTENBOROUGH

Key Questions

6.1 How many people can the earth support?

6.2 What factors influence the size of the human population?

6.3 How does a population's age structure affect its growth or decline?

6.4 How can we slow human population growth?

6.5 What are the major urban resource and environmental problems?

6.6 How does transportation affect urban environmental impacts?

6.7 How can we make cities more sustainable and livable?

Population pressure in Jaipur, India.
zanskar/Getty Images

Planet Earth: Population 7.4 Billion

It took about 200,000 years for the human population to reach an estimated 2 billion. It took less than 50 years to add the second 2 billion people (by about 1974), and 25 years to add the third 2 billion (by 1999). Sixteen years later, the earth had 7.4 billion people. In 2016, the three most populous countries, in order, were China with 1.38 billion people (Figure 6.1), India with 1.33 billion people, and the United States with 324 million people. The United Nations projects that the world's population will increase to 9.9 billion by 2050—an increase of 2.5 billion people.

Does it matter that there are now 7.4 billion people on the earth—almost 3 times as many as there were in 1950? Does it matter that each day, 246,000 more people show up for dinner and many of them will go hungry? Does it matter that there might be 2.5 billion more of us on the planet by 2050? Some say it does not matter and contend that we can develop new technologies that could easily support billions more people.

Many scientists disagree and argue that the current exponential growth of the human population (see Figure 1.12, p. 15) is unsustainable because as our population and economies grow, we use more of the earth's natural resources and our ecological footprints grow. As a result, we degrade the natural capital that keeps us alive and supports our lifestyles and economies.

According to *demographers*, or population experts, three major factors account for the rapid rise of the human population. *First*, the emergence of early and modern agriculture about 10,000 years ago increased food production. *Second*, additional technologies helped humans expand into almost all of the planet's climate zones and habitats (Figure 1.9, p. 11). *Third*, death rates dropped sharply with improved sanitation and health care and the development of antibiotics and vaccines to control infectious diseases.

Population experts have made low, medium, and high projections for the human population size at the end of this century, as shown in Figure 1.12 (p. 15). No one knows whether or for how long any of these population sizes are sustainable.

In this chapter, we examine population growth trends, the environmental impacts of the growing population, ways to slow human population growth, and ways to make the world's rapidly growing urban areas more sustainable and livable. ●

MACDUFF EVERTON/National Geographic Creative

FIGURE 6.1 This crowded street is located in China, where almost one-fifth of the world's people live.

6.1 HOW MANY PEOPLE CAN THE EARTH SUPPORT?

CONCEPT 6.1 The rapid growth of the human population and its impact on natural capital raises questions about how long the human population can keep growing.

Human Population Growth

For most of history, the human population grew slowly (see Figure 1.12, p. 15, left part of curve). However, it has grown rapidly for the last 200 years, resulting in the characteristic J-curve of exponential growth (Figure 1.12, right part of curve).

Demographers, or population experts, recognize three important trends related to the current size, growth rate, and distribution of the human population. *First*, the rate of population growth has slowed since 1960 (Figure 6.2), but the world's population is still growing at a rate of about 1.21%. This may not seem like much, but in 2016, this growth added about 89.8 million people to the population—an average of about 246,000 people every day.

Second, human population growth is unevenly distributed. About 96% of 89.8 million new arrivals on the planet in 2016 were added to the world's less-developed countries, where the population is growing 14 times faster than the population of the more-developed countries. At least 95% of the 2.5 billion people projected to be added to the world's population between 2016 and 2050 will be born in less-developed countries. Most of these countries are not equipped to deal with the pressures of rapid population growth.

Third, people have moved in large numbers from rural areas to urban areas. In 2016, about 54% of world's people lived in urban areas and this percentage is increasing.

Scientists and other analysts have long pondered the question: How long can the human population continue to grow while sidestepping many of the factors that sooner or later limit the growth of any population? These experts disagree over how many people the earth can support indefinitely (Science Focus 6.1). So far, advances in food production and health care have prevented sharp population declines, but there is extensive and growing evidence that human activities are depleting and degrading much of the earth's irreplaceable natural capital (Figure 6.3).

6.2 WHAT FACTORS INFLUENCE THE SIZE OF THE HUMAN POPULATION?

CONCEPT 6.2A Population size increases through births and immigration, and decreases through deaths and emigration.

CONCEPT 6.2B The key factor that determines the size of a human population is the average number of children born to the women in that population (*total fertility rate*).

The Human Population Can Grow, Decline, or Stabilize

The basics of global population change are quite simple. When there are more births than deaths, the human population increases. When there are more deaths than births, the population decreases. When the number of births equals the number of deaths, population size does not change.

The human population in a particular area grows or declines through the interplay of three factors: *births*

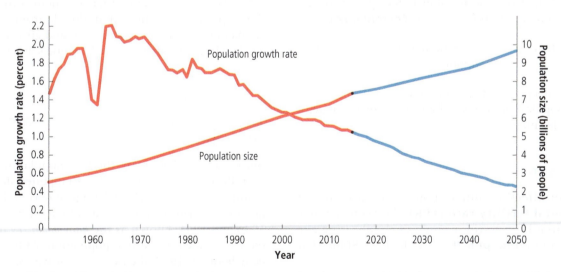

FIGURE 6.2 Global human population size compared with population growth rate, 1990–2015, with projections to 2050. ***Critical thinking:*** Why do you think that while the annual growth *rate* of world population has generally dropped since the 1960s, the population has continued to grow?

Compiled by the authors using data from United Nations Population Division, U.S. Census Bureau, and Population Reference Bureau.

How Long Can the Human Population Keep Growing?

Are there physical limits to human population growth and economic growth on a finite planet? Some say yes. Others say no.

This debate has been going on for more than 200 years. One current view is that we have already exceeded some of those limits, with too many people collectively degrading the earth's life-support system.

To some analysts, the key problem is the large and rapidly growing number of people in less-developed countries. To others, the key factor is *overconsumption* in affluent, more-developed countries because of their high rates of resource use per person. They debate over which is more important for shrinking the human ecological footprint: slowing population growth or reducing resource consumption. Some call for doing both.

Another view of population growth is that technology allows us to overcome the environmental limits faced by populations of other species. According to this view, these advances have increased the earth's carrying capacity for the human species. Some analysts point out that average life expectancy in most of the world has been steadily rising despite warnings that we are seriously degrading our life-support system.

These analysts argue that because of our technological ingenuity, there are few, if any, limits to human population growth and resource use per person. They believe that we can continue increasing economic growth and avoid serious damage to our life-support systems by making technological advances in areas such as food production and medicine, and by finding substitutes for resources that we are depleting. They see no need to slow the world's population growth or resource consumption.

Proponents of slowing or stopping population growth point out that in addition to degrading our life-support system, we are failing to provide the basic necessities for about 900 million people—one of eight on the planet—who struggle to survive on the equivalent of about $1.90 per day and the 2.1 billion people struggling to live on $3.10 per day or less. This raises a serious question: How will we meet the basic needs of the additional 2.5 billion people projected to be added between 2016 and 2050?

No one knows how close we are to environmental limits that some say eventually will reduce the size of the human population. These analysts call for us to confront this vital scientific, political, economic, and ethical issue.

CRITICAL THINKING

Do you think there are environmental limits to human population growth? If so, how close do you think we are to such limits? Explain.

(fertility), *deaths (mortality)*, and *migration*. We can calculate the **population change** of an area by subtracting the number of people leaving a population (through death and emigration) from the number entering it (through birth and immigration) during a year (**Concept 6.2A**).

Population change = (Births + Immigration) − (Deaths + Emigration)

When births plus immigration exceed deaths plus emigration, a population grows; when the reverse is true, a population declines.

Fertility Rates

A key factor affecting human population growth and size is the **total fertility rate (TFR)**: the average number of children born to the women of childbearing age in a population (**Concept 6.2B**). See the Case Study that follows.

Between 1955 and 2016, the global TFR dropped from 5 to 2.5. Those who support slowing the world's population growth view this as good news. However, to eventually halt population growth, the global TFR would have to drop to 2.1—the rate necessary for replacing both parents after considering infant mortality.

With a TFR of 4.7, Africa's population is growing more than twice as fast as any other continent. Africa's population is projected to more than double from 1.2 billion to 2.5 billion between 2016 and 2050 and grow to more than 4 billion by 2100. Africa is also the world's poorest continent.

CASE STUDY

The U.S. Population—Third Largest and Growing

Between 1900 and 2016, the U.S. population grew from 76 million to 324 million. This happened despite oscillations in the country's TFR (Figure 6.4) and population growth rate. During the period of high birth rates between 1946 and 1964, known as the *baby boom*, 79 million people were added to the U.S. population. At the peak of the baby boom in 1957, the average TFR was 3.7 children per woman. In most years since 1972, it has been at or below 2.1 children per woman, compared to a global TFR of 2.5.

The drop in the TFR has slowed the rate of population growth in the United States, but the country's

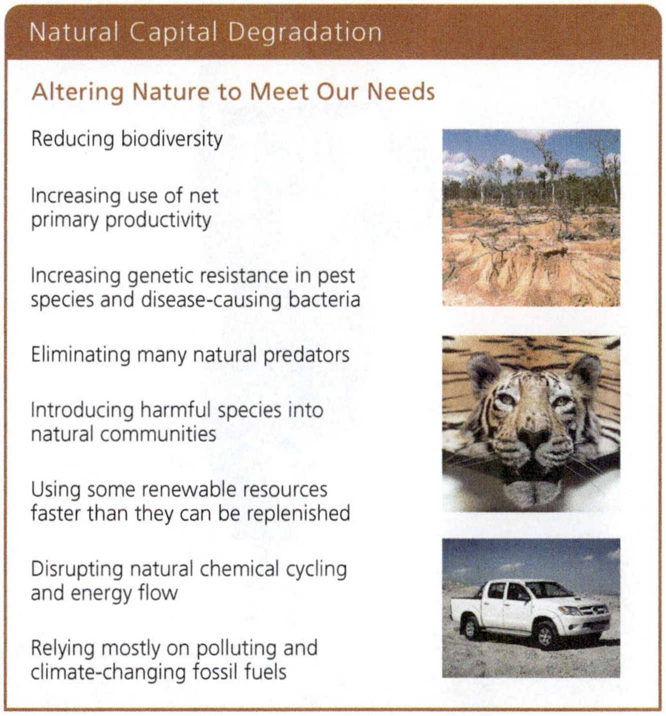

Natural Capital Degradation

Altering Nature to Meet Our Needs

Reducing biodiversity

Increasing use of net primary productivity

Increasing genetic resistance in pest species and disease-causing bacteria

Eliminating many natural predators

Introducing harmful species into natural communities

Using some renewable resources faster than they can be replenished

Disrupting natural chemical cycling and energy flow

Relying mostly on polluting and climate-changing fossil fuels

Top: Dirk Ercken/Shutterstock.com. Center: Fulcanelli/Shutterstock.com. Bottom: Werner Stoffberg/Shutterstock.com.

FIGURE 6.3 Human activities have altered the natural systems and ecosystem services that sustain our lives and economies in at least eight major ways to meet the increasing needs and wants of our growing population (**Concept 6.1**). *Critical thinking:* In your daily living, do you think you contribute directly or indirectly to any of these harmful environmental impacts? Which ones? Explain.

population is still growing. In 2016, about 3 million people were added to the U.S. population, according to the U.S. Census Bureau. About 2.2 million were added because there were more births than deaths and the rest were legal immigrants and refugees. Since 1820, the United States has admitted almost twice as many legal immigrants and refugees as all other countries combined. The United States also has an estimated 11 million illegal

immigrants. Since 2005 the flow of illegal immigrants into the country has been dropping, according to the Pew Research Center.

In addition to the fourfold increase in population growth since 1900, some amazing changes in lifestyles took place in the United States during the 20th century (Figure 6.5), which led to Americans living longer. Along with this came dramatic increases in per capita resource use and much larger total and per capita ecological footprints.

The U.S. Census Bureau projected that between 2016 and 2050, the U.S. population would likely grow from 324 million to 398 million—an increase of 74 million people. Because of a high per-person rate of resource use and the resulting waste and pollution, each addition to the U.S. population has an enormous environmental impact (see Figure 1.10, p. 13).

74 million
Projected increase in the U.S. population between 2016 and 2050

Factors That Affect Birth and Fertility Rates

Many factors affect a country's population growth rate and TFR. One is the *importance of children as a part of the labor force*, especially in less-developed countries. Many of the poor in those countries struggle to survive on less than $3.10 a day and some on less than $1.90 a day. Some of these couples have a large number of children to help haul drinking water, gather wood for heating and cooking, and grow or find food. Worldwide, one of every 10 children between ages 5 and 17 works to help the family survive (Figure 6.6).

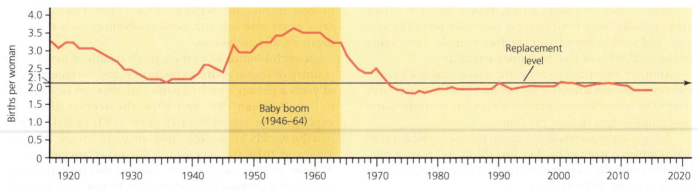

FIGURE 6.4 Total fertility rates for the United States between 1917 and 2016. *Critical thinking:* The U.S. fertility rate has declined and remained at or below replacement levels since 1972. So why is the population of the United States still growing?

Compiled by the authors using data from the Population Reference Bureau and U.S. Census Bureau.

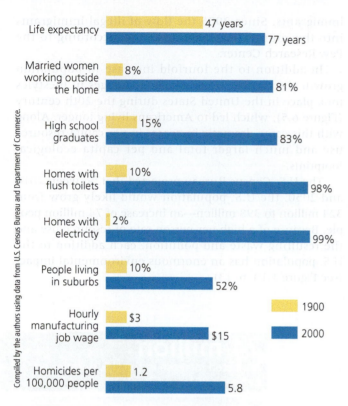

FIGURE 6.5 Some major changes took place in the United States between 1900 and 2000. *Critical thinking:* Which two of these changes do you think had the biggest impacts on the U.S. ecological footprint?

Bar chart categories (1900 in yellow, 2000 in blue):

- Life expectancy: 47 years / 77 years
- Married women working outside the home: 8% / 81%
- High school graduates: 15% / 83%
- Homes with flush toilets: 10% / 98%
- Homes with electricity: 2% / 99%
- People living in suburbs: 10% / 52%
- Hourly manufacturing job wage: $3 / $15
- Homicides per 100,000 people: 1.2 / 5.8

Legend: 1900 / 2000

Compiled by the authors using data from U.S. Census Bureau and Department of Commerce.

FIGURE 6.6 This young boy spends much of his day carrying bricks.

Zatletic/Dreamstime.com

Another economic factor is the *cost of raising and educating children*. Birth and fertility rates tend to be lower in more-developed countries, where raising children is much more costly because they do not enter the labor force until they are in their late teens or twenties. In the United States, the U.S. Department of Agriculture estimated in 2016 that the average cost of raising a child to the age of 18 was $245,000.

The *availability of pension systems* can influence the number of children that couples have, especially for poor people in less-developed countries. Pensions reduce a couple's need to have several children to ensure that they will have support in their old age.

Urbanization also plays a role. People living in urban areas usually have better access to family planning services and tend to have fewer children than do those living in the rural areas of less-developed countries.

Another important factor is the *educational and employment opportunities available for women*. Total fertility rates tend to be low when women have access to education and paid employment outside the home.

Average age at marriage (or, more precisely, the average age at which a woman has her first child) also plays a role. Women normally have fewer children when their average age at marriage is 25 or older.

Birth rates and TFRs are also affected by the *availability of reliable birth control methods* that allow women to control the number and spacing of their children.

Religious beliefs, traditions, and cultural norms also play a role. In some countries, these factors favor large families, because of opposition to abortion and some forms of birth control.

Factors That Affect Death Rates

The rapid growth of the world's population over the past 100 years is largely the result of declining death rates, especially in less-developed countries. More people in some of these countries are living longer and fewer infants are dying because of larger food supplies, improvements in food distribution, better nutrition, improved sanitation, safer water supplies, and medical advances such as immunizations and antibiotics.

A useful indicator of the overall health of people in a country or region is **life expectancy**: the average number of years a person born in a particular year can be expected to live. Between 1955 and 2016, the average global life expectancy increased from 48 years to 71. In 2016, Japan had the world's longest life expectancy of 83 years. Between 1900 and 2016, the average U.S. life expectancy rose from 47 years to 79 years. Research indicates that poverty, which reduces the average life span by 7 to 10 years, is the single most important factor that decreases life expectancy.

Another important indicator of overall health in a population is its **infant mortality rate**, the number of babies out of every 1,000 born who die before their first birthday. It is viewed as one of the best measures of a society's quality of life because it reflects a country's general level of nutrition and health care. A high infant mortality rate usually indicates insufficient food (*undernutrition*), poor nutrition (*malnutrition*, see Figure 1.13, p. 17), and a high incidence of infectious disease. Infant mortality also affects the TFR. In areas with low infant mortality rates, women tend to have fewer children because fewer of their children die at an early age.

Infant mortality rates in most countries have declined dramatically since 1965. Even so, every year more than 4 million infants (most of them in less-developed countries) die of *preventable* causes during their first year of life. This average of nearly 11,000 mostly unnecessary infant deaths per day is equivalent to 55 jet airliners, each loaded with 200 infants, crashing *every day* with no survivors.

Between 1900 and 2016, the U.S. infant mortality rate dropped from 165 to 5.8. This sharp decline was a major factor in the marked increase in U.S. average life expectancy during this period. However, 49 other nations (most in Europe) had lower infant mortality rates than the United States had in 2016.

Migration

A third factor in population change is **migration**: the movement of people into (*immigration*) and out of (*emigration*) specific geographic areas. Most people who migrate to another area within their country or to another country are seeking jobs and economic improvement. Others are driven to migrate by religious persecution, ethnic conflicts, political oppression, or war. There are also *environmental refugees*—people who have to leave their homes and sometimes their countries because of water or food shortages, soil erosion, flooding, or some other form of environmental degradation.

6.3 HOW DOES A POPULATION'S AGE STRUCTURE AFFECT ITS GROWTH OR DECLINE?

CONCEPT 6.3 The numbers of males and females in young, middle, and older age groups determine how fast a population grows or declines.

Age Structure

The **age structure** of a population is the numbers or percentages of males and females in young, middle, and older age groups in that population (**Concept 6.3**). Age structure is an important factor determining total fertility rates and whether the population of a country increases or declines.

Population experts construct a population *age-structure diagram* by plotting the percentages or numbers of males and females in the total population in each of three age categories: *pre-reproductive* (ages 0–14), consisting of individuals normally too young to have children; *reproductive* (ages 15–44), consisting of those normally able to have children; and *post-reproductive* (ages 45 and older), consisting of individuals normally too old to have children. Figure 6.7 presents generalized age-structure diagrams for countries with rapid, slow, zero, and negative population growth rates.

A country with a large percentage of its people younger than age 15 (represented by a wide base in Figure 6.7, far left) will experience rapid population growth unless death rates rise sharply. Because of this *demographic momentum*,

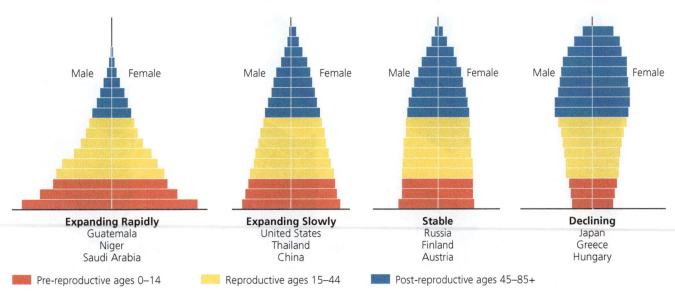

FIGURE 6.7 Generalized population age-structure diagrams for countries with rapid (1.5–3%), slow (0.3–1.4%), stable (0–0.2%), and negative (declining) population growth rates. ***Question:*** Which of these diagrams best represents the country where you live?

Compiled by the authors using data from U.S. Census Bureau and Population Reference Bureau.

the number of births in such a country will rise for several decades. This will occur even if women have an average of only one or two children each because the large number of girls entering their prime reproductive years. Most future human population growth will take place in less-developed countries because of their typically youthful age structure and rapid population growth rates.

The global population of seniors—people who are 65 and older—is projected to triple between 2016 and 2050, when one of every six people will be a senior (see the Case Study that follows).

CASE STUDY

The American Baby Boom

Changes in the distribution of a country's age groups have long-lasting economic and social impacts. For example, the American baby boom added 79 million people to the U.S. population between 1946 and 1964. Over time, this group looks like a bulge moving up through the country's age structure, as shown in Figure 6.8.

For decades, the baby-boom generation has strongly influenced the U.S. economy because they make up about 25% of the U.S. population. Baby boomers created the youth market in their teens and twenties and are now creating the late middle-age (50–60) and senior markets. In addition to their economic impact, the baby-boom generation plays an increasingly important role in deciding who is elected to public office and what laws are passed or weakened.

Since 2011, when the first baby boomers began turning 65, the number of Americans older than age 65 has grown at the rate of about 10,000 a day and will do so through 2030. Between 2015 and 2050, the number of Americans aged 65 and older is projected to grow from 48 million to 88 million. This process has been called the *graying of America*. As the number of working adults declines in proportion to the number of seniors, there may be political pressure from baby boomers to increase tax revenues to help support the growing senior population.

However, in 2015, according to the Census Bureau, the Millennial Generation—Americans born between1980 and 2005—overtook Baby Boomers to become the largest generation living in the United States, and eventually, this will change the political and economic power balance. This could lead to economic and political conflicts between older and younger Americans.

Aging Populations Can Decline Rapidly

The graying of the world's population is due largely to declining birth rates and medical advances that have extended life spans. The United Nations (UN) estimates that by 2050, the global number of people aged 60 and older will equal or exceed the number of people under age 15.

As the percentage of people aged 65 or older increases, more countries will begin experiencing population decline. If population decline is gradual, its harmful effects usually can be managed. However, some countries are experiencing rapid decline and feeling such effects more severely.

Japan has the world's highest percentage of elderly people (above age 65) and the world's lowest percentage of young people (below age 15). In 2016, Japan's population was 125 million. By 2050, its population is projected to be 101 million. As its population declines, there

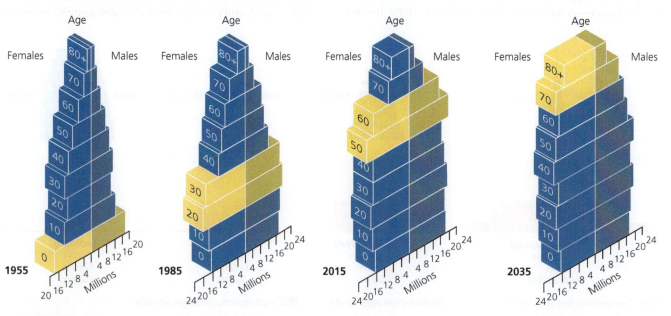

FIGURE 6.8 Age structure charts tracking the baby-boom generation in the United States, 1955, 1985, 2015, and 2035 (projected).

Compiled by the authors using data from U.S. Census Bureau and Population Reference Bureau.

will be fewer adults working and paying taxes to support an increasing elderly population. Because Japan discourages immigration, this could threaten its economic future. Figure 6.9 lists some of the problems associated with rapid population decline.

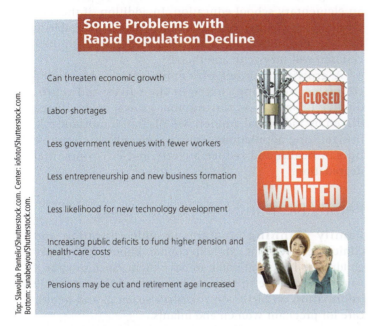

FIGURE 6.9 Rapid population decline can cause several problems. **Critical thinking:** Which two of these problems do you think are the most urgent?

6.4 HOW CAN WE SLOW HUMAN POPULATION GROWTH?

CONCEPT 6.4 We can slow human population growth by reducing poverty through economic development, elevating the status of women, and encouraging family planning.

Economic Development

There is controversy over whether we should slow population growth (Science Focus 6.1). Some analysts argue that we need to slow population growth in order to reduce environmental degradation of our life-support system. They have suggested several ways to do this, one of which is to reduce poverty through economic development.

Demographers have examined the birth and death rates of western European countries that became industrialized during the 19th century. Using such data, they developed a hypothesis on population change known as the **demographic transition**. It states that as countries become industrialized and economically developed, their per capita incomes rise, poverty declines, and their populations tend to grow more slowly. According to the hypothesis, this transition takes place in four stages, as shown in Figure 6.10.

Some analysts believe that most of the world's less-developed countries will make a demographic transition over the next few decades. They hypothesize that this

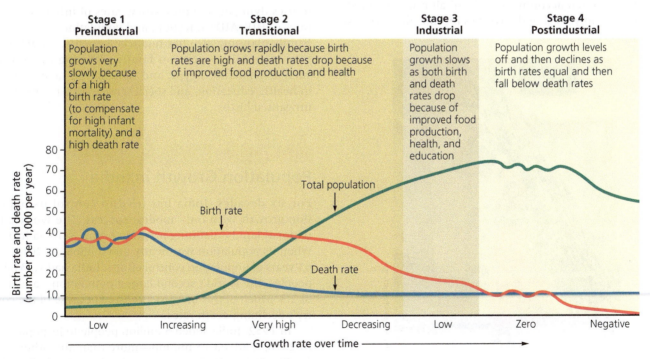

FIGURE 6.10 The *demographic transition*, which a country can experience as it becomes industrialized and more economically developed, can take place in four stages. **Critical thinking:** For each of the first three stages, what do you think would happen to a country that could not move past that stage?

transition will occur because newer technologies will help them to develop economically and to reduce poverty.

Other analysts fear that rapid population growth, extreme poverty, war, and increasing environmental degradation and resource depletion could leave countries with high population growth rates stuck in stage 2 of the demographic transition. This highlights the need to reduce poverty as a key to improving human health and stabilizing the population.

Educating and Empowering Women

A number of studies show that women tend to have fewer children if they are educated, control their own fertility, earn an income of their own, and live in societies that do not suppress their rights.

Only about 30% of the world's girls are enrolled in secondary education, and studies show that widespread education of girls is important for their future and for slowing population growth. In most societies, women have fewer rights and educational and economic opportunities than men have.

Women do almost all of the world's domestic work and childcare for little or no pay. They provide more unpaid health care (within their families) than do all of the world's organized health-care services combined. In rural areas of Africa, Latin America, and Asia, women do 60–80% of the work associated with growing food, hauling water, and gathering and hauling wood (Figure 6.11) and animal dung for use as fuel. As one Brazilian woman observed, "For poor women, the only holiday is when you are asleep."

While women account for 66% of all hours worked, they receive only 10% of the world's income and own just

2% of the world's land. They also make up 70% of the world's poor and 66% of the world's 775 million illiterate adults. Poor women who cannot read often have an average of five to seven children, compared to two or fewer children in societies where almost all women can read. This highlights the need to see that all children get at least an elementary school education. In addition, if the survival rates of children can be raised, parents will be able to have fewer children and feel confident that most of their children will survive to adulthood.

A growing number of women in less-developed countries are taking charge of their lives and reproductive behavior. As this number grows, such change driven by individual women will play an important role in stabilizing populations. This change will also improve human health and reduce poverty and environmental degradation.

Family Planning

Family planning programs provide education and clinical services that can help couples choose how many children to have and when to have them. Such programs vary from culture to culture, but most of them provide information on birth control, birth spacing, and health care for pregnant women and infants.

According to studies by the UN Population Division and other population agencies, family planning has been a major factor in reducing the global numbers of unintended pregnancies and births and abortions. In addition, family planning has reduced the number of mothers and fetuses dying during pregnancy, rates of infant mortality, rates of HIV/AIDS infection, and population growth rates. Family planning also has financial benefits. Studies show that each dollar spent on family planning in countries such as Thailand, Egypt, and Bangladesh saves $10–$16 in health, education, and social service costs by preventing unwanted births.

Population Growth in India

For six decades, India has tried to control its population growth with only modest success. The world's first national family planning program began in India in 1952, when its population was nearly 400 million. In 2016, after 63 years of population control efforts, India had 1.33 billion people—the world's second largest population—and a TFR of 2.3. Much of this increase occurred because of the country's declining death rates.

In 1952, India added 5 million people to its population. In 2016, it added 15 million—more than any other country. The United Nations projects that by 2029, India will be the world's most populous country, and that by 2050, it will have a population of 1.7 billion.

FIGURE 6.11 This woman in Nepal was bringing home firewood. Typically, she spends 2 hours a day, two or three times a week, gathering and hauling wood.

Iv Nikolny/Shutterstock.com

India has the world's fourth largest economy and a rapidly growing middle class. However, the country faces serious poverty, malnutrition, and environmental problems that could worsen as its population continues to grow rapidly. India is home to one-third of the world's poor (Figure 6.12). About one-fourth of all people in India's cities live in slums, and prosperity and progress have not touched hundreds of millions of Indians who live in rural villages.

Three factors help to account for larger families in India. *First*, most poor couples believe they need several children to work and care for them in their old age. *Second*, the strong cultural preference in India for male children means that some couples keep having children until they produce one or more boys. *Third*, even though 90% of Indian couples have access to at least one modern birth control method, only about 47% actually use one, according to the Population Reference Bureau.

India is undergoing rapid economic growth, which is expected to accelerate over the next few decades. This will help many people in India, but it will also put increasing pressure on India's and the earth's natural capital as rates of per capita resource use rise. India already faces serious soil erosion, overgrazing, water pollution, and air pollution problems. On the other hand, economic growth may help India to slow its population growth by accelerating its demographic transition.

FIGURE 6.12 Homeless people in Kolkata, India.

CASE STUDY

Slowing Population Growth in China

China is the world's most populous country, with 1.38 billion people in 2016 (Figure 6.1). According to the Population Reference Bureau and the United Nations Population Fund, China's population is projected to peak at around 1.4 billion by 2030 and then decline to 1.3 billion by 2050.

In the 1960s, China's large population was growing so rapidly that there was a serious threat of mass starvation and social upheaval. To avoid this, government officials took measures that eventually led to the establishment of the world's most extensive, intrusive, and strict family planning and birth control program.

The goal has been to sharply reduce population growth by promoting one-child families. The government provided contraceptives, sterilizations, and abortions for married couples. In addition, married couples pledging to have no more than one child received a number of benefits, including better housing, more food, free health care, salary bonuses, and preferential job opportunities for their child. Couples who broke their pledge lost such benefits.

Since this government-controlled program began in 1978, China has made impressive efforts to feed its people and bring its population growth under control. Between 1972 and 2016, the country reduced its TFR from 3.0 to 1.7. China's population is now growing more slowly than the U.S. population. Although China has avoided mass starvation, its strict population control program has been accused of violating human rights.

An unintended result of China's population control program is that because of the cultural preference for sons, many Chinese women have aborted female fetuses. This has reduced the female population, and as a result, about 30 million Chinese men are unable to find anyone to marry.

Since 1980, China has undergone rapid industrialization and economic growth. According to the Earth Policy Institute, between 1990 and 2010, this process reduced the number of people living in extreme poverty by almost 500 million. It has also helped at least 300 million Chinese—a number almost equal to the entire U.S. population—to become middle-class consumers. Over time, China's rapidly growing middle class will consume more resources per person, expanding China's ecological footprint within its own borders and in other parts of the world that provide it with resources. This will put a strain on China's and the earth's natural capital. Like India, China faces serious soil erosion, overgrazing, water pollution, and air pollution problems. And at least 400 million people still live in poverty in China's villages and cities (Figure 6.13).

Because of its one-child policy, in recent years, the average age of China's population has been increasing at one of the fastest rates ever recorded. In 2016, at least 137 million Chinese people over age 65—the largest number of people in this age group of all the world's countries. While China's population is not yet declining, the UN estimates that by 2030, the country is likely to have too few young workers (ages 15 to 64) to support its rapidly aging population. This graying of the Chinese population could lead to a declining work force, limited funds for

FIGURE 6.13 Old and new housing in heavily populated Shanghai, China.

supporting continued economic development, and fewer children and grandchildren to care for the growing number of elderly people. These concerns and other factors may slow China's economic growth.

Because of these concerns, in 2015, the Chinese government abandoned its one-child policy and replaced it with a two-child policy. Married couples can apply to the government for permission to have two children. However, the Chinese people have gotten used to small families. Because of the high cost of raising a second child, and because young women enjoy greatly increased educational and job opportunities, many married couples still choose to have only one child.

6.5 WHAT ARE THE MAJOR URBAN RESOURCE AND ENVIRONMENTAL PROBLEMS?

CONCEPT 6.5 Most cities are unsustainable because of high levels of resource use, waste, pollution, and poverty.

Three Important Urban Trends

In 2016, about 54% of the world's people, 82% of all Americans (see Case Study that follows), and 56% of China's population lived in urban areas.

54% Percentage of the world's people living in urban areas.

Urban areas grow in two ways—by *natural increase* when there are more births than deaths and by *immigration*, mostly from rural areas. People move from rural to urban areas in search of jobs, food, housing, educational opportunities, better health care, and entertainment. Some are driven to move by factors such as famine, loss of land for growing food, deteriorating environmental conditions, war, and religious, racial, and political conflicts

Population experts identify three major trends related to urban populations:

1. *The percentage of the global population that lives in urban areas has grown sharply and this trend is projected to continue.* Between 1850 and 2016, the percentage of the world's people living in urban areas increased from 2% to 54% and is likely to reach 67% by 2050. Between 2016 and 2050, the world's urban population

is projected to grow from 4.0 billion to 6.6 billion. The great majority of these 2.6 billion new urban dwellers will live in less-developed countries.

2. *The numbers and sizes of urban areas are increasing.* In 2016, there were 30 *megacities*—cities with 10 million or more people—22 of them in less-developed countries (Figure 6.14). Thirteen of these urban areas are *hypercities* with more than 20 million people. The largest hypercity is Tokyo, Japan, with 37.8 million—more than the entire population of Canada. By 2025, the number of megacities is expected to reach 37 with 21 of them in Asia. Some megacities and hypercities are merging into vast urban *megaregions,* each with more than 100 million people. The largest megaregion is the Hong Kong–Shenzhen–Guangzhou region in China with about 120 million people.

3. *Poverty is becoming increasingly urbanized, mostly in less-developed countries.* The United Nations estimates that at least 1 billion people live in the slums and shantytowns of most of the major cities in less-developed countries (see chapter-opening photo). This number may triple by 2050.

CONSIDER THIS …

THINKING ABOUT Urban Trends

If you could reverse one of the three urban trends discussed here, which one would it be? Explain.

CASE STUDY

Urbanization in the United States

Between 1800 and 2016, the percentage of the U.S. population living in urban areas increased from 5% to 82%. Figure 6.15 shows the major urban areas in the United States with more than 1 million people each. This population shift from rural to urban has occurred in three phases. First, *people migrated from rural areas to large central cities.* Second, *many people migrated from large central cities to nearby smaller cities and suburbs.* Currently, about half of all urban Americans live in the suburbs, nearly a third in central cities, and the rest in rural housing developments beyond suburbs. Third, *many people migrated from the North and East to the South and West.*

Since 1920, and especially since 1970, many of the worst urban environmental problems in the United States have been reduced significantly (Figure 6.5). Most people have better working and housing conditions and air and water quality have improved. Better sanitation, clean public water supplies, and expanded medical care have slashed death rates and incidences of sickness from infectious diseases. In addition, the concentration of most of the population in urban areas has helped to protect some of the country's biodiversity by reducing the destruction and degradation of wildlife habitat.

FIGURE 6.14 Megacities, or major urban areas with 10 million or more people, in 2015. *Question:* In order, what were the world's five most populous urban areas in 2015?

Compiled by the authors using data from National Geophysics Data Center, Demographia, National Oceanic and Atmospheric Administration, and United Nations Population Division.

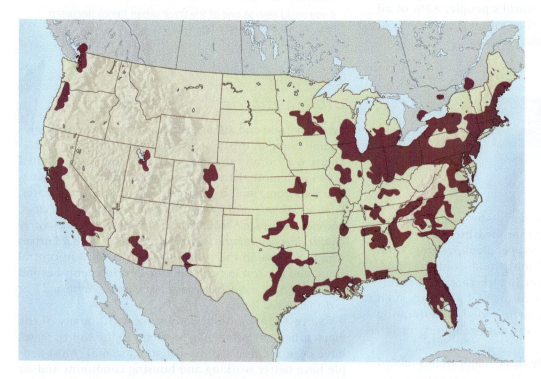

FIGURE 6.15 Urbanized areas (shaded) in the United States where cities, suburbs, and towns dominate the land area. *Critical thinking:* Why do you think many of the largest urban areas are located near water?

Compiled by the authors using data from National Geophysical Data Center/National Oceanic and Atmospheric Administration and U.S. Census Bureau.

However, a number of U.S. central cities—especially older ones—have deteriorating services and aging *infrastructures* (streets, bridges, dams, power lines, schools, water supply pipes, and sewers). For example, about 58,500 U.S. bridges that are used daily are in need of repair. If placed end-to-end, these bridges would stretch from Miami to New York City. Funds for repairing and upgrading urban infrastructure have declined in many urban areas as the flight of people and businesses to the suburbs and beyond has led to lower central city property tax revenues.

Urban Sprawl

In the United States and some other countries, **urban sprawl**—the growth of low-density development on the edges of cities and towns—is eliminating agricultural and wild lands around many cities (Figure 6.16). The result is a dispersed jumble of housing developments, shopping malls, parking lots, and office complexes that are loosely connected by multilane highways and freeways.

Urban sprawl is the product of ample affordable land, automobiles, government funding of highways, and lack of urban planning. Many people prefer living in suburbs. Compared to central cities, these areas provide lower-density living and access to single-family homes on larger lots. Often these areas also have newer public schools and lower crime rates. However, urban sprawl contributes to a number of environmental problems, as summarized in Figure 6.17.

CONSIDER THIS ...

THINKING ABOUT Urban Sprawl

Do you think the advantages of urban sprawl outweigh its disadvantages? Explain. Would you prefer to live in a central city, a suburb, or a rural housing development beyond the suburbs? Explain.

Urbanization Has Advantages

Urbanization has many benefits. Cities are centers of economic development, innovation, education, technological advances, social and cultural diversity, and jobs. Urban residents in many parts of the world tend to live longer than do rural residents and to have lower infant mortality and fertility rates. Urban residents usually have better access to medical care, family planning, education, and social services than do their rural counterparts.

Urban areas also have some environmental advantages. Recycling is more economically feasible because of the high concentrations of recyclable materials in urban areas. Satellite images show that urban areas containing 54% of the world's people occupy only 2.8% of the earth's land, excluding Antarctica. Concentrating people in urban areas preserves biodiversity by reducing the stress on wildlife habitats outside of urban areas. Heating and cooling multistory apartment and office buildings in central cities takes less energy per person than does heating and cooling single-family homes and smaller office buildings in the suburbs. Central-city dwellers also tend to drive less and rely more on mass transportation, car-pooling, walking, and bicycling.

1973

Courtesy of U.S. Geological Survey

2013

U.S. Department of the Interior/U.S. Geological Survey

FIGURE 6.16 *Urban sprawl* in and around the U.S. city of Las Vegas, Nevada, between 1973 and 2013—a trend that has continued since 2013. ***Critical thinking:*** What might be a limiting factor on population growth in Las Vegas?

Urban Sprawl

Land and Biodiversity

Loss of cropland

Loss and fragmentation of forests, grasslands, wetlands, and wildlife habitat

Water

Increased use and pollution of surface water and groundwater

Increased runoff and flooding

Energy, Air, and Climate

Increased energy use and waste

Increased emissions of carbon dioxide and other air pollutants

Economic Effects

Decline of downtown business districts

More unemployment in central cities

FIGURE 6.17 Some of the undesirable impacts of urban sprawl, or car-dependent development. ***Critical thinking:*** Which five of these effects do you think are the most harmful?

Left: Condor 36/Shutterstock.com.
Left center: Joseph Sohm/Shutterstock.com.
Right center: Ssuaphotos/Shutterstock.com.
Right: ronfromyork,2009/Shutterstock.Com.

Urbanization Has Disadvantages

Most urban areas are unsustainable systems. Although urban populations occupy only about 2.8% of the earth's land area, they consume about 75% of its resources and produce about 75% of the world's pollution and wastes. Because of high inputs of food, water, and other resources, and the resulting high waste outputs (Figure 6.18), most of the world's cities have huge ecological footprints that extend far beyond their boundaries. Here are a few reasons why:

Most Cities Lack Vegetation. In urban areas, most trees, shrubs, grasses, and other plants are cleared to make way for buildings, roads, parking lots, and housing developments. Thus, most cities do not benefit from the free ecosystem services provided by vegetation, including air purification, generation of oxygen, removal of atmospheric CO_2, control of soil erosion, and wildlife habitat.

CONSIDER THIS ...

CONNECTIONS Urban Living and Biodiversity Awareness

Recent studies reveal that urban dwellers tend to live most or all of their lives in artificial environments that isolate them from forests, grasslands, streams, and other natural areas. As a result, many urban residents are unaware of the importance of protecting not only the earth's increasingly threatened biodiversity but also its other forms of natural capital that support their lives and the cities in which they live.

Many Cities Have Water Problems. As urban areas grow, their water demands increase. This requires building expensive reservoirs or drilling deeper wells. This can deprive rural and wild areas of surface water and it can deplete groundwater supplies. In addition, projected climate change is expected to increase the melting of some mountaintop glaciers. Some of these glaciers might eventually disappear. Urban areas that depend on this melting ice for much of their annual water supplies will face severe water shortages.

Flooding also tends to be greater in cities built on floodplains near rivers or along low-lying coastlines. In most cities, buildings and paved surfaces cause precipitation to run off quickly and overload storm drains. Urban development has often destroyed or degraded large areas of wetlands that have served as natural sponges to help absorb excess storm water. Many coastal cities (Figure 6.14) will very likely face increased flooding in this century as sea levels rise because of projected climate change.

Cities Tend to Concentrate Pollution and Health Problems. Urban areas produce most of the world's air pollution, water pollution, and solid and hazardous wastes. Pollutant levels are generally higher because the pollution is produced in a confined area and cannot be dispersed and diluted as readily as pollution produced in rural areas. In addition, high population densities in urban areas can promote the spread of infectious diseases, especially if adequate drinking water and sewage systems are not in place.

Cities Have Excessive Noise. Because of the concentration of people and motor vehicles in cities, most urban dwellers are subjected to **noise pollution**: any unwanted, disturbing, or harmful sound. Noise pollution can interfere with and damage hearing. It can also cause stress, raise blood pressure, and hamper one's ability to concentrate and work efficiently. Noise levels are measured

Inputs

Energy
Food
Water
Raw materials
Manufactured goods
Money
Information

Outputs

Solid wastes
Waste heat
Air pollutants
Water pollutants
Greenhouse gases
Manufactured goods
Noise
Wealth
Ideas

FIGURE 6.18 Natural Capital Degradation: The typical city depends on nonurban areas for huge inputs of matter and energy resources, while it generates and concentrates large outputs of pollution, waste matter, and heat. ***Critical thinking:*** How would you apply the three **scientific principles of sustainability** to lessen some of these impacts?

in decibel-A (dbA) sound pressure units that vary with different human activities (Figure 6.19). Prolonged exposure to sound levels above 85 dbA causes permanent hearing damage. Just 1.5 minutes of exposure to 110 decibels or more can cause such damage.

We can reduce noise pollution and hearing damage by wearing earplugs or other protective devices, shielding workers and others from noisy activities or processes, moving noisy operations or machines away from people, and using *antinoise technologies* that cancel or muffle harmful noise with sounds that are not harmful.

Cities Affect Local Climates. Cities tend to be warmer, rainier, foggier, and cloudier than suburbs and nearby rural areas. In cities, the enormous amount of heat generated by cars, factories, furnaces, lights, air conditioners, and heat-absorbing dark roofs and streets creates an **urban heat island** surrounded by cooler suburban and rural areas. As urban areas grow and merge, these heat islands merge, which can reduce the natural dilution and cleansing of polluted air. The urban heat island effect can also increase dependence on air conditioning. This, in turn, leads to higher energy consumption, heat generation, greenhouse gas emissions, and other forms of air pollution.

CONSIDER THIS ...

THINKING ABOUT Disadvantages of Urbanization
Which two of these disadvantages of urbanization do you think are the most serious? Explain.

Poverty and Urban Living

Poverty is a way of life for many urban dwellers in less-developed countries. According to a UN study, there were 1 billion urban dwellers living in poverty in 2015.

Some poor people live in crowded slums (see chapter opening photo)—overcrowded neighborhoods dominated by dilapidated tenements where several people might live in a single room. Other poor people live in squatter settlements on the outskirts of cities. Some build shacks from corrugated metal, plastic sheets, scrap wood, cardboard, and other scavenged building materials. Others live in rusted shipping containers and junked cars.

Poor people living in shantytowns and squatter settlements, or on the streets (Figure 6.12), usually lack clean water supplies, sewers, and electricity. They are often subject to severe air and water pollution and hazardous wastes from nearby factories. Many of these settlements are in locations prone to landslides, flooding, or earthquakes. Some city governments regularly bulldoze squatter settlements and send police to drive the settlers out. The people usually move back in within a few days or weeks, or develop other shantytowns elsewhere.

Some governments have addressed these problems by granting settlers legal title to the land. They base this on evidence that poor people usually improve their living conditions once they know they have a permanent place to live.

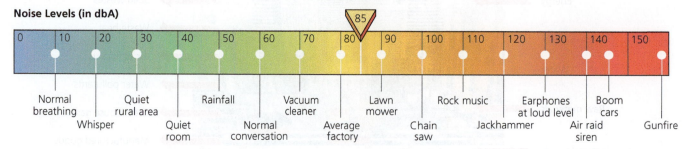

Permanent damage begins after 8-hour exposure

Noise Levels (in dbA)

| 0 | 10 | 20 | 30 | 40 | 50 | 60 | 70 | 80 | 85 | 90 | 100 | 110 | 120 | 130 | 140 | 150 |

Normal breathing — Whisper — Quiet rural area — Quiet room — Rainfall — Normal conversation — Vacuum cleaner — Average factory — Lawn mower — Chain saw — Rock music — Jackhammer — Earphones at loud level — Air raid siren — Boom cars — Gunfire

FIGURE 6.19 *Noise levels* (in decibel-A [dbA] sound pressure units) of some common sounds. **Question:** How often are your ears subjected to noise levels of 85 or more dbA?

Mexico City

With 20.1 million people, Mexico City is one of the world's hypercities (Figure 6.14). More than one-third of its residents live in slums called *favelas* or in squatter settlements without running water and electricity.

At least 3 million people have no sewage facilities. Their human waste is deposited in gutters, vacant lots, and open ditches every day, attracting armies of rats and flies. When the winds pick up dried excrement, a *fecal snow* blankets parts of the city. This bacteria-laden fallout leads to widespread salmonella and hepatitis infections, especially among children.

In 1992, the UN named Mexico City the most polluted city on the planet. Since then Mexico City has made progress in reducing the severity of some of its air pollution problems. In 2013, the Institute for Transportation and Development awarded the city its Sustainable Transportation Award for expanding its bus rapid-transit (BRT) system and its bike sharing program and bike lanes. Between 1992 and 2013, the percentage of days each year in which air pollution standards are violated fell from 50% to 20%.

The city government has moved refineries and factories out of the city, banned cars in its central zone, and required air pollution controls on all cars made after 1991. It has also phased out the use of leaded gasoline, expanded public transportation, and replaced some old buses, taxis, and delivery trucks with vehicles that produce fewer emissions.

Mexico City still has a long way to go as its human population increases along with its number of motor vehicles. However, its progress shows what a city can do to improve environmental quality once a community decides to act.

6.6 HOW DOES TRANSPORTATION AFFECT URBAN ENVIRONMENTAL IMPACTS?

CONCEPT 6.6 In some countries, many people live in widely dispersed urban areas and expand their ecological footprints by depending mostly on motor vehicles for their transportation.

Cities Can Grow Outward or Upward

If a city cannot spread outward, it must grow vertically—upward and below ground—so it occupies a small land area with a high population density. Most people living in *compact cities* such as Hong Kong, China, and Tokyo, Japan, get around by walking, biking, or using mass transit such as rail and bus systems. Some high-rise apartment buildings in these Asian cities contain everything from grocery stores to fitness centers that reduce the need for their residents to travel for food, entertainment, and other services.

In other parts of the world, a combination of plentiful land and networks of highways has produced *dispersed cities* whose residents depend on motor vehicles for most travel (**Concept 6.6**). Such cities are found on all continents and are especially prevalent in North America. The resulting urban sprawl can have a number of undesirable effects (Figure 6.17).

The United States is a prime example of a car-centered nation. With 4.4% of the world's people, the country has about 23% of the world's 1.1 billion motor vehicles, according to the U.S. Energy Information Agency (EIA). In its dispersed urban areas, U.S. passenger vehicles are used for 86% of all transportation and 76% of urban residents drive alone to work every day (up from 64% in 1980).

Pros and Cons of Motor Vehicles

Motor vehicles offer a convenient and comfortable way to get people around. In addition, much of the world's economy is built on producing motor vehicles and supplying fuel, roads, services, and vehicle maintenance and repairs.

Despite their important benefits, motor vehicles have harmful effects on people and the environment. Globally, automobile accidents kill more than 1.2 million people a year—an average of more than 3,300 deaths per day—and injure 50 million more. They also kill about 50 million wild animals and family pets every year.

Each year, motor vehicle accidents in the United States kill about 44,000 people and injure another 2 million, at least 300,000 of them severely. Car accidents have killed more Americans than have all the wars in the country's history.

Motor vehicles are the world's largest source of outdoor air pollution, which kills about 100,000 people per year in the United States, according to the Environmental Protection Agency. They are also the fastest-growing source of climate-changing CO_2 emissions. The average car in the United States emits about 1.8 metric tons (2.2 tons) of CO_2 each year. In addition, at least a third of the world's urban land and half of that in the United States is devoted to roads, parking lots, gasoline stations, and other automobile-related uses.

Widespread use of motor vehicles causes traffic congestion. If current trends continue, U.S. motorists will soon spend an average of 2 years of their lives in traffic jams. Traffic congestion in some cities in less-developed countries is much worse. Building more roads is not likely the answer because more roads usually encourage more people to drive.

Reducing Automobile Use

Some environmental scientists and economists suggest that we can reduce the harmful effects of automobile use by making drivers pay directly for most of the environmental and health costs they cause. This *user-pays* approach is a way to implement the *full-cost pricing* **principle of sustainability**.

One way to phase in such *full-cost pricing* is to charge a tax or fee on gasoline to cover the estimated harmful costs of driving. According to a study by the International Center for Technology Assessment, such a tax would amount to about $3.18 per liter ($12 per gallon) of gasoline in the United States. Automobile owners end up paying these harmful costs in the form of higher medical and health insurance bills. They also pay higher taxes to support federal, state, and local efforts to regulate and reduce air pollution from motor vehicles and for military defense of Persian Gulf oil supplies. However, most drivers do not relate these costs to their use of gasoline.

Gradually phasing in such a tax, as has been done in a number of European nations, would spur the use of motor vehicles that are more energy-efficient and mass transit. It would also reduce pollution, environmental degradation, and ocean acidification, and slow climate change.

Proponents of higher gasoline taxes urge governments to do two major things. *One is to* fund programs to educate people about the hidden harmful costs they are paying for gasoline. The other is to use gasoline tax revenues to help finance mass-transit systems, bike lanes, and sidewalks as alternatives to cars. This would also reduce taxes on income, wages, and wealth to offset the increased taxes on gasoline. In addition, such a *tax shift* would help make higher gasoline taxes more politically and economically acceptable.

Taxing gasoline heavily would be difficult in the United States, for three reasons. *First*, it faces strong opposition from people who feel they are already overtaxed, many of whom are largely unaware of the hidden costs they are paying for gasoline. Another opposition group is made up of the powerful transportation-related industries such as carmakers, oil and tire companies, road builders, and many real-estate developers. *Second*, the dispersed nature of most U.S. urban areas makes people dependent on cars, and higher taxes would be an economic burden for them. *Third*, mass-transit options, bike lanes, and sidewalks are not widely available in the United States. This is because most of the money from gasoline taxes is used for building and improving highways for motor vehicles.

Another user-pays approach to reduce automobile use and urban congestion is to raise parking fees and charge tolls on roads, tunnels, and bridges leading into cities—especially during peak traffic times. Densely populated Singapore is rarely congested because it auctions the rights to buy a car, and drivers are charged a fee every time they enter the city. Several European cities have also imposed stiff fees for motor vehicle use in their central cities, while others have banned the parking of cars on city streets and established networks of bike lanes. Shanghai, China, discourages car use by charging more than $9,000 for a license plate.

Car use in the United States is projected to decline in coming years because many Millennials are not driving or buying cars. In addition, there will be increased use of automated motor vehicles that can be summoned with a tap on a cell phone. They are projected to account for 66% of U.S. car use by 2030. This will also decrease the need to build parking lots and parking garages.

More than 300 European cities have *car-sharing* networks that provide short-term rental of cars. Network members reserve a car or contact the network and are directed to the closest car. In Berlin, Germany, car sharing has cut car ownership by 75%. According to the Worldwatch Institute, car sharing in Europe has reduced the average driver's CO_2 emissions by 40–50%. Car-sharing networks have sprouted in several U.S. cities and on some college campuses, and some large car-rental companies have begun renting cars by the hour.

Alternatives to Cars

There are several alternatives to motor vehicles. Figure 6.20 shows the transportation hierarchy in more-sustainable cities.

A widely used alternative for short distances is the *bicycle*, or pedal power. It is affordable, does not pollute, provides exercise, and requires little parking space. The use of bicycles with lightweight electric motors is also on the rise.

Bicycling accounts for at least a third of all urban trips in the Netherlands and in Copenhagen, Denmark, compared to less than 1% in the United States. About 25% of Americans polled would bike to work or school if safe bike lanes and secure bike storage were available. More than 700 cities in 50 countries, including 80 U.S. cities, have *bike-sharing systems* that allow individuals to rent bikes as needed from widely distributed stations. Portland, Oregon, is one of the country's most bicycle-friendly cities and has a widely used bus transit and light-rail system (Figure 6.21).

Each of the several alternatives to motor vehicles has advantages and disadvantages. Figures 6.22 through 6.25 summarize the pros and cons of using, respectively, bicycles, BRT systems, mass-transit rail systems (within urban areas), and high-speed rail systems (between urban areas).

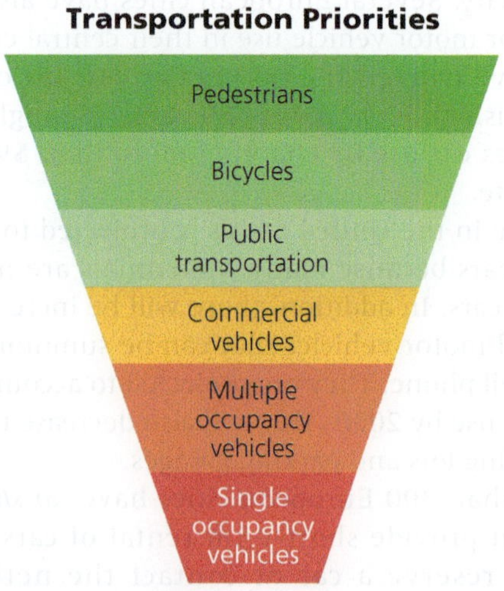
Transportation Priorities
Pedestrians
Bicycles
Public transportation
Commercial vehicles
Multiple occupancy vehicles
Single occupancy vehicles

FIGURE 6.20 Transportation priorities in more-sustainable cities. ***Critical thinking:*** Can you think of any drawbacks to using these priorities? Explain.

CONSIDER THIS ...

Learning From Nature

The kingfisher bird's long beak allows it to dive into water at a high speed without making a splash to catch fish. In Japan, designers increased the speed and reduced the noise from high-speed bullet trains by modeling the train's front end after the kingfisher's beak.

6.7 HOW CAN WE MAKE CITIES MORE SUSTAINABLE AND LIVABLE?

CONCEPT 6.7 An *eco-city* allows people to choose walking, biking, or mass transit for most transportation needs; recycle or reuse most of their wastes; grow much of their food; and protect biodiversity by preserving surrounding land.

Smart Growth

Smart growth is a set of policies and tools that encourages more environmentally sustainable urban development with less dependence on cars. It uses zoning laws and other tools to channel growth in order to reduce its ecological footprint.

Smart growth can discourage sprawl, reduce traffic, protect ecologically sensitive and important land and waterways, and develop neighborhoods that are more enjoyable places in which to live. Figure 6.26 lists popular smart growth tools that cities are using to control urban growth and prevent sprawl.

The Eco-City Concept: More Sustainable Cities

Many environmental scientists and urban planners call for us to make new and existing urban areas more sustainable and enjoyable places to live through good ecological design—an important way to decrease our harmful environmental impact and increase our beneficial environmental impact.

An important result of this trend is the **eco-city**, a people-oriented, not car-oriented city (**Concept 6.7**). Its residents are able to walk, bike, or use low-polluting mass transit for most of their travel. Its buildings, vehicles, and appliances meet high energy-efficiency standards. Trees and plants adapted to the local climate and soils are planted throughout the city to provide shade, beauty, and wildlife habitats, and to reduce air pollution, noise, and soil erosion.

In an eco-city, abandoned lots and industrial sites are cleaned up and used. Nearby forests, grasslands, wetlands, and farms are preserved. Much of the food that people eat comes from nearby organic farms, solar greenhouses, community gardens, and small gardens on rooftops, in yards, and in window boxes. Parks are easily available

FIGURE 6.21 Since 1986, widespread bicycle use and a light-rail system have helped to reduce car use in Portland, Oregon.

Ken Hawkins/Alamy Stock Photo

Trade-Offs

Bicycles

Advantages	Disadvantages
Are quiet and nonpolluting	Provide little protection in an accident
Take few resources to manufacture	Provide no protection from bad weather
Burn no fossil fuels	Are impractical for long trips
Require little parking space	Bike lanes and secure bike storage not yet widespread

Tyler Olson/Shutterstock.com

FIGURE 6.22 Bicycle use has advantages and disadvantages. *Critical thinking:* Which single advantage and which single disadvantage do you think are the most important?

Trade-Offs

Buses

Advantages	Disadvantages
Reduce car use and air pollution	Can lose money because they require affordable fares
Can be rerouted as needed	Can get caught in traffic and add to noise and pollution
Cheaper than heavy-rail system	Commit riders to transportation schedules

Isaak/Shutterstock.com

FIGURE 6.23 Bus rapid-transit (BRT) systems and conventional bus systems in urban areas have advantages and disadvantages. BRT systems make bus use more convenient by including fast express routes, allowing riders to pay at machines at each bus stop before they board a bus, and having three or four doors for quicker boarding. *Critical thinking:* Which single advantage and which single disadvantage do you think are the most important?

Trade-Offs

Mass-Transit Rail

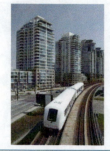

Advantages

- Uses less energy and produces less air pollution than cars do
- Uses less land than roads and parking lots use
- Causes fewer injuries and deaths than cars

Disadvantages

- Expensive to build and maintain
- Cost-effective only in densely populated areas
- Commits riders to transportation schedules

Steve Rosset/Shutterstock.com

FIGURE 6.24 Mass-transit rail systems in urban areas have advantages and disadvantages. They include *heavy-rail* systems (subways, elevated railways, and metro trains) and *light-rail* systems (streetcars, trolley cars, and tramways). ***Critical thinking:*** Which single advantage and which single disadvantage do you think are the most important?

Trade-Offs

Rapid Rail

Advantages

- Much more energy efficient per rider than cars and planes are
- Produces less air pollution than cars and planes
- Can reduce need for air travel, cars, roads, and parking areas

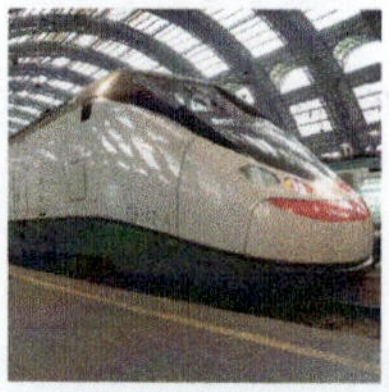

Disadvantages

- Costly to run and maintain
- Causes noise and vibration for nearby residents
- Adds some risk of collision at car crossings

Alfonso d'Agostino/Shutterstock.com

FIGURE 6.25 Rapid-rail systems between urban areas have advantages and disadvantages. Western Europe, Japan, and China have a number of high-speed bullet trains that travel between cities at up to 306 kilometers (190 miles) per hour. ***Critical thinking:*** Which single advantage and which single disadvantage do you think are the most important?

to everyone. People who design and live in eco-cities take seriously the advice that U.S. urban planner Lewis Mumford gave more than 3 decades ago: "Forget the damned motor car and build cities for lovers and friends."

The eco-city is not a futuristic dream, but a growing reality in a number of cities, including Portland, Oregon; Curitiba, Brazil (see the following Case Study); Bogotá, Colombia; Waitakere City, New Zealand; Stockholm, Sweden; Helsinki, Finland; Copenhagen, Denmark; Melbourne, Australia; Vancouver, Canada; Leicester, England; Neerlands, the Netherlands; Huangbaiyu and Tianjin Eco City in China; and in the United States,

Solutions

Smart Growth Tools

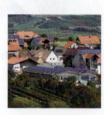

Limits and Regulations

- Limit building permits
- Draw urban growth boundaries
- Create greenbelts around cities

Zoning

- Promote mixed use of housing and small businesses
- Concentrate development along mass transportation routes

Planning

- Ecological land-use planning
- Environmental impact analysis
- Integrated regional planning

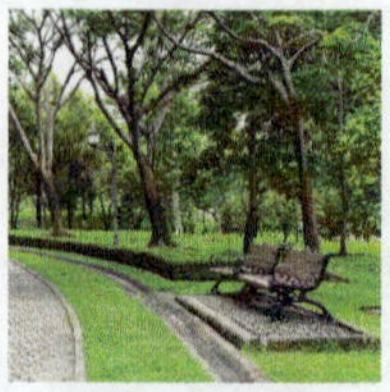

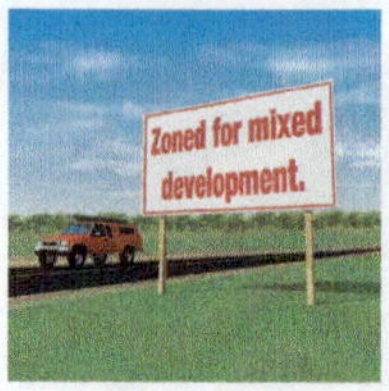

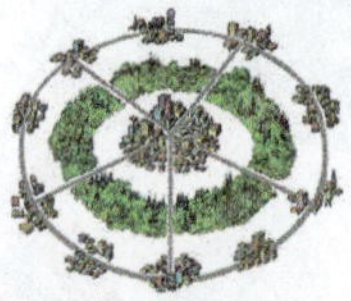

Protection

- Preserve open space
- Buy new open space
- Prohibit certain types of development

Taxes

- Tax land, not buildings
- Tax land on value of actual use instead of on highest value as developed land

Tax Breaks

- For owners agreeing not to allow certain types of development
- For cleaning up and developing abandoned urban sites

Revitalization and New Growth

- Revitalize existing towns and cities
- Build well-planned new towns and villages within cities

Top: Tungphoto/Shutterstock.com. Bottom: iStockphoto.com/Richard Schmidt-Zuper.

FIGURE 6.26 Smart growth tools can be used to prevent or control urban growth and sprawl. ***Critical thinking:*** Which five of these tools do you think would be best for preventing or controlling urban sprawl? Which, if any, of these tools are used in your community?

Davis, California; Olympia, Washington; and Chattanooga, Tennessee.

CASE STUDY

The Eco-City Concept in Curitiba, Brazil

An example of an eco-city is Curitiba, a city of 1.9 million people, known as the "ecological capital" of Brazil. In 1969, planners in this city decided to focus on an inexpensive and efficient mass-transit system rather than on the car.

Curitiba's BRT system efficiently moves large numbers of passengers, including 72% of the city's commuters. Each of the system's five major "spokes," connecting the city center to outlying districts (see map in Figure 6.27), has two

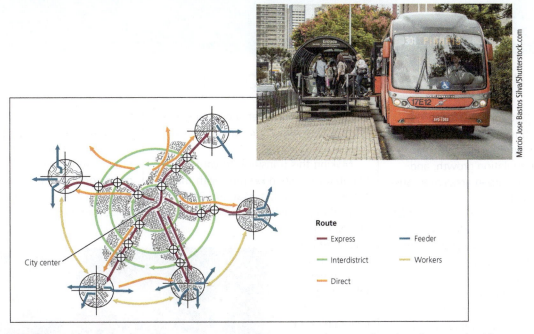

Route
— Express   — Feeder
— Interdistrict — Workers
— Direct

City center

FIGURE 6.27 Solutions: Curitiba's bus rapid-transit system has greatly reduced car use in this Brazilian city.

express lanes used only by buses. Double- and triple-length bus sections are coupled as needed to carry up to 300 passengers. Boarding is speeded up by the use of extra-wide bus doors and boarding platforms where passengers can pay before getting on the bus (photo in Figure 6.27).

Only high-rise apartment buildings are allowed near major bus routes, and the bottom two floors of each building must be devoted to stores—a practice that reduces the need for residents to travel. Cars are banned from 49 blocks in the center of the downtown area, which has a network of pedestrian walkways connected to bus stations, parks, and bicycle paths running throughout most of the city. As a result, Curitiba uses less energy per person and has lower emissions of greenhouse gases and other air pollutants and less traffic congestion than do most comparably sized cities.

Along the six streams that run within Curitiba's borders, the city removed most buildings and lined the streams with a series of interconnected parks. Volunteers have planted more than 1.5 million trees throughout the city, and no one can cut down a tree without a permit, which also requires that two trees must be planted for each one that is cut down.

Curitiba recycles roughly 70% of its paper and 60% of its metal, glass, and plastic. Recovered materials are sold mostly to the city's more than 500 major industries, which must meet strict pollution standards.

Curitiba's poor residents receive free medical and dental care, childcare, and job training, and 40 feeding centers are available for street children. People who live in areas not served by garbage trucks can collect garbage and exchange filled garbage bags for surplus food, bus tokens, and school supplies. The city uses old buses as roving classrooms to train its poor in basic job skills. Other retired buses have become health clinics, soup kitchens, and day-care centers that are free for low-income parents.

About 95% of Curitiba's citizens can read and write and 83% of its adults have at least a high school education. All school children study ecology. Polls show that 99% of the city's inhabitants would not want to live anywhere else.

Curitiba face challenges, as do all cities, mostly due to a fivefold increase in its population since 1965. Its once-clear streams are often overloaded with pollutants. The bus system is nearing capacity, and car ownership is on the rise. The city is considering building a light-rail system to relieve some of the pressure.

This internationally acclaimed model of urban planning and sustainability is the brainchild of architect and former college professor Jaime Lerner, who served as the city's mayor three times since 1969.

BIG IDEAS

- The human population and the global rate of resource use per person are both growing and putting an increasing strain on the earth's natural capital.

- We can slow human population growth by reducing poverty through economic development, elevating the status of women, and encouraging family planning.

- Most urban areas, home to more than half of the world's people, are unsustainable systems that can be made more sustainable and livable.

Population Growth, Urbanization, and Sustainability

In this chapter, we looked at the growth of the human population and urban areas, their environmental impacts, ways to slow population growth, and how we can make urban areas more sustainable and livable.

The three **scientific principles of sustainability**—reliance on solar energy, chemical cycling, and biodiversity—can guide us in dealing with the problems brought on by population growth and by urban growth. By employing solar and other renewable energy technologies more widely, we can cut pollution and emissions of climate-changing gases that are increasing as the population, resource use per person, and urban areas grow. By reusing and recycling more materials, we could cut our waste and reduce our ecological footprints. And in focusing on preserving biodiversity, we could help sustain the life-support system on which we and all other species depend, thereby increasing our beneficial environmental impact.

Making this transition toward sustainability is also in keeping with the three **principles of sustainability** derived from politics, economics, and ethics. Full-cost pricing requires that the harmful environmental costs of urbanization be included in the market prices of goods and services. To achieve this, people would have to work together to find win–win solutions to population and urban problems. By implementing these solutions, we could apply the ethical principle that calls for us leave the planet's life-support system in at least as good condition as we now enjoy and ideally in a more sustainable condition for future generations.

JeremyRichards/Shutterstock.com

Chapter Review

Core Case Study

1. Summarize the story of how the human population has surpassed 7.4 billion. List three factors that account for the rapid increase in the world's human population over the past 200 years.

Section 6.1

2. What is the key concept for this section? Summarize the three major population growth trends recognized by demographers. About how many people are added to the world's population each year? List eight major ways in which we have altered the earth's ecosystem services to meet our needs. Summarize the debate over whether and how long the human population can keep growing.

Section 6.2

3. What are the two key concepts for this section? List three variables that affect the growth and decline of human populations. How can we calculate the **population change** of an area? Define the **total fertility rate (TFR)**. How has the global TFR changed since 1955? Summarize the story of population growth in the United States. List six changes in lifestyles that have taken place in the United States during the 20th century, leading to a rise in per capita resource use.

4. List eight factors that affect birth rates and fertility rates. Define **life expectancy** and **infant mortality rate** and explain how they affect the population size of a country. What is **migration**? What factors can promote migration?

Section 6.3

5. What is the key concept for this section? What is the **age structure** of a population? Explain how age structure affects population growth and economic growth. What is demographic momentum? Describe the American baby boom and some of its economic and social effects. What are some problems related to rapid population decline due to an aging population?

Section 6.4

6. What is the key concept for this section? What is the **demographic transition** and what are its four stages? Explain how the reduction of poverty and empowerment of women can help countries to slow their population growth. What is **family planning** and how can it be used to help to stabilize populations? Describe India's efforts to control its population growth. Describe China's population control program and recent changes to it.

Section 6.5

7. What is the key concept for this section? What percentage of the world's people live in urban areas? List two ways in which urban areas grow. List three trends in global urban growth. Describe the three phases of urban growth in the United States. What is **urban sprawl**? List five factors that have promoted urban sprawl in the United States. List five undesirable effects of urban sprawl.

8. What are the major advantages and disadvantages of urbanization? Define **noise pollution**. Explain why

most urban areas are unsustainable systems. Describe the major aspects of poverty in urban areas. Summarize Mexico City's major urban and environmental problems and what government officials are doing about them.

Section 6.6

9. What is the key concept for this section? Distinguish between compact and dispersed cities, and give an example of each. What are the major advantages and disadvantages of using motor vehicles? List four ways to reduce dependence on motor vehicles. Explain why car use in the United States is projected to decline over the next two decades. List the major advantages and disadvantages of relying more on (a) bicycles, (b) bus rapid-transit systems, (c) mass-transit rail systems within urban areas, and (d) rapid-rail systems between urban areas.

Section 6.7

10. What is the key concept for this section? Define **smart growth** and explain its benefits. Define **eco-city** and describe the eco-city model. Give five examples of how Curitiba, Brazil, has attempted to become an eco-city. What are this chapter's *three big ideas*? Explain how eco-cities are applying the six **principles of sustainability** to become more sustainable urban areas.

Note: Key terms are in bold type.

Critical Thinking

1. Do you think that the global population of 7.4 billion is too large? Explain. If your answer was *yes*, what do you think should be done to slow human population growth? If your answer was *no*, do you believe that there is a population size that would be too big? Explain. Do you think that the population of the country where you live is too large? Explain.

2. If you could say hello to a new person every second without taking a break and working around the clock, how many years would it take you to greet the 89.9 million people who were added to the world's population in 2016? How many years would it take you to greet the 7.4 billion people living on the earth in 2016?

3. Identify a major local, national, or global environmental problem, and describe the role that population growth plays in this problem.

4. Some people think that our most important environmental goals should be to sharply reduce the rate of population growth in less-developed countries, where at least 92% of the world's population growth is expected to take place between now and 2050. Others argue that the most serious environmental problems stem from high levels of resource consumption per person in more-developed countries, which have much larger ecological footprints per person than do less-developed countries. What is your view on this issue? Explain.

5. Do you think that projected increases in the earth's population size and economic growth are sustainable? Explain. If not, how is this likely to affect your life? How will it affect the lives of any children or grandchildren you might have?

6. If you own a car or hope to own one, what conditions, if any, would encourage you to rely less on your car and to travel to school or work by bicycle, on foot, by mass transit, or by carpool?

7. Do you think the United States (or the country where you live) should develop a comprehensive and integrated mass-transit system over the next 20 years, including an efficient rapid-rail network for travel within and between its major cities? Explain. If so, how would you pay for such a system?

8. Consider the characteristics of an eco-city listed on pp. 126–128. How close to this eco-city model is the city in which you live or the city nearest to where you live? Pick what you think are the five most important characteristics of an eco-city and, for each of these characteristics, describe a way in which your city could attain it.

Doing Environmental Science

The campus where you go to school is something like an urban community. Choose five eco-city characteristics (pp. 126–128) and apply them to your campus. For each characteristic:

1. Create a scale of 1 to 10 in order to rate the campus on how well it does in having that characteristic. (For example, how well does it do in giving students options for getting around, other than by using a car? A rating of 1 could be *not at all*, while a rating of 10 could be *excellent*.)

2. Do some research and rate your campus for each characteristic.

3. Write an explanation of your research process and why you chose each rating.

Write a proposed plan for how the campus could improve its ratings.

Global Environment Watch Exercise

Go to your MindTap course to access the GREENR database. At the top of the page, do a "Basic Search" for *eco-city* or *green city*. Find an article about an eco-city and summarize the characteristics that make it a green city. Compare your summary to the description of eco-cities included in this chapter (pp. 126–128). What are some similarities between the two descriptions? What are some differences?

Data Analysis

The chart below shows selected population data for two different countries, A and B. Study the chart and answer the questions that follow.

	Country A	Country B
Population (millions)	144	82
Crude birth rate (number of live births per 1,000 people per year)	43	8
Crude death rate (number of deaths per 1,000 people per year)	18	10
Infant mortality rate (number of babies per 1,000 born who die in first year of life)	100	3.8
Total fertility rate (average number of children born to women during their childbearing years)	5.9	1.3
% of population under 15 years old	45	14
% of population older than 65 years	3	19
Average life expectancy at birth	47	79
% urban	44	75

1. Calculate the rates of natural increase (due to births and deaths, not counting immigration) for the populations of country A and country B. Based on these calculations and the data in the table, for each of the countries, suggest whether it is a more-developed country or a less-developed country and explain the reasons for your answers.

2. Describe where each of the two countries might be in the stages of demographic transition (Figure 6.10). Discuss factors that could hinder either country from progressing to later stages in the demographic transition.

3. Explain how the percentages of people under age 15 in each country could affect its per capita and total ecological footprints.

CENGAGE **brain**.com To access course materials, including Aplia homework, please visit www.cengagebrain.com.

WWW.CENGAGEBRAIN.COM • 133

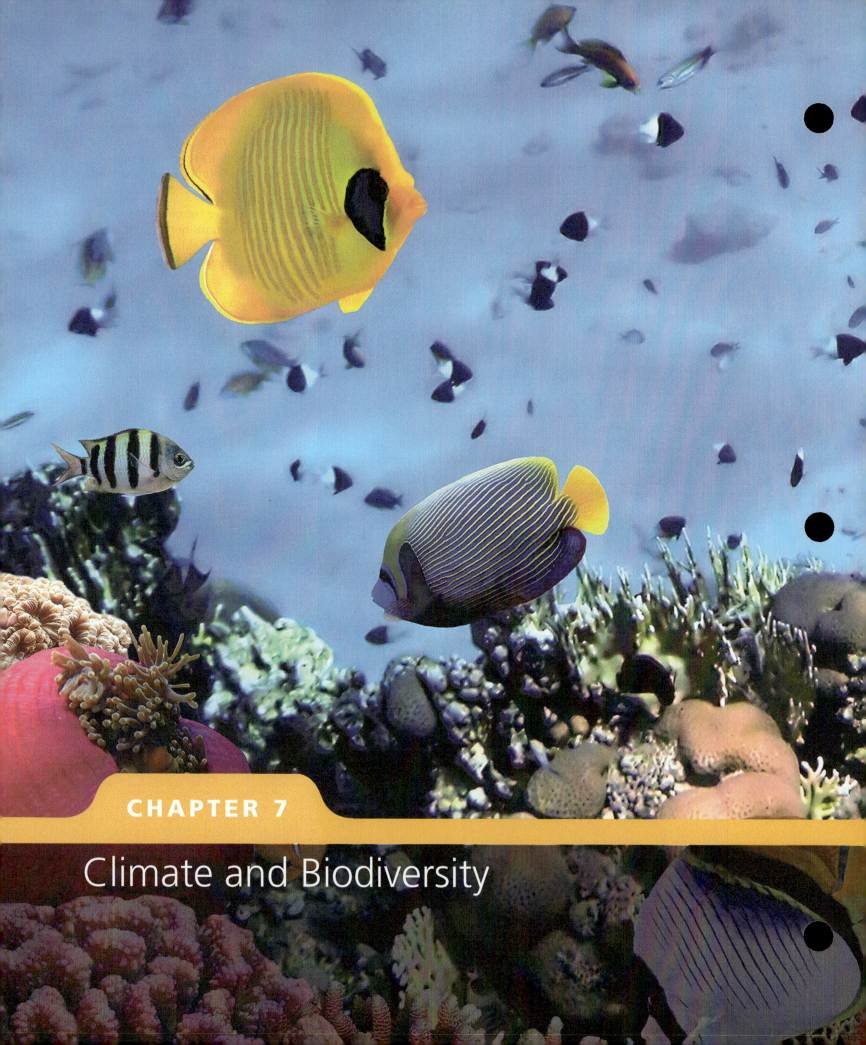

CHAPTER 7

Climate and Biodiversity

When we try to pick out anything by itself, we find it hitched to everything else in the universe.

JOHN MUIR

Key Questions

7.1 What factors influence climate?

7.2 What are the major types of terrestrial ecosystems and how are human activities affecting them?

7.3 What are the major types of marine aquatic systems and how are human activities affecting them?

7.4 What are the major types of freshwater systems and how are human activities affecting them?

Coral reef in Egypt's Red Sea.
Vlad61/Shutterstock.com

African Savanna

The earth has a great diversity of species and *habitats,* or places where these species can live. Some species live in *terrestrial,* or land, habitats—called biomes—such as grasslands, forests, and deserts. Others live in *aquatic,* or water-based, habitats. Examples are oceans, coral reefs (see chapter-opening photo), lakes, and rivers.

Why do grasslands grow on some areas of the earth's land while forests and deserts form in other areas? The answer lies largely in differences in *climate,* the average weather conditions in a given region over at least three decades to thousands of years. Differences in climate result mostly from long-term differences in weather, based primarily on average annual precipitation and temperature. These differences lead to three major types of climate—*tropical* (areas near the equator, receiving the most intense sunlight), *polar* (areas near the earth's poles, receiving the least intense sunlight), and *temperate* (areas between the tropical and polar regions).

Throughout these regions, we find different types of ecosystems, vegetation, and animals in land-based biomes adapted to the various climate conditions. For example, in tropical areas, we find a type of grassland called a *savanna.* This biome typically contains scattered trees and usually has warm temperatures year-round with alternating dry and wet seasons. Savannas in East Africa are home to *grazing* (primarily grass-eating) and *browsing* (twig- and leaf-nibbling) hoofed animals. They include wildebeests, gazelles, antelopes, zebras, elephants (Figure 7.1), and giraffes, as well as their predators such as lions, hyenas, and humans.

Archeological evidence indicates that our species emerged from African savannas and survived by gathering edible vegetation and hunting animals for food and clothing made from animal hides. After the last ice age, about 10,000 years ago, the earth's climate warmed and humans began their transition from hunter-gathers to farmers

growing food on the savanna and on other grasslands. Later, they cleared patches of forest to expand farmland and created villages and eventually towns and cities.

Today, vast areas of African savanna have been plowed up and converted to cropland or used for grazing livestock. Towns are also expanding there, and this trend will continue as the human population in Africa—the continent with the world's fastest population growth—increases. As a result, populations of elephants, lions, and other animals that roamed the savannas for millions of years have dwindled. Many of these animals will face extinction in the next few decades because of the loss of their habitats and because people kill them for food and their valuable parts such as the ivory tusks of elephants.

In this chapter, we explore the factors that determine climate, the nature of terrestrial and aquatic ecosystems, and the effects of human activities on these forms of natural capital. ●

FIGURE 7.1 Elephants on a tropical African savanna.

Amy Nichole Harris/Fotolia LLC

7.1 WHAT FACTORS INFLUENCE CLIMATE?

CONCEPT 7.1 Key factors that influence an area's climate are incoming solar energy, the earth's rotation, global patterns of air and water movement, gases in the atmosphere, and the earth's surface features.

The Earth Has Many Different Climates

It is important to understand the difference between weather and climate. **Weather** is a set of physical conditions of the lower atmosphere that includes temperature, precipitation, humidity, wind speed, cloud cover, and other factors that occur in a given area over a period of hours or days. The two most important factors in an area's weather are atmospheric temperature and precipitation.

While weather is the set of short-term atmospheric conditions over hours to days to years, **climate** is the pattern of atmospheric conditions in a given area over periods ranging from at least three decades to thousands of years. Weather often fluctuates daily, from one season

to another, and from one year to the next. However, climate tends to change slowly because it is the average of long-term atmospheric conditions over at least 30 years.

The key factors that influence an area's climate are incoming solar energy, the earth's rotation, global patterns of air and water movements, greenhouse gases in the atmosphere, and the earth's surface features (**Concept 7.1**). Scientists have used such factors and the temperature and precipitation of different parts of the world averaged over many decades to describe the various regions of the earth according to their climates. Figure 7.2 shows the earth's current major climate zones along with the major **ocean currents**—mass movements of surface ocean water driven by winds and shaped by landforms.

Some climate changes are caused by natural events such as changes in the input of energy from the sun, a change in the earth's orbit, extensive volcanic eruptions, and changes in ocean temperature and currents (**Concept 7.1**). Human activities, such as large inputs of carbon dioxide (CO_2) and other greenhouse gases that alter the earth's natural greenhouse effect (Figure 3.3, p. 48 and Science Focus 7.1), can lead to climate change.

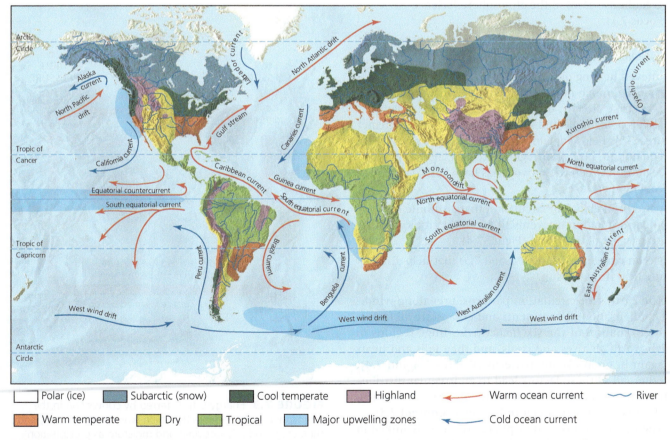

FIGURE 7.2 Natural Capital: Generalized map of the earth's climate zones, major ocean currents, and upwelling areas (where currents bring nutrients from the ocean bottom to the surface). *Question:* Based on this map, what is the general type of climate where you live?

Greenhouse Gases and Climate

Gases in the lower atmosphere affect its temperature and thus the earth's climates. As energy flows from the sun to the earth, some of it is reflected by the earth's surface back into the atmosphere. Molecules of certain gases in the atmosphere, including water vapor (H_2O), CO_2, methane (CH_4), and nitrous oxide (N_2O), absorb some of this solar energy and release a portion of it as infrared radiation (heat) that warms the lower atmosphere and the earth's surface. These gases are called **greenhouse gases**. They play a role in determining the lower atmosphere's average temperatures and therefore the earth's climates.

This natural warming of the lower atmosphere is called the **greenhouse effect** (see Figure 3.3, p. 48). Without this natural warming effect, the earth would be a very cold and mostly lifeless planet with an average temperature of near −18°C (0°F) instead of a much warmer 15°C (59°F).

Human activities such as producing and burning fossil fuels, clearing forests, and growing crops release large amounts of the greenhouse gases, carbon dioxide, CH_4, and N_2O, into the atmosphere. An enormous body of scientific evidence, combined with climate model projections, indicates that human activities

are emitting greenhouse gases into the atmosphere faster than natural processes such as the earth's carbon and nitrogen cycles (see Figures 3.16, p. 59, and 3.17, p. 60) can remove them. These emissions warm the earth's atmosphere, intensify the earth's natural greenhouse effect, and change the earth's climate, as discussed more fully in Chapter 15.

CRITICAL THINKING

How might your life change if human activities continue to enhance the earth's natural greenhouse effect?

Several factors help determine regional climates. They include

- greenhouse gases in the atmosphere (Science Focus 7.1).

- the cyclical movement of air in convection cells driven by solar energy (Figure 7.3).

- uneven heating of the earth's surface by the sun. Air is heated much more at the equator, where the sun's rays strike directly, than at the poles where sunlight strikes at an angle and spreads out over a much greater area. This helps explain why tropical regions near the equator are hot, polar regions are cold, and temperate regions between these two areas have both warm and cool temperatures.

- patterns of global air circulation that distribute heat and precipitation unevenly between the tropics and other parts of the world (Figure 7.4).

- Ocean currents (Figures 7.2 and 7.5) that help distribute heat from the sun.

The earth's air circulation patterns, prevailing winds, and configuration of continents and oceans are all factors in the formation of the six *Hadley cells*—huge regions in which warm air rises and cools, and then falls and heats up again in great cycling patterns (Figures 7.3 and 7.4). Together, all of these factors lead to an irregular distribution of climates and of the resulting deserts, grasslands, and forests, as shown in Figure 7.4, right (**Concept 7.1**).

The oceans and the atmosphere are strongly linked in two ways: ocean currents are driven partly by winds in the atmosphere and heat from the oceans affects atmospheric circulation. One example of the interactions between

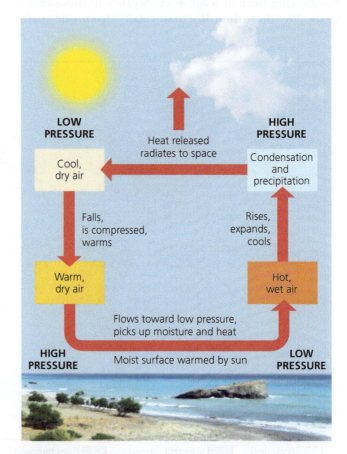

FIGURE 7.3 Energy is transferred by *convection* in the atmosphere—the process by which warm, wet air rises, and then cools and releases heat and moisture as precipitation (right side and top, center). Then the cooler, denser, and drier air sinks, warms up, and absorbs moisture as it flows across the earth's surface (left side and bottom) to begin the cycle again.

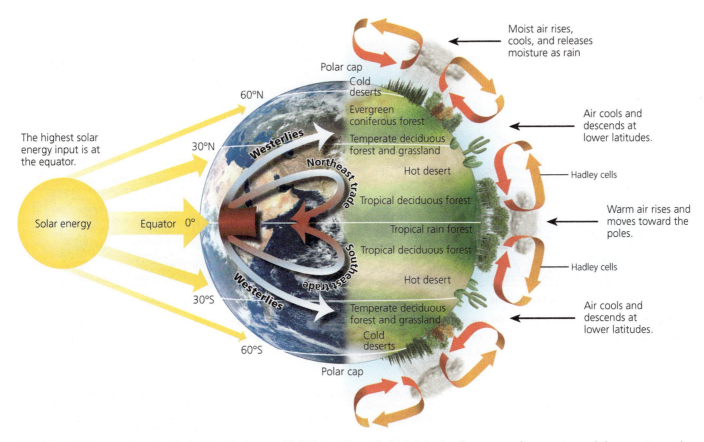

FIGURE 7.4 *Global air circulation:* As air rises and falls in Hadley cells (right), it also flows away from or toward the equator and is deflected to the east or west (left) by the rotation of the earth's axis, depending on where the cell is located. This creates global patterns of prevailing winds (westerlies, Northwest trades, and Southeast trades) that help distribute heat and moisture in the atmosphere, which leads to the earth's variety of forests, grasslands, and deserts (right).

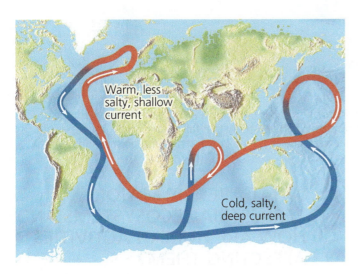

FIGURE 7.5 A connected loop of deep and shallow ocean currents transports warm and cool water to various parts of the earth.

the oceans and the atmosphere is the *El Niño–Southern Oscillation,* or *ENSO* (Figure 7.6). This large-scale weather phenomenon occurs every few years when prevailing winds in the tropical Pacific Ocean weaken and change

direction. The resulting above-average warming of Pacific waters alters the weather over at least two-thirds of the earth for 1 or 2 years by, for example, leading to one or two milder winters in some areas.

Earth's Surface Features Affect Local Climates

Various topographic features of the earth's surface can create local climatic conditions that differ from the general climate in some regions. For example, mountains interrupt the flow of prevailing surface winds and the movement of storms. When moist air blowing inland from an ocean reaches a mountain range, it is forced upward. As the air rises, it cools, expands, and loses most of its moisture as rain and snow that fall on the windward slope of the mountain.

As the drier air mass passes over the mountaintops, it flows down the leeward slopes (facing away from the wind) and warms up. This warmer air can hold more moisture, but it typically does not release much of it. Instead, the air tends to dry out plants and soil below. This process is called the **rain shadow effect** (Figure 7.7). Over many decades, it results in *semiarid* or *arid* conditions

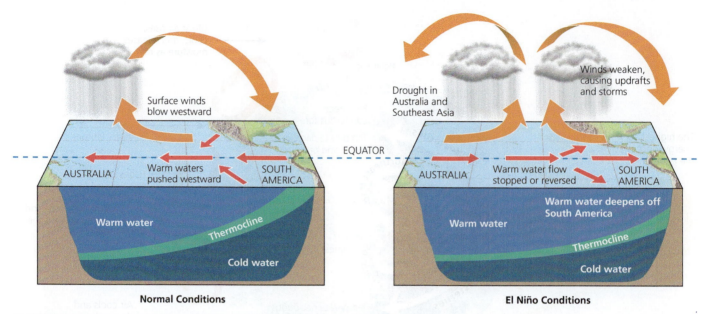

FIGURE 7.6 Normal prevailing or trade winds blowing east to west cause shore upwellings of cold, nutrient-rich bottom water in the tropical Pacific Ocean near the coast of Peru (left). Every few years, a shift in trade winds, known as the *El Niño–Southern Oscillation (ENSO),* disrupts this pattern for 1 to 2 years.

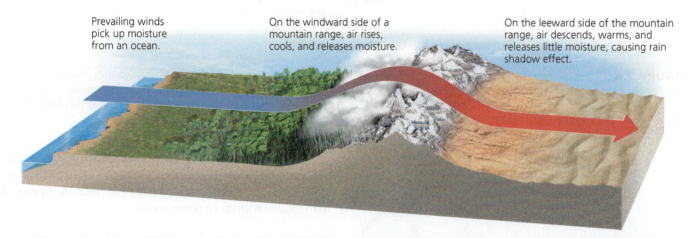

FIGURE 7.7 The *rain shadow effect* is a reduction of rainfall and loss of moisture from the landscape on the leeward side of a mountain. Warm, moist air in onshore winds loses most of its moisture as rain and snow that fall on the windward slopes of a mountain range. This leads to semiarid and arid conditions on the leeward side of the mountain range and on the land beyond.

on the leeward side of a high mountain range. Sometimes this effect leads to the formation of deserts such as Death Valley, a part of the Mojave Desert, which lies on the leeward side of mountains in the southwest United States.

Cities also create distinct microclimates. Bricks, concrete, asphalt, and other building materials absorb and hold heat, and buildings block wind. Motor vehicles and the heating and cooling systems of buildings release large quantities of heat and pollutants. As a result, cities on average tend to have more haze and smog, higher temperatures, and lower wind speeds than the surrounding countryside. These factors make them *heat islands.*

7.2 WHAT ARE THE MAJOR TERRESTRIAL ECOSYSTEMS AND HOW ARE HUMAN ACTIVITIES AFFECTING THEM?

CONCEPT 7.2A Desert, grassland, and forest biomes can be tropical, temperate, or cold depending on their climate and location.

CONCEPT 7.2B Human activities are disrupting ecosystem and economic services provided by many of the earth's deserts, grasslands, forests, and mountains.

Climate Affects Where Terrestrial Organisms Can Live

Differences in climate (Figure 7.2) help explain why one area of the earth's land surface is a desert, another a grassland, and another a forest. Different combinations of varying average annual precipitation and temperatures, along with global air circulation patterns and ocean currents, lead to the formation of tropical (hot), temperate (moderate), and polar (cold) deserts, grasslands, and forests, as summarized in Figure 7.8 (**Concept 7.2A**).

Climate and vegetation vary according to *latitude* and *elevation,* or height above sea level. If you climb a tall mountain, from its base to its summit, you can observe changes in plant life similar to those you would encounter in traveling from the equator to the earth's northern polar region.

Figure 7.9 shows how scientists have divided the world into several major **biomes**—large terrestrial regions, each characterized by a particular type of climate and a certain combination of dominant plant life. The variety of terrestrial biomes and aquatic systems is one of the four components of the earth's biodiversity (see Figure 4.2, p. 71)—a vital part of the earth's natural capital. Figure 4.4 (p. 73) shows how major biomes along the 39th parallel in the United States are related to different climates.

On maps such as the one in Figure 7.9, biomes are shown with sharp boundaries and uniform vegetation. In reality, biomes are not uniform. They consist of a patchwork of areas, each with somewhat different biological communities but with similarities typical of the biome. These patches occur because of the irregular distribution of the resources needed by plants and animals and because human activities have removed or altered the natural vegetation in many areas.

There are also differences along the transition zone (called the *ecotone*) between two different ecosystems or biomes. The ecotone contains habitats that are common to both ecosystems along with other habitats that are unique to the transition zone. The **edge effect** is the tendency for a transition zone to have greater species diversity and a higher density of organisms than are found in either of the individual ecosystems.

Types of Deserts

In a *desert,* annual precipitation is low and often scattered unevenly throughout the year. During the day, the baking sun warms the ground and evaporates water from plant

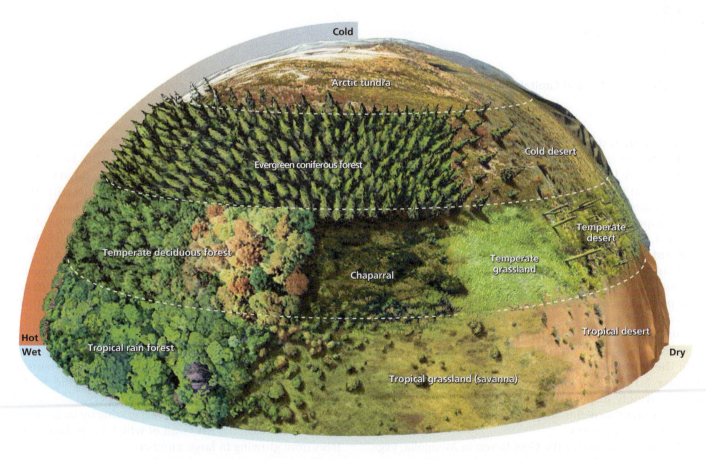

FIGURE 7.8 Natural Capital: Average precipitation and average temperature, acting together as limiting factors over a long time, help determine the type of desert, grassland, or forest in any particular area, and thus the types of plants, animals, and decomposers found in that area (assuming it has not been disturbed by human activities).

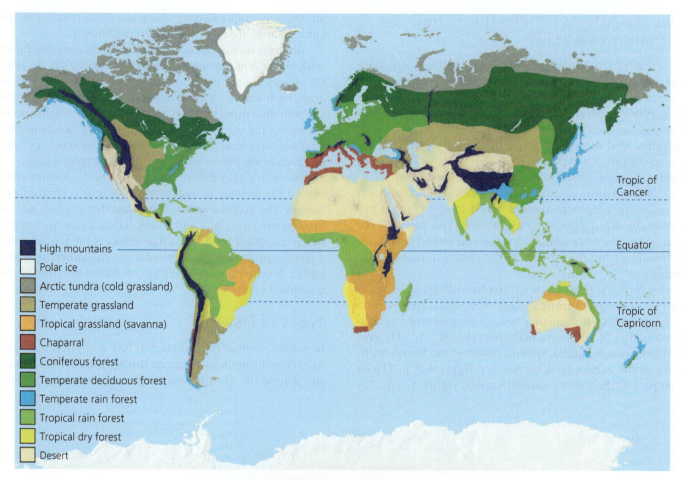

Legend:
- High mountains
- Polar ice
- Arctic tundra (cold grassland)
- Temperate grassland
- Tropical grassland (savanna)
- Chaparral
- Coniferous forest
- Temperate deciduous forest
- Temperate rain forest
- Tropical rain forest
- Tropical dry forest
- Desert

Tropic of Cancer

Equator

Tropic of Capricorn

FIGURE 7.9 Natural Capital: The earth's major *biomes* result primarily from differences in climate.

leaves and from the soil. At night, most of the heat stored in the ground radiates quickly into the atmosphere. This explains why in a desert, you may roast during the day but shiver at night.

A combination of low rainfall and varying average temperatures creates a variety of desert types—tropical, temperate, and cold (Figures 7.8 and 7.9 and **Concept 7.2A**). *Tropical deserts* (Figure 7.10, top photo) such as the Sahara and the Namib of Africa are hot and dry most of the year (Figure 7.10, top graph). They have few plants and a hard, windblown surface strewn with rocks and sand.

In *temperate deserts* (Figure 7.10, center photo), daytime temperatures are high in summer and low in winter and there is more precipitation than in tropical deserts (Figure 7.10, center graph). Their sparse vegetation is mostly widely dispersed, drought-resistant shrubs and cacti or other succulents adapted to the dry conditions and temperature variations.

In *cold deserts* such as the Gobi Desert in Mongolia, vegetation is sparse (Figure 7.10, bottom photo). Winters are cold, summers are warm or hot, and precipitation is low (Figure 7.10, bottom graph). In all types of deserts, plants

and animals have evolved adaptations that help them stay cool and get enough water to survive (Science Focus 7.2).

Desert ecosystems are vulnerable to disruption because they have slow plant growth, low species diversity, slow nutrient cycling due to low bacterial activity in the soils, and little water. It can take decades to centuries for desert soils to recover from disturbances such as off-road vehicle traffic, which can also destroy the habitats for a variety of animal species that live underground. The lack of vegetation, especially in tropical and polar deserts, also makes them vulnerable to heavy wind erosion from sandstorms.

Types of Grasslands

Grasslands occur primarily in the interiors of continents in areas that are too moist for deserts to form and too dry for forests to grow (Figure 7.9). Grasslands persist because of a combination of seasonal drought, grazing by large herbivores, and occasional fires—all of which keep shrubs and trees from growing in large numbers.

The three main types of grassland—tropical, temperate, and cold (arctic tundra)—result from combinations of low average precipitation and varying average temperatures

Tropical desert

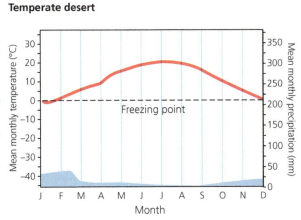

Temperate desert

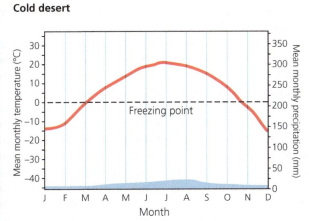

Cold desert

FIGURE 7.10 These climate graphs track the typical variations in annual temperature (red) and precipitation (blue) in tropical, temperate, and cold deserts. Top photo: a *tropical desert* in Morocco. Center photo: a *temperate desert* in southeastern California, with saguaro cactus, a prominent species in this ecosystem. Bottom photo: a *cold desert*, Mongolia's Gobi Desert. **Data analysis:** Which month of the year has the highest temperature and which month has the lowest rainfall for each of the three types of deserts?

Staying Alive in The Desert

Adaptations for survival in the desert have two themes: *beat the heat* and *every drop of water counts.*

Desert plants have evolved a number of adaptations based on such strategies. During long hot and dry spells, plants such as mesquite and creosote drop their leaves to survive in a dormant state. *Succulent (fleshy) plants* such as the saguaro cactus (Figure 7.10, center photo) have no leaves that can lose water to the atmosphere through *transpiration.* They reduce water loss by opening their pores only at night to take up CO_2. Succulents also store water and synthesize food in their expandable, fleshy tissue. The spines of these and many other desert plants guard them from being eaten by herbivores seeking the precious water they hold.

Some desert plants use deep roots to tap into groundwater. Others such as prickly pear and saguaro cacti use widely spread shallow roots to collect water after brief showers and they store it in their spongy tissues.

Other desert plants have wax-coated leaves that reduce water loss. Annual wildflowers and grasses store much of their biomass in seeds that remain inactive, sometimes for years, until they receive enough water to germinate. Shortly after a rain, these seeds germinate, grow, and carpet some deserts with dazzling arrays of colorful flowers that last several weeks.

Most desert animals are small. Some beat the heat by hiding in cool burrows or rocky crevices by day and coming out at night or in the early morning when it is cooler. Others become dormant during periods of extreme heat or drought. Larger animals such as camels can drink massive quantities of water when it is available and store it in their fat for use as needed. In addition, the camel's thick fur helps it keep cool because the air spaces in the fur insulate the camel's skin against the outside heat. In addition, camels do not sweat, which reduces their water loss through evaporation. Kangaroo rats never drink water. They get the water they need by breaking down fats in seeds that they consume.

Insects and reptiles such as rattlesnakes have thick outer coverings to minimize water loss through evaporation, and their wastes are dry feces and a dried concentrate of urine. Many spiders and insects get their water from dew or from the food they eat.

CRITICAL THINKING

What are three steps you would take to survive in the open desert if you had to?

(**Concept 7.2A**). One type of tropical grassland is *savanna* (**Core Case Study** and Figure 7.11, top photo). It contains widely scattered clumps of trees and usually has warm temperatures year-round with alternating dry and wet seasons (Figure 7.11, top graph).

Tropical savannas in East Africa (**Core Case Study**) are home to *grazing* (primarily grass-eating) and *browsing* (twig- and leaf-nibbling) hoofed animals, including wildebeests, gazelles, zebras, giraffes, and antelopes, as well as their predators such as lions, hyenas, and humans. Elephants eat a variety of foods including grass, tree leaves, tree bark, twigs, and shrubs and serve as keystone species (Science Focus 7.3). Herds of grazing and browsing animals migrate across the tropical savannas of East Africa to find water and food in response to seasonal and year-to-year variations in rainfall (Figure 7.11, blue areas in top graph) and food availability. Savanna plants, like those in deserts, are adapted to survive drought and extreme heat. Many have deep roots that can tap into groundwater.

In a *temperate grassland,* winters can be bitterly cold and summers are hot and dry. Annual precipitation is sparse and falls unevenly throughout the year (Figure 7.11, center graph). Because the aboveground parts of most of the grasses die and decompose each year, organic matter accumulates to produce deep, fertile topsoil. This topsoil is held in place by a thick network of intertwined roots.

If the topsoil is plowed, it can be blown away by high winds. This biome's grasses are adapted to droughts and to fires that burn the plant parts above the ground without harming the roots, from which new grass can grow. Many of the world's natural temperate grasslands have been converted to farmland, because their fertile soils are useful for growing crops (Figure 7.12) and grazing cattle.

Cold grasslands, or *arctic tundra,* lie south of the arctic polar ice cap (Figure 7.9). During most of the year, these treeless plains are bitterly cold (Figure 7.11, bottom graph), swept by frigid winds, and covered with ice and snow. Winters are long with few hours of daylight, and the scant precipitation falls primarily as snow.

A thick, spongy mat of low-growing plants lies under the snow. Trees and tall plants cannot survive in the cold and windy tundra because they would lose too much of their heat. Most of the annual growth of the tundra's plants occurs during the 7- to 8-week summer, when there is daylight almost around the clock.

One outcome of the extreme cold is the formation of **permafrost**, underground soil where captured water can stay frozen for more than two consecutive years. During the brief summer, the permafrost layer keeps melted snow and ice from draining into the ground. This forms shallow lakes, marshes, bogs, ponds, and other seasonal wetlands when snow and frozen surface soil melt. Hordes

Tropical grassland (savanna)

Temperate grassland (prairie)

Cold grassland (arctic tundra)

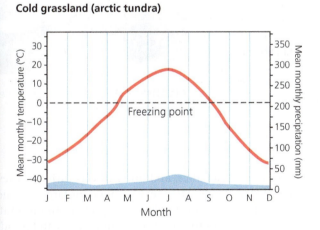

FIGURE 7.11 These climate graphs track the typical variations in annual temperature (red) and precipitation (blue) in tropical, temperate, and cold (arctic tundra) grasslands. Top photo: *savanna (tropical grassland)* in Kenya, Africa, with zebras grazing. Center photo: *prairie (temperate grassland)* in the U.S. state of Illinois. Bottom photo: *arctic tundra (cold grassland)* in Iceland in fall. ***Data analysis:*** Which month of the year has the highest temperature and which month has the lowest rainfall for each of the three types of grassland?

Revisiting the Savanna: Elephants as a Keystone Species

As in all biomes, the African savanna (**Core Case Study**) has food webs. Its food webs often include one or more keystone species that play a major role in maintaining the structure and functioning of the ecosystem.

Ecologists view elephants as a keystone species in the African savanna. They eat woody shrubs and young trees. This helps to keep the savanna from being overgrown by these woody plants and prevents the grasses from dying out. If this were to happen, antelopes, zebras, and other grass-eaters would leave the savanna in search of food and with them would go the carnivores such as lions and hyenas that feast on these grass-eaters. Elephants also dig for water during drought periods, creating or enlarging waterholes that are used by other animals. Without African elephants, the savanna food web would collapse and the savanna would become shrubland.

Conservation scientists classify the African elephant as *vulnerable* to extinction. In 1979, there were an estimated 1.3 million wild African elephants. Today, an estimated 400,000 remain in the wild. This sharp decline is due mostly to the illegal killing of elephants for their valuable ivory tusks (Figure 7.A, left) at an average rate of 96 animals a day in 2016. Since 1990, there has been an international ban of the sale of ivory and in some areas, elephants are protected as threatened or endangered species, but the illegal killing of elephants for their valuable ivory continues (Figure 7.A, right).

Another major threat to elephants is the loss and fragmentation of their habitats as human populations have expanded and taken over more land. Elephants are eating or trampling the crops of settlers who have moved into elephant habitat areas, and this has led to the killing of some elephants by farmers—a problem being addressed by conservation scientists (Individuals Matter 7.1). If the multiple threats are not curtailed, elephants may disappear from the savanna within your lifetime.

CRITICAL THINKING

Do you think African governments would be justified in setting aside large areas of elephant habitat and prohibiting development there? Explain your reasoning. What are other alternatives for preserving African elephants on the savanna?

Pearl Media/Shutterstock.com

Steve O.Taylor(GHF)/Nature Picture Library

FIGURE 7.A One reason African elephants are being threatened with extinction is their financially valuable ivory tusks.

of mosquitoes, black flies, and other insects thrive in these shallow surface pools. They serve as food for large colonies of birds (especially waterfowl) that migrate from the south to nest and breed in the tundra's summer bogs and ponds.

Animals in this biome survive the intense winter cold through adaptations such as thick coats of fur (arctic wolf, arctic fox, and musk oxen) or feathers (snowy owl) and living underground (arctic lemming). In the summer, caribou (often called reindeer) and other types of deer migrate to the tundra to graze on its vegetation.

Tundra is vulnerable to disruption. Because of the short growing season, tundra soil and vegetation recover very slowly from damage or disturbance. Human activities in the arctic tundra—primarily on and around oil and natural gas drilling sites, pipelines, mines, and military bases—leave scars that persist for centuries.

Types of Forests

Forests are lands that are dominated by trees. The three main types of forests—*tropical, temperate,* and *cold* (northern coniferous or boreal)—result from combinations of varying precipitation levels and varying average temperatures (**Concept 7.2A**) (Figures 7.8 and 7.9).

Tropical rain forests (Figure 7.13, top photo) are found near the equator (Figure 7.9), where hot, moisture-laden

Tuy Sereivathana: Elephant Protector

Since 1970, Cambodia's rain forest cover has dropped from over 70% of the country's land area to 3% primarily because of population growth, rapid development, illegal logging, and warfare. This severe forest loss forced elephants to search for food and water on farmlands. As a result, some poor farmers have killed elephants to protect their food supply.

Since 1995, Tuy Sereivathana, with a master's degree in forestry, has been on a mission to accomplish two goals. One is to double the population of Cambodia's endangered Asian elephants by 2030. The other is to show poor farmers that protecting elephants and other forms of wildlife can help them escape poverty.

Sereivathana has helped farmers set up nighttime lookouts for elephants. He taught them to scare elephants away by using foghorns and fireworks and to use solar-powered electric fences to mildly shock them. He has also encouraged farmers to stop growing watermelons and bananas, which elephants love, and to grow eggplant and chile peppers, which elephants shun.

Since 2005, mostly because of Sereivathana's efforts, no elephants have been killed in Cambodia over conflicts with humans. In 2010, Sereivathana was one of the six recipients of the Goldman Environmental Prize (often dubbed the "Nobel Prize for the environment"). In 2011, he was named a National Geographic Explorer.

Courtesy Tom Dusenbery

air rises and dumps its moisture (Figure 7.4). These lush forests have year-round, warm temperatures, high humidity, and almost daily heavy rainfall (Figure 7.13, top graph). This almost constant warm, wet climate is ideal for a wide variety of plants and animals.

Broadleaf evergreen plants that keep most of their leaves year-round dominate tropical rain forests. The tops of the trees form a dense *canopy* (Figure 7.13, top photo) that blocks most light from reaching the forest floor. Many of the relatively few plants living at the ground level have enormous leaves to capture what little sunlight filters down to them.

Some trees are draped with vines (called *lianas*) that reach for the treetops to gain access to sunlight. In the canopy, the vines grow from one tree to another, providing walkways for many species living there. When a large tree is cut down, its network of lianas can pull down other trees.

Tropical rain forests have a high net primary productivity (NPP; see Figure 3.14, p. 56). They are teeming with life and possess incredible biological diversity. Although tropical rain forests cover only about 2% of the earth's land surface, ecologists estimate that they contain at least 50% of the known terrestrial plant and animal species. A single tree in these forests may support several thousand different insect species. Plants from tropical rain forests are a source of a variety of chemicals, many of which have been used as blueprints for making most of the world's prescription drugs.

Rain forest species occupy a variety of specialized niches in distinct layers, which contribute to their high species diversity. Vegetation layers are structured, for the most part, according to the plants' needs for sunlight, as shown in Figure 7.14. Much of the animal life, particularly insects, bats, and birds, lives in the sunny canopy layer, with its abundant shelter and supplies of leaves, flowers, and fruits.

Dropped leaves, fallen trees, and dead animals decompose quickly in tropical rain forests because of the warm, moist conditions and the hordes of decomposers. About 90% of the nutrients released by this rapid decomposition are quickly taken up and stored by trees, vines, and other plants. Nutrients that are not taken up are soon leached from the thin topsoil by the frequent rainfall and plant litter does not build up on the ground. The resulting lack of fertile soil helps explain why rain forests are not good places to clear and grow crops or graze cattle on a sustainable basis.

At least half of all tropical rain forests have been destroyed or disturbed by human activities such as farming, and the pace of this destruction and degradation is increasing (see Chapter 3 Core Case Study, p. 46). Ecologists warn that without strong protective measures most of these forests, along with their rich biodiversity and other highly valuable ecosystem services, could be gone by the end of this century.

The second major type of forest is the temperate forest, the most common of which is the *temperate deciduous forest* (Figure 7.13, center photo). Such forests typically see warm summers, cold winters, and abundant precipitation—rain in summer and snow in winter months (Figure 7.13, center graph). They are dominated by a few species of *broadleaf deciduous trees* such as oak, hickory, maple, aspen, and birch. Animal species living in these forests include predators such as wolves, foxes, and wildcats. They feed on herbivores such as white-tailed deer,

FIGURE 7.12 Natural Capital Degradation: This intensively cultivated cropland is an example of the replacement of biologically diverse temperate grasslands (such as in the center photo of Figure 7.12) with a monoculture crop.

squirrels, rabbits, and mice. Warblers, robins, and other bird species live in these forests during the spring and summer, mating and raising their young.

In these forests, most of the trees' leaves, after developing their vibrant colors in the fall (Figure 7.13, center photo), drop off the trees. This allows the trees to survive the cold winters by becoming dormant. Each spring, they sprout new leaves and spend their summers growing and producing until the cold weather returns.

Because they have cooler temperatures and fewer decomposers than tropical forests have, temperate forests also have a slower rate of decomposition. As a result, they accumulate a thick layer of slowly decaying leaf litter, which becomes a storehouse of nutrients.

On a global basis, temperate forests have been degraded by various human activities, especially logging and urban expansion, more than any other terrestrial biome. However, within 100 to 200 years, forests of this type that have been cleared can return through secondary ecological succession (see Figure 5.9, p. 97).

Another type of temperate forest, the *coastal coniferous forests*, or *temperate rain* forests, is found in scattered coastal temperate areas with ample rainfall and moisture

from dense ocean fogs. These forests contain thick stands of large cone-bearing conifer trees such as giant sequoia (see Chapter 2 opening photo, pp. 26–27) that keep their leaves (needles) year-round.

Cold, or *northern coniferous forests* (Figure 7.13, bottom photo), also called *boreal forests* or *taigas*, are found south of arctic tundra (Figure 7.9). In this subarctic, cold, and moist climate, winters are long and extremely cold, with winter sunlight available only for 6 to 8 hours per day. Summers are short, with cool to warm temperatures (Figure 7.13, bottom graph), and the sun shines as long as 19 hours a day during mid-summer.

Most boreal forests are dominated by a few species of *coniferous* (cone-bearing) *evergreen trees* or *conifers* such as spruce, fir, cedar, hemlock, and pine that keep most of their leaves (or needles) year-round. Most of these species have small, needle-shaped, wax-coated leaves that can withstand the intense cold and drought of winter, when snow blankets the ground. Plant diversity is low because few species can survive the winters when soil moisture is frozen.

Beneath the stands of trees in these forests is a deep layer of partially decomposed conifer needles.

Tropical rain forest

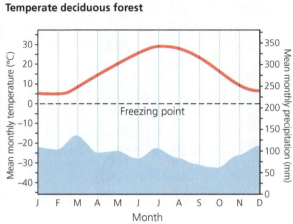

Temperate deciduous forest

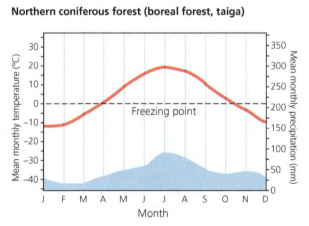

Northern coniferous forest (boreal forest, taiga)

FIGURE 7.13 These climate graphs track the typical variations in annual temperature (red) and precipitation (blue) in tropical, temperate, and cold (northern coniferous, or boreal) forests. Top photo: the closed canopy of a *tropical rain forest* in Costa Rica. Middle photo: a *temperate deciduous forest* near Hamburg, Germany in autumn. Bottom photo: a *northern coniferous forest* in Canada. **Data analysis:** Which month of the year has the highest temperature and which month has the lowest rainfall for each of the three types of forests?

FIGURE 7.14 Specialized plant and animal niches are *stratified*, or arranged roughly in layers, in a tropical rain forest. Filling such specialized niches enables many species to avoid or minimize competition for resources and results in the coexistence of a great variety of species.

Decomposition is slow because of low temperatures, the waxy coating on the needles, and high soil acidity. The decomposing conifer needles make the thin, nutrient-poor topsoil acidic, which prevents most other plants (except certain shrubs) from growing on the forest floor.

Year-round wildlife includes bears, wolves, moose, lynx, and many burrowing rodent species. Caribou spend winter in taiga and summer in the arctic tundra (Figure 7.11, bottom). During the brief summer, warblers and other insect-eating birds feed on flies, mosquitoes, and caterpillars.

Mountains Play Important Ecological Roles

Some of the world's most spectacular environments are high on *mountains*, steep or high-elevation lands that cover about one-fourth of the earth's land surface. About 1.2 billion people (16% of the world's population) live in mountain ranges or in their foothills, and 4 billion people (54% of the world's population) depend on mountain systems for all or some of their water. The soil on the steep slopes of mountains erodes easily when the vegetation holding them in place is removed by natural disturbances such as landslides and avalanches or by human activities such as timber cutting and agriculture. Many mountains are *islands of biodiversity* surrounded by a sea of lower-elevation landscapes that have been transformed by human activities.

Mountains play important ecological roles. They contain a large portion of the world's forests, which are habitats for much of the planet's terrestrial biodiversity. They often are habitats for *endemic species*—those that are found nowhere else on the earth. They also serve as sanctuaries for animals that are capable of migrating to higher altitudes and surviving in such environments. Every year, more of these animals are driven from lowland areas to mountain habitats by human activities and a warming climate.

CONNECTIONS Mountains and Climate

Mountains help regulate the earth's climate. Many mountain-tops are covered with glacial ice and snow that reflect some solar radiation back into space, which helps cool the earth. However, many mountain glaciers are melting, primarily because the atmosphere has warmed over recent decades. While glaciers reflect solar energy, the darker rocks exposed by melting glaciers absorb that energy. This helps warm the atmosphere above them, which melts more ice and warms the atmosphere more—in an escalating feedback loop.

Mountains also play a critical role in the hydrologic cycle (see Figure 3.15, p. 58) by serving as major store-houses of water. During winter, precipitation is stored as ice and snow. In the warmer weather of spring and summer, much of this snow and ice melts, releasing water to streams for use by wildlife and by humans for drinking and for irrigating crops. Because the atmosphere has warmed over the past 40 years, some mountaintop snow packs and glaciers have been melting earlier in the spring each year. This leads to lower food production in certain areas, because much of the water needed throughout the summer to irrigate crops is released too quickly and too early in the season.

Scientific measurements and climate models indicate that a large number of the world's mountaintop glaciers may disappear during this century if the atmosphere keeps getting warmer as projected. This could force many people to move from their homelands in search of new water supplies and places to grow their crops. Despite the ecological, economic, and cultural importance of mountain ecosystems, governments and many environmental organizations have not focused on protecting these areas.

Humans Have Disturbed Much of the Earth's Land

About 60% of the world's major terrestrial ecosystems are being degraded or used unsustainably, as the human ecological footprint gets bigger and spreads across the globe (see Figure 1.9, p. 11), according to the 2005 Millennium Ecosystem Assessment and later updates of such research. Figure 7.15 summarizes some of the most harmful human impacts on the world's deserts, grasslands, forests, and mountains (**Concept 7.2B**).

Natural Capital Degradation

Major Human Impacts on Terrestrial Ecosystems

Deserts	Grasslands	Forests	Mountains
Large desert cities	Conversion to cropland	Clearing for agriculture, livestock grazing, timber, and urban development	Agriculture
Destruction of soil and underground habitat by off-road vehicles	Release of CO_2 to atmosphere from burning grassland	Conversion of diverse forests to tree plantations	Timber and mineral extraction
Depletion of groundwater	Overgrazing by livestock	Damage from off-road vehicles	Hydroelectric dams and reservoirs
Land disturbance and pollution from mineral extraction	Oil production and off-road vehicles in arctic tundra	Pollution of forest streams	Air pollution blowing in from urban areas and power plants
			Soil damage from off-road vehicles

FIGURE 7.15 Human activities have had major impacts on the world's deserts, grasslands, forests, and mountains. *Critical thinking:* For each of these biomes, which two of the impacts listed do you think are the most harmful?

How long can we keep chipping away at these terrestrial forms of natural capital without threatening our economies and the long-term survival of our own and many other species? No one knows but there are increasing signs that we need to come to grips with this vital issue.

7.3 WHAT ARE THE MAJOR TYPES OF MARINE AQUATIC SYSTEMS AND HOW ARE HUMAN ACTIVITIES AFFECTING THEM?

CONCEPT 7.3 Oceans dominate the planet and provide vital ecosystem and economic services that are being disrupted by human activities.

Water Covers Most of the Planet

When viewed from outer space, the earth appears as a mostly blue planet with about 71% of its surface covered with ocean water. Although the *global ocean* is a single and continuous body of saltwater, geographers divide it into four oceans—the Arctic, Atlantic, Pacific, and Indian—separated by the continents. Together, the oceans hold almost 98% of the earth's water. Each of us is connected to, and utterly dependent on, the earth's global ocean through the water cycle (see Figure 3.15, p. 58).

71% of the earth is covered with ocean water.

The aquatic equivalents of biomes are called **aquatic life zones**—saltwater and freshwater portions of the biosphere that can support life. The distribution of many aquatic organisms is determined largely by the water's *salinity*—the amounts of various salts such as sodium chloride dissolved in a given volume of water. As a result, aquatic life zones are classified into two major types: **saltwater** or **marine life zones** (oceans and their bays, estuaries, coastal wetlands, shorelines, coral reefs, and mangrove forests) and **freshwater life zones** (lakes, rivers, streams, and inland wetlands).

In most aquatic systems, the key factors determining the types and numbers of organisms found in different areas of the ocean are *water temperature, dissolved oxygen content, availability of light,* and *availability of nutrients required*

for photosynthesis, such as carbon (as dissolved CO_2 gas), nitrogen (as NO_3^-), and phosphorus (mostly as PO_4^{3-}).

Oceans Provide Vital Ecosystem and Economic Services

Oceans dominate the planet and provide ecosystem and economic services (Figure 7.16) that help keep us and other species alive and that support our economies. They are also enormous reservoirs of biodiversity. Marine life is found in three major *life zones:* the coastal zone, the open sea, and the ocean bottom (Figure 7.17).

The **coastal zone** is the total area of warm, nutrient-rich, shallow coastal waters that make up less than 10% of the world's ocean area. It contains 90% of all marine species and is the site of most large commercial marine fisheries. This zone's aquatic systems include **estuaries**, the partially enclosed bodies of water where rivers meet the sea (Figure 7.18), and **coastal wetlands**, or land areas covered by coastal waters during all or part of the year. The latter include *coastal marshes* (Figure 7.19) and *mangrove forests* (Figure 7.20). Other important systems are *sea-grass beds* (Figure 7.21) and *coral reefs* (see chapter-opening photo and Science Focus 7.4).

These coastal aquatic systems provide important ecosystem and economic services. They help maintain water quality in tropical coastal zones by filtering toxic

Natural Capital

Marine Ecosystems

Ecosystem Services	Economic Services
Oxygen supplied through photosynthesis	Food
Water purication	Energy from waves and tides
Climate moderation	Pharmaceuticals
CO_2 absorption	Harbors and transportation routes
Nutrient cycling	
Reduced storm damage (mangroves, barrier islands, coastal wetlands)	Recreation and tourism
	Employment
Biodiversity: species and habitats	Minerals

FIGURE 7.16 Marine systems provide a number of important ecosystem and economic services (**Concept 7.3**). *Critical thinking:* Which two ecosystem services and which two economic services do you think are the most important? Why?

Top: Willyam Bradberry/Shutterstock.com. Bottom: James A. Harris/Shutterstock.com.

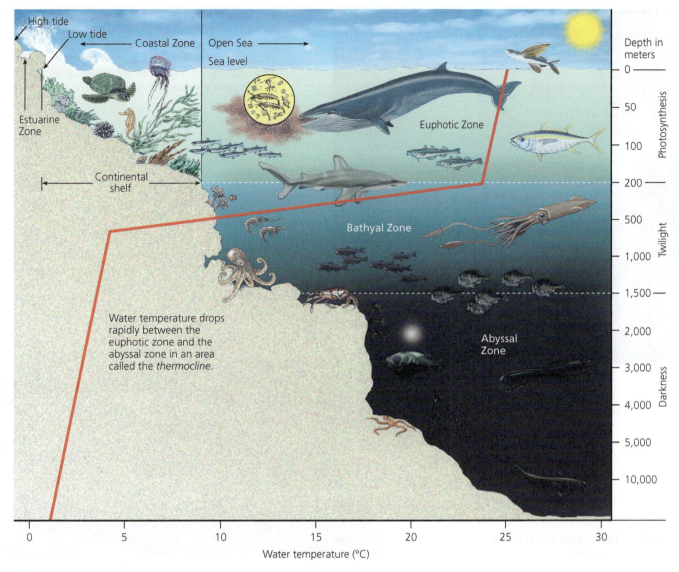

FIGURE 7.17 Major life zones and vertical zones (not drawn to scale) in an ocean. Actual depths of zones may vary. Available light determines the euphotic, bathyal, and abyssal zones. Temperature zones also vary with depth, shown here by the red line. *Critical thinking:* How is an ocean like a rain forest? (Hint: See Figure 7.14.)

pollutants, excess plant nutrients, and sediments and by absorbing other pollutants. They provide food, habitats, and nursery sites for a variety of aquatic and terrestrial species. Coastal wetlands also reduce storm damage and coastal erosion by absorbing waves and storing excess water produced by storms and tsunamis.

Open Sea and Ocean Floor

The sharp increase in water depth at the edge of the continental shelf separates the coastal zone from the vast volume of the ocean called the **open sea**. This aquatic life zone is divided into three *vertical zones* (Figure 7.17), or layers, primarily based on the degree of penetration of sunlight. Temperatures also change with depth (Figure 7.17, red line).

The *euphotic zone* is the brightly lit upper zone, where drifting phytoplankton carry out about 40% of the earth's photosynthetic activity. Large, fast-swimming predatory fishes such as swordfish, sharks, and bluefin tuna populate the euphotic zone.

The *bathyal zone* is the dimly lit middle zone that receives little sunlight and therefore does not contain photosynthesizing producers. Zooplankton and smaller fishes, many of which migrate to feed on the surface at night, are found in this zone.

The deepest zone, called the *abyssal zone*, is dark and cold. There is no sunlight to support photosynthesis, and this water has little dissolved oxygen. Nevertheless, the deep ocean floor is teeming with life because it contains enough nutrients to support a large number of species.

NASA/Landsat/Phil Degginger/Alamy Stock Photo

FIGURE 7.18 Satellite photo of an *estuary.* The Mississippi River carries sediment and plant nutrients from fertilizer runoff into the Gulf of Mexico. The excess plant nutrients create an algal bloom (green area) that upsets marine life by depleting dissolved oxygen near the Gulf's bottom.

FIGURE 7.19 Coastal marsh in the U.S. state of California.

jctdesign/Shutterstock.com

Manit Larpluechai/Dreamstime.com

FIGURE 7.20 Mangrove forest on the coast of Thailand.

FIGURE 7.21 Sea-grass beds, such as this one near the coast of San Clemente Island, California, support a variety of marine species.

James Forte/National Geographic Creative

Most of the zone's organisms in the deep waters and on the ocean floor get their food from showers of dead and decaying organisms—called *marine snow*—drifting down from upper, more lighted levels of the ocean.

NPP is quite low in the open sea (Figure 3.14, p. 56), except in upwelling areas, where currents bring up nutrients from the ocean bottom. However, because the open sea covers so much of the earth's surface, it makes the largest contribution to the earth's overall NPP.

Human Impacts on Marine Ecosystems

Certain human activities disrupt and degrade many of the ecosystem and economic services provided by marine aquatic systems. Areas most affected are coastal marshes, shorelines, mangrove forests, and coral reefs, as summarized in Figure 7.22 (**Concept 7.3**). We examine these harmful effects and possible ways to lessen them in Chapter 9.

The biggest threat to marine systems, according to many marine scientists, is climate change (as discussed in Chapter 15). The earth's average atmospheric temperature and the average temperature of the world's oceans have increased since 1980. This is raising the world's average sea level. One reason is that warmer ocean water expands. A second reason is that as the atmosphere has warmed land-based glaciers in Greenland and other parts of the world are slowly melting and

Coral Reefs

Coral reefs form in clear, warm coastal waters in tropical areas. These stunningly beautiful natural wonders (see chapter-opening photo) are among the world's oldest, most diverse, and most productive ecosystems.

Coral reefs are formed by massive colonies of tiny animals called *polyps* (close relatives of jellyfish). They slowly build reefs by secreting a protective crust of limestone (calcium carbonate) around their soft bodies. When the polyps die, their empty crusts remain behind as part of a platform for more reef growth. The resulting elaborate network of crevices, ledges, and holes serves as calcium carbonate "condominiums" for a variety of marine animals.

Coral reefs are the result of a mutually beneficial relationship between polyps and tiny single-celled algae called *zoo-xanthellae* that live in the tissues of the polyps. In this example of mutualism, the algae provide the polyps with food and oxygen through photosynthesis and help the corals produce calcium carbonate. Algae also give the reefs their stunning coloration. The polyps, in turn, provide the algae with a well-protected home and some of their nutrients.

Although shallow and deep-water coral reefs occupy only about 0.2% of the ocean floor, they provide important ecosystem and economic services. They act as natural barriers that help protect 15% of the world's coastlines from flooding and erosion caused by battering waves and storms. They also provide habitats, food, and spawning grounds for one-quarter to one-third of the organisms that live in the ocean, and they produce about one-tenth of the global fish catch. Through tourism and fishing, they provide goods and services worth about $40 billion a year.

Coral reefs are vulnerable to damage because they grow slowly and are disrupted easily. Runoff of soil and other materials from the land can cloud the water and block the sunlight that the algae in shallow reefs need for

FIGURE 7.B This bleached coral has lost most of its algae because of changes in the environment such as warming of the waters and deposition of sediments.

photosynthesis. In addition, the water in which shallow reefs live must have a temperature of 18–30°C (64–86°F) and cannot be too acidic. This explains why the two major long-term threats to coral reefs are *climate change,* which could raise the water temperature above tolerable limits in most reef areas, and *ocean acidification,* which could make it harder for polyps to build reefs and could even dissolve some of their calcium carbonate formations.

One result of stresses such as pollution and rising ocean water temperatures is *coral bleaching* (Figure 7.B). Such stressors can cause the colorful algae, upon which corals depend for food, to die off. Without food, the coral polyps die, leaving behind a white skeleton of calcium carbonate. Studies by the Global Coral Reef Monitoring Network and other scientist groups estimate that since the 1950s, some 45% to 53% of the world's shallow coral reefs have been destroyed

or degraded and that another 25% to 33% could be lost within 20 to 40 years. These centers of biodiversity are by far the most threatened marine ecosystems.

Coral bleaching and ocean acidification could have devastating effects on the biodiversity of coral reefs, on the food webs dependent on them, and on the ecosystem services they provide. Coral reef degradation and destruction will also have a severe impact on the approximately 500 million people who depend on them for their food or for income from fishing and tourism.

CRITICAL THINKING

How might the loss of most of the world's remaining tropical coral reefs affect your life and the lives of any children or grandchildren you might have? What are two things you could do to help reduce this loss?

Major Human Impacts on Marine Ecosystems and Coral Reefs

Marine Ecosystems   **Coral Reefs**

Marine Ecosystems	Coral Reefs
Half of coastal wetlands lost to agriculture and urban development	Ocean warming
Over one-fth of mangrove forests lost to agriculture, aquaculture, and development	Rising ocean acidity
	Rising sea levels
	Soil erosion
Beaches eroding due to development and rising sea levels	Algae growth from fertilizer runoff
Ocean-bottom habitats degraded by dredging and trawler shing	Bleaching
	Increased UV exposure
At least 20% of coral reefs severely damaged and 25–33% more threatened	Damage from anchors and from shing and diving

FIGURE 7.22 Human activities are having major harmful impacts on all marine ecosystems (left) and particularly on coral reefs (right) (**Concept 7.3**). *Critical thinking:* Which two of the threats to marine ecosystems do you think are the most serious? Why? Which two of the threats to coral reefs do you think are the most serious? Why?

Top left: Jorg Hackemann/Shutterstock.com. Top right: Rich Carey/Shutterstock.com.
Bottom left: Piotr Marcinski/Shutterstock.com. Bottom right: Rostislav Ageev/Shutterstock.com.

adding large quantities of water to the oceans. The rise in sea levels projected for this century would destroy shallow coral reefs and flood coastal marshes and other coastal ecosystems, as well as many coastal cities. In addition, because the oceans absorb much of the excess heat and CO_2 emitted into the atmosphere by human activities, they are warming and becoming more acidic, which will degrade marine biodiversity. We examine these harmful effects and possible ways to lessen them in Chapters 9 and 15.

7.4 WHAT ARE THE MAJOR TYPES OF FRESHWATER SYSTEMS AND HOW ARE HUMAN ACTIVITIES AFFECTING THEM?

CONCEPT 7.4 Freshwater lakes, rivers, and wetlands provide important ecosystem and economic services that are being disrupted by human activities.

Water Stands in Some Freshwater Systems and Flows in Others

Precipitation that does not sink into the ground or evaporate becomes **surface water**—freshwater that flows or is stored in bodies of water on the earth's surface. Surface water that flows into such bodies of water is called **runoff**. A **watershed**, or **drainage basin**, is the land area that delivers runoff, sediment, and dissolved substances to a stream, lake, or wetland.

Freshwater aquatic life zones include *standing* bodies of freshwater such as lakes, ponds, and inland wetlands and *flowing* systems such as streams and rivers. Although freshwater systems cover less than 2.5% of the earth's surface, they provide a number of important ecosystem and economic services (Figure 7.23).

Lakes are large natural bodies of standing freshwater formed when precipitation, runoff, streams, rivers, and groundwater seepage fill depressions in the earth's surface. Causes of such depressions include movements of glaciers, displacement of the earth's crust, and volcanic activity. Freshwater lakes vary greatly in size, depth, and nutrient content. Deep lakes normally consist of four distinct zones that are defined by their depth and distance from shore (Figure 7.24).

Ecologists classify lakes according to their nutrient content and primary productivity. Lakes that have a small supply of plant nutrients are called **oligotrophic lakes**. This type of lake (Figure 7.25) is often deep and has steep banks. Glaciers and mountain streams supply water to many of these lakes. They usually have crystal-clear water and small populations of phytoplankton and fish species, such as smallmouth bass and trout. Because of their low levels of nutrients, these lakes have a low NPP.

Over time, sediments, organic material, and inorganic nutrients wash into most oligotrophic lakes and plants grow and decompose to form bottom sediments.

Natural Capital

Freshwater Systems

Ecosystem Services	Economic Services
Climate moderation	Food
Nutrient cycling	Drinking water
Waste treatment	Irrigation water
Flood control	Hydroelectricity
Groundwater recharge	Transportation corridors
Habitats for many species	Recreation
Genetic resources and biodiversity	Employment
Scientific information	

FIGURE 7.23 Freshwater systems provide many important ecosystem and economic services (**Concept 7.4**). *Critical thinking:* Which two ecosystem services and which two economic services do you think are the most important? Why?

Top: Galyna Andrushko/Shutterstock.com. Bottom: Kletr/Shutterstock.com.

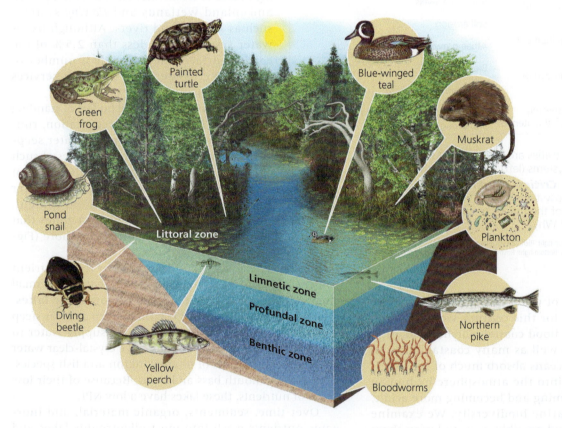

FIGURE 7.24 A typical deep temperate-zone lake has distinct zones of life. *Critical thinking:* How are deep lakes like tropical rain forests? (Hint: See Figure 7.14.)

Green frog

Painted turtle

Blue-winged teal

Muskrat

Pond snail

Littoral zone

Plankton

Diving beetle

Limnetic zone

Profundal zone

Benthic zone

Northern pike

Yellow perch

Bloodworms

FIGURE 7.25 Trillium Lake in the U.S. state of Oregon with a view of Mount Hood.

tusharkoley/Shutterstock.com

The process by which lakes gain nutrients is called **eutrophication**. A lake with a large supply of nutrients is called a **eutrophic lake** (Figure 7.26). Such lakes are typically shallow with murky brown or green water. Because of their high levels of nutrients, these lakes have a high NPP. Most lakes fall somewhere between the two extremes of nutrient enrichment.

Human inputs of nutrients through the atmosphere and from urban and agricultural areas within a lake's watershed can accelerate the eutrophication of the lake. This process is called **cultural eutrophication**.

Freshwater Streams and Rivers

In a watershed, water accumulates in small streams that join to form rivers. Collectively, streams and rivers carry huge amounts of water from highlands to lakes and oceans. Typically, a stream flows through three zones (Figure 7.27): the *source zone,* which contains *headwater* streams found in highlands and mountains; the *transition zone,* which contains wider, lower-elevation streams; and the *floodplain zone,* which contains rivers that empty into larger rivers or into the ocean.

As streams flow downhill, they shape the land through which they pass. Over millions of years, the friction of moving water has leveled mountains and cut deep canyons. Streams and rivers carry sand, gravel, and soil and deposit them as sediments in low-lying areas.

At its mouth, a river may divide into many channels as it flows through its **delta**—an area at the mouth of a river built up by deposited sediment and often containing estuaries (Figure 7.18) and coastal marshes (Figure 7.19). Deltas absorb and slow the velocity of floodwaters from

FIGURE 7.26 This eutrophic lake has received large flows of plant nutrients. As a result, its surface is covered with mats of algae.

Nicholas Rjabow/Dreamstime.com

coastal storms, hurricanes, and tsunamis and provide habitats for a wide variety of marine life.

Freshwater Inland Wetlands

Inland wetlands are lands located away from coastal areas that are covered with freshwater all or part of the time—excluding lakes, reservoirs, and streams. They include *marshes, swamps,* and *prairie potholes* (depressions carved out by ancient glaciers). Other examples are *floodplains,* which receive excess water from streams or rivers during heavy rains and floods.

Some wetlands are covered with water year-round and others remain under water for only a short time each year. The latter include prairie potholes, floodplain wetlands, and arctic tundra (see Figure 7.11, bottom).

Inland wetlands provide a number of free ecosystem and economic services. They

- filter and degrade toxic wastes and pollutants,
- reduce flooding and erosion by absorbing storm water and releasing it slowly and by absorbing overflows from streams and lakes,
- sustain stream flows during dry periods,

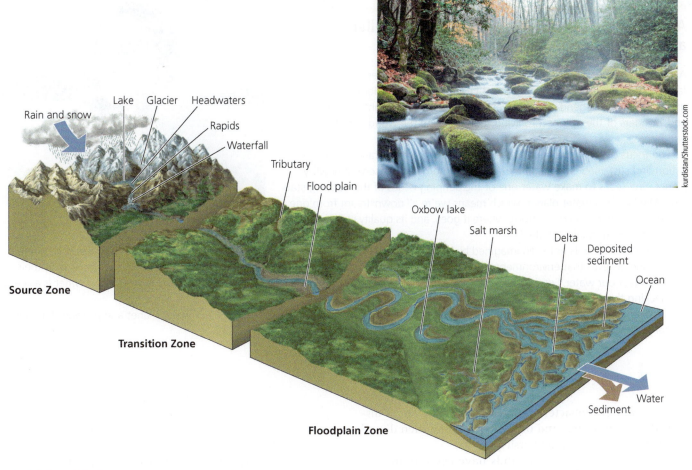

FIGURE 7.27 Three zones in the downhill flow of water—the source zone (see photo), transition zone, and floodplain zone.

- recharge groundwater aquifers,
- maintain biodiversity by providing habitats for a variety of species,
- supply valuable products such as fishes and shellfish, blueberries, cranberries, and wild rice, and
- provide recreation for birdwatchers, nature photographers, boaters, anglers, and waterfowl hunters.

Human Activities Are Disrupting and Degrading Freshwater Systems

Human activities are disrupting and degrading many of the ecosystem and economic services provided by freshwater rivers, lakes, and wetlands (**Concept 7.4**) in four major ways. *First,* dams and canals restrict the flows of about 40% of the world's 237 largest rivers. This alters or destroys terrestrial and aquatic wildlife habitats along these rivers and in their coastal deltas and estuaries by reducing water flow and the flow of sediments to river deltas. *Second,* flood control levees and dikes built along rivers disconnect the rivers from their floodplains and destroy aquatic habitats by altering or degrading the functions of adjoining wetlands.

Third, cities and farms add pollutants and excess plant nutrients to nearby streams, rivers, and lakes. For example, runoff of nutrients into a lake (cultural eutrophication, Figure 7.26) causes explosions in the populations

Alexandra Cousteau: Environmental Storyteller and National Geographic Explorer

Alexandra Cousteau is proud of her heritage as granddaughter of Captain Jacques-Yves Cousteau and daughter of Philippe Cousteau. Her father and grandfather were legendary underwater explorers who brought the mysteries and wonders of the oceans into living rooms around the world with their films and books.

The focus of Alexandra's work is on advocating the importance of conservation and sustainable management of water in order to preserve a healthy planet. She seeks to make water one of the defining issues of this century, stating: "We live on a water planet, which means we're all downstream from one another. Where water comes from, where it goes, and its quality is intricately connected to our quality of life."

She is utilizing tools not even imagined by her grandfather—those of social networking and other modes of mobile communication. She believes that environmental advocates can use such new media tools and technology to inform people about how their actions affect our water. For example, she imagines a day in the future when knowing the quality and quantity of our water is as easy as checking the weather on our smart phones.

Alexandra's nonprofit Blue Legacy International harnesses technology to tell the stories of our water planet and provides film and digital resources to allow others to explore and understand water issues.

© Bill Zelman

of algae and cyanobacteria, which deplete the lake's dissolved oxygen. Fishes and other species may then die off and reduce the lake's biodiversity.

Fourth, many inland wetlands have been drained or filled to grow crops or have been covered with concrete, asphalt, and buildings. More than half of the inland wetlands estimated to have existed in the continental United States during the 1600s are gone—most of them drained to grow crops. This loss of natural capital has been an important factor in increasing flood damage in parts of the United States. Many other countries have suffered similar losses. For example, 80% of all inland wetlands in Germany and France have been destroyed.

Many scientists and other individuals are devoting their lives to understanding aquatic systems and learning how we can use them more sustainably (Individuals Matter 7.1).

BIG IDEAS

- Differences in climate, based primarily on long-term differences in average temperature and precipitation, largely determine the types and locations of the earth's deserts, grasslands, and forests.

- Saltwater and freshwater aquatic systems cover almost three-fourths of the earth's surface, and oceans dominate the planet.

- The earth's terrestrial and aquatic systems provide important ecosystem and economic services that are being degraded and disrupted by human activities.

Tying It All Together

Tropical African Savanna and Sustainability

This chapter's Core Case Study began with questions about the earth's diversity of terrestrial ecosystems and how they form. We examined the difference between weather and climate and how climate is a major factor in the formation and distribution of these biomes—the world's deserts, grasslands, and forests—as well as the life forms that live in those systems. In particular, we focused on the savanna, a grassland biome that is threatened by expansion of the human population.

We also discussed the influence of climate on terrestrial biodiversity in the formation of biomes as well as the life forms that live in those systems. These relationships are also in keeping with the three **scientific principles of sustainability**. The earth's dynamic climate system helps distribute heat from solar energy and recycle the earth's nutrients. This in turn helps generate and support the biodiversity found in the earth's various biomes.

Finally, we looked at aquatic life zones and at how human activities are degrading the vital ecosystem and economic services provided by terrestrial and aquatic systems. Scientists call for much more research on the components and workings of the world's terrestrial and

aquatic systems, on how they are interconnected, and on which systems are in the greatest danger of being disrupted by human activities.

Chapter Review

Core Case Study

1. Describe the African savanna and explain why it serves as an example of how differences in climate lead to the formation of different types of ecosystems.

Section 7.1

2. What is the key concept for this section? Define and distinguish between **weather** and **climate**. Define **ocean currents**. Describe three major factors that determine how air circulates in the lower atmosphere. Explain how varying combinations of temperature and precipitation, along with global air circulation and ocean currents, lead to the formation of various types of forests, grasslands, and deserts.

3. Define and give three examples of a **greenhouse gas**. What is the **greenhouse effect** and why is it important to the earth's life and climate? What is the **rain shadow effect** and how can it lead to the formation of deserts? Why do cities tend to have more haze and smog, higher temperatures, and lower wind speeds than the surrounding countryside?

Section 7.2

4. What are the two key concepts for this section? What is a **biome**? Explain why there are three major types

of each of the major biomes (deserts, grasslands, and forests). Explain why biomes are not uniform. What is an ecotone and what is the **edge effect**?

5. Explain how the three major types of deserts differ in their climate and vegetation. Why are desert ecosystems vulnerable to long-term damage? How do desert plants and animals survive? Explain how the three major types of grasslands differ in their climate and vegetation. What is a savanna? Explain how savanna animals survive seasonal variations in rainfall (**Core Case Study**). Why is the elephant an important component of the African savanna? Describe Tuy Sereivathana's efforts to prevent elephants from becoming extinct in Cambodia. Why have many of the world's temperate grasslands disappeared? Describe Arctic tundra and define **permafrost**? Explain how the three major types of forests differ in their climate and vegetation. Why is biodiversity so high in tropical rain forests? Why do most soils in tropical rain forests hold few plant nutrients. Why do temperate deciduous forests typically have a thick layer of decaying litter? What are coastal coniferous or temperate rain forests? How do most species of coniferous evergreen trees survive the cold winters in boreal forests? What important ecological roles do mountains play?

6. About what percentage of the world's major terrestrial ecosystems are being degraded or used unsustainably. Summarize the ways in which human activities have affected the world's deserts, grasslands, forests, and mountains.

Section 7.3

7. What is the key concept for this section? What percentage of the earth's surface is covered with ocean water? What is an **aquatic life zone**? Distinguish between a **saltwater (marine) life zone** and a **freshwater life zone**, and give two examples of each. List four factors that determine the types and numbers of organisms found in the layers of aquatic life zones.

8. What major ecosystem and economic services are provided by marine systems? What are the three major life zones in an ocean? Define **coastal zone** and distinguish between **estuaries** and **coastal wetlands.** Explain why each has a high NPP. Explain the ecological and economic importance of coastal marshes, mangrove forests, and sea-grass beds. Explain how coral reefs are formed and how climate change and ocean acidification are threatening them. Define **open sea** and describe its three major zones. List five human activities that pose major threats

to marine systems and eight human activities that threaten coral reefs.

Section 7.4

9. What is the key concept for this section? Define **surface water**, **runoff**, and **watershed (drainage basin)**. What major ecosystem and economic services do freshwater systems provide? What is a **lake**? What four zones are found in deep lakes? What is **eutrophication?** Distinguish between **oligotrophic** and **eutrophic lakes**. What is **cultural eutrophication**? Describe the three zones that a stream passes through as it flows from highlands to lower elevations. What is a **delta**? Define and give three examples of **inland wetlands** and list the ecological and economic services provided by such wetlands. What are four ways in which human activities are disrupting and degrading freshwater systems. How is Alexandra Cousteau attempting to educate people about the importance of aquatic systems?

10. What are this chapter's *three big ideas*? Explain how terrestrial and aquatic systems are living examples of the **scientific principles of sustainability** in action.

Note: Key terms are in bold type.

Critical Thinking

1. Why is the African savanna (Core Case Study) a good example of the three **scientific principles of sustainability** in action? For each of these principles, give an example of how it applies to the African savanna and explain how it is being violated by human activities that now affect the savanna.

2. For each of the following, decide whether it represents a likely trend in weather or in climate: **(a)** an increase in the number of thunderstorms in your area from one summer to the next; **(b)** a decrease of 20% in the depth of a mountain snowpack between 1975 and 2017; **(c)** a rise in the average winter temperatures in a particular area over a decade; and **(d)** an increase in the earth's average global temperature since 1980.

3. Why do most animals in a tropical rain forest live in its trees?

4. How might the distribution of the world's forests, grasslands, and deserts shown in Figure 7.9 differ if the prevailing winds shown in Figure 7.4 did not exist?

5. Which biomes are best suited for **(a)** raising crops and **(b)** grazing livestock? Use the three

scientific principles of sustainability to come up with three guidelines for growing crops and grazing livestock more sustainably in these biomes.

6. You are a defense attorney arguing in court for sparing a tropical rain forest from being cut down. Give your three best arguments for the defense of this ecosystem. Do the same for sparing a threatened coral reef. If you had to choose between protecting a tropical rain forest and a coral reef, which one would you select? Explain.

7. Why is ocean acidification considered serious environmental problem? If acidity levels in the ocean rise sharply during your lifetime, how might this affect you? Can you think of ways in which you might be contributing to this problem? What could you do to reduce your impact?

8. Suppose you have a friend who owns property that includes a freshwater wetland and the friend tells you she is planning to fill the wetland to make more room for her lawn and garden. What would you say to this friend?

Doing Environmental Science

Find a natural ecosystem near where you live or go to school, either a terrestrial ecosystem such as a forest or an aquatic system such as a lake or wetland. Study and write a description of the system, including its dominant vegetation and any animal life that you are aware of. Also, note how any human disturbances have changed the system. Compare your notes with those of your classmates.

Global Environment Watch Exercise

Go to your MindTap course and access the GREENR database. Starting on the home page, under "Browse Issues and Topics," click on *Environment and Ecology*, and then select *Coral Reefs* and use the topic portal to find information on **(a)** trends in the global rate of coral reef destruction; **(b)** what areas of the world are seeing rising rates of coral reef destruction and what areas are seeing falling rates; and **(c)** what is being done to protect coral reefs in various areas. Write a report on your findings.

Data Analysis

In this chapter, you learned how long-term variations in average temperatures and average precipitation play a major role in determining the types of deserts, forests, and grasslands found in different parts of the world. Below are typical annual climate graphs for a tropical grassland (savanna) in Africa (**Core Case Study**) and a temperate grassland in the Midwestern United States.

1. In what month (or months) does the most precipitation fall in each of these areas?

2. What are the driest months in each of these areas?

3. What is the coldest month in the tropical grassland?

4. What is the warmest month in the temperate grassland?

Tropical grassland (savanna)

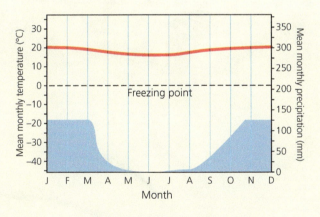

Temperate grassland (prairie)

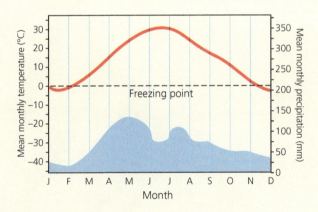

CHAPTER 8

Sustaining Biodiversity: Saving Species

The last word in ignorance is the person who says of an animal or plant: "What good is it?" … If the land mechanism as a whole is good, then every part of it is good, whether we understand it or not.

ALDO LEOPOLD

Endangered wild Siberian tiger.
Volodymyr Burdiak/Shutterstock.com

Key Questions

8.1 What role do humans play in the loss of species and ecosystem services?

8.2 Why should we try to sustain wild species and the ecosystem services they provide?

8.3 How do humans accelerate species extinction and degradation of ecosystem services?

8.4 How can we sustain wild species and the ecosystem services they provide?

Where Have All the European Honeybees Gone?

In meadows, forests, farm fields, and gardens around the world, industrious honeybees (Figure 8.1) flit from one flowering plant to another. They are collecting nectar and pollen that they take back to their hives. They feed young honeybees the protein-rich pollen and adults feed on the honey made from the collected nectar and stored in the hive.

Honeybees provide us one of nature's most important ecosystem services: *pollination.* It involves a transfer of pollen stuck on their bodies from the male to female reproductive organs of the same flower or among different flowers. This fertilization enables the flower to produce fruit and seed to grow new plants. Honeybees pollinate many plant species and some of our most important food crops, including many vegetables, fruits, and tree nuts such as almonds. European honeybees pollinate about 71% of the fruit and vegetable crops that provide 90% of the world's food and a third of the U.S. food supply.

Nature relies on the earth's free pollination service provided by a diversity of bees and other wild pollinators. In contrast, farmers practicing industrialized agriculture on vast croplands and orchards rely mostly on this single honeybee species to pollinate their crops. Many U.S. growers rent European honeybees from commercial beekeepers that truck about 2.7 million hives to farms across the country to pollinate different crops.

However, European honeybee populations have been in decline since the 1980s because of a variety of factors including exposure to new parasites, viruses, and fungal diseases. During the past decade, a new threat has emerged. Massive numbers of European bees in the United States and some European countries have been disappearing from their colonies, especially during winter. Since 2006 this phenomenon, named **colony collapse disorder (CCD)**, has affected 23% to 43% of the European honeybee colonies in the United States. This was well above the historical loss rates of 10% to 15%. Researchers are looking for the causes and for ways to reverse this decline of European honeybee populations.

Many farmers believe that we need the industrialized honeybee pollination system to grow enough food. However, many ecologists view such heavy dependence on a single bee species as a potentially dangerous violation of the earth's biodiversity **principle of sustainability**. They warn that this dependence could put food supplies at risk if European honeybee populations continue declining. If this occurs, food prices will rise. Ecologists call for more reliance on the free crop pollination services provided by a variety of wild bee species and other pollinators.

The honeybee crisis is a classic case of how the decline of a species can threaten vital ecosystem and economic services. Scientists project that during this century, human activities, especially those that contribute to habitat loss and climate change, are likely to play a key role in the extinction of one-fourth to one-half of the world's plant and animal species. Many scientists view this threat as one of the most serious and long-lasting environmental and economic problems we face. In this chapter, we discuss the causes of this problem and possible ways to deal with it.

FIGURE 8.1 European honeybee sipping nectar from a flower.

DARLYNE A. MURAWSKI/National Geographic Creative

8.1 WHAT ROLE DO HUMANS PLAY IN THE LOSS OF SPECIES AND ECOSYSTEM SERVICES?

CONCEPT 8.1 Species are becoming extinct at least 1,000 times faster than the historical rate, and by the end of this century, the extinction rate is projected to be 10,000 times higher.

Extinctions Are Natural but Sometimes They Increase Sharply

When a species is no longer found anywhere on the earth, it has experienced **biological extinction**. Extinction is a natural process and has occurred at a low rate throughout most of the earth's history. This natural rate is known as the **background extinction rate**. Scientists estimate that the background rate typically amounted to a loss of about 1 species per year for every 1 million species living on the earth. If the earth has 10 million species, this would result in 10 natural extinctions per year.

However, extinction does not always happen at a constant rate. The extinction of many species in a relatively short period of geologic time is called a **mass extinction**. Geologic, fossil, and other records indicate that the earth has experienced five mass extinctions, during which 50% to 90% of all species present at that time went extinct (Figure 4.11, p. 83) over thousands of years.

The causes of past mass extinctions are poorly understood but probably involved global changes in environmental conditions. Examples are sustained and significant global warming or cooling, large changes in sea levels and ocean water acidity, and catastrophes such as multiple large-scale volcanic eruptions and large asteroids or comets hitting the planet.

Scientific evidence indicates that after each mass extinction, the earth's overall biodiversity eventually returned to equal or higher levels (Figure 4.11, p. 83). However, each recovery took several million years.

Scientific evidence indicates that extinction rates have increased as the human population has grown, spread over most of the globe, and created large and growing ecological footprints (Figure 1.9, p. 11). In the words of biodiversity expert Edward O. Wilson (see Individuals Matter 4.1, p. 73), "The natural world is everywhere disappearing before our eyes—cut to pieces, mowed down, plowed under, gobbled up, replaced by human artifacts."

Scientists estimate that the current annual extinction rate is at least 1,000 times the natural background extinction rate (**Concept 8.1**). Assuming there are 10 million species on the earth, this means that today we are losing an estimated 10,000 species per year.

Biodiversity researchers project that during this century, the extinction rate is likely to rise to at least 10,000 times the background rate—mostly because of habitat loss and degradation, climate change, ocean acidification, and other environmentally harmful effects of human activities (**Concept 8.1**). At this rate, if there were 10 million species on the earth, then about 100,000 species would be expected to disappear each year. By the end of this century, most of the big carnivorous cats, including cheetahs, tigers (see chapter-opening photo), and lions, will probably exist only in zoos and small wildlife sanctuaries. Most elephants, rhinoceroses, gorillas, chimpanzees, and orangutans will likely disappear from the wild.

Why does this matter? According to biodiversity researchers, including Edward O. Wilson and Stuart Pimm, at this extinction rate, an estimated 20% to 50% of the world's roughly 2 million identified animal and plant species could vanish from the wild by the end of this century, along with many of the millions of unidentified species and some of world's ecosystem services. If these estimates are correct (see Science Focus 8.1), the earth is entering a *sixth mass extinction* caused primarily by human activities. Unlike previous mass extinctions, much of this mass extinction is projected to take place over the course of a human lifetime instead of over many thousands of years (Figure 4.11, p. 83). Conservation scientists view this potential massive loss of biodiversity and ecosystem services mostly within the span of a human lifetime as one of our most important and long-lasting environmental problems. As extinctions increase, the planet's life-sustaining biodiversity reaches a tipping point that causes ecosystems to collapse, which can lead to more extinctions in a runaway or positive feedback loop.

20–50% Range for percentage of the earth's known species that could disappear this century primarily because of human activities.

Wilson, Pimm, and other extinction experts consider a projected extinction rate of 10,000 times the background extinction rate to be low, for two reasons. *First*, both the rate of extinction and the resulting threats to ecosystem services are likely to increase sharply during the next 50–100 years because of the harmful environmental impacts of the rapidly growing human population and its growing per capita use of resources.

Second, we are eliminating, fragmenting, or degrading many biologically diverse environments—including tropical forests, coral reefs, wetlands, and estuaries—that serve as potential sites for the emergence of new species. Thus, in addition to greatly increasing the rate of extinction, we may be limiting the long-term recovery of biodiversity by eliminating places where new species can evolve. In other words, we are also creating a *speciation crisis*.

Estimating Extinction Rates

Scientists who estimate past extinction rates and project future extinction rates face three problems. *First,* because the natural extinction of a species typically takes a very long time, it is difficult to document. *Second,* scientists have identified only about 2 million of the world's estimated 7 to 10 million and perhaps as many as 100 million species. *Third,* scientists know little about the ecological roles of most of the species that have been identified.

One approach to estimating future extinction rates is to study records documenting past rates at which easily observable mammals and birds have become extinct. Most of these extinctions have occurred since humans began to dominate the planet about 10,000 years ago. This information can be compared with fossil records of extinctions that occurred before that time.

Another approach is to observe how reductions in habitat area affect extinction rates. The *species–area relationship,* studied by Edward O. Wilson (see Individuals Matter 4.1, p. 73) and Robert MacArthur, suggests that, on average, a 90% loss of land habitat in a given area can cause the extinction of about 50% of the species living in that area. Thus, we can base extinction rate estimates on the rates of habitat destruction and degradation, which is increasing around the world.

Scientists also use mathematical models to estimate the risk of a particular species becoming endangered or extinct within a certain period and run them on computers. These models include factors such as trends in population size, past and projected changes in habitat availability, interactions with other species, and genetic factors.

Researchers are working hard to get more and better data and to improve the models they use in order to make better estimates of extinction rates and to project the effects of such extinctions on vital ecosystem services such as pollination (**Core Case Study**). These scientists contend that our need for better data and models should not delay our acting now to keep from hastening extinctions and the accompanying losses of ecosystem services through human activities.

CRITICAL THINKING

Does the fact that extinction rates can only be estimated make them unreliable? Why or why not?

Biologists Philip Levin, Donald Levin, and others warn that, while our activities are likely to reduce the speciation rates and population sizes for some species, they could increase the speciation rates and population sizes for other rapidly reproducing species such as weeds and rats and species of insects such as cockroaches. Rapidly expanding populations of such species could reduce the populations of various other species, further accelerating their extinction and threatening key ecosystem services.

Endangered and Threatened Species Are Ecological Smoke Alarms

Biologists classify species that are heading toward biological extinction as either endangered or threatened. An **endangered species** has so few individual survivors that the species could soon become extinct. A **threatened species** has enough remaining individuals to survive in the short term, but because of declining numbers, it is likely to become endangered in the near future. Some species have characteristics that increase their chances of becoming extinct (Figure 8.2).

The International Union for Conservation of Nature (IUCN) has been monitoring the status of the world's species for 50 years and each year publishes a Red List that identifies species that are critically endangered, endangered, or threatened. Between 1996 and 2016, the total number of species in these three categories increased by 96%.

Figure 8.3 shows 4 of the nearly 23,000 species on the 2016 Red List. The actual number of species in trouble is very likely much higher. As biodiversity expert Edward O. Wilson puts it, "The first animal species to go are the big, the slow, the tasty, and those with valuable parts such as tusks and skins."

8.2 WHY SHOULD WE TRY TO SUSTAIN WILD SPECIES AND THE ECOSYSTEM SERVICES THEY PROVIDE?

CONCEPT 8.2 We should avoid speeding up the extinction of wild species because of the ecosystem and economic services they provide because it can take millions of years for nature to recover from large-scale extinctions and because many people believe that species have a right to exist regardless of their usefulness to us.

Species Are a Vital Part of the Earth's Natural Capital

According to the World Wildlife Fund (WWF), only 61,000 orangutans (Figure 8.4) remain in the wild. Most of them live among the trees in the dense tropical forests of Borneo, Asia's largest island.

Characteristic	Examples
Low reproductive rate (K-strategist)	Blue whale, giant panda, rhinoceros
Specialized niche	Blue whale, giant panda, Everglades kite
Narrow distribution	Elephant seal, desert pupfish
Feeds at high trophic level	Bengal tiger, bald eagle, grizzly bear
Fixed migratory patterns	Blue whale, whooping crane, sea turtle
Rare	African violet, some orchids
Commercially valuable	Snow leopard, tiger, elephant, rhinoceros, rare plants and birds
Require large territories	California condor, grizzly bear, Florida panther

FIGURE 8.2 Certain characteristics can put a species in greater danger of becoming extinct.

These highly intelligent animals are disappearing at an estimated rate of 1,000–2,000 per year. A key reason is that much of their tropical forest habitat is being cleared for plantations that grow oil palms. They are a source of palm oil. This vegetable oil is used in numerous products such as cookies and cosmetics and is used to produce bio-diesel fuel for motor vehicles. Another reason for their decline is smuggling. An illegally smuggled, live orangutan sells for thousands of dollars on the black market. Because of their low birth rate, orangutans have a hard time increasing their numbers. Without urgent protective action, the endangered orangutan may disappear in the wild within the next two decades.

Orangutans are considered keystone species in the ecosystems they inhabit. The dispersal of fruit and plant seeds in their wastes throughout their tropical rain forest habitat is an important ecosystem service. If orangutans disappear, many rain forest plants and other animals that consume them may be threatened.

Does it matter that orangutans—or any species, for that matter—may disappear in the wild largely due to human activities? New species eventually evolve to take the places of species lost through background and mass extinctions, so why should we care if we greatly speed up the global extinction rate over the next 50 to 100 years? According to biologists, there are four major reasons why we should prevent our activities from causing or hastening the extinction of other species.

First, many people believe that wild species, such as orangutans, have a right to exist, regardless of their usefulness to us (**Concept 8.2**).

a. Mexican gray wolf: About 97 in the forests of Arizona and New Mexico

b. California condor: 410 in the southwestern United States (up from 9 in 1986)

c. Whooping crane: 442 in North America

d. Sumatran tiger: Less than 400 on the Indonesian island of Sumatra

FIGURE 8.3 Endangered Natural Capital: These four critically endangered species are threatened with extinction, largely because of human activities. The number below each photo indicates the estimated total number of individuals of that species remaining in the wild.

FIGURE 8.4 Natural Capital Degradation: These endangered orangutans depend on a rapidly disappearing tropical forest habitat in Borneo. **Critical thinking:** What difference will it make if human activities hasten the extinction of the orangutan?

Seatraveler/Dreamstime.com

Second, the world's species provide vital *ecosystem services* (see Figure 1.3, p. 7) that help keep us alive and support our economies (**Concept 8.2**).

Third, many species contribute to *economic services* on which we depend (**Concept 8.2**). Various plant species provide economic value as sources of food, fuel, lumber, paper, and medicines (Figure 8.5).

For example, *bioprospectors* search tropical forests and other ecosystems to find plants and animals that scientists can use to make medicinal drugs—an example of *learning from nature.* According to a United Nations University report, 62% of all cancer drugs have been derived from the discoveries of bioprospectors. Less than 0.5% of the world's known plant species have been studied for their medicinal properties. **GREEN CAREER: Bioprospecting.**

CONSIDER THIS ...

Learning from Nature

Scientist Richard Wrangham is identifying medicinal compounds useful for humans by observing which plants chimpanzees eat to heal themselves when they are ill.

Another economic benefit from preserving species and their habitats is the revenues from *ecotourism.* This rapidly growing industry specializes in environmentally responsible travel to natural areas and generates more than $1 million per minute in tourist expenditures, worldwide. Conservation biologist Michael Soulé estimates that a male lion living to age 7 generates about $515,000 through ecotourism in Kenya but only about $10,000 if it is killed for its skin.

Pacific yew
Taxus brevifolia,
Pacific Northwest
Ovarian cancer

Rosy periwinkle
Cathranthus roseus,
Madagascar
Hodgkin's disease,
lymphocytic leukemia

Rauvolfia
Rauvolfia sepentina,
Southeast Asia
Anxiety, high
blood pressure

Foxglove
Digitalis purpurea,
Europe
Digitalis for heart failure

Cinchona
Cinchona ledogeriana,
South America
Quinine for malaria treatment

Neem tree
Azadirachta indica,
India
Treatment of many
diseases, insecticide,
spermicide

FIGURE 8.5 Natural Capital: These plant species are examples of *nature's pharmacy*. Once the active ingredients in the plants have been identified, scientists can usually produce them synthetically. The active ingredients in 9 of the 10 leading prescription drugs originally came from wild organism.

A *fourth* major reason for not hastening extinctions through our activities is that it will take 5 to 10 million years for natural speciation to replace the species we are likely to be lost during this century.

8.3 HOW DO HUMANS ACCELERATE SPECIES EXTINCTION AND DEGRADATION OF ECOSYSTEM SERVICES?

CONCEPT 8.3 The greatest threats to species and ecosystem services are loss or degradation of habitat, harmful invasive species, human population growth, pollution, climate change, and overexploitation.

Habitat Destruction Poses the Greatest Threat: Remember HIPPCO

Biodiversity researchers summarize the most important direct causes of the extinction of species and threats to ecosystem services using the acronym **HIPPCO**: **H**abitat destruction, degradation, and fragmentation; **I**nvasive (nonnative) species; **P**opulation growth and the resulting increased use of resources; **P**ollution; **C**limate change; and **O**verexploitation (**Concept 8.3**).

According to biodiversity researchers, the greatest threat to wild species is habitat loss (Figure 8.6), degradation, and fragmentation. Specifically, deforestation in tropical areas (see Figure 3.1, p. 46) is the greatest threat to species and to the ecosystem services they provide. The next largest habitat threats are the destruction and degradation of coastal wetlands and coral reefs (see Science Focus 7.4, p. 156), the plowing of grasslands for planting of crops (see Figure 7.12, p. 148), and the pollution of streams, lakes, and oceans.

Island species—many of them found nowhere else on earth—are especially vulnerable to extinction when their habitats are destroyed, degraded, or fragmented, because they have nowhere else to go. This is why the Hawaiian Islands are America's "extinction capital"—with 63% of their species at risk.

Habitat fragmentation occurs when a large, intact area of habitat such as a forest or natural grassland is divided into smaller, isolated patches or *habitat islands* (Figure 8.7)—typically by roads, logging operations, crop fields, and urban development. Fragmentation can divide populations of a species into increasingly isolated small groups that are more vulnerable to predators, competitor species, diseases, and catastrophic events such as storms and fires. In addition, habitat fragmentation creates barriers that limit the abilities of some species to disperse and colonize areas, locate adequate food supplies, and find mates.

Beneficial and Harmful Nonnative Species

The introduction of many nonnative species to the United States has been beneficial. According to a study by ecologist David Pimentel, nonnative species such as corn, wheat,

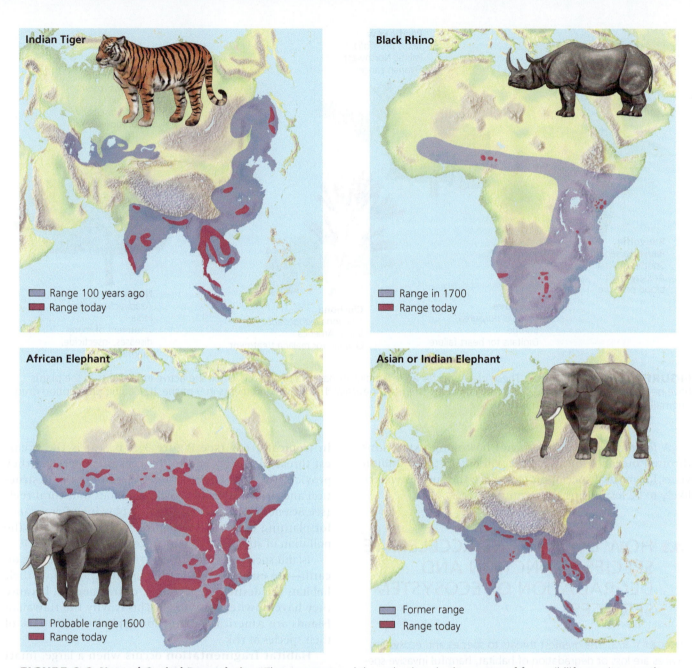

FIGURE 8.6 Natural Capital Degradation: These maps reveal the reductions in the ranges of four wildlife species, mostly as the result of severe habitat loss and fragmentation and illegal hunting for some of their valuable body parts. ***Critical thinking:*** Would you support expanding these ranges even though this would reduce the land available for human habitation and farming? Explain.

(Compiled by the authors using data from International Union for Conservation of Nature and World Wildlife Fund)

rice, and other food crops, as well as some species of cattle, poultry, and other livestock, provide more than 98% of the U.S. food supply. Similarly, nonnative tree species are grown in about 85% of the world's tree plantations. In the 1600s, English settlers brought highly beneficial European honeybees (**Core Case Study**) to North America to provide honey. Today they pollinate one-third of the crops grown in the United States.

A problem can occur when an introduced species does not face the natural predators, competitors, parasites,

viruses, bacteria, or fungi that controlled its populations in its native habitat. This can allow such nonnative species to out-compete populations of many native species for food, disrupt ecosystem services, transmit diseases, and lead to economic losses. Such nonnative species are viewed as harmful *invasive species*. The spread of such species into ecosystems is the second largest cause of extinctions and loss of ecosystem services (**Concept 8.3**).

Figure 8.8 shows some of the 7,100 or more invasive species that, after being deliberately or accidentally

FIGURE 8.7 The fragmentation of landscapes reduces biodiversity by eliminating or degrading grassland and forest wildlife habitats and degrading ecosystem services.

Deliberately Introduced Species

Purple loosestrife

African honeybee ("Killer bee")

Kudzu

Nutria

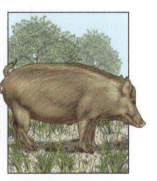

European wild boar (Feral pig)

Accidentally Introduced Species

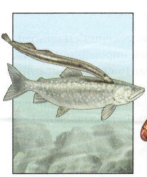

Sea lamprey (attached to lake trout)

Red fire ant

Burmese python

Formosan termite

Zebra mussel

FIGURE 8.8 Some of the estimated 7,100 harmful invasive species that have been deliberately or accidentally introduced into the United States.

introduced into the United States, have caused ecological and economic harm. According to the U.S. Fish and Wildlife Service (USFWS), about 42% of the species listed as endangered in the United States and 95% of those in the U.S. state of Hawaii are on the list because of threats from invasive species.

In the 1930s, the extremely aggressive red fire ant (Figure 8.8) was accidentally introduced into the United States probably on shiploads of lumber or coffee imported from South America. These ants have no natural predators in the southern United States where they have spread rapidly by land and by water because they can float. They have also invaded other countries, including China, Taiwan, Malaysia, and Australia.

When these ants invade an area, they can displace up to 90% of native ant populations, which provide important ecosystem services such as enrichment of topsoil, dispersal of plant seeds, and control of pest species such as flies, bedbugs, and cockroaches. Step on a red fire ant mound and as many as 100,000 ants may swarm out of their nest to attack you with painful, burning stings. Fire ants have killed deer fawns, ground-nesting birds, baby sea turtles, newborn calves, pets, and at least 80 people who were allergic to their venom.

Widespread insecticide spraying in the 1950s and 1960s temporarily reduced red fire ant populations. However, the insecticides also reduced populations of many native ant species. Widespread insecticide use also promoted genetic resistance to the insecticides in red fire ants through natural selection. The introduction of a species of tiny parasitic fly has shown some success in controlling red fire ant populations, but more research is needed to understand the long-term impacts of this biological remedy.

The Burmese python is an example of what can happen when nonnative species escape or are released into the wild and become invasive species. Large numbers of these snakes are imported from Asia for sale as pets. Some buyers, after learning that these reptiles do not make good pets, let them go in the wetlands of Florida's Everglades.

The Burmese python (Figure 8.9) can live 20 to 25 years and grow as long as 5 meters (16 feet), weigh as much 77 kilograms (170 pounds), and be as big around as a telephone pole. They have huge appetites, seizing prey with their sharp teeth, wrapping around them, and squeezing them to death before feeding on them. They feed at night and eat a variety of birds and mammals such as rabbits, foxes, raccoons, and white-tailed deer. Occasionally the pythons eat other reptiles, including young American alligators—a keystone species in the Everglades ecosystem (see Chapter 4, Case Study, p. 75). The pythons have also been known to eat pet cats, dogs, small farm animals, and geese. Research indicates that predation by these snakes is altering the complex food webs and ecosystem services of the Everglades.

According to wildlife scientists, the Burmese python population in Florida's wetlands cannot be controlled.

They are hard to find and kill or capture and they reproduce rapidly. Trapping and moving the snakes from one area to another has not worked because they are able to return to the areas where they are captured. Another concern is that the Burmese python could spread to other swampy wetlands in the southern half of the United States.

Some invasive species, such as *kudzu* (Figure 8.8), have been deliberately introduced into ecosystems. In the 1930s, this plant was imported from Japan and planted in the southeastern United States to control soil erosion.

Kudzu does control erosion, but it grows so rapidly that it can engulf hillsides, gardens, trees, stream banks, cars (Figure 8.10), and anything else in its path. Dig it up or burn it, and it still keeps spreading. It can grow in sunlight or shade and is very difficult to kill, even with herbicides that can contaminate water supplies. This plant—sometimes called "the vine that ate the South"—has spread throughout much of the southeastern United States. As the climate gets warmer, it could spread to the north.

Invasive species also affect aquatic systems and are blamed for about two-thirds of all fish extinctions in the United States since 1990. The Great Lakes of North America have been invaded by at least 180 nonnative species and the number keeps rising. One of the biggest threats is the fish-killing sea lamprey (see Figure 5.5, p. 94), which has depleted some Great Lakes populations of important sport fish species such as lake trout.

Another aquatic invader is a thumbnail-sized mollusk called the *zebra mussel* (Figure 8.8), which reproduces rapidly and has no known natural enemies in the Great Lakes. It has displaced other mussel species and depleted the food supply for some native aquatic species. Zebra mussels have also caused massive economic damages by clogging irrigation pipes, shutting down cooling water intake pipes for power plants and city water supplies, jamming ship rudders, and growing in large masses on boat hulls, piers, and other exposed aquatic surfaces. Zebra mussels have spread into a number of rivers and, in 2016, had been reported in at least 24 U.S. states.

Controlling Invasive Species

Once a harmful nonnative species becomes established in an ecosystem, removing it is almost impossible. Americans pay more than $160 billion a year to eradicate or control an increasing number of invasive species—without much success. Thus, the best way to limit the harmful impacts of nonnative species is to prevent them from being introduced into ecosystems.

Scientists suggest several ways to do this, including:

- Greatly increasing research to identify the characteristics of successful invaders, the types of ecosystems vulnerable to invaders, and the natural predators, parasites, bacteria, and viruses that could be used to control populations of established invaders.

FIGURE 8.9 University of Florida researchers hold a 4.6-meter-long (15-foot-long), 74-kilogram (162-pound) Burmese python captured in Everglades National Park shortly after it had eaten a 1.8-meter (6-foot) long American alligator. *Critical thinking:* What would happen if we just allowed pythons to take over this ecosystem and accepted whatever changes might occur? Would this present a problem? Explain.

Dan Callister/Alamy Stock Photo

- Increasing ground surveys and drone satellite observations to track invasive plant and animal species and developing better models for predicting how they will spread and what harmful effects they could have.

- Identifying major harmful invader species and establishing international treaties banning their transfer from one country to another, as is now done for endangered species, while stepping up inspection of imported goods to enforce such bans.

- Educating the public about the effects of releasing exotic plants and pets into the environment where they live.

Figure 8.11 shows some of the things you can do to help prevent or slow the spread of harmful invasive species.

$2.7 million
Estimated global cost per minute of harm from invasive species.

Population Growth, Resource Use, Pollution, and Climate Change Contribute to Species Extinctions

Past and projected *human population growth* (Figure 1.12, p. 15) and rising rates of *resource use per person* have greatly expanded the human ecological footprint (see Figure 1.9, p. 11). People have eliminated, degraded, and fragmented

FIGURE 8.10 Kudzu has grown over this car in the U.S. state of Georgia.

What Can You Do?

Controlling Invasive Species

- Do not buy wild plants and animals or remove them from natural areas.

- Do not release wild pets in natural areas.

- Do not dump aquarium contents or unused shing bait into waterways or storm drains.

- When camping, use only local firewood.

- Brush or clean pet dogs, hiking boots, mountain bikes, canoes, boats, motors, shing tackle, and other gear before entering or leaving wild areas.

FIGURE 8.11 Individuals Matter: Some ways to prevent or slow the spread of harmful invasive species. *Critical thinking:* Which two of these actions do you think are the most important ones to take? Why? Which of these actions do you plan to take?

FIGURE 8.12 *Bioaccumulation and biomagnification:* DDT is a fat-soluble chemical that can accumulate in the fatty tissues of animals. In a food chain or food web, the accumulated DDT is biologically magnified in the bodies of animals at each higher trophic level, such as the food chain in the U.S. state of New York, illustrated here. *Critical thinking:* How does this story demonstrate the value of pollution prevention?

vast areas of wildlife habitat as they have spread out over the planet and increased their use of resources. This has caused the extinction of many species (**Concept 8.3**).

Pollution of the air, water, and soil by human activities also threatens some species with extinction. According to the USFWS, each year, pesticides kill about one-fifth of the European honeybee colonies that pollinate almost one-third of all U.S. food crops (**Core Case Study** and Science Focus 8.2). The USFWS estimates that pesticides also kill more than 67 million birds and 6 to 14 million fish each year. They also threaten about 20% of the country's endangered and threatened species.

During the 1950s and 1960s, populations of fish-eating birds such as ospreys, brown pelicans, and bald eagles plummeted because of the widespread use of a pesticide called DDT to kill mosquitoes. The concentration of a chemical derived from the pesticide DDT remained in the environment and was taken up and accumulated in the tissues of organisms, a process called **bioaccumulation**. The chemical became more concentrated as it moved up through food chains and webs, a process called **biomagnification** (Figure 8.12). It made the eggshells of these top predator birds so fragile that they could not reproduce successfully. Populations of other predatory birds also declined sharply. They included the prairie falcon, sparrow hawk, and peregrine falcon, which helped control populations of rabbits and ground squirrels. Since the U.S. ban on DDT in 1972, most of these bird species have made a comeback—an example of the effectiveness of pollution prevention.

According to Conservation International, failure to sharply reduce greenhouse gas emissions could help drive one-quarter to one-half of all land animals and plants to extinction by the end of this century. For

DDT in fish-eating birds (ospreys)
25 ppm

DDT in large fish (needlefish)
2 ppm

DDT in small fish (minnows)
0.5 ppm

DDT in zooplankton
0.04 ppm

DDT in water
0.000003 ppm, or 3 ppt

FIGURE 8.13 On floating ice in the Arctic sea, this polar bear has killed a bearded seal, one of its major sources of food. *Critical thinking:* Do you think it matters that the polar bear may become extinct in the wild during this century primarily because of human activities? Explain.

Vladimir Seliverstov/Dreamstime.com

example, scientific studies indicate that the polar bear is threatened because higher temperatures are melting sea ice in its Arctic habitat. Shrinkage of this floating ice makes it harder for polar bears to find seals, their favorite prey (Figure 8.13). According to the IUCN and the U.S. Geological Survey, the Arctic polar bear population is likely to decline by 30–35% by 2050 due to loss of habitat and prey.

The Killing, Capturing, and Selling of Wild Species Threatens Biodiversity

Some protected species are illegally killed for their valuable parts or are captured and sold live to collectors. Organized crime has moved into illegal wildlife smuggling of parts of species and live members of species because of the huge profits involved. Few of the smugglers are caught or punished.

It is highly profitable for poachers and organized crime to kill highly endangered eastern mountain gorillas (of which there are about 880 left in the wild) and endangered giant pandas (1,864 left in the wild in China as of 2016) for their valuable pelts. Four of the five rhino species are critically endangered, mostly because so many have been illegally killed for their very valuable horns (Figure 8.14).

The illegal killing of elephants, especially African elephants (Figure 7.A, p. 146) for their valuable ivory tusks has increased in recent years, despite an international ban on the trade of ivory. China has the largest market for illegal ivory, followed by the United States. Elephants are being killed at a rate of 30,000 a year, according to WildAid.

Honeybee Losses: A Search for Causes

Over the past 50 years, the European honeybee population in the United States has been cut in half. Since colony collapse disorder (CCD) emerged in 2006 (**Core Case Study**) commercial beekeepers in the United States have lost 23–43% of their hives on average each year. Scientific research has found several possible reasons for this decline. They include parasites, viruses, insecticides, stress, and diet.

Parasites such as the varroa mite feed on the blood of adult honeybees and their larvae, weakening their immune systems and shortening their lives. Varroa mites have killed millions of honeybees since first appearing in the United States in 1987—probably among bees imported from South America.

Several *viruses* are known to affect the winter survival of European honeybees. One example is the tobacco ringspot virus that can infect honeybees feeding on pollen containing the virus. The virus is thought to attack the bees' nervous systems. The virus has also been detected in varroa mites, which may help spread the virus as they feed on honeybees.

As honeybees forage for nectar, they are exposed to *insecticides* sprayed on crops and can carry these chemicals back to the hives. Some research indicates that widely used insecticides called *neonicotinoids* may play a role in CCD by disrupting the nervous systems of bees and decreasing their ability to find their way back to their hives. These chemicals can also disrupt the immune systems of bees and make them vulnerable to the harmful effects of other threats.

Stress from honeybees that are transported across the United States (Figure 8.A) can also play a role. Overworking and overstressing honeybees by moving them around the country can weaken their immune systems and make them more vulnerable to death from parasites, viruses, fungi, and insecticides.

Another factor is *diet*. In natural ecosystems, honeybees gather nectar and pollen from a variety of flowering plants. However, industrial worker honeybees feed mostly on pollen or nectar from one crop or a small number of crops that may lack the nutrients they need. In winter, bees in hives where most of the honey has been removed for sale are often fed sugar or high fructose corn syrup that provide calories but not enough protein for good health.

The growing consensus among bee researchers is that the decline of European honeybee populations and CCD occur because of a combination of these factors. These annual bee deaths raise the costs for beekeepers and farmers who use their services and could put many of them out of business if the problem continues. This could lead to higher food prices.

CRITICAL THINKING

Can you think of some ways in which commercial beekeepers could lessen one or more of the threats described here? Explain.

Cristi111/Dreamstime.com

FIGURE 8.A European honeybee hive boxes in an acacia orchard. Each year, commercial beekeepers rent and deliver several million hives by truck to farmers throughout the United States.

CONSIDER THIS ...

CONNECTIONS Drones, Elephants, and Poachers

Researchers are using small drones with cameras connected to smart phones to track and monitor wildlife species such as endangered elephants and rhinos in Africa, tigers in Nepal, and orangutans in Sumatra. Drones with infrared cameras can find illegal poachers at night, expose their locations to wildlife rangers, and deter them by using bright strobe lights.

Since 1900, the number of the world's wild tigers has declined by 99%, mostly because of habitat loss (Figure 8.6) and illegal poaching. More than half of the world's remaining 3,890 wild tigers are in India, which is doing more than other countries to protect them by establishing tiger reserves.

The Indian, or Bengal, tiger is at risk because its fur can be used to make expensive coats. The bones and penis

FIGURE 8.14 This critically endangered northern white rhinoceros was killed illegally in South Africa for its two horns. This species is now extinct in the wild. ***Critical thinking:*** What would you say if you could talk to the poacher who illegally killed this animal?

Avalon/Photoshot License/Alamy Stock Photo

of a single tiger can fetch thousands of dollars on the black market. According to the WWF, without emergency action to curtail poaching and to preserve tiger habitat, few if any tigers, including the Sumatran tiger (Figure 8.3d), will be left in the wild by 2022. During the past 100 years, the number of cheetahs—the world's fastest land animal—has dropped about 100,000 to 7,100 mainly because of habitat loss and poachers killing them for their coats.

CONSIDER THIS …

THINKING ABOUT Tigers

Would it matter to you if all of the world's wild tigers were to disappear? Explain. List two steps you could take to help protect the world's remaining wild tigers from extinction.

Around the globe, the legal and illegal trade in wild species for use as pets is a huge and highly profitable business. However, many owners of wild pets do not know that, for every live animal captured and sold in the legal and illegal pet market, many others are killed or die in transit. According to the IUCN, more than 60 bird species, mostly parrots, are endangered or threatened because of the wild bird trade (see the Case Study that follows).

Buyers of wild animals for pets might also be unaware that some imported exotic animals carry diseases such as Hantavirus, Ebola virus, Asian bird flu, herpes B virus (carried by most adult macaques), and salmonella (from pets such as hamsters, turtles, and iguanas). These diseases can easily spread from pets to their owners and then to other people.

Other wild species whose populations are depleted because of the pet trade include many amphibians (see Chapter 4, Core Case Study, p. 70), reptiles, and tropical fishes taken mostly from the coral reefs of Indonesia and the Philippines. Some divers catch tropical fish by using plastic squeeze bottles of poisonous cyanide to stun them. For each fish caught alive, many more die. In addition, the cyanide solution kills the polyps, the tiny animals that create the reef.

Some exotic plants, especially orchids and cacti (see Figure 7.10, center, p. 143), are endangered because they are removed and sold, often illegally, for thousands of dollars to collectors to decorate houses, offices, and landscapes. According to the USFWS, collectors of exotic birds will pay thousands of dollars for an endangered hyacinth macaw (Figure 8.15) smuggled out of Brazil (Individuals Matter 8.1). However, during its lifetime, a single hyacinth macaw left in the wild could attract an estimated $165,000 in ecotourism revenues.

Rising Demand for Bushmeat Threatens Some African Species

For centuries, indigenous people in much of West and Central Africa have sustainably hunted wildlife for *bushmeat* as a source of food. In the last three decades,

Roy Toft/National Geographic Creative

FIGURE 8.15 Many species of wildlife such as this endangered hyacinth macaw in Mato Grosso, Brazil, are sources of beauty and pleasure. Habitat loss and illegal capture in the wild by pet traders endanger this species.

INDIVIDUALS MATTER 8.1

Juliana Machado Ferreira: Conservation Biologist and National Geographic Explorer

© REBECCA DROBIS

Every year, poachers illegally remove about 38 million wild animals from their natural habitats in Brazil. Some of these animals stay in Brazil and others end up in the United States, Europe, and other parts of the world. Juliana Machado Ferreira is a conservation biologist with a PhD in genetics who fights this illegal removal of wildlife in her native country of Brazil.

She founded FREELAND Brasil to help combat this highly profitable illegal trade. Many people in Brazil keep parrots, macaws, songbirds, monkeys, and other wild animals in their homes as pets and believe that it is harmless cultural tradition. Her organization educates the public about the harmful ecological effects of removing birds and other species from wild for the amusement of people who take them.

Ferreira has used her knowledge of genetics to develop molecular markers that can help identify the origins of illegal birds seized by police so that the birds can be returned to the areas where they lived. Ferreira is also trying to get the Brazilian government to pass and enforce strict laws against illegal wildlife trafficking.

In 2014, she was selected as a National Geographic Explorer. When asked what people can do to help save wild species she says: "Do n[ot] regard wild animals as pets."

bushmeat hunting in some areas has skyrocketed. Some hunters have tried to provide food for rapidly growing populations. Others make a living by supplying restaurants in major cities with exotic meats from gorillas (Figure 8.16) and other species. Logging roads in once-inaccessible forests have made such hunting much easier.

Bushmeat hunting has driven at least one species— Miss Waldron's red colobus monkey—to complete extinction. It is also a factor in the reduction of some populations of orangutans (Figure 8.4), gorillas, chimpanzees, elephants, and hippopotamuses. Another problem is that butchering and eating some forms of bushmeat has helped spread fatal diseases such as HIV/AIDS and the Ebola virus from animals to humans.

The U.S. Agency for International Development (USAID) is trying to reduce unsustainable hunting for bushmeat in some areas of Africa by introducing alternative sources of food, including farmed fish. They are also showing villagers how to breed large rodents such as cane rats as a source of protein.

FIGURE 8.16 *Bushmeat* such as this severed head of an endangered lowland gorilla in the Congo is consumed as a source of protein by local people in parts of West and Central Africa and is sold in national and international marketplaces and served in some restaurants where wealthy patrons regard gorilla meat as a source of status and power. **Critical thinking:** How, if at all, is this different from killing a cow for food?

CASE STUDY

A Disturbing Message from the Birds

Approximately 70% of the world's more than 10,000 known bird species are declining in numbers. Much of this decline is related to human activities, summarized by HIPPCO. According to the IUCN 2014 Red List of Endangered Species, roughly one of every eight (13%) of all bird species is threatened with extinction mostly by *habitat loss, degradation, and fragmentation* (the H in HIPPCO)—primarily in tropical forests.

According to a 2014 *State of the Birds study*, almost one-third of the more than 800 bird species in the United States are endangered (Figure 8.3b and c), threatened, or in decline, mostly because of habitat loss and degradation. Sharp declines in bird populations have occurred among songbird species that migrate long distances. These birds nest deep in North American woods in the summer and spend their winters in Central or South America or on the Caribbean Islands. Research indicated that the primary causes of these population declines were habitat loss and fragmentation of the birds' breeding habitats in North America and Central and South America.

The second greatest threat to birds is the intentional or accidental introduction of nonnative species that become *invasive species* and affect about 28% of the world's threatened birds. Such invasive species (the I in HIPPCO) include the brown tree snake and mongooses. In the United States, feral cats and pet cats kill at least 1.4 billion birds each year, according to a study by Peter Mara of the Smithsonian Conservation Biology Institute.

Çağan Hakkı Sekercioğlu: Protector of Birds and National Geographic Emerging Explorer

Çağan Sekercioğlu, is a bird expert, tropical biologist, accomplished wildlife photographer, and a National Geographic Explorer. He has seen over 64% of the planet's known bird species in 75 countries, developed a global database on bird ecology, and has become an expert on the causes and consequences of bird extinctions around the world.

Recently he has focused his research on monitoring the effects of habitat loss on birds in Costa Rica's forests and agricultural areas and on the effects of climate change on birds. He notes that climate change is driving birds into higher elevations in mountain ranges.

Sekercioğlu also works hard to address threats to birds and other forms of wildlife in Turkey, his home country. In 2007, he founded KuzeyDoğa. It is an award-winning ecological research and community-based conservation organization devoted to conserving and protecting the wildlife of northeastern Turkey. He also developed a plan for Turkey's first protected wildlife corridor. In 2011, he was named as Turkey's Scientist of the Year.

Based on his extensive research Sekercioğlu estimates that the percentage of the world's known bird species that are endangered could approximately double from 13% in 2013 to 25% by the end of this century. He says, "My ultimate goal is to prevent extinctions and consequent collapses of critical ecosystem processes while making sure that human communities benefit from conservation as much as the wildlife they help conserve…. I don't see conservation as people versus nature, I see it as a collaboration."

Population growth, the first P in HIPPCO, also threatens some bird species, as more people spread out over the landscape each year and increase their use of timber, food, and other resources, which results in disturbance or destruction of bird habitats. According to bird expert Daniel Klem, Jr., about 600 million birds die each year from collisions with windows in the United States and Canada.

Pollution, the second P in HIPPCO, also threatens birds. Countless birds are exposed to oil spills, insecticides, and herbicides.

Another rapidly growing threat to birds is *climate change*, the C in HIPPCO. A study done for the WWF found that the effects of climate change, such as more intense heat waves and flooding, are causing declines of some bird populations in every part of the globe. Such losses are expected to increase sharply during this century.

Overexploitation (the O in HIPPCO) is also a major threat to bird populations. Fifty-two of the world's 388 parrot species (Figure 8.15) are threatened, partly because so many parrots are captured, often illegally, for sale as pets, usually to buyers in Europe and the United States.

Biodiversity scientists view this decline of bird species with alarm. One reason is that birds are excellent *indicator species* because they live in every climate and biome, respond quickly to environmental changes in their habitats, and are relatively easy to track and count. To these scientists, the decline of many bird species indicates widespread environmental degradation.

A second reason for alarm is that birds perform critically important economic and ecosystem services throughout the world. For example, many birds play specialized roles in free pollination and seed dispersal, especially in tropical areas. Extinctions of these bird species could lead to extinctions of plants that depend on the birds for pollination. Then, some specialized animals that feed mostly on these plants might also become extinct. Such a *cascade of extinctions*, in turn, could affect human food supplies. Biodiversity scientists (Individuals Matter 8.2) urge us to listen more carefully to what birds are telling us about the state of the environment, for the birds' sake, as well as for ours.

8.4 HOW CAN WE SUSTAIN WILD SPECIES AND THE ECOSYSTEM SERVICES THEY PROVIDE?

CONCEPT 8.4 We can reduce species extinction and sustain ecosystem services by establishing and enforcing national environmental laws and international treaties and by creating and protecting wildlife sanctuaries.

Treaties and Laws Can Help Protect Species

Some governments are working to reduce species extinction and sustain ecosystem services (see the Case Study that follows) by establishing and enforcing international treaties and conventions, as well as national environmental laws (**Concept 8.4**).

One important international agreement is the 1975 *Convention on International Trade in Endangered Species of Wild Fauna and Flora (CITES)*. This treaty, signed by

181 countries, bans the hunting, capturing, and selling of threatened or endangered species. It lists 931 species that are in danger of extinction and that cannot be commercially traded as live specimens or for their parts or products. It restricts the international trade of roughly 5,600 animal species and 30,000 plant species at risk of becoming threatened.

CITES has helped reduce the international trade of many threatened animals, including elephants, crocodiles, cheetahs, and chimpanzees. The treaty has also raised public awareness about the illegal trade of wildlife and poaching.

However, CITES is limited because enforcement varies from country to country and convicted violators often pay only small fines. Also, member countries can exempt themselves from protecting any listed species. In addition, much of the highly profitable illegal trade in wildlife and wildlife products goes on in countries that have not signed the treaty.

Another important international treaty is the *Convention on Biological Diversity (CBD)*, ratified or accepted by 196 countries. It legally commits participating governments to reducing the global rate of biodiversity loss and to sharing the benefits from the use of the world's genetic resources. It also aims to prevent or control the spread of ecologically harmful invasive species.

This convention is a landmark in international law because it focuses on ecosystems rather than on individual species. However, implementation has been slow because some key countries (including the United States, as of 2016) have not ratified it. The law also lacks severe penalties or other enforcement mechanisms.

CASE STUDY

The U.S. Endangered Species Act

The United States enacted the **Endangered Species Act (ESA)** in 1973 and has amended it several times. The act is designed to identify and protect endangered species in the United States and abroad (**Concept 8.4**). The ESA creates recovery programs for the species it lists. Its goal is to help the populations of protected species recover to levels where legal protection is no longer needed. When that happens, a species can be taken off the list, or delisted.

Under the ESA, the National Marine Fisheries Service (NMFS) is responsible for identifying and listing endangered and threatened ocean species, while the USFWS is to identify and list all other endangered and threatened species. Any decision by either agency to list or delist a species must be based on biological factors alone, without consideration of economic or political factors. However, the two agencies can use economic factors in deciding whether and how to protect endangered habitat and in developing recovery plans for listed species.

The ESA forbids federal agencies (except the Defense Department) to carry out, fund, or authorize projects that would jeopardize any endangered or threatened species or destroy or modify its critical habitat. The law also makes it illegal for Americans to sell or buy any product made from an endangered or threatened species or to hunt, kill, collect, or injure such species in the United States.

For offenses committed on private lands, fines as high as $100,000 and prison terms of up to 1 year can be imposed to ensure protection of the habitats of endangered species. Although this provision has rarely been used, it has been controversial because at least 90% of the listed species live totally or partially on private land. Since 1982, the ESA has been amended to give private landowners various economic incentives to help save endangered species living on their lands.

The ESA requires that all commercial shipments of wildlife and wildlife products enter or leave the country through 1 of 17 designated airports and ocean ports. The 120 full-time USFWS inspectors can inspect only a small fraction of the more than 200 million wild animals brought legally into the United States annually. Each year, tens of millions of wild animals are also brought in illegally, but few illegal shipments of endangered or threatened animals or plants are confiscated. In addition, many violators are not prosecuted and convicted violators often pay only a small fine.

Between 1973 and March 2017, the number of U.S. species on the official endangered and threatened species lists increased from 92 to 1,652, with 70% of the listed species having active recovery plans. According to a study by the Nature Conservancy, about 33% of the country's species are at risk of extinction, and 15% of all species are at high risk—far more than the current number officially listed as threatened or endangered.

Since 1995, there have been numerous efforts to weaken the ESA and to reduce its already meager annual budget. Opponents of the act contend that it puts the rights and welfare of endangered plants and animals above those of people. Some critics would do away with this act. They call it an expensive failure because only 30 species had recovered enough to be removed from the endangered list by January 2017.

Most biologists view the act as one of the world's most successful environmental laws, for several reasons. *First,* species are listed only when they face serious danger of extinction. ESA supporters argue that this is similar to a hospital emergency room set up to take only the most desperate cases, often with little hope for recovery. Such a facility could not be expected to save all or even most of its patients.

Second, according to federal data, the conditions of more than half of the listed species are stable or improving, 90% are recovering at rates specified by their recovery plans, and 99% of the listed species are still surviving. A hospital emergency room having similar results would be considered an astounding success story.

Third, it takes many decades for a species to reach the point where it is in danger of extinction. Thus, it takes many decades to bring a species back to the point where it can be removed from the endangered list.

Fourth, the small federal budget for protecting endangered species has been flat or declining in recent years. To ESA supporters, it is amazing that the federal agencies responsible for enforcing the act have managed to stabilize or improve the conditions of 99% of the listed species on such a small budget.

A national poll conducted by the CBD found that two out of three Americans want the ESA strengthened or left alone. However, some members of Congress continue efforts to weaken or do away with the law.

A U.S. National Academy of Sciences study recommended three major changes in the law to make it more scientifically sound and effective:

- Greatly increase the funding for implementing the act.

- Put greater emphasis on developing recovery plans more quickly.

- When a species is first listed, establish the core of its habitat as critical for its survival and give that area the maximum protection.

Wildlife Refuges and Other Protected Areas

In 1903, President Theodore Roosevelt (Figure 1.16, p. 20) established the first U.S. federal wildlife refuge at Pelican Island, Florida, to help protect the brown pelican and other birds from extinction (Figure 8.17). In 2009, the brown pelican was removed from the U.S. Endangered Species list, thanks to Roosevelt's early protection. By 2016, there were more than 560 refuges in the National Wildlife Refuge System. Each year, more than 47 million Americans visit these refuges to hunt, fish, hike, and watch birds and other wildlife.

More than three-fourths of the refuges serve as wetland sanctuaries that are vital for protecting migratory waterfowl. At least one-fourth of all U.S. endangered and threatened species have habitats in the refuge system, and some refuges have been set aside specifically for certain endangered species (**Concept 8.4**). Such areas have aided the recovery of populations of Florida's Key deer, the brown pelican, and the trumpeter swan.

GOOD NEWS

Despite their benefits, activities that are harmful to wildlife, such as mining, oil drilling, and use of off-road vehicles, take place in nearly 60% of the nation's wildlife refuges, according to a General Accounting Office study. Biodiversity researchers urge the U.S. government to set aside more refuges and to increase the meager budget for the refuge system.

Seed Banks and Botanical Gardens

Recent research indicates that between 60,000 and 100,000 species of the world's plants—roughly one-fourth of all known plant species—are in danger of extinction. *Seed banks* are refrigerated, low-humidity storage environments that are used to preserve the seeds of endangered and other plant species. More than 1,400 seed banks around the world collectively hold about 3 million samples.

Seed banks vary in quality, are expensive to operate, and are difficult to protect against destruction by fire or other mishaps. However, the Svalbard Global Seed Vault, an underground facility on a remote island in the Arctic, is expected to withstand natural and human-caused disasters. It will eventually contain seeds from 2 million of the world's plant species. In addition, some species cannot be preserved in seed banks.

The world's 1,600 *botanical gardens* contain living plants that represent almost one-third of the world's known plant species. However, they contain only about 3% of the world's rare and threatened plant species and have too little space and funding to preserve most of those species.

Zoos, Aquariums, and Wildlife Farms

Zoos, aquariums, game parks, and animal research centers preserve some individuals of critically endangered animal species. The long-term goal is to reintroduce such species into protected wild habitats.

Two techniques for preserving endangered terrestrial species are egg pulling and captive breeding. *Egg pulling* involves collecting eggs laid in the wild by critically endangered bird species and then hatching them in zoos or research centers. In *captive breeding*, some or all of the wild individuals of a critically endangered species are collected for breeding in captivity, with the aim of reintroducing their offspring into the wild. Captive breeding has been used to save the peregrine falcon and the California condor (Figure 8.3b).

Several other techniques are used to increase the populations of captive species. They include *artificial insemination embryo transfer* (the surgical implantation of eggs of one species into a surrogate mother of another species) and *cross fostering* (in which the young of a rare species are raised by parents of a similar species). Scientists also match individuals for mating by using DNA analysis along with computer databases that hold information on family lineages of endangered zoo animals—a computer dating service for zoo animals.

The ultimate goal of captive breeding programs is to build populations to a level where they can be reintroduced into the wild. Successes include the black-footed ferret, the golden lion tamarin (a highly endangered monkey species), the Arabian oryx, and the California condor. However, most reintroductions fail because of a lack of suitable habitat, an inability of the individuals bred in captivity to survive in the wild, renewed overhunting or poaching, or pollution and other hazards in the environment.

FIGURE 8.17 The Pelican Island National Wildlife Refuge in Florida was America's first National Wildlife Refuge.

George Gentry/U.S. Fish and Wildlife Service; Inset: Chuck Wagner/Shutterstock.com

One problem for captive breeding programs is that a captive population of an endangered species must typically number 100–500 individuals to avoid extinction resulting from accidents, diseases, or the loss of genetic diversity through inbreeding. Recent genetic research indicates that 10,000 or more individuals are needed for an endangered species to maintain its capacity for biological evolution. Zoos and research centers do not have the funding or space to house such large populations.

Public aquariums that exhibit unusual and attractive species of fish and marine animals such as seals and dolphins help educate the public about the need to protect such species. However, mostly due to limited funds,

public aquariums have not served as effective gene banks for endangered marine species, especially marine mammals that need large volumes of water in which to live.

We can take pressure off some endangered or threatened species by raising individuals of these species on *farms* for commercial sale. In Florida, American alligators are raised on farms for their meat and hides (see Case Study, p. 75). Butterfly farms established to raise and protect endangered butterfly species flourish in Papua New Guinea, where many such species are threatened by development activities. These farms are also used to educate visitors about the need to protect butterfly species.

Protecting Species and Ecosystem Services Raises Difficult Questions

Efforts to prevent the extinction of wild species and the accompanying losses of ecosystem services require the use of financial and human resources that are limited. This raises some challenging questions:

- Should we focus on protecting species or should we focus more on protecting ecosystems and the ecosystem services they provide?

- How do we allocate limited resources between these two priorities?

- How do we decide which species should get the most attention in our efforts to protect as many species as possible? For example, should we focus on protecting the most threatened species or on protecting keystone species?

- Protecting species that are appealing to humans, such as panda bears and orangutans (Figure 8.4), can increase public awareness of the need for wildlife conservation. Is this more important than focusing on the ecological importance of species when deciding which ones to protect?

- How do we determine which habitat areas are the most critical to protect?

Conservation biologists continually struggle to deal with these questions. Because of limited funds, they must decide which species will get priority.

Figure 8.18 lists some guidelines that you can follow to play your part to help protect species and to increase your beneficial environmental impact.

What Can You Do?

Protecting Species

- Do not buy furs, ivory products, or other items made from endangered or threatened animal species.

- Do not buy wood or wood products from tropical or old-growth forests

- Do not buy pet animals or plants taken from the wild

- Tell friends and relatives what you're doing about this problem.

FIGURE 8.18 Individuals Matter: You can help prevent the extinction of species. *Critical thinking:* Which two of these actions do you think are the most important ones to take? Why?

BIG IDEAS

- We are hastening the extinction of wild species and degrading the ecosystem services they provide by destroying and degrading natural habitats, introducing harmful invasive species, and increasing human population growth, pollution, climate change, and overexploitation.

- We should avoid causing or hastening the extinction of wild species because of the ecosystem and economic services they provide and because their existence should not depend primarily on their usefulness to us.

- We can work to prevent the extinction of species and to protect overall biodiversity and ecosystem services by establishing and enforcing environmental laws and treaties and by creating and protecting wildlife sanctuaries.

Tying It All Together

Honeybees and Sustainability

In this chapter, we learned about the human activities that are hastening the extinction of many species and about how we might curtail those activities. We learned that as many as half of the world's wild species could go extinct during this century, largely because of human activities that threaten many species and some of the vital ecosystem services they provide. For example, populations of honeybees, vital for pollinating crops that supply much of our food, have been declining for a variety of reasons

(**Core Case Study**), many of them related to human activities. One of the key reasons for such problems is that most people are unaware of the highly valuable ecosystem and economic services provided by the earth's species.

Acting to prevent the extinction of species from human activities implements two of the three **scientific principles of sustainability**. It preserves not only the earth's biodiversity but also the vital ecosystem services that sustain us, including chemical cycling. It

Malwina Szweda/Shutterstock.com

also implements the ethical **principle of sustainability** that call for us to leave the earth in a condition as good as or better than what we inherited.

Chapter Review

Core Case Study

1. What economic and ecological services do honeybees provide? How are human activities contributing to the decline of many populations of European honeybees? What is **colony collapse disorder (CCD)**?

Section 8.1

2. What is the key concept for this section? Define and distinguish between **biological extinction** and **mass extinction**. What is the **background extinction rate**? Explain how scientists estimate extinction rates and describe the challenges they face in doing so. What percentage of the world's identified species could become extinct primarily because of human activities during this century? Give two reasons why many extinction experts think that projected extinction rates are probably on the low side. Distinguish between **endangered species** and **threatened species** and give an example of each. List four characteristics that make some species especially vulnerable to extinction.

Section 8.2

3. What is the key concept for this section? What are four reasons for trying to avoid hastening the extinction of wild species? Describe two economic and two ecological benefits of species diversity. Explain how saving other species and the ecosystem services they provide can help us save our own species and our cultures and economies.

Section 8.3

4. What is the key concept for this section? What is **HIPPCO**? What is the greatest threat to wild species? What is **habitat fragmentation**? Describe the major effects of habitat loss and fragmentation. Why are island species especially vulnerable to extinction? What are habitat islands?

5. Give two examples of the benefits gained by the introduction of nonnative species. Give two examples of the harmful effects of nonnative species that have been introduced deliberately and two examples of the same for accidentally introduced species. Explain why prevention is the best way to reduce threats from invasive species and list four proposed ways to implement this strategy.

6. Summarize the roles of population growth, overconsumption, pollution, and climate change in the extinction of wild species. Explain how concentrations of pesticides such as DDT can accumulate to high levels in food webs. Define **bioaccumulation** and **biomagnification**. List five possible causes of the decline of European honeybee populations in the United States. Give three examples of species that are threatened by poaching. Why are wild tigers likely to disappear within a few decades? What is the connection between infectious diseases in humans and the pet trade?

7. List the major threats to the world's bird populations and give two reasons for protecting bird species from

extinction. Describe the threat to some forms of wildlife from the growing practice of hunting for bushmeat. Summarize environmental scientist Çağan Sekercioğlu's contributions to our understanding of the ecological importance of birds and threats to their extinction.

Section 8.4

8. What is the key concept for this section? Name two international treaties that are used to help protect species. What is the U.S. **Endangered Species Act**? How successful has it been, and why is it controversial?

9. Summarize the roles and limitations of wildlife refuges, seed banks, botanical gardens, zoos,

aquariums, and wildlife farms in protecting some species. Describe the role of captive breeding in efforts to prevent species extinction and give an example of success in returning a nearly extinct species to the wild. What are five important questions related to protecting biodiversity?

10. What are this chapter's *three big ideas*? Explain how preventing the extinction of honeybees and other species implements two of the three **scientific principles of sustainability**.

_____

Note: Key terms are in bold type.

Critical Thinking

1. What are three aspects of your lifestyle that might directly or indirectly contribute to decline in European honeybee populations and the endangerment of other pollinator species (**Core Case Study**)?

2. Give your response to the following statement: "Eventually, all species become extinct. So it does not really matter that the world's remaining tiger species or a tropical forest plant are endangered mostly because of human activities." Be honest about your reaction, and give arguments to support your position.

3. Do you accept the ethical position that each species has the inherent right to survive without human interference, regardless of whether it serves any useful purpose for humans? Explain. Would you extend this right to the *Anopheles* mosquito, which transmits malaria, and to harmful infectious bacteria? Explain. If your answer is no, where would you draw the line?

4. Wildlife ecologist and environmental philosopher Aldo Leopold wrote this with respect to preventing the extinction of wild species: "To keep every cog and wheel is the first precaution of intelligent tinkering." Explain how this statement relates to the material in this chapter.

5. What would you do if red fire ants invaded your yard and house? Explain your reasoning behind your course of action. How might your actions affect other species or the ecosystem you are dealing with?

6. How do you think your daily habits might contribute directly or indirectly to the extinction of some bird species? List three things that you think should be done to reduce the rate of extinction of bird species?

7. Which of the following statements best describes your feelings toward wildlife?
   a. As long as it stays in its space, wildlife is okay.
   b. As long as I do not need its space, wildlife is okay.
   c. I have the right to use wildlife habitat to meet my own needs.
   d. When you have seen one redwood tree, elephant, or some other form of wildlife, you have seen them all, so preserve a few of each species in a zoo or wildlife park and do not worry about protecting the rest.
   e. All wildlife species should be protected in their current ranges.
   f. We should do whatever we can to expand the current ranges of wildlife species wherever possible.

8. How might your lifestyle change if human activities were to contribute to the extinction of 20–50% of the world's identified species during this century? How might this affect the lives of any children or grandchildren you eventually might have? List two aspects of your lifestyle that contribute to this threat to the earth's natural capital.

Doing Environmental Science

Identify examples of habitat destruction or degradation in the area in which you live or go to school. Try to determine and record any harmful effects that these activities have had on the populations of one wild plant and one animal species. (Name each of these species and describe how they have been affected.) Do some research on the Internet and/or in a library on *wildlife management plans*, and then develop a management plan for restoring the habitats and species you have studied. Try to determine whether trade-offs are necessary with regard to the human activities you have observed and account for these trade-offs in your management plan. Compare your plan with those of your classmates.

Global Environment Watch Exercise

Go to your MindTap course to access the GREENR database. Starting on the home page under "Browse Issues and Topics," click on *Environment and Ecology,* then select *Extinction,* and scroll to statistics on the portal page. Click on "Known Causes of Animal Extinction since 1600." You will find four general categories of causes. Thinking about history from 1600 through today, how do you think humans have changed their impact on species in each of these categories? Has the impact increased or decreased over this period? Give specific examples of changes in this timeframe to support your answers.

Data Analysis

Examine the following data released by the World Resources Institute and answer these questions:

1. Complete the table by filling in the last column. For example, to calculate this value for Costa Rica, divide the number of threatened breeding bird species by the total number of known breeding bird species and multiply the answer by 100 to get the percentage.

2. Arrange the countries from largest to smallest according to total land area. Does there appear to be any correlation between the size of country and the percentage of threatened breeding bird species? Explain your reasoning.

Country	Total Land Area in Square Kilometers (Square Miles)	Protected Area as Percent of Total Land Area (2003)	Total Number of Known Breeding Bird Species (1992–2002)	Number of Threatened Breeding Bird Species (2002)	Threatened Breeding Bird Species as Percent of Total Number of Known Breeding Bird Species
Afghanistan	647,668 (250,000)	0.3	181	11	
Cambodia	181,088 (69,900)	23.7	183	19	
China	9,599,445 (3,705,386)	7.8	218	74	
Costa Rica	51,114 (19,730)	23.4	279	13	
Haiti	27,756 (10,714)	0.3	62	14	
India	3,288,570 (1,269,388)	5.2	458	72	
Rwanda	26,344 (10,169)	7.7	200	9	
United States	9,633,915 (3,718,691)	15.8	508	55	

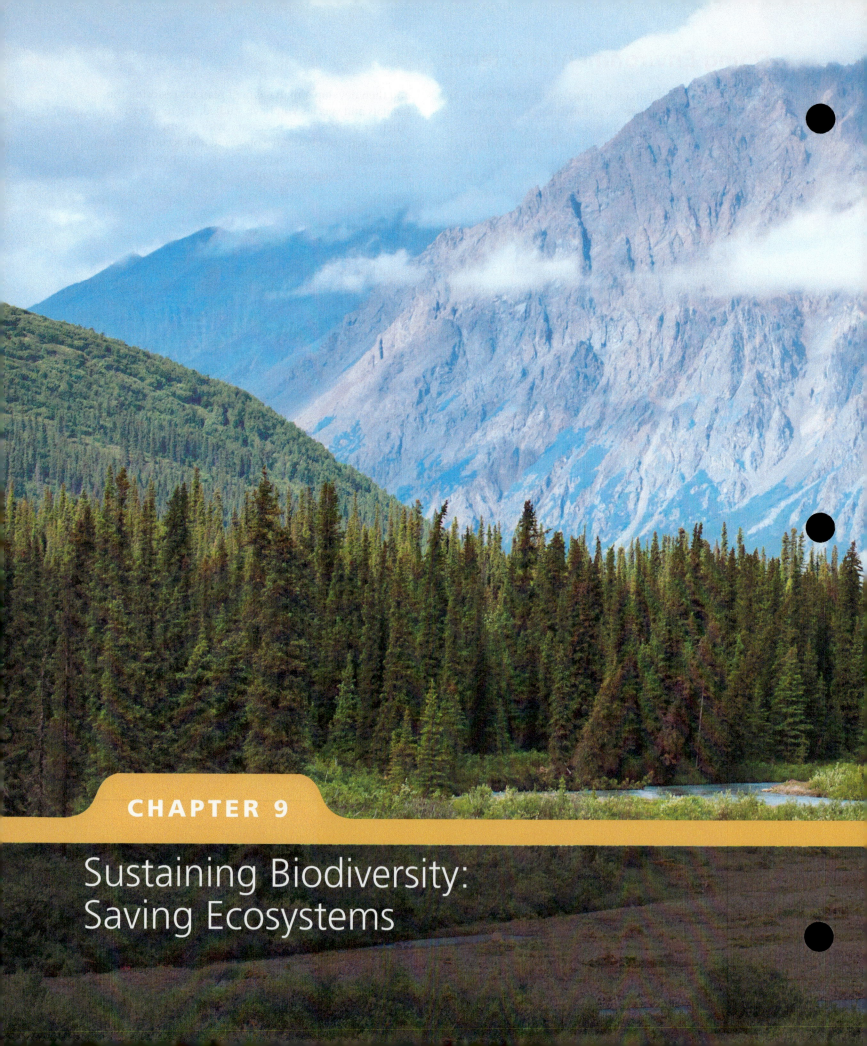

CHAPTER 9

Sustaining Biodiversity:
Saving Ecosystems

There is no solution, I assure you, to save Earth's biodiversity other than preservation of natural environments in reserves large enough to maintain wild populations sustainably.

EDWARD O. WILSON

Key Questions

9.1 What are the major threats to forest ecosystems?

9.2 How can we manage and sustain forests?

9.3 How can we manage and sustain grasslands?

9.4 How can we sustain terrestrial biodiversity?

9.5 How can we sustain aquatic biodiversity?

Denali National Park, Alaska (USA).
Jerryway/Dreamstime.com

Costa Rica—A Global Conservation Leader

Tropical forests once covered Central America's Costa Rica, which is smaller in area than the U.S. state of West Virginia. Between 1963 and 1983, politically powerful ranching families cleared much of the country's forests in order to graze cattle.

Despite such widespread forest loss, tiny Costa Rica is a superpower of biodiversity, with an estimated 500,000 plant and animal species. A single park in Costa Rica is home to more bird species than are found in all of North America.

This oasis of biodiversity results mostly from two factors. One is the country's tropical location. It lies between two oceans and has both coastal and mountainous regions that provide a variety of microclimates and habitats for wildlife. The other is the government's strong commitment to biodiversity conservation.

Costa Rica established a system of nature reserves and national parks (Figure 9.1) in the mid-1970s. By 2016, this system included more than 27% of its land—6% of it reserved for indigenous peoples. Costa Rica has increased its beneficial environmental impact by devoting a larger proportion of its land to biodiversity conservation than any other country—an example of the biodiversity **principle of sustainability**.

To reduce *deforestation*—the clearing and loss of forests—the government eliminated subsidies for converting forestland to grazing lands. Instead, it pays landowners to maintain or restore tree cover.

The strategy has worked. Costa Rica has gone from having one of the world's highest deforestation rates to having one of the lowest. Over three decades, the land area covered by forests grew from 20% to 50%.

Ecologists warn that human population growth, economic development, and poverty put increasing pressure on the earth's ecosystems and on the ecosystem services they provide. According to a recent joint report by two United Nations environmental bodies: "Unless radical and creative action is taken to conserve the earth's biodiversity, many local and regional ecosystems that help to support human lives and livelihoods are at risk of collapsing."

This chapter describes the threats to the earth's terrestrial and aquatic ecosystems and ways to sustain these vital ecosystems and the ecosystem services they provide. ●

FIGURE 9.1 La Fortuna Falls is located in a tropical rain forest in Costa Rica's Arenal Volcano National Park.

gary yim/Shutterstock.com

9.1 WHAT ARE THE MAJOR THREATS TO FOREST ECOSYSTEMS?

CONCEPT 9.1A Forests provide ecosystem services far greater in value than the value of wood and other raw materials they provide.

CONCEPT 9.1B Unsustainable cutting and burning of forests and climate change are the chief threats to forest ecosystems.

Forests Provide Economic and Ecosystem Services

Forests provide highly valuable economic and ecosystem services (Figure 9.2 and **Concept 9.1A**). Forests support biodiversity by providing habitats for about two-thirds of the earth's terrestrial species. They are also home to more than 300 million people. About 1 billion people living in extreme poverty depend on forests for their survival.

Forests also play a role in maintaining human health. Certain chemicals found in tropical forest plants are used as blueprints for making most of the world's prescription drugs (Figure 8.5, p. 173).

Forests have tremendous economic value (Figure 9.2, right). Scientists and economists have estimated the economic value of major ecosystem services that the world's forests and other ecosystems provide (Science Focus 9.1).

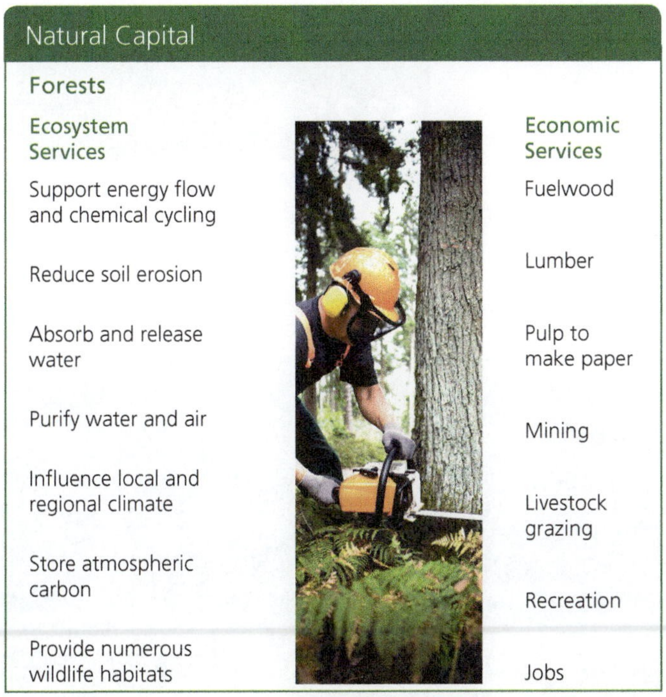

Forests Vary in Age and Structure

Scientists divide the earth's forests into major types, based on their age and structure. *Forest structure* refers to the distribution of vegetation, both horizontally and vertically, as well as the types, sizes, and shapes of that vegetation. An **old-growth forest**, or **primary forest**, is an uncut or regrown forest that has not been seriously disturbed by human activities or natural disasters for 200 years or more (Figure 9.3). Old-growth forests are reservoirs of biodiversity because they provide ecological niches for a multitude of wildlife species (see Figure 7.14, p. 150).

A **second-growth forest** is a stand of trees resulting from secondary ecological succession (see Figure 5.9, p. 97). Such forests develop after the trees in an area have been removed by human activities, such as clear-cutting for timber or conversion to cropland, or by natural forces such as fires and hurricanes.

A **tree plantation** (Figure 9.4) is a managed forest that contains only one or two species of trees that are all of the same age. It is also known as a *tree farm* or *commercial forest*. Old-growth or second-growth forests are often cleared for timber and then replaced by tree plantations. Trees grown in plantations are usually harvested by clear-cutting as soon as they become commercially valuable. The land is then replanted and clear-cut again in a regular cycle (Figure 9.4).

When managed carefully, tree plantations can produce wood at a rapid rate and could supply most of the wood used for industrial purposes, including papermaking and construction. This would help protect the world's remaining old-growth and second-growth forests, as long as they are not cleared to make room for tree plantations.

Natural Capital

Forests

Ecosystem Services	Economic Services
Support energy flow and chemical cycling	Fuelwood
Reduce soil erosion	Lumber
Absorb and release water	Pulp to make paper
Purify water and air	Mining
Influence local and regional climate	Livestock grazing
Store atmospheric carbon	Recreation
Provide numerous wildlife habitats	Jobs

Val Thoerner/Shutterstock.com

FIGURE 9.2 Forests provide many important ecosystem and economic services (**Concept 9.1A**). *Critical thinking:* Which two ecosystem services and which two economic services do you think are the most important?

Aleksander Bolbot/Shutterstock.com

FIGURE 9.3 Old-growth forest in Poland.

Putting a Price Tag on Nature's Ecosystem Services

Currently, forests and other ecosystems are valued mostly for their economic services (Figure 9.2, right). In 2014, a team of ecologists, economists, and geographers, led by ecological economist Robert Costanza, estimated the monetary worth of 17 ecosystem services (Figure 9.2, left) provided by 16 of the earth's biomes. Examples are waste treatment, erosion control, climate regulation, nutrient cycling, food production, and recreation.

Their conservative estimate was that the global monetary value of these services is at least $125 trillion per year—much more than the $73.2 trillion that the entire world spent on goods and services in 2015. This means that every year, the earth provides you and every other person in the world with ecosystem services worth of at least $17,123, on average. On a global basis, the top five ecosystem services

are waste treatment ($22.5 trillion per year), recreation ($20.6 trillion), erosion control ($16.2 trillion), food production ($14.8 trillion), and nutrient cycling ($11.1 trillion).

The researchers also estimated that since 1997, the world has been losing ecosystem services with an estimated value of about $20.2 trillion a year. This annual loss is more than the $18.1 trillion Gross National Product (GNP) of the United States in 2015.

According to their research, the world's forests provide us with ecosystem services worth at least $15.6 trillion per year. This is hundreds of times more than the economic value of lumber, paper, and other wood products that forests provide us. The researchers pointed out that their estimates were very conservative.

We can draw four important conclusions from this and related studies:

(1) the earth's ecosystem services are essential for all humans and their economies; **(2)** the economic value of these services is huge; **(3)** these ecosystem services will be an ongoing source of ecological income, as long as they are used sustainably, and **(4)** we need to use the full-cost pricing **principle of sustainability**, to include the huge economic values of these irreplaceable ecosystem services in the prices of goods and services provided by the earth's ecosystems.

CRITICAL THINKING

Some people believe that we should not try to put economic values on the world's irreplaceable ecosystem services because their value is infinite. Do you agree with this view? Explain. What is the alternative?

The downside of tree plantations is that they contain only one or two tree species. As a result, they are much less biologically diverse and sustainable than old-growth and second-growth forests are. In addition, repeated cycles of cutting and replanting can eventually deplete the nutrients in the area's topsoil. This can hinder the regrowth of any type of forest on such land.

$125 trillion Conservative estimate of the annual value of nature's ecosystem services.

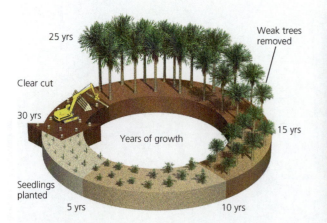

FIGURE 9.4 Oil palm tree plantation. A large area of diverse tropical forest was cleared and planted with this monoculture of oil palm trees.

Ways to Harvest Trees

Because of the immense economic value of forests, the harvesting of wood for timber and to make paper is one of the world's major industries. The first step in harvesting trees is to build roads for access and timber removal. Even carefully designed logging roads can have a number of harmful effects (Figure 9.5), including topsoil erosion, sediment runoff into waterways, habitat loss, and biodiversity loss. Logging roads also increase chances of invasion by disease-causing organisms and nonnative pests, as well as disturbances from human activities like farming and ranching.

Loggers use a variety of methods to harvest trees. With *selective cutting*, loggers cut intermediate-aged and mature trees singly or in small groups, leaving the forest largely intact (Figure 9.6a). However, loggers often remove all the trees from an area in what is called a *clear-cut* (Figures 9.6b and 9.7). Clear-cutting is the most efficient and sometimes the most cost-effective method for harvesting trees. It also provides profits in the shortest time for landowners and timber companies. However, clear-cutting can be harmful to forest ecosystems. Wholesale removal of trees leads to soil erosion, increased sediment pollution of nearby waterways, and decreased biodiversity.

A variation of clear-cutting that produces a more sustainable timber yield without widespread destruction is *strip cutting* (Figure 9.6c). It involves clear-cutting a strip of trees along the contour of the land within a corridor narrow enough to allow the natural forest to grow back within a few years. After one strip grows back, loggers cut another strip next to the first, and so on.

Fires and Forest Ecosystems

Two types of fires can affect forest ecosystems. *Surface fires* (Figure 9.8, left) usually burn only undergrowth and leaf litter on the forest floor. They kill seedlings and small trees, but spare most mature trees and allow most wild animals to escape.

Occasional surface fires have a number of ecological benefits. They:

- burn away flammable material such as dry brush and help prevent fires that are more disruptive;
- free valuable plant nutrients trapped in slowly decomposing litter and undergrowth;
- release seeds from the cones of tree species such as lodgepole pines and stimulate the germination of other seeds such as those of the giant sequoia;
- help control the presence of destructive insects and tree diseases.

Another type of fire, called a *crown fire* (Figure 9.8, right), is an extremely hot fire that leaps from treetop to treetop, burning whole trees. Such fires usually occur in forests that have not experienced surface fires for several decades. This absence of fire allows dead wood, leaves, and other flammable ground litter to build up. These rapidly burning fires can destroy most vegetation, kill wildlife, increase topsoil erosion, and burn or damage buildings and homes.

Almost Half of the World's Old-Growth Forests Have Been Cut Down

Deforestation is the temporary or permanent removal of large expanses of forest for agriculture, settlements, or other uses. Surveys by the World Resources Institute (WRI) indicate that during the past 8,000 years, deforestation has eliminated almost half of the earth's old-growth forest cover. Most of this loss occurred in the past 65 years.

According to the WRI, if current deforestation rates continue, about 40% of the world's remaining intact forests will be logged or converted to other uses within two decades. Clearing large areas of forests, especially old-growth forests, has important short-term economic benefits (Figure 9.2, right column), but it also has a number of harmful environmental effects (Figure 9.9), including severe erosion and loss of topsoil (Figure 9.10).

FIGURE 9.5 Natural Capital Degradation: Building roads into previously inaccessible forests is the first step in harvesting timber, but it also paves the way to fragmentation, destruction, and degradation of forest ecosystems.

a. Selective cutting

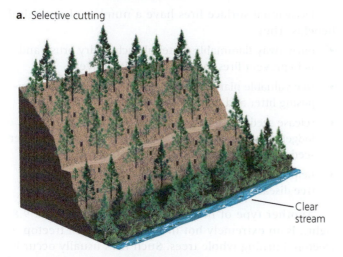

Clear stream

b. Clear-cutting

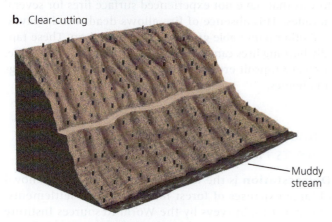

Muddy stream

c. Strip cutting

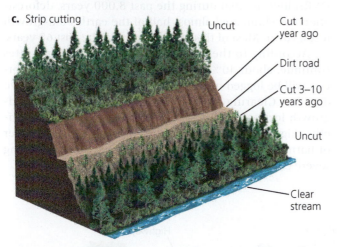

Uncut

Cut 1 year ago

Dirt road

Cut 3–10 years ago

Uncut

Clear stream

FIGURE 9.6 Three major ways to harvest trees. ***Critical thinking:*** If you were cutting trees in a forest you owned, which method would you choose and why?

Forest cover has increased because of the spread of commercial tree plantations in some areas and natural regrowth in other areas. In addition, a global reforestation program sponsored by the United Nations Environment Programme (UNEP) seeks to plant billions of trees throughout much of the world—many of them in tree

FIGURE 9.7 Clear-cut forest.

plantations. China now leads the world in new forest cover, mostly due to its plantations of fast-growing trees. Other countries that have increased their forest cover are Costa Rica (**Core Case Study**), the Philippines, Russia, and the United States (see the following Case Study).

CASE STUDY

Many Cleared Forests in the United States Have Grown Back

Forests cover about 30% of the U.S. land area. They provide habitats for more than 80% of the country's wildlife species and contain about two-thirds of the nation's surface water.

Today, forests in the United States (including tree plantations) cover more area than they did in 1920. The primary reason is that many of the old-growth forests that were cleared or partially cleared between 1620 and 1920 have grown back naturally through secondary ecological succession (Figure 9.11).

There are now diverse second- or third-growth forests in every region of the United States except in much of the West. Environmental writer Bill McKibben has cited this forest regrowth in the United States—especially in the East—as "the great environmental success story of the United States, and in some ways, the whole world." Protected forests make up about 40% of the country's total forest area, mostly in the *National Forest System*, which

FIGURE 9.8 Surface fires (left) usually burn only undergrowth and leaf litter on a forest floor. They can help prevent more destructive crown fires (right) by removing flammable ground material.

Left: David J. Moorhead/The University of Georgia; Right: age fotostock/Alamy Stock Photo

FIGURE 9.9 Deforestation has some harmful environmental effects that can reduce biodiversity and degrade the ecosystem services provided by forests (Figure 9.2, left).

consists of 155 national forests managed by the U.S. Forest Service but owned jointly by the citizens of the United States. However, since the mid-1960s, a large area of the nation's remaining old-growth and diverse second-growth forests has been cut down and replaced with biologically simplified tree plantations.

Tropical Forests Are Disappearing Rapidly

Tropical forests (see Figure 7.13, top, p. 149) cover about 6% of the earth's land area—roughly the area of the continental United States. Climatic and biological data indicate that mature tropical forests once covered at least twice the

area they do now. Most loss of the world's tropical forests has taken place since 1950 (see Chapter 3, Core Case Study, p. 46). According to the Earth Policy Institute and Global Forest Watch, between 2000 and 2014, the world lost the equivalent of more than 50 soccer fields of tropical forest every minute, mainly due to deforestation, and similar rates continue today.

Satellite scans and ground-level surveys indicate that large areas of tropical forests are being cut rapidly in parts of Africa, Southeast Asia, and South America—especially in Brazil's vast Amazon Basin, which has more than 40% of the world's remaining tropical forests. About 20% of the tropical forest in the Amazon Basin has been cleared and nearly that much more has been degraded by human activities.

Currently, tropical forests absorb and store about one-third of the world's terrestrial carbon emissions as part of the carbon cycle. Thus, in reducing these forests, we reduce their absorption of carbon dioxide (CO_2) and contribute to atmospheric warming and climate change. Burning and clearing tropical forests also adds CO_2 to the atmosphere, accounting for 10–15% of global greenhouse gas emissions.

Water evaporating from trees and vegetation in tropical rain forests plays a major role in determining the amount of rainfall there. Removing large areas of trees can lead to drier conditions that dehydrate the topsoil by exposing it to sunlight and allowing it to be blown away. This makes it difficult for a forest to grow back in the area, which is often replaced by tropical grassland or savanna. Scientists project that if current burning and deforestation

FIGURE 9.10 Natural Capital Degradation: Severe soil erosion and desertification caused by clearing an area of tropical forest followed by livestock overgrazing.

rates continue, 20–30% of the Amazon Basin could become savanna by 2080.

Studies indicate that at least half of the world's known species of terrestrial plants, animals, and insects live in tropical forests. Because of their specialized niches (Figure 7.14, p. 150), many of these species are vulnerable to extinction when their forest habitats are destroyed or degraded. The UN Food and Agriculture Organization (FAO) warns that at the current global rate of tropical deforestation, as much as half of the world's remaining old-growth tropical forests will be gone or severely degraded by the end of this century (**Concept 9.1B**).

Tropical deforestation results from a number of underlying and direct causes. Underlying causes, such as pressures from population growth and poverty, push subsistence farmers and the landless poor into tropical forests, where they cut or burn trees for firewood or try to grow enough food to survive. Government subsidies can accelerate other direct causes such as large-scale logging and livestock overgrazing (Figure 9.10) by reducing the costs of these enterprises.

The major direct causes of tropical deforestation vary by location. Tropical forests in the Amazon and other South American countries are cleared or burned primarily for cattle grazing and large soybean plantations (see Figure 1.6, p. 8). In Indonesia, Malaysia, and other areas of Southeast Asia, large oil palm plantations (Figure 9.4) are replacing tropical forests. In Africa, the primary direct cause of deforestation is people clearing plots for small-scale farming and harvesting wood for fuel.

Tropical forest degradation follows a predictable cycle in most cases. First, access roads are into the forest interior (Figure 9.5), often by international timber corporations. Then loggers selectively cut down the largest and best

a. 1620

b. 1920

c. 2000

FIGURE 9.11 In 1620, **(a)** when European settlers were moving to North America, forests covered more than half of the current land area of the continental United States. By 1920, **(b)** most of these forests had been decimated. In 2000, **(c)** secondary and commercial forests covered about a third of U.S. land in the lower 48 states.

trees (Figure 9.6a). When loggers cut one tree, many other trees often fall because of their shallow roots and the network of vines connecting the trees in the forest's canopy.

After trees are harvested, logging companies often sell the land to ranchers who burn any remaining timber to clear the land for cattle grazing. Within a few years, the land typically is overgrazed. Then the ranchers sell the degraded land to settlers for small-scale farming or to farmers who plow it up to plant large crops such as soybeans

(see Figure 1.6, p. 8). After a few years of crop growing and erosion from rain, the topsoil is depleted of nutrients. Then the farmers and settlers move on to newly cleared land to repeat this environmentally destructive process.

CONSIDER THIS ...

> **THINKING ABOUT** Tropical Forests
>
> Why should you care if most of the world's remaining tropical forests are burned, cleared, or converted to savanna within your lifetime? What are three ways in which this might affect your life or the lives of any children and grandchildren that you eventually might have?

9.2 HOW CAN WE MANAGE AND SUSTAIN FORESTS?

CONCEPT 9.2 We can sustain forests by emphasizing the economic value of their ecosystem services, halting government subsidies that hasten their destruction, protecting old-growth forests, harvesting trees no faster than they are replenished, and planting trees to reestablish forests.

Managing Forests More Sustainably

Biodiversity researchers and a growing number of foresters have called for more sustainable forest management (Figure 9.12) (**Concept 9.2**). They recognize that sustaining a forest's ecosystem services is the key to maintaining its economic services over time. For example, loggers could be encouraged or required to make greater use of more sustainable selective cutting (Figure 9.6a) and strip cutting (Figure 9.6c) to harvest tropical trees instead of clear-cutting the forests (Figure 9.6b). To reduce damage to neighboring trees when cutting and removing individual trees, loggers can first cut the canopy vines (lianas) that connect them.

Many economists urge governments to begin making a shift to more sustainable forest management. They recommend phasing out government subsidies and tax breaks that encourage forest degradation and deforestation and replacing them with forest-sustaining economic rewards. Doing so would likely lead to higher prices on unsustainably produced timber and wood products, in keeping with the full-cost pricing **principle of sustainability**. Costa Rica (**Core Case Study**) is taking a lead in using this approach. Governments can also encourage tree-planting programs to help restore degraded forests. **GREEN CAREER: Sustainable forestry**

One important tool is the certification of sustainably grown timber and of sustainably produced forest products. The nonprofit Forest Stewardship Council (FSC) oversees the certification of forestry operations that meet certain

Solutions

More Sustainable Forestry

- Include ecosystem services of forests in estimates of their economic value
- Identify and protect highly diverse forest areas
- Stop logging in old-growth forests
- Stop clear-cutting on steep slopes
- Reduce road-building in forests and rely more on selective and strip cutting
- Leave most standing dead trees and larger fallen trees for wildlife habitat and nutrient cycling
- Put tree plantations only on deforested and degraded land
- Certify timber grown by sustainable methods

FIGURE 9.12 Ways to grow and harvest trees more sustainably (**Concept 9.2**). ***Critical thinking:*** Which three of these methods of more sustainable forestry do you think are the best methods? Why?

sustainable forest standards. To become certified, operators must demonstrate that they do not cut trees at a rate that exceeds long-term forest regeneration in a given area. They must maintain roads and use harvesting systems in ways that limit ecological damage. They must also prevent unreasonable damage to topsoil and leave in place some downed wood and standing dead trees to provide wildlife habitat. The paper used in this book was produced with the use of sustainably grown timber, as certified by the FSC, and contains recycled paper fibers.

Improving Management of Forest Fires

The U.S. Forest Service and the National Advertising Council launched the Smokey Bear campaign in the 1940s to educate the public about the dangers of forest fires. The campaign has helped prevent numerous forest fires, saved many lives, and prevented billions of dollars in losses of trees, wildlife, and human-built structures. However, the campaign also convinced much of the public that all forest fires are bad and should be prevented or put out. This is not the case.

In fact, ecologists warn that trying to prevent all forest fires can actually make forests more vulnerable to fires. Because such efforts can lead to the accumulation of highly flammable underbrush, they can raise the likelihood of destructive crown fires (Figure 9.8, right).

Ecologists and forest fire experts recommend several strategies for limiting the harmful effects of forest fires:

- Use carefully planned and controlled fires, called *prescribed burns,* to remove flammable small trees and underbrush in the highest-risk forest areas.
- Allow some fires on public lands to burn underbrush and smaller trees, as long as the fires do not threaten human structures or human lives.
- Protect houses and other buildings in fire-prone areas by thinning trees and other vegetation in a zone around them and eliminating the use of highly flammable construction materials such as wood shingles.

Reducing the Demand for Harvested Trees

According to the Worldwatch Institute and to forestry analysts, *up to 60% of the wood consumed in the United States is wasted unnecessarily.* This waste results from inefficient use of construction materials, excessive packaging, overuse of junk mail, inadequate paper recycling, and the failure to reuse or find substitutes for wooden shipping containers.

One way to reduce demand for harvested trees is to produce tree-free paper. For example, China uses rice straw and other agricultural residues to make some of its paper. Most of the small amount of tree-free paper produced in the United States is made from the fibers of a rapidly growing woody annual plant called *kenaf.* Kenaf and other non-tree fiber sources yield more paper pulp per area of land than tree farms do and can be grown using fewer pesticides and herbicides. Another way to reduce the demand for tree cutting is to reduce the use of throwaway paper products made from trees. Reusable plates and cups, cloth napkins and handkerchiefs, and cloth bags can replace such products.

More than 2 billion people in less-developed countries use fuelwood (see Figure 6.11, p. 116) and charcoal made from wood for heating and cooking. Most of these countries are experiencing fuelwood shortages because people have been harvesting trees for fuelwood and other forest products 10–20 times faster than new trees are being planted.

For example, Haiti, a country with 11 million people, was once a tropical paradise, 60% of it covered with forests. Now it is an ecological and economic disaster. Largely because its trees were cut for fuelwood and to make charcoal, less than 2% of its land is now covered with trees (Figure 9.13). With the trees gone, soils have eroded away in many areas, making it much more difficult to grow crops.

The U.S. Agency for International Development funded the planting of 60 million trees over more than two decades in Haiti. However, local people cut most of them down for firewood and to make charcoal before they could grow into mature trees. This unsustainable use of natural capital and failure to follow the biodiversity and chemical cycling **principles of sustainability** have played a role in the downward spiral of environmental degradation, poverty, disease, social injustice, crime, and violence in Haiti.

FIGURE 9.13 Natural Capital Degradation: Haiti's deforested brown landscape (left) contrasts sharply with the heavily forested green landscape of its neighboring country, the Dominican Republic.

JAMES P. BLAIR/National Geographic Creative

One way to reduce the severity of the fuelwood crisis in less-developed countries is to establish small plantations of fast-growing fuelwood trees and shrubs around farms and in community woodlots. Another option is to use stoves that burn renewable biomass, such as sun-dried roots of various gourds and squash plants, or methane produced from crop and animal wastes. Doing this would also reduce the large number of deaths caused by indoor air pollution from open fires and poorly designed stoves.

Reducing Tropical Deforestation

Analysts have suggested various ways to protect tropical forests and use them more sustainably (Figure 9.14).

At the international level, *debt-for-nature swaps* can make it financially attractive for countries to protect their tropical forests and use them more sustainably. Under the terms of such swaps, participating countries agree to set aside and protect forest reserves in return for foreign aid or debt relief. In a similar strategy, called *conservation concessions*, governments or private conservation organizations pay nations for agreeing to preserve their natural resources such as forests.

National governments can take steps to reduce deforestation (**Core Case Study**). Between 2005 and 2013, Brazil cut its deforestation rate by 80% by cracking down on illegal logging and setting aside a large conservation reserve in the Amazon Basin. Governments can also end subsidies that fund the construction of logging roads and instead subsidize sustainable forestry and tree planting programs.

The late Wangari Maathai, the first environmentalist to be awarded the Nobel Peace Prize, promoted tree planting

Sustaining Tropical Forests

Prevention

Protect the most diverse and endangered areas

Educate settlers about sustainable agriculture and forestry

Subsidize only sustainable forest use

Protect forests through debt-for-nature swaps and conservation concessions

Certify sustainably grown timber

Reduce poverty and slow population growth

Sustainably grown timber

Restoration

Encourage regrowth through secondary succession

Rehabilitate degraded areas

Concentrate farming and ranching in already-cleared areas

FIGURE 9.14 Ways to protect tropical forests and to use them more sustainably (**Concept 9.2**). *Critical thinking:* Which three of these solutions do you think are the best ones? Why?

Top: STILLFX/Shutterstock.com. Center: Manfred Mielke/USDA Forest Service Bugwood.org

in her native country of Kenya and throughout the world in what became the Green Belt Movement. Her efforts inspired the UNEP to implement a global effort to plant at least 1 billion trees a year beginning in 2006. By 2012, the year Maathai died, about 12.6 billion trees had been planted in 193 countries. [GOOD NEWS]

Consumers can reduce the demand for unsustainable and illegal logging in tropical forests. They can choose to buy only wood and wood products that have been certified as sustainably produced by the FSC and other organizations, including the Rainforest Alliance and the Sustainability Action Network.

People who live in tropical forests, many of them poor farmers trying to feed their families, are looking for ways to grow the food they need without having to cut and burn trees, and several organizations are assisting them. In 1997, Florence Reed founded Sustainable Harvest International, a nonprofit organization dedicated to helping poor farmers learn how to grow nutritious food on the same land year after year and to raise their incomes without having to clear and burn more forests. By 2014, it had helped more than 10,000 people in 100 or more farming communities learn how to grow crops more sustainably, have a more nutritious diet, and increase their average income from around $475 to $5,000 per year. [GOOD NEWS]

9.3 HOW CAN WE MANAGE AND SUSTAIN GRASSLANDS?

CONCEPT 9.3 We can sustain the productivity of grasslands by controlling the numbers and distribution of grazing livestock and by restoring degraded grasslands.

Some Grasslands Are Overgrazed

Grasslands cover about one-fourth of the earth's landmass. They provide many important ecosystem services, including soil formation, erosion control, chemical cycling, storage of atmospheric CO_2 in biomass, and maintenance of biodiversity.

After forests, grasslands are the ecosystems most widely used and altered by human activities. Only about 5% of the original grasslands in the U.S. remain. Most of the other 95% have been converted to cropland. **Rangelands** are unfenced grasslands in temperate and tropical climates that supply *forage,* or vegetation for grazing (grass-eating) and browsing (shrub-eating) animals. Livestock also graze in **pastures**, which are managed grasslands or fenced meadows often planted with domesticated grasses or other forage crops such as alfalfa and clover.

Cattle, sheep, and goats graze on about 42% of the world's natural grasslands. This could increase to 70% by the end of this century according to the UN Millennium Assessment—a four-year study by 1,360 experts from 95 countries.

Blades of rangeland grass grow from the base, not at the tip as broadleaf plants do. As long as only the upper portion of the blade is eaten and the lower portion remains, rangeland grass is a renewable resource that can be grazed repeatedly. Moderate levels of grazing are healthy for grasslands, because removal of mature vegetation stimulates rapid regrowth and encourages greater plant diversity.

Overgrazing occurs when too many animals graze for too long, damaging or killing the grasses (Figure 9.15, left) and exceeding the area's carrying capacity for grazing. Overgrazing reduces grass cover, exposes the topsoil to erosion by water and wind, and compacts the soil, which reduces its capacity to hold water. Overgrazing also promotes the invasion of rangeland by species such as sagebrush, mesquite, cactus, and cheatgrass, which cattle will not eat. The FAO has estimated that overgrazing by livestock has reduced productivity on as much as 20% of the world's rangeland.

Managing Rangelands More Sustainably

Managing rangelands more sustainably and preventing overgrazing typically involve controlling how many animals are allowed to graze in a given area and for how long (**Concept 9.3**). One way to prevent overgrazing is to use *rotational grazing,* in which small groups of cattle are confined by portable fencing to one area for a few days and then moved to a new location. Another way to discourage unwanted

FIGURE 9.15 Natural Capital Degradation: To the left of the fence is overgrazed rangeland. The land to the right of the fence is lightly grazed.

USDA, Natural Resources Conservation Service

vegetation in some areas is through controlled, short-term trampling by large numbers of livestock such as sheep, goats, and cattle that destroy the invasive plants' root systems.

Cattle prefer to graze around natural water sources, especially along streams or rivers lined by strips of vegetation known as *riparian zones*. Overgrazing can destroy the vegetation in riparian zones (Figure 9.16, left). Ranchers can protect such land through rotational grazing and by fencing off damaged areas, which leads to eventual restoration through natural ecological succession (Figure 9.16, right).

9.4 HOW CAN WE SUSTAIN TERRESTRIAL BIODIVERSITY?

CONCEPT 9.4A We can establish and protect wilderness areas, parks, and nature preserves.

CONCEPT 9.4B We can identify and protect biological hotspots that are highly threatened centers of biodiversity.

CONCEPT 9.4C We can protect important ecosystem services, restore damaged ecosystems, and share areas that we dominate with other species.

Strategies for Sustaining Terrestrial Biodiversity

Since the 1960s, a number of strategies have been used to help sustain biodiversity, including:

- Protecting species from extinction, as discussed in Chapter 8
- Setting aside wilderness areas that are protected from harmful human activities
- Establishing parks and nature reserves in which people and nature can interact with some restrictions
- Identifying and protecting *biodiversity hotspots* that contain a high diversity of species, which are under severe threat of extinction from human activities

FIGURE 9.16 Natural Capital Restoration: In the mid-1980s, cattle had degraded the vegetation and soil on this stream bank along the San Pedro River in the U.S. state of Arizona (left). Within 10 years, the area was restored through secondary ecological succession (right) after grazing and off-road vehicle use were banned (**Concept 9.3**).

Left: U.S. Bureau of Land Management; Right: U.S. Bureau of Land Management

- Shifting new development to lands that have already been cleared or degraded

- Avoiding the destruction of forests and grasslands by increasing crop productivity on existing cropland

- Protecting important ecosystem services

- Rehabilitating and partially restoring damaged ecosystems

- Sharing areas that we dominate with other species

According to most ecologists and conservation biologists, the best way to preserve terrestrial biodiversity is to create a worldwide network of areas that are strictly or partially protected from harmful human activities. Currently, less than 13% of the earth's land area (excluding Antarctica) is protected to some degree in more than 177,000 wilderness area, parks, nature reserves, and wildlife refuges. However, no more than 6% of the earth's land is strictly protected from potentially harmful human activities. In other words, we have reserved 94% of the earth's land for human use. According to biodiversity expert Edward O. Wilson, unless we learn more about its importance for biodiversity and act quickly to protect much more of it, we will lose many of the world's species and the vital ecosystem services they provide within this century—an irreversible loss of natural capital.

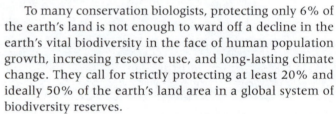

6% Percentage of the earth's land that is strictly protected from potentially harmful human activities.

To many conservation biologists, protecting only 6% of the earth's land is not enough to ward off a decline in the earth's vital biodiversity in the face of human population growth, increasing resource use, and long-lasting climate change. They call for strictly protecting at least 20% and ideally 50% of the earth's land area in a global system of biodiversity reserves.

Many developers and resource extractors oppose protecting even 13% of the earth's land (the current amount), arguing that these protected areas might contain valuable resources that would provide economic benefits. In contrast, ecologists and conservation biologists view protected areas as islands of biodiversity and ecosystem services that help sustain all life and economies indefinitely and that serve as centers of future evolution.

Establishing Wilderness Areas

One way to protect existing wildlands from human exploitation is to designate them as **wilderness**—areas essentially undisturbed by humans and protected by federal law from harmful human activities (**Concept 9.4a**). Theodore Roosevelt (see Figure 1.15, p. 19), the first U.S. president to set aside protected areas, summarized his thoughts on what to do with wilderness: "Leave it as it is. You cannot improve it."

Most developers and resource extractors oppose establishing protected wilderness areas because they contain resources that could provide short-term economic benefits. Ecologists and conservation biologists take a longer view. To them, wilderness areas are islands of biodiversity and ecosystem services needed to support life and human economies now and in the future and to serve as centers

FIGURE 9.17 Diablo Lake in the wilderness area of North Cascades National Park in the U.S. state of Washington.

tusharkoley/Shutterstock.com

for future evolution in response to changes in environmental conditions such as climate change.

In 1964, the U.S. Congress passed the Wilderness Act, which allowed the government to protect undeveloped tracts of U.S. public land as part of the National Wilderness Preservation System (Figure 9.17). The country's area of protected wilderness grew by nearly 12-fold between 1964 and 2015. Even so, only 5% of all U.S. land is protected as wilderness—more than 54% of it in Alaska. Only about 2.7% of the land of the lower 48 states is protected as wilderness, most of it in the West.

Establishing Parks and Other Nature Reserves

According to the International Union for the Conservation of Nature (IUCN), there are more than 6,600 major national parks located in more than 120 countries (see chapter-opening photo). However, most of these parks are too small to sustain many large animal species and many are "paper parks" that receive little protection, especially in less-developed countries. Many parks also suffer from invasions by harmful nonnative species that out-compete and reduce the populations of native species. Some national parks are so popular that large numbers of visitors are degrading the natural features that make them attractive (see the Case Study that follows).

Stresses on U.S. Public Parks

The U.S. National Park System, established in 1912, includes 59 major national parks, sometimes called the country's crown jewels (see chapter-opening photo). The U.S. national park system also has 353 monuments, recreational areas, battlefields, and historic sites. States, counties, and cities also operate public parks.

Popularity threatens many U.S. national parks. Between 1960 and 2015, the number of recreational visitors to U.S. national parks more than tripled, reaching about 307 million. In some U.S. parks and on other public lands, dirt bikes, dune buggies, jet skis, snowmobiles, and other off-road vehicles have become a problem. These recreational vehicles destroy or damage fragile vegetation, disturb wildlife, and degrade the experience for many other visitors.

Many U.S. national parks have become threatened islands of biodiversity surrounded by commercial development. The parks' wildlife and recreational values are threatened by nearby activities such as mining, logging, livestock grazing, water diversion, operation of coal-fired power plants, and urban development. The National Park Service reports that air pollution, mainly caused by coal-fired power plants and dense vehicle traffic, impairs scenic views more than 90% of the time in many of its parks.

A number of parks also suffer damage from the migration or deliberate introduction of harmful nonnative species. European wild boars, imported into the state of North Carolina in 1912 for hunting, now threaten vegetation in parts of the popular Great Smoky Mountains National Park. Nonnative mountain goats in Washington State's Olympic National Park trample and destroy the root systems of native vegetation and accelerate soil erosion.

Native species—some of which are threatened or endangered—are hunted or illegally removed from almost half of all U.S. national parks. However, the endangered gray wolf, a keystone species, was successfully reintroduced into Yellowstone National Park after a 50-year absence (Science Focus 9.2).

Designing and Managing Nature Reserves

In establishing nature reserves, the size and design of the reserve is important. Research indicates that large nature reserves typically sustain more species and provide greater habitat diversity than do small reserves. Research also indicates that in some areas, several well-placed medium-size reserves may better protect a variety of habitats and sustain more biodiversity than a single large reserve can.

Establishing protected *habitat corridors* between isolated reserves can benefit species and allow migration by vertebrates that need large ranges. Corridors also allow some species to move to areas that are more favorable if climate change alters their existing areas.

On the other hand, corridors can threaten isolated populations by allowing movement of fire, disease, pests, and invasive species between reserves. They can also expose migrating species to natural predators, human hunters, and pollution. Some research suggests that the benefits of corridors outweigh their potential harmful effects, especially as the climate changes.

Conservation biologists suggest using the *buffer zone concept* whenever possible in designing and managing nature reserves. Establishing a buffer zone means strictly protecting the inner core of a reserve, usually by establishing one or more outer buffer zones in which local people can extract resources sustainably without harming the inner core (see Case Study that follows). By 2015, the United Nations had used this concept to create a global network of 669 *biosphere reserves* in 120 countries. However, most biosphere reserves fall short of these design ideals and receive too little funding for their protection and management.

CASE STUDY

Identifying and Protecting Biodiversity in Costa Rica

For several decades, Costa Rica (**Core Case Study**) has been using government and private research agencies to identify the plants and animals that make it one of the world's most biologically diverse countries (Figure 9.18). The government has consolidated the country's parks and reserves into several large conservation areas, or *megareserves*, designed with the goal of sustaining about 80% of the country's biodiversity (Figure 9.19).

Each reserve contains a protected inner core surrounded by two buffer zones that local and indigenous people can use for sustainable logging, crop farming, cattle grazing, hunting, fishing, and ecotourism. Instead of shutting local people out of reserve areas, this approach enlists local people as partners in protecting a reserve from activities such as illegal logging and poaching. It is an application of the biodiversity and win–win **principles of sustainability**.

In addition to its ecological benefits, this strategy has paid off financially. Today, Costa Rica's largest source of income is its $3-billion-a-year tourism industry, almost two-thirds of which involves ecotourism.

There are some potential threats to Costa Rica's conservation efforts. One is the clearing of forests to grow pineapples in plantations for export to China. Ecotourism helps fund Costa Rica's parks and conservation efforts and reduces exploitation of conservation areas by providing income for local people in visited areas, but excessive numbers of ecotourists can degrade sensitive areas.

The Ecosystem Approach for Sustaining Terrestrial Biodiversity: A Five-Point Plan

Most wildlife biologists and conservationists believe that the best way to keep from hastening the extinction of wild species through human activities is to protect

Reintroducing the Gray Wolf to Yellowstone National Park

FIGURE 9.A After becoming almost extinct in much of the western United States, the *gray wolf* was listed and protected as an endangered species in 1974.

In the 1800s, at least 350,000 gray wolves (Figure 9.A) roamed over 75% of America's lower 48 states, especially in the West. They survived mostly by preying on abundant bison, elk, caribou, and deer. Between 1850 and 1900, most of them were shot, trapped, or poisoned by ranchers, hunters, and government employees. This drove the gray wolf to near extinction in the lower 48 states.

Ecologists recognize the important roles that this keystone predator species once played in the Yellowstone National Park region. The wolves culled herds of bison, elk, moose, and mule deer, and kept down coyote populations. By leaving some of their kills partially uneaten, they provided meat for scavengers such as ravens, bald eagles, ermines, grizzly bears, and foxes.

When the number of gray wolves declined, herds of plant-browsing elk, moose, and mule deer expanded and over-browsed the willow and aspen trees growing near streams and rivers. This led to increased soil erosion and declining populations of other wildlife species such as beaver, which eat willow and aspen. This in turn affected species that depended on wetlands created by dam-building beavers.

In 1974, the gray wolf was listed as an endangered species in the lower 48 states. In 1987, the U.S. Fish and Wildlife Service (USFWS) proposed reintroducing gray wolves into the Yellowstone National Park to try to help stabilize the ecosystem. The proposal brought angry protests from area ranchers who feared the wolves would leave the park and attack their cattle and sheep and from hunters who feared the wolves would kill too many big-game animals. Mining and logging companies also objected, fearing that the government would halt their operations on wolf-populated federal lands.

In 1996, USFWS officials captured 41 gray wolves in Canada and northwest Montana and relocated them in Yellowstone National Park. Scientists estimate that the long-term carrying capacity of the park is 110 to 150 gray wolves. In 2016, the park had 99 wolves in 10 packs.

Scientists have been using radio collars to track some of the wolves and study the ecological effects of reintroducing the wolves. Their research indicates that the return of this keystone predator has decreased populations of elk, the wolves' primary food source.

The leftovers of elk killed by wolves have again become an important food source for scavengers.

However, a study led by U.S. Geological Survey scientist Matthew Kauffman indicated that the aspen were not recovering despite a 60% decline in elk numbers. Declining populations of elk were also supposed to allow for the return of willow trees along streams. Research indicates that willows have only partly recovered.

The wolves have cut in half the Yellowstone population of coyotes—the top predators in the absence of wolves. This has reduced coyote attacks on cattle from area ranches and has led to larger populations of small animals such as ground squirrels, mice, and gophers, which are hunted by coyotes, eagles, and hawks.

Overall, this experiment has had some important ecological benefits for the Yellowstone ecosystem but more research is needed. The focus has been on the gray wolf, but other factors, such as drought and the rise of bear and cougar populations, may play a role in the observed ecological changes and need to be examined.

The wolf reintroduction has also produced economic benefits for the region. One of the main attractions of the park for many visitors is the hope of spotting wolves chasing their prey across its vast meadows.

CRITICAL THINKING

If the gray wolf population in the park were to reach its estimated carrying capacity of 110 to 150 wolves, would you support a program to kill wolves to maintain this population level? Explain. Can you think of other alternatives?

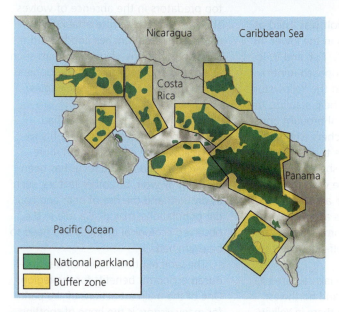

FIGURE 9.18 This scarlet macaw parrot is one of more than half a million species found in Costa Rica.

FIGURE 9.19 Solutions: Costa Rica has created several *megareserves*. Green areas are protected natural parklands and yellow areas are the surrounding buffer zones.

National parkland

Buffer zone

threatened habitats and their ecosystem services. This *ecosystem approach* would generally employ the following five-point plan:

1. Map the world's terrestrial ecosystems and create an inventory of the species contained in each of them, along with the ecosystem services they provide.

2. Identify terrestrial ecosystems that are resilient and can recover if not overwhelmed by harmful human activities, along with ecosystems that are fragile and need protection.

3. Protect the most endangered terrestrial ecosystems and species, with emphasis on protecting plant biodiversity and ecosystem services.

4. Restore as many degraded ecosystems as possible.

5. Make development *biodiversity-friendly* by providing significant financial incentives (such as tax breaks and subsidies) and technical help to private landowners who agree to help protect endangered ecosystems.

Protecting Biodiversity Hotspots and Ecosystem Services

The ecosystem approach calls for identifying and taking emergency action to protect the earth's **biodiversity hotspots**. They are areas rich in highly endangered species found nowhere else and threatened by human activities (**Concept 9.4b**). These areas have suffered serious ecological disruption, mainly due to rapid human population growth and the resulting pressure on natural resources and ecosystem services.

Figure 9.20 shows 34 terrestrial biodiversity hotspots biologists have identified. According to the IUCN, these areas cover only about 2% of the earth's land surface but are home for the majority of the world's endangered or critically endangered species, as well as for 1.2 billion people. However, only about 5% of the total area of these hotspots is protected with government funding and law enforcement.

Another way to sustain the earth's biodiversity is to identify and protect areas where vital ecosystem services (see the orange boxed labels in Figure 1.3, p. 7) are being impaired. Scientists call for identifying highly stressed areas with high poverty levels where most people heavily depend on ecosystem services for survival.

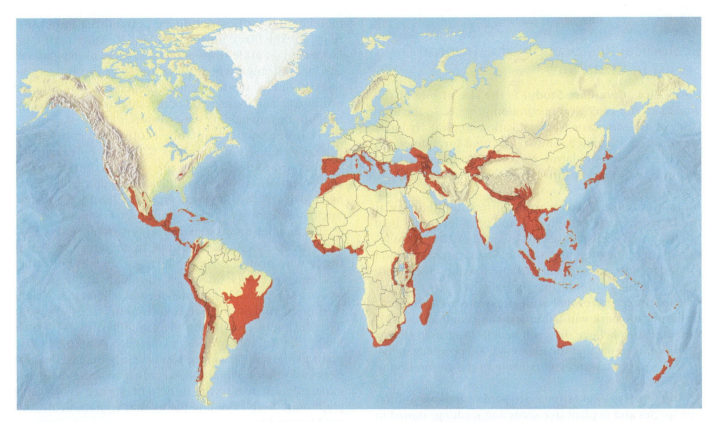

FIGURE 9.20 *Endangered natural capital:* Biologists have identified these 34 biodiversity hotspots. Compare this map with the global map of the human ecological footprint, shown in Figure 1.9 (p. 11). ***Critical thinking:*** Why do you think so many hotspots are located near coastal areas?

Compiled by the authors using data from the Center for Applied Biodiversity Science at Conservation International.

Restoring Damaged Ecosystems

Almost every natural place on the earth has been impacted to some degree by human activities, most often in harmful ways. We can partially reverse much of the damage through **ecological restoration**: the process of repairing damage to ecosystems caused by human activities. Examples include replanting forests (see the Case Study that follows), reintroducing keystone native species (Science Focus 9.2), removing harmful invasive species, freeing river flows by removing dams, and restoring grasslands, coral reefs, wetlands, and stream banks (Figure 9.16, right). This is an important way to expand our beneficial environmental impact.

By studying how natural ecosystems recover through ecological succession, scientists have found ways to restore degraded ecosystems, including the following four:

- *Restoration:* returning a degraded habitat or ecosystem to a condition as similar as possible to its original one.

- *Rehabilitation:* turning a degraded ecosystem into a functional or useful ecosystem without trying to restore it to its original condition. Examples include

removing pollutants from abandoned industrial sites and replanting trees to reduce soil erosion in clear-cut forests.

- *Replacement:* replacing a degraded ecosystem with another type of ecosystem. For example, a degraded forest could be replaced by a productive pasture or tree plantation.

- *Creating artificial ecosystems:* for example, artificial wetlands have been created in some areas to help reduce flooding and to treat sewage.

Researchers have suggested the following four-step strategy for carrying out most forms of ecological restoration and rehabilitation.

1. Identify the causes of the degradation, such as pollution, farming, overgrazing, mining, or invasive species.

2. Stop the degradation by eliminating or sharply reducing these factors.

3. Reintroduce keystone species to help restore natural ecological processes, as was done with gray wolves in the Yellowstone ecosystem (Science Focus 9.2).

4. Protect the area from further degradation to allow natural recovery (Figure 9.16, right).

Ecological Restoration of a Tropical Dry Forest in Costa Rica

Costa Rica (**Core Case Study**) is the site of one of the world's largest ecological restoration projects. In the lowlands of its Guanacaste National Park, a tropical dry forest was burned, degraded, and fragmented for conversion to cattle ranches and farms. Now it is being restored and reconnected to a rain forest on nearby mountain slopes. **[GOOD NEWS]**

Daniel Janzen, professor of conservation biology at the University of Pennsylvania and a leader in the field of restoration ecology, used his own MacArthur Foundation grant money to purchase the Guanacaste forestland for designation as a national park. He has also raised more than $10 million for restoring the park.

Janzen recognizes that ecological restoration and protection of the park will fail unless the people in the surrounding area believe they will benefit from such efforts. His vision is to see that the nearly 40,000 people who live near the park play an essential role in the restoration of the forest.

In the park, local farmers are paid to remove nonnative species and to plant tree seeds and seedlings started in Janzen's laboratory. Local grade school, high school, and university students and citizens' groups study the park's ecology during field trips. The park's location near the Pan American Highway makes it an ideal area for ecotourism, which stimulates the local economy.

This project also serves as a training ground in tropical forest restoration for scientists from around the world. Research scientists working on the project give guest classroom lectures and lead field trips. Janzen believes that education, awareness, and involvement—not guards and fences—are the best ways to protect largely intact ecosystems from unsustainable use. This is an application of the biodiversity and win–win **principles of sustainability**.

Sharing Ecosystems with Other Species

Humans dominate most of the world's ecosystems, which is a cause of species extinction and loss of ecosystem services. Ecologist Michael L. Rosenzweig calls for humans to share some of the spaces they dominate with other species. He calls this approach **reconciliation ecology**. It focuses on establishing and maintaining new habitats to conserve species diversity in places where people live, work, or play.

By encouraging sustainable forms of ecotourism, people can protect local wildlife and ecosystems and provide economic resources for their communities. In the Central American country of Belize, for instance, conservation biologist Robert Horwich helped establish a local sanctuary for the black howler monkey. He convinced local farmers to set aside strips of forest to serve as habitats and corridors through which these monkeys can travel. The reserve, run by a local women's cooperative, has attracted ecotourists and biologists. Local residents receive income for housing and guiding these visitors.

Without proper controls, ecotourism can lead to degradation of popular sites if they are overrun by visitors or are degraded by the construction of nearby hotels and other tourist facilities. However, when managed properly, ecotourism can be a useful form of reconciliation ecology.

Reconciliation ecology is also a way to protect vital ecosystem services. For example, some people are learning how to protect insect pollinators, such as butterflies and honeybees (see Chapter 8, Core Case Study, p. 168), which are vulnerable to pesticides and habitat loss. Neighborhoods and municipal governments are doing this by reducing or eliminating the use of pesticides on their lawns, fields, golf courses, and parks. People can also plant gardens of flowering plants as a source of food for bees and other pollinators. According to some honeybee experts, gardeners trying to help bees must avoid using glyphosate herbicides and plants that contain neonicotinoid insecticides.

People have also worked together to protect bluebirds within human-dominated habitats. In such areas, bluebird populations have declined because most of their nesting trees have been cut down. Specially designed boxes have provided artificial nesting places for bluebirds. Their widespread use has allowed populations of this species to grow. Figure 9.21 lists some **[GOOD NEWS]** ways in which you can help sustain the earth's terrestrial biodiversity.

What Can You Do?

Sustaining Terrestrial Biodiversity

- Plant trees and take care of them
- Recycle paper and buy recycled paper products
- Buy sustainably produced wood and wood products and wood substitutes such as recycled plastic furniture and decking
- Help restore a degraded forest or grassland
- Landscape your yard with a diversity of native plants

FIGURE 9.21 Individuals Matter: Ways to help sustain terrestrial biodiversity. **Critical thinking:** Which two of these actions do you think are the most important ones to take? Why? Which of these things do you already do?

9.5 HOW CAN WE SUSTAIN AQUATIC BIODIVERSITY?

CONCEPT 9.5A Aquatic species and the ecosystem and economic services they provide are threatened by habitat loss, invasive species, pollution, climate change, and overexploitation.

CONCEPT 9.5B We can help sustain aquatic biodiversity and increase our beneficial environmental impact by establishing protected sanctuaries, managing coastal development, reducing water pollution, and preventing overfishing.

Human Activities Are Threatening Aquatic Biodiversity

Human activities have destroyed or degraded a large portion of the world's coastal wetlands, coral reefs, mangroves, and even the ocean floor. They also have disrupted many of the world's freshwater ecosystems.

Sea-bottom habitats are being degraded and destroyed by impacts from dredging operations and trawler fishing boats. Like giant submerged bulldozers, thousands of trawler fishing boats drag huge nets weighted down with chains and steel plates over the ocean floor to harvest a few species of bottom fish and shellfish (Figure 9.22). Each year, thousands of trawlers scrape and disturb an area of ocean floor many times larger than the annual

global total area of forests that are clear-cut. Some marine scientists call bottom trawling the largest human-caused disturbance in the biosphere.

During this century, rising sea levels, mainly caused by projected climate change, are likely to destroy many coral reefs (see Science Focus 7.4, p. 156) and flood some low-lying islands along with their protective coastal mangrove forests.

Ocean waters are growing warmer, mostly due to heat absorbed from a warmer atmosphere. A WRI study estimated that 75% of the world's shallow coral reefs are at risk of being destroyed, mainly by a combination of warmer waters, overfishing, pollution, and ocean acidification (Science Focus 9.3). The latter decreases the carbonate ions in ocean water that corals need to build their calcium carbonate skeletons. Today, coral reefs, on average, are exposed to the warmest and most acidic ocean waters of the past 400,000 years.

Habitat disruption is also a problem in freshwater aquatic zones. The main causes are the building of dams and excessive water withdrawal from rivers for irrigation and urban water supplies. These activities destroy aquatic habitats, degrade water flows, and disrupt freshwater biodiversity. Globally, the extinction rate for freshwater species is five times the rate for terrestrial species according to the latest IUCN Red List.

Another problem that threatens aquatic biodiversity is the deliberate or accidental introduction of hundreds of

FIGURE 9.22 Natural Capital Degradation: An area of ocean bottom before (left) and after (right) a trawler net scraped it. *Critical thinking:* What land activities are comparable to this?

Ocean Acidification: The Other CO_2 Problem

By burning an increasingly large amount of carbon-containing fossil fuels, especially since 1950, we have added CO_2 to the lower atmosphere faster than it can be removed by the carbon cycle (see Figure 3.16, p. 59). At least 90% of the world's climate scientists agree that this increase in CO_2 levels is raising the temperature of the earth's lower atmosphere and changing the earth's climate. Extensive research indicates that if we continue to increase CO_2 levels in the atmosphere, it will likely lead to severe disruption the earth's climate during this century, as we discuss in Chapter 15.

Another serious environmental problem related to CO_2 emissions is **ocean acidification**. The oceans have helped reduce atmospheric warming and climate change by absorbing about one-fourth of the excess CO_2 that human activities have added to the atmosphere. When this absorbed CO_2 combines with ocean water, it forms carbonic acid (H_2CO_3), a weak acid also found in carbonated drinks. This process increases the level of hydrogen ions (H^+) in ocean water and makes the water less basic (lower pH). This also decreases the level of carbonate ions (CO_3^{2-}) in the water because these

ions react with hydrogen ions (H^+) to form bicarbonate ions (HCO_3^-).

The problem is that many aquatic species—including phytoplankton, corals, sea snails, crabs, and oysters—use carbonate ions to produce calcium carbonate ($CaCO_3$), the main component of their shells and bones. In less basic waters, carbonate ion concentrations drop (Figure 9.B) and shell-building species and coral reefs grow more slowly. When the hydrogen ion concentration in surrounding seawater reaches high enough levels, the calcium carbonate in the shells and bones of these organisms begins to dissolve.

According to a 2013 study by more than 540 of the world's experts on ocean acidification, the average acidity of surface ocean water has risen by 30% (actually a 30% decrease in average basicity) since 1800 and by the end of this century could increase by 170%. According to the report, the oceans are acidifying "faster than at any time during the last 300 million years." The report also warned that this would reduce the ability of the oceans to help slow the rate of climate change by absorbing CO_2 from the atmosphere.

According to most marine scientists, the only way to slow these changes is through a quick and sharp reduction in the use of fossil fuels around the world, which would lessen the massive inputs of CO_2 into the air and from there into the ocean. This can be done by sharply reducing energy waste and shifting from a dependence on carbon-containing fossil fuels to greater reliance on the sun, wind, and heat stored in the earth's interior (geothermal energy) over the next several decades. (We discuss this in Chapters 13 and 15.) We can also slow the rise of acidity levels in ocean waters by protecting and restoring mangrove forests, sea grasses, and coastal wetlands, because these aquatic systems take up and store some of the atmospheric CO_2 that is at the heart of this problem.

CRITICAL THINKING

How might widespread losses of some forms of marine aquatic life due to ocean acidification affect life on land? How might it affect your life? (Hint: Think food webs.)

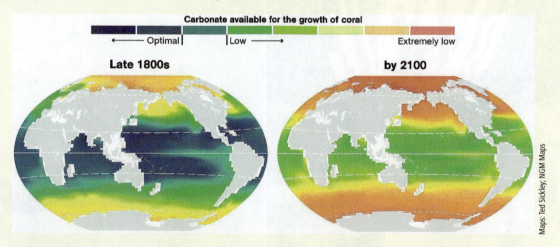

FIGURE 9.B Calcium carbonate levels in ocean waters, calculated from historical data (left), and projected for 2100 (right). Colors shifting from blue to red indicate waters becoming less basic. **Data analysis:** In what two areas of the globe has the change between these two maps been the greatest?

Sources: Andrew G. Dickson, Scripps Institution of Oceanography, U.C. San Diego, and Sarah Cooley, Woods Hole Oceanographic Institution. Used by permission from National Geographic

Cigdem Cooper/Shutterstock.com

FIGURE 9.23 The *common lionfish* has invaded the eastern coastal waters of North America, where it has few if any predators. One scientist described it as "an almost perfectly designed invasive species."

harmful invasive species (Figure 9.23) into coastal waters, wetlands, and lakes throughout the world. According to the USFWS, bioinvaders are blamed for about two-thirds of all fish extinctions in the United States since 1900 and have caused huge economic losses.

Overfishing: Gone Fishing, Fish Gone

Fish and fish products provide about 20% of the world's animal protein for billions of people. A **fishery** is a concentration of a wild aquatic species suitable for commercial harvesting in a given ocean area or inland body of water.

Today, 4.4 million fishing boats hunt for and harvest fish from the world's oceans. Industrial fishing fleets use a variety of methods to harvest marine fishes and shellfish. They include global satellite positioning equipment, sonar fish-finding devices, huge nets, long fishing lines, spotter planes and drones, and refrigerated factory ships that can process and freeze their enormous catches. Figure 9.24 shows the major methods used in the commercial harvesting of marine fishes and shellfish. These highly efficient fleets help supply the growing demand for seafood, but critics say they are overfishing many species, reducing marine biodiversity, and degrading important marine ecosystem services.

According to the Woods Hole National Fisheries Service, 57% of the world's fisheries have been fully exploited and another 30% are overexploited or depleted. Such overharvesting has led to the collapse of some of the world's major fisheries (Figure 9.25).

87% Percentage of the world's commercial fisheries that have been fully exploited or overfished.

One result of the efficient global hunt for fish is that larger individuals of commercially valuable wild species—including cod, marlin, swordfish, and tuna—are becoming scarce. A study conducted by Canadian scientists found that 90% or more of these and other large, predatory, open-ocean fishes disappeared between 1950 and 2006, a trend that is increasing. Another effect of overfishing is that when larger predatory species dwindle, rapidly reproducing invasive species such as jellyfish (see Case Study that follows) can take over and disrupt ocean food webs.

The decline in commercially valuable large fish species has led the fishing industry to begin working its way down to lower tropic levels in marine food webs by shifting to smaller marine species known as forage fish. Examples include anchovies, herring, sardines, and shrimp-like krill. Ninety percent of this catch is converted to fishmeal and fish oil and fed to farmed fish. Scientists warn that this reduces the food supply for larger species and makes it harder for them to rebound from overfishing. The result will likely be further disruption of marine ecosystems and their ecosystem services.

CASE STUDY

Upsetting Marine Ecosystems: Jellyfish Invasions

Jellyfish are often found in large swarms or *blooms* of thousands, even millions of individuals. In recent years, the number of these blooms has been rising. Often they are as big as 5 to 6 city blocks in diameter.

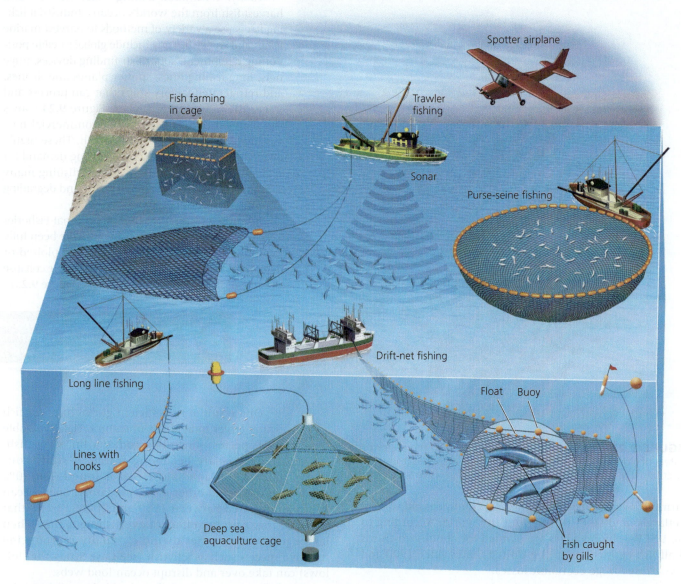

FIGURE 9.24 Major commercial fishing methods used to harvest various marine species (along with methods used to raise fish through aquaculture).

These blooms cause beach closings, disrupt commercial fishing operations by clogging or tearing fishnets, wipe out coastal fish farms, and shut down ship engines. They can also close down coal-burning and nuclear power plants by blocking their cooling water intakes.

Jellyfish reproduce rapidly and certain human activities are helping boost their populations. This includes depleting populations of their major predators, reducing populations of fish species that compete with jellyfish for food, warming and making the oceans more acidic, creating numerous large oxygen-depleted zones, and spreading jellyfish throughout the world as hitchhikers on ships.

According to Chinese oceanographer Wei Hao and other marine scientists, the startling growth of jellyfish populations threatens to upset marine food webs and ecosystem services and turn some of the world's most productive ocean areas into jellyfish empires. Once jellyfish take over a marine ecosystem, they might dominate it for millions of years.

Protecting and Sustaining Marine Biodiversity

Protecting marine biodiversity is difficult for several reasons:

- The human ecological footprint is expanding so rapidly that it is difficult to monitor the human impact on marine diversity.

- Much of the damage to the oceans and other bodies of water is not visible to most people.

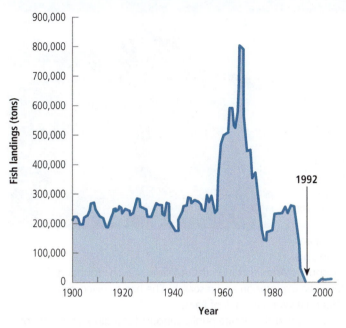

FIGURE 9.25 Natural Capital Degradation: The collapse of Newfoundland's Atlantic cod fishery. ***Data analysis:*** By roughly what percentage did the catch of Atlantic cod drop between the peak catch in 1960 and 1970?

Compiled by the authors using data from Millennium Ecosystem Assessment.

global network of fully protected *marine reserves,* some of which already exist. These areas are declared off-limits to commercial fishing, dredging, mining, and waste disposal in order to enable their ecosystems to recover and flourish.

When patrolled and protected, marine reserves work and they work quickly. Studies show that in fully protected marine reserves, on average, commercially valuable fish populations double, fish size grows by almost a third, fish reproduction triples, and species diversity increases by almost one-fourth. These improvements can happen within 2–4 years after strict protection begins.

Despite the importance of protected reserves, only 1.2% of the world's oceans are strictly protected, compared to 5% of the world's land. In other words, 98.8% of the world's oceans are not effectively protected from harmful human activities. In addition, many existing reserves are fully protected only on paper because of shortages of funding and a need for more trained staff to manage and monitor them.

98.8% Percentage of the world's oceans without effective protection from harmful human activities.

- Many people incorrectly view the seas as an inexhaustible resource that can absorb an almost infinite amount of waste and pollution and still produce all the seafood and other products that we want.

- Most of the world's ocean area lies outside the legal jurisdiction of any country. Thus, much of it is an open-access resource, subject to overexploitation. This is a classic example of the tragedy of the commons (see Chapter 1, p. 12).

Nevertheless, there are several ways to protect and sustain marine biodiversity, thereby increasing our beneficial environmental impact (**Concept 9.5**). For example, we can *protect endangered and threatened aquatic species,* as discussed in Chapter 8, and we can restore and sustain streams, wetlands, and other aquatic systems.

We can also *establish protected marine sanctuaries.* Since 1986, the IUCN has helped several nations establish a global system of *marine protected areas (MPAs)*—areas of ocean partially protected from human activities. According to the U.S. National Ocean Service, there are more than 5,800 MPAs worldwide (more than 1,800 in U.S. waters), covering about 2.8% of the world's ocean surface, and their numbers are growing. However, most MPAs allow dredging, trawler fishing, and other ecologically harmful resource extraction activities. In addition, many of them are too small to be effective in protecting larger species.

Many scientists and policymakers call for protecting and sustaining entire marine ecosystems within a

Many marine scientists want to set aside 10–30% of the world's oceans as fully protected marine reserves—an important way to increase our beneficial environmental impact. Sylvia Earle, one of the world's leading marine scientists, is spearheading this effort (Individuals Matter 9.1).

CONSIDER THIS ...

THINKING ABOUT Marine Reserves

Do you support setting aside at least 30% of the world's oceans as fully protected marine reserves? Explain. How would this affect your life and the lives of any children and grandchildren you might have? How would you fund this protection?

Restoration Can Protect Marine Biodiversity

A dramatic example of marine system restoration is Japan's attempt to restore its largest coral reef—90% of it dead—by seeding it with new corals. Divers drill holes into the dead reefs and insert ceramic discs holding sprigs of fledgling coral. Figure 9.26 shows how protection has helped restore coral reefs near Kanton Island, an atoll located in the South Pacific.

Many scientists support efforts to restore aquatic systems, but they warn that these projects could fail if the

Sylvia Earle—Advocate for the Oceans

Sylvia Earle is one of the world's most respected oceanographers and is a National Geographic Society Explorer. She has taken a leading role in helping us understand the world's oceans and to protect them. *Time* magazine named her the first Hero for the Planet and the U.S. Library of Congress calls her "a living legend."

Earle has led more than 100 ocean research expeditions and has spent more than 7,000 hours underwater, either diving or descending in research submarines to study ocean life. She has focused her research on the ecology and conservation of marine ecosystems, with an emphasis on developing deep-sea exploration technology.

She is the author of more than 175 publications and has been a participant in numerous radio and television productions. During her long career, Earle has also served as the Chief Scientist of the U.S. National Oceanic and Atmospheric Administration (NOAA) and she has founded three companies devoted to developing submarines and other devices for deep-sea exploration and research. She has received more than 100 major international and national honors, including a place in the National Women's Hall of Fame.

These days, Earle is leading a campaign called *Mission Blue* to finance research and to ignite public support for a global network of marine protected areas, which she dubs "hope spots." Her goal is to help save and restore the oceans, which she calls "the blue heart of the planet." She says, "There is still time, but not a lot, to turn things around. . . . This mostly blue planet has kept us alive. It's time for us to return the favor."

problems that caused their degradation are not addressed. They also call for more emphasis on *preventing aquatic ecosystem degradation*, which is far less expensive and more effective than restoration efforts.

For example, a study by IUCN and scientists from the Nature Conservancy concluded that the world's shallow coral reefs and mangrove forests could survive currently projected climate change if we reduce other threats such as overfishing and pollution. However, while some shallow coral species may be able to adapt to warmer temperatures, they may not have enough time to do this unless we act now to slow down the rate of ocean warming.

Ecosystem Approach to Sustaining Aquatic Biodiversity

Edward O. Wilson (see Individuals Matter 4.1, p. 73) and other biodiversity experts have proposed the following priorities for an ecosystem approach to sustaining aquatic biodiversity and ecosystem services:

- Map and inventory the world's aquatic biodiversity.
- Identify and preserve the world's aquatic biodiversity hotspots and areas where deterioration of ecosystem services threatens people and many forms of aquatic life.

- Create large and fully protected marine reserves to allow damaged marine ecosystems to recover and fish stocks to be replenished.
- Protect and restore the world's lakes and river systems, which are among the world's most threatened ecosystems but emphasize pollution prevention, because ecological restorations are expensive and have a high failure rate.
- Initiate worldwide ecological restoration projects in systems such as coral reefs and inland and coastal wetlands.
- Find ways to raise the incomes of people who live on or near protected waters so that they can become partners in the protection and sustainable use of aquatic ecosystems.

There is growing evidence that the current harmful effects of human activities on both terrestrial and aquatic biodiversity and ecosystem services could be reversed over the next two decades. Doing this will require implementing an ecosystem approach to sustaining both terrestrial and aquatic ecosystems. According to Edward O. Wilson, such a conservation strategy would cost about $30 billion per year—an amount that could be provided by a tax of one penny per cup of coffee consumed in the world each year.

FIGURE 9.26 Recovery of a coral reef in a protected area near Kanton Island in the South Pacific.

BRIAN J. SKERRY/National Geographic Creative

CONSIDER THIS ...

THINKING ABOUT The Cost of Sustaining Ecosystems

Would you be willing to pay an extra penny for each cup of coffee you buy to help pay for sustaining ecosystems and biodiversity? Can you think of other things for which you would pay a little more to support this effort?

This strategy for protecting the earth's vital biodiversity and increasing our beneficial environmental impact will not be implemented without political pressure on elected officials from individual citizens and groups. It will also require cooperation among scientists, engineers, and business and government leaders in applying the win–win **principle of sustainability**.

A key part of this strategy will be for individuals to "vote with their wallets" by trying to buy only products and services that do not have harmful impacts on terrestrial and aquatic biodiversity.

BIG IDEAS

- The economic values of the ecosystem services provided by the world's ecosystems are far greater than the value of raw materials obtained from those systems.

- We can sustain terrestrial biodiversity and ecosystem services and increase our beneficial environmental impact by protecting severely threatened areas, protecting remaining undisturbed areas, restoring damaged ecosystems, and sharing much of the land we dominate with other species.

- We can sustain aquatic biodiversity and increase our beneficial environmental impact by establishing protected marine sanctuaries, managing coastal development, reducing water pollution, and preventing overfishing.

Tying It All Together

Sustaining Costa Rica's Biodiversity

Eduardo Rivero/Shutterstock.com

In this chapter, we looked at how human activities are destroying or degrading much of the earth's terrestrial and aquatic biodiversity. We discussed the importance of preserving what remains of diverse and highly endangered biodiversity hotspots and sustaining the earth's ecosystem services. We also saw how we can reduce this destruction and degradation by using the earth's resources more sustainably and by employing restoration ecology and reconciliation ecology. The **Core Case Study** introduced much of this by reporting on what Costa Rica is doing to protect and restore its precious biodiversity.

Preserving terrestrial and aquatic biodiversity involves applying the three **scientific principles of sustainability**. First, it means respecting biodiversity and understanding the value of sustaining it. Also, if we rely less on fossil fuels and more on direct solar energy and its indirect forms, such as wind and flowing water, we will generate less pollution and interfere less with chemical cycling and other forms of natural capital that sustain biodiversity and our own lives and economies.

We can also apply the economic, political, and ethical **principles of sustainability** to help preserve biodiversity. By placing economic value on ecosystem services and including such values in the prices of goods and services, we would more clearly acknowledge their importance. By working together to find win–win solutions to problems of environmental degradation, we would benefit both the earth and its people. The search for such solutions could be guided by an ethical responsibility to sustain biodiversity and ecosystem services for current and future generations.

Chapter Review

Core Case Study

1. Summarize the story of Costa Rica's efforts to preserve its rich biodiversity.

Section 9.1

2. What are the two key concepts for this section? What major ecological and economic benefits do forests provide? Describe the efforts of scientists and economists to put a price tag on the major ecosystem services provided by forests and other ecosystems. Distinguish among **old-growth (primary) forests**, **second-growth forests**, and **tree plantations** (tree farms or commercial forests). Explain how increased reliance on tree plantations can reduce overall forest biodiversity and degrade forest topsoil.

3. Explain how building roads into previously inaccessible forests can harm the forests. Distinguish among selective cutting, clear-cutting, and strip cutting in the harvesting of trees. What are two types of forest fires? What are some benefits of occasional surface fires?

4. What is **deforestation**? List some major harmful environmental effects of deforestation. Summarize the story of reforestation in the United States. Summarize the trends in tropical deforestation. What are four major causes of tropical deforestation? Explain how widespread tropical deforestation can convert a tropical forest to tropical grassland (savanna).

Section 9.2

5. What is the key concept for this section? What is certified sustainably grown timber? List four ways to manage forests more sustainably. What are three ways to reduce the harm caused by forest fires to forests and to people? What are two ways to reduce the need to harvest trees for making paper? Describe the global fuelwood crisis. Describe deforestation in Haiti.

What are five ways to protect tropical forests and use them more sustainably?

Section 9.3

6. What is the key concept for this section? Distinguish between **rangelands** and **pastures**. What is **overgrazing** and what are its harmful environmental effects? What are three ways to reduce overgrazing and use rangelands more sustainably?

Section 9.4

7. What are the three key concepts for this section? What are nine strategies for sustaining terrestrial biodiversity? What percentage of the earth's land is strictly protected from potentially harmful human activities and what percentage do conservation biologists call for? What is **wilderness** and why is it important, according to conservation biologists? Summarize the history of wilderness protection in the United States. What are the major environmental threats to national parks in the world and in the United States? Describe some of the ecological effects of reintroducing the gray wolf to Yellowstone National Park. What is the buffer zone concept? How has Costa Rica applied this approach?

8. Summarize the five-point strategy recommended by biologists for protecting terrestrial ecosystems. What is a **biodiversity hotspot** and why is it important to protect such areas? Explain the importance of protecting ecosystem services and list three ways to do this. Define **ecological restoration**. What are four approaches to restoration? Summarize the science-based, four-step strategy for carrying out

ecological restoration and rehabilitation. Describe the ecological restoration of Guanacaste National Park in Costa Rica. Define and give three examples of **reconciliation ecology**.

Section 9.5

9. What is the key concept for this section? Summarize the threats to aquatic biodiversity resulting from human activities. What is **ocean acidification** and why is it a major threat to aquatic biodiversity? Define **fishery** and summarize the threats to marine fisheries. Describe the major industrial fish harvesting methods. What percentage of the world's commercial fisheries has been fully exploited overfished? Explain how jellyfish can upset marine ecosystems and how human activities affect jellyfish populations. Why is it difficult to protect marine biodiversity? What is a marine protected area? What is a marine reserve? What are three ways to protect marine biodiversity? What percentage of the world's oceans is strictly protected from harmful human activities in marine reserves? Summarize the contributions of Sylvia Earle to the protection of aquatic biodiversity. Describe the role of restoration in protecting marine biodiversity. What are six ways to apply the ecosystem approach to protecting aquatic biodiversity?

10. What are this chapter's *three big ideas*? Explain the relationship between preserving biodiversity as it is done in Costa Rica and the six **principles of sustainability**.

Note: Key terms are in bold type.

Critical Thinking

1. Why do you think Costa Rica (**Core Case Study**) has set aside a much larger percentage of its land for biodiversity conservation than the United States has? Should the United States reserve more of its land for this purpose? Explain.

2. In the early 1990s, Miguel Sanchez, a subsistence farmer in Costa Rica, was offered $600,000 by a hotel developer for a piece of land that he and his family had been using sustainably for many years. An area under rapid development surrounded the land, which contained an old-growth rain forest and a black sand beach. Sanchez refused the offer. Explain how Sanchez's decision was an application of the ethical **principle of sustainability**. What would you have done if you were Sanchez? Explain.

3. Should more-developed countries provide at least half of the money needed to help preserve the remaining tropical forests in less-developed countries? Explain. Do you think that the long-term economic and ecological benefits of doing this would outweigh the short-term economic costs? Explain.

4. Are you in favor of establishing more wilderness areas in the United States (or in the country where you live)? Explain. What might be some drawbacks of doing this?

5. You are a defense attorney arguing in court for preserving an old-growth forest that developers want to clear for a suburban development. Give your three strongest arguments for preserving this ecosystem. How would you counter the argument that preserving the forest would harm the economy by causing a loss of jobs in the timber industry?

6. What do you think are the three greatest threats to aquatic biodiversity and aquatic ecosystem services? For each of them, explain your thinking. Imagine that you are a national official in charge of setting policy for preserving aquatic biodiversity and outline a plan for dealing specifically with these threats.

7. Some scientists consider ocean acidification to be one of the most serious environmental and economic threats that the world faces. How do you think you might be contributing to ocean acidification in your daily life? What are three things you could do to help reduce the threat of ocean acidification?

8. How might your life and the lives of any children or grandchildren you might have be affected if we fail to control the spread of jellyfish populations? What are three things you could do to help prevent this from happening?

Doing Environmental Science

Pick an area near where you live or go to school that hosts a variety of plants and animals. It could be a yard, an abandoned lot, a park, a forest, or some part of your campus. Visit this area at least three times and make a survey of the plants and animals that you find there, including any trees, shrubs, groundcover plants, insects, reptiles, amphibians, birds, and mammals. Also, take a small sample of the topsoil and find out what organisms are living there. (Be careful to get permission from whoever owns or manages the land before doing any digging.) Using guidebooks and other resources to help identify different species, record your findings and categorize them into the general types of organisms listed above. Then do some research to find out about the ecosystem services that some or all of these organisms provide. Try to find and record five of these services. Finally, do some research to find a range of values that economists have assigned to these ecosystem services at the global level. Write a report summarizing your findings.

Global Environment Watch Exercise

Go to your MindTap course to access the GREENR database. Starting on the home page, under "Browse Issues and Topics," click on *Environment and Ecology*, then select *Forests and Deforestation* to enter the portal. Go to the Statistics heading and click "View All." On this page, click on "Share of Tropical Deforestation." Choose one of these countries and research the deforestation in this country further (tip: use the World Map feature). Write a report on your findings and include possible solutions for this deforestation problem. Solutions may include those legislated by governments, as well as those being tried by private individuals or companies.

Ecological Footprint Analysis

A *fishprint* provides a measure of a country's fish harvest in terms of area. The unit of area used in fishprint analysis is the global hectare (gha), a unit weighted to reflect the relative ecological productivity of the area fished. When compared with the fishing area's *sustainable biocapacity* (its ability to provide a stable supply of fish year after year, expressed in terms of yield per area), its fishprint indicates whether the country's annual fishing harvest is sustainable. The fishprint and biocapacity are calculated using the following formulas:

Fishprint in (gha) = metric tons of fish harvested per year/productivity in metric tons per hectare × weighting factor

Biocapacity in (gha) = sustained yield of fish in metric tons per year/productivity in metric tons per hectare × weighting factor

The following graph shows the earth's total fishprint and biocapacity between 1950 and 2000. Study it and answer the following questions:

1. Based on the graph,
   a. In what year did the global fishprint begin to exceed the biological capacity of the world's oceans?
   b. By how much did the global fishprint exceed the biological capacity of the world's oceans in 2000?

2. Assume a country harvests 18 million metric tons of fish annually from an ocean area with an average productivity of 1.3 metric tons per hectare and a weighting factor of 2.68. What is the annual fishprint of that country?

3. If biologists determine that this country's sustained yield of fish is 17 million metric tons per year,
   a. What is the country's sustainable biological capacity?
   b. Is the county's annual fishing harvest sustainable?
   c. To what extent, as a percentage, is the country undershooting or overshooting its biological capacity?

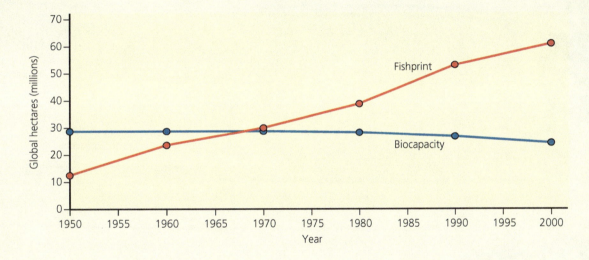

CHAPTER 10

Food Production and
The Environment

There are two spiritual dangers in not owning a farm. One is the danger of supposing that breakfast comes from the grocery, and the other that heat comes from the furnace.

ALDO LEOPOLD

Key Questions

10.1 What is food security and why is it difficult to attain?

10.2 How is food produced?

10.3 What are the environmental effects of industrialized food production?

10.4 How can we protect crops from pests more sustainably?

10.5 How can we produce food more sustainably?

10.6 How can we improve food security?

The Rodale Research Center in Kutztown, Pennsylvania conducts research on organic and other forms of agriculture.

JIM RICHARDSON/National Geographic Creative

225

Growing Power—An Urban Food Oasis

A *food desert* is an urban area where people have little or no easy access to grocery stores or other sources of nutritious food. An estimated 23.5 million Americans, including 6.5 million children, live in such neighborhoods. They tend to rely on convenience stores and fast food restaurants for their meals, which typically consist of high-calorie, highly processed foods that can lead to obesity, diabetes, and heart disease.

Will Allen (Figure 10.1) was one of six children of a sharecropper and grew up on a farm in Maryland. He left the farm life for college and a professional basketball career, followed by a successful corporate marketing career. In 1993, Allen had decided to return to his roots. He bought the last working farm within the city limits of Milwaukee, Wisconsin, and, in time, created a food oasis in a food desert.

On this small urban plot, Allen developed Growing Power, Inc. As an ecologically based farm, it showcases forms of agriculture that apply all three **scientific principles of sustainability** (see inside back cover of this book). Allen's farm is powered partly by solar electricity and solar hot water systems and uses several greenhouses to capture solar energy for growing food year-round. The farm produces an amazing diversity of crops with about 150 varieties of organic produce. It also produces organically raised chickens, turkeys, goats, fish, and honeybees. In addition, the farm's nutrients are recycled in creative ways. For example, wastes from the farmed fish provide nutrients for some of the crops.

The farm's products are sold locally at Growing Power farm stands throughout the region and to restaurants. Allen worked with the city of Milwaukee to establish the Farm-to-City Market Basket program through which people can sign up for weekly deliveries of organic produce at modest prices.

In addition, Growing Power runs an education program in which schoolchildren visit the farm and learn about where their food comes from. Allen also trains about 1,000 people every year who want to learn organic farming methods. The farm partnered with the city of Milwaukee to create 150 new green jobs for unemployed and low-income workers, who build greenhouses and grow food organically. Growing Power has expanded, opening an urban farm in a neighborhood of Chicago, Illinois, and setting up training sites in five other states.

For his creative and energetic efforts, Allen has won several prestigious awards. However, he is most proud of the fact that his urban farm helps feed more than 10,000 people every year and puts people to work raising good food.

In this chapter, you will learn about how food is produced, the environmental effects of food production, and how we can produce food more sustainably. ●

FIGURE 10.1 In 1996, Will Allen founded Growing Power—an urban farm in Milwaukee, Wisconsin.

Growing Power/Flickr

10.1 WHAT IS FOOD SECURITY AND WHY IS IT DIFFICULT TO ATTAIN?

CONCEPT 10.1A Many people in less-developed countries have health problems from not getting enough food, while many people in more-developed countries suffer health problems from eating too much food.

CONCEPT 10.1B The greatest obstacles to providing enough food for everyone are poverty, war, bad weather, climate change, and the harmful environmental effects of industrialized food production.

Poverty Is the Root Cause of Food Insecurity

Food security is the condition under which people have access to enough safe and nutritious food for a healthy and active lifestyle. More than 1 billion people work in agriculture to grow crops on about 38% of the earth's ice-free land. They produce more than enough food to meet the basic nutritional needs of every person on the earth. Despite this food surplus, one of every nine people in the world—about 800 million in all—is not getting enough to eat. These people face **food insecurity** by having to live with chronic hunger and poor nutrition that threaten their ability to lead healthy and active lifestyles (**Concept 10.1A**).

Most agricultural experts agree that *the root cause of food insecurity is poverty,* which prevents poor people from growing or buying enough nutritious food to live healthy and active lives. This is not surprising given that in 2014, one of every three people, or 2.6 billion people, struggled to live on the equivalent of $3.10 a day and 900 million people struggled to live on the equivalent of less than $1.90 a day, according to the World Bank. Other obstacles to food security are war, corruption, bad weather (such as prolonged drought, flooding, and heat waves), and climate change (**Concept 10.1B**).

Each day, there are about 246,000 more people at the world's dinner tables and many of them will have little or no food on their plates. By 2050, there will likely be at least 2.5 billion more people to feed. Most of these newcomers will be born in the major cities of less-developed countries. A critical question is how will we feed the projected 9.9 billion people in 2050 without causing serious harm to the environment? We explore possible answers to this question throughout this chapter.

Chronic Hunger and Malnutrition

To maintain good health and resist disease, individuals need large amounts of *macronutrients* (such as carbohydrates, proteins, and fats) and smaller amounts of *micronutrients*—vitamins, such as A, B, C, and E, and minerals, such as iron, iodine, and calcium.

People who cannot grow or buy enough food to meet their basic energy needs suffer from **chronic undernutrition**, or **hunger**, which threatens their ability to lead healthy and productive lives (**Concept 10.1A**). Most of the world's hungry people can afford only a low-protein, high-carbohydrate, vegetarian diet consisting mostly of grains such as wheat, rice, and corn. In other words, they live low on the food chain (Figure 10.2).

In more-developed countries, people living in food deserts (**Core Case Study**) have a similar problem, except that their diet is heavy on cheap food loaded with fats, sugar, and salt. In both cases, people often suffer from **chronic malnutrition**, a condition in which they do not get enough protein and other key nutrients. This can weaken them, make them more vulnerable to disease, and hinder the normal physical and mental development of children.

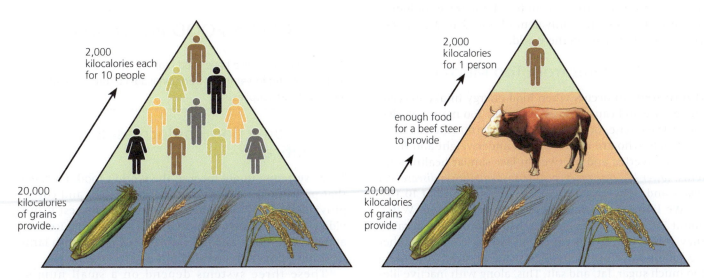

FIGURE 10.2 The poor cannot afford to eat meat and survive by eating lower on the food chain on a diet of grain.

According to the United Nations Food and Agriculture Organization (FAO), in 2015 there were about 793 million chronically undernourished and malnourished people in the world. According to the FAO, at least 3.1 million children younger than age 5 died from chronic hunger and malnutrition in 2013 (the latest year for which data are available). Globally, the number and percentage of people suffering from chronic hunger has been declining since 1992 but there is still a long way to go.

Lack of Vitamins and Minerals

About 2 billion people, most of them in less-developed countries, suffer from a deficiency of one or more vitamins and minerals, usually *vitamin A, iron,* and *iodine* (**Concept 10.1A**). According to the World Health Organization (WHO), at least 250,000 children younger than age 6, most of them in less-developed countries, go blind every year from a lack of vitamin A. Within a year, more than half of them die. Providing children with adequate vitamin A and zinc could save an estimated 145,000 lives per year.

Having too little *iron* (Fe) in the blood is a condition called *anemia*. It causes fatigue, makes infection more likely, and increases a woman's chances of dying from hemorrhage in childbirth. According to the WHO, one of every five people in the world—mostly women and children in less-developed countries—suffers from iron deficiency.

The chemical element *iodine* (I) is essential for proper functioning of the thyroid gland, which produces hormones that control the body's rate of metabolism. Chronic lack of iodine can cause stunted growth, mental retardation, and goiter—a severely swollen thyroid gland that can lead to deafness (Figure 10.3). According to the United Nations (UN), some 600 million people (almost twice the current U.S. population) suffer from goiter, most of them in less-developed countries. In addition, 26 million children suffer irreversible brain damage every year from lack of iodine. The FAO and the WHO estimate that eliminating this serious health problem by adding traces of iodine to salt would cost the equivalent of only 2 to 3 cents per year for every person in the world.

Health Problems from Too Much Food

Overnutrition occurs when food energy intake exceeds energy use and causes excess body fat. Too many calories, too little exercise, or both can cause overnutrition.

People who are underfed and underweight and those who are overfed and overweight face similar health problems: *lower life expectancy, greater susceptibility to disease and illness,* and *lower productivity and life quality* (**Concept 10.1A**).

We live in a world where, according to the WHO, about 793 million people face health problems because they do not get enough nutritious food to eat and another 2.1 billion have health problems caused mostly by eating too much sugar, fat, and salt. This, along with inactive lifestyles, causes them to become overweight or obese.

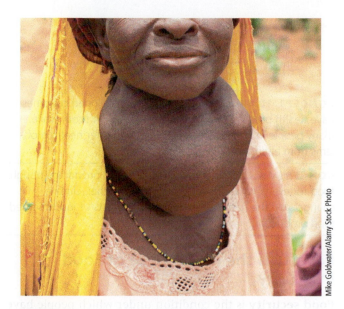

FIGURE 10.3 This woman suffers from goiter, an enlargement of the thyroid gland, caused by a lack of iodine in her diet.

Mike Goldwater/Alamy Stock Photo

According to 2015 statistics from the U.S. Centers for Disease Control and Prevention (CDC), about 72% of adults over age 20 and 33% of all children are overweight or obese. A study by Columbia University and the Robert Wood Johnson Foundation found that obesity plays an important role in nearly one in five deaths in the United States from heart disease, stroke, type 2 diabetes, and some forms of cancer.

72% Percentage of U.S. adults over age 20 who are obese (38%) or overweight (34%).

10.2 HOW IS FOOD PRODUCED?

CONCEPT 10.2 We have used high-input industrialized agriculture and lower-input traditional agriculture to greatly increase food supplies.

Food Production Has Increased Dramatically

Three systems supply most of the world's food. *Croplands* that produce grains—primarily rice, wheat, and corn—provide about 77% of the world's food. The rest is provided by *rangelands, pastures,* and *feedlots* that produce meat and meat products and *fisheries* and *aquaculture* (fish farming) that supply fish and shellfish.

These three systems depend on a small number of plant and animal species. At least half of the world's people survive primarily by eating three different grain crops—*rice, wheat,* and *corn*—because they cannot afford to eat meat. Only a few species of mammals and fish provide most of the world's meat and seafood.

Such food specialization puts humans in a vulnerable position. If any of the small number of crop strains, livestock breeds, and fish and shellfish species that people depend on were to become depleted, the consequences could be dire. Such depletion could be caused by plant or livestock diseases, environmental degradation, or climate change. This food specialization violates the biodiversity **principle of sustainability**, which calls for depending on a variety of food sources as an ecological insurance policy for dealing with changing environmental conditions.

Despite such genetic vulnerability, since 1960, there has been a staggering increase in global food production from all three of the major food production systems (**Concept 10.2**). Three major technological advances have been especially important: **(1)** the development of **irrigation**—a mix of methods by which water is supplied to crops by artificial means; **(2) synthetic fertilizers**—manufactured chemicals that contain nutrients such as nitrogen, phosphorus, potassium, calcium, and several others; and **(3) synthetic pesticides**—chemicals manufactured to kill or control populations of organisms that interfere with crop production.

Industrialized Agriculture

There are two major types of agriculture: industrialized agriculture and traditional. **Industrialized agriculture,** or **high-input agriculture**, uses motorized equipment (Figure 10.4) along with large amounts of financial capital, fossil fuels, water, commercial inorganic fertilizers, and pesticides. Industrialized agriculture produces a single crop at a time on a plot of land, a practice known as **monoculture.** The major goal of industrialized agriculture is to steadily increase each crop's **yield**—the amount of food produced per unit of land. Industrialized agriculture is practiced on 25% of all cropland, mostly in more-developed countries, and produces about 80% of the world's food (**Concept 10.2**).

Plantation agriculture is a form of industrialized agriculture used primarily in less-developed tropical countries. It involves growing *cash crops* such as bananas, coffee, vegetables, soybeans (mostly to feed livestock; see Figure 1.6, p. 8), sugarcane (for sugar and ethanol fuel), and palm oil (for cooking oil and biodiesel fuel). These crops are grown on large monoculture plantations, mostly for export to more-developed countries.

Traditional Agriculture

Traditional, low-input agriculture provides about 20% of the world's food crops on about 75% of its cultivated land, mostly in less-developed countries. It takes two basic forms.

Traditional subsistence agriculture combines energy from the sun with the labor of humans and draft animals to produce enough crops for a farm family's survival, with little left over to sell or store as a reserve for hard times. In **traditional intensive agriculture**, farmers try to obtain higher crop yields by increasing their inputs of human and draft-animal labor, animal manure for fertilizer, and water. With good weather, farmers can produce enough food to feed their families and have some left over to sell for income.

Some traditional farmers cultivate a single crop, but many grow several crops on the same plot simultaneously, a practice known as **polyculture**. This method relies on solar energy and natural fertilizers such as animal manure. The various crops mature at different times. This provides food year-round and keeps topsoil covered to reduce erosion from wind and water. Polyculture also lessens the need for fertilizer and water because root systems at different depths in the soil capture nutrients and moisture efficiently.

Polyculture is an application of the biodiversity **principle of sustainability**. Crop diversity helps protect and replenish the soil and reduces the chance of losing most or all of the year's food supply to pests, bad weather, and other misfortunes. Research shows that, on average, low-input polyculture produces higher average yields than high-input industrialized monoculture, while using less energy and fewer resources, and provides more food security for small landowners. For example, ecologists Peter Reich and David Tilman found that carefully controlled polyculture plots with 16 different species of plants consistently out-produced plots with 9, 4, or only 1 type of plant species.

CONSIDER THIS ...

Learning from Nature

Scientists are studying natural biodiversity to learn how to grow crops using polyculture. The idea is to grow stable crop systems, less vulnerable to environmental threats than monoculture crops are, and to increase yields.

Such research explains why some analysts argue for greatly expanding the use of polyculture to produce food more sustainably. The Growing Power farm (**Core Case Study**) practices polyculture by growing a variety of crops in inexpensive greenhouses—an application of the solar energy and biodiversity **principles of sustainability**.

Organic Agriculture

A fast-growing sector of U.S. and world food production is **organic agriculture.** Organic crops are grown without the use of synthetic pesticides, synthetic inorganic fertilizers, and genetically engineered varieties. Animals are raised on 100% organic feed without the

FIGURE 10.4 This farmer, harvesting a wheat crop in the Midwestern United States, relies on expensive heavy equipment and uses large amounts of seed, manufactured inorganic fertilizer and pesticides, and fossil fuels to produce the crop.

Brenda Carson/Shutterstock.com

use of antibiotics or growth hormones. Growing Power (**Core Case Study**) has become a well-known model for such food production. Figure 10.5 compares organic agriculture with industrialized agriculture.

In the United States, by law, a label of *100 percent organic* (or *USDA Certified Organic*) means that a product is produced only by organic methods, contains all organic ingredients, and has undergone a certification process. Products labeled *"organic"* must contain at least 95% organic ingredients. Those labeled *"made with organic ingredients"* must contain at least 70% organic ingredients. The food label *natural* has no requirement for organic ingredients. Will Allen and other Growing Power farmers (**Core Case Study**) are learning how to use sustainable farming methods to get higher yields of a variety of organic crops at affordable prices.

Green Revolutions Have Increased Yields

Farmers have two ways to produce more food: farm more land or increase yields from existing cropland. Since 1950, most of the increase in global grain production has been the result of increasing crop yields through industrialized agriculture.

This process, called the **green revolution**, involves three steps. *First,* develop and plant monocultures of selectively bred or genetically engineered varieties of key grain crops such as rice, wheat, and corn. *Second,* produce high yields by using large inputs of water, synthetic inorganic fertilizers, and pesticides. *Third,* increase the number of crops grown per year on a plot of land through *multiple cropping*.

In the *first green revolution*, which occurred between 1950 and 1970, this high-input approach dramatically

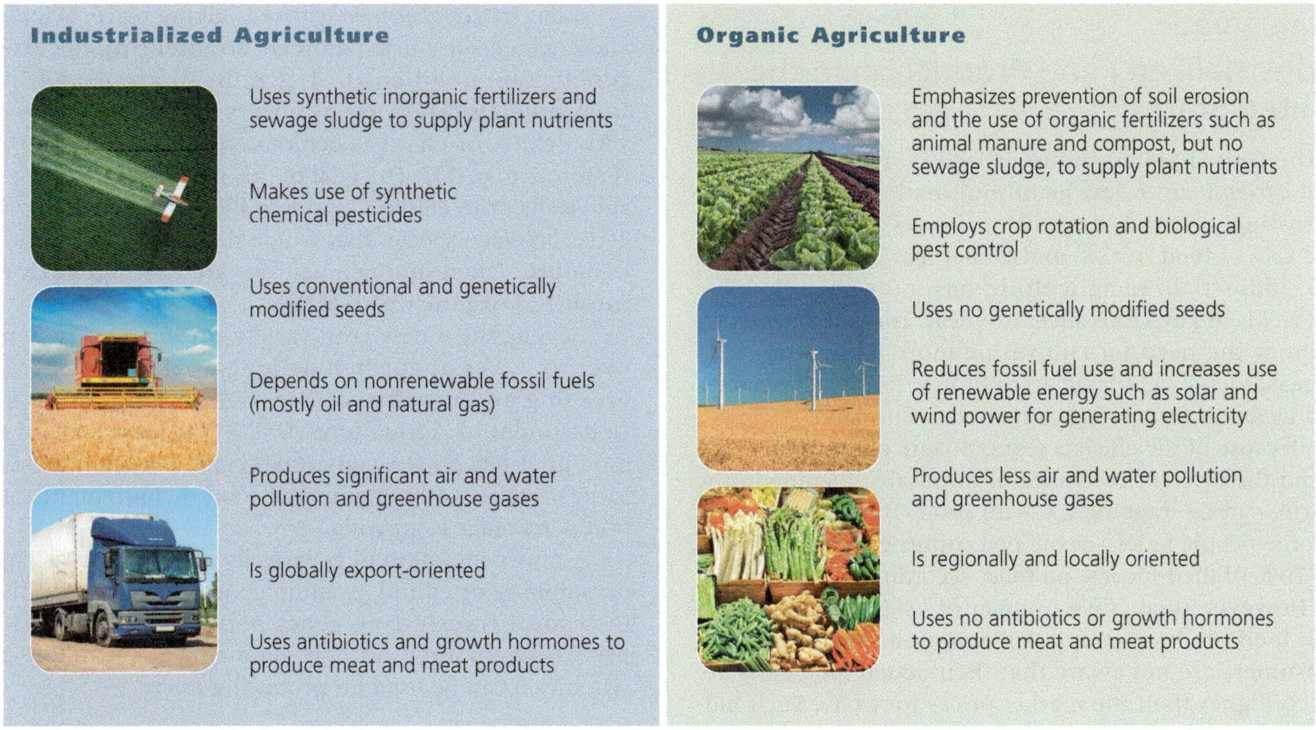

Industrialized Agriculture
Uses synthetic inorganic fertilizers and sewage sludge to supply plant nutrients
Makes use of synthetic chemical pesticides
Uses conventional and genetically modified seeds
Depends on nonrenewable fossil fuels (mostly oil and natural gas)
Produces significant air and water pollution and greenhouse gases
Is globally export-oriented
Uses antibiotics and growth hormones to produce meat and meat products
Organic Agriculture
Emphasizes prevention of soil erosion and the use of organic fertilizers such as animal manure and compost, but no sewage sludge, to supply plant nutrients
Employs crop rotation and biological pest control
Uses no genetically modified seeds
Reduces fossil fuel use and increases use of renewable energy such as solar and wind power for generating electricity
Produces less air and water pollution and greenhouse gases
Is regionally and locally oriented
Uses no antibiotics or growth hormones to produce meat and meat products

FIGURE 10.5 Major differences between industrialized agriculture and organic agriculture.

Left top: B Brown/Shutterstock.com. Left center: ZoranOrcik/Shutterstock.com. Left bottom: Art Konovalov/Shutterstock.com. Right top: Noam Armonn/Shutterstock.com. Right center: Varina C/Shutterstock.com. Right bottom: Adisa/Shutterstock.com.

Grain production (millions of metric tons)
2,500
2,000
1,500
1,000
500
0
1950
1960
1970
1980
1990
2000
2010
2020
Year
Per capita grain production (kilograms per person)
400
350
300
250
200
150
1950
1960
1970
1980
1990
2000
2010
2020
Year

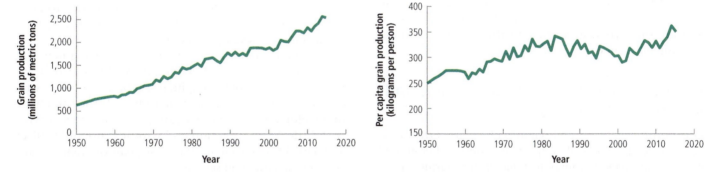

FIGURE 10.6 Growth in worldwide grain production (left) of wheat, corn, and rice and per capita grain production (right) between 1950 and 2016. ***Critical thinking:*** Why do you think per capita grain production has grown less consistently than total grain production?

Compiled by the authors using data from U.S. Department of Agriculture, Worldwatch Institute, UN Food and Agriculture Organization, and Earth Policy Institute

raised crop yields in most of the world's more-developed countries, especially the United States (see the Case Study that follows). In the *second green revolution,* which began in 1967, fast-growing varieties of rice and wheat, specially bred for tropical and subtropical climates, were introduced into middle-income, less-developed countries, such as India, China, and Brazil.

Largely because of the two green revolutions, between 1950 and 2015, world grain production (Figure 10.6, left) increased by 313% and per capita grain production (Figure 10.6, right) grew by 37% between 1950 and 2015. The world's three largest grain-producing countries—China, India, and the United States—produce almost half of the world's grains. However, the rate of growth in crop yields has slowed from an average of 2.2% per decade before 1990 to 1.2% on average since then.

People directly consume about half of the world's grain production. Most of the rest is fed to livestock and thus is consumed by people who can afford to eat meat and meat products.

Industrialized Food Production in the United States

In the United States, industrialized farming has evolved into *agribusiness*. A few giant multinational corporations increasingly control the growing, processing, distribution, and sale of food in U.S. and global markets. Since 1960, U.S. industrialized agriculture has more than doubled the yields of key crops such as wheat, corn, and soybeans without the need for cultivating more land. Such yield increases have kept large areas of U.S. forests, grasslands, and wetlands from being converted to farmland.

Because of the efficiency of U.S. agriculture, Americans spend the lowest percentage of disposable income in the world—an average of 9%—on food. By contrast, low-income people in less-developed countries typically spend 50–70% of their income on food, according to the FAO.

However, there are a number of *hidden costs* related to U.S. food production and consumption. Most American consumers are not aware that their actual food costs are much higher than the market prices they pay. Such hidden costs include the costs of pollution, environmental degradation, and higher health insurance bills related to the harmful environmental and health effects of industrialized agriculture. Other hidden costs include the taxes to pay for *farm subsidies,* or government payments intended to help farmers stay in business and increase their yields. Most of these subsidies go to producers of corn, wheat, soybeans, and rice. U.S. farm subsidies amount to about $20 billion a year, according to the U.S. government accountability office.

Genetic Revolutions: Crossbreeding and Genetic Engineering

For centuries, farmers and scientists have used *crossbreeding* to develop genetically improved varieties of crops and livestock animals. Through *artificial selection,* farmers have developed genetically improved varieties of crops and livestock animals. For example, a tasty but small species of tomato might be crossbred with a larger species of tomato to produce a larger, tasty tomato species. Such selective breeding in this first *gene revolution* has yielded amazing results. For example, ancient ears of corn were about the size of your little finger, and wild tomatoes were once the size of grapes, but most of the large varieties used now were selectively bred for specific desirable traits.

GOOD NEWS

Traditional crossbreeding is a slow process. It often takes 15 years or more to produce a commercially valuable new crop variety and it can combine traits only from species that are genetically similar. Typically, resulting varieties remain useful for only 5 to 10 years before pests and diseases reduce their yields. Important advances are still being made with this method.

Today, a *second gene revolution* is taking place. Scientists and engineers are using **genetic engineering** to develop genetically modified (GM) strains of crops and livestock animals. They use a process called *gene splicing* to add, delete, or change segments of an organism's DNA (Figure 2.6, p. 34). The goal of this process is to add desirable traits or to eliminate undesirable ones by transferring genes between species that would not normally interbreed in nature. The resulting organisms are called **genetically modified organisms (GMOs)**.

Developing a new crop variety through genetic engineering takes about half as long as traditional crossbreeding and usually costs less. According to the U.S. Department of Agriculture (USDA), at least 80% of the food products on U.S. supermarket shelves contain some form of genetically engineered food or ingredients, and that percentage is growing.

A new generation of genetically altered crops is based on snipping or editing existing genes at precise locations instead of transferring genes between species. Scientists are evaluating this new gene-editing technique and whether it can be used for producing crops.

Controversy over Genetically Engineered Foods

The first genetically engineered crops were planted in 1996. Their use has increased in the United States and several other countries. However, globally this is an emerging and controversial technology, which some European countries have banned. It is also too costly for use by poor farmers in less-developed countries.

Bioengineers have talked about developing new GM varieties of crops that have higher yields and that are resistant to heat, cold, drought, insect pests, parasites, viral diseases, herbicides, and salty or acidic soil. They also hope to develop crop plants that can grow faster and survive with little or no irrigation and with less use of fertilizer and pesticides. Accomplishing these goals can reduce hunger and increase food security.

However, critics have raised some concerns about the widespread use of GM crops and foods. One concern is that while many people are consuming GM foods daily, we know too little about their long-term health effects.

Critics also point out that if GM crops or seeds released into the environment caused some long-term harmful genetic or ecological effects, as some scientists project, those organisms could not be recalled. In addition, genes in plant pollen from genetically engineered crops have been known to spread to non–genetically engineered species. This could result in hybrids with wild crop varieties, which could reduce the natural genetic biodiversity of the wild strains. This could in turn reduce the gene pool from which new species can evolve or be engineered—a violation of the biodiversity **principle of sustainability.**

Some 64 countries require that food labels identify GM foods. Polls indicate that 90% of U.S. consumers want to have such information clearly listed on food labels but for decades food producers have opposed doing this. In 2016, the U.S. Congress passed a law requiring such labels. However, it allows food manufacturers to use digital bar codes to enable buyers to read the labels. Many consumers are opposed to having to use a smartphone to scan a bar code to get this information because one-third of the U.S. population—many of them low-income, minority, and elderly—do not have smartphones and most shoppers would not take the time to do this for all of the items they buy. Critics view this industry-supported approach as an attempt to make it difficult for consumers to get such information.

In 2015, an advisory panel of experts for the U.S. National Academies of Science and Engineering concluded that genetically engineered food does not appear to pose serious risks to human health or the environment based on analysis of more than 1,000 studies and testimonies from 80 witnesses in a series of public meetings. However, the report noted that GM crops have not increased the ability to feed the world because the crops have not substantially increased crop yields, as proponents promised.

The report also pointed that while GM crops have decreased the use of insecticides, some herbicide-resistant GM crops have led to increased herbicide use and to herbicide-resistant superweeds. This has forced farmers to spend more money increasing their use of herbicides or switching to stronger herbicides.

The Ecological Society of America and various critics of genetically engineered crops call for more controlled field experiments and testing to better understand the long-term ecological and health risks of using GM crops. They also want stricter regulation of this rapidly growing technology.

CONSIDER THIS ...

THINKING ABOUT GM Crops

Are you for or against the widespread use of genetically engineered crops and foods? Explain.

Meat Consumption Has Grown

Meat and animal products such as eggs and milk are sources of high-quality protein and represent the world's second major food-producing system. Between 1950 and 2016, global meat production grew more than sixfold. According to the FAO global meat consumption is likely to more than double again by 2050 as incomes rise and millions of people in rapidly developing countries move up the food chain and consume more meat and meat products. For example, meat consumption in China increased more than 10-fold between 1975 and 2015.

About half of the world's meat comes from livestock grazing on grass in unfenced rangelands and enclosed pastures. The other half is produced through an industrialized factory farm system. This involves raising large numbers of animals bred to gain weight quickly, mostly in *feedlots* (Figure 10.7) or in crowded pens and cages in huge buildings. These operations take place in *concentrated animal feeding operations (CAFOs), also called factory farms* (Figure 10.8)

In CAFOs, animals are fed grain, soybeans, fishmeal, or fish oil, and some of this feed contains growth hormones and antibiotics to accelerate livestock growth. Because of the crowding and runoff of animal wastes from feedlots, CAFOs can have serious impacts on the air and water. These impacts are examined later in this chapter.

Fish and Shellfish Production Has Risen

The world's third major food-producing system consists of fisheries and aquaculture. A **fishery** is a concentration of particular aquatic species suitable for commercial harvesting in a given ocean area or inland body of water. Industrial fishing fleets harvest use a variety of methods (Figure 9.24, p. 216) to harvest most of the world's marine catch of wild fish. Fish and shellfish are also produced through **aquaculture** or **fish farming** (Figure 10.9). It is the practice of raising fish in freshwater ponds, lakes, reservoirs, and rice paddies, and in underwater cages in coastal and deeper ocean waters.

Aquaculture is the world's fastest growing type of food production. Between 1950 and 2015, global seafood production of wild and farmed fish increased ninefold (**Concept 10.2**) while the global wild catch leveled off and declined. In 2015, aquaculture accounted for nearly half of the world's fish and shellfish production, and the rest were caught mostly by industrial fishing fleets (Figure 10.9). According to the FAO, about 87% of the world's commercial ocean fisheries are being overfished (30%) or harvested at full capacity (57%).

Most of the world's aquaculture involves raising species that feed on algae or other plants—mainly carp in China and India, catfish in the United States, and tilapia and shellfish in a number of countries. However, the farming of meat-eating species such as shrimp and salmon is growing rapidly, especially in more-developed countries. Such species are often fed fishmeal and fish oil produced from other fish and their wastes.

10.3 WHAT ARE THE ENVIRONMENTAL EFFECTS OF INDUSTRIALIZED FOOD PRODUCTION?

CONCEPT 10.3 Future food production may be limited by soil erosion and degradation, desertification, irrigation water shortages, air and water pollution, climate change, and loss of biodiversity.

FIGURE 10.7 *Industrialized beef production:* On this cattle feedlot in Arizona, thousands of cattle are fattened on grain for a few months before being slaughtered.

PETE MCBRIDE/National Geographic Creative

FIGURE 10.8 Concentrated chicken feeding operation in Iowa (USA). Such operations can house up to 100,000 chickens.

Scott Sinklier/Passage/Getty Images

Industrialized Food Production Requires Huge Energy Inputs

The industrialization of food production and increased crop yields has been made possible by using fossil fuels—mostly oil and natural gas—to run farm machinery and fishing vessels, to pump irrigation water for crops, and to produce synthetic pesticides and synthetic inorganic fertilizers. Fossil fuels are also used to process food and transport it long distances within and between countries. Altogether, agriculture accounts for about 17% of all energy used in the United States more than any other industry. In the United States, food items travel an average of 2,400 kilometers (1,300 miles) from farm to plate. Burning such large quantities of fossil fuels disturbs land, pollutes the air and water, and contributes to climate change.

FIGURE 10.9 *Aquaculture:* Shrimp farms on the southern coast of Thailand.

puwanai/Shutterstock.com

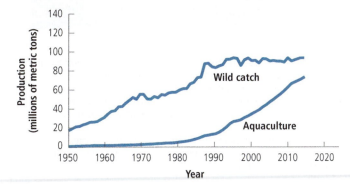

FIGURE 10.10 World seafood production, including both wild catch (marine and inland) and aquaculture, grew between 1950 and 2014, with the wild catch generally leveling off since 1996 and aquaculture production rising sharply since 1990. ***Data analysis:*** In about what year did aquaculture surpass the 1980 wild catch?

Compiled by the authors using data from UN Food and Agriculture Organization, Worldwatch Institute, and Earth Policy Institute.

According to a study led by ecological economist Peter Tyedmers, the world's fishing fleets use about 12.5 units of energy for every 1 unit of food energy from seafood on the table. When we consider the energy used to grow, store, process, package, transport, refrigerate, and cook all plant and animal food, it takes about 10 units of fossil fuel energy to put 1 unit of food energy on the table in the United States. In other words, today's food production systems result a large net energy loss.

On the other hand, the amount of energy per calorie used to produce crops in the United States has declined by about 50% since the 1970s. One factor in this decline is that the amount of energy used to produce synthetic nitrogen fertilizer has dropped sharply. Another reason for the decline is the rising use of conservation tillage, which sharply reduces energy use and the harmful environmental effects of plowing.

Producing Food Has Major Environmental Impacts

Industrialized food production has allowed farmers to use less land to produce more food. This has reduced the need to convert forests and grasslands to cropland, thereby destroying the wildlife habitats provided by these ecosystems.

However, many analysts point out that industrialized agriculture has greater overall harmful environmental impacts (Figure 10.11) than any other human activity. These impacts may limit future food production (**Concept 12.3**).

According to a study by 27 experts assembled by the United Nations Environment Programme (UNEP), agriculture uses massive amounts of the world's resources and pollutes the air and water. It uses about 70% of the world's freshwater, produces about 60% of all water pollution, degrades and erodes topsoil, emits about 25% of the world's greenhouse gas emissions, and uses about 38% of the world's ice-free land. As a result, many analysts view today's industrialized agriculture as environmentally and economically unsustainable. However, proponents of industrialized agriculture argue that its benefits outweigh its harmful effects.

Topsoil Erosion Is a Serious Problem

Topsoil is the fertile top layer of many soils (Figure 3.9, p. 52). It is one of the most important components of the earth's natural capital because all terrestrial life depends directly or indirectly on this potentially renewable resource. Topsoil stores and purifies water and supplies most of the nutrients needed for plant growth. It recycles these nutrients endlessly as long as they are not removed faster than natural processes replenish them. Organisms living in topsoil remove and store carbon dioxide (CO_2) from the atmosphere, thereby helping control the earth's climate as part of the carbon cycle.

A major environmental problem related to agriculture is **soil erosion**—the movement of soil components, especially surface litter and topsoil from one place to another by the actions of wind and water. Some topsoil erosion is natural, but much of it is caused by clearing forests and grasslands for agriculture, plowing the soil to plant new crops each year, and leaving the soil exposed during part of the year.

Flowing water, the largest cause of erosion, carries away particles of exposed topsoil that have been loosened by rainfall (Figure 10.12, left). Severe erosion of this type leads to the formation of gullies (Figure 10.12, right). Wind also loosens and blows particles of topsoil away, especially in areas with a dry climate and relatively flat and exposed land (Figure 10.13).

In undisturbed, vegetated ecosystems, the roots of plants help anchor the topsoil and prevent some erosion. However, topsoil can erode when soil-holding grasses, trees, and other vegetation are removed through activities such as farming (see Figure 7.12, p. 148), clear-cut logging (see Figure 9.7, p. 198), and overgrazing (see Figure 9.15, p. 205).

Natural Capital Degradation

Food Production

Biodiversity Loss	Soil	Water	Air Pollution	Human Health
Conversion of grasslands, forests, and wetlands to crops or rangeland	Erosion	Aquifer depletion	Emissions of greenhouse gases CO_2 from fossil fuel use, N_2O from inorganic fertilizer use, and methane (CH_4) from cattle	Nitrates in drinking water (blue baby)
Fish kills from pesticide runoff	Loss of fertility	Increased runoff, sediment pollution, and flooding from cleared land		Pesticide residues in water, food, and air
	Salinization			
Killing of wild predators to protect livestock	Waterlogging	Pollution from pesticides	Other air pollutants from fossil fuel use and pesticide sprays	Livestock wastes in drinking and swimming water
	Desertification	Algal blooms and fish kills caused by runoff of fertilizers and farm wastes		
Loss of agrobiodiversity replaced by monoculture strains				Bacterial contamination of meat

FIGURE 10.11 Food production has a number of harmful environmental effects (**Concept 10.3**). *Critical thinking:* Which item in each of these categories do you think is the most harmful?

Left: Orientaly/Shutterstock.com. Left center: pacopi/Shutterstock.com. Center: Tim McCabe/USDA Natural Resources Conservation Service. Right center: Mikhail Malyshev/Shutterstock.com. Right: B Brown/Shutterstock.com.

FIGURE 10.12 Natural Capital Degradation: Flowing water from rainfall is the leading cause of topsoil erosion as seen on this farm in the U.S. state of Tennessee (left). Severe water erosion can become gully erosion, which has damaged this cropland in western Iowa (right).

FIGURE 10.13 Wind is an important cause of topsoil erosion in dry areas that are not covered by vegetation such as this bare crop field in the U.S. state of Iowa.

A joint survey by the UNEP and the World Resources Institute indicated that topsoil is eroding faster than it forms on about one-third of the world's cropland (Figure 10.14).

Erosion of topsoil has three major harmful effects. One is *loss of soil fertility* through depletion of plant nutrients in topsoil (see Figure 3.9, p. 52). A second effect is *water pollution* in surface waters where eroded topsoil ends up as sediment, which can kill fish and shellfish and clog irrigation ditches, boat channels, reservoirs, and lakes. Additional water pollution occurs when the eroded sediment contains pesticide residues that can be ingested by aquatic organisms and in some cases biomagnified within food webs (see Figure 8.12, p. 178). Third, erosion releases carbon stored in the soil by vegetation into the atmosphere as CO_2, which contributes to atmospheric warming and climate change.

The rise of industrialized agriculture has exposed irreplaceable topsoil to erosion by water and wind and reduced the plant nutrient content of topsoil in many areas. This erosion of soil nutrients from topsoil, and from synthetic chemical fertilizers added to the soil, sends the nutrients on a one-way trip to crops and then to nearby bodies of surface water, which often become overloaded with plant nutrients. The continuing disruption of the nitrogen and phosphorus cycles (see Figures 3.17 and 3.18, pp. 60 and 61), due to the loss of topsoil and depletion of its key nutrients, is another factor that could eventually make industrialized agriculture unsustainable (**Concept 10.3**).

Soil pollution is also a problem in parts of the world. Some of the chemicals emitted into the atmosphere by industrial and power plants and by motor vehicles can pollute soil and water used to irrigate soil. Some pesticides can also contaminate soil.

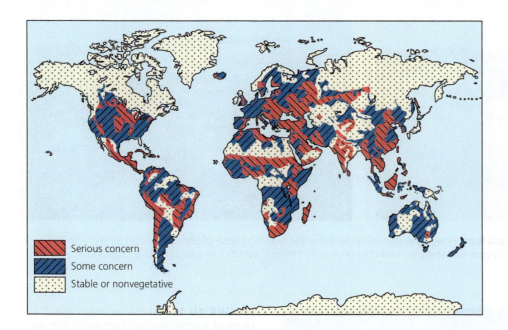

FIGURE 10.14 Natural Capital Degradation: Topsoil erosion is a serious problem in some parts of the world. *Critical thinking:* Can you see any geographical pattern associated with this problem?

Compiled by the authors using data from the UN Environment Programme and the World Resources Institute.

Serious concern

Some concern

Stable or nonvegetative

A recent study by China's environment ministry found that an estimated 19% of China's arable (farmable) land is contaminated, especially with toxic metals such as cadmium, arsenic, and nickel. The ministry also estimated that 2.5% of the country's cropland is too contaminated to grow food safely. China, with 19% of the world's people and only 7% of the world's arable land, cannot afford to lose 2.5% of its cropland.

Desertification

Drylands in regions with arid and semiarid climates occupy about 40% of the world's land area and are home to some 2 billion people. A major threat to food security in some of these areas is **desertification**—the process in which the productive potential of topsoil falls by 10% or more because of a combination of prolonged drought and human activities such as overgrazing, deforestation, and excessive plowing, which expose topsoil to erosion.

Desertification can be *moderate* (with a 10–25% drop in productivity), *severe* (with a drop of 25–50%), or *very severe* (with a drop of more than 50%), usually resulting in large gullies and sand dunes. Desertification decreases soil productivity but only in extreme cases does it lead to what we call a desert.

Over thousands of years, the earth's deserts have expanded and contracted, primarily because of climate change. However, human uses of land, especially for agricultural purposes, have increased desertification in some parts of the world.

Excessive Irrigation Pollutes Soil and Water

Irrigation accounts for about 70% of the water that humanity uses. Currently, the 16% of the world's cropland that is irrigated produces about 44% of the world's food.

However, irrigation has a downside. Most irrigation water is a dilute solution of various salts, such as sodium chloride, that are picked up as the water flows over or through soil and rocks. Irrigation water that is not absorbed into the topsoil evaporates and leaves behind a thin crust of dissolved mineral salts in the topsoil. Repeated applications of irrigation water in dry climates lead to the gradual accumulation of salts in the upper soil layers—a soil degradation process called **soil salinization**. It stunts crop growth, lowers crop yields, and can eventually kill plants and ruin the land.

The FAO estimates that severe soil salinization has reduced yields on at least 10% of the world's irrigated cropland, and that by 2020, 30% of the world's farmable land will be salty. Salinization affects almost one-fourth of irrigated cropland in the United States, especially in western states (Figure 10.15).

Another problem with irrigation is **waterlogging**, in which water accumulates underground and gradually raises the water table. This can occur when farmers apply large amounts of irrigation water in an effort to reduce salinization by leaching salts deeper into the soil. Waterlogging lowers the productivity of crop plants and kills them after prolonged exposure by depriving them of the oxygen they need to survive. At least 10% of the world's irrigated land suffers from this worsening problem, according to the FAO.

Industrialized Agriculture Contributes to Pollution and Climate Change

Eroded topsoil flows as sediments into streams, lakes, and wetlands where it can smother fish and shellfish and clog irrigation ditches, boat channels, reservoirs, and lakes. This problem gets worse when the eroded sediment contains pesticide residues that can be ingested by aquatic

FIGURE 10.15 Natural Capital Degradation: White alkaline salts have displaced crops that once grew on this heavily irrigated land in the U.S. state of Colorado.

USDA Natural Resources Conservation Service

organisms and in some cases biomagnified within food webs (see Figure 8.12, p. 178).

Farmers contribute to water pollution through *over-fertilizing* their fields. Globally, the use of fertilizers has grown 45-fold since 1940. Nitrates in fertilizers can percolate down through the soil into aquifers where it can contaminate groundwater used for drinking. According to the FAO, fully one-third of all water pollution from the runoff of nitrogen and phosphorus is due to excessive use of synthetic fertilizers.

CONSIDER THIS ...

CONNECTIONS Meat Production and Ocean Dead Zones

Huge amounts of synthetic inorganic fertilizers are used in the Midwestern United States to produce corn for animal feed and ethanol fuel for cars. Much of this fertilizer runs off cropland and eventually goes into the Mississippi River. The added nitrate and phosphate nutrients overfertilize coastal waters in the Gulf of Mexico, where the river flows into the ocean. Each year, this creates a huge oxygen-depleted "dead zone" that threatens one-fifth of the nation's seafood yield. In other words, growing corn in the Midwest, largely to feed cattle and fuel cars, degrades aquatic biodiversity and seafood production in the Gulf of Mexico.

Agricultural activities also pollute the air. Clearing and burning forests to raise crops or livestock adds dust and smoke into the air. Applying fertilizer and pesticides spreads particles and various chemicals into the air. Agriculture also accounts for more than a quarter of all human-generated emissions of CO_2. The emissions of this and other greenhouse gases warm the atmosphere and contribute to climate change, which can affect crop productivity and food security.

Industrialized Food Production Reduces Biodiversity

Natural biodiversity and some ecosystem services are threatened when forests are cleared and when grasslands are plowed up and replaced with croplands used to produce food (**Concept 10.3**).

For example, one of the fastest-growing threats to world's biodiversity is happening in Brazil. Large areas of tropical forest in its Amazon Basin and in cerrado—a *huge* tropical grassland region south of the Amazon Basin—are being lost. This land is being burned or cleared for cattle ranches and for large plantations of soybeans grown for cattle feed. Biodiversity is threatened in these and many other areas because tropical forests and grasslands host many more varieties of organisms than agricultural land.

A related problem is the increasing loss of **agrobiodiversity**—the genetic variety of animal and plant species used on farms to produce food. Scientists estimate that since 1900, we have lost 75% of the genetic diversity of agricultural crops. For example, India once planted 30,000 varieties of rice. Now more than 75% of its rice production comes from only 10 varieties. Soon, most of its production might come from just one or two varieties. In the United States, about 97% of the food plant varieties available to farmers in the 1940s no longer exist, except perhaps in small amounts in seed banks and in occasional home gardens.

Ecologists warn that farming practices that reduce agrobiodiversity are rapidly shrinking the world's genetic "library" of plant varieties, which are critical for increasing food yields through crossbreeding and genetic engineering. This failure to preserve agrobiodiversity is a serious violation of the biodiversity **principle of sustainability** that could reduce the sustainability of food production (**Concept 10.3**).

must be regularly collected. Unless this is done, seed banks become seed morgues.

Limits to Expanding Green Revolutions

So far, several factors have limited the success of the green revolutions and may limit them even more in the future (**Concept 10.3**). For example, without large inputs of water and synthetic inorganic fertilizers and pesticides, most green revolution and genetically engineered crop varieties produce yields that are no higher (and are sometimes lower) than those from traditional strains. Climate change and the growing world population also limit the success of green revolutions, as does cost. The high inputs that sustain green revolutions cost too much for most subsistence farmers in less-developed countries.

Scientists point out that there comes a point where yields stop increasing because of the inability of crop plants to take up nutrients from additional fertilizer and irrigation water. This helps explain the slowdown in the rate of growth in global grain yields since 1990.

Can we expand the green revolutions by irrigating more cropland? Since 1978, the amount of irrigated land per person has been declining, and it is projected to fall much more by 2050. One reason for this per person decline is population growth, which is projected to add 2.5 billion more people between 2015 and 2050. Other factors are limited availability of irrigation water, soil salinization, and the fact that most of the world's farmers cannot afford to irrigate their crops.

Climate change is expected to reduce yields of crops such as wheat, rice, and corn during this century. In addition, mountain glaciers that provide irrigation and drinking water for many millions of people in China, India, and South America are melting and this will lessen the area of crops that can be irrigated. During this century, fertile croplands in coastal areas, including many of the major rice-growing floodplains and river deltas in Asia, are likely to be flooded by rising sea levels resulting from climate change.

Can we increase the food supply by cultivating more land? Clearing more tropical forests and irrigating arid land could more than double the area of the world's cropland. However, massive clearing of forests and irrigation of arid land would decrease biodiversity, speed up climate change and its harmful effects, and increase soil erosion. In addition, much of this land has poor soil fertility, steep slopes, or both, and cultivating such land would be expensive and probably not ecologically sustainable.

Industrialized Meat Production Harms the Environment

Proponents of industrialized meat production point out that it has increased meat supplies, reduced overgrazing, and kept food prices down. However, feedlots and CAFOs

FIGURE 10.16 Svalbard Global Seed Vault.

JIM RICHARDSON/National Geographic Creative

Efforts are underway to save individual plants and seeds from endangered varieties of crops and wild plant species important to the world's food supply. About 1,400 refrigerated seed banks store individual plants and seeds. They are also stored in agricultural research centers and botanical gardens scattered around the world.

The world's most secure seed bank is the underground Svalbard Global Seed Vault, also called the "doomsday seed vault," carved into the permafrost on a frozen Norwegian island near the North Pole (Figure 10.16). It is being stocked with duplicates of much of the world's seed collections. This provides insurance against irreversible losses of stored seeds at other sites that are subject to power failures, fires, storms, and wars.

However, the seeds of many plants cannot be stored successfully in seed banks. In addition, stored seeds must be planted and germinated periodically and new seeds produce widespread harmful health and environmental effects. Analysts point out that meat produced by industrialized agriculture is artificially cheap because most of its harmful environmental and health costs are not included in the market prices of meat and meat products. This is a violation of the full-cost pricing **principle of sustainability** (see inside back cover of this book).

A major problem with feedlots and CAFOs is that huge amounts of water are used to irrigate the grain crops that feed the livestock. According to waterfootprint.org, producing a quarter-pound hamburger requires 1,810 liters (478 gallons) of water—the equivalent of 15 to 20 showers for the average person. Industrialized meat production also uses large amounts of energy (mostly from oil), which helps make it one of the chief sources of air and water pollution and greenhouse gas emissions.

Another growing problem is the use of antibiotics in industrialized livestock production facilities. According to the U.S. Food and Drug Administration (FDA), about 80% of all antibiotics sold in the United States (and 50% of those in the world) are added to animal feed. This is done to try to prevent the spread of diseases in crowded feedlots and CAFOs and to promote the growth of the animals before they are slaughtered. According to FDA data and several studies, this plays a role in the rise of genetic resistance among many disease-causing bacteria (see Figure 4.9, p. 79). Such resistance can reduce the effectiveness of some antibiotics used to treat humans for bacterial infections, and it can promote the development of new, more genetically resistant infectious disease organisms.

Finally, according to the USDA, animal waste produced by the American meat industry amounts to about 67 times the amount of waste produced by the country's human population. Ideally, the manure slurry from CAFOs should be returned to the soil as a nutrient-rich fertilizer in keeping with the chemical cycling **principle of sustainability** (see inside back cover of this book). However, it is often so contaminated with residues of antibiotics and pesticides that it is unfit for use as a fertilizer.

Despite such potential contamination, up to half of the manure slurry from CAFOs in the United States is applied to fields and creates severe odor problems for people living nearby. Much of the other half of CAFO animal waste is pumped into lagoons which eventually leak and pollute nearby surface and groundwater, produce foul odors, and emit large quantities of climate-changing greenhouse gases into the atmosphere. Figure 10.17 summarizes the advantages and disadvantages of industrialized meat production.

When livestock feed on grass in rangelands and pastures, environmental impacts can still be high, especially when forests are cut down or burned to make way for grazing land, as is done in Brazil's Amazon forests. According to an FAO report, overgrazing and erosion by livestock has degraded about 20% of the world's grasslands and pastures. The same report estimated that rangeland

Trade-Offs	
Animal Feedlots and CAFOs	
Advantages	Disadvantages
Increased meat production	Animals unnaturally confined and crowded
Higher profits	Large inputs of grain, fishmeal, water, and fossil fuels
Less land use	Greenhouse gas (CO_2 and CH_4) emissions
Reduced overgrazing	Concentration of animal wastes that can pollute water
Reduced soil erosion	Use of antibiotics can increase genetic resistance to microbes in humans
Protection of biodiversity	

FIGURE 10.17 Use of animal feedlots and concentrated animal feeding operations has advantages and disadvantages. *Critical thinking:* Which single advantage and which single disadvantage do you think are the most important? Why?

Top: Mikhail Malyshev/Shutterstock.com. Bottom: Maria Dryfhout/Shutterstock.com.

grazing and industrialized livestock production has caused about 55% of all topsoil erosion and sediment pollution.

In addition, grass-fed cows emit more of the powerful greenhouse gas methane than do grain-fed cows. Thus, expanding grass-fed production could increase the agricultural contributions to deforestation and climate change.

Aquaculture Can Harm Aquatic Ecosystems

Aquaculture produces about 50% of the world's seafood, according to the FAO, and by 2030 could produce 62% of all seafood. Figure 10.18 lists the major benefits and drawbacks of aquaculture. Some analysts warn that the harmful environmental effects of aquaculture could limit its future production potential (**Concept 10.3**) unless efforts are made to make it more sustainable.

A major environmental problem associated with aquaculture is that about a third of the wild fish from marine fisheries is converted to fishmeal and fish oil and fed to farmed fish. This is contributing to the depletion of many populations of wild fish that are crucial to marine food webs—a serious threat to marine biodiversity and ecosystem services.

Another problem is that some fishmeal and fish oil fed to farm-raised fish is contaminated with long-lived toxins such as PCBs and dioxins that are picked up from the ocean floor. Fish farms, especially those that raise carnivorous fish such as salmon and tuna, also produce large amounts of wastes that can contain pesticides and antibiotics used on fish farms. These chemicals can contaminate

Aquaculture

Advantages	Disadvantages
High efficiency	Use of fish oil and fishmeal on fish farms depletes wild fisheries
High yield	
	Large waste output
Reduces over-harvesting of fisheries	Loss of mangrove forests and estuaries
Jobs and profits	Dense populations vulnerable to disease

FIGURE 10.18 Use of aquaculture has advantages and disadvantages. *Critical thinking:* Which single advantage and which single disadvantage do you think are the most important? Why?

Top: Vladislav Gajic/Shutterstock.com. Bottom: FeellFree/Shutterstock.com.

Cathy Keifer/Shutterstock.com

FIGURE 10.19 Natural Capital: This ferocious-looking wolf spider with a grasshopper in its mouth is one of many important insect predators that can be killed by some pesticides.

farm-raised fish and people who eat such fish. Aquaculture producers contend that the concentrations of these chemicals are not high enough to threaten human health, but some scientists disagree.

Another problem is that farmed fish can escape their pens and mix with wild fish and possibly disrupt the gene pools of wild populations or become invasive species. Aquaculture can also destroy or degrade aquatic ecosystems, particularly mangrove forests (see Figure 7.20, p. 154) that are cleared for coastal fish farms. This loss of mangrove forest decreases biodiversity and valuable ecosystem services such as natural flood control in coastal areas that are expected to experience severe flooding because of rising sea level caused by climate change.

10.4 HOW CAN WE PROTECT CROPS FROM PESTS MORE SUSTAINABLY?

CONCEPT 10.4 We can sharply cut pesticide use without decreasing crop yields by using a mix of cultivation techniques, biological pest controls, and small amounts of selected chemical pesticides as a last resort (integrated pest management).

Nature Helps Control Many Pests

A **pest** is any species that interferes with human welfare by competing with us for food, invading our homes,

lawns, or gardens, destroying building materials, spreading disease, invading ecosystems, or simply being a nuisance. Worldwide, only about 100 species of plants (weeds), animals (mostly insects), fungi, and microbes cause most of the damage to crops.

In natural ecosystems and in many polyculture crop fields, *natural enemies* (predators, parasites, and disease organisms) control the populations of most potential pest species. This free ecosystem service is an important part of the earth's natural capital. For example, biologists estimate that about 40,000 known species of spiders kill far more crop-eating insects every year than humans do by using insecticides. Most species of spiders, including the wolf spider (Figure 10.19), do not harm humans.

When we clear forests and grasslands, plant monoculture crops, and douse fields with chemicals that kill pests, we upset many of these natural population checks and balances that are in keeping with the biodiversity **principle of sustainability** (see inside back cover of this book). Society must then devise ways to protect monoculture crops, tree plantations, lawns, and golf courses from insects, weeds, and other pests that nature helps control at no charge.

Synthetic Pesticides Are an Option

Scientists and engineers have developed a variety of synthetic **pesticides**. Common types of synthetic pesticides include *insecticides* (insect killers), *herbicides* (weed killers), *fungicides* (fungus killers), and *rodenticides* (rat and mouse killers).

Since 1950, synthetic pesticide use has increased more than 50-fold and most of today's pesticides are 10 to 100 times more toxic to pests than those used in the 1950s. Some synthetic pesticides, called *broad-spectrum agents,* are toxic to beneficial species as well as to pests. Examples are organochlorine compounds (such as DDT), organophosphates (such as malathion and parathion), carbamates, pyrethroids, and neonicotinoids (which have been linked to the serious decline in honeybee populations).

50-fold
Increase in synthetic pesticide use since 1950.

Other synthetic pesticides, called *selective,* or *narrow-spectrum, agents,* are each effective against a narrowly defined group of organisms. Examples are fungicides, and glyphosate, a widely used herbicide that kills weeds without hurting crops such as corn and soybeans.

Pesticides vary in their *persistence,* the length of time they remain deadly in the environment. Some, such as DDT and related compounds, remain in the environment for years and can be biologically magnified in food chains and webs (see Figure 8.12, p. 178). Others, such as organophosphates, are active for days or weeks and are not biologically magnified but can be highly toxic to humans.

About one-fourth of the pesticides used in the United States are aimed at ridding houses, gardens, lawns, parks, and golf courses of insects and other species that people view as pests. According to the U.S. Environmental Protection Agency (EPA), the amount of synthetic pesticides used on the average U.S. homeowner's lawn is 10 times the amount (per unit of land area) typically used on U.S. croplands.

Benefits of Synthetic Pesticides

The use of synthetic pesticides has its advantages and disadvantages. Proponents contend that the benefits of pesticides (Figure 10.20, left) outweigh their harmful effects (Figure 10.20, right). They point to the following benefits:

- *They have saved human lives.* Since 1945, DDT and other insecticides probably have prevented the premature deaths of at least 7 million people (some say as many as 500 million) from insect-transmitted diseases such as malaria (carried by the *Anopheles* mosquito), bubonic plague (carried by rat fleas), and typhus (carried by body lice and fleas).

- *They can increase food supplies* by reducing food losses due to pests for some crops.

- *They can help farmers control soil erosion and build soil fertility.* In conventional no-till farming, farmers apply herbicides instead of weeding the soil by plowing. This dramatically reduces soil erosion and soil nutrient depletion.

- *They can help farmers reduce costs and increase profits.* The costs of using pesticides can be regained, at least in the near term, through higher crop yields.

- *They work fast.* Pesticides control most pests quickly, have a long shelf life, and are easily shipped and applied.

- *Newer pesticides are safer to use and more effective than many older ones.*

Problems with Synthetic Pesticides

Opponents of widespread use of synthetic pesticides contend that the harmful effects of these chemicals (Figure 10.20, left) outweigh their benefits (Figure 10.20, right). They cite several problems.

- *They accelerate the development of genetic resistance to pesticides in pest organisms* (Figure 10.21). Since 2010, according to the WHO, 60 countries have reported genetic resistance to at least one class of insecticide, with 49 countries reporting resistance to two or more classes. Superweeds that are resistant to herbicides have also spread.

- *They can put farmers on a financial treadmill.* Farmers can find themselves having to pay more and more for a chemical pest control program that can become less and less effective as pests develop genetic resistance to pesticides.

- *Some insecticides kill natural predators and parasites that help control pest populations.* About 100 of the 300 most destructive insect pests in the United States were

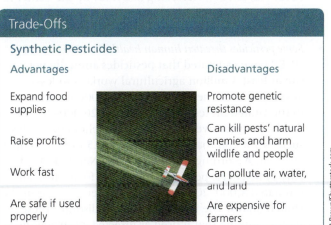

Trade-Offs

Synthetic Pesticides

Advantages	Disadvantages
Expand food supplies	Promote genetic resistance
Raise profits	Can kill pests' natural enemies and harm wildlife and people
Work fast	Can pollute air, water, and land
Are safe if used properly	Are expensive for farmers

B Brown/Shutterstock.com

FIGURE 10.20 Use of synthetic pesticides has advantages and disadvantages. ***Critical thinking:*** Which single advantage and which single disadvantage do you think are the most important? Why?

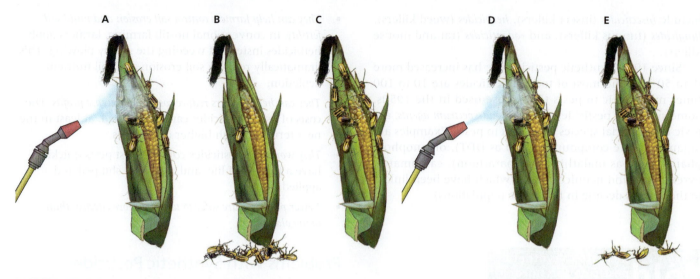

FIGURE 10.21 When a pesticide is sprayed on a crop **(a)**, a few pest insects resist it and survive **(b)**. The survivors reproduce and pass on their trait for resistance to the pesticide **(c)**. When the crop is sprayed again **(d)**, more insects resist and survive it and continue reproducing **(e)**. The pesticide has now become ineffective and the farmer must look for a stronger pesticide.

minor pests until widespread use of insecticides wiped out many of their natural predators. (See Case Study that follows.)

- *Pesticides are usually applied inefficiently and pollute the environment.* According to the USDA, about 98–99.9% of the insecticides and more than 95% of the herbicides applied by aerial spraying or ground spraying do not reach the target pests. They end up in the air, surface water, groundwater, bottom sediments, food, and nontarget organisms, including humans.

- *Some pesticides harm wildlife.* According to the USDA and the U.S. Fish and Wildlife Service, each year, some of the pesticides applied to cropland poison honeybee colonies on which we depend for pollination of many food crops (see Chapter 8, Core Case Study, p. 168, and Science Focus 8.2, p. 180). According to a study by the Center for Biological Diversity, pesticides menace about a third of all endangered and threatened species in the United States.

- *Some pesticides threaten human health.* The WHO and UNEP have estimated that pesticides annually poison at least 3 million agricultural workers in less-developed countries and at least 300,000 workers in the United States. They also cause 20,000–40,000 deaths per year, worldwide and household pesticides such as ant and roach sprays sicken 2.5 million people per year. According to studies by the National Academy of Sciences, pesticide residues in food cause an estimated 4,000–20,000 cases of cancer per year in the United States. The pesticide industry disputes these claims, arguing that if used as directed, pesticides do not remain in the environment at levels high enough to cause serious environmental or health problems. Figure 10.22 lists some ways to reduce your exposure to residues of synthetic pesticides.

What Can You Do?

Reducing Exposure to Pesticides

- Grow some of your food using organic methods
- Buy certified organic food
- Wash and scrub all fresh fruits and vegetables
- Eat less meat, no meat, or certified organically produced meat
- Before cooking, trim the fat from meat

FIGURE 10.22 Individuals Matter: You can reduce your exposure to pesticides. *Critical thinking:* Which three of these steps do you think are the most important ones to take? Why?

CONSIDER THIS ...

CONNECTIONS Pesticides and Food Choices

According to the Environmental Working Group (EWG), a research organization, you could reduce your pesticide intake by up to 90% by eating only 100% USDA Certified Organic versions of 12 types of fruits and vegetables that tend to have the highest pesticide residues. In 2016, these foods, which the EWG calls the "dirty dozen," were strawberries, apples, nectarines, peaches, celery, grapes, cherries, spinach, tomatoes, sweet bell peppers, cherry tomatoes, and cucumbers.

CASE STUDY

Ecological Surprises: The Law of Unintended Consequences

Malaria once infected 9 of every 10 people in North Borneo, now known as the eastern Malaysian state of Sabah. In 1955, the WHO sprayed the island with dieldrin (a DDT relative) to kill malaria-carrying mosquitoes. The program was so successful that the dreaded disease was nearly eliminated.

Then unexpected things began to happen. The dieldrin also killed other insects, including flies and cockroaches living in houses, which made the islanders happy. Next, small insect-eating lizards living in the houses died after gorging themselves on dieldrin-contaminated insects. Then cats began dying after feeding on the lizards. In the absence of cats, rats flourished in and around the villages. When the residents became threatened by sylvatic plague carried by rat fleas, the WHO parachuted healthy cats onto the island to help control the rats. Operation Cat Drop worked.

Then the villagers' roofs began to fall in. The dieldrin had killed wasps and other insects that fed on a type of caterpillar that was not affected by the insecticide. With most of its predators eliminated, the caterpillar population exploded, munching its way through its favorite food: the leaves used in thatch roofs.

Ultimately, this story ended well. Both malaria and the unexpected effects of the spraying program were brought under control. Nevertheless, this chain of unintended and unforeseen events reminds us that whenever we intervene in nature and affect organisms that interact with one another, we need to ask, "Now what will happen?"

Pesticides Have Not Consistently Reduced U.S. Crop Losses to Pests

Largely because of genetic resistance and the loss of many natural predators, synthetic pesticides have not always succeeded in reducing U.S. crop losses. David Pimentel, an expert on insect ecology, evaluated data from more than 300 agricultural scientists and economists. He reached three major conclusions. He found that between 1942 and 1997, estimated crop losses from insects almost doubled from 7% to 13%, despite a 10-fold increase in the use of synthetic insecticides. He also estimated that alternative pest management practices could cut the use of synthetic pesticides by half on 40 major U.S. crops without reducing crop yields (**Concept 10.4**).

The pesticide industry disputes these findings. However, numerous studies and experience support them. For example, Sweden has cut its pesticide use in half with almost no decrease in crop yields.

Regulating Synthetic Pesticide Use

More than 20,000 different pesticide products are used in the United States. Three federal agencies, the EPA, the USDA, and the FDA, regulate the use of these pesticides under the Federal Insecticide, Fungicide, and Rodenticide Act (FIFRA), first passed in 1947 and amended in 1972. Critics argue that that FIFRA has not been well enforced and the EPA says that the U.S. Congress has not provided them with enough funds to carry out the complex and lengthy process of evaluating pesticides for toxicity.

In 1996, Congress passed the Food Quality Protection Act, mostly because of growing scientific evidence and citizen pressure concerning the effects of small amounts of pesticides on children. This act requires the EPA to reduce the allowed levels of pesticide residues in food by a factor of 10 when there is inadequate information on the potentially harmful effects on children. Some scientists call for reducing the levels by a factor of 100.

Between 1972 and 2016, the EPA used FIFRA to ban or severely restrict the use of 64 active pesticide ingredients, including DDT and most other chlorinated hydrocarbon insecticides. However, according to studies by the National Academy of Sciences, federal laws regulating pesticide use generally are inadequate and poorly enforced. A 2015 study by the U.S. General Accounting Office found that the FDA tests less than one-tenth of 1% of all imported fruits and vegetables. The FDA also does not test foods for some of the pesticide residues that are strictly regulated by the EPA.

In what environmental scientists call a *circle of poison*, or the *boomerang effect*, residues of synthetic pesticides that have been banned or not registered in one country but exported to other countries can return to the exporting countries on imported food. Winds can also carry persistent pesticides from one country to another.

In 2000, more than 100 countries developed an international agreement to ban or phase out the use of 12 especially hazardous persistent organic pollutants (POPs)—9 of them persistent hydrocarbon pesticides such as DDT and other chemically similar pesticides. By 2015, the initial list of 12 chemicals had been expanded to 25. In 2004, the POPs treaty went into effect. By 2016, it had been signed or ratified by 180 countries, not including the United States.

Alternatives to Synthetic Pesticides

Many scientists urge us to greatly increase the use of biological, ecological, and other alternative methods for controlling pests and diseases that affect crops and human health (**Concept 10.4**). Here are some of these alternatives:

- *Fool the pest.* A variety of *cultivation practices* can be used to fake out pests. Examples include rotating the types of crops planted in a field each year and adjusting planting times so that major insect pests either starve or are eaten by their natural predators.

- *Provide homes for pest enemies.* Farmers can increase the use of polyculture, which uses plant diversity to reduce losses to pests by providing habitats for the predators of pest species.

- *Implant genetic resistance.* Use genetic engineering to speed up the development of pest- and disease-resistant crop strains.

- *Bring in natural enemies.* Use *biological control* by importing natural predators (Figures 10.19 and 10.23), parasites, and disease-causing bacteria and viruses to help regulate pest populations. This approach is nontoxic to other species and is usually less costly than applying pesticides. However, some biological control agents are difficult to mass-produce and are often slower acting and more difficult to apply than synthetic pesticides are. Sometimes the agents can multiply and become pests themselves.

- *Use insect scents.* Trace amounts of *sex attractants* (called *pheromones*) can be used to lure pests into traps or to attract their natural predators into crop fields. Each of these chemicals attracts only one species. They have little chance of causing genetic resistance and are not harmful to nontarget species. However, they are costly and time-consuming to produce.

- *Use insect hormones.* Hormones produced by animals control their developmental processes at *different* stages of life. Scientists have learned how to identify and use hormones that disrupt an insect's normal life cycle, thereby preventing it from reaching maturity and reproducing. Use of insect hormones has some of the same advantages and disadvantages as use of sex attractants has. In addition, they take weeks to kill an insect, are often ineffective with large infestations of insects, and sometimes break down before they can act.

- *Use natural methods to control weeds.* Farmers can control weeds by methods such as crop rotation, mechanical cultivation, hand weeding, and the use of cover crops and mulches.

FIGURE 10.23 Natural Capital: In this example of biological pest control, a wasp is parasitizing a gypsy moth caterpillar.

Scott Bauer/USDA Agricultural Research Service

Integrated Pest Management

Many pest control experts and farmers believe that the best way to control crop pests is through **integrated pest management (IPM)**, a program in which each crop and its pests are evaluated as parts of an ecosystem (**Concept 10.4**). The overall aim of IPM is to reduce crop damage to an economically tolerable level with minimal use of synthetic pesticides.

When farmers detect an economically damaging level of pests in a field, they start with using biological methods (natural predators, parasites, and disease organisms) and cultivation controls (such as altering planting time and growing different crops on fields from year to year to disrupt pests). They apply small amounts of synthetic insecticides only when insect or weed populations reach a threshold where the potential cost of pest damage to crops outweighs the cost of applying the pesticide.

IPM works. In Sweden and Denmark, farmers have used it to cut their synthetic pesticide use by more than half. In Cuba, where organic farming is used almost exclusively, farmers make extensive use of IPM. In Brazil, IPM has reduced pesticide use on soybeans by as much as 90%.

According to the U.S. National Academy of Sciences, a well-designed IPM program can reduce synthetic pesticide use and pest control costs by 50–65%, without reducing crop yields and food quality. IPM can also reduce inputs of fertilizer and irrigation water and slow the development of genetic resistance. IPM is an important example of pollution prevention that reduces risks to wildlife and human health and applies the biodiversity **principle of sustainability**.

Despite its potential, IPM has some drawbacks. It requires expert knowledge about each pest situation and takes more time than does relying solely on synthetic pesticides. Methods developed for a crop in one area might not apply to areas with even slightly different growing conditions. Initial costs may be higher, although long-term costs typically are lower than those of using conventional pesticides. Widespread use of IPM has been hindered in the United States and other countries by government subsidies that support use of synthetic pesticides, as well as by opposition from pesticide manufacturers, and a shortage of IPM experts. **GREEN CAREER: Integrated pest management**

A growing number of scientists urge the USDA to use a three-point strategy to promote IPM in the United States. *First,* add a small sales tax on synthetic pesticides and use the revenue to fund IPM research and education. *Second,* set up a federally supported IPM demonstration project on at least one farm in every county in the United States. *Third,* train USDA field personnel and county farm agents in IPM so they can help farmers use this alternative.

Several UN agencies and the World Bank have joined to establish an IPM facility. Its goal is to promote the use of IPM by disseminating information and establishing networks among researchers, farmers, and agricultural extension agents involved in IPM.

GOOD NEWS

10.5 HOW CAN WE PRODUCE FOOD MORE SUSTAINABLY?

CONCEPT 10.5 We can produce food more sustainably by conserving topsoil, producing meat more efficiently, reducing the harmful effects of aquaculture, and eliminating government subsidies that promote environmentally harmful types of agriculture.

Protecting Topsoil

Land used for food production must have fertile topsoil (Figure 3.9, p. 52), which takes hundreds of years to form. Thus, sharply reducing topsoil erosion is a key component of more sustainable agriculture and an important way to increase our beneficial environmental impact.

Soil conservation involves using a variety of methods to reduce topsoil erosion and restore soil fertility, mostly by keeping the land covered with vegetation. For example, *terracing* involves converting steeply sloped land into a series of broad, nearly level terraces that run across the land's contours (Figure 10.24a). Each terrace retains water for crops and reduces topsoil erosion by controlling runoff.

On less steeply sloped land, *contour planting* (Figure 10.24b) can be used to reduce topsoil erosion. It involves plowing and planting crops in rows across the slope of the land rather than up and down. Each row acts as a small dam to help hold topsoil by slowing runoff. Similarly, *strip-cropping* (Figure 10.24b) helps reduce erosion and restore soil fertility with alternating strips of a row crop (such as corn or cotton) and another crop that completely covers the soil, called a *cover crop* (such as alfalfa, clover, oats, or rye). The cover crop traps topsoil that erodes from the row crop and catches and reduces water runoff.

Alley cropping, or *agroforestry* (Figure 10.24c), is another way to slow the erosion of topsoil and maintain soil fertility. One or more crops, usually legumes or other crops that add nitrogen to the soil, are planted together in alleys between orchard trees or fruit-bearing shrubs, which provide shade. This reduces water loss by evaporation and helps retain and slowly release soil moisture.

Farmers can also establish *windbreaks*, or *shelterbelts*, of trees around crop fields to reduce wind erosion (Figure 10.24d). The trees retain soil moisture, supply wood for fuel, and provide habitats for birds and insects that help with pest control and pollination.

Another way to greatly reduce topsoil erosion is to eliminate or minimize the plowing and tilling of topsoil and to leave crop residues on the ground. Such *conservation-tillage farming* uses specialized machines that inject seeds and fertilizer directly through crop residues into minimally disturbed topsoil. Weeds are controlled with herbicides.

This type of farming increases crop yields and greatly reduces soil erosion and water pollution from sediment and fertilizer runoff. It also helps farmers survive prolonged drought by holding more moisture in the soil. However, one drawback is that the greater use of herbicides promotes the growth of herbicide-resistant weeds that force farmers to use larger doses of weed killers or, in some cases, to return to plowing. However, organic no-till methods being developed at the Rodale Research Institute (see chapter-opening photo) could be herbicide-free and will not require plowing.

According to government surveys, farmers used conservation tillage methods on about 35% of U.S. cropland. Worldwide, it is used on only about 10% of all cropland, but its use is increasing.

Another way to conserve topsoil is to grow plants without using soil. Some producers are raising crops in greenhouses, using a system called *hydroponics*. At the Growing Power farm (**Core Case Study**), Will Allen has developed such a system for raising salad greens and fish together. Wastewater from the fish tanks flows into hydroponic troughs where it nourishes the plants. The plant roots filter the water, which is then returned to the fish runs. This closed-loop, chemical-free *aquaponic system* conserves soil, water, and energy while supporting more than 100,000 tilapia and perch, which are sold in local markets along with the salad greens.

Still another way to conserve topsoil is to retire the estimated one-tenth of the world's highly erodible cropland. The goal is to identify *erosion hotspots*, stop cultivating them, and plant them with grasses or trees, at least until their topsoil has been renewed.

Some countries, such as the United States, have paid farmers to set aside considerable areas of cropland for conservation purposes. Under the 1985 Food Security Act (Farm Act), more than 400,000 farmers participating in the Conservation Reserve Program received subsidy payments for taking highly erodible land out of production and replanting it with grass or trees for 10–15 years. Since 1985, these efforts have cut total topsoil losses on U.S. cropland by 40%.

CONSIDER THIS ...

CONNECTIONS Corn, Ethanol, and Soil Conservation

In recent years, some U.S. farmers took erodible land out of the conservation reserve in order to receive more generous government subsidies for planting corn (which removes nitrogen from the soil and reduces the ability of soil to store carbon by removing CO_2 from the atmosphere) to make ethanol for use as a motor vehicle fuel. This led to mounting political pressure to abandon or sharply cut back on the nation's highly successful topsoil conservation reserve program.

a.

b.

c.

d.

Lim Yong Hian/Shutterstock.com

Ron Nichols/USDA Natural Resources Conservation Service

inga spence/Alamy Stock Photo

Fedorov Oleksiy/Shutterstock.com

FIGURE 10.24 Soil conservation methods include **(a)** terracing; **(b)** contour planting and strip cropping; **(c)** alley cropping; and **(d)** windbreaks between crop fields.

Restoring Soil Fertility

Another way to protect soil is to restore some of the lost plant nutrients that have been washed, blown, or leached out of topsoil, or that have been removed by repeated crop harvesting. To restore topsoil, farmers can use **organic fertilizer** derived from plant and animal materials or synthetic inorganic fertilizer made of inorganic compounds that contain *nitrogen, phosphorus,* and *potassium* along with trace amounts of other plant nutrients.

There are several types of *organic fertilizers*. One is **animal manure**: the dung and urine of cattle, horses, poultry, and other farm animals. Manure improves topsoil structure, adds organic nitrogen, and stimulates the

growth of beneficial soil bacteria and fungi. Another type, called **green manure**, consists of freshly cut or growing green vegetation that is plowed into the topsoil to increase the organic matter and humus available to the next crop. A third type is **compost**, produced when microorganisms break down organic matter such as leaves, crop residues, food wastes, paper, and wood in the presence of oxygen.

The Growing Power farm (**Core Case Study**) depends greatly on its large piles of compost. Will Allen invites local grocers and restaurant owners to send their food wastes to add to the pile. To make this compost, Allen uses millions of red wiggler worms, which reproduce rapidly and eat their own weight in food wastes every day, converting the wastes to plant nutrients. The composting process generates a considerable amount of heat, which is used to help warm the farm's greenhouses during cold months.

Another form of organic fertilizer is *biochar*. It is a form of charcoal made from often discarded woody materials through a process called *pyrolysis* that heat them at low temperatures in containers that limit oxygen input until the materials become charcoal. The biochar can be buried to enrich topsoil. It has the added benefit of removing CO_2 from the atmosphere, thereby helping slow climate change.

We degrade soils when we plant crops such as corn and cotton on the same land several years in a row, a practice that can deplete nutrients—especially nitrogen—in the topsoil. One way to reduce such losses is through **crop rotation**, in which a farmer plants a series of different crops in the same area from season to season. For example, where a nitrogen-depleting crop is grown one year, the farmer can plant the same area the following year with a crop such as legumes, which add nitrogen to the soil.

Many farmers, especially those in more-developed countries, rely on synthetic inorganic fertilizers. Their use accounts for about 25% of the world's crop yield. While these fertilizers can replace depleted inorganic nutrients, they do not replace organic matter. Completely restoring topsoil nutrients requires both inorganic and organic fertilizers.

Reducing Soil Salinization and Desertification

We know how to prevent and deal with soil salinization, as summarized in Figure 10.25. The problem is that most of these solutions are costly.

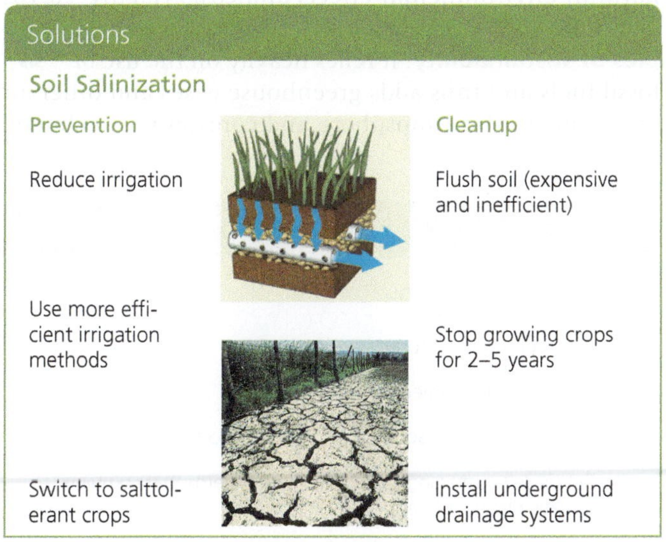

Solutions	
Soil Salinization	
Prevention	**Cleanup**
Reduce irrigation	Flush soil (expensive and inefficient)
Use more efficient irrigation methods	Stop growing crops for 2–5 years
Switch to salttolerant crops	Install underground drainage systems

FIGURE 10.25 Ways to prevent and ways to clean up soil salinization. **Critical thinking:** Which two of these solutions do you think are the best ones? Why?

USDA Natural Resources Conservation Service

Reducing desertification is not easy. We cannot control the timing and location of prolonged droughts caused by changes in weather and climate patterns. However, we can reduce population growth, overgrazing, deforestation, and destructive forms of planting and irrigation in dryland areas, which have left much land vulnerable to topsoil erosion and thus desertification. We can also work to decrease the human contribution to projected climate change, which could increase the severity of droughts in larger areas of the world during this century.

It is possible to restore land suffering from desertification by planting trees and other plants that anchor topsoil and hold water. We can also grow trees and crops together (alley cropping, Figure 10.24c) and establish windbreaks around farm fields (Figure 10.24d).

Reducing the Environmental Effects of Meat Production

The production of meat and dairy products has a huge environmental impact. Meat consumption is the largest factor in the growing ecological footprints of individuals in affluent nations.

Some types of meat are produced more efficiently than others (Figure 12.26). For example, producing a pound of beef requires more than three times the amount of grain needed to produce a pound of pork. A more sustainable form of meat production and consumption would involve shifting from less grain-efficient forms of animal protein, such as beef, pork, and carnivorous fish produced by aquaculture, to more grain-efficient forms, such as poultry and plant-eating farmed fish.

Insects are another form of protein. Would you consider trying a beetle salad, a caterpillar stew, or a handful of crunchy fried ants for a snack? If you are thinking *yuck*, you are not alone. Yet, at least 2,000 species of insects provide nutrients to more than 2 billion people, according to the FAO. In countries such as Thailand, Australia, and Mexico, insects are deep-fried with spices, made into flours and tasty sauces, baked into breads, and cooked in stews. Insects can be rich in protein, fiber, and healthy fats. They

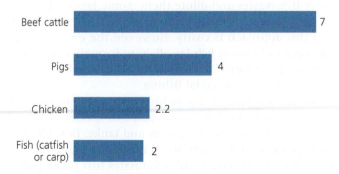

FIGURE 10.26 Kilograms of grain required for each kilogram of body weight added for each type of animal.

Compiled by the authors using data from U.S. Department of Agriculture.

can provide vital micronutrients such as calcium, iron, B vitamins, and zinc.

A growing number of people who can afford to eat meat have one or two meatless days per week. Others go further and eliminate most or all meat from their diets. They replace the meat with a balanced vegetarian diet of fruits, vegetables, and protein-rich foods such as peas, beans, and lentils. According to one estimate, if all Americans picked one day per week to have no meat, the reduction in greenhouse gas emissions would be equivalent to taking 30–40 million cars off the road for a year.

CONSIDER THIS ...

THINKING ABOUT Meat Consumption

If you do not do this already, would you be willing to live lower on the food chain by eating much less meat, or no meat at all? Explain.

Practicing More Sustainable Aquaculture

Various organizations have established guidelines, standards, and certifications to encourage more sustainable aquaculture and fishing practices. The Aquaculture Stewardship Council (ASC), for example, has developed aquaculture sustainability standards, although it has certified only about 4.6% of the world's aquaculture production operations. The Marine Stewardship Council (MSC) performs a similar program for wild-caught fisheries. Like the organic certification, the programs have associated labels for food products that help consumers purchase more sustainable options.

Scientists and producers are working on ways to make aquaculture more sustainable and to reduce its harmful environmental effects. One approach is open-ocean aquaculture, which involves raising large carnivorous fish in underwater pens—some as large as a high school gymnasium. They located far offshore (see Figure 9.24, p. 216), where rapid currents can sweep away fish wastes and dilute them. Some farmed fish can escape from such operations and breed with wild fish, and this approach is costly. However, the environmental impact of raising fish far offshore is smaller than that of raising fish near shore and much smaller than that of industrialized commercial fishing.

Other fish farmers are reducing coastal damage from aquaculture by raising shrimp and fish species in inland facilities using freshwater ponds and tanks. In such *recirculating aquaculture systems,* the water used to raise the fish is continually recycled. This eliminates fish waste pollution of aquatic systems and reduces the need for antibiotics and other chemicals used to combat disease among farmed fish. It also eliminates the problem of farmed

fish escaping into natural aquatic systems. The Growing Power aquaponic system (Core Case Study) captures its fish wastes and converts them to fertilizer used to grow salad greens.

Making aquaculture more sustainable will require some changes for producers and consumers. One change is for more consumers to choose fish species such as carp, tilapia, and catfish that eat algae and other vegetation rather than fish oil and fishmeal produced from other fish. Raising carnivorous fishes such as salmon, trout, tuna, grouper, and cod contributes to overfishing and population crashes of species and is unsustainable. Aquaculture producers can avoid this problem by raising herbivorous fishes as long as they do not try to increase yields by feeding fishmeal to such species, as many of them are doing.

Fish farmers can also emphasize *polyaquaculture,* which has been part of aquaculture for centuries, especially in Southeast Asia. Polyaquaculture operations raise fish or shrimp along with algae, seaweeds, and shellfish in coastal lagoons, ponds, and tanks. The wastes of the fish or shrimp feed the other species. Polyaquaculture applies the chemical cycling and biodiversity principles of sustainability. GREEN CAREER: Sustainable aquaculture

Figure 10.27 lists some ways to make aquaculture more sustainable and to reduce its harmful environmental effects.

Shift to More Sustainable Food Production

Modern industrialized food production yields large amounts of food at affordable prices. However, to a growing number of analysts, it is unsustainable because of its harmful environmental effects and costs (Figure 10.10) and because it violates the three scientific principles of sustainability. It relies heavily on the use of fossil fuels and thus adds greenhouse gases and other air pollutants to the atmosphere and contributes to climate

Solutions

More Sustainable Aquaculture

- Protect mangrove forests and estuaries
- Improve management of wastes
- Reduce escape of aquaculture species into the wild
- Set up self-sustaining polyaquaculture systems that combine aquatic plants, fish, and shellfish
- Certify sustainable forms of aquaculture

FIGURE 10.27 Ways to make aquaculture more sustainable and reduce its harmful effects. *Critical thinking:* Which two of these solutions do you think are the best ones? Why?

More Sustainable Food Production

More	Less
High-yield polyculture	Soil erosion
Organic fertilizers	Soil salinization
Biological pest control	Water pollution
Integrated pest management	Aquifer depletion
Efficient irrigation	Overgrazing
Perennial crops	Overfishing
Crop rotation	Loss of biodiversity and agrobiodiversity
Water-efficient crops	Fossil fuel use
Soil conservation	Greenhouse gas emissions
Subsidies for sustainable farming	Subsidies for unsustainable farming

FIGURE 10.28 More sustainable, low-input food production has a number of major components. *Critical thinking:* For each list in this diagram (left and right), which two items do you think are the most important? Why?

Top: Marko5/Shutterstock.com. Center: Anhong/Dreamstime.com. Bottom: Pacopi/Shutterstock.com.

Organic Farming

Advantages	Disadvantages
Reduces soil erosion	More use of manure can cause more surface and groundwater pollution
Retains more water in soil during drought years	Large-scale composting can generate greenhouse gases
Improves soil fertility	Lower crop yields in large-scale production
Uses less energy and emits less CO_2	Can require plowing for weed control leading to more erosion
Eliminates water pollution from pesticides and synthetic fertilizers	Higher cost can lead to higher prices
Benefits birds, bats, bees, and other wildlife	

FIGURE 10.29 Organic farming has advantages and disadvantages. *Critical thinking:* Which single advantage and which single disadvantage do you think are the most important? Why?

Bottom: Marbury67/Dreamstime.com. Top: Robert Kneschke/Shutterstock.com

change. It also reduces biodiversity and agrobiodiversity and interferes with the cycling of plant nutrients. These harmful effects are hidden from consumers because most of them are not included in the market prices of food—a violation of the full-cost pricing **principle of sustainability** (see inside back cover of this book).

A more sustainable food production system would have several major components (Figure 10.28; **Concept 10.5**). One component is USDA 100% Certified Organic Agriculture. Like any technology, organic farming has advantages and disadvantages (Figure 10.29). However, many experts, including Scientists the Rodale Institute Center (see chapter-opening photo), have been conducting research on organic farming since 1981. According to these and number of other scientists, research shows that the advantages of organic farming outweigh its disadvantages. By eating USDA 100% certified organic foods, consumers also reduce their exposure to pesticide residues and to bacteria-resistant antibiotics that can be found in conventional foods. They also increase their intake of beneficial antioxidants by almost 70%.

Sustainable agriculture would rely less on conventional monoculture and more on organic polyculture. Research shows that, on average, low-input polyculture produces higher yields per unit of land than does high-input industrialized monoculture. It also uses less energy and fewer resources and provides more food security for small landowners. Ecologist David Tilman (Individuals Matter 10.1) has been instrumental in demonstrating the benefits of polyculture. The Growing Power farm (**Core Case Study**) practices polyculture by growing a variety of crops in several greenhouses.

Of particular interest to some scientists is the idea of using polyculture to grow *perennial crops*—crops that grow back year after year on their own (Science Focus 10.2).

Another key to developing more sustainable agriculture is to shift from using fossil fuels to relying more on renewable energy for food production—an important application of the solar energy **principle of sustainability** that has been well demonstrated by the Growing Power farm (**Core Case Study**). To produce the electricity and fuels needed for food production, farmers can make greater use of renewable solar energy, wind, flowing water, and biofuels produced from farm wastes in tanks called *biogas digesters*. Most proponents of more sustainable agriculture call for using more environmentally sustainable forms of both high-yield polyculture and high-yield monoculture.

Agricultural experts such as Jonathan Foley argue that industrialized farming could play a major role in shifting to more sustainable food production. Foley notes that farmers

David Tilman—Polyculture Researcher

One of the world's most prominent ecologists and agricultural experts is David Tilman, a professor at the University of Minnesota. Since 1981, he has conducted more than 150 long-term controlled experiments on a university-owned grassland. For example, he applied certain mixes of fertilizers and water on experimental plots and observed how the plants responded. He then compared the results from the experimental plots with those from control plots that did not receive the experimental treatments. He also used varying mixes of plant species in such experiments, focusing on the benefits of polyculture.

With ecologist Peter Reich, Tilman found that carefully controlled polyculture plots consistently out-produced monoculture plots. Such research explains why some analysts argue for greatly expanding the use of polyculture to produce food more sustainably.

Tilman's findings also support the scientific idea that biodiversity can make ecosystems more stable and sustainable. A diverse forest, for example, suffers damage during pest infestations, but there are enough different species to withstand the damage. Some species are wiped out, but not the whole forest. A vast field of corn or wheat, on the other hand, is highly vulnerable to heat waves, drought, crop disease, and pests.

For his important research efforts, Tilman has received several awards. In 2010, he was awarded the prestigious Heineken Prize for Environmental Sciences for his important contributions to the science of ecology.

are already finding ways to apply pesticides and fertilizers in smaller amounts and more precisely, using computerized tractors and remote sensing and Global Positioning System (GPS) technology. Fertilizers can also be mixed and tailored to different soil conditions to help minimize runoff into waterways, and irrigation can be done much more efficiently. Such methods could increase production on the current total area of farmland by 50–60% with a lower harmful environmental impact, according to Foley.

Proponents of more sustainable food production systems say that education is an important key to producing and consuming food more sustainably. They seek to inform people, especially young consumers, about where their food really comes from, how it is produced, and what the environmentally harmful effects of food production are. They also call for economic policies that reward more sustainable agriculture. A major part of such policies would be to shift subsidies from unsustainable to more sustainable food production so that food should be good for people and for the planet, an application of the ethical **principle of sustainability**. Figure 10.30 lists ways in which you can promote more sustainable food production.

10.6 HOW CAN WE IMPROVE FOOD SECURITY?

CONCEPT 10.6 We can improve food security by subsidizing environmentally sustainable food production, producing food more sustainably, reducing poverty and chronic malnutrition, relying more on locally grown food, and cutting food waste.

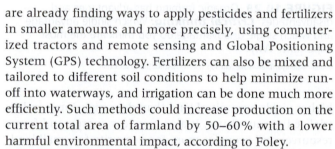

What Can You Do?

More Sustainable Food Production

- Eat less meat, no meat, or organically certified meat
- Choose sustainably produced herbivorous fish
- Use organic farming to grow some of your food
- Buy certified organic food
- Eat locally grown food
- Compost food wastes
- Cut food waste

FIGURE 10.30 Individuals Matter: Ways to promote more sustainable food production (**Concept 10.6**). **Critical thinking:** Which three of these actions do you think are the most important ones to take? Why?

Use Government Policies to Improve Food Production and Security

Agriculture is a financially risky business. Whether farmers have a good or bad year depends on factors over which they have little control, including weather, crop prices, pests and diseases, interest rates on loans, and global food markets.

Governments use two main approaches to influence food production with hopes of strengthening food security. First, they can *control food prices* by putting a legally mandated upper limit on them to keep them artificially

Perennial Polyculture and The Land Institute

Some scientists call for greater reliance on polycultures of perennial crops as a component of more sustainable agriculture. Such crops can live for several years without having to be replanted and are better adapted to regional soil and climate conditions than are most annual crops.

More than three decades ago, plant geneticist Wes Jackson cofounded the Land Institute in the U.S. state of Kansas. One of the institute's goals has been to grow a diverse mixture (polyculture) of edible perennial plants to supplement traditional annual monoculture crops and to help reduce the latter's harmful environmental effects.

Perennial crops, which live for several years, help farmers copy nature by better using and conserving natural resources—sunlight, soil, and water. Because there is no need to till (plow)

the soil and replant perennials each year, this approach produces much less topsoil erosion and water pollution. It also reduces the need for irrigation because the deep roots of such perennials retain more water than do the shorter roots of annuals (Figure 10.A). Often, there is a reduced need for chemical fertilizers and pesticides, and thus little or no pollution from these sources. Perennial crops also remove and store more carbon from the atmosphere and growing them requires less energy than does growing annual crops in conventional monocultures.

CRITICAL THINKING

Why do you think large seed companies generally oppose this form of more sustainable agriculture?

FIGURE 10.A The roots of an annual wheat crop plant (left) are much shorter than those of big bluestem (right), a tallgrass prairie perennial plant.

Photo Courtesy of The Land Institute

---

low. This makes consumers happy but makes it harder for farmers to make a living.

Second, they can *provide subsidies* by giving farmers price supports, tax breaks, and other financial support to help them stay in business and to encourage them to increase food production. In the United States, most subsidies go to industrialized food production, often in support of environmentally harmful practices.

Some opponents of subsidies call for ending them. They point to New Zealand, which ended farm subsidies in 1984. After the shock wore off, innovation took over and production of some foods such as milk quadrupled. Brazil has also ended most of its farm subsidies. Some analysts call for replacing traditional subsidies for farmers with subsidies that promote more environmentally sustainable farming practices.

Similarly, government subsidies to fishing fleets can promote overfishing and the reduction of aquatic biodiversity. For example, several governments give the highly destructive bottom-trawling industry (see Figure 9.22, p. 213) millions of dollars per year in subsidies, which is the main reason fishers who use this environmentally destructive practice can stay in business. Many analysts call for replacing those harmful subsidies with

subsidies that would promote more sustainable fishing and aquaculture.

Some analysts call for countries to use laws and regulations to guide people in producing food more sustainably. Such laws could set standards for sustainable production, cut subsidies that promote unsustainable practices, regulate industrial farm pollution, and set goals for agricultural greenhouse gas reductions. Some have argued for levying more taxes on pesticide and fertilizer use and emissions of climate-changing methane. The proceeds could be used to subsidize organic farming, IPM, and sustainable food production on farms (**Concept 10.6**).

Other Government and Private Programs Are Increasing Food Security

Government and private programs aimed at reducing poverty can improve food security. For example, some programs provide small loans at low interest rates to poor people to help them start businesses or buy land to grow their own food.

Some analysts urge governments to establish programs focused on saving children from the harmful health effects of poverty. Studies by the United Nations Children's Fund

Jennifer Burney: Environmental Scientist and National Geographic Explorer

Environmental scientist and National Geographic Explorer Jennifer Burney notes that subsistence farmers represent the majority of the world's poorest people and need to boost their productivity for better standards of living and health. She is trying to help such farmers in Africa to grow, distribute, and cook their food using resources like water, fertilizer, and energy as efficiently as possible.

For example, to deal with the problem of seasonal water shortages in many parts of sub-Saharan Africa, Burney has helped organizations connect two technologies—solar irrigation systems and drip irrigation—that could serve as a solution. Drip irrigation systems sip water and drip it directly onto plant roots instead of pumping and dumping it. Solar-powered pumps work without the need for batteries or fuel. On sunny days, when crops need water more, the solar panels speed the pumping; on cloudy days, when there is less evaporation, the pumping slows down. Thus, only the amount of water that is needed is pumped on most days. This has allowed farmers to grow fruits and vegetables on a larger scale and to improve their incomes and food security.

UC San Diego

(UNICEF) indicate that one-half to two-thirds of nutrition-related childhood deaths could be prevented at an average annual cost of $5 to $10 per child. This involves simple measures such as immunizing more children against childhood diseases, preventing dehydration due to diarrhea by giving infants a mixture of sugar and salt in their water, and combating blindness by giving children an inexpensive vitamin A capsule twice a year.

Many private, mostly nonprofit, organizations are also working to help individuals, communities, and nations improve their food security and produce food more sustainably. For example, Growing Power's Will Allen (**Core Case Study**) argues that instead of trying to transfer complex technologies such as genetic engineering to less-developed countries, we should be helping them develop simple, sustainable, local food production and distribution systems that would give them more control over their food security.

Sustainable agriculturalists and National Geographic Explorers Cid Simones and Paola Segura work with small farmers to show them how to grow food more sustainably on small plots in the tropical forests of Brazil. They train one family at a time. In return, each family must teach five other families and thus help spread more sustainable farming methods. Another person who is working toward this goal in Africa is National Geographic Explorer Jennifer Burney (Individuals Matter 10.2).

Buying Locally Grown Food and Cutting Food Waste

One way to increase food security is to grow more of our food locally or regionally, ideally with USDA 100%

certified organic farming practices. A growing number of consumers are becoming "locavores" who try to buy as much of their food as possible from local and regional producers in farmers' markets, which provide access to fresher seasonal foods, many of them grown organically.

In addition, many people participate in *community-supported agriculture (CSA)* programs. In these programs, people buy shares of a local farmer's crops and receive a box of fruits or vegetables on a regular basis during the growing season. Growing Power (**Core Case Study**) runs such a program for inner-city residents. For many of these people, the organically grown food they get from the urban farm greatly improves their diets and increases their chances of living longer and healthier lives.

By buying food locally, people support local economies and farm families. Buying locally also reduces fossil fuel energy costs for food producers, as well as the greenhouse gas emissions from storing and transporting food products over long distances. There are limits to this benefit, however. Food scientists point out that the largest share of carbon footprint for most foods is in production. Thus, for example, an apple grown through high-input agriculture and trucked across North America could have a larger footprint than an apple grown through low-input farming and sent on a ship from South America.

An increase in the demand for locally grown food could result in more small, diversified farms that produce organic, minimally processed food from plants and animals. Such ecofarming could be one of this century's challenging new careers for many young people. **GREEN CAREER: Small-scale sustainable agriculture**

People can also waste less food. According to Jonathan Foley, 25% of the world's food calories are lost or wasted. In poor countries with unreliable food storage and transportation, much is lost before it gets to consumers. In wealthy countries, much waste occurs in restaurants, homes, and supermarkets. According to studies by the EPA and the Natural Resources Defense Council, Americans throw away 30–40% of the country's food supply each year while 49 million Americans experience chronic hunger.

Growing More Food in Urban Areas

Sustainable agriculture entrepreneurs and ordinary citizens who live in urban areas could grow more of their own food, as the Growing Power farm has shown (**Core Case Study**). According to the USDA, approximately 15% of the world's food is grown in urban areas, and this percentage could easily be doubled.

Increasingly, people are sharing garden space, labor, and produce in community gardens in vacant lots. People are planting gardens and raising chickens in backyards, growing dwarf fruit trees in large containers of soil, and raising vegetables in containers on rooftops, balconies, and patios. One study estimates that converting 10% of American lawns into food-producing gardens would supply one-third of the country's fresh produce.

Many urban schools, colleges, and universities are benefitting from having gardens on school grounds. Not only do the students have a ready source of fresh produce, but they also learn about where their food comes from and how to grow their own food more sustainably.

In the future, much of our food might be grown in cities within specially designed high-rise buildings. Such a building would have rooftop solar panels for generating electricity, and it could capture and recycle rainwater for irrigating its wide diversity of crops. Sloped glass facing south would bring in sunlight, and excess heat collected in this way could be stored in tanks containing water or sand underneath the building for use as needed. This approach would put into practice the three **scientific principles of sustainability** (see inside back cover of this book).

BIG IDEAS

- About 793 million people have health problems because they do not get enough to eat and 2.1 billion people face health problems from eating too much.

- Modern industrialized agriculture has a greater harmful impact on the environment than any other human activity.

- More sustainable forms of food production could greatly reduce the harmful environmental and health impacts of industrialized food production systems.

Tying It All Together

Growing Power and Sustainability

This chapter began with a look at how Growing Power, an ecologically based urban farm (**Core Case Study**), is providing a diversity of good food to people living in a food desert. Its founder Will Allen, in demonstrating how organic food can be grown more sustainably at affordable prices, is showing how to make the transition to more sustainable food production while applying the three **scientific principles of sustainability**. Modern industrialized agriculture, aquaculture, and other forms of industrialized food production violate all of these principles.

Making the transition to more sustainable food production means relying more on solar and other forms of renewable energy and less on fossil fuels. It also means sustaining chemical cycling by conserving topsoil and

returning uncontaminated crop residues and animal wastes to the soil. It involves working to sustain natural, agricultural, and aquatic biodiversity by relying on a greater variety of crop and animal strains and seafood, produced by certified organic methods and sold locally in grocery stores and farmers' markets (see photo). Controlling pest populations through broader use of conventional and perennial polyculture and integrated pest management will also help sustain biodiversity.

Such efforts will be enhanced if we can slow the growth of the human population and sharply reduce our waste of food and other resources. Governments could help these efforts by replacing environmentally harmful agricultural and fishing subsidies and tax breaks with more environmentally beneficial ones.

Finally, the transition to more sustainable food production would be accelerated for the benefit of the environment as well as current and future generations if we could find ways to include the harmful environmental and health costs of food production in the market prices of food, in keeping with the economic, political, and ethical **principles of sustainability**.

Chapter Review

Core Case Study

1. What is a food desert? Summarize the benefits that the Growing Power farm has brought to its community. How does the farm showcase the three **scientific principles of sustainability**?

Section 10.1

2. What are the two key concepts for this section? Define **food security** and **food insecurity**. What is the root cause of food insecurity? Distinguish between **chronic undernutrition (hunger)** and **chronic malnutrition** and describe their harmful effects. Describe the effects of diet deficiencies in vitamin A, iron, and iodine. What is **overnutrition** and what are its harmful effects?

Section 10.2

3. What is the key concept for this section? What three systems supply most of the world's food? Define **irrigation**, **synthetic fertilizers**, and **synthetic pesticides**. Define and distinguish

among **industrialized agriculture (high-input agriculture)**, **plantation agriculture**, **traditional subsistence agriculture**, and **traditional intensive agriculture**. What is **monoculture**? Define **yield**. Define **polyculture** and summarize its benefits. Define **organic agriculture** and compare its main components with those of conventional industrialized agriculture. What is a **green revolution**? Distinguish between the first and second green revolutions. Summarize the story of industrialized food production in the United States.

4. Define **genetic engineering** and distinguish between it and crossbreeding through artificial selection. Describe the first genetic revolution based on crossbreeding. Describe the second gene revolution based on genetic engineering. What is a **genetically modified organism (GMO)**? Summarize the controversy over genetically engineered foods. Summarize the growth of industrialized meat production. What are feedlots and CAFOs? What is a **fishery**? What is **aquaculture (fish farming)**?

Section 10.3

5. What is the key concept for this section? Explain why industrialized food production requires large inputs of energy. Why does it result in a net energy loss? Describe the harmful environmental effects of industrialized agriculture on biodiversity, soil, water, air, human health, and resource use. What is **topsoil** and why is it one of our most important resources? What is **soil erosion** and what are its three major harmful environmental effects? What is **desertification** and what are its harmful environmental effects? How much of the water used by humanity is for irrigation? Define **soil salinization** and **waterlogging** and explain why they are harmful. What is soil pollution and what are two of its causes?

6. Summarize industrialized agriculture's contribution to climate change, water pollution, air pollution, and climate change. Explain how synthetic fertilizer use has increased and list two effects of overfertilization. Explain how industrialized food production systems have caused losses in biodiversity. What is **agrobiodiversity** and how is it being affected by industrialized food production? What factors can limit green revolutions? Compare the benefits and harmful effects of industrialized meat production. Explain the connection between feeding livestock and the formation of ocean dead zones. Compare the benefits and harmful effects of aquaculture.

Section 10.4

7. What is the key concept for this section? What is a **pest**? How much has pesticide use grown since 1950? Summarize the advantages and disadvantages of using synthetic pesticides. List three ways to reduce your exposure to pesticides. How effective have pesticides been in reducing U.S. crop losses to pesticides? Describe the use and effectiveness of laws and treaties to help protect humans from the harmful effects of pesticides. List seven alternatives to conventional pesticides. Define **integrated pest management (IPM)** and list its advantages.

Section 10.5

8. What is the key concept for this section? What is **soil conservation**? Describe six ways to reduce top-soil erosion. What is no-till farming and what are its advantages? What is hydroponics? Distinguish among **organic fertilizer**, **animal manure**, **green manure**, and **compost**. What is biochar? Define **crop rotation** and explain how it can help restore topsoil fertility? What are some ways to prevent and some ways to clean up soil salinization? How can we reduce desertification? What are some ways to make meat production and consumption more sustainable? Describe three ways to make aquaculture more sustainable. What are the advantages and disadvantages of organic agriculture? What are four important components of a more sustainable food production system? List the advantages of relying more on organic polyculture and perennial crops. Describe ecologist David Tilman's research on the importance of polyculture. What are five strategies that could help farmers and consumers to shift to more sustainable food production? What are three important ways in which individual consumers can help promote more sustainable food production?

Section 10.6

9. What is the key concept for this section? What are the two main approaches used by governments to influence food production? How have governments used subsidies to influence food production and what have been some of their effects? What are two ways in which private organizations are improving food security? Explain three of the benefits of buying locally grown food. When it is not a good choice? How can urban farming help increase food security? Describe the system used by Jennifer Burney to help people grow crops in parts of sub-Saharan Africa.

10. What are the three big ideas of this chapter? Explain how making the transition to more sustainable food production such as that promoted by the Growing Power farm (**Core Case Study**) will involve applying the six **principles of sustainability**.

_____

Note: Key terms are in bold type.

Critical Thinking

1. Suppose you got a job with Growing Power, Inc. (**Core Case Study**) and were given the assignment to turn an abandoned suburban shopping center and its large parking lot into an organic farm. Write up a plan for how you would accomplish this.

2. Food producers can now produce more than enough food to feed everyone on the planet a healthy diet. Given this fact, why do you think that about 793 million people are chronically undernourished or malnourished? Assume you are in charge of solving this problem, and write a plan for how you will accomplish it.

3. Explain why you support or oppose greatly increasing the use of (a) genetically modified food production and (b) organic perennial polyculture.

4. Suppose you work for a farmer and are given the assignment of deciding whether to use no-till agriculture on the farmer's fields or to continue using conventional plowing and weed control methods. Compare the advantages and disadvantages of each and decide how you will advise your boss. Write up a report and provide evidence to support your arguments.

5. You are the head of a major agricultural agency in the area where you live. Weigh the advantages and disadvantages of using synthetic pesticides and explain why you would support or oppose the increased use of such pesticides as a way to help farmers raise their yields. What are the alternatives?

6. If the mosquito population in the area where you live were proven to be carrying malaria or some dangerous viral disease, would you want to spray DDT in your yard, inside your home, or all through the local area to reduce this risk? Explain. What are the alternatives?

7. Do you think that the advantages of organic agriculture outweigh its disadvantages? Explain. Do you eat or grow organic foods? If so, explain your reasoning for making this choice. If not, explain your reasoning for some of the food choices you do make.

8. According to physicist Albert Einstein, "Nothing will benefit human health and increase the chances of survival of life on Earth as much as the evolution to a vegetarian diet." Explain your interpretation of this statement. Are you willing to eat less meat or no meat? Explain.

Doing Environmental Science

For 1 week, weigh the food that is purchased in your home and the food that is thrown out. Also, keep track of the types of food you eat, using categories like fruits, vegetables, meats, and dairy. Record and compare these numbers and other data from day to day. Develop a plan for cutting your household food waste in half. Consider making a similar study for your school cafeteria and reporting the results and your recommendations to school officials.

Global Environment Watch Exercise

Go to your MindTap course to access the GREENR database. Starting on the home page under "Browse Issues and Topics," click on *Resource Management,* and then select *Soil Erosion.* Use this portal to search for information on causes of soil erosion and how it affects soil fertility. Write a report on your findings. If you were to overhear a group of farmers complaining about how much money they must spend on fertilizers, what suggestions would you give them for saving money? Include your answer to this question, along with your reasoning, in your report.

Ecological Footprint Analysis

The following table gives the world's fish harvest and population data.

1. Use the world fish harvest and population data in the table to calculate the per capita fish consumption for 1990–2014 in kilograms per person. (*Hints:* 1 million metric tons equals 1 billion kilograms; the human population data are expressed in billions; and per capita consumption can be calculated directly by dividing the total amount consumed by a population figure for any year.)

2. Did per capita fish consumption generally increase or decrease between 1990 and 2014?

3. In what years did per capita fish consumption decrease?

WORLD FISH HARVEST

Years	Fish Catch (million metric tons)	Aquaculture (million metric tons)	Total (million metric tons)	World Population (in billions)	Per Capita Fish Consumption (kilograms/person)
1990	84.8	13.1	97.9	5.27	
1991	83.7	13.7	97.4	5.36	
1992	85.2	15.4	100.6	5.44	
1993	86.6	17.8	104.4	5.52	
1994	92.1	20.8	112.9	5.60	
1995	92.4	24.4	116.8	5.68	
1996	93.8	26.6	120.4	5.76	
1997	94.3	28.6	122.9	5.84	
1998	87.6	30.5	118.1	5.92	
1999	93.7	33.4	127.1	6.00	
2000	95.5	35.5	131.0	6.07	
2001	92.8	37.8	130.6	6.15	
2002	93.0	40.0	133.0	6.22	
2003	90.2	42.3	132.5	6.31	
2004	94.6	45.9	140.5	6.39	
2005	94.2	48.5	142.7	6.46	
2006	92.0	51.7	143.7	6.54	
2007	90.1	52.1	142.2	6.61	
2008	89.7	52.5	142.3	6.69	
2009	90.0	55.7	145.7	6.82	
2010	89.0	59.0	148.0	6.90	
2011	93.5	62.7	156.2	7.00	
2012	90.2	66.5	156.7	7.05	
2013	92.7	70.3	163.0	7.18	
2014	93.4	73.8	167.2	7.27	

Compiled by the authors using data from UN Food and Agriculture Organization and Earth Policy Institute.

CENGAGE**brain**.com To access course materials, including Aplia homework, please visit www.cengagebrain.com.

CHAPTER 11

Water Resources and Water Pollution

Our liquid planet glows like a soft blue sapphire in the hard-edged darkness of space. There is nothing else like it in the solar system. It is because of water.
JOHN TODD

Key Questions

11.1 Will we have enough usable water?

11.2 How can we increase freshwater supplies?

11.3 How can we use freshwater more sustainably?

11.4 How can we reduce water pollution?

Brown pelican severely oiled by the 2010 BP *Deepwater Horizon* oil well rupture in the Gulf of Mexico.

Joel Sartore/National Geographic Stock

The Gulf of Mexico's Annual Dead Zone

The Mississippi River basin (Figure 11.1, top) lies within 31 states and contains almost two-thirds of the continental U.S. land area. With more than half of all U.S. croplands, it is one of the world's most productive agricultural regions. Water drains off the land from farms, cities, factories, and sewage treatment plants throughout the huge Mississippi river basin into the Mississippi River and its tributaries. This water contains sediments and other pollutants that end up in the Gulf of Mexico (Figure 11.1, bottom photo)—a major supplier of the country's fish and shellfish.

Each spring and summer, huge quantities of nitrogen and phosphorus plant nutrients—mostly nitrates and phosphates from crop fertilizers—flow into the Mississippi River, end up in the northern Gulf of Mexico, and overfertilize the coastal waters of the U.S. states of Mississippi, Louisiana, and Texas. This excess of plant nutrients leads to an explosive growth of phytoplankton (mostly algae) that eventually die, sink to the bottom, and are decomposed by hordes of oxygen-consuming bacteria. This depletes most of the dissolved oxygen in the Gulf's bottom layer of water.

The resulting massive volume of water with a low dissolved-oxygen content (below 2 parts per million) is called a *dead zone* because it contains little or no animal marine life. Its low dissolved-oxygen levels (Figure 11.1, bottom) drive away faster-swimming fish and other marine organisms and suffocate bottom-dwelling fish, crabs, oysters, and shrimp that cannot move to less polluted areas. Large amounts of sediment, mostly from soil eroded from the Mississippi River basin, can kill bottom-dwelling forms of animal aquatic life. The dead zone appears each spring and grows until fall when storms churn the water and redistribute dissolved oxygen to the Gulf bottom.

The size of the Gulf of Mexico's annual dead zone varies with the

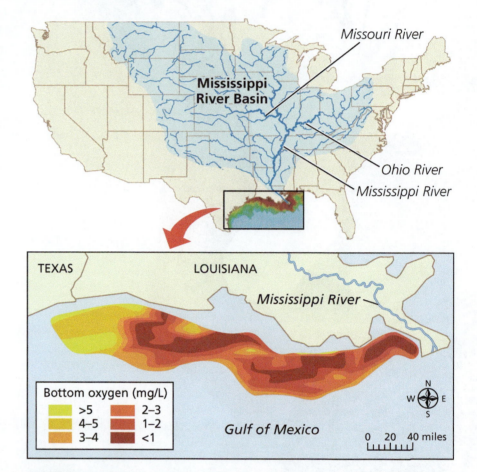

FIGURE 11.1 Water containing sediments, dissolved nitrate fertilizers, and other pollutants drains from the Mississippi River basin (top) into the Mississippi River and from there into the northern Gulf of Mexico (bottom). This creates a dead zone with low levels of dissolved oxygen (1–3 ppm), indicated by the dark and light red shaded areas in the bottom figure for 2015.

(Compiled by the authors using data from NOAA)

amount of water flowing into the Mississippi River. In years with ample rainfall and snowmelt, such as 2003, it covered an area as large as the state of Massachusetts—27,300 square kilometers (10,600 square miles). In years with less rainfall, such as 2015, it covered a smaller area of 16,760 square kilometers (6,474 square miles)—slightly less than the combined areas of the states of Connecticut and Rhode Island.

The annual Gulf of Mexico dead zone is one of 400 dead zones found throughout the world, 200 of them in the United States. Thus, producing crops to feed livestock and people and producing ethanol from corn to fuel cars in the vast Mississippi basin end up disrupting coastal aquatic life and seafood production in the Gulf of Mexico.

In this chapter, we look at the world's supply of freshwater that sustains life and at how the world's freshwater supplies and oceans are threatened with pollution. ●

11.1 WILL WE HAVE ENOUGH USABLE WATER?

CONCEPT 11.1A We are using available freshwater unsustainably by extracting it faster than nature can replace it and by wasting, polluting, and underpricing this irreplaceable natural resource.

CONCEPT 11.1B Freshwater supplies are not evenly distributed, and one of every ten people on the planet does not have adequate access to clean water.

We Are Managing Freshwater Poorly

We live on a planet that is unique because of a precious layer of water—most of it saltwater—covering about 71% of its surface. You could survive for several weeks without food, but only a few days without **freshwater**, or water that contains very low levels of dissolved salts.

It takes huge amounts of water to produce food and most of the other things that we use to meet our daily needs and wants. Water also plays a key role in determining the earth's climates and in removing and diluting some of the pollutants and wastes that we produce.

Despite its importance, freshwater is one of our most poorly managed resources. We waste it, pollute it, and do not value it highly enough. As a result, it is available at too low a cost to billions of consumers, and this encourages waste and pollution of this resource, for which we have no substitute (**Concept 11.1A**).

Access to freshwater is a *global health issue*. The World Health Organization (WHO) has estimated that every year, an average of about 669,000 people die from waterborne infectious diseases because they lack access to safe drinking water.

Access to freshwater is also an *economic issue* because water is vital for producing food and energy and reducing poverty. According to the WHO, about 57% of the world's people have water piped to their homes. The rest have to find and carry it from distant sources or wells. This daily task usually falls to women and children (Figure 11.2).

Access to freshwater is also a *national and global security issue* because of increasing tensions within and between some nations over access to limited freshwater resources they share.

Finally, water is an *environmental issue*. Excessive withdrawal of freshwater from rivers and aquifers has resulted in falling water tables, dwindling river flows, shrinking lakes, and disappearing wetlands. This, in combination with water pollution in many areas of the world, has degraded water quality. It has also reduced fish populations, hastened the extinction of some aquatic species, and degraded aquatic ecosystem services.

Most of the Earth's Water Is Not Available

Only *0.024%* of the planet's enormous water supply is readily available to people as liquid freshwater. This water is found in accessible underground deposits and in lakes, rivers, and streams. The rest is in the salty oceans (about 96.5% of the earth's volume of liquid water), in frozen polar ice caps and glaciers (1.7%), and in deep underground aquifers (1.7%).

Fortunately, the earth's freshwater supply is continually recycled, purified, and distributed in the earth's *hydrologic cycle* (see Figure 3.15, p. 58). However, this vital ecosystem service begins to fail when we overload it with water pollutants or withdraw freshwater from underground and surface water supplies faster than natural processes replenish it.

Research also indicates that atmospheric warming is altering the water cycle by evaporating more water into the atmosphere. As a result, wet places will get wetter with more frequent and heavier flooding and dry places will get drier with more intense drought.

Most people have paid little attention to their effects on the water cycle mostly because they think of the earth's freshwater as a free and infinite resource. As a result, we have placed little or no economic value on the irreplaceable ecosystem services that water provides, a serious violation of the full-cost pricing **principle of sustainability** (see Inside Back Cover).

On a global basis, there is plenty of freshwater, but it is not distributed evenly. For example, Canada, with only 0.5% of the world's population, has 20% of its liquid freshwater, while China, with 19% of the world's people, has only 6.5% of the supply.

Groundwater and Surface Water

Much of the earth's water is stored underground. Some precipitation soaks into the ground and sinks downward through spaces in soil, gravel, and rock until an impenetrable layer of rock or clay stops it. The freshwater in these underground spaces is called **groundwater,** a key component of the earth's natural capital (Figure 11.3).

The spaces in soil and rock close to the earth's surface hold little moisture. However, below a certain depth, in the **zone of saturation**, these spaces are completely filled with freshwater. The top of this groundwater zone is the **water table**. The water table rises in wet weather. It falls in dry weather or when we remove groundwater faster than nature can replenish it.

Deeper down are geological layers called **aquifers**, caverns and porous layers of sand, gravel, or rock through which groundwater flows. Most aquifers are like large, elongated sponges through which groundwater

FIGURE 11.2 Each day these women carry water to their village in a dry area of India.

ShivJi Joshi/National Geographic Creative

seeps—typically moving only a meter or so (about 3 feet) per year and rarely more than 0.3 meter (1 foot) per day. Watertight layers of rock or clay below aquifers keep the freshwater from escaping deeper into the earth. We use pumps to bring groundwater to the surface for irrigating crops, supplying households, and meeting the needs of industries.

Most aquifers are replenished naturally by precipitation that sinks downward through exposed soil and rock. Others are recharged from nearby lakes, rivers, and streams. However, aquifers recharge slowly and in urban areas, so much of the landscape has been built on or paved over that freshwater can no longer penetrate the ground to recharge aquifers. Some aquifers, called *deep aquifers*, cannot be recharged or take many thousands of years to recharge. On a human time scale, deep aquifers are nonrenewable one-time deposits of freshwater.

Another crucial resource is **surface water**, the freshwater from rain and melted snow that is stored in lakes, reservoirs, wetlands, streams, and rivers. Precipitation that does not soak into the ground or return to the atmosphere by evaporation is called **surface runoff**. The land from which surface runoff drains into a particular stream, lake, wetland, or other body of water is called its **watershed**, or **drainage basin**.

Water Use Is Increasing

Two-thirds of the annual surface runoff of freshwater into rivers and streams is lost in seasonal floods and is not available for human use. The remaining one-third is **reliable surface runoff**, which is regarded as a stable source of freshwater from year to year.

During the past century, the human population tripled, global water withdrawals increased sevenfold, and per capita withdrawals quadrupled. As a result, we now withdraw an estimated 34% of the world's reliable runoff. This is a global average. In the arid American Southwest, up to 70% of the reliable runoff is withdrawn for human purposes, mostly for irrigation. Some water experts project that because of population growth, rising rates of water use per person, longer dry periods in some areas, and unnecessary water waste, the world's people are likely to be withdrawing up to 90% of the world's reliable freshwater runoff by 2025.

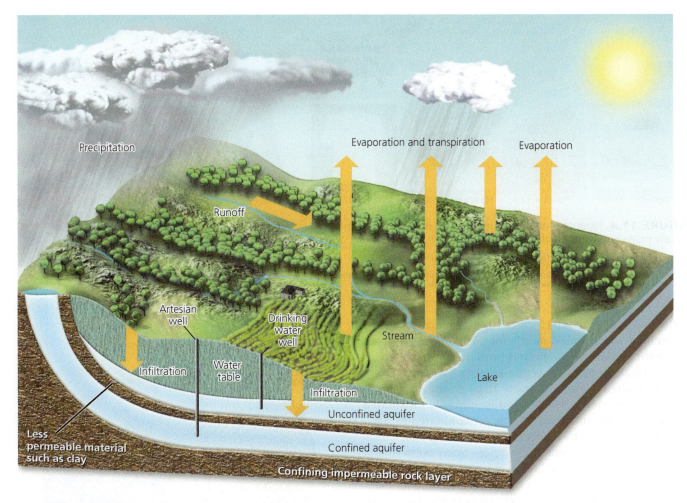

FIGURE 11.3 Natural Capital: Much of the water that falls in precipitation seeps into the ground to become groundwater, stored in aquifers.

Worldwide, people use 70% of the freshwater withdrawn each year from rivers, lakes, and aquifers to irrigate cropland and raise livestock. In arid regions, up to 90% of the available water supply is used for food production. Industry uses roughly another 20% of the water withdrawn globally each year. Cities and residences use the remaining 10%.

Your **water footprint** is a rough measure of the total volume of freshwater you use directly or indirectly. Your daily water footprint includes the freshwater you use directly (for example, to drink, bathe, or flush a toilet) and the water used indirectly to produce the food you eat, the energy you use, and the products you buy.

Freshwater that is not directly consumed but is used to produce food and other products is called **virtual water**. It makes up a large part of the water footprints, of individuals, especially in more-developed countries. Agriculture accounts for about 92% of humanity's water footprint. Producing and delivering a typical hamburger, for example, takes about 1,800 liters (480 gallons or about 12 bathtubs) of freshwater—most of it used to grow grain that is fed to cattle. Figure 11.4 shows one way to measure the amounts of virtual water used for producing and delivering products.

For some water-short countries, it makes sense to save real freshwater by importing virtual water through food imports, instead of producing food domestically. Such countries include Egypt and other Middle Eastern nations in dry climates with little freshwater. Large exporters of virtual water—mostly in the form of wheat, corn, soybeans, alfalfa, and other foods—are the European Union, the United States, Canada, Brazil, and Australia.

Freshwater Resources in the United States

According to the U.S. Geological Survey (USGS), the average American directly uses between 300 and 379 liters (80 to 100 gallons) of freshwater a day—enough water to fill at least two typical bathtubs full of water. (The average

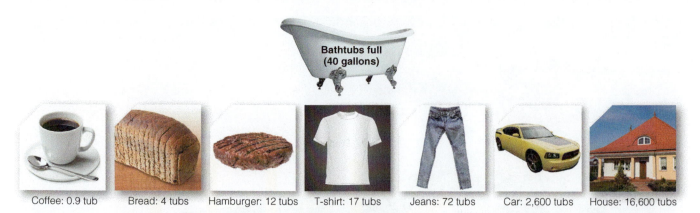

Coffee: 0.9 tub | Bread: 4 tubs | Hamburger: 12 tubs | T-shirt: 17 tubs | Jeans: 72 tubs | Car: 2,600 tubs | House: 16,600 tubs

FIGURE 11.4 Producing and delivering a single one of each of the products listed here requires the equivalent of nearly one and usually many bathtubs full of freshwater, called *virtual water. Note:* 1 bathtub = 151 liters (40 gallons).

(Compiled by the authors using data from UN Food and Agriculture Organization, UNESCO-IHE Institute for Water Education, World Water Council, and Water Footprint Network)

Bathtub: Baloncici/Shutterstock.com. Coffee: Aleksandra Nadeina/Shutterstock.com. Bread: Alexander Kalina/Shutterstock.com. Hamburger: Joe Belanger/Shutterstock.com. T-shirt: grmarc/Shutterstock.com. Jeans: Eyes wide/Shutterstock.com. Car: L Barnwell/Shutterstock.com. House: Rafal Olechowski/Shutterstock.com

bathtub contains about 151 liters or 40 gallons of water.) Household water is used mostly for flushing toilets, washing clothes, taking showers, and running faucets, or is lost through leaking pipes, faucets, and other fixtures.

The United States has more than enough freshwater to meet its needs. However, it is unevenly distributed (**Concept 11.1B**) and much of it is contaminated by agricultural and industrial practices. The eastern states usually have ample precipitation, whereas many western and southwestern states have little (Figure 11.5).

In the eastern United States, most water is used for manufacturing and for cooling power plants (with most of the water heated and returned to its source). In many parts of this region, the most serious water problems are flooding, occasional water shortages because of drought, and pollution.

In the arid and semiarid regions of the western half of the United States, irrigation counts for as much as 85% of freshwater use. The major water problem is a shortage of freshwater runoff caused by low precipitation (Figure 11.5), high evaporation, and recurring prolonged drought.

The U.S. Department of the Interior has identified *water scarcity hotspots* in 17 western states (Figure 11.6). In these areas, there is increasing competition for scarce freshwater to support growing urban areas, irrigation, recreation, and wildlife. This competition for freshwater could trigger intense political and legal conflicts between states and between rural and urban areas within states. In addition, Columbia University climate researchers led by Richard Seager used well-tested climate models to project that the southwestern United States is very likely to have long periods of extreme drought throughout most of the rest of this century.

Freshwater Shortages Will Grow

Freshwater scarcity stress is a comparison of the amount of freshwater available with the amount used by humans. The main causes of freshwater scarcity in a particular

Average annual precipitation (centimeters)

Less than 41   81–122
41–81   More than 122

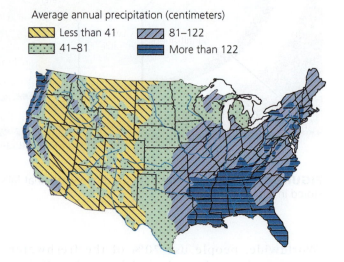

FIGURE 11.5 Long-term average annual precipitation in the continental United States.

(Compiled by the authors using data from U.S. Water Resources Council and U.S. Geological Survey)

area are a dry climate, drought, too many people using a water supply, and wasteful use of water. Many of the world's rivers and their basins (see Case Study that follows) suffer from various levels of freshwater scarcity stress (Figure 11.7). By 2050, as many as 60 countries—most of them in Asia—are likely to suffer from freshwater scarcity stress.

Currently, about 30% of the earth's land—an area roughly five times the size of the United States—experiences severe drought. By 2059, as much as 45% of the earth's land could experience *extreme drought* from a combination of natural drought cycles and projected climate change, according to a study by climate researcher David Rind and his colleagues.

FIGURE 11.6 Water scarcity hotspots in 17 western states that, by 2025, could face intense conflicts over scarce water needed for urban growth, irrigation, recreation, and wildlife.

Highly likely conflict potential
Substantial conflict potential
Moderate conflict potential
Unmet rural water needs

(Compiled by the authors using data from U.S. Department of the Interior and U.S. Geological Survey)

In 276 of the world's water basins, two or more countries share the available freshwater supplies. However, countries in only 118 of those basins have water-sharing agreements. Consequently, conflicts over shared freshwater resources are likely to occur more as populations grow, demand for water increases, and supplies shrink in many parts of the world.

In 2015, the United Nations (UN) and the WHO reported that 783 million people—nearly 2.4 times the U.S. population—did not have regular access to clean water for drinking, cooking, and washing, mostly because of poverty (**Concept 11.1B**). The report also noted that more than 2 billion people gained access to clean water between 1990 and 2012. Still, many analysts view the likelihood of expanding water shortages in many parts of the world as one of the most serious environmental, health, and economic problems that society faces.

783 million Number of people without regular access to clean water.

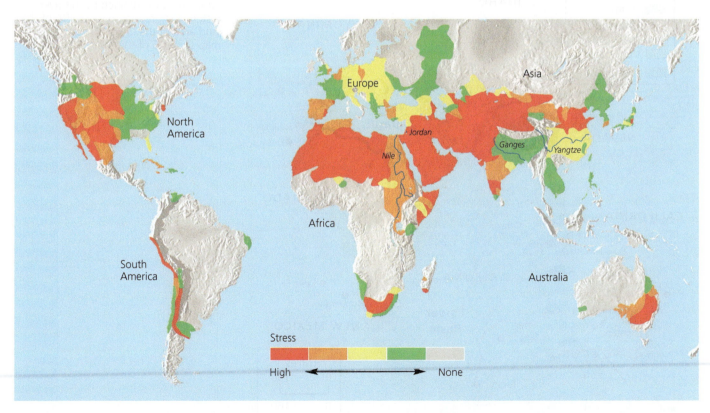

FIGURE 11.7 Natural Capital Degradation: The world's major river basins differ in their degree of *freshwater scarcity stress* (**Concept 11.1B**). *Critical thinking:* Do you live in a freshwater-stressed area? If so, what signs of stress have you noticed? In what ways, if any, has it affected your life?

(Compiled by the authors using data from World Commission on Water Use for the 21st Century, UN Food and Agriculture Organization, and World Water Council)

The Colorado River

The Colorado River flows 2,300 kilometers (1,400 miles) through seven arid states in the southwestern United States to the Gulf of California (Figure 11.8). Most of its water comes from snowmelt in the Rocky Mountains. During the past 100 years, this once free-flowing river has been tamed by a gigantic plumbing system consisting of 14 major dams and reservoirs.

This system of dams and reservoirs provides water and electricity from hydroelectric plants for roughly 40 million people in seven states—about one of every eight people in the United States. The river's water is used to produce about 15% of the nation's crops and livestock. The system supplies water to some of the nation's driest and hottest cities such as Las Vegas, Nevada, Phoenix, Arizona, and San Diego and Los Angeles, California. Take away this tamed river, and these cities would become largely uninhabitable desert areas and California's Imperial Valley would not be able to produce half of the country's fruits and vegetables.

So much water is withdrawn from the Colorado River to grow crops and support cities in a desert-like climate that very little of it reaches the sea. Since 1999, the river's watershed has also experienced severe **drought**, a prolonged period in which precipitation is lower and evaporation is higher than normal.

There are three major problems associated with the use of freshwater from this river. *First*, the Colorado River basin includes some of the driest lands in the United States and Mexico. *Second*, long-standing legal agreements between Mexico and the affected western states allocated more freshwater for human use than the river can supply—even in rare years when there is no drought. These pacts allocated no water for protecting aquatic and terrestrial wildlife. *Third*, since 1960, because of drought, damming, and heavy withdrawals, the river has rarely flowed all the way to the Gulf of California and this has degraded the river's aquatic ecosystems and dried up its *delta*, the wetland area at its mouth, destroying the delta's estuary ecosystem. This overuse of the Colorado River system illustrates the challenges of managing how to share water resources among different governments and groups of people living in dry climates.

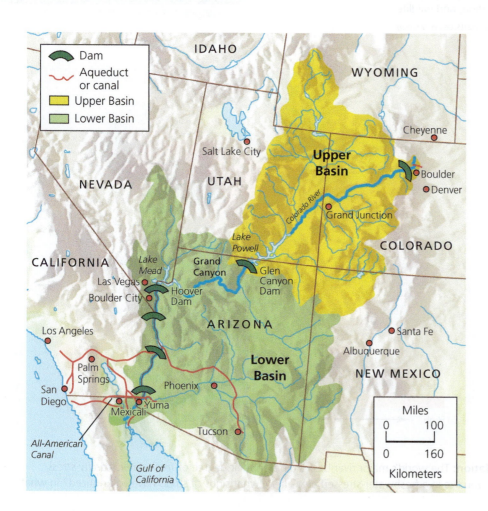

FIGURE 11.8 The *Colorado River basin:* The area drained by this river system is more than one-twelfth of the land area of the lower 48 states. This map shows 6 of the river's 14 dams.

11.2 HOW CAN WE INCREASE FRESHWATER SUPPLIES?

CONCEPT 11.2A Groundwater used to supply cities and grow food is being pumped from aquifers in some areas faster than it is renewed by precipitation.

CONCEPT 11.2B Large dam-and-reservoir systems and water transfer projects have greatly expanded water supplies in some areas but have also disrupted ecosystems and displaced people.

CONCEPT 11.2C We can convert salty ocean water to freshwater, but the cost is high, and the resulting salty brine must be disposed of without harming aquatic or terrestrial ecosystems.

Aquifers Are Being Depleted

Aquifers provide drinking water for nearly half of the world's people, while surface water provides drinking water for the other half. In the United States, aquifers supply almost all drinking water in rural areas (20% in urban areas) and 43% of the country's irrigation water, according to the USGS. Most aquifers are renewable resources unless the groundwater they contain is removed faster than it is replenished from rainfall and snowmelt. This practice is referred to as *overpumping* (see the Case Study that follows). Relying more on groundwater has advantages and disadvantages (Figure 11.9).

Trade-Offs

Withdrawing Groundwater

Advantages	Disadvantages
Useful for drinking and irrigation	Aquifer depletion from overpumping
Exists almost everywhere	Sinking of land (subsidence) from overpumping
Renewable if not overpumped or contaminated	Some deeper wells are nonrenewable
Cheaper to extract than most surface waters	Pollution of aquifers lasts decades or centuries

FIGURE 11.9 Withdrawing groundwater from aquifers has advantages and disadvantages. *Critical thinking:* Which two advantages and which two disadvantages do you think are the most important? Why?

Top: Ulrich Mueller/Shutterstock.com

Test wells and satellite data indicate that water tables are falling in many areas of the world because of overpumping (**Concept 11.2A**). The world's three largest grain producers—China, the United States, and India—as well as Mexico, Saudi Arabia, Iran, Iraq, Egypt, Pakistan, Spain, and other countries are overpumping many of their aquifers. Much of the Middle East is facing a growing water and food crisis and increasing tensions among its nations, brought on mostly by falling water tables, rapid population growth, and disagreements over access to shared water supplies from the region's rivers.

For decades, Saudi Arabia has pumped freshwater from a nonrenewable deep aquifer to irrigate crops such as wheat grown on desert land (Figure 11.10). It was also used to fill fountains and swimming pools, which lose a great deal of water through evaporation into the dry desert air. In 2008, Saudi Arabia announced that irrigated wheat production had largely depleted this major deep aquifer. In 2016, the country stopped producing wheat and imported grain (virtual water) to help feed its 32 million people.

CASE STUDY

Overpumping the Ogallala Aquifer

In the United States, groundwater is being withdrawn from aquifers, on average, four times faster than it is being replenished, according to the USGS (**Concept 11.2A**). Figure 11.11 shows the areas of greatest aquifer depletion in the continental United States. One of the most serious overpumping of groundwater is occurring in the lower half of the Ogallala Aquifer. It is one of the world's largest aquifers, stretching beneath eight states from southern South Dakota to Texas (blowup section of Figure 11.11).

The Ogallala Aquifer supplies one-third of all the groundwater used in the United States. This aquifer has helped make the Great Plains one of world's most productive irrigated agricultural regions. However, the Ogallala is a deposit of liquid natural capital with a slow rate of recharge.

In parts of the southern half of the Ogallala, groundwater is being pumped out 10–40 times faster than the natural recharge rate. This has lowered water tables and raised pumping costs, especially in parts of Texas (blowup map in Figure 11.11). The overpumping of this aquifer, along with urban development and restricted access to Colorado River water, has decreased irrigated cropland in Texas, Arizona, Colorado, and California. It has also increased competition for water among farmers, ranchers, and growing urban areas.

Government *subsidies*—payments or tax breaks designed to increase crop production—have encouraged farmers to grow water-thirsty crops in dry areas, which has accelerated depletion of the Ogallala Aquifer.

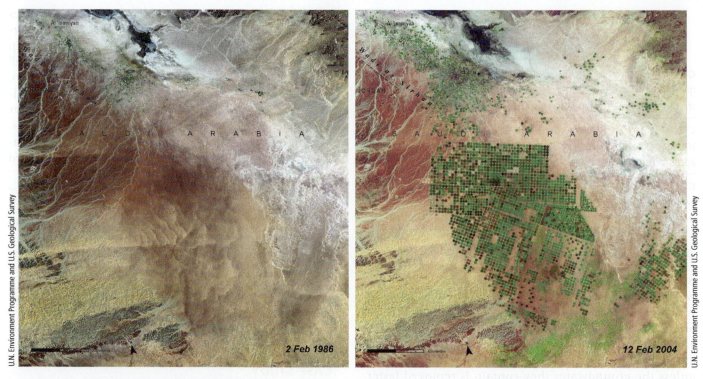

FIGURE 11.10 Natural Capital Degradation: Satellite photos of farmland irrigated by groundwater pumped from a deep aquifer in a vast desert region of Saudi Arabia between 1986 (left) and 2004 (right). Irrigated areas appear as green dots (each representing a circular spray system) and brown dots show areas where wells have gone dry and the land has returned to desert. Since 2004, many more wells have been abandoned.

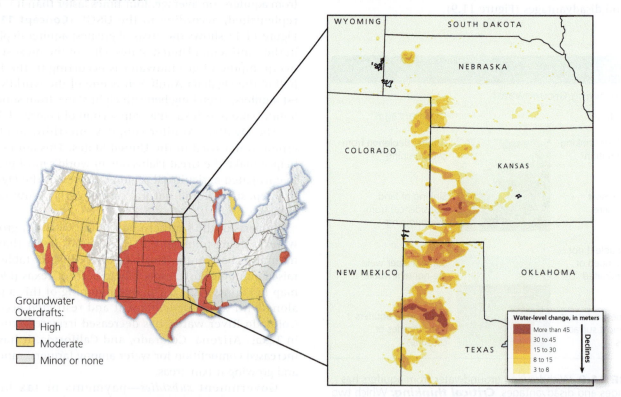

FIGURE 11.11 Natural Capital Degradation: Areas of greatest aquifer depletion from groundwater overdraft in the continental United States, with blowup (right) showing where water levels in the Ogallala Aquifer have dropped sharply at its southern end beneath parts of Kansas, Oklahoma, Texas, and New Mexico. *Questions:* Do you depend on any of these aquifers for your drinking water? If so, what is the level of severity of overdraft where you live?

(Compiled by the authors using data from U.S. Water Resources Council and U.S. Geological Survey)

In particular, corn—a very thirsty crop—has been planted widely on fields watered by the Ogallala. Serious aquifer depletion is also taking place in California's semiarid Central Valley (the long red area in the California portion of Figure 11.11), which supplies half of the country's fruits and vegetables.

Harmful Effects of Overpumping Aquifers

Overpumping aquifers contributes to limits on food production, rising food prices, and widening gaps between the rich and the poor in some areas. As water tables drop, the energy and financial costs of pumping water from lower depths rise sharply. Farmers must drill deeper wells, buy larger pumps, and use more electricity to run the pumps. Farmers who cannot afford to do this often lose their land. This forces them to work for wealthier farmers or migrate to cities in search of other work.

Withdrawing large amounts of groundwater can sometimes cause the sand and rock that is held in place by water pressure in aquifers to collapse. This can cause the land above the aquifer to *subside* or sink, a phenomenon known as *subsidence*. Extreme and sudden subsidence, sometimes referred to as a *sinkhole,* can swallow cars and houses. Once an aquifer becomes compressed by subsidence, recharge is impossible. Subsidence can also damage roadways, water and sewer lines, and foundations beneath buildings.

Since 1925, overpumping of an aquifer to irrigate crops in California's San Joaquin Valley has caused half of the valley's land to subside by more than 0.3 meter (1 foot) and, in one area, by more than 8.5 meters (28 feet) (Figure 11.12). Mexico City and parts of Beijing, China, also suffer from severe subsidence problems.

Overpumping groundwater in coastal areas, where many of the world's largest cities are found, can pull saltwater into freshwater aquifers. The resulting contaminated groundwater is undrinkable and cannot be used for irrigation. This problem is especially serious in coastal areas of the U.S. states of California, Texas, Florida, Georgia, South Carolina, and New Jersey, as well as in Turkey, Thailand, and the Philippines.

Figure 11.13 lists ways to prevent or slow the problem of aquifer depletion by using this potentially renewable resource more sustainably.

Large Dams Have Benefits and Drawbacks

A **dam** is a structure built across a river to control its flow. Usually, dammed water creates an artificial lake, or **reservoir**, behind the dam (Figure 2.10, p. 37). The purpose of a dam-and-reservoir system is to capture and store the surface runoff from a river's watershed, and release it as needed to control floods, generate

FIGURE 11.12 This pole shows subsidence from overpumping of an aquifer for irrigation in California's San Joaquin Central Valley between 1925 and 1977. In 1925, this area's land surface was near the top of the pole. Since 1977, this problem has gotten worse.

electricity (hydropower), and supply freshwater for irrigation and for use in towns and cities. Reservoirs also provide water for recreational activities such as swimming, fishing, and boating.

The world's 45,000 large dams capture and store 14% of the earth's surface runoff. They provide water for almost half of all irrigated cropland, and supply more than half the electricity used in 65 countries. Dams have increased the annual reliable runoff available for human use by nearly 33%.

However, dam-and-reservoir systems have drawbacks. For example, the world's reservoirs have displaced 40–80 million people from their homes and flooded large areas of mostly productive land. Dams have impaired some of the important ecosystem services that rivers provide.

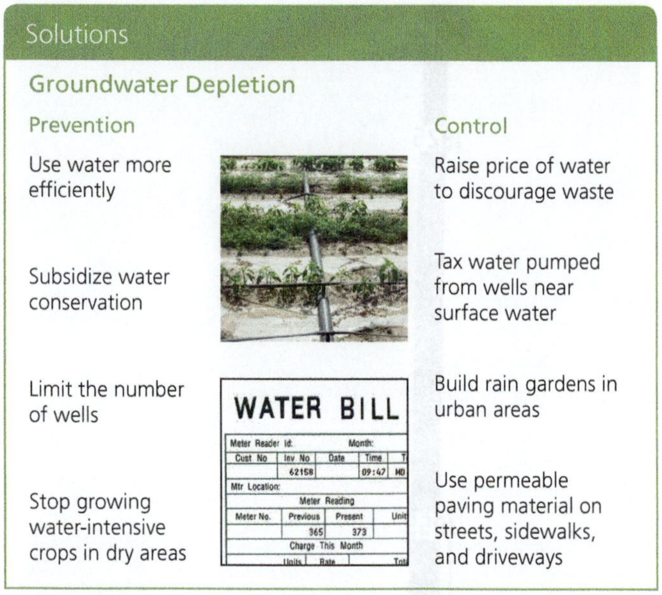

Groundwater Depletion

Prevention	Control
Use water more efficiently	Raise price of water to discourage waste
Subsidize water conservation	Tax water pumped from wells near surface water
Limit the number of wells	Build rain gardens in urban areas
Stop growing water-intensive crops in dry areas	Use permeable paving material on streets, sidewalks, and driveways

FIGURE 11.13 Ways to prevent or slow groundwater depletion by using freshwater more sustainably. *Critical thinking:* Which two of these solutions do you think are the best ones? Why?

Top: iStock.com/anhong. Bottom: Banol2007/Dreamstime.com.

According to a study by the World Wildlife Fund (WWF) only 21 of the planet's 177 longest rivers consistently run all the way to the sea before running dry (see Case Study that follows). The WWF study also estimated that about one out of five of the world's freshwater fish and plant species are either extinct or endangered primarily because dams and water withdrawals have sharply decreased the river flows. This helps explain why estimated extinction rates for freshwater life are four to six times larger than for marine or terrestrial species.

Reservoirs also have limited lifespans. Within 50 years, the reservoirs behind dams typically fill up with sediments (mud and silt), which make them useless for storing water or producing electricity. In the Colorado River system (Figure 11.8) the equivalent of roughly 20,000 dump-truck loads of silt are deposited on the bottoms of Lake Powell and Lake Mead every day. Sometime during this century these two vital reservoirs will very likely be too full of silt to function as designed. About 85% of all U.S. dam-and-reservoir systems will be 50 years old or older by 2025. Some aging dams are already being removed because their reservoirs have filled with silt. Figure 11.14 summarizes the major advantages and disadvantages of dam-and-reservoir systems.

If climate change occurs as projected during this century, water shortages will intensify in many parts of the world. For example, mountain snows that feed the Colorado River system (Figure 11.8) will melt faster and earlier, making less freshwater available to the river system when it is needed for irrigation during hot and dry summer months. Agricultural production would drop sharply and the region's major desert cities such as Las Vegas, Nevada and Phoenix, Arizona, would be challenged to survive.

Nearly 3 billion people in South America, China, India, and other parts of Asia depend on river flows fed by mountain glaciers. These glaciers serve as freshwater savings accounts. They store precipitation as ice and snow in wet periods, and during dry seasons much of the ice and snow melts and slowly releases water for use on farms and in cities. In 2015, according to the World Glacier Monitoring Service, many of these mountain glaciers had been shrinking for 24 consecutive years, mostly due to a warming atmosphere.

CASE STUDY

How Dams Can Kill a Delta

Since 1905, the amount of water flowing to the mouth of the Colorado River (Figure 11.8) has dropped dramatically. In most years since 1960, the river has dwindled to a small, sluggish stream by the time it reaches the Gulf of California.

The Colorado River once emptied into a vast delta, which had hosted forests, lagoons, and marshes rich in plant and animal life and supported a thriving coastal fishery for hundreds of years. Since the damming of the Colorado—within one human lifetime—this biologically diverse delta ecosystem has collapsed and is now covered mostly by mud flats and desert.

Historically, about 80% of the water withdrawn from the Colorado has been used to irrigate crops and raise cattle. This is because the U.S. government (taxpayers) paid for the dams and reservoirs and has supplied many farmers and ranchers with water at low prices. These subsidies have led to the wasteful use of irrigation water for growing thirsty crops such as rice, cotton, and alfalfa in dry areas.

Water experts call for the seven states using the Colorado River to enact and enforce strict water conservation measures. They also call for sharply decreasing or phasing out state and federal government subsidies for agriculture in this region. The goal would be to shift water-thirsty crops to less arid areas and to severely restrict the watering of golf courses and lawns in the desert areas of the Colorado River basin. They suggest that the best way to implement such solutions is to sharply raise the historically low price of the river's freshwater over the next decade—an application of the full-cost pricing **principle of sustainability**.

Water Transfers Have Benefits and Drawbacks

In some heavily populated dry areas of the world, governments have tried to solve water shortage problems by using canals and pipelines to transfer water from water-rich to water-poor areas. If you consume lettuce in the

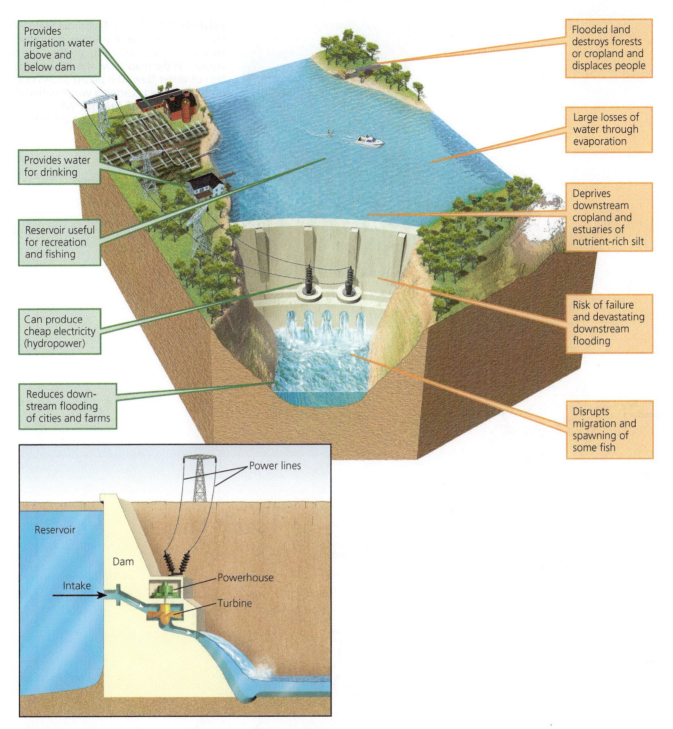

Provides irrigation water above and below dam

Provides water for drinking

Reservoir useful for recreation and fishing

Can produce cheap electricity (hydropower)

Reduces down-stream flooding of cities and farms

Flooded land destroys forests or cropland and displaces people

Large losses of water through evaporation

Deprives downstream cropland and estuaries of nutrient-rich silt

Risk of failure and devastating downstream flooding

Disrupts migration and spawning of some fish

Power lines

Reservoir

Dam

Intake

Powerhouse

Turbine

FIGURE 11.14 **Trade-offs:** Use of large dams and reservoirs has its advantages (green) and disadvantages (orange) (**Concept 11.2B**). *Critical thinking:* Which single advantage and which single disadvantage do you think are the most important? Why?

United States, chances are it was grown in the arid Central Valley of California, partly with the use of irrigation water from snow melting off the tops of the High Sierra Mountains of northeastern California.

The *California State Water Project* (Figure 11.15) is one of the world's largest freshwater transfer projects. It uses a maze of giant dams, pumps, and lined canals, or *aqueducts*

(see photo in Figure 11.15), to transfer freshwater from the mountains of northern California to heavily populated cities and agricultural regions of water-poor central and southern California.

This massive water transfer has yielded many benefits. California's heavily irrigated Central Valley supplies half of the nation's fruits and vegetables, and the arid cities

FIGURE 11.15 The California State Water Project transfers huge volumes of freshwater from one watershed to another. The red arrows on the map show the general direction of water flow. The photo shows one of the aqueducts carrying water within the system. ***Critical thinking:*** What effects might this system have on the areas from which the water is taken?

of San Diego and Los Angeles have grown and flourished because of the water transfer.

However, the project has also reduced the flow of the Sacramento River, threatening fisheries and reducing the flushing action that helps cleanse the San Francisco Bay of pollutants. As a result, the bay has suffered from pollution and the flow of freshwater to its coastal marshes and other ecosystems has dropped. These factors have placed stress on the wildlife species that depend on the bay's many ecosystems (**Concept 11.2B**).

The federal government and the state of California have subsidized this water transfer project. These subsidies have promoted inefficient use of large volumes of water to irrigate thirsty crops like lettuce, alfalfa, and almonds, in desert-like areas. In central California, agriculture consumes three-fourths of the water that is transferred. Much of this water is wasted through inefficient irrigation systems. Studies show that making irrigation just 10% more efficient would provide all the water needed for domestic and industrial uses in southern California.

According to several studies, projected climate change during this century will make matters worse in California by reducing surface water availability. California depends on *snowpacks* of densely packed, slowly melting snow in the Sierra Nevada Mountains, for more than 60% of its freshwater during the hot, dry summer, according to the Sierra Nevada Nature Conservancy. Projected atmospheric warming could shrink the snowpacks by as much as 40% by 2050 and by as much as 90% by the end of this century. Shrinking snowpacks will sharply reduce the amount of freshwater available for northern California residents and ecosystems, as well as for transfer to arid and semiarid central and southern California.

CASE STUDY

The Aral Sea Disaster: An Example of Unintended Effects

The shrinking of the Aral Sea (Figure 11.16) is the result of a freshwater transfer project in central Asia. Starting in 1960, enormous amounts of irrigation water were diverted from the two rivers that supply water to the Aral Sea. The goal was to create one of the world's largest irrigated areas, mostly for raising cotton and rice. The irrigation canal, the world's longest, stretches more than 1,300 kilometers (800 miles)—roughly the distance between the two U.S. cities of Boston, Massachusetts, and Chicago, Illinois.

This project, coupled with drought and high evaporation rates due to the area's hot and dry climate, has caused a regional ecological and economic disaster. Since 1961, the sea's salinity has risen sevenfold and the average level of its water has dropped by an amount roughly equal to the height of a six-story building. The Southern Aral Sea has lost more than 90% of its volume of water and most of its lake bottom is now a white salt desert (Figure 11.16, right photo). Water withdrawals for agriculture have reduced the two rivers feeding the sea to mere trickles.

About 85% of the area's wetlands have been eliminated. The sea's greatly increased salt concentration—three times saltier than ocean water—has caused the presumed local extinction of 26 of the area's 32 native fish species. This has devastated the area's fishing industry, which once provided work for more than 60,000 people. Fishing villages and boats once located on the sea's coastline now sit abandoned in a salty desert.

Winds pick up the sand and salty dust and blow it onto fields as far as 500 kilometers (310 miles) away. As the salt spreads, it pollutes water and kills wildlife, crops, and other vegetation. Aral Sea dust settling on glaciers in the Himalayas is causing them to melt at a faster-than-normal rate.

1976

2016

FIGURE 11.16 Natural Capital Degradation: The *Aral Sea*, straddling the borders of Kazakhstan and Uzbekistan, was one of the world's largest saline lakes. These satellite photos show the sea in 1976 (left) and in 2016 (right). ***Critical thinking:*** What do you think should be done to help prevent further shrinkage of the Aral Sea?

The shrinkage of the Aral Sea has also altered the area's climate. The shrunken sea no longer acts as a thermal buffer to moderate the heat of summer and the extreme cold of winter. Now there is less rain, summers are hotter and drier, winters are colder, and the growing season is shorter. The combination of such climate change and severe salinization has reduced crop yields by 20–50% on almost one-third of the area's cropland—the opposite of the project's intended effects.

Since 1999, the UN, the World Bank, and the five countries surrounding the lake have worked to improve irrigation efficiency. They have also partially replaced thirsty crops with other crops that require less irrigation water. Because of a dike built to block the flow of water from the Northern Aral Sea into the southern sea, the level of the northern sea has risen by 2 meters (7 feet), its salinity has dropped, dissolved oxygen levels are up, and it supports a healthy fishery.

However, the formerly much larger southern sea is still shrinking. By 2016, its eastern lobe was essentially gone (Figure 11.16, right photo). The European Space Agency projects that the rest of the Southern Aral Sea could dry up completely by 2020.

Removing Salt from Seawater to Provide Freshwater

Desalination is the process of removing dissolved salt from ocean water or from brackish (slightly salty) water in aquifers or lakes. It is another way to increase supplies of freshwater (**Concept 11.2C**).

Currently, the two most widely used methods for desalinating water are distillation and reverse osmosis. *Distillation* involves heating saltwater until it evaporates (leaving behind salts in solid form) and condenses as freshwater. *Reverse osmosis* (or *microfiltration*) uses high pressure to force saltwater through a membrane filter with pores small enough to remove the salt and other impurities. The world's more than 18,400 desalination plants, including more than 300 in the United States, supply less than 1% of freshwater used in the United States and in the world.

Three major problems hinder the widespread use of desalination.

1. Desalination is costly because removing salt from seawater requires a lot of energy. Using fossil fuels to produce that energy increases emissions of climate-changing CO_2 and other air pollutants.

2. Pumping large volumes of seawater through pipes requires chemicals to sterilize the water and prevent algae growth. This kills many marine organisms and requires large inputs of energy and money.

3. Desalination produces huge quantities of wastewater that are much saltier than ocean water and that require proper disposal. Dumping the salty waste into coastal ocean waters increases the salinity of

those waters, which can threaten food resources and aquatic life—especially if it is dumped near coral reefs, marshes, or mangrove forests. Disposing the salty wastes on land can contaminate groundwater and surface water (**Concept 11.2C**).

Currently, desalination is practical only for water-short countries and cities that can afford its high cost. However, scientists and engineers are working to develop better and more affordable desalination technologies.

11.3 HOW CAN WE USE FRESHWATER MORE SUSTAINABLY?

CONCEPT 11.3 We can use freshwater more sustainably by cutting water waste, raising water prices, slowing population growth, and protecting aquifers, forests, and other ecosystems that store freshwater.

Cutting Water Waste

According to water resource expert Mohamed El-Ashry of the World Resources Institute, about 66% of the freshwater used in the world and about 50% of the freshwater used in the United States is lost through evaporation, leaks, and inefficient use. El-Ashry estimates that it is economically and technically feasible to reduce such losses to 15%, thereby meeting most of the world's freshwater needs.

Why do we have such large losses of freshwater? According to water resource experts, there are two major reasons. First, the cost of freshwater to most users is low—a violation of the full-cost pricing **principle of sustainability**. This gives users little or no financial incentive to invest in water-saving technologies.

Higher prices for freshwater encourage water conservation but make it difficult for low-income farmers and city dwellers to buy enough water to meet their needs. When South Africa raised water prices, it dealt with this problem by establishing *lifeline rates*, which give each household a set amount of free or low-priced water to meet basic needs. When users exceed this amount, they pay increasingly higher prices as their water use increases. This is a *user-pays* approach.

The second major cause of unnecessary losses of freshwater is a lack of government subsidies for improving the efficiency of water use. Many hydrologists and economists call for replacing current water subsidies that encourage water waste with subsidies that would encourage using water more efficiently. Understandably, farmers and industries that receive subsidies that keep their water prices low have vigorously opposed efforts to eliminate or reduce them. However, the environmental and economic benefits of shifting from subsidies that promote water waste to subsidies that promote water conservation far outweigh the environmental and economic harm from not making such a shift, according to many water resource experts and economists.

Improving Efficiency in Irrigation

Only about 60% of the world's irrigation water reaches crops, which means that most irrigation systems are highly inefficient. The most inefficient irrigation system is *flood irrigation,* in which water is pumped from a groundwater or surface water source through unlined ditches where it flows by gravity to the crops (Figure 11.17, left). This method delivers far more water than is needed for crop growth and typically about 45% of this water is lost through evaporation, seepage, and runoff.

The *center pivot spray irrigation system* (Figure 11.17, right), which uses pumps to spray water on crops, allows about 80% of the water to reach crops. An improved system that sprays the water closer to the ground puts 90–95% of the water where crops need it.

Drip, or *trickle irrigation* (Figure 11.17, center), is the most efficient way to deliver small amounts of water precisely to crops. It consists of a network of perforated plastic tubing installed at or below the ground level. Small pinholes in the tubing deliver drops of water at a slow and steady rate, close to the roots of individual plants and allow 90–95% of the water to reach the crops.

Drip irrigation is used on less than 4% of the irrigated crop fields in the world and in the United States, largely because most drip irrigation systems are costly. This percentage rises to 13% in the U.S. state of California, 66% in Israel, and 90% in Cyprus. If freshwater were priced closer to the value of the ecosystem services it provides, and if government subsidies for inefficient use of water were reduced or eliminated, drip irrigation could be used to irrigate most of the world's crops.

According to the UN, reducing the current global withdrawal of water for irrigation by just 10% would save enough water to grow crops and meet the estimated additional water demands of the earth's cities and industries through 2025. This would also reduce the need for costly desalination. Figure 11.18 summarizes several ways to

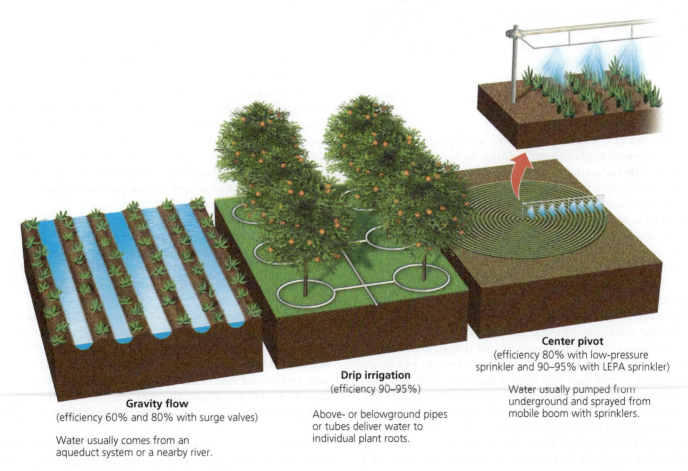

Center pivot
(efficiency 80% with low-pressure sprinkler and 90–95% with LEPA sprinkler)

Water usually pumped from underground and sprayed from mobile boom with sprinklers.

Drip irrigation
(efficiency 90–95%)

Above- or belowground pipes or tubes deliver water to individual plant roots.

Gravity flow
(efficiency 60% and 80% with surge valves)

Water usually comes from an aqueduct system or a nearby river.

FIGURE 11.17 Traditional irrigation methods rely on gravity and flowing water (left). Newer systems such as center-pivot, sprinkler irrigation (right), and drip irrigation (center) are far more efficient.

Reducing Irrigation Water Losses

- Avoid growing thirsty crops in dry areas
- Import water-intensive crops and meat
- Encourage organic farming and polyculture to retain soil moisture
- Monitor soil moisture to add water only when necessary
- Expand use of drip irrigation and other efficient methods
- Irrigate at night to reduce evaporation
- Line canals that bring water to irrigation ditches
- Irrigate with treated wastewater

FIGURE 11.18 Ways to reduce freshwater losses in irrigation. *Critical thinking:* Which two of these solutions do you think are the best ones? Why?

reduce water losses in crop irrigation. Since 1950, water-short Israel has used many of these techniques to slash irrigation water losses by 84% while irrigating 44% more land. Israel treats and purifies 30% of its municipal sewage water and uses it to provide more than 50% of the freshwater used agriculture, households, and industry and plans to increase this to 80% by 2025. Israel also uses desalination to provide nearly half of its water. In addition, the government gradually eliminated most water subsidies to raise Israel's price of irrigation water, which is now one of the highest in the world.

Cutting Freshwater Waste in Industries and Homes

Producers of chemicals, paper, oil, coal, primary metals, and processed foods consume almost 90% of the freshwater used by industries in the United States. Some of these industries purify wastewater and recycle it to reduce their water use and water treatment costs. For example, more than 95% of the water used to make steel can be recycled. Even so, most industrial processes could be redesigned to use much less water. **GREEN CAREER: Water conservation specialist**

Flushing toilets with freshwater—most of it clean enough to drink—is the largest use of domestic freshwater in the United States and accounts for about one-fourth of home water use in the United States. Since 1992, U.S. government standards have required that new toilets use no more than 6.1 liters (1.6 gallons) of water per flush. Even at this rate, just two flushes use more than the daily amount of water available per person for all uses to many of the world's poor who live in arid regions.

Other water-saving and money-saving appliances are widely available. Low-flow showerheads can save large amounts of water and money by cutting the flow of a shower in half. Front-loading clothes washers use 30% less water than top-loading machines use. According

to the American Water Works Association, the typical American household could cut its water use and water bills by using water-saving appliances and stopping water leaks.

According to UN studies, 30–60% of the water supplied in nearly all of the world's major cities in less-developed countries is lost, primarily through leakage from water mains, pipes, pumps, and valves. Water experts say that fixing these leaks should be a high priority for water-short countries, because it would increase water supplies and cost less than building dams or importing water.

Even in advanced industrialized countries such as the United States, losses to leakage average 10–30%. However, leakage losses have been reduced to about 3% in Copenhagen, Denmark, and to 5% in Fukuoka, Japan.

CONSIDER THIS ...

CONNECTIONS Water Leaks and Water Bills

Any water leak wastes freshwater and raises water bills. You can detect a silent toilet water leak by adding a few drops of food coloring to the toilet tank and waiting 5 minutes. If the color shows up in the bowl, you have a leak. A faucet leaking water at the rate of 1 drop per second can about waste 10,000 liters (2,650 gallons) per year. This also represents money going down the drain.

Many homeowners and businesses in water-short areas are using drip irrigation to cut water losses. Some use smart sprinkler systems with moisture sensors that have cut water use on lawns by up to 40%. Others copy nature by replacing green lawns with a mix of native plants that need little or no watering. Such water-thrifty landscaping saves money by reducing water use by 30–85% and by sharply reducing labor, fertilizer, and fuel requirements. It also can help landowners reduce polluted runoff, air pollution, and yard wastes.

This example of reconciliation ecology (see Chapter 9, p. 212) also provides habitats and food for threatened honeybee, butterfly, and songbird species. It is an application of the biodiversity **principle of sustainability**, as well as a good way to make a beneficial environmental impact.

Water used in homes can be reused. **Gray water** is used water from bathtubs, showers, sinks, dishwashers, and clothes washers. About 50–75% of a household's gray water could be stored in a holding tank and reused to irrigate lawns and nonedible plants, flush toilets, and wash cars. Such efforts mimic the way nature recycles water, and thus follow the chemical cycling **principle of sustainability**.

Large-scale harvesting of rainwater in urban areas can increase water supplies and reduce flooding by reducing storm flows. In Singapore, for example, most urban runoff of water is collected and deposited in reservoirs.

The relatively low cost of water in most communities is one of the major causes of excessive water use and waste. About one-fifth of all U.S. public water systems do not use water meters, which can help track water use and reveal water leaks. These public water systems charge a single, low annual rate for almost unlimited use of high-quality water.

When the U.S. city of Boulder, Colorado, introduced water meters, water use per person dropped by 40%. In some cities in Brazil, people buy *smart cards*, each of which contains a certain number of water credits that entitle their owners to measured amounts of freshwater. Brazilian officials say this approach saves water and typically reduces household water bills by 40%. Figure 11.19 lists various ways to use water more efficiently in industries, homes, and businesses (**Concept 11.3**).

In 2015, the state of California had been experiencing drought for four years and projected climate change is likely to make the state hotter and drier throughout much of this century. To deal with these problems, California has raised water prices, with heavy users paying more. Many Californians are replacing their grass lawns with water-saving ground cover or native vegetation adapted to dry conditions. Others are installing more efficient toilets and showerheads and are showering and washing clothes less frequently. In 2015, California's urban residents reached a 25% water use reduction goal set by the state government.

Finding more sustainable ways to use freshwater is the subject of some major research efforts. One group that is working on this problem is the Global Water Policy Project, founded by the renowned water supply expert and National Geographic Explorer Sandra Postel (Individuals Matter 11.1).

Solutions

Reducing Water Losses

- Redesign manufacturing processes to use less water
- Recycle water in industry
- Fix water leaks
- Landscape yards with plants that require little water
- Use drip irrigation on gardens and lawns
- Use water-saving showerheads, faucets, appliances, and toilets (or waterless composting toilets)
- Collect and reuse gray water in and around houses, apartments, and office buildings
- Raise water prices and use meters, especially in dry urban areas

FIGURE 11.19 Ways to reduce freshwater losses in industries, homes, and businesses (**Concept 11.3**). *Critical thinking:* Which three of these solutions do you think are the best ones? Why?

What Can You Do?

Water Use and Waste

- Use water-saving toilets, showerheads, and faucets
- Take short showers instead of baths
- Turn off sink faucets while brushing teeth, shaving, or washing
- Wash only full loads of clothes or use the lowest possible water-level setting for smaller loads
- Repair water leaks
- Wash your car from a bucket of soapy water, use gray water, and use the hose for rinsing only
- If you use a commercial car wash, try to find one that recycles its water
- Replace your lawn with native plants that need little if any watering
- Water lawns and gardens only in the early morning or evening and use gray water
- Use drip irrigation and mulch for gardens and flowerbeds

FIGURE 11.20 Individuals Matter: You can reduce your use and waste of freshwater. *Critical thinking:* Which of these steps have you taken? Which would you like to take?

Each of us can reduce our water footprints by using less water and using it much more efficiently (Figure 11.20).

Using Less Water to Remove Wastes

Currently, large amounts of freshwater clean enough to drink are used to flush away industrial, animal, and household wastes. According to the UN Food and Agriculture Organization (FAO), if current growth trends in population and water use continue, within 40 years, the equivalent of the world's entire annual reliable flow of river water will be needed just to dilute and transport such wastes.

Recycling and reusing gray water from homes and businesses and wastewater from sewage plants could save much of this freshwater. In Singapore, all sewage water is treated at reclamation plants for reuse by industry. U.S. cities such as Las Vegas, Nevada, and Los Angeles, California, are beginning to clean up and reuse some of their wastewater. However, less than 10% of the water in the United States is recycled, cleaned up, and reused. Sharply raising this percentage would be a way to apply the chemical cycling **principle of sustainability**.

Another way to keep freshwater out of the waste stream is to rely more on waterless composting toilets. These devices convert human fecal matter to a small amount of dry and odorless soil-like humus material that can be removed from a composting chamber and returned to the soil as fertilizer. One of the authors (Miller) used a composting toilet for over a decade with no problems, while living and working deep in the woods in an a small passive solar home and office used for evaluating solutions to water, energy, and other environmental problems.

Sandra Postel: National Geographic Explorer and Freshwater Conservationist

Sandra Postel is one of the world's most respected authorities on water issues. In 1994, she founded the Global Water Policy Project, a research and education organization that promotes more sustainable use of the earth's finite freshwater supply. Postel has authored or coauthored several influential books and written dozens of articles about using water more sustainably.

In her quest to educate people about water supply issues, Postel has appeared on several television news shows, taken part in many environmental documentary films (including the BBC's *Planet Earth*), and addressed the European Parliament. In 2010, she was appointed Freshwater Fellow of the National Geographic Society, where she serves as the lead water expert for the society's freshwater conservation efforts.

Postel is also co-director of *Change the Course*, a national freshwater conservation and restoration campaign being piloted in the Colorado River Basin. In 2002, Postel was named one of the "Scientific American 50" for her contributions to science and technology.

Mark Thiessen/National Geographic Creative

Reducing Flooding

Some areas have too little freshwater. Other areas sometimes have too much water because of natural flooding by streams, caused mostly by heavy rain or rapidly melting snow. A flood happens when freshwater in a stream or river overflows its normal channel and spills into the adjacent area, called a **floodplain**.

Human activities have contributed to flooding in several ways. First, in efforts to reduce the threat of flooding on floodplains, some rivers have been narrowed and straightened, banked by protective dikes and *levees* (long mounds of earth along their banks), and dammed to create reservoirs that store and release water as needed. However, such measures can lead to greatly increased flood damage when heavy snowmelt or prolonged rains overwhelm dikes and levees.

A second activity that makes flooding more likely is the removal of water-absorbing vegetation, especially on hillsides (Figure 11.21). Once the trees on a hillside have been cut for timber, fuelwood, livestock grazing, or farming, water from precipitation rushes down the naked slope, eroding precious topsoil. This practice increases flooding and pollution in local streams.

A third human activity that increases the severity of flooding is the draining of wetlands that naturally absorb floodwaters. These areas often end up being covered with pavement and buildings, which leads to greatly increase runoff. The increased runoff contributes to flooding and pollution of surface water.

A fourth human-related factor that will likely increase flooding is a rise in sea levels, which is projected to occur during this century mostly because of climate change related to human activities. Climate change models project that, by 2075, as many as 150 million people living in the world's largest coastal cities could be flooded out by rising sea levels.

According to many scientists, people can reduce flooding and water pollution by relying less on engineered devices such as dams and levees and more on nature's systems. By preserving existing wetlands and restoring degraded wetlands that lie in floodplains, we can take advantage of the natural flood control they provide. These and other ways to reduce our contribution to flooding are listed in Figure 11.22.

11.4 HOW CAN WE REDUCE WATER POLLUTION?

CONCEPT 11.4 Reducing water pollution requires that we prevent it, work with nature to treat sewage, cut resource use and waste, reduce poverty, and slow population growth.

Point and Nonpoint Sources of Water Pollution

Water pollution is any change in water quality that can harm living organisms or make the water unfit for human uses such as drinking, irrigation, and recreation. Water pollution can come from single (point) sources or from larger and dispersed (nonpoint) sources. **Point sources**

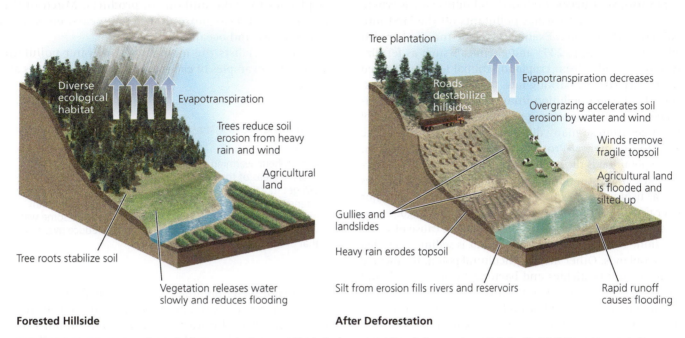

Forested Hillside

Diverse ecological habitat

Evapotranspiration

Trees reduce soil erosion from heavy rain and wind

Agricultural land

Tree roots stabilize soil

Vegetation releases water slowly and reduces flooding

After Deforestation

Tree plantation

Roads destabilize hillsides

Evapotranspiration decreases

Overgrazing accelerates soil erosion by water and wind

Winds remove fragile topsoil

Agricultural land is flooded and silted up

Gullies and landslides

Heavy rain erodes topsoil

Silt from erosion fills rivers and reservoirs

Rapid runoff causes flooding

FIGURE 11.21 Natural Capital Degradation: A hillside before and after deforestation. ***Critical thinking:*** How might a drought in this area make these conditions even worse?

Solutions

Reducing Flood Damage

Prevention

Preserve forests in watersheds

Preserve and restore wetlands on floodplains

Tax development on floodplains

Increase use of flood-plains for sustainable agriculture and forestry

Control

Strengthen and deepen streams (channelization)

Build levees or floodwalls along streams

Build dams

FIGURE 11.22 Methods for reducing the harmful effects of flooding. ***Critical thinking:*** Which two of these solutions do you think are the best ones? Why?

Top: allensima/Shutterstock.com. Bottom: Zeljko Radojko/Shutterstock.com.

FIGURE 11.23 Point source of water pollution from an industrial plant.

discharge pollutants into bodies of surface water at specific locations through drain pipes, ditches, or sewer lines. Examples include factories (Figure 11.23), sewage treatment plants (which remove some, but not all, pollutants), underground mines, oil wells, and oil tankers.

Point sources are relatively easy to identify, monitor, and regulate. Most more-developed countries have laws that help control point-source discharges of harmful chemicals into aquatic systems. In most of the less-developed countries, there is little control of such discharges.

Nonpoint sources are broad and diffuse areas where rainfall or snowmelt washes pollutants off the land into bodies of surface water. Examples include runoff of eroded soil (Figure 10.12, p. 237) and chemicals such as fertilizers and pesticides from cropland, animal feedlots, logged forests, city streets, parking lots, lawns, and golf courses. Controlling water pollution from nonpoint sources is challenging. It is difficult and expensive to identify and control discharges from so many diffuse sources. According to the U.S. Environmental Protection Agency (EPA), nonpoint source pollution is the main reason why 40% of all U.S. rivers, lakes, and estuaries are still not clean enough for uses such as fishing and swimming, despite the passage of major water pollution control laws more than 40 years ago.

Agricultural activities are by far the leading cause of water pollution. The most common pollutant is sediment eroded from croplands. Other major agricultural pollutants include fertilizers and pesticides and bacteria from livestock and food-processing wastes. Point sources from *industrial facilities* that emit a variety of harmful chemicals are the second largest source of water pollution. *Mining* is the third largest source. Surface mining disturbs the land, which in turn leads to major erosion of sediments and runoff of toxic chemicals.

Another form of water pollution originates from the widespread use of human-made materials such as plastics to make millions of products. Much of this plastic, which lasts more than 1,000 years, ends up in rivers, lakes, and oceans.

Table 11.1 lists the major types of water pollutants along with examples of each and their harmful effects and sources.

A major water pollution problem is exposure to infectious bacteria, viruses, and parasites that are transferred into water from the wastes of 2.5 billion humans without access to toilets and other forms of waste disposal, and from the wastes of other animals. Drinking water contaminated with these biological pollutants can cause painful, debilitating, and often life-threatening diseases,

TABLE 11.1 Types, Effects, and Sources of Major Water Pollutants

Type/Effects	Examples	Major Sources
Infectious agents *Cause diseases*	Bacteria, viruses, protozoa, and parasites	Human and animal wastes
Oxygen-demanding wastes *Deplete dissolved oxygen needed by aquatic species*	Biodegradable animal wastes and plant debris	Sewage, animal feedlots, food-processing facilities, paper mills
Plant nutrients *Cause excessive growth of algae and other species*	Nitrates (NO_3^-) and phosphates (PO_4^{3-})	Sewage, animal wastes, inorganic fertilizers
Organic chemicals *Add toxins to aquatic systems*	Oil, gasoline, plastics, pesticides, cleaning solvents	Industry, farms, households
Inorganic chemicals *Add toxins to aquatic systems*	Acids, bases, salts, and metal compounds	Industry, households, surface runoff
Sediments *Disrupt photosynthesis, food webs, and other processes*	Soil, silt	Land erosion
Heavy metals *Cause cancer, disrupt immune and endocrine systems*	Lead, mercury, arsenic	Unlined landfills, household chemicals, mining refuse, and industrial discharges
Thermal *Make some species vulnerable to disease*	Heat	Electric power and industrial plants

including typhoid fever, cholera, hepatitis B, giardiasis, and cryptosporidium. The WHO estimates that each year, more than 1.6 million people die from largely preventable waterborne infectious diseases that they get by drinking contaminated water or by not having enough clean water to keep clean.

Pollution of Rivers and Streams

Because they are flowing, rivers and streams can recover from moderate levels of biodegradable wastes. The wastes are diluted by the flowing water and are broken down by bacteria. However, this natural recovery process does not work when a stream is overloaded with biodegradable pollutants or when drought, damming, or water diversion reduce its flow (**Concept 11.4**). In addition, this process does not eliminate slowly biodegradable and nondegradable pollutants.

In a flowing stream, the breakdown of biodegradable wastes by bacteria depletes dissolved oxygen and creates an *oxygen sag curve* (Figure 11.24). This reduces or eliminates populations of organisms with high oxygen requirements until the stream is cleansed of oxygen-demanding wastes.

Wastewater is any water that contains sewage, other wastes, or polluting chemicals from homes and industries. Wastewater can be treated to remove or reduce pollutants. Laws enacted in the 1970s to control water pollution have greatly increased the number of facilities that treat wastewater in the United States and in most other more-developed countries. Environmental laws also require industries to reduce or eliminate their point-source discharges of harmful chemicals into surface waters.

One success story is the cleanup of the U.S. state of Ohio's Cuyahoga River. It was so polluted that it caught fire several times and, in 1969, was photographed while burning as it flowed through the city of Cleveland toward Lake Erie. The highly publicized image of this burning river prompted elected officials to enact laws to limit the discharge of industrial wastes into the river and to provide funds for upgrading sewage treatment facilities. Today, the river is cleaner, is no longer flammable, and is widely used by boaters and anglers. This accomplishment illustrates the power of bottom-up pressure by citizens who prodded elected officials to change a severely polluted river into an economically and ecologically valuable public resource. GOOD NEWS

In most less-developed countries, stream pollution from discharges of untreated sewage, industrial wastes,

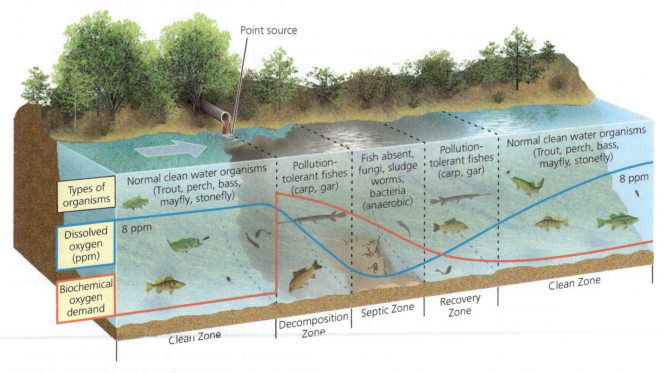

FIGURE 11.24 Natural Capital: A stream can dilute and decay degradable, oxygen-demanding wastes, and it can also dilute heated water. This figure shows the oxygen sag curve (blue) and the curve of oxygen demand (red). Streams recover from oxygen-demanding wastes and from the injection of heated water if they are given enough time and are not overloaded. ***Critical thinking:*** What would be the effect of putting another discharge pipe emitting biodegradable waste to the right of the one in this picture?

and discarded trash is a serious and growing threat. According to the Global Water Policy Project, most cities in less-developed countries discharge 80–90% of their untreated sewage directly into rivers, streams, and lakes. These bodies of water are often used for drinking, bathing, and washing clothes.

80–90% Percentage of the raw sewage in most cities in less-developed countries that is discharged directly into rivers, streams, and lakes.

According to the World Commission on Water for the 21st Century, half of the world's 500 major rivers are heavily polluted, with most of these polluted waterways running through less-developed countries. Most of these countries cannot afford to build waste treatment plants and do not have, or do not enforce, laws for controlling water pollution.

Industrial wastes and sewage pollute more than two-thirds of India's water resources as well as 54 of the 78 rivers and streams monitored in China. According to the Ministry of Environmental Protection, some 380 million Chinese people drink unsafe water and nearly half of China's rivers carry water that is too toxic to touch, much less drink.

Pollution of Lakes and Reservoirs

Large lakes and artificial reservoirs are generally less effective at diluting pollutants than streams are, for two reasons. *First,* lakes and reservoirs often contain layers (see Figure 7.24, p. 158) that undergo little vertical mixing. *Second,* lakes and reservoirs have low flow rates or no flow at all. The flushing and changing of water in lakes and large artificial reservoirs can take from 1 to 100 years, compared with several days to several weeks for streams.

These two factors make lakes and reservoirs more vulnerable than streams to contamination. Sources include runoff or discharges of plant nutrients, oil, pesticides, and nondegradable toxic substances, such as lead, mercury, and arsenic. Many toxic chemicals also enter lakes and reservoirs from the atmosphere.

Eutrophication is natural nutrient enrichment of a shallow lake, a coastal area at the mouth of a river (**Core Case Study**), or a slow-moving stream. It is caused mostly by the natural runoff of plant nutrients such as nitrates and phosphates from land bordering such bodies of water. An *oligotrophic lake* is low in nutrients and its water is clear (see Figure 7.25, p. 159). Over time, some lakes become more eutrophic (see Figure 7.26, p. 160) as nutrients are added from natural and human sources in the surrounding watersheds.

Human activities increase the input of plant nutrients into lakes near urban or agricultural areas. These inputs involve mostly nitrate- and phosphate-containing effluents from various sources. Sources include fertilized farmland, animal feedlots, urban streets and parking lots, fertilized lawns, mining sites, and municipal sewage treatment plants. Some nitrogen also reaches lakes by deposition from the atmosphere. This process of accelerated eutrophication from human activities is called **cultural eutrophication**.

During hot weather or drought, this nutrient overload can produce dense growths, or "blooms," of organisms such as algae and cyanobacteria (see Figure 7.26, p. 160). When the algae die, swelling populations of oxygen-consuming bacteria decompose them. These bacteria deplete dissolved oxygen in the surface layer of water near the shore and in the bottom layer of a lake or coastal area (Figure 11.1). This lack of oxygen can kill fish, shellfish, and other aerobic aquatic animals that cannot move to safer waters. If excess nutrients continue to flow into a lake, bacteria that do not require oxygen take over and produce gaseous products such as smelly, highly toxic hydrogen sulfide and flammable methane.

According to the EPA, about one-third of the 100,000 medium to large lakes and 85% of the large lakes near major U.S. population centers have some degree of cultural eutrophication. The International Water Association estimates that more than half of the lakes in China suffer from cultural eutrophication.

There are several ways to *prevent* or *reduce* cultural eutrophication. Advanced (but expensive) waste treatment processes can remove nitrates and phosphates from wastewater before it enters a body of water. Other preventive approaches include banning or limiting the use of phosphates in household detergents and other cleaning agents. Soil conservation (Figure 10.24, p. 248) and land-use control can also reduce nutrient runoff.

CONSIDER THIS ...

Learning from Nature

Scientists are finding ways to mimic the earth's natural cycling of nutrients by adding nutrients removed from wastewater to the soil (unless they contain toxic chemicals) instead of dumping them into waterways.

Removing weeds, adding herbicides and algaecides, and pumping air into lakes and reservoirs to prevent oxygen depletion can clean up lakes suffering from cultural eutrophication. However, these methods are expensive and energy-intensive. Most lakes can recover from cultural eutrophication if excessive inputs of plant nutrients are stopped.

Groundwater Pollution

Groundwater pollution is a serious but hidden global threat to human health. Common pollutants such as fertilizers, pesticides, gasoline, oil, and organic solvents can seep into groundwater from numerous sources (Figure 11.25). People who dump or spill gasoline, oil, and organic solvents onto the ground can also contaminate groundwater.

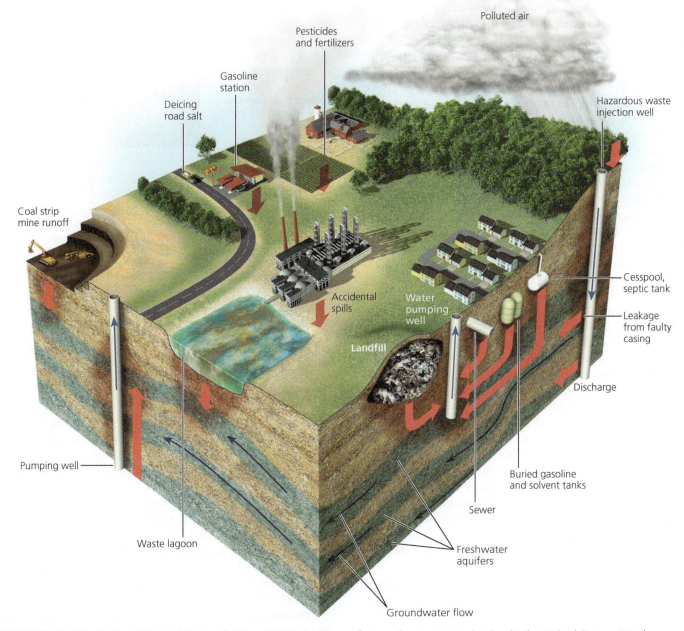

FIGURE 11.25 **Natural Capital Degradation:** Principal sources of groundwater contamination in the United States. Another source in coastal areas is saltwater intrusion from excessive groundwater withdrawal. (Figure is not drawn to scale.) ***Critical thinking:*** What are three sources shown in this figure that might be affecting groundwater in your area?

Hydraulic fracturing or *fracking* used to extract oil and natural gas from thousands of wells is a new and growing potential threat to groundwater in parts of the United States. Groundwater contamination can result from leaky pipes and pipe fittings in oil and natural gas wells and from contaminated wastewater brought to the surface and often stored in deep underground in hazardous waste wells. (Fracking is discussed further in Chapter 13.)

Removing contaminants from groundwater is difficult and costly. Groundwater flows so slowly that contaminants are not effectively diluted and dispersed. In addition, groundwater usually has much lower concentrations of dissolved oxygen (which helps decompose many contaminants) and smaller populations of decomposing bacteria. The usually cold temperature of groundwater also slows down chemical reactions that decompose wastes.

Thus, it can take decades to thousands of years for contaminated groundwater to cleanse itself of slowly degradable wastes (such as DDT). On a human time scale, nondegradable wastes (such as toxic lead and arsenic) remain in the water permanently.

On a global scale, we do not know much about groundwater pollution because few countries go to the great expense of locating, tracking, and testing aquifers. However, the results of scientific studies in scattered parts of the world are alarming.

Groundwater provides about 70% of China's drinking water. According to the Chinese Ministry of Land and Resources, about 90% of China's shallow groundwater is polluted with chemicals such as toxic heavy metals, organic solvents, nitrates, petrochemicals, and pesticides. About 37% of this groundwater is so polluted that it cannot even be treated for use as drinking water. Every year, according to the WHO and the World Bank, contaminated drinking water in China sickens an estimated 190 million and kills about 60,000 people.

In the United States, an EPA survey of 26,000 industrial waste ponds and lagoons found that one-third of them had no liners to prevent toxic liquid wastes from seeping into aquifers. In addition, almost two-thirds of the country's liquid hazardous wastes are injected into the ground in disposal wells (Figure 11.25). Leaking injection pipes and seals in such wells can contaminate aquifers used as sources of drinking water.

By the end of 2016, the EPA had cleaned up about 461,000 of the more than 532,000 underground tanks in the United States that were leaking gasoline, diesel fuel, home heating oil, or toxic solvents into groundwater. During this century, scientists expect many of the millions of such tanks that have been installed around the world to corrode, leak, and become a major global health problem. Determining the extent of a leak from a single underground tank can cost \$25,000–\$250,000, and cleanup costs can be even higher.

If toxic chemicals reach an aquifer, effective cleanup is often not possible or is too costly. Although there are ways to clean up contaminated groundwater (Figure 11.26, right), such methods are expensive. Cleaning up a single contaminated aquifer can cost anywhere from 10 million to several hundred million dollars depending on the size of the aquifer and the types of contaminants. Thus, preventing groundwater contamination (Figure 11.26, left) is the most effective way to deal with this serious water pollution problem.

Purifying Drinking Water

Most more-developed countries have laws establishing drinking water standards. However, most less-developed countries do not have such laws or, if they have them, they do not enforce them.

In more-developed countries, surface water withdrawn for use as drinking water is typically stored in a reservoir

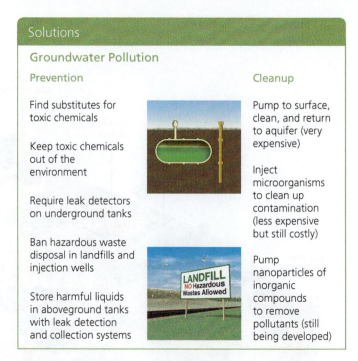

Solutions

Groundwater Pollution

Prevention	Cleanup
Find substitutes for toxic chemicals	Pump to surface, clean, and return to aquifer (very expensive)
Keep toxic chemicals out of the environment	Inject microorganisms to clean up contamination (less expensive but still costly)
Require leak detectors on underground tanks	
Ban hazardous waste disposal in landfills and injection wells	Pump nanoparticles of inorganic compounds to remove pollutants (still being developed)
Store harmful liquids in aboveground tanks with leak detection and collection systems	

FIGURE 11.26 We can clean up contaminated groundwater, but prevention is the only effective approach. **Critical thinking:** Which two of these preventive solutions do you think are the most important? Why?

for several days. This improves clarity and taste by increasing dissolved oxygen content and allowing suspended matter to settle. The water is then pumped to a purification plant and treated to meet government drinking water standards.

In areas with pure groundwater or surface water sources, little treatment is necessary. Several major U.S. cities, including New York City, Boston, Seattle, and Portland, Oregon, have avoided building expensive water treatment facilities. Instead, they have invested in protecting the forests and wetlands in the watersheds that provide their water supplies.

Technology exists that can convert sewer water into pure drinking water. One process begins with microfiltration to remove bacteria and suspended solids. The wastewater then undergoes reverse osmosis to remove minerals, viruses, and various organic compounds. Finally, hydrogen peroxide and ultraviolet (UV) light remove additional organic compounds. In a world where people will face increasing shortages of drinking water, wastewater purification is likely to become a major growth business. **GREEN CAREER: Wastewater purification**

There are also simpler ways to purify drinking water. In tropical countries that lack centralized water treatment systems, the WHO urges people to purify drinking water by exposing a clear plastic bottle filled with contaminated water to intense sunlight. The sun's heat and UV rays can kill infectious microbes in as little as 3 hours. Painting one

FIGURE 11.27 The *LifeStraw*™ is a personal water purification device that gives many poor people access to safe drinking water. Here, four young men in Uganda demonstrate its use. ***Critical thinking:*** Do you think the development of such devices should make prevention of water pollution less of a priority? Explain.

Vestergaard Frandsen

side of the bottle black can improve heat absorption in this simple solar disinfection method, which applies the solar energy **principle of sustainability**. Where this simple measure has been used, the incidence of dangerous childhood diarrhea has decreased by 30–40%. Researchers found that they can speed up this disinfection process by adding lime juice to the water bottles.

Danish inventor Torben Vestergaard Frandsen has developed the *LifeStraw*™, an inexpensive, portable water filter that eliminates many viruses and parasites from water that is drawn through it (Figure 11.27). It has been particularly useful in Africa, where aid agencies are distributing it. Another option being used by more and more people around the world is bottled water, which has created or worsened some environmental problems (see Case Study that follows).

CASE STUDY

Is Bottled Water a Good Option?

Bottled water can be a useful but expensive option in countries and areas where people do not have access to safe and clean drinking water. However, experts say that the United States has some of the world's cleanest drinking water. Municipal water systems in the United States are required to test their water regularly for a number of pollutants and to make the results available to citizens. Yet, about two-thirds of all Americans worry about getting sick from tap water contaminants and many drink high-priced bottled water or install expensive water purification systems.

Studies by the Natural Resources Defense Council (NRDC) reveal that in the United States, a bottle of water costs between 240 and 10,000 times as much as the same volume of tap water. Water expert Peter Gleick has estimated that more than 40% of the expensive bottled water that Americans drink is bottled tap water. A 4-year study by the NRDC concluded that most bottled water is of good quality but they found bacteria and synthetic organic chemicals in 23 of the 123 brands tested. Bottled water is less regulated than tap water and the EPA contamination standards that apply to public water supplies do not apply to bottled water.

Use of bottled water also causes environmental problems. In the United States, according to the Container Recycling Institute, more than 67 million plastic water bottles are discarded every day—enough bottles in a year, if lined up end-to-end, to wrap around the planet at its equator about 280 times. Most water bottles are made of recyclable PET plastic, but in the United States only about 38% of these bottles get recycled. Many of the billions of discarded bottles end up in landfills, where they can remain for hundreds of years. In addition, millions get scattered on the land and end up in rivers, lakes, and oceans. By contrast, in Germany most bottled water is sold in returnable and reusable glass bottles.

It takes huge amounts of energy to manufacture bottled water and to transport it across countries and around the world, as well as to refrigerate much of it in stores. Toxic gases and liquids are released during the manufacture of plastic water bottles, and greenhouse gases and other air pollutants are emitted by the fossil fuels burned to make them and to deliver bottled water to suppliers. According to the Pacific Institute, the oil used to pump, process, bottle, transport, and refrigerate the bottled water used annually in the United States would be enough to fuel 3 million cars for a year. In addition, withdrawing groundwater for bottling is helping deplete some aquifers.

Because of these harmful environmental impacts and the high cost of bottled water, there is a growing *back-to-the-tap* movement. From San Francisco to New York to Paris, city governments, restaurants, schools, religious groups, and many consumers are refusing to buy bottled water. In 2015, San Francisco became the first city to ban the sale of plastic water bottles. Violators can face fines of up to $1,000.

As part of this movement, people are refilling portable bottles with tap water and using simple filters to improve the taste and color of water where necessary. Some health officials suggest that before drinking expensive bottled water or buying costly home water purifiers, consumers have their water tested by local health departments or private labs (but not by companies trying to sell water purification equipment).

Using Laws to Protect Drinking Water Quality

About 54 countries, most of them in North America and Europe, have legal standards for safe drinking water. For example, the U.S. Safe Drinking Water Act of 1975 (amended in 1996) requires the EPA to establish national drinking water standards, called *maximum contaminant levels,* for any pollutants that could have adverse effects on human health. Currently, this act strictly limits the levels of 91 potential contaminants in U.S. tap water. However, in most less-developed countries, such laws do not exist or are not enforced.

Health scientists call for strengthening the U.S. Safe Drinking Water Act. However, various industries have pressured elected officials to weaken the Safe Drinking Water Act, because complying with it increases their costs.

One proposal is to eliminate national testing of drinking water and requirements for public notification about violations of drinking water standards. Another proposal would allow states to give providers of drinking water a permanent right to violate the standard for a given contaminant if they claim they cannot afford to meet the standard. Some critics also call for greatly reducing the EPA's already-low budget for enforcing the Safe Drinking Water Act.

CASE STUDY

Lead in Drinking Water

In 2014, many people in older and poorer neighborhoods of Flint, Michigan—a city of nearly 100,000 people with 40% living in poverty—were exposed to potentially dangerous levels of lead in their tap water. The problem began when, in an effort to save money, Flint officials began withdrawing drinking water from the Flint River instead of from Lake Huron.

Officials did not take into account that there were at least 20,000 lead pipes connecting the city's main line water pipes (that do not contain lead) to homes, many of them in older and poorer neighborhoods. They failed to add chemicals to reduce the leaching of lead from the pipes exposed to the more corrosive water supply.

As a result, lead began leaching into the water supply for many homes in older and poorer neighborhoods. Research shows that prolonged exposure to high levels of lead (well beyond the EPA's 15 parts per billion (ppb) action level standard for lead in drinking water) is harmful to the developing brains and nervous systems of children.

After Flint changed its drinking water source, the percentage of children citywide with elevated blood levels of lead increased from 2.4% to 4.9% and to 15.7% in its poorest neighborhood. This meant that roughly 1 out of 6 children in Flint and 1 out of 3 in its poorest neighborhood were exposed to blood levels of lead that the Center for Disease Control and Prevention (CDC) uses as a standard to identify children who need medical help. In 1986, the EPA banned the use of lead water pipes, fittings, solder, and other plumbing material in new homes. However, this regulation did not cover the millions of lead pipes and plumbing material in older homes in Flint and elsewhere in the United States.

After a highly publicized outcry, officials switched the city's water supply source back to Lake Huron but the health threats to children remain. The ultimate solution is to replace all of the lead pipes connecting homes to the city's main line water pipes. This will be costly for the city and homeowners and will take decades.

Public health officials say that Flint is "just the tip of the iceberg." Investigations revealed that in 2015 almost 2,000, or 20%, of U.S. water systems tested have failed to meet the EPA's standard of 15 ppb for lead in drinking water. Many of them are small water utilities that provide water for about 4 million Americans. Because of a lack of money, many of these utilities skip water safety tests or fail to treat water contaminated with lead or other pollutants. The EPA and health officials have urged all U.S. residents to have their drinking water tested for lead.

Solutions to the country's lead poisoning problem include replacing all of the lead pipes leading to and existing within homes at a cost of at least $55 billion and providing free lead testing for all children ages 1–6. Meanwhile in 2012 Congress, under pressure from the lead industry, slashed funding for lead screening of children, lead removal from older homes, and CDC lead testing and research.

Ocean Pollution

We should care about the oceans because they keep us alive. Oceans help provide and recycle the planet's freshwater through the water cycle (see Figure 3.15, p. 58). They also affect weather and climate, help regulate the earth's temperature, and absorb some of the massive amounts of carbon dioxide that human activities emit into the atmosphere.

As oceanographer and explorer Sylvia A. Earle (see Individuals Matter 9.1, p. 218) reminds us: "Even if you never have the chance to see or touch the ocean, the ocean touches you with every breath you take, every drop of water you drink, every bite you consume. Everyone, everywhere is inextricably connected to and utterly dependent upon the existence of the sea." Despite its importance, we treat the ocean as the world's largest dump for the massive and growing amount of wastes and pollutants that we produce.

Coastal areas such as wetlands, estuaries, coral reefs, and mangrove swamps receive the largest inputs of pollutants and wastes (Figure 11.28). About 40% of the world's people (53% in the United States) live on or near coastlines, and coastal populations are projected to double by 2050. This explains why 80% of marine pollution originates on land.

According to a study by the UN Environment Programme (UNEP), in coastal areas of less-developed counties, 80–90% of city sewage is dumped into oceans without treatment. This often overwhelms the ability of the coastal waters to degrade the wastes. For example, many areas of China's coastline are so choked with algae growing on the nutrients provided by sewage that some scientists have concluded that large areas of China's coastal waters can no longer sustain marine ecosystems. Dumping biodegradable wastes and plant wastes into coastal waters instead of recycling these vital plant nutrients to the soil violates the chemical cycling **principle of sustainability**.

In deeper waters, the oceans can dilute, disperse, and degrade large amounts of raw sewage and other types of degradable pollutants. Some scientists suggest that it is safer to dump sewage sludge, toxic mining wastes, and most other harmful wastes into the deep ocean than to bury them on land or burn them in incinerators. Other scientists disagree and point out that dumping harmful wastes into the ocean would delay urgently needed pollution prevention measures and promote further degradation of this vital part of the earth's life-support system.

Recent studies of some U.S. coastal waters have found huge colonies of viruses thriving in raw sewage and in effluents from sewage treatment plants (which do not remove viruses) and leaking septic tanks. According to one study, one-fourth of the people using coastal beaches in the United States develop ear infections, sore throats, eye irritations, respiratory diseases, or gastrointestinal diseases from swimming in seawater containing infectious viruses and bacteria.

Scientists also point to the underreported problem of pollution from cruise ships. A cruise liner can carry as many as 6,300 passengers and 2,400 crewmembers, and it can generate as much waste (toxic chemicals, garbage, sewage, and waste oil) as a small city. Many cruise ships dump these wastes at sea. In U.S. waters, such dumping is illegal, but some ships continue dumping secretively, usually at night. Some environmentally aware vacationers are refusing to go on cruise ships that do not have sophisticated systems for dealing with the wastes they produce.

Runoff of sewage and agricultural wastes into coastal waters introduces large quantities of nitrate (NO_3^-) and phosphate (PO_4^{3-}) plant nutrients that can cause explosive growths of algae and lead to dead zones (**Core Case Study** and Figure 11.1). These *harmful algal blooms*—known as red, brown, or green toxic tides—can release waterborne and airborne toxins that poison seafood, damage fisheries, kill some fish-eating birds, and reduce tourism. Each year, harmful algal blooms poison about 60,000 Americans who eat shellfish contaminated by the algae.

Harmful algal blooms occur annually at least 400 *oxygen-depleted zones* around the world mostly in temperate coastal waters and in large bodies of water with restricted outflows, such as the Baltic and Black seas. The largest such zone in U.S. coastal waters forms each year in the northern Gulf of Mexico (**Core Case Study**). A study by Luan Weixin, of China's Dalain Maritime University, found that nitrates and phosphates have seriously contaminated about half of China's shallow coastal waters.

Ocean Pollution from Oil

Crude petroleum (oil as it comes out of the ground) reaches the ocean from natural seeps on the ocean floor and from human activities. The most visible human sources of ocean oil pollution are tanker accidents, such as the huge *Exxon Valdez* oil spill in the U.S. state of Alaska in 1989. Other sources are blowouts at offshore oil drilling rigs, such as that from the BP Deepwater Horizon rig in the Gulf of Mexico in 2010 (Figure 11.29)—the worst-ever oil spill in U.S. waters, with cleanup, damages, and fines amounting to $61.6 billion (see chapter-opening photo).

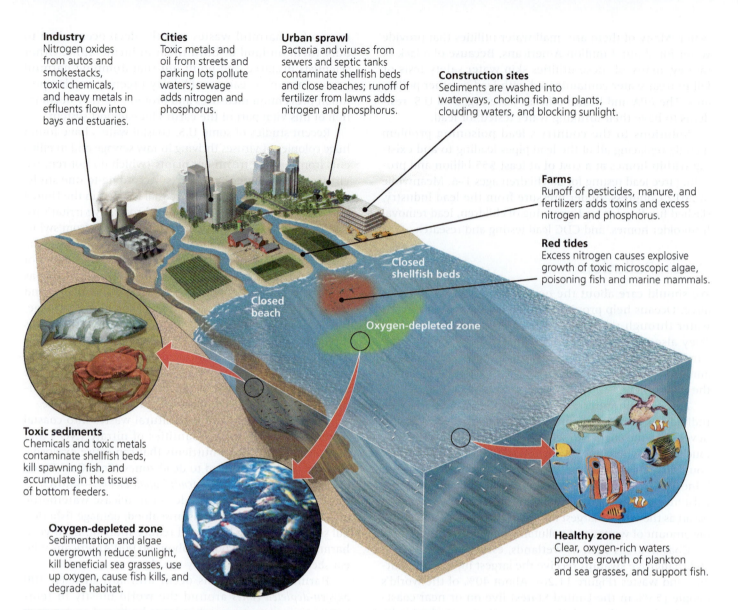

Industry
Nitrogen oxides from autos and smokestacks, toxic chemicals, and heavy metals in effluents flow into bays and estuaries.

Cities
Toxic metals and oil from streets and parking lots pollute waters; sewage adds nitrogen and phosphorus.

Urban sprawl
Bacteria and viruses from sewers and septic tanks contaminate shellfish beds and close beaches; runoff of fertilizer from lawns adds nitrogen and phosphorus.

Construction sites
Sediments are washed into waterways, choking fish and plants, clouding waters, and blocking sunlight.

Farms
Runoff of pesticides, manure, and fertilizers adds toxins and excess nitrogen and phosphorus.

Red tides
Excess nitrogen causes explosive growth of toxic microscopic algae, poisoning fish and marine mammals.

Closed shellfish beds

Closed beach

Oxygen-depleted zone

Toxic sediments
Chemicals and toxic metals contaminate shellfish beds, kill spawning fish, and accumulate in the tissues of bottom feeders.

Oxygen-depleted zone
Sedimentation and algae overgrowth reduce sunlight, kill beneficial sea grasses, use up oxygen, cause fish kills, and degrade habitat.

Healthy zone
Clear, oxygen-rich waters promote growth of plankton and sea grasses, and support fish.

FIGURE 11.28 Natural Capital Degradation: Residential areas, factories, and farms all contribute to the pollution of coastal waters. ***Critical thinking:*** What do you think are the three worst pollution problems shown here? For each one, how does it affect two or more of the ecosystem and economic services listed in Figure 7.16 (p. 152)?

However, studies show that the largest source of ocean oil pollution is urban and industrial runoff from land. Most of this comes from leaks in pipelines, refineries, and other oil-handling and storage facilities and from oil and oil products that are intentionally dumped or accidentally spilled or leaked onto the land or into sewers by homeowners and industries.

Chemicals in oil called volatile organic hydrocarbons in oil kill many aquatic organisms immediately upon contact. Other chemicals in oil form tarlike globs that float on the surface and coat the feathers of seabirds (see chapter-opening photo) and the fur of marine mammals. The oil destroys their natural heat insulation and buoyancy, causing many of them to drown or die from loss of body heat.

Heavy oil components that sink to the ocean floor or wash into estuaries and coastal wetlands can smother bottom-dwelling organisms such as crabs, oysters, mussels, and clams, or make them unfit for human consumption. Some oil spills have killed coral reefs.

Research shows that populations of many forms of marine life can recover from exposure to large amounts of *crude oil* in warm waters with rapid currents within about 3 years. However, in cold and calm waters, recovery can take decades. Recovery from exposure to spills of *refined oil* such as gasoline and diesel fuel, especially in estuaries and salt marshes, can take 10–20 years or longer. Oil slicks that wash onto beaches can have a serious economic impact on coastal residents, who lose income normally gained from fishing and tourist activities.

Small oil spills can be partially cleaned up by mechanical means, including floating booms, skimmer boats, and absorbent devices such as giant pillows filled with feathers

FIGURE 11.29 The *Deepwater Horizon* drilling platform, located 64 kilometers (40 miles) off the coast of Louisiana, exploded, burned, and sank in the Gulf of Mexico on April 20, 2010. The ruptured wellhead on the ocean floor released 3.1 million barrels (130 million gallons) of crude oil over a 3-month period before it was capped.

U.S. Coast Guard

or hair. However, scientists estimate that current cleanup methods can recover typically no more than 15% of the oil from a major spill.

Preventing oil pollution is the most effective and least costly approach. One of the best ways to prevent oil tanker spills is to use tankers with double hulls. Stricter safety standards and inspections could reduce oil well blowouts at sea. In addition, businesses, institutions, and citizens living in coastal areas must take care to prevent leaks and spillage of even the smallest amounts of oil and oil products such as paint thinners and gasoline. Figure 11.30 lists ways to prevent pollution of coastal waters and ways to reduce it.

Reducing Water Pollution from Nonpoint Sources

Most nonpoint sources of water pollution come from agricultural practices. To reduce this type of pollution, farmers keep cropland covered with vegetation and use conservation tillage and other methods (see Chapter 10, pp. 247–248) to reduce soil erosion. They can also minimize the amount of fertilizer that runs off into surface water by using fertilizers that release plant nutrients slowly, using no fertilizer on steeply sloped land, and planting buffer zones of vegetation between cultivated fields and nearby surface waters.

The annual formation of the dead zone in the Gulf of Mexico (**Core Case Study**) will be difficult to prevent because of the importance of the Mississippi River basin for growing crops. However, nutrient inputs can be reduced with widespread use of the fertilizer management practices listed above.

Organic farming (see Figure 10.5, p. 231) and other forms of more sustainable food production (see Figure 10.28, p. 251) can also prevent water pollution caused by nutrient overload because they use little if any synthetic inorganic fertilizers and pesticides. Farmers can reduce pesticide runoff by applying pesticides only when needed and by relying more on integrated pest management (see Chapter 10, p. 246). In addition, they can control runoff and infiltration of manure from animal feedlots by planting buffer zones and by locating feedlots, pastures, and animal waste storage sites away from steeply sloped land, surface water, and flood zones.

Reducing Water Pollution from Point Sources

The Federal Water Pollution Control Act of 1972 (renamed the Clean Water Act when it was amended in 1977) and the 1987 Water Quality Act form the basis of U.S. efforts to control pollution of the country's surface waters. The Clean Water Act sets standards for allowed levels of 100 key water pollutants. The Act requires polluting industries to get permits that limit how much of these various pollutants they can discharge into aquatic systems.

The EPA has also been experimenting with a *discharge trading policy,* which uses market forces to reduce water pollution in the United States. Under this program, a permit holder can pollute at higher levels than allowed in its permit if it buys credits from permit holders who are polluting below their allowed levels.

Solutions

Coastal Water Pollution

Prevention		Cleanup
Separate sewage and storm water lines	Ban dumping of wastes and sewage by ships in coastal waters	Improve oil-spill cleanup capabilities
Require secondary treatment of coastal sewage	Strictly regulate coastal development, oil drilling, and oil shipping	Use nanoparticles on sewage and oil spills to dissolve the oil or sewage (still under development)
Use wetlands and other natural methods to treat sewage	Require double hulls for oil tankers	

FIGURE 11.30 Methods for preventing excessive pollution of coastal waters and methods for cleaning it up (**Concept 11.4**). ***Critical thinking:*** Which two of these solutions do you think are the most important? Why?

notion that reducing protection of our waters will somehow ignite the economy, we will shortchange our health, environment, and economy."

According to the EPA, the Clean Water Act of 1972 and other water quality acts led to numerous improvements in U.S. water quality, including the following:

- 60% of all tested U.S. streams, lakes, and estuaries can be used safely for fishing and swimming, compared to 33% in 1972.

- Sewage treatment plants serve 75% of the U.S. population.

- Annual losses of U.S. wetlands that naturally absorb and purify water have been reduced by 80% since 1992.

These are impressive achievements, given the increases in the U.S. population and its per capita consumption of water and other resources since 1972.

Environmental scientists and economists warn that the effectiveness of such a system depends on how low the cap on total pollution levels in any given area is set and on how regularly the cap is lowered. They also warn that discharge trading could allow water pollutants to build up to dangerous levels in areas where credits are bought. Neither adequate scrutiny of the cap levels nor gradual lowering of caps is a part of the current EPA discharge trading system.

Some scientists call for strengthening the U.S. Clean Water Act. Suggested improvements include:

- Shifting the focus of the law from end-of-pipe removal of specific pollutants to water pollution prevention.

- Greatly increased monitoring for violations of the law, with much larger mandatory fines for violators.

- Regulating irrigation water quality.

- Expanding the rights of citizens to bring lawsuits to ensure that water pollution laws are enforced.

Many people oppose these proposals, contending that the Clean Water Act's regulations are already too restrictive and costly. Some state and local officials argue that in many communities, it is unnecessary and too expensive to test all the water for pollutants as required by federal law.

Some members of Congress, under pressure from regulated industries, go further and want to seriously weaken or repeal the Clean Water Act and other government environmental regulations, arguing that such regulations hinder economic growth and prevent job growth. In 2012, William K. Reilly, former head of the EPA who also served as co-chairman of a presidential commission on offshore drilling, said: "If we buy into the misguided

Sewage Treatment

In rural and suburban areas with suitable soils, sewage from each house usually is discharged into a **septic tank** with a large drainage field. In such a system, household sewage and wastewater is pumped into a settling tank, where grease and oil rise to the top, and solids, called *sludge,* fall to the bottom and are decomposed by bacteria. The partially treated wastewater that results is discharged into a large drainage (absorption) field through small holes in perforated pipes embedded in porous gravel or crushed stone just below the soil's surface. As these wastes drain from the pipes and sink downward, the soil filters out some potential pollutants and soil bacteria decompose biodegradable materials. About one-fourth of all homes in the United States are served by septic tanks. They work well as long as they are not overloaded and the sludge in the tank is regularly pumped out.

In urban areas in the United States and other more-developed countries, most waterborne wastes from homes, businesses, and storm runoff flow through a network of sewer pipes to *sewage treatment plants*. Raw sewage reaching a treatment plant typically undergoes one or two levels of wastewater treatment. The first is **primary sewage treatment,** a *physical* process that uses screens and a grit tank to remove large floating objects and to allow solids such as sand and rock to settle out. Then the waste stream flows into a primary settling tank where suspended solids settle out as sludge (Figure 11.31, left).

The second level is **secondary sewage treatment**—a *biological* process in which aerobic bacteria remove as much as 90% of dissolved and biodegradable, oxygen-demanding organic wastes (Figure 11.31, right).

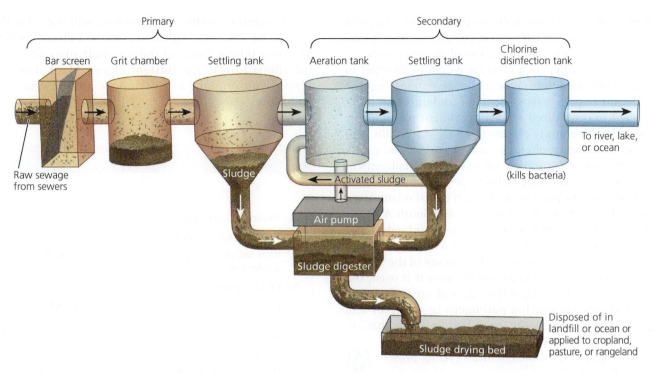

Primary | Secondary

Bar screen | Grit chamber | Settling tank | Aeration tank | Settling tank | Chlorine disinfection tank

Raw sewage from sewers

Sludge

Activated sludge

Air pump

Sludge digester

Sludge drying bed

To river, lake, or ocean

(kills bacteria)

Disposed of in landfill or ocean or applied to cropland, pasture, or rangeland

FIGURE 11.31 Solutions: Primary and secondary sewage treatment systems help reduce water pollution. ***Critical thinking:*** What do you think should be done with the sludge produced by sewage treatment plants?

A combination of primary and secondary treatment removes 95–97% of the suspended solids and oxygen-demanding organic wastes, 70% of most toxic metal compounds and degradable synthetic organic chemicals, 70% of the phosphorus, and 50% of the nitrogen. However, the treatment process removes only a tiny fraction of persistent and potentially toxic organic substances found in some pesticides and in discarded medicines. In addition, and it does not kill viruses, bacteria, and other disease-causing agents.

Before discharge, water from sewage treatment plants usually undergoes *bleaching*, to remove water coloration, and *disinfection* to kill disease-carrying bacteria and some (but not all) viruses. The usual method for accomplishing this is *chlorination*. However, chlorine can react with organic materials in water to form small amounts of chlorinated hydrocarbons. Some of the chemicals cause cancers in test animals, can increase the risk of miscarriages, and can damage the human nervous, immune, and endocrine systems. Use of other disinfectants, such as ozone and UV light, is increasing, but they cost more and their effects do not last as long as those of chlorination.

Improving Sewage Treatment

A serious problem is that EPA regulations allow sludge from waste treatment plants to be applied as fertilizer to cropland, public parks, and other land used by the public. However, the EPA allows industries and homeowners to send wastewater that contains toxic heavy metals, organic chemicals, pesticides, and pharmaceuticals to sewage treatment plants and many of these harmful chemicals end up in the toxic sludge that is applied to these lands.

Environmental and health scientists suggest several ways to prevent pollutants from ending up in sewage sludge that is applied to land and in wastewater from sewage treatment plants. One is to require industries and businesses to remove toxic and hazardous wastes from water sent to municipal sewage treatment plants, which would help implement the full-cost pricing **principle of sustainability** by increasing the cost of creating waste and pollution. Industries could also be encouraged or subsidized to reduce or eliminate their use and waste of toxic chemicals, which would reduce their expense in complying with water pollution control laws.

Another approach is to view the nitrates and phosphates in the sludge and wastewater in sewage treatment plants as ecologically and financially valuable soil nutrients that should be removed from sewage, sold as assets to help pay for their removal, and recycled to the land as is done in Sweden and The Netherlands. This practice implements the chemical cycling **principle of sustainability.**

Another option is to require or encourage more households, apartment buildings, and offices to eliminate sewage outputs by switching to waterless, odorless composting toilet systems, to be installed, maintained, and managed by professionals. Unlike a conventional toilet, a composting toilet does not use water, and waste is not flushed away. Instead, naturally occurring aerobic bacteria break down the waste with the aid of air and heat. Composting toilets do not require the use of any chemicals.

On a larger scale, such systems would be cheaper to install and maintain than current sewage systems, because they do not require vast systems of underground pipes connected to centralized sewage treatment plants. They also save large amounts of water, reduce water bills, and decrease the amount of energy used to pump and purify water. This more environmentally sustainable replacement for the conventional toilet and sewage treatment plants is now being used in parts of more than a dozen countries, including China, India, Mexico, Syria, and South Africa.

A Swedish entrepreneur has developed a biodegradable single-use plastic bag that can be used as a toilet in urban slums and in other areas where many of the world's people do not have access to toilets. After it is used, the bag is knotted and buried. A thin layer of urea in this bag kills the disease-producing pathogens in the feces and helps break down the waste into plant nutrients that are then recycled. This is a simple and inexpensive, low-tech application of the chemical cycling **principle of sustainability**.

Some communities are also using unconventional, but highly effective, *ecological sewage treatment systems,* which work with nature (Science Focus 11.1).

Preventing Water Pollution

It is encouraging that since 1970, most of the world's more-developed countries have enacted laws and regulations that have significantly reduced point-source water pollution. These improvements were largely the result of *bottom-up* political pressure on elected officials from individuals and groups. On the other hand, little has been done to reduce water pollution in most of the less-developed countries.

To many environmental and health scientists, the next step is to increase efforts to reduce and prevent water pollution in both more- and less-developed countries as an important way to improve human health and increase our beneficial environmental impact. They would begin by asking the question: *How can we avoid producing water pollutants in the first place?* (**Concept 11.4**) Figure 11.32 lists ways to achieve this goal over the next several decades.

This shift to pollution prevention will not take place unless citizens put political pressure on elected officials and also take actions to reduce their own daily contributions to water pollution. Figure 11.33 lists some actions you can take to help reduce and prevent water pollution.

Solutions

Water Pollution

- Prevent groundwater contamination
- Reduce nonpoint runoff
- Work with nature to treat sewage and reuse treated wastewater
- Find substitutes for toxic pollutants
- Practice the four Rs of resource use (refuse, reduce, reuse, recycle)
- Reduce air pollution
- Reduce poverty
- Slow population growth

FIGURE 11.32 Ways to prevent or reduce water pollution (**Concept 11.4**). *Critical thinking:* Which two of these solutions do you think are the most important? Why?

What Can You Do?

Reducing Water Pollution

- Fertilize garden and yard plants with manure or compost instead of commercial inorganic fertilizer
- Minimize use of pesticides, especially near bodies of water
- Prevent yard wastes from entering storm drains
- Do not use water fresheners in toilets
- Do not flush unwanted medicines down the toilet
- Do not pour pesticides, paints, solvents, oil, antifreeze, or other harmful chemicals down the drain or onto the ground

FIGURE 11.33 Individuals Matter: You can help reduce water pollution. *Critical thinking:* Which three of these steps do you think are the most important ones to take? Why?

Treating Sewage by Learning from Nature

Some communities and individuals are seeking better ways to purify sewage by learning how nature purifies water using sunlight, plants, aquatic organisms, and natural filtration by soil, sand, and gravel (**Concept 11.4**). Biologist John Todd has developed an ecological approach to treating sewage, called the Living Machine (Figure 11.A).

This purification process begins when sewage flows into a passive solar greenhouse or outdoors site containing rows of large open tanks. The tanks are populated by an increasingly complex series of organisms. In the first set of tanks, algae and microorganisms decompose organic wastes, with sunlight speeding up the process. Water hyacinths, cattails, bulrushes, and other aquatic plants growing in the tanks take up the resulting nutrients.

After flowing though several of these natural purification tanks, the water passes into an artificial marsh, made of sand, gravel, and bulrushes, which filters out algae and remaining organic waste. Some of the plants also remove toxic metals such as lead and mercury and secrete natural antibiotic compounds that kill pathogens.

Next, the water flows into aquarium tanks, where snails and zooplankton consume microorganisms and are in turn consumed by crayfish, tilapia, and other fish that can be eaten or sold as bait. After 10 days, the clear water flows into a second artificial marsh for final filtering and cleansing. The water can be made pure enough to drink by treating it with UV light or by passing the water through an ozone generator, usually under water and out of sight, as part of an attractive pond or wetland habitat.

Operating costs are about the same as those of a conventional sewage treatment plant. These systems are widely used on a small scale. However, they have been difficult to maintain on a scale large enough to handle the typical variety of chemicals in the sewage wastes from large urban areas.

More than 800 cities and towns around the world (150 in the United States) mimic nature by using natural or artificially created wetlands to treat sewage as a lower-cost alternative to expensive waste treatment plants. For example, in Arcata, California—a coastal town of almost 18,000 people—scientists and workers created some 65 hectares (160 acres) of wetlands between the town and the adjacent Humboldt Bay. The marshes and ponds, developed on land that was once a dump, act as a natural waste treatment plant. The cost of the project was less than half the estimated price of a conventional treatment plant.

This system returns purified water to Humboldt Bay, and the sludge that is removed is processed for use as fertilizer. The marshes and ponds also serve as an Audubon Society bird sanctuary, which provides habitats for thousands of seabirds, otters, and other marine animals. The town celebrates its natural sewage treatment system with an annual "Flush with Pride" festival.

Ocean Arks International

FIGURE 11.A Solutions: The Solar Sewage Treatment Plant in the U.S. city of Providence, Rhode Island, is an ecological wastewater purification system, also called a *living machine*. Biologist John Todd is demonstrating this ecological process he invented for purifying wastewater by using the sun and a series of tanks containing living organisms.

This approach and the living machine system developed by John Todd apply all three **scientific principles of sustainability**: using solar energy, employing natural processes to remove and recycle nutrients and other chemicals, and relying on a diversity of organisms and natural processes.

CRITICAL THINKING

Can you think of any disadvantages of using such a nature-based system instead of a conventional sewage treatment plant? Do you think any such disadvantages outweigh the advantages? Why or why not?

BIG IDEAS

- One of the major global environmental problems is the deepening shortage of freshwater in many parts of the world.

- We can use water more sustainably by reducing water use, using water more efficiently, cutting water losses, raising water prices, and protecting aquifers, forests, and other ecosystems that store and release water.

- Reducing water pollution requires that we prevent it, work with nature in treating sewage, and cut resource use and waste.

Tying It All Together

Dead Zones and Sustainability

NO SWIMMING

Bull's-Eye Arts/Shutterstock.com

The Core Case Study that opens this chapter explains how fertilizers from

farmland in the Mississippi River Basin end up over-fertilizing an area of the Gulf of Mexico. This form of water pollution creates a huge dead zone by depleting dissolved oxygen, which reduces seafood production and biodiversity in these coastal waters.

The chapter also discussed how we are using water resources unsustainably and how we can use this irreplaceable resource more sustainably. It also discussed various types of water pollutants and how we can prevent and reduce water pollution.

The three **scientific principles of sustainability** can guide us in using the water more sustainably and in reducing and preventing water pollution.

We can use solar energy to desalinate water expand freshwater supplies, and purify much of the water we use, which will reduce water waste and water pollution. We can promote chemical cycling by not disrupting the natural water cycle, converting sewer water into freshwater, and treating sewage in wetland-based sewage treatment systems (Science Focus 11.1). Finally, by preventing the pollution and degradation of aquatic systems and their bordering terrestrial systems, we help preserve biodiversity—a key component of the life-support system that we depend on for maintaining water supplies and water quality.

Chapter Review

Core Case Study

1. Describe the nature and causes of the annual dead (oxygen-depleted) zone in the Gulf of Mexico.

Section 11.1

2. What are the two key concepts for this section? Define **freshwater**. Explain why access to water is a health issue, an economic issue, a national and global security issue, and an environmental issue. What percentage of the earth's freshwater is available to us? Explain how water is recycled by the hydrologic cycle and how human activities can interfere with this cycle. Define **groundwater**, **zone of saturation**, **water table**, **aquifer**, **surface water**, **surface run-off**, and **watershed basin**.

3. What is **reliable surface runoff?** What percentage of the world's reliable runoff are we using and what percentage are we likely to be using by 2025? How is most of the world's water used? What is a **water footprint**? Define and give two examples of **virtual water**. Describe the availability and use of freshwater resources in the United States and the water shortages that could occur during this century. Define freshwater scarcity stress, and explain how widespread it is

and how serious it may be by 2050. What percentage of the earth's land suffers from severe drought today and how might this change by 2059? How many people in the world lack regular access to clean water today? Summarize the importance of the Colorado River basin in the United States and how human activities are stressing this system. What are three major problems resulting from the way people are using water from the Colorado River basin?

Section 11.2

4. What are the three key concepts for this section? What are the advantages and disadvantages of withdrawing groundwater? Summarize the problem of groundwater depletion in the world, in Saudi Arabia, and in the United States (especially in the Ogallala Aquifer). List three problems that result from the overpumping of aquifers. List some ways to prevent or slow groundwater depletion.

5. What is a **dam**? What is a **reservoir**? What are the advantages and disadvantages of using large dams and reservoirs? How might climate change affect water supplies in the Colorado River system and in areas that depend on mountain glaciers for much

of their water? Explain how dams on the Colorado River have affected its delta. List the pros and cons of the California Water Project. Describe the environmental and health disaster caused by the Aral Sea water transfer project. Define **desalination** and distinguish between distillation and reverse osmosis as methods for desalinating water. What three problems can limit the use of desalination?

Section 11.3

6. What is the key concept for this section? What percentage of available freshwater is lost through inefficient use and other causes in the world and in the United States? What are two major reasons for those losses? Describe three major irrigation methods and list ways to reduce water losses in irrigation. List four ways to reduce water waste in industries and homes and three ways to use less water to remove wastes. What is **gray water**? Describe Sandra Postel's efforts to educate people about water supply issues. List four ways in which you can reduce your use and waste of water. What is a **floodplain?** List four human activities that increase the threat of flooding and four ways to reduce our contribution to flooding.

Section 11.4

7. What is the key concept for this section? What is **water pollution**? Define and distinguish between **point sources** and **nonpoint sources** of water pollution and give an example of each. Summarize the relationship between atmospheric warming and water pollution. List seven major types of water pollutants and three diseases that can be transmitted to humans through polluted water. How many people die each year from water-borne infections diseases?

8. Explain how streams can cleanse themselves of oxygen-demanding wastes and how these cleansing processes can be overwhelmed. What is **wastewater?** Describe the state of stream pollution in more-developed and less-developed countries. Give two reasons why lakes and reservoirs cannot cleanse themselves of pollutants very well. Define

and distinguish between **eutrophication** and **cultural eutrophication**. List three ways to prevent or reduce cultural eutrophication. What are the major sources of groundwater contamination in the United States? Explain why groundwater cannot cleanse itself very well. List three ways to prevent or clean up groundwater contamination. List some ways to purify drinking water. Describe the purification of drinking water in more-developed and less-developed countries. Describe environmental problems caused by the widespread use of bottled water. How are laws used to protect drinking water in the United States? List three ways to strengthen the U.S. Safe Drinking Water Act. Describe how lead threatens drinking water in the United States.

9. Why should we care about the oceans? How are coastal waters and deeper ocean waters most often polluted? What causes algal blooms and what are their harmful effects? What are the effects of oil pollution of the oceans and what can be done to reduce such pollution? List four ways to prevent and reduce pollution of coastal waters.

10. List ways to reduce water pollution from **(a)** nonpoint sources and **(b)** point sources. Describe the U.S. experience with reducing point-source water pollution and list ways to improve such efforts. What is a **septic tank** and how does it work? Explain how **primary sewage treatment** and **secondary sewage treatment** are used to treat wastewater. What are three ways to improve conventional sewage treatment? What is a waterless composting toilet system? Describe John Todd's use of living machines to treat sewage by working with nature. Explain how wetlands can be used to treat sewage. List six ways to prevent and reduce water pollution. List five things you can do to reduce water pollution. What are this chapter's *three big ideas*? Explain how the three **scientific principles of sustainability** can guide us in using water resources more sustainably and in reducing and preventing water pollution.

Note: Key terms are in bold type.

Critical Thinking

1. How might you be contributing directly or indirectly to the annual dead zone that forms in the Gulf of Mexico (**Core Case Study**)? What are three things you could do to reduce your impact?

2. What do you think are the three most important priorities for dealing with the water resource problems of the Colorado River basin? Explain your thinking.

3. Explain why you are for or against **(a)** raising the price of water while providing lower lifeline rates for poor consumers, **(b)** withdrawing government subsidies that provide farmers with water at low cost, and **(c)** providing government subsidies to farmers for improving irrigation efficiency.

4. Calculate how many liters (and gallons) of water are lost in 1 month by a toilet or faucet that leaks 2 drops

of water per second. (One liter of water equals about 3,500 drops and 1 liter equals 0.265 gallon.) How many bathtubs (each containing about 151 liters or 40 gallons) could be filled with this lost water?

5. List the three most important ways in which you could cut your water waste. Which, if any, of these measures do you already take?

6. How might you be contributing directly or indirectly to groundwater pollution? What are three things you could do to reduce your contribution?

7. When you flush your toilet, where does the wastewater go? Trace the actual flow of this water in your community from your toilet through sewers to a wastewater treatment plant (or to a septic system)

Doing Environmental Science

Do some research on the water resources in your community and write a report answering the following questions:

a. What are the principle sources of your community's drinking water?

b. How is your drinking water treated?

c. What are your community's principal nonpoint sources of contamination of surface water and groundwater?

Global Environment Watch Exercise

Go to your MindTap course to access the GREENR database. Using the "Basic Search" box at the top of the page, search for *Ogallala Aquifer*. Research articles that quantify how much the aquifer has declined and list the three areas over the aquifer where the decline is the worst. Look for projections on how much more the aquifer could decline

and from there to the environment. Try to visit a local sewage treatment plant to see what it does with wastewater. Compare the processes it uses with those shown in Figure 11.31. What happens to the sludge produced by this plant? What improvements, if any, would you suggest for this plant?

8. Congratulations! You are in charge of the world. What are three actions you would take to **(a)** sharply reduce point-source water pollution in more-developed countries, **(b)** sharply reduce nonpoint-source water pollution throughout the world, **(c)** sharply reduce groundwater pollution throughout the world, and **(d)** provide safe drinking water for the poor and for other people in less-developed countries?

d. What problems related to drinking water, if any, have arisen in your community? What actions, if any, has your local government taken to solve such problems?

e. Is groundwater contamination a problem? If so, where, and what has been done about the problem?

in the future and take notes on this. Find information on the causes of this decline and determine which are the three largest causes. Learn what is being done to address each of these causes and write a report explaining the causes, projections, and possible ways to slow the decline of the Ogallala Aquifer.

Data Analysis

In 2006, scientists assessed the overall condition of the estuaries on the western coasts of the U.S. states of Oregon and Washington. To do so, they took measurements of various characteristics of the water, including dissolved oxygen (DO), in selected locations within the estuaries. The concentration of DO for each site was measured in terms of milligrams (mg) of oxygen per liter (L) of water sampled. The scientists used the following DO concentration ranges and quality categories to rate their water samples: water with greater than 5 mg/L of DO was considered *good* for supporting aquatic life; water with 2–5 mg/L of DO was rated as *fair*; and water with less than 2 mg/L of DO was rated as *poor*.

The graph below shows measurements taken in bottom water at 242 locations. Each triangular mark represents one or more measurements. The x-axis on this graph represents DO concentrations in mg/L. The y-axis represents percentages of the total area of estuaries studied (estuarine area).

To read this graph, pick one of the triangles and observe the values on the x- and y-axes. For example, note that the circled triangle lines up approximately with the 5-mg/L mark on the x-axis and with a value of about 34% on the y-axis. This means that waters at this particular measurement station (or stations), along with about 34% of the total area being studied, are estimated to have a 5% or lower DO concentration.

Use this information, along with the graph, to answer the following questions:

1. Half of the estuarine area has waters falling below a certain DO concentration level, and the other half has levels above that level. What is that level, in mg/L?

2. Give your estimate of the highest DO concentration measured and your estimate of the lowest concentration.

3. Approximately what percentage of the estuarine area studied is considered to have poor DO levels? About what percentage has fair DO levels, and about what percentage has good DO levels?

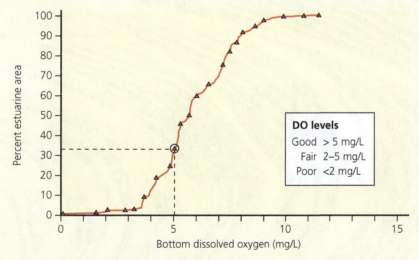

Concentrations of dissolved oxygen in bottom waters of estuaries in Washington and Oregon.

CHAPTER 12

Geology and Nonrenewable Mineral Resources

Key Questions

12.1 What are the earth's major geological processes and what are mineral resources?

12.2 How long might supplies of nonrenewable mineral resources last?

12.3 What are the environmental effects of using nonrenewable mineral resources?

12.4 How can we use mineral resources more sustainably?

12.5 What are the earth's major geological hazards?

Open-pit copper mine in Utah. It is almost 5 kilometers (3 miles) wide and 1,200 meters (4,000 feet) deep, and is getting deeper.

Lee Prince/Shutterstock.com

The Real Cost of Gold

Mineral resources are extracted from the earth's crust through a variety of processes called **mining**. They are processed into an amazing variety of products that make life easier and provide economic benefits and jobs. However, extracting minerals from the ground and using them to manufacture products results in a number of harmful environmental effects.

For example, gold mining often involves digging up massive amounts of rock (Figure 12.1) containing only small concentrations of gold. Many newlyweds would be surprised to know that mining enough gold to make their wedding rings produces roughly enough mining waste to equal the total weight of more than three mid-size cars. This waste is usually left piled near the mine site and can pollute the air and nearby surface water.

About 90% of the world's gold mining operations extract the gold by spraying a solution of highly toxic cyanide salts onto piles of crushed rock. The solution reacts with the gold and then drains off the rocks, pulling some gold with it, into settling ponds (Figure 12.1, foreground). After the solution is recirculated a number of times, the gold is removed from the ponds.

Until sunlight breaks down the cyanide, the settling ponds are extremely toxic to birds and mammals that go to them in search of water. These ponds can also leak or overflow, posing threats to underground drinking water supplies and to fish and other organisms in nearby lakes and streams. Special liners in the settling ponds can help prevent leaks, but some have failed. According to the U.S. Environmental Protection Agency (EPA), all such liners are likely to leak, eventually.

In 2000, snow and heavy rains washed out an earthen dam on one end of a cyanide leach pond at a gold mine in Romania. The dam's collapse released large amounts of water laced with cyanide and toxic metals into the Tisza and Danube Rivers, which flow through parts of Romania, Hungary, and Yugoslavia. Several hundred thousand people living along these rivers were told not to fish or to drink or withdraw water from them or from wells along the rivers. Businesses located there were shut down. Thousands of fish and other aquatic animals and plants were killed. This accident and another one that occurred in January 2001 could have been prevented if the mining company had

installed a stronger containment dam and a backup collection pond to prevent leakage into nearby surface water.

After extracting the gold in a mine, some mining companies have declared bankruptcy. This has allowed them to walk away from cleaning up their mining operations, leaving behind holding ponds full of cyanide-laden water. In addition, millions of poor miners have illegally cleared areas of tropical forest in parts of Asia, Africa, and Latin America. These small-scale, unregulated mining operations also have severe effects on human health and the environment.

In 2016, the world's five top gold producing countries were, in order, China, Australia, Russia, the United States, and Peru. These countries vary in how they deal with the environmental impacts of gold mining. In this chapter, we look at the earth's dynamic geologic processes, the valuable minerals such as gold that some of these processes produce, and the potential supplies of these resources. We will also study the environmental impacts of extracting and processing these resources and how people can use these resources more sustainably. ●

FIGURE 12.1 Gold mine in the Black Hills of the U.S. state of South Dakota with cyanide leach piles and settling ponds in foreground.

Larry Mayer/Getty Images

12.1 WHAT ARE THE EARTH'S MAJOR GEOLOGICAL PROCESSES AND WHAT ARE MINERAL RESOURCES?

CONCEPT 12.1A Dynamic processes within the earth and on its surface produce the mineral resources we depend on.

CONCEPT 12.1B Mineral resources are nonrenewable because it takes millions of years for the earth's rock cycle to produce or renew them.

The Earth Is a Dynamic Planet

Geology is the scientific study of dynamic processes that take place on the earth's surface and in its interior. Geology helps scientists identify potential natural resources and understand natural geological hazards, such as volcanoes and earthquakes.

Scientific evidence indicates that the earth formed about 4.6 billion years ago. As the primitive earth cooled over millions of years, its interior separated into three major layers: the *core*, the *mantle*, and the *crust* (Figure 12.2).

The **core** is the earth's innermost zone. It is composed primarily of iron. It is extremely hot and has a solid inner part, surrounded by a thick layer of *molten rock*, or hot liquid rock, and semisolid material. Surrounding the core is a thick zone called the **mantle**. The outer part of the mantle is solid rock. Beneath it is the **asthenosphere**—a volume of hot, partly melted rock that flows.

Tremendous heat within the core and mantle generates *convection cells*, or *currents*, that slowly move

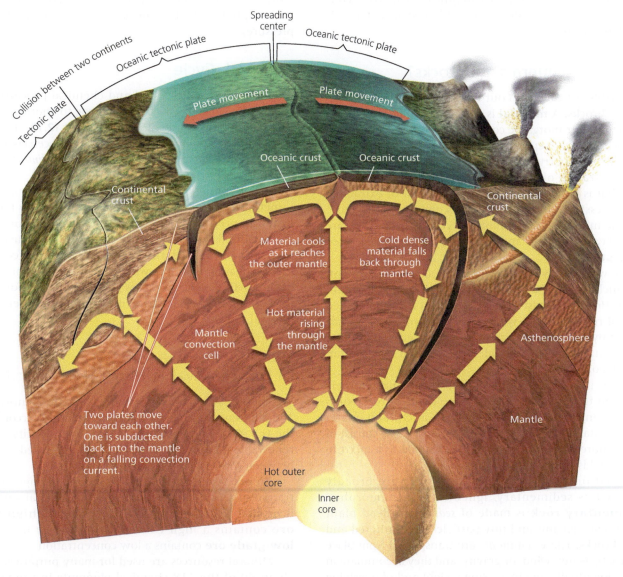

FIGURE 12.2 The earth has a core, mantle, and crust. Within the core and mantle, dynamic forces have major effects on what happens in the crust and on the surface.

large volumes of rock and heat in loops within the mantle like gigantic conveyer belts (Figure 12.2). The innermost material heats, rises, and begins to cool. As it cools, it becomes denser and sinks back down toward the core where it is reheated and rises again in a convection loop. Some of the molten rock in the asthenosphere flows upward into the crust, where it is called *magma*. Magma that emerges onto the earth's surface is called *lava*. These convection cells move rocks and minerals and transfer heat and energy throughout the earth's core and mantle.

The outermost and thinnest zone of solid material is the earth's **crust**. It consists of the continental crust and the oceanic crust. The *continental crust* underlies the continents and includes the continental shelves extending into the ocean. The *oceanic crust* underlies the ocean basins and makes up 71% of the earth's crust. The combination of the crust and the rigid, outermost part of the mantle is called the **lithosphere**. This zone contains the mineral resources on which society depends (**Concept 12.1A**).

What Are Minerals and Rocks?

The earth's crust beneath our feet consists mostly of minerals and rocks. A **mineral** is a naturally occurring chemical element or inorganic compound that exists as a solid with a regularly repeating internal arrangement of its atoms or ions (a *crystalline solid*). A **mineral resource** is a concentration of one or more minerals in the earth's crust that is large enough to cover the cost of extracting and processing it into raw materials and useful products. Because minerals take millions of year to form, they are *nonrenewable resources,* and their supplies can be depleted (**Concept 12.1B**).

A few minerals consist of a single chemical element. They include gold (**Core Case Study** and Figure 2.3, right, p. 31). However, most of the more than 2,000 identified mineral resources that we use are inorganic compounds formed by various combinations of elements. Examples include quartz (silicon dioxide, or SiO_2) and salt (sodium chloride, or NaCl).

Rock is a solid combination of one or more minerals found in the earth's crust. Some kinds of rock, such as limestone (calcium carbonate, or $CaCO_3$) and quartz (SiO_2) contain only one mineral, but most rocks consist of two or more minerals. Granite, for example, is a mixture of mica, feldspar, and quartz crystals.

Based on the way they form, rocks are broadly classified as sedimentary, igneous, or metamorphic. **Sedimentary rock** is made of *sediments*—dead plant and animal remains and tiny particles of weathered and eroded rocks. These sediments are transported from place to place by water, wind, or gravity, and they accumulate in layers. Eventually, the increasing weight and of overlying layers, along with certain chemical processes, transform the underlying sedimentary layers to rock. Examples include *sandstone* and *shale* (formed primarily from sand and silt, respectively), *dolomite* and *limestone* (formed from the compacted shells, skeletons, and other remains of dead aquatic organisms), and *lignite* and *bituminous coal* (derived from compacted plant remains).

Igneous rock forms below or on the earth's surface under intense heat and pressure when magma wells up from the earth's mantle and then cools and hardens. Examples include *granite* (formed underground) and *basalt* (formed aboveground). Igneous rock forms the bulk of the earth's crust but is usually buried beneath sedimentary rock.

Metamorphic rock forms when an existing rock is subjected to high temperatures (which may cause it to melt partially), high pressures, chemically active fluids, or a combination of these agents. Examples include *slate* (formed when shale and mudstone are heated) and *marble* (produced when limestone is exposed to heat and pressure).

The Rock Cycle

The interaction of physical and chemical processes that change the earth's rocks from one type to another is called the **rock cycle** (Figure 12.3 and **Concept 12.1B**). Rocks are recycled over millions of years by three processes—*erosion, melting,* and *metamorphism*—that produce *sedimentary, igneous,* and *metamorphic* rocks, respectively.

In these processes, rocks are broken down, buried, and sometimes melted and fused together into new forms by heat and pressure. They are also cooled and sometimes recrystallized within the earth's interior and crust. Some rock is then uplifted and exposed at the surface where erosion again breaks it down. The rock cycle is the slowest of the earth's cyclic processes and plays the major role in the formation of concentrated deposits of nonrenewable mineral resources.

We Depend on a Variety of Nonrenewable Mineral Resources

We have learned how to find and extract more than 100 different minerals from the earth's crust. According to the U.S. Geological Survey (USGS), the quantity of nonrenewable minerals extracted globally increased threefold between 1995 and 2016.

An **ore** is rock that contains a large enough concentration of a particular mineral—often a metal—to make it profitable for mining and processing. A **high-grade ore** contains a high concentration of the mineral. A **low-grade ore** contains a low concentration.

Mineral resources are used for many purposes. Today, about 60 of the 118 chemical elements in the periodic

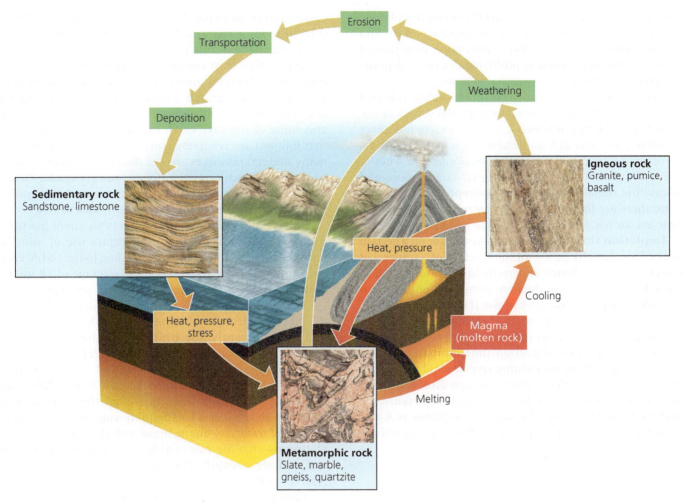

FIGURE 12.3 Natural Capital: The rock cycle is the slowest of the earth's cyclical processes.

Left: Dwight Smith/Shutterstock.com. Center: LesPalenik/Shutterstock.com. Right: Bragin Alexey/Shutterstock.com.

table are used for making computer chips. *Aluminum* (Al) is used as a structural material in beverage cans, motor vehicles, aircraft, and buildings. *Steel,* an essential material used in buildings, machinery, and motor vehicles, is a mixture (or *alloy*) of iron (Fe) and other elements that give it certain properties. *Manganese* (Mn), *cobalt* (Co), and *chromium* (Cr) are widely used in steel alloys. *Copper* (Cu), a good conductor of electricity, is used to make electrical and communications wiring and plumbing pipes. *Gold* (Au) is a component of electrical equipment, tooth fillings, jewelry, coins, and some medical implants.

There are several widely used nonmetallic mineral resources. *Sand,* which is mostly SiO_2, is used to make glass, bricks, and concrete for the construction of roads and buildings. *Gravel* is used for roadbeds and to make concrete. Another common nonmetallic mineral is *limestone,* which is crushed to make concrete and cement. Still another is *phosphate,* used to make inorganic fertilizers and certain detergents.

12.2 HOW LONG MIGHT SUPPLIES OF NONRENEWABLE MINERAL RESOURCES LAST?

CONCEPT 12.2A Nonrenewable mineral resources exist in finite amounts and can become economically depleted when it costs more than it is worth to find, extract, and process the remaining deposits.

CONCEPT 12.2B There are several ways to extend supplies of mineral resources, but each of them is limited by economic and environmental factors.

Supplies of Nonrenewable Mineral Resources Can Be Economically Depleted

Most published estimates of the supply of a given nonrenewable mineral resource refer to its **reserves:** identified deposits from which we can extract the mineral profitably at current prices. Reserves can be expanded when we find new, profitable deposits or when higher prices or improved mining technologies make it profitable to extract deposits that previously were too expensive to remove.

The future supply of any nonrenewable mineral resource depends on the actual or potential supply of the mineral and the rate at which it is used. Society has never completely run out of a nonrenewable mineral resource. However, a mineral becomes *economically depleted* when it costs more than it is worth to find, extract, transport, and process the remaining deposits (**Concept 12.2A**). At that point, there are five choices: *recycle or reuse existing supplies, waste less, use less, find a substitute,* or *do without.*

Depletion time is the time it takes to use up a certain proportion—usually 80%—of the reserves of a mineral at a given rate of use. When experts disagree about depletion times, it is often because they are using different assumptions about supplies and rates of use (Figure 12.4).

The shortest depletion-time estimate assumes no recycling or reuse and no increase in reserves (curve A, Figure 12.4). A longer depletion-time estimate assumes that recycling will stretch existing reserves and that better mining technology, higher prices, or new discoveries will increase the reserve (curve B). The longest depletion-time estimate (curve C) makes the same assumptions as A and B, and also assumes that people will reuse and reduce

consumption to expand the reserve further. Finding a substitute for a resource leads to a new set of depletion curves for the new mineral.

The earth's crust contains abundant deposits of non-renewable mineral resources such as iron and aluminum. However, concentrated deposits of important mineral resources such as manganese, chromium, cobalt, platinum, and some of the rare earth metals (see the Case Study that follows) are relatively scarce. In addition, deposits of many mineral resources are not distributed evenly among countries. Five nations—the United States, Canada, Russia, South Africa, and Australia—supply most of the nonrenewable mineral resources that modern societies use.

Since 1900, and especially since 1950, there has been a sharp rise in the total and per capita use of mineral resources in the United States. According to the USGS, each person in the United States uses an average of 18 metric tons (about 20 tons) of mineral resources per year.

The United States has economically depleted some of its once-rich deposits of metals such as lead, aluminum, and iron. Currently, the United States imports all of its supplies of 24 key nonrenewable mineral resources and relies on imports for more than 50% of 43 other minerals. Most of these mineral imports come from reliable and politically stable countries. However, there are serious concerns about access to adequate supplies of four *strategic metal resources*—manganese, cobalt, chromium, and platinum—that are essential for the country's economic and military strength. The United States has little or no reserves of these metals.

Lithium (Li), the world's lightest metal, is a vital component of lithium-ion batteries, which are used in cell phones, iPads, laptop computers, electric cars, and a growing number of other products. The problem is that some countries, including the United States, do not have large supplies of lithium. Bolivia has about 50% of these reserves, whereas the United States holds only about 3%.

Japan, China, South Korea, and the United Arab Emirates have been buying up access to global lithium reserves to ensure their ability to sell lithium and lithium-ion batteries to the rest of the world. Within a few decades, the United States may be heavily dependent on expensive imports of lithium and lithium-ion batteries.

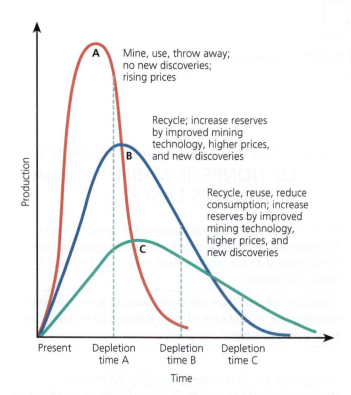

FIGURE 12.4 *Natural capital depletion:* Each of these *depletion curves* for a mineral resource is based on a different set of assumptions. Dashed vertical lines represent the times at which 80% depletion occurs.

CASE STUDY

The Crucial Importance of Rare Earth Metals

Some mineral resources are familiar, such as gold, copper, aluminum, sand, and gravel. Less well known are the *rare earth metals and oxides,* which are crucial to many technologies that support modern lifestyles and economies.

The 17 rare earth metals, also known as *rare earths,* include scandium, yttrium, and 15 lanthanide chemical elements, including lanthanum. Because of their superior magnetic strength and other unique properties, these elements and their compounds are important for a number of widely used technologies.

Rare earths are used to make liquid crystal display (LCD) flat screens for computers and television sets, energy-efficient light-emitting diode (LED) light bulbs, solar cells, fiber-optic cables, smart phones, and digital cameras. They are also part of batteries and motors for electric and hybrid-electric cars (Figure 12.5), solar cells, catalytic converters in car exhaust systems, jet engines, and the powerful magnets in wind turbine generators. Rare earths also go into missile guidance systems, jet engines, smart bombs, aircraft electronics, and satellites.

Without affordable supplies of these metals, industrialized nations could not develop current versions of cleaner energy technology and other high-tech products that will be major sources of economic growth during this century. Many nations also need these metals to maintain their military strength.

Most rare earth elements are not actually rare, but they are hard to find in concentrations high enough to extract and process at an affordable price. According to the USGS, in 2014, China had roughly 25% of the world's known rare earth reserves. Brazil had the second largest share with 17%, and the United States, with the fifth largest share, had just 1.4% of the global reserves.

In 2016, China produced about 95% of the world's rare earth metals and oxides. Australia and Chile are increasing their shares of global production. China holds the lead, partly because it does not strictly regulate the environmentally disruptive mining and processing of rare earths. This means that Chinese companies have lower production costs than do companies in countries with stricter regulations. However, since 2012 a glut of rare earths has caused global prices to fall, shutting down some mines in China and elsewhere.

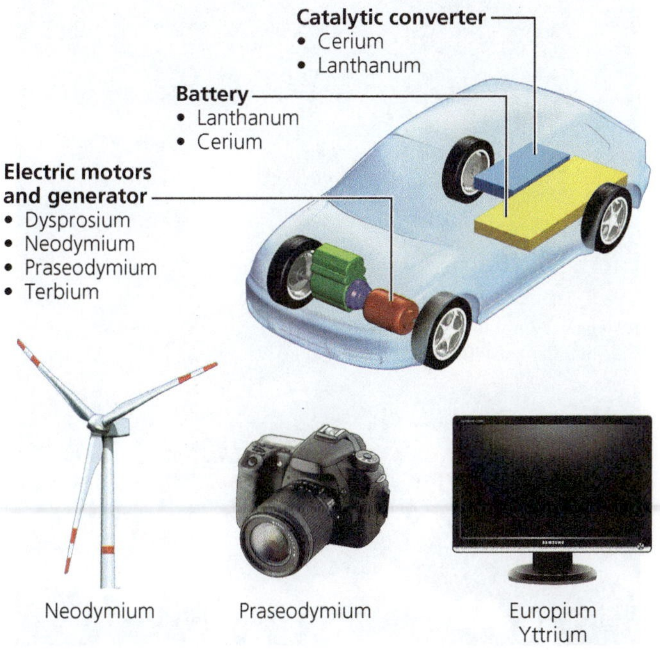

The United States and Japan are heavily dependent on rare earths and their oxides. Japan has no rare earth reserves. In the United States, the only rare earth mine, located in California, was once the world's largest supplier of rare earth metals. However, it closed down because of the expense of meeting pollution regulations, and because China had driven the prices of rare earth metals down to a point where the mine was too costly to operate. In 2015, the company that owns the mine declared bankruptcy.

One way to increase supplies of rare earths is to extract and recycle them from the massive amounts of electronic wastes that are being produced. So far, however, less than 1% of rare earth metals are recovered and recycled. Another approach is to find substitutes for rare earth metals. In 2016, Honda produced a hybrid electric car engine with magnets that do not need heavy rare-earth metals and that are 10% cheaper and 8% lighter.

Market Prices Affect Supplies of Mineral Resources

Geological processes determine the quantity and location of a nonrenewable mineral resource in the earth's crust, but economics determines what part of the known supply is extracted and used. According to standard economic theory, in a competitive market system when a resource becomes scarce, its price rises. Higher prices can encourage exploration for new deposits, stimulate development of better mining technology, and make it profitable to mine lower-grade ores. They can also promote resource conservation and a search for substitutes.

CONSIDER THIS ...

CONNECTIONS High Metal Prices and Thievery

Resource scarcity can also promote thievery. For example, copper prices have risen sharply in recent years. As a result, in several U.S. communities, people have been stealing copper to sell it. They strip abandoned houses of copper pipe and wiring and steal outdoor central air conditioning units for their copper coils. They also steal wiring from beneath city streets and copper piping from farm irrigation systems. In 2015, thieves stole copper wiring from New York City's subway system, temporarily shutting down two of the city's busiest lines.

According to some economists, this price effect may no longer apply completely in most of the more-developed countries. Governments in such countries often use subsidies, tax breaks, and import tariffs to help control the supply, demand, and prices of key mineral resources. In the United States, for instance, mining companies get various

Catalytic converter
• Cerium
• Lanthanum

Battery
• Lanthanum
• Cerium

Electric motors and generator
• Dysprosium
• Neodymium
• Praseodymium
• Terbium

Neodymium

Praseodymium

Europium
Yttrium

FIGURE 12.5 Manufacturers of all-electric and hybrid-electric cars use a variety of rare earth metals.

types of government subsidies, including *depletion allow-ances*—which allow the companies to deduct the costs of developing and extracting mineral resources from their taxable incomes. These allowances amount to 5–22% of their gross income gained from selling the mineral resources.

Generally, the mining industry maintains that they need subsidies and tax breaks to keep the prices of minerals low for consumers. They also claim that, without these subsidies and tax breaks, they might move their operations to other countries where they would not have to pay taxes or comply with strict mining and pollution control regulations.

Expanding Mineral Reserves

Some analysts contend that we can increase supplies of some minerals by extracting them from lower-grade ores. They point to the development of new earth-moving equipment, improved techniques for removing impurities from ores, and other technological advances in mineral extraction and processing that can make lower-grade ores accessible, sometimes at lower costs. For example, in 1900, the copper ore mined in the United States was typically about 5% copper by weight. Today, it is typically about 0.5%, yet copper costs less (when prices are adjusted for inflation).

Several factors can limit the mining of lower-grade ores (**Concept 12.2B**). For example, it requires mining and processing larger volumes of ore, which takes much more energy and costs more. Another factor is the dwindling supplies of freshwater needed for the mining and processing of some minerals, especially in dry areas. A third limiting factor is the growing environmental impacts of land disruption, along with waste material and pollution produced during mining and processing.

One way to improve mining technology and reduce its environmental impact is to use a biological approach, sometimes called *biomining*. Miners use natural or genetically engineered bacteria to remove desired metals from ores through wells bored into the deposits. This leaves the surrounding environment undisturbed and reduces the air and water pollution associated with removing the metal from metal ores. On the down side, biomining is slow. It can take decades to remove the same amount of material that conventional methods can remove within months or years. So far, biomining methods are economically feasible only for low-grade ores for which conventional techniques are too expensive.

Minerals from the Ocean

Most of the minerals found in seawater occur in such low concentrations that removing them takes more energy and money than they are worth. Currently, only magnesium, bromine, and sodium chloride are abundant enough to be extracted profitably from seawater. On the other hand, sediments along the shallow continental shelf and adjacent shorelines contain significant deposits of minerals such as sand, gravel, phosphates, copper, iron, tungsten, silver, titanium, platinum, and diamonds.

Another potential ocean source of some minerals is *hydrothermal ore deposits* that form wherever superheated, mineral-rich water shoots out of vents in volcanic regions of the ocean floor. As the hot water mixes with cold seawater, black particles of various metal sulfides precipitate out and accumulate as chimney-like structures, called *black* smokers or *hydrothermal vents*, near the hot water vents (Figure 12.6). These deposits are especially rich in minerals such as copper, lead, zinc, silver, gold, and some of the rare earth metals. Exotic communities of marine life—including giant clams, six-foot tubeworms, and eyeless shrimp—live in the dark depths around these black smokers.

Because of the rapidly rising prices of many of these metals, interest in deep-sea mining is growing. Companies from Australia, the United States, and China have been exploring the possibility of mining black smokers in several areas. The United Nations International Seabed Authority, established to manage seafloor mining in international waters, began issuing mining permits in 2011.

According to some analysts, seafloor mining is less environmentally harmful than mining on land. Other scientists, however, are concerned that the sediment stirred up by such mining could harm or kill organisms that feed

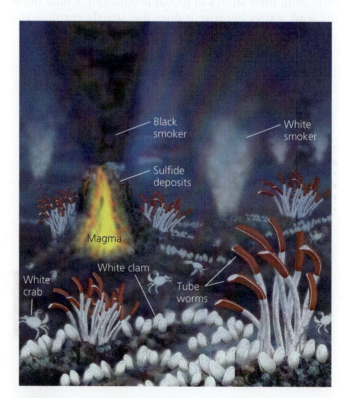

FIGURE 12.6 Natural Capital: Hydrothermal deposits, or black smokers, are rich in various minerals.

by filtering seawater. Supporters of seafloor mining say that the number of potential mining sites, and thus the overall environmental impact, will be quite small.

Another possible source of metals from the ocean is potato-size *manganese nodules* that cover large areas of the Pacific Ocean floor and smaller areas of the Atlantic and Indian Ocean floors. They also contain some low concentrations of various rare earth minerals. These nodules could be sucked up by giant vacuum pipes or scooped up by underwater mining machines.

To date, mining on the ocean floor has been hindered by the high costs involved, the potential threat to marine ecosystems, and arguments over shared rights to the minerals in deep ocean areas that do not belong to any specific country.

12.3 WHAT ARE THE ENVIRONMENTAL EFFECTS OF USING NONRENEWABLE MINERAL RESOURCES?

CONCEPT 12.3 Extracting minerals from the earth's crust and converting them to useful products can disturb the land, erode soils, produce large amounts of solid waste, and pollute the air, water, and soil.

Extracting Minerals Can Have Harmful Environmental Effects

Every metal product has a *life cycle* that includes mining the mineral, processing it, manufacturing the product, and disposal or recycling of the product (Figure 12.7). This process requires large amounts of energy and water, and can disturb the land, erode soils, and produce large amounts of pollution and mining wastes (**Concept 12.3**).

The environmental impacts of mining a metal ore are determined partly by the ore's percentage of metal

content, or *grade*. The more accessible higher-grade ores are usually exploited first. Mining lower-grade ores takes more money, energy, water, and other resources, and leads to more land disruption, mining waste, and pollution.

Several different mining techniques are used to remove mineral deposits. Shallow mineral deposits are removed by **surface mining**, in which vegetation, soil, and rock overlying a mineral deposit are cleared away. This waste material, called **overburden**, is usually deposited in piles called **spoils** (Figure 12.8). Surface mining is used to extract about 90% of the nonfuel mineral resources and 60% of the coal used in the United States.

Different types of surface mining can be used, depending on two factors: the resource being sought and the local topography. In **open-pit mining**, machines are used to dig very large holes and remove metal ores containing copper (see chapter-opening photo), gold (**Core Case Study**), or other metals, or sand, gravel, or stone.

> **9 million** Number of people who could sit in Bingham Copper Mine (see chapter-opening photo) if it were a stadium.

Strip mining involves extracting of mineral deposits that lie in large horizontal beds close to the earth's surface. **Area strip mining** is used on flat terrain. In this type of mining, a gigantic earthmover strips away the overburden. Then a power shovel—which can be as tall as a 20-story building—removes the mineral deposit such as gold (Figure 12.9) or an energy resource such as coal. The resulting trench is filled with overburden, and a new cut is made parallel to the previous one. This process is repeated over the entire site.

Contour strip mining (Figure 12.10) is used mostly to mine coal and various mineral resources on hilly or mountainous terrain. Huge power shovels and bulldozers cut a series of terraces into the side of a hill. Then, earthmovers remove the overburden, an excavator or power

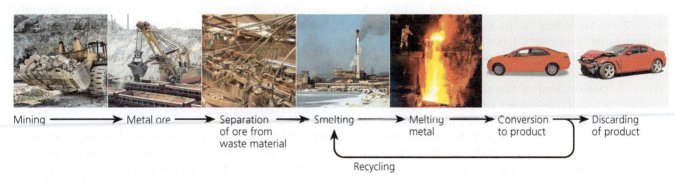

Mining ⟶ Metal ore ⟶ Separation of ore from waste material ⟶ Smelting ⟶ Melting metal ⟶ Conversion to product ⟶ Discarding of product

Recycling

FIGURE 12.7 Each metal product that we use has a *life cycle*.

Left: kaband/Shutterstock.com. Second to left: Andrey N Bannov/Shutterstock.com. Center left: Vladimir Melnik/Shutterstock.com. Center: mares/Shutterstock.com. Center right: zhu difeng/Shutterstock.com. Second to right: Michael Shake/Shutterstock.com. Right: Pakhnyushchy/Shutterstock.com.

FIGURE 12.8 Natural Capital Degradation: This spoils pile in Zielitz, Germany, is made up of waste material from the mining of potassium salts used to make fertilizers.

shovel extracts the coal, and the overburden from each new terrace is dumped onto the one below. Unless the land is restored, what is left are a series of spoils banks and a highly erodible bank of soil and rock called a *highwall*.

Another surface mining method is **mountaintop removal,** in which explosives are used to remove the top of a mountain to expose seams of coal that are then extracted (Figure 12.11). Enormous machines plow the waste rock and dirt into valleys below the mountaintops. This destroys forests, buries mountain streams, and increases the risk of flooding. Wastewater and toxic sludge, produced when the coal is processed, are often stored in pits behind dams in these valleys. Some dams have overflowed or collapsed and released toxic substances such as arsenic and mercury.

In the United States, more than 500 mountaintops in West Virginia and other Appalachian states have been removed to extract coal (Figure 12.11 and Individuals Matter, 12.1). According to the U.S. EPA, the resulting spoils have buried more than 1,100 kilometers (700 miles) of streams—a total roughly equal to the distance between the two U.S. cities of New York and Chicago.

The U.S. Department of the Interior estimates that at least 500,000 surface-mined sites dot the U.S. landscape, mostly in the West. Such sites can be cleaned up and restored, but it is costly. The U.S. Surface Mining Control and Reclamation Act of 1977 requires the restoration of surface-mined sites. The government (taxpayers) is required to restore mines abandoned before 1977 and companies are required to restore active mines and mines abandoned after 1977. However, the program is greatly underfunded and many mines have not been reclaimed. Many coal companies have gone bankrupt because of the decline of coal use in the United States. They likely will not be able to meet their legal obligation to restore mining sites, which will have to be restored using taxpayer dollars.

Deep deposits of minerals are removed by **subsurface mining**, in which underground mineral resources are removed through tunnels and shafts. This method is used to remove metal ores and coal too deep to be extracted by surface mining. Miners dig a deep, vertical shaft and blast open subsurface tunnels and chambers to reach the deposit. Then they use machinery to remove the resource and transport it to the earth's surface.

Subsurface mining disturbs less than one-tenth as much land as surface mining and usually produces less waste material. However, the environmental damage is significant and subsurface mines can lead to other hazards such as cave-ins, explosions, and fires for miners. Miners often get lung diseases caused by prolonged inhalation of mineral or coal dust in subsurface mines. Another problem is *subsidence*—the collapse of land above some underground mines. It can damage houses, crack sewer lines, break gas mains, and disrupt groundwater systems. Subsurface mining also requires large amounts of water and energy to process and transport the mined material.

Collectively, surface and subsurface mining operations produce three-fourths of all U.S. solid waste and cause major water and air pollution. For example, *acid mine drainage* occurs when rainwater that seeps through an underground mine or a spoils pile from a surface mine carries sulfuric acid (H_2SO_4, produced when aerobic bacteria act on minerals in the spoils) to nearby streams and groundwater. This is one of the problems often associated with gold mining (Core Case Study).

According to the EPA, mining has polluted mountain streams in 40% of the western watersheds in the United States. It also accounts for half of the country's emissions of toxic chemicals into the atmosphere. In fact, the mining industry produces more of such toxic emissions than any other U.S. industry.

Mining can be even more harmful to the environment in countries where environmental regulations are lacking or not reliably enforced. In China, for instance, the mining and processing of rare earth metals and oxides has stripped land of its vegetation and topsoil. It also has polluted the air, acidified streams, and left toxic and radioactive waste piles.

Processing Ores Has Harmful Environmental Effects

Ore extracted by mining typically has two components: the ore mineral that contains the desired metal and waste material. Removing the waste material from ores produces **tailings**—rock wastes that are left in piles or put into ponds where they settle out. Particles of toxic metals in tailings piles can be blown by the wind, washed out

FIGURE 12.9 **Natural Capital Degradation:** Area strip mining for gold in Yukon Territory, Canada.

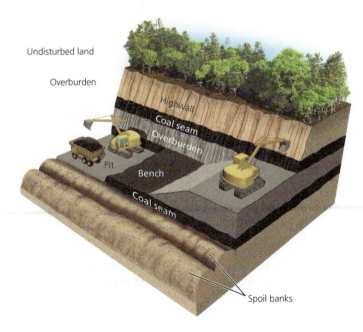

Undisturbed land

Overburden

Highwall

Coal seam

Overburden

Pit

Bench

Coal seam

Spoil banks

FIGURE 12.10 **Natural Capital Degradation:** Contour strip mining is used in hilly or mountainous terrain.

by rain, or leak from holding ponds, and thus they can contaminate surface water and groundwater.

After the waste material is removed, heat or chemical solvents are used to extract the metals from mineral ores. Heating ores to release metals is called **smelting** (Figure 12.7). Without effective pollution control equipment, a smelter emits large quantities of air pollutants. These pollutants include sulfur dioxide and suspended toxic particles that can damage vegetation and acidify soils in the surrounding area. Smelters also cause water pollution and produce liquid and solid hazardous wastes that require safe disposal. A 2012 study found that lead

FIGURE 12.11 Natural Capital Degradation: Mountaintop removal coal mining near Whitesville, West Virginia.

Jim West/AGE Fotostock

smelting is the world's second most toxic industry after the recycling of lead-acid batteries.

Using chemicals to extract metals from their ores can also create numerous problems, as we saw in the case of using cyanide to remove gold (**Core Case Study**). Even on a smaller scale, this is the case. For example, millions of poverty-stricken miners in less-developed countries have gone into tropical forests in search of gold (Figure 12.12). They have cleared trees to get access to gold, and such illegal deforestation has increased rapidly, especially in parts of the Amazon Basin. They dig up and sift through tons of soil to find flecks of gold. The miners use toxic mercury to separate gold from its ore. They heat the mixture of gold and mercury to vaporize the mercury and leave the gold, causing dangerous air and water pollution. They leave

behind land stripped of its vegetation and topsoil loaded with mercury. Many of these miners and villagers living near the mines eventually inhale toxic mercury vapor, drink mercury-laden water, or eat fish contaminated with mercury.

12.4 HOW CAN WE USE MINERAL RESOURCES MORE SUSTAINABLY?

CONCEPT 12.4 We can try to find substitutes for scarce mineral resources, reduce mineral resource waste, and recycle and reuse minerals.

Maria Gunnoe: Fighting to Save Mountains

Goldman Environmental Foundation

In the 1800s, Maria Gunnoe's Cherokee ancestors arrived in what is now Boone County, West Virginia. Her grandfather bought the land where she now lives. In 2000, miners blew up the mountaintop above that land to extract underlying coal (Figure 12.11). Gunnoe's land now sits near a 10-story-high pile of mine waste.

With the soil and vegetation gone, rains running off the mountain ridge flooded her land seven times since 2000. They covered her land with toxic sludge, contaminated her water well and the soil she used to grow food, and washed out two small bridges that linked her to the only road out. Gunnoe had to hike to and from the road for years. When she complained to the coal company officials, they called the floods "acts of God."

Gunnoe, a mother of two, refused to leave her land and decided to fight the powerful coal companies and to try to end mountaintop coal mining. Mining companies and miners who worried about losing their jobs viewed her as an enemy. She and her children received death threats and two family dogs were killed. People tried to run her off the road and fired shots around her house. For several years, Gunnoe wore a bulletproof vest when she went outside, but she kept up the fight.

With only a high school education, Gunnoe educated herself about complex mining and water pollution regulations and harmful chemicals found in streams and groundwater near some mining sites. She organized communities, argued in court, and testified before Congress contending that the "valley fills" from mountaintop mining violate the federal Clean Water Act by burying streams and destroying aquatic and animal habitats. The EPA agreed and fined certain coal companies multiple times. She also pressured the president and federal government to ban mountaintop mining because of its harmful health effects.

In 2009, this courageous and inspiring woman won a Goldman Environmental Prize—considered the Nobel Prize equivalent for grassroots environmental leaders around the world.

Finding Substitutes for Scarce Mineral Resources

Some analysts believe that even if supplies of key minerals become too expensive or too scarce due to unsustainable use, human ingenuity will find substitutes. They point to the current *materials revolution* in which silicon and other materials are replacing some metals for common uses. They also point out the possibilities of finding substitutes for scarce minerals through nanotechnology (Science Focus 12.1) and other emerging technologies.

For example, fiber-optic glass cables that transmit pulses of light are replacing copper and aluminum wires in telephone cables, and nanowires may eventually replace fiber-optic glass cables. High-strength plastics and materials, strengthened by lightweight carbon, hemp, and glass fibers, are beginning to transform the automobile and aerospace industries. These new materials do not need painting (which reduces pollution and costs) and can be molded into any shape. Use of such materials in manufacturing motor vehicles and planes could greatly increase vehicle fuel efficiency by reducing vehicle weights.

CONSIDER THIS ...

Learning from Nature

Without using toxic chemicals, spiders rapidly build their webs by producing threads of silk that are strong enough to capture insects flying at high speeds. Learning how spiders do this could revolutionize the production of high-strength fibers with a very low environmental impact.

We can also find substitutes for rare earths (**Concept 12.4**). For example, electric car battery makers are beginning to switch from making nickel-metal-hydride batteries, which require the rare earth metal lanthanum, to manufacturing lighter-weight lithium-ion batteries, which researchers are trying to improve (Individuals Matter 12.2).

Despite its potential, resource substitution is not a cure-all. For example, platinum is currently unrivaled as a catalyst and is used in industrial processes to speed up chemical reactions, and chromium is an essential ingredient of stainless steel. Finding acceptable and affordable substitutes for such scarce resources may not always be possible.

Using Mineral Resources More Sustainably

Figure 12.13 lists several ways to use mineral resources more sustainably (**Concept 12.4**). One strategy is to focus on recycling and reuse of nonrenewable mineral

FIGURE 12.12 Illegal gold mining on the banks of the Pra River in Ghana, Africa.

RANDY OLSON/National Geographic Creative

resources, especially valuable or scarce metals such as gold (**Core Case Study**), iron, copper, aluminum, and platinum. Recycling, an application of the chemical cycling **principle of sustainability**, has a much lower environmental impact than that of mining and processing metals from ores.

For example, recycling aluminum beverage cans and scrap aluminum produces 95% less air pollution and 97% less water pollution, and uses 95% less energy, than mining and processing aluminum ore. We can also extract and recycle valuable gold (**Core Case Study**) from discarded cell phones. Cleaning up and reusing items instead of recycling them has an even lower environmental impact.

GOOD NEWS

12.5 WHAT ARE THE EARTH'S MAJOR GEOLOGICAL HAZARDS?

CONCEPT 12.5 Dynamic processes move matter within the earth and on its surface and can cause volcanic eruptions, earthquakes, tsunamis, erosion, and landslides.

The Earth beneath Your Feet Is Moving

We tend to think of the earth's crust as solid and unmoving. However, according to geologists, the flows of energy and heated material within the earth's convection cells

The Nanotechnology Revolution

Nanotechnology uses science and engineering to manipulate and create materials out of atoms and molecules at the ultra-small scale of less than 100 nanometers. The diameter of the period at the end of this sentence is about a half million nanometers.

Currently, nanomaterials are used in more than 1,300 consumer products and that number is growing. Such products include certain batteries, stain-resistant and wrinkle-free clothes, self-cleaning glass surfaces, self-cleaning sinks and toilets, sunscreens, waterproof coatings for cell phones, some cosmetics, some foods, and food containers that release nanosilver ions to kill bacteria, molds, and fungi.

Graphene, a nanomaterial made from single layers of carbon atoms (Figure 12.A), is the world's thinnest and strongest material and is light, flexible, and stretchable. A single layer of graphene is 150,000 times lighter than a human hair. A single layer stretched over a coffee mug could support the weight of a car. It is also a better conductor of electricity than copper and conducts heat better than any known material.

Nanotechnologists envision innovations such as a supercomputer smaller than a grain of rice, thin and flexible solar cell films that could be attached to or painted onto almost any surface, and biocomposite materials that would make our bones and tendons super strong. Some nanomolecules could be specifically designed to seek out and kill cancer cells or to eliminate the need for allergy shots. Scientists are working on a wearable graphene patch that would help diabetics manage their blood glucose levels. It would measure blood sugar levels in sweat and maintain acceptable levels by delivering a dose of a diabetes drug through the skin without the use of needles. Nanotechnology allows us make materials from the bottom up, using atoms of abundant elements (primarily hydrogen, oxygen, nitrogen, carbon, silicon, and aluminum) as substitutes for scarcer elements, such as copper, cobalt, and tin.

Nanotechnology has many potential environmental benefits. Designing and building products on the molecular level would greatly reduce the need to mine many materials. It also requires less matter and energy and would reduce waste production. We may be able to use nanoparticles to remove industrial pollutants from contaminated air, soil, and groundwater. Nanofilters might someday be used to desalinate and purify seawater at an affordable cost, thereby increasing drinking water supplies. **GREEN CAREER: Environmental nanotechnology**

What is the catch? Because of their tiny size, nanoparticles are potentially more toxic than many conventional materials. Laboratory studies involving mice and other test animals reveal that nanoparticles can be inhaled deeply into the lungs and absorbed into the bloodstream. This can result in lung damage similar to that caused by mesothelioma, a deadly cancer resulting from the inhalation of asbestos particles. Nanoparticles can also penetrate cell membranes, including those in the brain, and move across the placenta from a mother to her fetus.

A panel of experts from the U.S. National Academy of Sciences has said that the U.S. government is not doing enough to evaluate the potential health and environmental risks of using nano-materials. For example, the U.S. Food and Drug Administration does not maintain a list of the food products and cosmetics

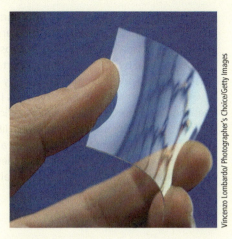

FIGURE 12.A Graphene, which consists of layers of carbon atoms linked together in a hexagonal lattice, is a revolutionary new material.

Vincenzo Lombardo/ Photographer's Choice/Getty Images

that contain nanomaterials. By contrast, the European Union takes a precautionary approach to the use of nanomaterials, requiring that manufacturers demonstrate the safety of their products before they can enter the marketplace.

Nanotechnology could transform the way we make and use products. However, before unleashing nanotechnology more broadly, many analysts call for greatly increasing research on the potential harmful health effects of nanoparticles and for developing regulations to control its growing applications until we know more about such effects. Many are also calling for the labeling of all products that contain nanoparticles.

CRITICAL THINKING

Do you think the potential benefits of nanotechnology products outweigh their potentially harmful effects? Explain.

Yu-Guo Guo: Designer of Nanotechnology Batteries and National Geographic Explorer

Yu-Guo Guo is a professor of chemistry and a nanotechnology researcher at the Chinese Academy of Sciences in Beijing. He has invented nanomaterials that can be used to make lithium-ion battery packs smaller, more powerful, and less costly, which makes them more useful for powering electric cars and electric bicycles. This is an important scientific advance, because the battery pack is the most important and expensive part of any electric vehicle.

Guo's innovative use of nanomaterials has greatly increased the power of lithium-ion batteries by enabling electric current to flow more efficiently through what he calls "3-D conducting nanonetworks." With this promising technology, lithium-ion battery packs in electric vehicles can be fully charged in just a few minutes, as quickly and easy as filling a car with gas. They also have twice the energy storage capacity of today's batteries, and thus will extend the range of electric vehicles by enabling them to run longer. Guo is also interested in developing nanomaterials for use in solar cells and fuel cells that could be used to generate electricity and to power vehicles.

© Jinsong Hu

> **Solutions**
>
> **Sustainable Use of Nonrenewable Minerals**
>
> - Reuse or recycle metal products whenever possible
> - Redesign manufacturing processes to use less mineral resources
> - Reduce mining subsidies
> - Increase subsidies for reuse, recycling, and finding substitutes

FIGURE 12.13 We can use nonrenewable mineral resources more sustainably (**Concept 12.4**). *Critical thinking:* Which two of these solutions do you think are the most important? Why?

(Figure 12.2) are so powerful that they have caused the lithosphere to break up into many rigid plates, called **tectonic plates**, which move extremely slowly atop the asthenosphere (Figure 12.14). Currently, there are seven major plates and dozens of smaller ones.

These gigantic plates are somewhat like the world's largest and slowest-moving surfboards on which we ride without noticing their movement. Their typical speed is about the rate at which your fingernails grow. Throughout the earth's history, landmasses have split apart and joined as tectonic plates shifted around atop the earth's asthenosphere (Figure 4.C, p. 81). The slow movement of the continents across the earth's surface is called **continental drift**.

Much of the geological activity at the earth's surface takes place at the boundaries between tectonic plates as they separate, collide, or grind along against each other (Figure 12.14, bottom and Figure 12.15). The tremendous forces produced at these plate boundaries can cause mountains or deep rifts to form, earthquakes to shake parts of the crust, and volcanoes to erupt. Scientist Bob Ballard (Individuals Matter 12.3) has played a key role in helping us understand more about tectonic plates. In the process, he and others discovered mineral-rich black smokers (Figure 12.6).

Volcanoes Release Molten Rock from the Earth's Interior

An active **volcano** occurs where magma rising in a plume through the lithosphere reaches the earth's surface through a central vent or a long crack, called a *fissure* (Figure 12.16). Magma that reaches the earth's surface is called *lava* and often builds into a cone.

Many volcanoes form along the boundaries of the earth's tectonic plates when one plate slides under or moves away from another plate (Figure 12.14, bottom). A volcanic eruption releases large chunks of lava rock, glowing hot ash, liquid lava, and gases (including water vapor, carbon dioxide, and sulfur dioxide) into the environment (**Concept 12.5**). Eruptions can be explosive and extremely destructive, causing loss of life and obliterating ecosystems and human communities. They can also be slow and much less destructive with lava gurgling up and spreading slowly across the land or sea floor. It is this slower form of eruption that builds the cone-shaped mountains so commonly associated with volcanoes, as well as layers of rock made of cooled lava on the earth's surface.

While volcanic eruptions can be destructive, they can also form majestic mountains and lakes, and the weathering of lava contributes to fertile soils. Hundreds of volcanoes have erupted on the ocean floor, building cones

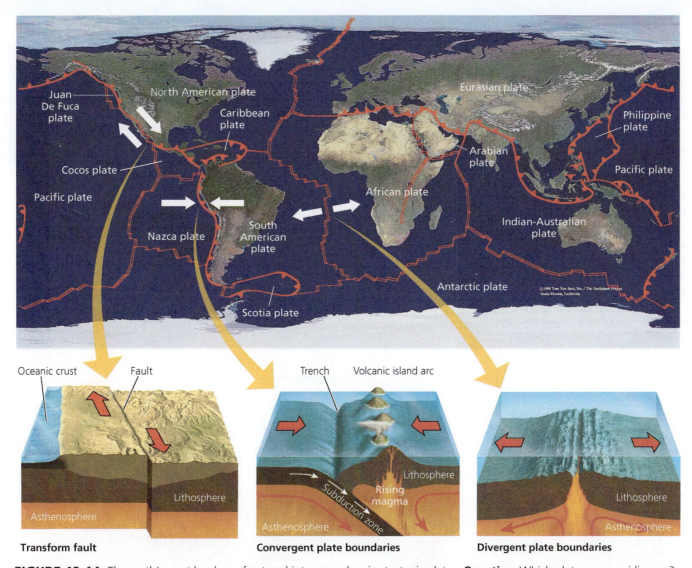

FIGURE 12.14 The earth's crust has been fractured into several major tectonic plates. *Question:* Which plate are you riding on?

Oceanic crust | Fault

Transform fault

Trench | Volcanic island arc

Convergent plate boundaries

Divergent plate boundaries

FIGURE 12.15 The North American Plate and the Pacific Plate (see map) slide very slowly against each other in opposite directions along the San Andreas fault (see photo), which runs almost the full length of California. It has been the site of many earthquakes of varying magnitudes.

Kevin Schafer/AGE Fotostock

Robert Ballard, Ocean Explorer

In 1977, oceanographer Bob Ballard and a team of scientists piloted a submersible to the bottom of the ocean near the Galapagos Islands. There they made a shocking discovery: jets of boiling water coming from beneath the ocean floor had formed large chimney structures that were teaming with life. The chimneys, or hydrothermal vents, formed from the chemicals in the hot water jets that were mixing with seawater (Figure 12.6). When the hot magma cools, it contracts and cracks. Seawater penetrates the cracks, sinks down below the earth's surface, and interacts with the hot, sometimes molten, rock. The water pulls chemicals and minerals from the rock and brings them back to the surface when it emerges through the hydrothermal vents. It is these minerals that give the ocean its chemical makeup and make it salty.

Another major discovery related to the hydrothermal vents was of their biological communities. The vents were far too deep for sunlight to reach. Until this discovery of giant tubeworms, enormous clams, and eye-less shrimp and fish, all living at extreme depths, pressure, and temperatures, scientists thought all life depended on sunlight. The discovery of these organisms proved that some life can survive without the sun's light. Rather than using sunlight to photosynthesize, the organisms rely on sulfur chemicals produced around the hydrothermal sites as their energy source.

Ballard is a National Geographic Explorer and is the founder of JASON, an organization that brings real-world science to students all around the world through a science curriculum guided by practicing scientists.

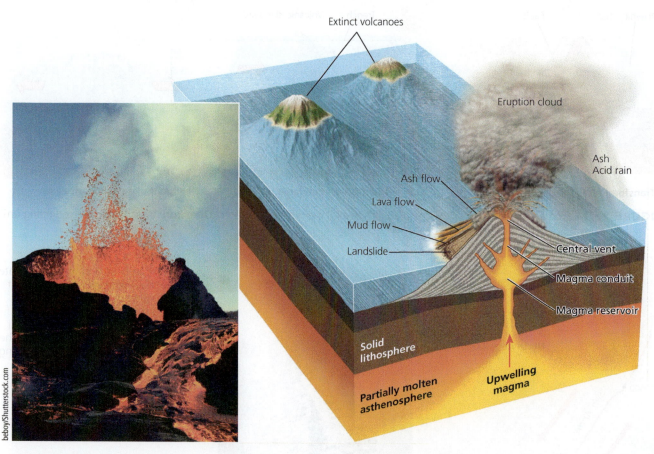

FIGURE 12.16 Sometimes, the internal pressure in a volcano is high enough to cause lava, ash, and gases to be ejected into the atmosphere (photo inset) or to flow over land, causing considerable damage.

that have reached the ocean's surface, eventually to form islands that have become suitable for human settlement, such as the Hawaiian Islands.

We can reduce the loss of human life and some of the property damage caused by volcanic eruptions by using historical records and geological measurements to identify high-risk areas, so that people can avoid living in those areas. Scientists and engineers are also developing monitoring devices that warn us when volcanoes are likely to erupt. In some areas that are prone to volcanic activity, evacuation plans have been developed.

Earthquakes Are Geological Rock-and-Roll Events

Forces inside the earth's mantle put tremendous stress on rock within the crust. Such stresses can be great enough to cause sudden breakage and shifting of the rock, producing a *fault*, or fracture in the earth's crust (Figure 12.15, right). When a fault forms, or when there is abrupt movement on an existing fault, energy that has accumulated over time is released in the form of vibrations, called *seismic waves*, which move in all directions through the surrounding rock—an event called an **earthquake** (Figure 12.17 and **Concept 12.5**). Most earthquakes occur at the boundaries of tectonic plates (Figure 12.14).

Seismic waves move upward and outward from the earthquake's center, or *focus*, like ripples in a pool of water. Scientists measure the severity of an earthquake by the *magnitude* of its seismic waves. The magnitude is a measure of ground motion (shaking) caused by the earthquake, as indicated by the *amplitude*, or size of the seismic waves when they reach a recording instrument, called a *seismograph*.

Scientists use the *Richter scale*, on which each unit has an amplitude 10 times greater than the next smaller unit. Scientists who study earthquakes rate them as *insignificant* (less than 4.0 on the Richter scale), *minor* (4.0–4.9), *damaging* (5.0–5.9), *destructive* (6.0–6.9), *major* (7.0–7.9),

and *great* (over 8.0). The largest-ever recorded earthquake occurred in Chile on May 22, 1960, and measured 9.5 on the Richter scale. Each year, scientists record the magnitudes of more than 1 million earthquakes, most of which are too small to feel.

The primary effects of earthquakes include shaking and sometimes a permanent vertical or horizontal displacement of a part of the crust. These effects can have serious consequences for people and for buildings, bridges, freeway overpasses, dams, and pipelines. A major earthquake is a very large rock-and-roll geological event.

One way to reduce the loss of life and property damage from earthquakes is to examine historical records and make geological measurements to locate active fault zones. Scientists can then map high-risk areas (Figure 12.18) and establish building codes that regulate the placement and design of buildings in such areas. Then people can evaluate the risk and factor it into their decisions about where to live. In addition, engineers know how to make homes, other buildings, bridges, and freeways more earthquake resistant, although this is costly.

Earthquakes on the Ocean Floor Can Cause Tsunamis

A **tsunami** is a series of large waves generated when part of the ocean floor suddenly rises or drops (Figure 12.19). Most large tsunamis are caused when certain types of faults in the ocean floor move up or down because of a large underwater earthquake. Other causes are landslides generated by earthquakes and volcanic eruptions (**Concept 12.5**).

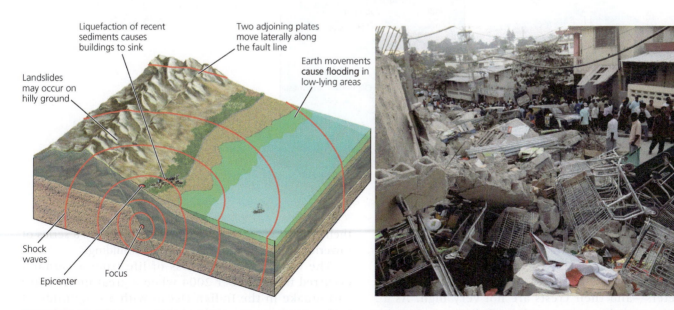

Liquefaction of recent sediments causes buildings to sink

Two adjoining plates move laterally along the fault line

Earth movements **cause flooding** in low-lying areas

Landslides may occur on hilly ground

Shock waves

Epicenter

Focus

AP Images/Jorge Cruz

FIGURE 12.17 An *earthquake* (left) is one of nature's most powerful events. The photo shows damage from a 2010 earthquake in Port-au-Prince, Haiti.

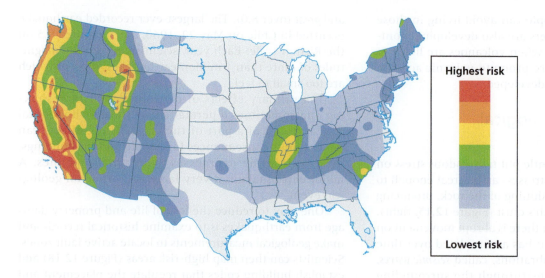

FIGURE 12.18 Comparison of degrees of earthquake risk across the continental United States. **Question:** What is the earthquake risk where you live or go to school?

Compiled by the authors using data from the U.S. Geological Survey

Highest risk

Lowest risk

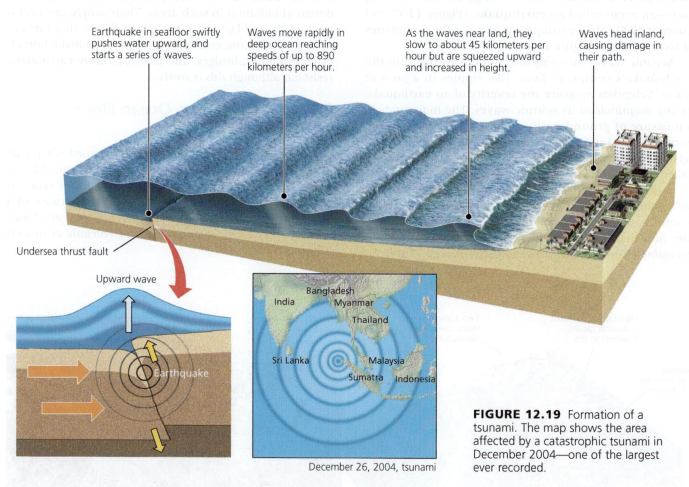

Earthquake in seafloor swiftly pushes water upward, and starts a series of waves.

Waves move rapidly in deep ocean reaching speeds of up to 890 kilometers per hour.

As the waves near land, they slow to about 45 kilometers per hour but are squeezed upward and increased in height.

Waves head inland, causing damage in their path.

Undersea thrust fault

Upward wave

Earthquake

Bangladesh
India
Myanmar
Thailand
Sri Lanka
Malaysia
Sumatra
Indonesia

December 26, 2004, tsunami

FIGURE 12.19 Formation of a tsunami. The map shows the area affected by a catastrophic tsunami in December 2004—one of the largest ever recorded.

Tsunamis are often called *tidal waves,* although they have nothing to do with tides. They can travel far across the ocean at the speed of a jet plane. In deep water, the waves are very far apart—sometimes hundreds of kilometers—and their crests are not very high. As a tsunami approaches a coast with its shallower waters, it slows down, its wave crests squeeze closer together, and

their heights grow rapidly. It can hit a coast as a series of towering walls of water that can level buildings.

The largest recorded loss of life from a tsunami occurred in December 2004 when a great underwater earthquake in the Indian Ocean with a magnitude of 9.15 caused a tsunami that killed around 230,000 people. It also devastated many coastal areas of Indonesia

FIGURE 12.20 The Banda Aceh Shore near Gleebruk, Indonesia on June 23, 2004 (left), and on December 28, 2004 (right), after it was struck by a tsunami.

New York Public Library/Science Source

(Figures 12.19 and 12.20), Thailand, Sri Lanka, South India, and eastern Africa. It displaced about 1.7 million people (1.3 million of them in India and Indonesia) and destroyed or damaged about 470,000 buildings and houses. There were no recording devices in place to provide an early warning of this tsunami.

In 2011, a large tsunami caused by a powerful earthquake off the coast of Japan generated 3-story high waves that killed almost 19,000 people, displaced more than 300,000 people, and destroyed or damaged 125,000 buildings. It also heavily damaged three nuclear reactors, which then released dangerous radioactivity into the surrounding environment. This nuclear accident led Japan to shut down all of its nuclear reactors.

In some areas, scientists and engineers have built networks of ocean buoys and pressure recorders on the ocean floor to collect data that can be relayed to tsunami emergency warning centers. However, these networks are far from complete.

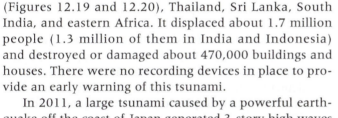

BIG IDEAS

- Dynamic forces that move matter within the earth and on its surface recycle the earth's rocks, form deposits of mineral resources, and cause volcanic eruptions, earthquakes, and tsunamis.

- The available supply of a mineral resource depends on how much of it is in the earth's crust, how fast we use it, the mining technology used to obtain it, its market prices, and the harmful environmental effects of removing and using it.

- We can use mineral resources more sustainably by trying to find substitutes for scarce resources, reducing resource waste, and reusing and recycling nonrenewable minerals.

Tying It All Together

The Real Cost of Gold and Sustainability

In this chapter's Core Case Study, we considered the harmful effects of gold mining as an example of the impacts of our extraction and use of mineral resources. We saw that these effects make gold much more costly, in terms of environmental and human health costs, than is reflected in the price of gold.

In this chapter, we looked at technological developments that could help us expand supplies of mineral resources and to use them more sustainably. For example, if we develop it safely, we could use nanotechnology to make new materials that could replace scarce mineral resources and greatly reduce the environmental impacts of mining and processing such resources. For example, we might use graphene to produce more efficient and affordable solar cells to generate electricity—an application of the solar energy **principle of sustainability**.

We can also use mineral resources more sustainably by reusing and recycling them, and by reducing unnecessary resource use and waste—applying the chemical cycling **principle of sustainability**. In addition, industries can mimic nature by using a diversity of ways to reduce the harmful environmental impacts of mining and processing mineral resources, thus applying the biodiversity **principle of sustainability**.

Matt Benoit/Shutterstock.com

Chapter Review

Core Case Study

1. What is **mining**? Explain why the real cost of gold is more than what most people pay for it. What are some examples of costs not accounted for?

Section 12.1

2. What are the two key concepts for this section? Define **geology**. Define and distinguish among the earth's **core**, **mantle**, **asthenosphere**, **crust**, and **lithosphere**. Define **mineral**, **mineral resource**, and **rock**. Define and distinguish among **sedimentary rock**, **igneous rock**, and **metamorphic rock** and give an example of each. Define the **rock cycle** and explain its importance. Define **ore** and distinguish between a **high-grade ore** and a **low-grade ore**. List five important nonrenewable mineral resources and their uses.

Section 12.2

3. What are the two key concepts for this section? What are the **reserves** of a mineral resource and how can they be expanded? What two factors determine the future supply of a nonrenewable mineral resource? Explain how the supply of a nonrenewable mineral resource can be economically depleted and list the five choices we have when this occurs. What is **depletion time** and what factors affect it?

4. Which five nations supply most of the world's nonrenewable mineral resources? How dependent is the United States on other countries for important nonrenewable mineral resources? Explain why uneven distribution of lithium reserves among various countries is a concern, especially for the United States. Summarize the importance of rare earth metals. Explain the concern over U.S. access to rare earth mineral resources and how the country could solve this problem. Describe the conventional view of the relationship between the supply of a mineral resource and its market price. Explain why some economists believe this relationship no longer applies. Summarize the pro and cons of providing subsidies and tax breaks for mining companies in the United States and in other countries.

5. Summarize the opportunities and limitations of expanding mineral supplies by mining lower-grade ores. What are the advantages and disadvantages of biomining? Describe the opportunities and possible problems that could result from deep-sea mining.

Section 12.3

6. What is the key concept for this section? Summarize the life cycle of a metal product. What is the relationship between the grade of a metal ore and the environmental effects of mining the ore?

7. What is **surface mining**? Define **overburden** and **spoils**. Define **open-pit mining** and **strip mining**, and distinguish among **area strip mining**, **contour strip mining**, and **mountaintop removal mining**. Describe three harmful environmental effects of surface mining. Explain why not all surface mining areas are cleaned up and restored after the mining is done. What is **subsurface mining**? Summarize the harmful environmental and health effects of subsurface mining. Define **tailings** and explain why they can be hazardous. What is **smelting** and what are its major harmful environmental effects?

Section 12.4

8. What is the key concept for this section? Give two examples of promising substitutes for key mineral resources. What is **nanotechnology** and what are some of its potential environmental and other benefits? What are some problems that could arise from the widespread use of nanotechnology? Describe Yu-Guo Guo's scientific contributions to nanotechnology. Describe the potential of using graphene as a new resource. Explain the benefits of recycling and reusing valuable metals. List five ways to use nonrenewable mineral resources more sustainably.

Section 12.5

9. What is the key concept for this section? What are **tectonic plates**, and what typically happens when they collide, move apart, or grind against one another? What is **continental drift?** What is a **volcano** and what are the major effects of a volcanic eruption? What is an **earthquake** and what are its major effects? What is a **tsunami** and what are its major effects?

10. What are the three big ideas of this chapter? Explain how we can apply the three **scientific principles of sustainability** to obtain and use gold and other nonrenewable mineral resources in more sustainable ways.

Note: Key terms are in bold type.

Critical Thinking

1. Do you think that the benefits we get from gold— its uses in jewelry, dentistry, electronics, and other uses—are worth the real cost of gold (**Core Case Study**)? If so, explain your reasoning. If not, explain your argument for cutting back on or putting a stop to the mining of gold.

2. You are an igneous rock. Describe what you experience as you move through the rock cycle. Repeat this exercise, assuming you are a sedimentary rock and again assuming you are a metamorphic rock.

3. What are three ways in which you benefit from the rock cycle?

4. Suppose your country's supply of rare earth metals was cut off tomorrow. How would this affect your life? Give at least three examples. How would you adjust to these changes? Explain.

5. Use the second law of thermodynamics (see Chapter 2, p. 38) to analyze the scientific and economic feasibility of each of the following processes:

   a. Extracting certain minerals from seawater
   b. Mining increasingly lower-grade deposits of minerals
   c. Continuing to mine, use, and recycle minerals at increasing rates

6. Suppose you were told that mining deep-ocean mineral resources would mean severely degrading ocean bottom habitats and life forms such as giant tubeworms and giant clams. Do you think that such information should be used to prevent ocean bottom mining? Explain.

7. List three ways in which a nanotechnology revolution could benefit you and three ways in which it could harm you. Do you think the benefits outweigh the harms? Explain.

8. What are three ways to reduce the harmful environmental impacts of the mining and processing of nonrenewable mineral resources? What are three aspects of your lifestyle that contribute to these harmful impacts?

Doing Environmental Science

Do research to determine which mineral resources go into the manufacture of each of the following items and how much of each of these resources are required to make each item: **(a)** a cell phone, **(b)** a wide-screen TV, and **(c)** a large pickup truck. Pick three of the lesser-known mineral materials that you have learned about in this exercise and do more research to find out where in the world most of that material comes from. For each of the three minerals

you chose, try to find out what kinds of environmental effects have resulted from the mining of the mineral in at least one of the places where it is mined. You might also find out about steps that have been taken to deal with those effects. Write a report summarizing all of your findings.

Global Environment Watch Exercise

Go to your MindTap course to access the GREENR database. Using the "Basic Search" box at the top of the page, search for an article that deals with rare earth metal supplies. Summarize the conclusions expressed in the article. Is there scientific information cited in the article to support the author's conclusions? Give specific examples. Do you think there are any types of supporting scientific data

not mentioned in the article that would strengthen the author's conclusions? For example, would you add statistical data to support a point, or data in a graph indicating possible cause-and-effect relationships? Be specific and give reasons for your suggestions.

Data Analysis

Rare earth metals are widely used in a variety of important products. According to the USGS, China has about 50% of the world's reserves of rare earth metals. Use this information to answer the following questions.

1. In 2014, China had 55 million metric tons of rare earth metals in its reserves and produced 95,000 metric tons of these metals. At this rate of production how long will China's rare earth reserves last?

2. In 2014, the global demand for rare earth metals was about 133,600 metric tons. At this annual rate of use, if China were to produce all of the world's rare earth metals, how long would its reserves last?

3. The annual global demand for rare earth metals is projected to rise to at least 185,000 metric tons by 2020. At this rate, if China were to produce all of the world's rare earth metals, how long would its reserves last?

Energy Resources

Key Questions

13.1 What is net energy and why is it important?

13.2 What are the advantages and disadvantages of using fossil fuels?

13.3 What are the advantages and disadvantages of using nuclear power?

13.4 Why is energy efficiency an important energy resource?

13.5 What are the advantages and disadvantages of using renewable energy resources?

13.6 How can we make the transition to a more sustainable energy future?

Wind turbines and solar cell panels

Vaclav Volrab/Shutterstock.com

Using Hydrofracking to Produce Oil and Natural Gas

Geologists have known for decades about vast deposits of oil and natural gas that are dispersed and trapped between compressed layers of shale rock formations. These deposits are found deep underground in many areas of the United States, including North Dakota, Texas, and Pennsylvania.

For years, it cost too much to extract such oil (called *tight oil*) and natural gas from shale rock. This changed in the late 1990s when oil and gas producers combined two existing extraction technologies (Figure 13.1). One is **horizontal drilling**, which involves drilling a vertical well deep into the earth, turning the flexible shaft of the drill, and then drilling horizontally to gain access to multiple oil and natural gas deposits held tightly between layers in shale rock formations. Usually, wells are drilled vertically for 1.6 to 2.4 kilometers (1 to 1.5 miles) and then horizontally for up to 1.6 kilometers (1 mile). Two or three horizontally drilled wells can often produce as much oil as 20 vertical wells, which reduces the area of land damaged by drilling operations.

The second technology, called **hydraulic fracturing** (also called **hydrofracking** or **fracking**), is then used to free the oil and natural gas trapped in the shale rock. High-pressure pumps force a mixture, or *slurry*, of water, sand, and a cocktail of chemicals through holes in the underground well pipe to fracture the shale rock and create cracks. The sand becomes wedged into the cracks and props them open. When the pressure is released a mixture of the oil or natural gas and about half of the slurry flows out of the cracks and is pumped to the surface through the well pipe (Figure 13.1).

The returning slurry contains a mix of naturally occurring salts, toxic heavy metals, and radioactive materials leached from the rock, along with some potentially harmful drilling chemicals that oil and natural gas companies are not required to identify. The hazardous slurry is injected under high pressure into deep underground hazardous waste wells (the most widely used option), sent to sewage treatment plants that often cannot handle the wastes, stored in open air holding ponds that can leak or collapse, or cleaned up and reused in the fracking process—the best but least used option because of its high cost.

Energy companies drill a well horizontally, frack it several times, and then drill a new well and repeat the process. The use of these two extraction technologies in at least 25 states between 1990 and 2015 has brought about a new era of increased oil and natural gas production in the United States. It could last as long as market prices of oil and natural gas are high enough to make these drilling operations profitable. Like any technology, this approach has advantages and disadvantages that we discuss later in this chapter.

In this chapter, we explore and compare the benefits and drawbacks of using *nonrenewable resources*, such as oil, natural gas, coal, and nuclear power, versus making improvements in energy efficiency and using *renewable resources* such as wind, solar energy, flowing water, and the earth's internal heat. ●

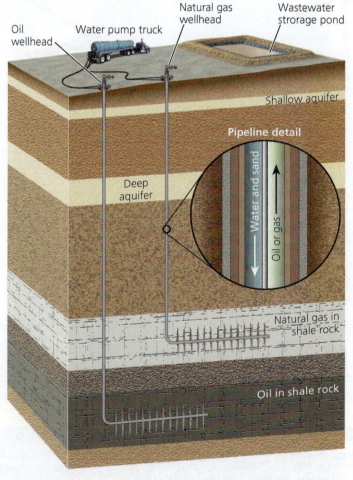

FIGURE 13.1 *Fracking:* Horizontal drilling and hydraulic fracturing are used to release large amounts of oil and natural gas that are tightly held in underground shale rock formations.

13.1 WHAT IS NET ENERGY AND WHY IS IT IMPORTANT?

CONCEPT 13.1A About 90% of the commercial energy used in the world comes from nonrenewable energy resources (mostly oil, natural gas, and coal) and 10% comes from renewable energy resources.

CONCEPT 13.1B Energy resources vary greatly in their *net energy*—the amount of energy available from a resource minus the amount of energy needed to make it available.

Where Does the Energy We Use Come From?

The energy that heats the earth and makes life possible comes from the sun—in keeping with the solar energy **principle of sustainability**. Without this free and essentially inexhaustible input of solar energy, the earth's average temperature would be −240°C (−400°F) and life as we know it would not exist.

To supplement the sun's life-sustaining energy, we use *commercial energy* produced from a variety of nonrenewable and renewable energy resources and sold in the marketplace to supplement the sun, which provides 99% of the earth's energy. *Nonrenewable energy resources* exist in fixed amounts that took millions of years to form. They include fossil fuels (oil, natural gas, coal) and the nuclei of certain elements (nuclear energy). *Renewable energy resources* are replenished by natural processes and include energy from the sun, wind, flowing water (hydropower), biomass (energy stored in plants), and heat in the earth's interior (geothermal energy).

In 2015, 90% of the commercial energy used in the world and in the United States came from nonrenewable

resources (mostly oil, coal, and natural gas) and 10% came from renewable resources (Figure 13.2) (**Concept 13.1A**).

90% Percentage of commercial energy used in the world and in the United States that comes from nonrenewable energy (mostly fossil fuels).

Net Energy: It Takes Energy to Get Energy

Producing high-quality energy from any energy resource requires an input of high-quality energy. For example, before oil can be used, it must be located, pumped from beneath the ground or ocean floor, transferred to a refinery, converted to gasoline and other fuels, and delivered to consumers. Each of these steps uses high-quality energy, obtained mostly by burning fossil fuels, especially gasoline and diesel fuel produced from oil. Because of the second law of thermodynamics (Chapter 2, p. 38), some of the high-quality energy used in each step is degraded to lower-quality energy that typically flows into the environment as heat.

Net energy is the amount of high-quality energy available from an energy resource minus the high-quality energy needed to make the energy available (**Concept 13.1B**).

Net Energy = energy output − energy input

Suppose that it takes about 9 units of high-quality energy to produce 10 units of high-quality energy from an energy resource. Then the net energy for the resource is 1 unit of energy (10 − 9 = 1), a low value. Net energy is like the net profit earned by a business after it deducts its expenses. If a

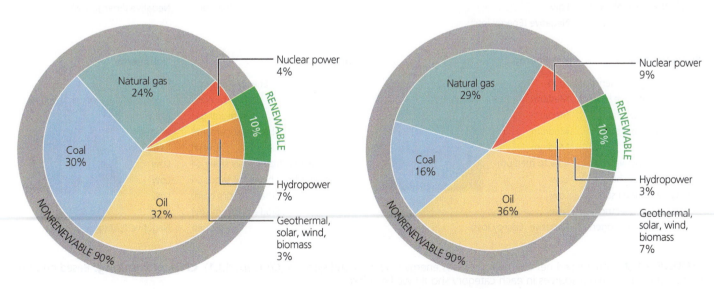

FIGURE 13.2 Energy use by source throughout the world (left) and in the United States (right) in 2015.

(Compiled by the authors using data from British Petroleum, U.S. Energy Information Administration (EIA), and International Energy Agency (IEA).)

business has $1 million in sales and $900,000 in expenses, its net profit is $100,000.

Figure 13.3 shows the generalized net energy for major energy resources and systems. Many analysts view net energy as an important measure for evaluating the long-term economic usefulness of different energy resources.

Some Energy Resources Need Subsidies

Resources with a low net energy are costly to bring to market. This makes it difficult for such energy resources to compete in the marketplace against energy resources with higher net energies unless they receive subsidies and tax breaks from the government (taxpayers) or other outside sources.

For example, electricity produced by nuclear power has a low net energy. This is because large amounts of high-quality energy are needed for each step in the *nuclear power fuel cycle:* to extract and process uranium ore, upgrade it to nuclear fuel, build and operate nuclear power plants, dismantle each nuclear plant after its useful life (typically 40–60 years), and safely store for thousands of years the high-level radioactive wastes created in operating and dismantling each plant.

The low net energy and the resulting high cost for the entire nuclear fuel cycle (further discussed later in this chapter) is one reason why governments throughout the world heavily subsidize nuclear-generated electricity to make it available to consumers at an affordable price. However, such subsidies hide the true costs of the nuclear power fuel cycle and thus violate the full-cost pricing **principle of sustainability** (see inside back cover).

13.2 WHAT ARE THE ADVANTAGES AND DISADVANTAGES OF USING FOSSIL FUELS?

CONCEPT 13.2 Oil, natural gas, and coal are currently abundant and relatively inexpensive, but using them causes air and water pollution, and releases climate-changing greenhouse gases to the atmosphere.

We Depend Heavily on Oil

Crude oil, or **petroleum**, is a black, gooey liquid containing a mixture of combustible hydrocarbons

Electricity	Net Energy
Energy efficiency	High
Hydropower	High
Wind	High
Coal	High
Natural gas	Medium
Geothermal energy	Medium
Solar cells	Low to medium
Nuclear fuel cycle	Low
Hydrogen	Negative (Energy loss)

High-Temperature Industrial Heat	Net Energy
Energy efficiency (cogeneration)	High
Coal	High
Natural gas	Medium
Oil	Medium
Heavy shale oil	Low
Heavy oil from oil sands	Low
Direct solar (concentrated)	Low
Hydrogen	Negative (Energy loss)

Space Heating	Net Energy
Energy efficiency	High
Passive solar	Medium
Natural gas	Medium
Geothermal energy	Medium
Oil	Medium
Active solar	Low to medium
Heavy shale oil	Low
Heavy oil from oil sands	Low
Electricity	Low
Hydrogen	Negative (Energy loss)

Transportation	Net Energy
Energy efficiency	High
Gasoline	High
Natural gas	Medium
Ethanol (from sugarcane)	Medium
Diesel	Medium
Gasoline from heavy shale oil	Low
Gasoline from heavy oil sands oil	Low
Ethanol (from corn)	Low
Biodiesel (from soy)	Low
Hydrogen	Negative (Energy loss)

FIGURE 13.3 Generalized net energy for various energy resources and systems (**Concept 13.1**). ***Critical thinking:*** Based only on these data, which two resources in each category should we be using?

(Compiled by the authors using data from the U.S. Department of Energy; U.S. Department of Agriculture; Colorado Energy Research Institute, *Net Energy Analysis,* 1976; Howard T. Odum and Elisabeth C. Odum, *Energy Basis for Man and Nature,* 3rd ed., New York: McGraw-Hill, 1981; and Charles A. S. Hall and Kent A. Klitgaard, *Energy and the Wealth of Nations,* New York: Springer, 2012.)
Top left: racorn/Shutterstock.com. Bottom left: Donald Aitken/National Renewable Energy Laboratory. Top right: Serdar Tibet/Shutterstock.com. Bottom right: Michel Stevelmans/Shutterstock.com.

(compounds containing hydrogen and carbon atoms) along with small amounts of sulfur, oxygen, and nitrogen impurities. It is also known as *conventional* or *light crude oil*. Crude oil forms over millions of years from the decayed remains of ancient organisms, which become crushed beneath layers of rock and are subjected to intense pressure and heat. Oil is the most widely commercial energy resource in the world (Figure 13.2, left) and the most widely used energy resource in the United States (Figure 13.2, right).

Deposits of conventional crude oil are typically trapped beneath layers of nonporous rock in the earth's crust on land or under the seafloor. The crude oil in such deposits is dispersed in microscopic pores and cracks in underground rock formations, somewhat like water saturating a sponge.

Geologists identify potential oil deposits by using large machines to pound the earth and send shock waves deep underground and measuring how long it takes the waves to be reflected back. They feed this information into computers to produce *three-dimensional seismic maps* that show the locations and sizes of various underground rock formations, including those containing oil.

Once they identify a potential site, oil companies drill holes and remove rock cores to determine if there is enough oil to be extracted profitably. If there is enough oil, wells are drilled and the oil—drawn out of the rock pores by gravity—flows to the bottom of the well and is pumped to the surface.

After about a decade of pumping, the pressure in a well usually drops and its rate of crude oil production starts to decline. This point in time is referred to as **peak production** for the well. The same decline in production can take place in an oil field with numerous wells.

Crude oil from a well cannot be used as it is. It is transported to a refinery by pipeline, truck, rail, or ship (oil tanker). There the crude oil is heated to separate it into various fuels and other components with different boiling points (Figure 13.4) in a process called **refining**. Refining requires a large amount of high-quality energy and decreases the oil's net energy. About 2% of the products of refining, called **petrochemicals**, are used as raw materials to make industrial organic chemicals, cleaning fluids, pesticides, plastics, synthetic fibers, paints, medicines, cosmetics, and many other products.

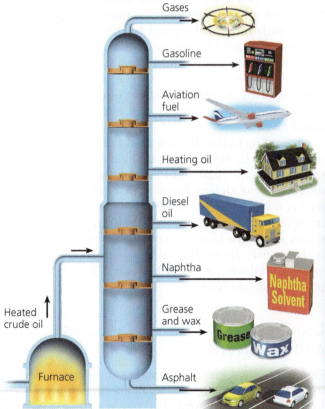

Lowest Boiling Point

Gases
Gasoline
Aviation fuel
Heating oil
Diesel oil
Naphtha
Grease and wax
Asphalt

Heated crude oil

Furnace

Highest Boiling Point

Natalia Bratslavsky/Shutterstock.com

FIGURE 13.4 When crude oil is refined, many of its components are removed at various levels of a distillation column, depending on their boiling points. The most volatile components with the lowest boiling points are removed at the top of the column, which can be as tall as a nine-story building. The photo shows an oil refinery in Texas.

Is the World Running Out of Crude Oil?

We rely heavily on crude oil. In 2016, the world used 34.7 billion barrels of crude oil. (One barrel of oil contains 159 liters or 42 gallons of oil.) Laid end to end, these barrels of crude oil would stretch to about 31.7 million kilometers (19.7 million miles)—far enough to reach to the moon and back about 41 times.

How much crude oil exists on the earth? No one knows the actual amount but geologists have estimated the amounts in identified oil deposits. **Proven oil reserves** are known deposits from which crude oil can be extracted profitably at current prices using current technology. Proven oil reserves are not fixed. They grow when new deposits or found or when new extraction technologies such as horizontal drilling and hydraulic fracking (**Core Case Study**) make it possible and profitable to produce oil from deposits that were once too costly to tap.

The net energy of crude oil is still high in places like Saudi Arabia where huge deposits located near the earth's surface yield oil at a cost of $3 to $10 a barrel. However, since 1999, oil producers have had to spend more money and use more energy to dig deeper wells on land and at sea and to extract and transport oil from more remote and challenging areas such as the Arctic. As a result, oil has a medium net energy. There is no global shortage of oil but there is an increasing shortage of cheap oil, as the world's easy-to-reach concentrated crude oil deposits are being depleted.

The 12 countries that make up the Organization of Petroleum Exporting Countries (OPEC) have about 81% of the world's proven crude oil reserves and thus are likely to control most of the world's conventional oil supplies for decades. OPEC's member countries are Algeria, Angola, Ecuador, Iran, Iraq, Kuwait, Libya, Nigeria, Qatar, Saudi Arabia, United Arab Emirates, and Venezuela. However, the recent increase in U.S. oil production (**Core Case Study**) has weakened the ability of OPEC nations to control global oil prices.

81% Percentage of the world's proven oil reserves held by OPEC's 12 countries.

In 2015, the three countries with the world's largest proven crude oil reserves were Venezuela, Saudi Arabia, and Canada (including heavy oil from oil sands, covered later in chapter). The three largest producers of crude oil in 2015 were the United States, Saudi Arabia, and Russia. In 2015, the world's three largest consumers of crude oil—the United States, China, and Japan—had, in order, only about 3.2%, 1.1%, and 0.003% of the world's proven crude oil reserves.

Oil Production and Consumption in the United States

In 2015, the United States got 81% of its commercial energy from fossil fuels, with the largest percentage coming from crude oil (Figure 13.2, right).

Since 1982, the United States has imported some of the oil it uses because its oil consumption has exceeded its domestic production. In 2015, the United States imported 24% of its crude oil, compared to 60% in 2005, mostly because of rising domestic production of tight oil from shale rock (**Core Case Study**). In 2015, the five largest suppliers of imported oil for the United States, in order, were Canada, Saudi Arabia, Venezuela, Mexico, and Columbia.

Can the United States continue reducing its dependence on oil imports by increasing its oil production? Some say "yes" and project that domestic oil production will increase dramatically over the next few decades—especially from oil found in layers of shale rock (**Core Case Study**).

Other analysts question the long-term availability of this source of oil for two reasons. *First*, drilling a horizontal oil well costs more than drilling a vertical well. Thus, if oil prices go too low (less than $50–$60 a barrel), using horizontal drilling and fracking to develop new wells is not profitable and the number of new wells drops sharply, as happened in the United States between 2014 and 2016, unless the process can be made more efficient and less costly. *Second*, the output of oil from shale rock drops off about twice as fast as it does in most conventional oil fields.

According to the International Energy Agency (IEA), oil produced from shale rock in the United States is likely to peak around 2020 and then decline for two to three decades as the richest deposits are depleted. If this projection is correct, the current U.S. boom in oil production from shale rock is a temporary bubble, not a long-term source of oil. The long-term problem for the United States is that it uses about 20% of the world's oil production, while it produces 13%, and has only 3.2% of the world's proven crude oil reserves.

Using Crude Oil Has Advantages and Disadvantages

Figure 13.5 lists major advantages and disadvantages of using conventional light oil as an energy resource. A critical problem is that burning oil or any carbon-containing fossil fuel releases the greenhouse gas CO_2 into the atmosphere. According to decades of research and at least 90% of the world's top climate scientists, this plays an important role in warming the atmosphere and changing the world's climate, as discussed in Chapter 15.

Trade-Offs

Conventional Oil

Advantages	Disadvantages
Ample supply for several decades	Water pollution from oil spills and leaks
Net energy is medium but decreasing	Environmental costs not included in market price
Low land disruption	Releases CO_2 and other air pollutants when burned
Efficient distribution system	Vulnerable to international supply interruptions

Richard Goldberg/Shutterstock.com

FIGURE 13.5 Using conventional light oil as an energy resource has advantages and disadvantages. *Critical thinking:* Which single advantage and which single disadvantage do you think are the most important? Why? Do the advantages outweigh the disadvantages? Explain.

Heavy Oil Has a High Environmental Impact

An alternative to conventional or light crude oil is heavy crude oil that is thicker and stickier. Two sources of heavy oil are oil shale rock and oil sands.

Heavy oil extracted from oil shale rock is called **shale oil**. It is *dispersed within* bodies of shale rock compared to lighter oil that is *trapped between* layers of shale rock (**Core Case Study**). Producing shale oil involves mining, crushing, and heating oil shale rock (Figure 13.6, left) to extract a mixture of hydrocarbons called *kerogen* that can be distilled to produce shale oil (Figure 13.6, right). Before the thick shale oil is pumped through a pressurized pipeline to a refinery, it must be heated to increase its flow rate and processed to remove sulfur, nitrogen, and other impurities, which reduces its net energy.

About 72% of the world's estimated reserves of oil shale rock are buried deep in rock formations found mostly under government-owned land in the U.S. states of Colorado, Wyoming, and Utah. The potential supply is huge but its net energy is low (Figure 13.3), which means that these deposits are too costly to develop. The process also has a large harmful environmental impact, including the production of rock waste, wastewater, possible leaks of the extracted kerogen into groundwater, wastewater, and high water use. If the production price drops or the price of conventional oil rises sharply, and if its harmful environmental effects can be reduced, shale oil could become an important energy source. Otherwise, it will remain in the ground.

U.S. Department of Energy

FIGURE 13.6 Heavy shale oil (right) can be extracted from oil shale rock (left).

A growing source of heavy oil is **oil sands**, or **tar sands.** Oil sands are a mixture of clay, sand, water, and *bitumen*—a thick, sticky, tar-like heavy oil with a high sulfur content. Much of the world's proven reserves of heavy oil in oil sands are found under sandy soil in a vast area of remote boreal forest in Alberta, Canada. If we include its conventional light oil and its heavy oil from oil sands, Canada has the world's third largest proven oil reserves. The United States also has large deposits of oil sands in Utah, Wyoming, and Colorado.

The two big drawbacks to producing oil from oil sands are its low net energy (Figure 13.2) and its major harmful impacts on the land (Figure 13.7), air, water, wildlife, and climate. It takes two to four tons of oil sands and two to five barrels of water to produce one barrel of this heavy synthetic oil. The process also emits large quantities of air pollutants and 20% more climate-changing CO_2 than does the production of conventional crude oil, according to a 2015 study by the U.S. Department of Energy (DOE). A 2016 study by Shao-Meng Li and other Canadian scientists found that the production of oil from tar sands in Canada is also a major source of fine-particle (aerosol) air pollution, which can increase the risk of heart and lung disease.

Producing this heavy oil is expensive. Between 2014 and July 2017, global oil prices of light crude oil were well below the average cost of producing heavy oil from oil sands and new production dropped. Figure 13.8 lists the major advantages and disadvantages of producing heavy oil from oil shale rock and oil sands.

Natural Gas Is an Important Energy Resource

Natural gas contains a mixture of gases of which 50–90% is methane (CH_4). It also contains smaller amounts of heavier gaseous hydrocarbons such as propane (C_3H_8) and butane (C_4H_{10}), and small amounts of highly toxic

FIGURE 13.7 Oil sands surface mining operation in Alberta, Canada, involves clearing boreal forest and strip mining the land to remove the oil sands.

Christopher Kolaczan /Shutterstock.com

Trade-Offs	
Heavy Oil from Oil Sands	
Advantages	Disadvantages
Large potential supplies	Low net energy
	Expensive
Easily transported within and between countries	Releases CO_2 and other air pollutants
	Severe land disruption
Efficient distribution system in place	Water pollution and high water use

Christopher Kolaczan /Shutterstock.com

FIGURE 13.8 Using heavy oil from oil sands and from oil shale rock as an energy resource has advantages and disadvantages (**Concept 13.2**). *Critical thinking:* Which single advantage and which single disadvantage do you think are the most important? Why? Do the advantages outweigh the disadvantages? Explain.

hydrogen sulfide (H_2S). Natural gas has a medium net energy (Figure 13.3). This fuel is widely used for cooking, heating, and industrial processes. It is also used to fuel fleets of trucks and cars and to power turbines that produce electricity in power plants.

This versatility and the use of horizontal drilling and fracking (**Core Case Study**) help explain why natural gas provided about 29% of the energy (Figure 13.3, right) and 34% of the electricity consumed in the United States in 2016 (up from 17% in 2003). Natural gas power plants cost less and take much less time to build than do coal-powered and nuclear power plants. Natural gas also burns cleaner than oil and much cleaner than coal. When burned completely, it emits about 30% less CO_2 than oil and about 50% less than coal for the same amount of energy. This explains why natural gas production and consumption has grown and is projected to grow over the next few decades. Increased natural gas production in the United States has reduced its market price.

Natural gas is often found in deposits above deposits of crude oil. It also exists in tightly held deposits between layers of shale rock and can be extracted through horizontal drilling and fracking (**Core Case Study**).

When a natural gas deposit is tapped, propane and butane gases can be liquefied and removed as **liquefied petroleum gas (LPG)**. LPG is stored in pressurized tanks and used mostly in rural areas not served by natural gas pipelines. The rest of the natural gas (mostly methane) is purified and pumped into pressurized pipelines for distribution across land areas. The U.S. exports natural gas, mostly by pipeline, to Mexico and Canada.

Natural gas can also be transported across oceans, by converting it to **liquefied natural gas (LNG)** at a high pressure and a very low temperature. This highly flammable liquid is transported in refrigerated tanker ships. At its destination port, it is heated and converted back to the gaseous state and then distributed by pipeline. LNG has a lower net energy than natural gas because more than a third of its energy content is used to liquefy it, process it, deliver it to users by ship, and convert it back to natural gas. By 2020, the DOE projects that the United States will be the world's third largest exporter of LNG.

In 2015, the three countries with world's largest proven natural gas reserves in order were Iran, Russia, and Qatar; the three largest producers of natural gas were the United States, Russia, and China; and the three largest consumers were the United States, Russia, and Iran.

Currently, the United States does not have to rely on natural gas imports because domestic production has been increasing rapidly, mostly because of the growing use of horizontal drilling and fracking to extract natural gas from shale rock beds (**Core Case Study**). The increased supply has reduced the price of natural gas, which has accelerated a shift from coal to natural gas for generating electricity in the United States.

However, natural gas production from shale rock tends to peak and drop off much faster than does production from conventional natural gas wells. In addition, extracting and producing natural gas from shale rock reduces its net energy and, without effective regulation, can increase the harmful environmental impacts of production (Science Focus 13.1).

Natural Gas and Climate Change

Some see increased use of natural gas as a way to slow climate change because it emits much less CO_2 per unit of energy than coal. Critics cite two problems with this. First, abundant and cheap natural gas can reduce the use of environmentally harmful coal but its low price can also slow the shift to energy efficiency and renewable energy sources (as discussed later in this chapter). The other problem is that methane (CH_4) is a much more potent greenhouse gas per molecule than CO_2. Research indicates that natural gas drilling, production, and distribution processes leak large quantities of CH_4 into the atmosphere. Unless these leaks are plugged, greater reliance on natural gas could hasten rather than slow climate change.

Figure 13.10 lists the advantages and disadvantages of using conventional natural gas as an energy resource.

Coal Is a Plentiful but Dirty Fuel

Coal is a solid fossil fuel formed from the remains of land plants that were buried and exposed to intense heat and pressure for 300–400 million years. Figure 13.11 identifies the major types of coal that have formed from this process.

Coal is burned to produce electricity in power plants (Figure 13.12). In 2016, coal generated about 40% of the world's electricity, 33% of the electricity used in the United States (down from 30% in 2003), 75% in China, and 62% in India. Coal is also burned in industrial plants to make steel, cement, and other products.

In 2015, the three countries with the largest proven reserves of coal in order were the United States, Russia, and China; the three largest producers of coal were China, the United States, and Australia; and the three largest consumers were China, India, and the United States. In 2015, China consumed roughly as much coal as the rest of the world combined.

Coal is the world's most abundant fossil fuel. However, coal is by far the dirtiest of all fossil fuels because mining it degrades land (see Figure 12.11, p. 312) and burning it pollutes the air and water. Coal is mostly carbon but contains small amounts of sulfur, which are converted to the air pollutant sulfur dioxide (SO_2) when the coal burns. These emissions contribute to acid precipitation and to serious human health problems.

Burning coal also releases large amounts of black carbon particulates, or soot, and much smaller, fine particles of air pollutants such as toxic mercury. The fine particles can get into our lungs and with prolonged exposure can cause illnesses (such as emphysema and lung cancer), and contribute to heart attacks and strokes. According to a study by the Clean Air Task Force, fine-particle pollution in the United States, mostly from the older coal-burning power plants without the latest air pollution control technology, kills at least 13,000 people a year. In China,

Environmental Effects of Natural Gas Production and Fracking in the United States

The U.S. Energy Information Administration projects that over the next one or two decades at least 100,000 more natural gas wells will be drilled and fracked in the United States (**Core Case Study**).

Scientific studies indicate that without more monitoring and regulation of the entire natural gas production and distribution process (including fracking), greatly increased production of natural gas (and oil) from shale rock (Figure 13.1) could have several harmful environmental effects:

- Fracking requires enormous volumes of water—10 to 100 times more than in a conventional vertical well. In water-short areas, this could deplete aquifers, degrade aquatic habitats, and reduce the availability of water for irrigation and other purposes.
- Each fracked well produces huge volumes of hazardous wastewater that are pumped to the surface with the released natural gas (or oil).
- Providing the massive amounts of sand used for fracking is destroying large areas of prime midwestern farmland in states such as Illinois, Wisconsin, and Minnesota.
- To prevent leaks of natural gas (or oil) from conventional and fracked wells and from the hazardous wastewater that returns to the surface, steel pipe called casing is inserted into the drilled well and cement is pumped into a small space between the steel pipe and surrounding rock (Figure 13.1). Failure of this well casing and poor cementing can release methane (or oil), and contaminants in the wastewater into groundwater and drinking water wells far above the fracking site. When the contaminated water is drawn from a tap, this natural gas can catch fire (Figure 13.9). Experience shows that cementing the steel well casing is the weakest link in producing natural gas

(and oil) and in terms of preventing leaks from active and abandoned wells.

- According to recent studies by the National Academy of Sciences and the U.S. Geological Survey, one of the causes of hundreds of small and a few medium strength earthquakes in 13 states, especially Oklahoma, in recent years has been the shifting of bedrock resulting from the high-pressure injection of fracking wastewater into a growing number of deep underground hazardous waste wells. Geologists are concerned that such earthquakes could release hazardous wastewater into aquifers and cause breaks in the steel linings and cement seals of oil and gas well pipes.

In 2016, a five-year EPA study concluded that the major threats posed by fracking include potential contamination of groundwater from well casing and cement seal failures, poor management of the contaminated wastewater resulting from the process, and chemical spills at drilling sites.

Currently, there is little protection from the harmful environmental impacts of natural gas fracking on people and the environment. This is because, under political pressure from natural gas suppliers, the U.S. Congress in 2005 excluded natural gas fracking from EPA regulations under 7 major federal environmental laws. There is concern that without stricter regulation and monitoring, drilling another 100,000 natural gas wells during

FIGURE 13.9 Natural gas fizzing from this faucet in a Pennsylvania home can be lit like a natural gas stove burner. This began to happen in the area after an energy company drilled a fracking well, but the company denies responsibility. The homeowners have to keep their windows partly open year-round to keep the lethal and explosive gas from building up in the house.

MARK THIESSEN/National Geographic creative

the next 10 to 20 years could increase the risk of air and water pollution from the natural gas production process and lead to a political backlash against this technology. To avoid this, some analysts call for the following measures:

- eliminate all exemptions from environmental laws for the natural gas industry,
- put a cap on the number of hazardous waste injection wells in an area with natural gas (or oil) wells, as Oklahoma has had to do,
- increase the monitoring and regulation of the entire natural gas production and distribution system, and
- repair existing natural gas leaks.

However, there is strong political opposition from the natural gas industry to these policy changes.

CRITICAL THINKING

How might your life be affected if the policy changes listed above are not implemented?

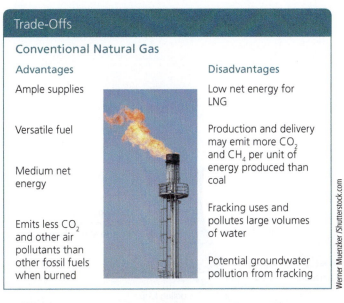

Werner Muenzker /Shutterstock.com

FIGURE 13.10 Use of conventional natural gas as an energy resource has advantages and disadvantages. **Critical thinking:** Which single advantage and which single disadvantage do you think are the most important? Why? Do you think that the advantages outweigh the disadvantages? Explain.

outdoor air pollution from the burning of coal contributes to 336,000 premature deaths per year (an average of 921 deaths per day), according to a study by Teng Fei at Tsinghua University.

Coal-burning power and industrial plants are among the largest emitters of CO_2 (Figure 13.13), which contributes to atmospheric warming and climate change and ocean acidification (Science Focus 9.2, p. 214). Because coal is mostly carbon, coal combustion emits about twice as much CO_2 per unit of energy as natural gas and produces about 42% of global CO_2 emissions. China leads the world in such emissions, followed by the United States. Coal use in China produces more greenhouse gas emissions than those from all of the coal, natural gas, and oil consumed in the United States. The consulting group HIS Energy projects that China's coal production will not peak until 2026. Coal combustion also emits trace amounts of radioactive materials as well as toxic mercury into the atmosphere and from there into lakes, where it can accumulate to high levels in fish consumed by humans.

Because of air pollution laws, many coal-burning plants in more-developed nations employ scrubbers to remove some of these pollutants before they leave the smokestacks. This reduces air pollution but produces a dust-like material called *coal ash* (Figure 13.12), which can contain harmful, indestructible chemical elements such as arsenic, lead, mercury, cadmium, and radioactive radium. This ash must be stored safely, essentially forever. However, political pressure by the U.S. coal industry has kept coal ash from being classified as a hazardous waste.

Since the 1980s, there has been a sharp drop in the number of coal mining jobs in the United States. This is largely the result of two factors. One is automation, which involves a shift to mining technology that relies heavily on machinery instead of miners and reduces the cost of producing coal for coal mining companies. The other factor is that the use of coal to produce electricity in the United States is decreasing because natural gas is abundant and offers a cleaner and cheaper way to produce electricity. In some areas, wind farms can also produce electricity at a lower cost than coal. Thus, the driving force for the decrease in coal use and coal mining jobs in the United States is economics, not environmental regulations.

Figure 13.14 lists the advantages and disadvantages of using coal as an energy resource. Scientists and many energy experts call for reducing energy waste and shifting from using abundant coal to less environmentally harmful natural gas and renewable energy resources over the next several decades to reduce air pollution and help slow

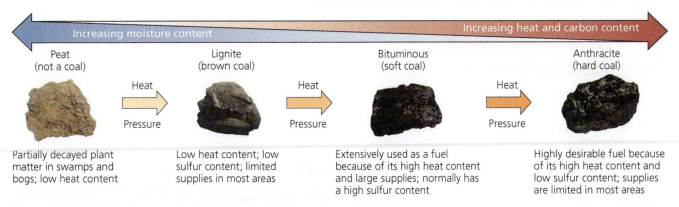

Peat (not a coal)		Lignite (brown coal)		Bituminous (soft coal)		Anthracite (hard coal)
Partially decayed plant matter in swamps and bogs; low heat content	Heat / Pressure	Low heat content; low sulfur content; limited supplies in most areas	Heat / Pressure	Extensively used as a fuel because of its high heat content and large supplies; normally has a high sulfur content	Heat / Pressure	Highly desirable fuel because of its high heat content and low sulfur content; supplies are limited in most areas

Increasing moisture content → Increasing heat and carbon content

FIGURE 13.11 Over millions of years, several different types of coal have formed. Peat is a soil material made of moist, partially decomposed organic matter, similar to coal. It is not classified as a coal, although it is used as a fuel. These different major types of coal vary in the amounts of heat, carbon dioxide, and sulfur dioxide released per unit of mass when they are burned.

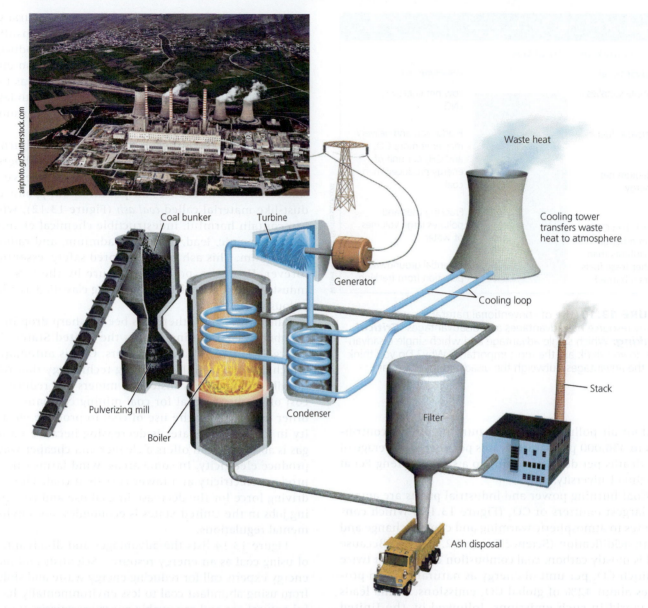

airphoto.gr/Shutterstock.com

FIGURE 13.12 This power plant burns pulverized coal to boil water and produce steam that spins a turbine to produce electricity. About 65% of the energy released when coal is burned in such a plant is wasted and ends up as heat that flows into the atmosphere or into the water used to cool the plant. ***Critical thinking:*** Does the electricity that you use come from a coal-burning power plant?

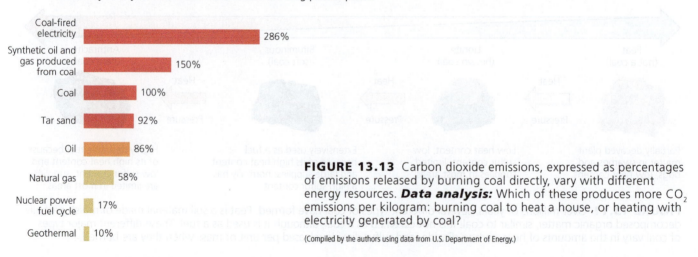

Coal-fired electricity	286%
Synthetic oil and gas produced from coal	150%
Coal	100%
Tar sand	92%
Oil	86%
Natural gas	58%
Nuclear power fuel cycle	17%
Geothermal	10%

FIGURE 13.13 Carbon dioxide emissions, expressed as percentages of emissions released by burning coal directly, vary with different energy resources. ***Data analysis:*** Which of these produces more CO_2 emissions per kilogram: burning coal to heat a house, or heating with electricity generated by coal?

(Compiled by the authors using data from U.S. Department of Energy.)

Trade-Offs

Coal

Advantages	Disadvantages
Ample supplies in many countries	Severe land disturbance and water pollution
Medium to high net energy	Fine particle and toxic mercury emissions threaten human health
Low cost when environmental costs are not included	Emits large amounts of CO_2 and other air pollutants when produced and burned

FIGURE 13.14 Using coal as an energy resource has advantages and disadvantages. ***Critical thinking:*** Which single advantage and which single disadvantage do you think are the most important? Why? Do you think that the advantages outweigh the disadvantages? Explain.

climate change. However, climate scientists estimate that to do this, we would have to leave 82% of the world's current coal reserves and 92% of U.S. coal reserves in the ground. This is a controversial and difficult economic, political, and ethical challenge for countries such as the United States, China, and India with large coal reserves.

13.3 WHAT ARE THE ADVANTAGES AND DISADVANTAGES OF USING NUCLEAR POWER?

CONCEPT 13.3 Nuclear power has a low environmental impact and a low accident risk, but its use has been limited by a low net energy, high costs, fear of accidents, long-lived radioactive wastes, and its role in the spread of nuclear weapons technology.

How Does a Nuclear Fission Reactor Work?

To evaluate the advantages and disadvantages of nonrenewable nuclear power, we need to know how a nuclear power plant and its accompanying nuclear fuel cycle work. A nuclear power plant is complex and costly system designed to perform a relatively simple task: boil water to produce steam that spins a turbine and generates electricity.

What makes nuclear power complex and costly is the use of a controlled nuclear fission reaction to provide the heat needed to boil the water. **Nuclear fission** occurs when a neutron is used to split the nuclei of certain isotopes with large mass numbers (such as uranium-235) into two or more lighter and smaller nuclei. Each fission releases neutrons, which cause more nuclei to fission. The resulting cascade of fissions produces a *chain reaction* that releases an enormous amount of energy in a short time (Figure 13.15). The heat released by the chain reaction of fissions inside the reactor of a nuclear power plant is used to convert water into steam, which spins a turbine that generates electricity. Most nuclear-generated electricity is produced by light-water reactors (LWRs, see Figure 13.16).

The fuel for a nuclear reactor is made from uranium ore mined from the earth's crust. After it is mined, the ore must be enriched to increase the concentration of its fissionable material (uranium-235) from 1% to about 5%. The enriched uranium-235 is processed into small pellets of uranium dioxide. Each pellet, about the size of an eraser on a pencil, contains as much energy as a ton of coal. Large numbers of pellets are packed into closed pipes, called *fuel rods*. The rods are bundled together in *fuel assemblies* and placed in the core of a nuclear reactor.

Plant operators move *control rods* in and out of the reactor core to absorb more or fewer neutrons in the nuclear fission chain reaction and regulate how much power is produced. A *coolant,* usually water, circulates through the reactor's core to remove heat and prevent the fuel rods and other reactor components from melting and releasing massive amounts of radioactivity into the environment. An emergency core cooling system can flood the reactor core with water to help prevent such meltdowns of the reactor's highly radioactive core from a loss of cooling water.

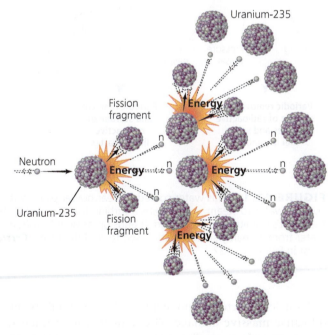

FIGURE 13.15 A nuclear fission chain reaction releases a large amount of energy inside the reactor of a nuclear power plant.

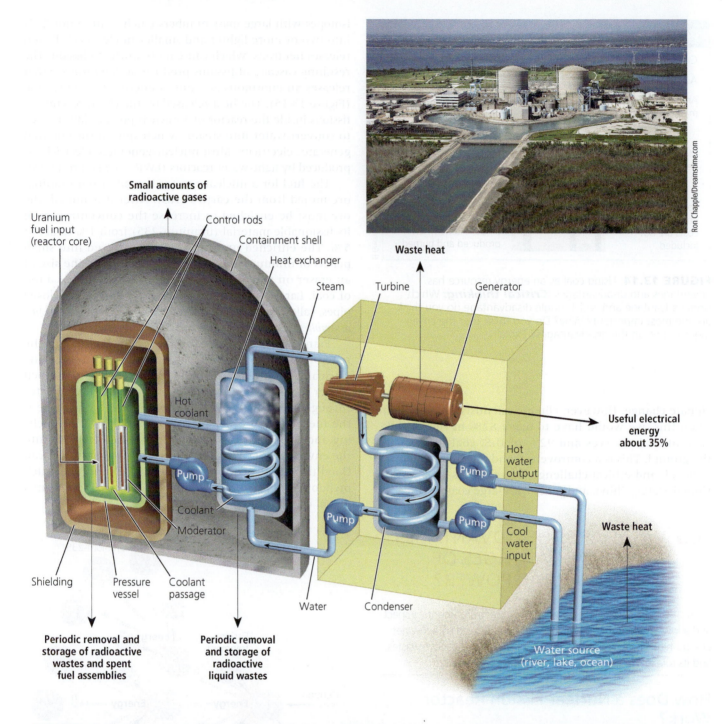

Small amounts of radioactive gases

Uranium fuel input (reactor core)

Control rods

Containment shell

Heat exchanger

Steam

Turbine

Generator

Waste heat

Hot coolant

Useful electrical energy about 35%

Pump

Pump

Coolant

Hot water output

Moderator

Pump

Pump

Shielding

Pressure vessel

Coolant passage

Water

Condenser

Cool water input

Waste heat

Periodic removal and storage of radioactive wastes and spent fuel assemblies

Periodic removal and storage of radioactive liquid wastes

Water source (river, lake, ocean)

Ron Chapple/Dreamstime.com

FIGURE 13.16 This water-cooled nuclear power plant, with a pressurized water reactor, produces intense heat that is used to convert water to steam, which spins a turbine that generates electricity. About 65% of the energy released by the nuclear fission of the plant's uranium fuel is wasted. It ends up as heat that flows into the atmosphere, usually through large cooling towers or into water from a nearby source that is used to cool the plant. *Critical thinking:* How does this plant differ from the coal-burning plant in Figure 13.12?

A nuclear reactor cannot explode like an atomic bomb and cause massive damage. The danger from a nuclear reactor comes from smaller explosions or a core meltdown because of a loss of coolant water that can release radioactive materials into the environment.

A *containment shell* made of thick, steel-reinforced concrete surrounds the reactor core. It is designed to keep radioactive materials from escaping into the environment if an internal explosion or a core meltdown occurs. The shell also protects the core from external threats such as tornadoes and plane crashes. These essential safety features and the 10 or more years it typically takes to build a nuclear power plant reduce the overall net energy of nuclear power and helps explain why a new nuclear power plant can cost $9 billion to $11 billion.

In 2016, the world's four leading producers of nuclear power were, in order, the United States, France, Russia, and China. France generates 75% of its electricity and the United States generates 20% of its electricity using nuclear power.

What Is the Nuclear Fuel Cycle?

Building and running a nuclear power plant is only one part of the **nuclear fuel cycle** (Figure 13.17). This cycle includes the **(1)** mining of uranium, **(2)** processing and enriching the uranium to make the nuclear fuel, **(3)** using it in a reactor, **(4)** safely storing the highly radioactive spent fuel rods for thousands of years until their radioactivity falls to safe levels, and **(5)** retiring the worn-out plant by taking it apart and storing its high- and moderate-level radioactive parts safely for thousands of years.

As long as a reactor is operating safely, the power plant itself has a low environmental impact and little risk of an accident. However, considering the entire nuclear fuel cycle, the potential environmental impact increases. In evaluating the safety, economic feasibility, net energy, and overall environmental impact of nuclear power, energy experts and economists caution us to look at the entire nuclear fuel cycle, not just the power plant itself. Figure 13.18 lists the major advantages and disadvantages of using the nuclear fuel cycle to produce electricity (**Concept 13.3**).

A major problem with nuclear power is the high cost of building the plant and operating the nuclear fuel cycle, which leads to a low net energy. As a result, nuclear power cannot compete in the marketplace with other energy resources such as natural gas, wind, and solar cells unless it receives large government subsidies and tax breaks. An increasing number of existing nuclear plants in the United States are being closed down because electricity can be produced more cheaply by burning natural gas and in some areas by wind and solar energy.

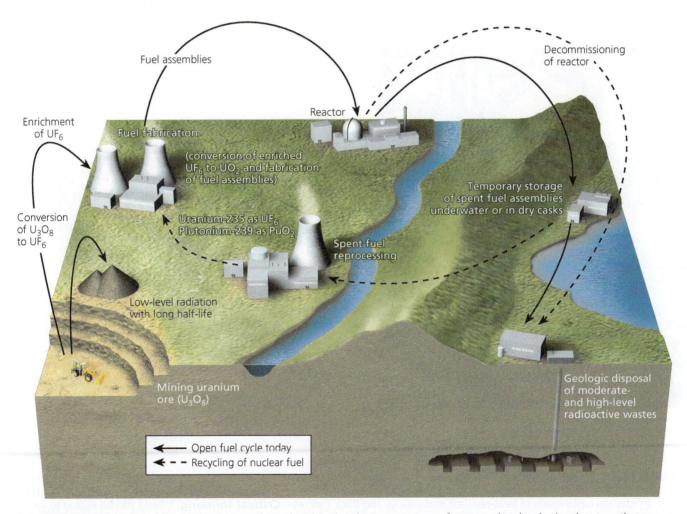

FIGURE 13.17 Using nuclear power to produce electricity involves a sequence of steps and technologies that together are called the *nuclear fuel cycle*. **Critical thinking:** Do you think the market price of nuclear-generated electricity should include all the costs of the nuclear fuel cycle, in keeping with the full-cost pricing **principle of sustainability**? Explain.

Trade-Offs

Conventional Nuclear Fuel Cycle

Advantages	Disadvantages
Low environmental impact (without accidents)	Low net energy
	High overall cost
Emits 1/6 as much CO_2 as coal	Produces long-lived, harmful radioactive wastes
Low risk of accidents in modern plants	Promotes spread of nuclear weapons

FIGURE 13.18 Using the nuclear fuel cycle (Figure 13.17) to produce electricity has advantages and disadvantages. **Critical thinking:** Which single advantage and which single disadvantage do you think are the most important? Why? Do you think that the advantages outweigh the disadvantages? Explain.

Dealing with Radioactive Nuclear Wastes

After 3 to 4 years, the enriched uranium fuel in a typical nuclear reactor becomes *spent*, or useless, and must be replaced. The spent-fuel rods are so thermally hot and highly radioactive that they cannot be thrown away. Researchers have found that 10 years after being removed from a reactor, a single spent-fuel rod assembly can still emit enough radiation to kill a person standing 1 meter (39 inches) away in less than 3 minutes.

After spent-fuel rod assemblies are removed from reactors, they are stored in *water-filled pools* (Figure 13.19, left). After several years of cooling and decay of some of their radioactivity, the rods can be transferred to *dry casks* made of heat-resistant metal alloys and concrete and filled with inert helium gas (Figure 13.19, right). These casks are licensed for 20 years and could last for 100 or more years. Even so, this is a tiny fraction of the thousands of years required to safely store this high-level radioactive waste.

Spent nuclear fuel rods can also be processed to remove their radioactive plutonium, which can then be used as

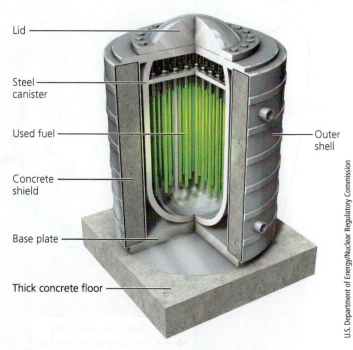

Lid
Steel canister
Used fuel
Outer shell
Concrete shield
Base plate
Thick concrete floor

FIGURE 13.19 After 3 or 4 years in a reactor, spent-fuel rods are removed and stored in a deep pool of water contained in a steel-lined concrete basin (left) for cooling. After about 5 years of cooling, the fuel rods can be stored upright on concrete pads (right) in sealed dry-storage casks made of heat-resistant metal alloys and thick concrete. **Critical thinking:** Would you be willing to live within a block or two of these casks or have them transported through the area where you live in the event that they were transferred to a long-term storage site? Explain. What are the alternatives?

nuclear fuel or for making nuclear weapons, thus closing the nuclear fuel cycle (Figure 13.17). This reprocessing reduces the storage time for the remaining wastes from up to 240,000 years to about 10,000 years. (For comparison, modern humans appeared about 200,000 years ago.)

However, reprocessing is costly. It also produces bomb-grade plutonium that nations could use to make nuclear weapons, as India did in 1974. This is mainly why the U.S. government, after spending billions of dollars, abandoned this nuclear fuel recycling approach in 1977. In recent years, France, Russia, Japan, India, the United Kingdom, and China have reprocessed some of their nuclear fuel.

Most scientists and engineers agree in principle that underground burial is the safest way to store high-level radioactive wastes. Between 1987 and 2009, the DOE spent $12 billion on research and testing for a long-term underground nuclear waste storage site in the Yucca Mountain desert region of Nevada. In 2010, this taxpayer-funded project was abandoned for scientific, economic, and political reasons. A government panel is looking for alternative solutions and sites and some call for completing the Yucca Mountain site. Meanwhile highly radioactive nuclear wastes are accumulating, with about 78% stored in pools and 22% stored in dry casks (Figure 13.19, right), which secure the wastes for 100 years at most. Thus, after 60 years, the United States has not come up with a scientifically, economically, and politically acceptable solution for storing its high-level radioactive wastes for thousands of years.

Another costly radioactive waste problem arises when a nuclear power plant reaches the end of its useful life after 40 to 60 years. At that point, the plant must be *decommissioned* or retired. Globally, 285 of the world's 448 commercial nuclear reactors operating in 2016 will have to be decommissioned by 2025.

Scientists and engineers have proposed three ways to decommission nuclear power plants: **(1)** remove and store the highly radioactive parts in a permanent, secure repository (which does not yet exist), **(2)** install a physical barrier around the plant and set up full-time security for 30 to 100 years before dismantling the plant and storing its radioactive parts in a repository, and **(3)** enclose the entire plant in a concrete and steel-reinforced tomb, called a containment structure.

The last option was used with a reactor at the Chernobyl nuclear power plant in Ukraine after it exploded and nearly melted down in 1986, due to a combination of poor reactor design and human operator error. The explosion and the radiation released over large areas killed hundreds and perhaps thousands of people and contaminated a large area of land with long-lasting radioactive fallout. A few years after the containment structure was built, it began to crumble and leak radioactive wastes. The structure is being rebuilt at great cost and, in 2016, was surrounded by a gigantic hangar-like structure that is unlikely to last more than a century. Thirty years after the accident, an area surrounding the reactor roughly the size of Rhode Island remains off limits because of radioactive fallout.

The high cost of retiring nuclear plants adds to the enormous cost of the nuclear power fuel cycle and reduces its already low net energy. Even if all the nuclear power plants in the world were shut down tomorrow, their high-level radioactive wastes and components would need to be safely stored for thousands of years.

Can Nuclear Power Slow Climate Change?

Supporters of nuclear power claim that increased use of this energy resource could greatly reduce CO_2 emissions that contribute to climate change because, they argue, nuclear power is a carbon-free energy resource. Scientists point out that this is only partially correct. While nuclear plants are operating, they do not emit CO_2 but the 10-year process of building a plant and every other step in the nuclear fuel cycle (Figure 13.17) involves CO_2 emissions. Such emissions are much lower than those from coal-burning power plants, but they still contribute to atmospheric warming and climate change. In other words, the nuclear power fuel cycle is not a carbon-free source of electricity.

Nuclear Power's Future Is Uncertain

After more than 60 years of development, a huge financial investment, and enormous government subsidies, 448 commercial nuclear reactors in 30 countries produced only 4% of the world's commercial energy and 11% of its electricity in 2016. In the United States, 99 government-licensed commercial nuclear power reactors in 30 states produced about 8% of the country's overall energy and 20% of its electricity in 2016.

Globally in 2016, 61 new nuclear reactors were under construction (20 of them in China). Another 156 reactors are planned (most of them in China) but even if completed after a decade or two, they will not replace the world's 285 aging reactors scheduled for decommissioning by 2025. This helps explain why the global production of electricity from nuclear power is the world's slowest-growing form of commercial energy (Figure 13.20) and is projected to grow little between 2016 and 2035, according to the IEA. In the United States, electricity produced from nuclear power has not grown since 2000 and is not expected to grow between 2016 and 2035 because of the costs involved and because electricity can be produced much more quickly and cheaply from natural gas, wind, and solar cells.

There is controversy over the future of nuclear power. Critics argue that its three most serious problems are its high cost, low net energy of the nuclear power fuel cycle, and its contribution to the spread of technology that can be used to make nuclear weapons. They contend that the

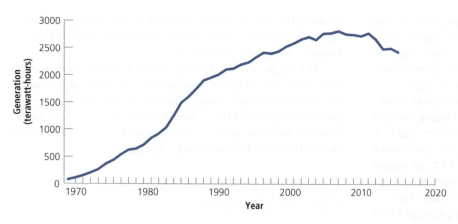

FIGURE 13.20 Global electricity generation from nuclear power, 1970–2015. **Data analysis:** By what percentage did electricity produced by nuclear power decrease between 2006 and 2015?

(Compiled by the authors using data from the International Energy Agency, BP, Worldwatch Institute, and Earth Policy Institute)

nuclear power industry could not exist without high levels of financial support from governments and taxpayers, because of the high cost of ensuring safety and the low net energy of the nuclear fuel cycle (Figure 13.18).

For example, the U.S. government has provided large research and development subsidies, tax breaks, and loan guarantees to the nuclear power industry (with taxpayers taking on the risk of any debt defaults) for more than 50 years. In addition, the government provides accident insurance guarantees (under the Price-Anderson Act passed by Congress in 1957), because insurance companies have refused to fully insure any U.S. nuclear reactor against the effects of a catastrophic accident.

According to the nonpartisan Congressional Research Service, since 1948, the U.S. government has spent more than $95 billion (in 2011 dollars) on nuclear energy research and development (R & D). This is more than four times the amount spent on R & D for solar, wind, geothermal, biomass, biofuels, and hydropower combined. Many people question the need for continuing such taxpayer support for nuclear power, especially since its energy output has not grown for several decades. It is unlikely to grow significantly during the next several decades because electricity can be produced more cheaply from natural gas, wind power, and solar cells according to the IEA and the DOE.

A serious national and global security concern related to commercial nuclear power is the spread of nuclear weapons technology. The United States and eight other countries have been selling commercial and experimental nuclear reactors and uranium fuel-enrichment and waste reprocessing technology to other countries for decades. According to energy expert John Holdren, the 60 countries that have nuclear weapons or the knowledge to develop them have gained most of such information by using civilian nuclear power technology. Some critics view this threat to global security as the single most important reason for not building more nuclear power plants that use the fissionable isotopes uranium-235 or plutonium-239 as a fuel or that produce plutonium-239, all of which can be used to make nuclear weapons.

Because of the multiple built-in safety features, the risk of exposure to radioactivity from nuclear power plants in the United States and in most other more-developed countries is very low. However, between 1952 and 2016, several serious nuclear accidents have occurred, including some that involved explosions and partial or complete reactor core meltdowns. They include major accidents at Pennsylvania's Three Mile Island nuclear plant in 1979, Ukraine's Chernobyl nuclear power plant in 1986, and Japan's Fukushima Daiichi nuclear power plant in 2011. These and other accidents have dampened public, government, and investor confidence in nuclear power. For example, the 2011 nuclear power accident in Japan prompted Germany, Switzerland, and Belgium to announce plans for phasing out nuclear power.

Proponents of nuclear power argue that governments should continue funding research, development, and pilot-plant testing of potentially safer and less costly new types of reactors. The nuclear industry claims that hundreds of new *advanced light-water reactors (ALWRs)* could be built in just a few years. ALWRs have built-in safety features designed to make meltdowns and releases of radioactive emissions almost impossible and thus do not need expensive automatic cooling. The industry is also evaluating the development of smaller modular light-water reactors— about the size of a school bus—that could be built in a factory, delivered to a site, and installed underground. As of 2016, no commercial versions of any of the proposed new-generation reactors have been built or evaluated.

Some scientists call for replacing today's uranium-based reactors with new ones to be fueled by thorium. They contend that such reactors would be much less costly and safer because they cannot melt down. In addition, the nuclear waste they produce does not contain fissionable isotopes that can be used to make nuclear weapons. China plans to explore this option.

Is Nuclear Fusion the Answer?

Some proponents of nuclear power hope to develop nuclear fusion. In **nuclear fusion** the nuclei of two isotopes of a light element such as hydrogen are forced together at extremely high temperatures until they fuse to form a heavier nucleus, releasing energy in the process (Figure 13.21).

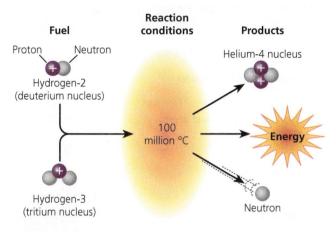

FIGURE 13.21 The nuclear fusion of two isotopes of a light element such as hydrogen releases a huge amount of energy.

With nuclear fusion, there would be no risk of a meltdown or of a release of large amounts of radioactive materials, and little risk of the additional spread of nuclear weapons. Fossil fuels would not be needed to produce electricity, thereby eliminating most of the earth's air pollution and climate-changing CO_2 emissions from power plants. Fusion power also might be used to destroy toxic wastes and to supply electricity for desalinating water and for decomposing water to produce clean burning hydrogen fuel.

However, in the United States, after more than 50 years of research and a $25 billion investment (mostly by the government), controlled nuclear fusion is still in its infancy. None of the tested approaches have produced more energy than they used.

In 2006, the United States, China, Russia, Japan, South Korea, India, and the European Union agreed to spend at least $12.8 billion in a joint effort to build a large-scale experimental nuclear fusion reactor by 2026. The goal is to determine if nuclear fusion can have a high net energy at an affordable cost. By 2014, the estimated cost of this project had doubled and it was far behind schedule. Unless there is an unexpected scientific breakthrough, some skeptics say, "Nuclear fusion is the power of the future and always will be."

13.4 WHY IS ENERGY EFFICIENCY AN IMPORTANT ENERGY RESOURCE?

CONCEPT 13.4A Increasing energy efficiency and reducing energy waste could save at least a third of the energy used in the world and up to 43% of the energy used in the United States.

CONCEPT 13.4B We have a variety of technologies for significantly increasing the energy efficiency of industrial operations, motor vehicles, appliances, and buildings.

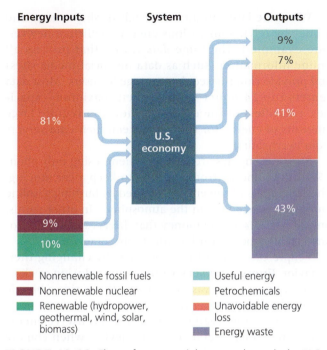

FIGURE 13.22 Flow of commercial energy through the U.S. economy. Only 16% of the country's high-quality energy ends up performing useful tasks. ***Critical thinking:*** What are two examples of unnecessary energy waste?

(Compiled by the authors using data from U.S. Department of Energy.)

We Waste a Lot of Energy and Money

Improving energy efficiency and conserving energy are key strategies reducing energy waste. **Energy efficiency** is a measure of how much useful work we can get from each unit of energy. Improving energy efficiency means using less energy to provide the same amount of work. People can do this by using more fuel-efficient cars, light bulbs (such as LED bulbs), appliances, computers, and industrial processes.

No energy-using device operates at 100% efficiency because some energy is always lost to the environment as low-quality heat, as required by the second law of thermodynamics (Chapter 2, p. 38). About 84% of all commercial energy used in the United States is wasted (Figure 13.22). About 41% of this energy unavoidably ends up as low-quality waste heat lost to the environment because of the degradation of energy quality imposed by the second law of thermodynamics. The other 43% is wasted unnecessarily, mainly due to the inefficiency of industrial motors, most motor vehicles, power plants, light bulbs, and numerous other energy-consuming devices (**Concept 13.4A**). This wasted energy is the country's largest untapped source of energy and reducing this huge waste of energy would save consumers money.

43% Percentage of energy used in the United States that is unnecessarily wasted.

We waste large amounts of high-quality energy and money by relying on various energy-inefficient devices. One example is the huge *data centers* that process all online information (such as data on social media sites) and provide cloud-based data storage for users. Most data centers run 24 hours a day at their maximum capacities, regardless of the demand. They also require large amounts of energy for cooling to keep their data servers from overheating.

Most car engines are also inefficient, using only about 25% of the energy from gasoline to keep a car moving. The other 75% of the energy released by burning gasoline ends up as waste heat in the atmosphere. In other words, only about 25% of the money that drivers spend on gasoline actually goes toward getting them somewhere.

People can also cut energy waste by changing their behavior. **Energy conservation** means reducing or eliminating unnecessary waste of energy. If you ride a bicycle to school or work rather than driving a car, you are practicing energy conservation. You can also conserve energy by turning off lights and electronic devices when you are done using them.

Many people waste energy and money by living and working in poorly insulated and leaky houses and buildings that require extra heating in cold weather and extra cooling in hot weather. Three of every four Americans commute to work, mostly in energy-inefficient vehicles, and only 5% of commuters use more energy-efficient mass transit.

Improving energy efficiency and conserving energy have numerous economic, health, and environmental benefits (Figure 13.23). In 2017, energy efficiency jobs in the United States numbered 2.2 million, more than 12 times the number of jobs provided by the U.S. coal industry. To most energy analysts, *they are the quickest, cleanest, and cheapest ways to provide more energy, reduce pollution and environmental degradation, and slow climate change and ocean acidification* (**Concept 13.4B**).

Improving Energy Efficiency in Industries and Utilities

Industry accounts for about 30% of the world's energy consumption and 33% of U.S. energy consumption. Industries that use the most energy are those that produce petroleum, chemicals, cement, steel, aluminum, and paper and wood products.

Industries and utility companies can save energy by using **cogeneration** to produce two useful forms of energy from the same fuel source. For example, the steam used for generating electricity in an industrial or power plant can be captured and used to heat the plant or nearby buildings. The energy efficiency of these systems is 60–80%, compared to 25–35% for coal-fired and nuclear power plants. Denmark uses cogeneration to produce 53% of its electricity compared to 12% in the United States.

Solutions

Improving Energy Efficiency

- Prolongs fossil fuel supplies

- Reduces oil imports and improves energy security

- Very high net energy

- Low cost

- Reduces pollution and environmental degradation

- Buys time to phase in renewable energy

- Creates local jobs

FIGURE 13.23 Improving energy efficiency and conserving energy can have important benefits. *Critical thinking:* Which two of these benefits do you think are the most important? Why?

Top: Dmitriy Raykin/Shutterstock.com. Center: V. J. Matthew/Shutterstock.com. Bottom: andrea lehmkuhl/Shutterstock.com.

Industries can also save energy and money by using more energy-efficient variable-speed electric motors. In contrast, standard electric motors run at full speed with their output throttled to match the task. This is somewhat like using one foot to push the gas pedal to the floorboard of your car and putting your other foot on the brake pedal to control its speed.

Recycling materials such as steel and other metals can also save industries energy and money and reduce their harmful environmental impact. For example, producing steel from recycled scrap iron uses 75% less high-quality energy than does producing steel from virgin iron ore, and it emits 40% less CO_2. Industries can also save energy by using energy-efficient LED lighting and shutting off computers, printers, and lights when they are not being used.

A growing number of major corporations are saving money by improving energy efficiency. For example, between 1990 and 2014, Dow Chemical Company, which operates 165 manufacturing plants in 37 countries, saved $27 billion in a program to improve energy efficiency. Ford Motor Company saves $1 million a year by turning off computers that are not in use.

GOOD NEWS

Building a Smarter and More Energy-Efficient Electrical Grid

In the United States, electricity is delivered to consumers through an *electrical grid* consisting of a network of transmission and distribution lines. The U.S. electrical grid system was designed more than 100 years ago and is inefficient and outdated. As former U.S. energy secretary Bill Richardson observed, "We're a major superpower with a third-world electrical grid system."

There is increasing pressure to convert and expand the current U.S. electrical grid system into a *smart grid*. This new grid will be digitally controlled, transmit ultra-high-voltage, and have superefficient transmission lines. It will be less vulnerable to power outages because it will quickly adjust for a major power loss in one area by automatically rerouting available electricity from other parts of the country. A national network of wind farms and solar cell power plants connected to a smart grid would make the sun and wind reliable sources of electricity around the clock. Smart electricity meters will allow consumers to save money by reducing electricity use during times when rates are high.

According to the DOE, building such a nationwide grid will cost up to $800 million over the next 20 years. However, it will save the U.S. economy at least $2 trillion during that period. So far, Congress has not authorized significant funding for this vital component of the country's energy and economic future.

Making Transportation More Energy-Efficient

In 1975, the U.S. Congress established Corporate Average Fuel Economy (CAFE) standards to improve the average fuel economy of new cars and light trucks, vans, and sport utility vehicles (SUVs) in the United States. Between 1973 and 2015, these standards increased the average fuel economy for such vehicles from 5 kilometers per liter, or *kpl* (11.9 miles per gallon, or *mpg*) to 10.6 kpl (24.9 mpg). The average government fuel-economy goal for such vehicles is 23.3 kpl (54.5 mpg) by 2025, which according to the EPA would provide $100 billion of benefits from reduced air pollution while lowering carbon dioxide emissions and reducing oil imports. However, since 2017 automakers have been pressuring the EPA and the U.S. Congress to decrease these fuel efficiency standards.

Energy experts project that by 2040, all new cars and light trucks sold in the United States could get more than 43 kpl (100 mpg), using available technology (**Concept 13.4B**). Achieving this level of fuel efficiency is an important way to reduce energy waste, save consumers' money, cut air pollution, and slow climate change and ocean acidification.

However, many consumers buy energy-inefficient sport-utility vehicles (SUVs), minivans, and pickup trucks, which are more profitable for automakers, especially when gasoline prices fall. One reason is that most consumers do not realize that gasoline costs them much more than what they pay at the pump. A number of *hidden costs* are not included in the market price of gasoline. Such costs include government subsidies and tax breaks for oil companies, car manufacturers, and road builders. They also include costs related to pollution control and cleanup and higher medical bills and health insurance premiums resulting from illnesses caused by pollution related to the production and use of motor vehicles.

The International Center for Technology Assessment estimates that the hidden costs of gasoline for U.S. consumers amount to $3.18 per liter ($12 per gallon). If people were more aware of these costs, they might be more motivated to save money and improve environmental quality by purchasing fuel-efficient vehicles.

One way to include more of these hidden costs in the market price is through higher gasoline taxes— an application of the full-cost pricing **principle of sustainability**. However, higher gas taxes are politically unpopular, especially in the United States. Some analysts call for increasing U.S. gasoline taxes while reducing payroll and income taxes to offset any additional financial burden to consumers. Another way for governments to encourage higher energy efficiency is to give consumers significant tax breaks or other economic incentives to encourage them to buy more fuel-efficient vehicles.

Other ways to save energy and money in transportation include building or expanding mass transit systems within cities, constructing high-speed rail lines between cities (as is done in Japan, much of Europe, and China), and carrying more freight by rail instead of in heavy trucks. Another approach is to encourage bicycle use by building bike lanes along highways and city streets.

Switching to Energy-Efficient Vehicles

Energy-efficient vehicles are available (**Concept 13.4B**). One example is the gasoline-electric *hybrid car* (Figure 13.24, left). It has a small gasoline-powered engine and a battery-powered electric motor that provides the energy needed for acceleration and hill climbing. The most efficient models get a combined city/highway mileage of up to 23 kpl (54 mpg) and emit about 65% less CO_2 per kilometer driven than do comparable conventional cars.

Another option is the *plug-in hybrid electric vehicle* (Figure 13.24, right). Available models can travel 21–33 kilometers (13–21 miles) on electricity alone. Then the small gasoline motor kicks in, recharges the battery, and extends the driving range to 600 kilometers (370 miles) or more. The battery can be plugged into a conventional 110-volt outlet and be fully charged in 6 to 8 hours, or faster using a 220-volt outlet.

According to a DOE study, replacing most of the current U.S. vehicle fleet with plug-in hybrid vehicles over 3 decades would cut U.S. oil consumption by 70–90%,

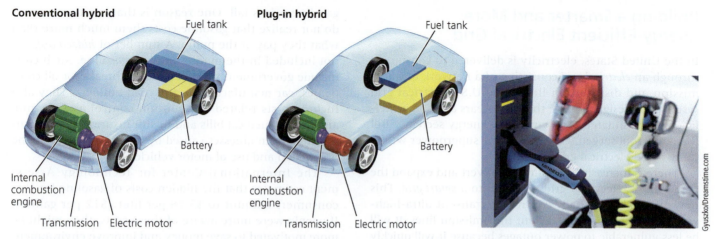

Conventional hybrid

Fuel tank

Internal combustion engine

Transmission   Electric motor

Battery

Plug-in hybrid

Fuel tank

Internal combustion engine

Transmission   Electric motor

Battery

FIGURE 13.24 Solutions: A *conventional gasoline-electric hybrid* vehicle (left) is powered mostly by a small internal combustion engine with an assist from a strong battery. A *plug-in hybrid electric* vehicle (center) typically has a smaller internal combustion engine with a second and more powerful battery that can be plugged into a 110-volt or 220-volt outlet and recharged (right). An all-electric vehicle (not shown) runs completely on a rechargeable battery. ***Critical thinking:*** Would you buy one of these vehicles? Explain.

eliminate the need for any costly oil imports, save consumers money, and reduce CO_2 emissions by 27%. Recharging the batteries in these cars mostly by electricity generated by renewable resources such as wind turbines, solar cells, or hydroelectric power would cut U.S. emissions of CO_2 by 80–90%. This would slow climate change and ocean acidification and save thousands of lives by reducing air pollution from motor vehicles and coal-burning power plants.

Another option is an *all-electric vehicle* that runs on a battery only. In 2015, the typical all-electric vehicle could go 161 kilometers (100 miles) on a single battery charge. By 2017, some all-electric vehicles had a range of about 322 kilometers (200 miles) per charge.

The problem for the average consumer is that the prices of hybrid, plug-in hybrid, and all-electric cars are high because of the high cost of their batteries. The key to greatly increasing the use of hybrid, plug-in hybrid, and all-electric motor vehicles is greatly increased research and development of better and more affordable batteries (Science Focus 13.2). Building a network of recharging stations within and between communities will also increase the use of battery-powered vehicles. If the federal government reduces fuel efficiency standards, carmakers may reduce their introduction of electric cars, which help them meet higher fuel-efficiency standards.

Another potential alternative fuel resource is the **hydrogen fuel cell,** which could be used to power all-electric vehicles. This device uses hydrogen gas (H_2) as a fuel to produce electricity when it reacts with oxygen gas (O_2) in the atmosphere and emits harmless water vapor. A fuel cell is much more energy efficient than an internal combustion engine, has no moving parts, and requires little maintenance. The H_2 fuel is usually produced by passing electricity through water, or it can be produced from methane stored in a vehicle. Two major problems

with fuel cells are that they are expensive and that H_2 has a negative net energy, which means that it takes more energy to produce it than it can provide, as discussed later in this chapter. **GREEN CAREER: Fuel-cell technology**

CONSIDER THIS ...

LEARNING FROM NATURE

Blue-green algae use sunlight and an enzyme to produce hydrogen from water. Scientists are evaluating this as a way to produce hydrogen fuel for motor vehicles and for home heating without the use of costly high-temperature processes or electricity. If successful, this would sharply reduce emissions of CO_2 and other air pollutants.

Reducing the weight of a vehicle is another way to improve fuel efficiency. Car bodies can be made of *ultralight* and *ultrastrong* composite materials such as fiberglass, carbon fiber, hemp-fiber, and graphene (Science Focus 12.1, p. 315). They are also safer in a crash than cars with conventional bodies. The current cost of making such car bodies is high, but technological innovations and mass production would likely bring the cost down.

Energy conservation can also play a role. Since cars are the biggest energy user for most Americans, shifting to a more fuel-efficient car that gets at least 17 kpl (40 mpg) is one of the best ways to save money and reduce one's harmful environmental impact.

Designing Buildings That Save Energy and Money

Worldwide, buildings are responsible for more than 40% of global energy use and up to one-third of the world's greenhouse gas emissions, according to a 2015 study by

The Search for Better Batteries

The major obstacle standing in the way of wider use of plug-in hybrid-electric and all-electric vehicles is the need for an affordable, small, lightweight, and easily rechargeable car battery that can store enough energy to power a vehicle for long-distance trips.

Lithium-ion batteries are light (because lithium is the lightest solid chemical element) and can pack a lot of energy into a small volume. Many of them are hooked together and used to power hybrid, plug-in hybrid, and all-electric motor vehicles. However, they take a long time to recharge, and must be replaced every few years. Lithium-ion batteries are also expensive but prices dropped by 50% between 2010 and 2015.

Researchers at the Massachusetts Institute of Technology (MIT) have developed a new type of lithium-ion battery using nanotechnology (Science Focus 12.1, p. 315). It is less expensive and can be charged 40 times faster than the batteries that are used to power many of today's hybrid vehicles. Researchers at the University of California, Irvine, are using nanowires to develop a lithium-ion battery with an average life span of 300–400 years compared to the current life span of 3 years.

Scientists have also developed *ultracapacitors,* which are small mechanical batteries consisting of two metal surfaces separated by an electric insulator. They quickly store and release large amounts of electrical energy to supply the power needed for quick acceleration and hill climbing. They can be recharged in minutes, can hold a charge much longer than conventional batteries, and do not have to be replaced as frequently as conventional batteries.

If any one or a combination of these or other new battery technologies can be mass-produced at an affordable cost, plug-in hybrid and all-electric vehicles could take over the car and truck market within a few decades. This would greatly reduce three of the world's most serious environmental problems: air pollution, climate-changing CO_2 emissions, and ocean acidification. **GREEN CAREER: Battery engineer**

CRITICAL THINKING

How might your life change if better and cheaper battery technologies become a reality?

---

the United Nations Environment Program. Green architecture is a rapidly growing new field. It focuses on building design that is energy efficient, resource efficient, and cost efficient (**Concept 13.4B**). Green architecture makes use of natural lighting, direct solar heating, insulated windows, smart thermostats, and energy-efficient appliances and lighting. It also focuses on using nontoxic building materials and recycled materials from deconstructed buildings. Some green buildings use solar hot water heaters and get all or most of their electricity from solar cells.

Some homes and urban buildings have *living roofs,* or *green roofs,* (Figure 13.25) covered with specially designed soil and vegetation that is watered with a smart-controlled drip irrigation system. Living roofs reduce the costs of cooling and heating a building by absorbing heat from the summer sun and helping insulate the structure and retain heat in the winter. Roofs painted white on houses, factories, and other buildings reflect solar energy and decrease air conditioning use and costs.

A key goal of many green buildings is to produce as much energy as they use each year—a concept known as zero net energy. Other goals are net zero net water, and zero net carbon.

The three best tools for reducing the waste of energy and money in houses are a thermal imaging camera, which can reveal heat flowing out of a house (Figure 13.26), insulation, and a caulking gun, used to seal air leaks, thereby reducing heat losses. *Superinsulation* is important in energy-efficient design. A house can be so heavily insulated and airtight that heat from direct sunlight, appliances, and human bodies can warm it with little or no need for a backup heating system, even in extremely cold climates. Superinsulated houses in Sweden use 90% less energy for heating and cooling than do typical American homes of the same size.

One example of superinsulation is straw-bale construction in which the walls of a house are built of straw bales that are covered on the inside and outside with adobe (Figure 13.27). Such walls can have insulating values of two to six times those of conventional walls.

The World Green Building Council and the U.S. Green Building Council's Leadership in Energy and Environmental Design (LEED) have developed standards for certifying that a building meets certain energy efficiency and environmental standards. **GREEN CAREERS: Sustainable environmental design and architecture**

Saving Energy and Money in Existing Buildings

Here are ways to save energy and money in existing homes and other buildings (**Concept 13.4B**):

- *Use an energy audit to detect air leaks.*
- *Use energy-efficient windows.*

FIGURE 13.25 Green roof on Chicago's City Hall.

- *Seal leaky heating and cooling ducts in attics and unheated basements.*
- *Insulate the building and plug air leaks.* About one-third of the heated air in typical U.S. homes and other buildings escapes through holes, cracks, and single-pane windows (Figure 13.26). During hot weather, these windows and cracks let in heat, which increases the need for air conditioning.
- *Heat interior spaces more efficiently.* In order, the most energy-efficient ways to heat indoor space are: superinsulation (including plugging leaks); a

geothermal heat pump that transfers heat stored underground into a home; passive solar heating; a high-efficiency, conventional heat pump (in warm climates only); and a high-efficiency natural gas furnace.

- *Heat water more efficiently.* One option is to use a roof-mounted solar hot water heater. Another option is a *tankless instant water heater* about the size of a suitcase. These heaters burn natural gas or LPG (but not an electric heater, which is inefficient) to deliver hot water as it is needed instead of keeping water in a

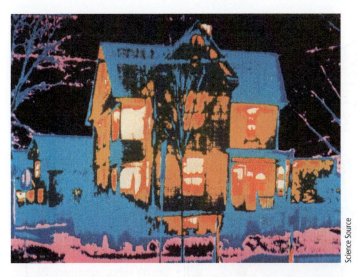

FIGURE 13.26 A *thermogram,* or infrared photo, of a house that is poorly insulated and sealed and loses heat (reddish areas) and wastes money. **Critical thinking:** How do you think the place where you live would compare to this house in terms of heat loss?

large tank hot all the time. They have been used for decades in many European countries.

- *Use energy-efficient appliances.* A refrigerator with its freezer located in its bottom uses about half as much energy as one with the freezer on top or on the side, which allows dense cold air to flow out quickly when the door is opened. Microwave ovens use 25–50% less electricity than electric stoves. Front-loading clothes washers use 55% less energy and 30% less water than top-loading models use and cut operating costs in half.

- *Use energy-efficient computers.* According to the EPA, if all computers sold in the United States met its Energy Star requirements, consumers would save $1.8 billion a year in energy costs and reduce greenhouse gas emissions by an amount equal to that of taking about 2 million cars off the road.

- *Use energy-efficient lighting.* The DOE estimates that switching to energy-efficient LED bulbs over the next 20 years in the United States would save consumers large amounts of money and eliminate the need to build 40 new power plants. In recent years, the cost of LED bulbs has fallen by 90%. They last 25 times longer than traditional incandescent bulbs and 2.5 times longer than compact fluorescent bulbs.

- *Stop using the standby mode.* Consumers can reduce their energy use and their monthly power bills by plugging their standby electronic devices into a smart power strip that cuts off power to a device when it detects that the device has been turned off.

Figure 13.28 lists ways that you can save energy and money in the place where you live.

Why Are We Wasting So Much Energy and Money?

Considering its impressive array of benefits (Figure 13.23), why is there so little emphasis on improving energy efficiency and conserving energy? One reason is that energy resources such as fossil fuels and nuclear power are artificially cheap because of the government subsidies and tax breaks they receive and because their market prices do not

FIGURE 13.27 Solutions: Energy-efficient, straw-bale house in Crested Butte, Colorado, during construction (left) and after it was completed (right). **Question:** Would you like to live in such a house? Explain.

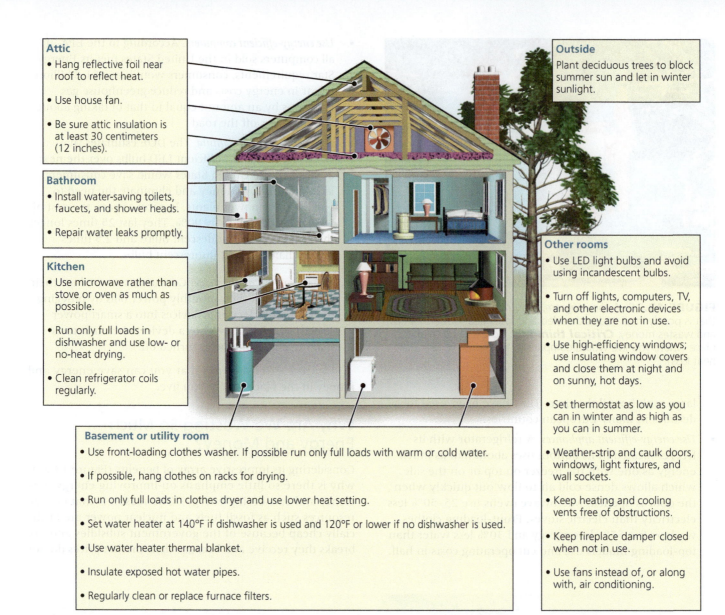

Attic
- Hang reflective foil near roof to reflect heat.
- Use house fan.
- Be sure attic insulation is at least 30 centimeters (12 inches).

Bathroom
- Install water-saving toilets, faucets, and shower heads.
- Repair water leaks promptly.

Kitchen
- Use microwave rather than stove or oven as much as possible.
- Run only full loads in dishwasher and use low- or no-heat drying.
- Clean refrigerator coils regularly.

Basement or utility room
- Use front-loading clothes washer. If possible run only full loads with warm or cold water.
- If possible, hang clothes on racks for drying.
- Run only full loads in clothes dryer and use lower heat setting.
- Set water heater at 140°F if dishwasher is used and 120°F or lower if no dishwasher is used.
- Use water heater thermal blanket.
- Insulate exposed hot water pipes.
- Regularly clean or replace furnace filters.

Outside
Plant deciduous trees to block summer sun and let in winter sunlight.

Other rooms
- Use LED light bulbs and avoid using incandescent bulbs.
- Turn off lights, computers, TV, and other electronic devices when they are not in use.
- Use high-efficiency windows; use insulating window covers and close them at night and on sunny, hot days.
- Set thermostat as low as you can in winter and as high as you can in summer.
- Weather-strip and caulk doors, windows, light fixtures, and wall sockets.
- Keep heating and cooling vents free of obstructions.
- Keep fireplace damper closed when not in use.
- Use fans instead of, or along with, air conditioning.

FIGURE 13.28 Individuals Matter: You can save energy and money where you live. ***Critical thinking:*** Which of these things do you already do? Which ones could you do tomorrow?

include the harmful environmental and health costs of producing and using them.

Another reason for continuing energy waste is that there are little significant government tax breaks, rebates, low-interest long-term loans, and other economic incentives for individuals and businesses to invest in energy efficiency. A third reason is that most governments and utility companies have not put a high priority on educating the public about the money-saving and environmental advantages of improving energy efficiency and conserving energy.

There has been some progress. In 2016, the American Council for an Energy Efficient Economy rated the energy efficiency performance of world's 23 top energy-consuming countries in terms of buildings, industry, and transportation. In order, the three most energy-efficient countries were Germany, Italy, and Japan. The United States ranked eighth.

13.5 WHAT ARE THE ADVANTAGES AND DISADVANTAGES OF USING RENEWABLE ENERGY RESOURCES?

CONCEPT 13.5 We can use a mix of renewable energy resources to meet our energy needs while drastically reducing pollution, greenhouse gas emissions, climate change, and ocean acidification.

Relying More on Renewable Energy

We can use renewable energy from the sun, wind, flowing water, biomass, and heat from the earth's interior (geothermal energy) to produce electricity. Studies show that with increased and consistent government backing in the form of research and development funds, subsidies, and tax breaks, renewable energy could provide 20% of the world's electricity by 2025 and 50% by 2050. The IEA projects that between 2016 and 2040, 60% of all new production of electric power will come from renewable energy—primarily wind and solar power. The IEA also projected that by 2040, renewable forms of energy will be competitive with other sources of energy without subsidies in most parts of the world. The U.S. National Renewable Energy Laboratory (NREL) projects that, with a crash program, the United States could get 50% of its electricity from renewable energy sources (mostly wind and solar) by 2050 because of decreasing costs of producing electricity from these two sources. Between 2008 and 2015, the cost of producing electricity from wind dropped by 41% and the cost of solar power electricity dropped by 64%, according to the DOE. This helps explain why polls show that more than 70% of U.S. consumers want the U.S. government to invest in renewable energy.

According to the IEA, solar and wind are the world's fastest-growing energy resources and nuclear energy is the slowest (Figure 13.20). China has the world's largest installed capacity for electricity from wind power and solar cells and plans to become the largest user and seller of wind turbines and solar cells—projected to be two of the world's fastest growing businesses over the next few decades. China's goal is to greatly expand its production of electricity from renewable wind, sun, and flowing water (hydropower) to help reduce its use of coal and the resulting outdoor air pollution that kills about 1.2 million of its citizens each year.

In 2015, China was the world's largest investor in renewable energy with an investment of $102 billion compared to $44 billion by the United States. However, in 2016 China reduced its targets for wind and solar energy through 2020 because a number of its solar power plants and wind farms do not have access to the country's slowing expanding electricity grid.

If renewable energy is so great, why does it provide only 10% of the world's energy (Figure 13.2, left) and 10% of the energy used in the United States (Figure 13.2, right)? There are several reasons. *First*, people tend to think that solar and wind energy are too diffuse, too intermittent and unreliable, and too expensive to use on a large scale. These perceptions are out of date.

Second, since 1950, U.S. government tax breaks, subsidies, and funding for research and development of renewable energy resources have been much lower than those for fossil fuels and nuclear power. According to the IEA, global subsidies for fossil fuels are nearly 10 times more than subsidies for renewable energy.

Third, U.S. government subsidies and tax breaks for renewable energy have been increasing but Congress requires them to be renewed every few years, which hinders investments in renewable energy, and changing political alignments may put such federal government subsidies in doubt. In contrast, billions of dollars of annual subsidies for fossil fuels and nuclear power have essentially been guaranteed for many decades, due in large part to political pressure from these industries.

Fourth, prices for nonrenewable fossil fuels and nuclear power do not include most of the harmful environmental and human health costs of producing and using them. As a result, they are partially shielded from free-market competition with cleaner renewable sources of energy.

Heating Buildings and Water with Solar Energy

A building that has enough access to sunlight can get all or most of its heat through a **passive solar heating system** (Figure 13.29, left, and Figure 13.30). Such a system absorbs and stores heat from the sun directly within a well-insulated and airtight structure. Water tanks, walls, and

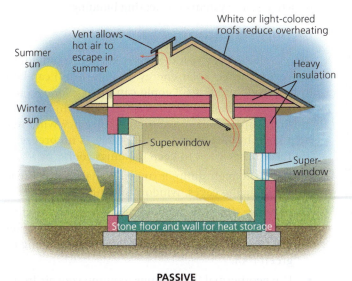

PASSIVE

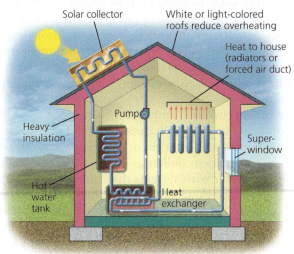

ACTIVE

FIGURE 13.29 Solutions: Passive (left) and active (right) solar home heating systems.

FIGURE 13.30 This passive solar home in Golden, Colorado, collects and stores incoming solar energy to provide much of its heat in a climate with cold winters. Notice the solar hot water heating panels in the yard. Some passive solar houses have sunrooms (see inset photo) to help collect incoming solar energy.

Alan Ford/National Renewable Energy Laboratory

floors of concrete, adobe, brick, or stone can store much of the collected solar energy as heat that is slowly released.

An **active solar heating system** (Figure 13.29, right) captures energy from the sun by pumping a heat-absorbing fluid such as water or an antifreeze solution through special collectors. The collectors are usually mounted on a roof or on special racks that face the sun. Some of the collected heat is used directly. The rest can be stored in large insulated containers filled with gravel, water, clay, or a heat-absorbing chemical, and released as needed.

Rooftop active solar collectors are used to heat water in many homes and apartment buildings. One in ten houses and apartment buildings in China use the sun to provide hot water with systems that cost the equivalent of $200. Once the initial cost is paid, the water is heated free. According to the UN Development Programme, solar water heaters could provide half of the world's hot water.

Passive and active solar systems can be used to heat new homes in areas with adequate sunlight, as long as trees or other buildings do not block solar access. Figure 13.31 shows the availability of solar energy in the continental United States and Canada. Figure 13.32 lists the major advantages and disadvantages of using passive or active solar systems for heating buildings.

Cooling Buildings Naturally

Direct solar energy works against keeping a building cool. However, indirect solar energy (mainly wind) can be used to help cool buildings. People can open windows to take advantage of cooling breezes and use fans to keep the air moving. Superinsulation and high-efficiency windows can help keep hot air outside when there is no breeze. Here are three other ways to keep buildings cool:

- Block the high summer sun with shade trees, broad overhanging eaves, window awnings, or shades.

- In warm climates, use a light-colored roof to reflect as much as 90% of the sun's heat (compared to only 10–15% for a dark-colored roof), or use a living or green roof.

- Use geothermal heat pumps to pump cool air from underground into a building during summer.

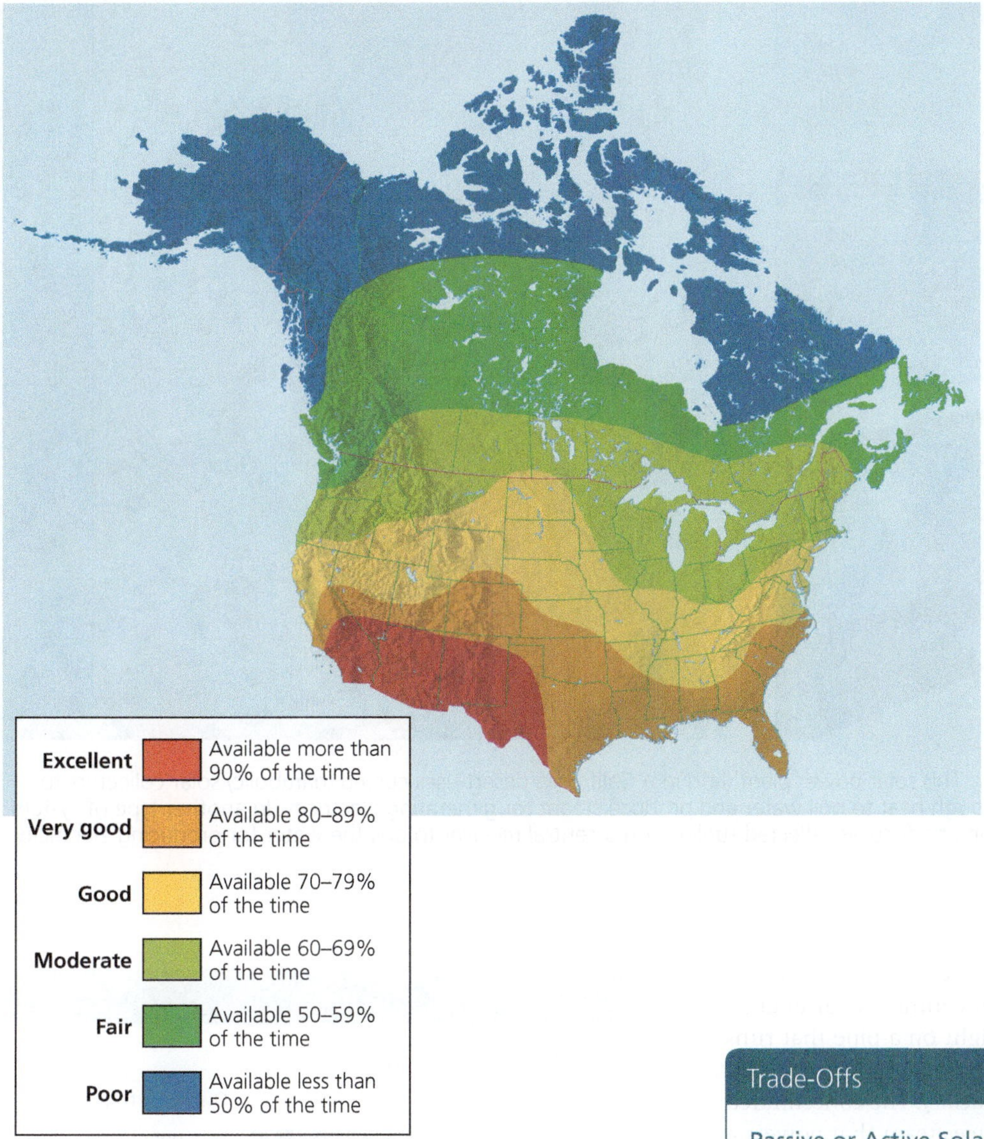

FIGURE 13.31 Availability of direct solar energy in the continental United States and Canada.

(Compiled by the authors using data from the U.S. Geological Society and U.S. Department of Energy.)

Excellent		Available more than 90% of the time
Very good		Available 80–89% of the time
Good		Available 70–79% of the time
Moderate		Available 60–69% of the time
Fair		Available 50–59% of the time
Poor		Available less than 50% of the time

CONSIDER THIS ...

LEARNING FROM NATURE

Some species of African termites stay cool in a hot climate by building tall mounds that allow air to circulate through them. Engineers have used this design lesson from nature to cool buildings naturally, reduce energy use, and save money. One example is Canada's Manitoba Hydro Place office building—North America's most energy-efficient building.

Concentrating Sunlight to Produce High-Temperature Heat and Electricity

One problem with direct solar energy is that it is dispersed. **Solar thermal systems**, also known as *concentrating solar power* (CSP) systems, collect and concentrate solar energy and use it to boil water and produce steam for generating electricity. These systems can be used in deserts and other open areas with ample sunlight.

Trade-Offs

Passive or Active Solar Heating

Advantages	Disadvantages
Medium net energy	Need access to sun 60% of time during daylight
Very low emissions of CO_2 and other air pollutants	Sun can be blocked by trees and other structures
Very low land disturbance	High installation and maintenance costs for active systems
Moderate cost (passive)	Need backup system for cloudy days

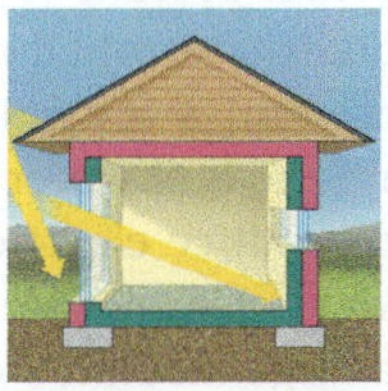

FIGURE 13.32 Heating a house with a passive or active solar energy system has advantages and disadvantages. *Critical thinking:* Which single advantage and which single disadvantage do you think are the most important? Why? Do you think that the advantages outweigh the disadvantages?

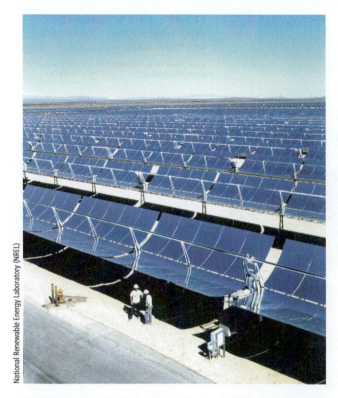

National Renewable Energy Laboratory (NREL)

Sandia National Laboratories/National Renewable Energy Laboratory

FIGURE 13.33 *Solar thermal power:* This solar power plant (left) in a California desert uses curved (parabolic) solar collectors to concentrate solar energy to provide enough heat to boil water and produce steam for generating electricity. In another type of system (right), an array of mirrors tracks the sun and focuses reflected sunlight on a central receiver to boil the water for producing electricity.

One such system uses rows of highly curved mirrors, called parabolic troughs, to collect and concentrate solar energy. Each trough focuses incoming sunlight on a pipe that runs through its center and is filled with synthetic oil (Figure 13.33, left) that goes to a power producing facility. The concentrated heat is used to boil water and produce steam that powers a turbine that drives a generator to produce electricity.

Another solar thermal system (Figure 13.33, right) uses an array of computer-controlled mirrors to track the sun and focus its energy on a central power tower. The concentrated heat is used to boil water and produce steam that drives turbines to produce electricity. Heat produced by either of these systems can be used to melt a certain type of salt stored in a large insulated container. The heat stored in this molten salt can be released as needed to produce electricity at night or on cloudy days.

In 2014, the world's largest solar thermal plant using mirrors opened in California's Mojave Desert. This $2.2 billion plant has 350,000 mirrors focused on three 40-story power towers. It can produce enough electricity to power 140,000 homes and eliminate annual CO_2 emissions equivalent to taking 88,000 cars off the road.

Because solar thermal systems have a low net energy, they require large government subsidies or tax breaks to be competitive in the marketplace. Figure 13.34 lists the major advantages and disadvantages of solar thermal systems.

Trade-Offs

Solar Thermal Systems

Advantages	Disadvantages
High potential for growth	Low net energy and high costs
No direct emissions of CO_2 and other air pollutants	Needs backup or storage system on cloudy days
Lower costs with natural gas turbine backup	Requires high water use
Source of new jobs	Can disrupt desert ecosystems

FIGURE 13.34 Using solar energy to generate high-temperature heat and electricity has advantages and disadvantages (**Concept 13.5**). ***Critical thinking:*** Which single advantage and which single disadvantage do you think are the most important? Why? Do you think that the advantages outweigh the disadvantages?

Bottom: National Renewable Energy Laboratory (NREL). Top: Sandia National Laboratories/National Renewable Energy Laboratory.

FIGURE 13.35 Solutions: Simple solar oven.

People can concentrate solar energy on a smaller scale. In some sunny areas, people use inexpensive *solar cookers* to focus and concentrate sunlight for boiling and sterilizing water and cooking food (Figure 13.35). Solar cookers can replace wood and charcoal fires and reduce indoor air pollution, a major killer of many people in less-developed nations. They also reduce deforestation by lowering the need for firewood and charcoal made from firewood.

GOOD NEWS

Using Solar Cells to Produce Electricity

Solar energy can be converted directly into electrical energy using **photovoltaic (PV) cells**, commonly called **solar cells**. Solar cells are the world's fastest growing technology for producing electricity. Between 2001 and 2015, the cost per watt of electricity produced by solar cells fell by 83% and the cost is expected to keep falling because the prices are driven by advances in technology.

Most solar cells are thin transparent wafers of purified silicon (Si) or polycrystalline silicon with trace amounts of metals that allow them to produce electricity when sunlight strikes them. Solar cells are wired together in a panel and many panels can be connected to produce electricity for a house or a large solar power plant (see Figure 13.36 and chapter-opening photo). Such systems can be connected to electrical grids or to batteries that store the electrical energy until it is needed. Large solar-cell power plants are operating in Germany, Spain, Portugal, South Korea, China, and the southwestern United States.

Arrays of solar cells can be mounted on rooftops or incorporated into almost any type of roofing material. Nanotechnology and other emerging technologies will likely allow the manufacturing of solar cells in paper-thin, rigid or flexible sheets (see Figure 12.A, p. 315) that can be printed like newspapers and attached to or embedded in a variety of surfaces such as outdoor walls, windows, and clothing (to recharge mobile phones). Solar power providers in Japan, Great Britain, India, Italy, and Australia are putting floating arrays of solar cell panels on the surfaces of lakes, reservoirs, ponds, and canals. Engineers are also developing dirt and water-repellent coatings to keep solar panels and collectors clean without having to use water. **GREEN CAREER: solar-cell technology**

Nearly 1.3 billion people, most of them in rural villages in less-developed countries, are not connected to an electrical grid. A growing number of these people are using rooftop solar panels (Figure 13.37) to power energy-efficient LED lamps that replace inefficient kerosene lamps that pollute indoor air. Expanding off-grid solar cell systems to additional rural villages will help hundreds of millions of people lift themselves out of poverty and reduce their exposure to deadly indoor air pollution.

Solar cells have no moving parts, need no water for cooling, and operate safely and quietly. Solar cells do not emit greenhouse gases or other air pollutants, but they are not a carbon-free option because fossil fuels are used to produce and transport the panels. However, these emissions per unit of electricity are much lower than emissions released by using fossil fuels or the nuclear power fuel cycle to produce electricity. Conventional solar cells also contain toxic materials that must be recovered when the cells wear out after 20–25 years of use, or when they are replaced by new systems.

One problem with solar cells is their low energy efficiency. They typically convert only 20% of the incoming solar energy into electricity, although their efficiency is rapidly improving. In 2014, researchers at Germany's Fraunhofer Institute for Solar Energy Systems developed a solar cell with an efficiency of 45%—compared to an efficiency of 35% for fossil fuel and nuclear power plants. They are working to scale up this prototype cell for commercial use. Figure 13.38 lists the major advantages and disadvantages of using solar cells.

Some businesses and homeowners are spreading the cost of rooftop solar power systems over decades by including them in their mortgages. Others are leasing solar-cell systems from companies that install and maintain them.

Some communities and neighborhoods use community solar or shared solar systems to provide electricity for individuals who rent or live in condominiums, or who have their access to sunlight blocked by buildings or trees. Customers buy the power from a centrally located small solar cell power plant with the power delivered by the local utility.

Producing electricity from solar cells is the world's fastest growing way to produce electricity (Figure 13.39). It is expected to continue growing at a rapid rate because solar energy is unlimited and available throughout the world. It is also a technology, not a depletable fuel like coal or natural gas whose prices are controlled by available supplies. **GREEN CAREER: solar-cell technology**

FIGURE 13.36 *Solar cell power plant:* Huge arrays of solar cells can be connected to produce electricity.

Ollyy/Shutterstock.com

If pushed hard and supported with government subsidies equivalent to or greater than fossil fuel subsidies, solar energy could supply up as much as 23% of U.S. electricity by 2050, according to projections by the NREL. After 2050, solar electricity is likely to become one of the top sources of electricity for the United States and much of the world. If this happens, it will represent a global application of the solar energy **principle of sustainability**.

Producing Electricity from Falling and Flowing Water

Hydropower is any technology that uses the kinetic energy of flowing or falling water to produce electricity. This renewable energy resource is an indirect form of solar

FIGURE 13.37 Solutions: A solar cell panel provides electricity for lighting this hut in rural West Bengal, India. ***Critical thinking:*** Do you think your government should provide aid to poor countries for obtaining solar-cell systems? Explain.

Jim Welc/National Renewable Energy Laboratory

Solar Cells

Advantages	Disadvantages
Little or no direct emissions of CO_2 and other air pollutants	Need access to sun
	Need electricity storage system or backup
Easy to install, move around, and expand as needed	Low net energy but likely to improve
Competitive cost for newer cells	Solar-cell power plants could disrupt desert ecosystems

FIGURE 13.38 Using solar cells to produce electricity has advantages and disadvantages (**Concept 13.5**). ***Critical thinking:*** Which single advantage and which single disadvantage do you think are the most important? Why? Do you think that the advantages outweigh the disadvantages? Why?

Top: Martin D. Vonka/Shutterstock.com. Bottom: pedrosala/Shutterstock.com.

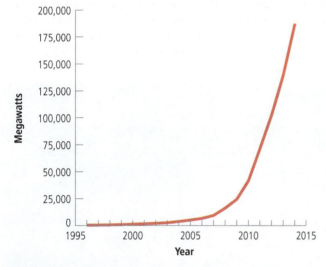

FIGURE 13.39 Global installed electricity capacity of solar cells, 1996–2015. ***Data analysis:*** By what factor and percent did the capacity of installed solar cell increase between 1996 and 2015?

(Compiled by the authors using data from U.S. Energy Information Administration, International Energy Agency, Worldwatch Institute, and Earth Policy Institute.)

energy because it depends on heat from the sun evaporating surface water as part of the earth's solar-powered water cycle (see Figure 3.15, p. 58).

Hydropower is the world's most widely used renewable energy resource. In 2015, it produced more than 16% of the world's electricity in 159 countries, according to the IEA. In 2015, the world's top four consumers of hydropower were in order China, Canada, Brazil, and the United States. That year, hydropower supplied about 6% of the electricity used in the United States and about half of the electricity used on the West Coast, mostly in Washington and California.

The most common way to harness hydropower is to build a high dam across a large river to create a reservoir. Some of the water stored in the reservoir is allowed to flow through large pipes at controlled rates, turning blades on a turbine that produces electricity (see Figure 11.14, p. 273) that is distributed by the electrical grid.

According to the United Nations, only 13% of the world's potential for hydropower has been developed. Countries with the greatest hydropower potential include China, India, and several countries in South America and Central Africa. China, with the world's largest hydropower output, plans to double its output during the next decade and is also building or funding more than 200 hydropower dams around the world.

Hydropower is the least expensive renewable energy resource. Building a dam is costly, but once a dam is up and running, its source of energy—flowing water—is free and is usually renewed annually by snow and rainfall. Despite their potential, some analysts expect the use of large-scale hydropower plants to fall slowly over the next several decades as many existing reservoirs fill with silt and become useless faster than new systems are built.

There is also growing concern over emissions of methane, a potent greenhouse gas, from the decomposition of submerged vegetation in hydropower plant reservoirs, especially in warm climates. Scientists at Brazil's National Institute for Space Research estimate that the world's largest dams altogether are the single largest human-caused source of climate-changing methane. The electricity output of hydropower plants may also drop if atmospheric temperatures continue to rise and melt mountain glaciers that are a primary source of water for these plants.

It is unlikely that large new hydroelectric dams will be built in the United States because most of the best sites already have dams and because of controversy over controlling river flows and the high cost of building new dams. However, the turbines at many existing U.S. hydropower dams could be modernized and upgraded to increase their output of electricity. Figure 13.40 lists the major advantages and disadvantages of using large-scale hydropower plants to produce electricity.

Another way to produce electricity from flowing water is to tap into the energy from *ocean tides* and *waves*. In some coastal bays and estuaries, water levels can rise or fall by 6 meters (20 feet) or more between daily high and low tides. Dams can be built across the mouths of such bays and estuaries to capture the energy in these flows for hydropower. According to energy experts, tidal power will make only a minor contribution to the world's future electricity production because sites with large tidal flows are rare.

Trade-Offs

Large-Scale Hydropower

Advantages	Disadvantages
High net energy	Large land disturbance and displacement of people
Large untapped potential	
Low-cost electricity	High CH_4 emissions from rapid biomass decay in shallow tropical reservoirs
Low emissions of CO_2 and other air pollutants in temperate areas	
	Disrupts downstream aquatic ecosystems

FIGURE 13.40 Advantages and disadvantages of using large dams and reservoirs to produce electricity (**Concept 13.5**). *Critical thinking:* Which single advantage and which single disadvantage do you think are the most important? Why? Do you think that the advantages outweigh the disadvantages? Why?

For decades, scientists and engineers have been trying to produce electricity by tapping wave energy along seacoasts where there are almost continuous waves. However, production of electricity from tidal and wave systems is limited because of a lack of suitable sites,

citizen opposition at some sites, high costs, and damage to equipment from saltwater corrosion and storms.

China is building a pilot plant to evaluate the feasibility of producing electricity by using the difference in temperature between warm surface water and cold deep water in parts of the world's tropical oceans to generate a flow of electrons. The United States experimented with this approach, called ocean thermal-energy conversion (OTEC), in the 1980s, but abandoned it because of its high cost.

Using Wind to Produce Electricity

Land near the earth's Equator absorbs more solar energy than does land near its poles. This uneven heating of the earth's surface and its atmosphere, combined with the earth's rotation, causes prevailing winds to blow (Figure 7.4, p. 139). Because wind is an indirect form of solar energy, using it is a way to apply the solar energy **principle of sustainability**.

The kinetic energy of wind can be captured and converted to electrical energy by *wind turbines*. As a turbine's blades spin, they drive an electric generator, which produces electricity (Figure 13.41, left). Groups of wind turbines called *wind farms* transmit electrical energy to electrical grids. Wind farms can operate both on land (chapter-opening photo) and at sea (Figure 13.41, right).

Today's wind turbines can be as tall as a 60-story building and have blades as long as 70 meters (230 feet)—the

Wind turbine

FIGURE 13.41 Wind turbines convert the kinetic energy in wind to electricity, another form of kinetic energy (moving electrons). Wind power is an indirect form of solar energy.

combined length of six school buses. This height allows them to tap into stronger and more constant winds at higher altitudes and to produce more electricity at a lower cost. A typical wind turbine can generate enough electricity to power more than 1,000 homes.

Harvard University researcher Xi Lu estimates that wind power has the potential to produce 40 times the world's current demand for electricity. Most of world's wind farms have been built on land in parts of Europe, China, and the United States. However, the frontier for wind energy is offshore wind farms (Figure 13.41, right) because winds are generally much stronger and steadier over coastal waters than over land. When located far offshore, wind farms are not visible from the land. Building offshore costs more but avoids the need for negotiations among multiple landowners over the locations of turbines and electrical transmission lines. Offshore wind farms have been built off the coasts of China, Japan, and 10 European countries. The map in Figure 13.42 shows the potential availability of land- and ocean-based wind in the United States.

Since 1990, wind power has been the world's second fastest-growing source of electricity (Figure 13.43) after solar cells (Figure 13.39). In 2015, the United States led the world in producing electricity from wind, followed by China, Germany, and Spain. Many of China's wind turbines sit idle because they are not near major cities and are not connected to the country's electrical grid.

In 2015, 315,000 wind turbines in more than 85 countries produced about 3.7% of the world's electricity. The IEA projects that by 2050 this percentage could grow to 18% of the world's electricity. Around the world, more than 400,000 people are employed in the production, installation, and maintenance of wind turbines and such job numbers will grow as wind power continues its rapid expansion. Over the next decade, wind-farm technician is projected to be one of the fastest growing occupations in the United States.

In 2015, wind power produced 45% of Denmark's electricity and the country plans to increase this to 85% by 2035. By 2015, wind turbines in the United States produced 4.7% of the country's electricity, an amount equal to that produced by 64 large nuclear plant reactors. Texas leads the nation in wind energy production, followed by Iowa, California, and Oklahoma. In 2015, Iowa, Illinois, Nebraska, Kansas, and parts of Texas wind farms generated electricity at a lower cost, without subsidies, than any other technology, according to a 2016 study by researchers at the University of Texas at Austin.

The United States is a superpower of wind energy. A study published in the *Proceedings of the U.S. National*

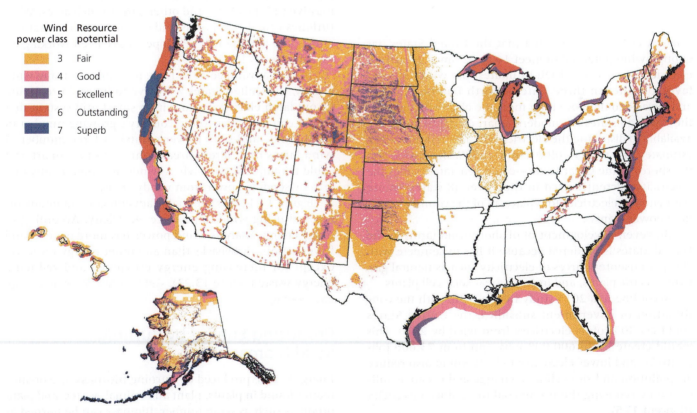

Wind power class	Resource potential
3	Fair
4	Good
5	Excellent
6	Outstanding
7	Superb

FIGURE 13.42 Potential supply of land- and ocean-based wind energy in the United States.

(Compiled by the authors using data from U.S. Geological Survey and U.S. Department of Energy.)

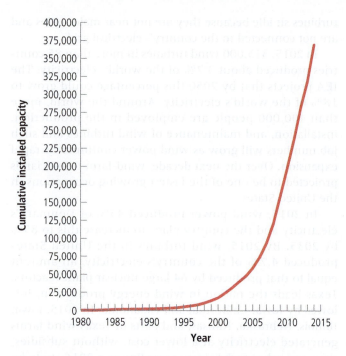

FIGURE 13.43 Global installed capacity for generation of electricity by wind energy, 1980–2014. **Data analysis:** In 2014, the world's installed capacity for generating electricity by wind power was about how many times more than it was in 2005?

(Compiled by the authors using data from Global Wind Energy Council, European Wind Energy Association, American Wind Energy Association, Worldwatch Institute, World Wind Energy Association, and Earth Policy Institute.)

Academy of Sciences estimated that the United States has enough wind potential to meet 16 to 22 times its current electricity needs. The DOE estimates that wind farms at favorable sites in three states—North Dakota, Kansas, and Texas—could more than meet the electricity needs of the lower 48 states if a modern smart electrical grid were available to distribute the electricity. In addition, the NREL estimates that winds—off Atlantic and Pacific coasts, and the shores of the Great Lakes—could generate 9 times the electricity currently used in the lower 48 states—more than enough electricity to replace all of the country's coal-fired power plants.

However, development of offshore wind farms in the United States is hindered because it has to compete with cheaper onshore sources of electricity such as natural gas, hydroelectric plants, and in some areas solar cell plants.

According to a 2015 study by the DOE, with the continuation of government subsidies, the United States could get 30% of its electricity from wind by 2030. This would create up to 600,000 jobs—up from 73,000 jobs in 2015—and lower electricity bills. It would also reduce air pollution and slow climate change and ocean acidification by reducing the use of coal to produce electricity (**Concept 13.5**).

Wind is abundant, widely distributed, and inexhaustible, and wind power is mostly carbon-free and pollution-free. A wind farm can be built in 9 to 12 months and expanded as needed. Although wind farms can cover large areas of land, the turbines themselves occupy only a small portion of land.

Many U.S. landowners in favorable wind areas are investing in wind farms. Landowners typically receive $3,000 to $10,000 a year in royalties, for each wind turbine on their land. An acre of land in northern Iowa planted in corn can produce about $1,000 worth of ethanol fuel a year. The same site used for a single wind turbine can produce $300,000 worth of electricity per year and the area around the turbine towers can still be used to grow crops and graze cattle.

Since 1990, prices for wind-generated electricity in the United States (and in other countries) have been falling sharply. Prices are expected to keep falling because of technological improvements in wind turbine design that will increase their energy efficiency and reduce mass production and maintenance costs. The DOE and the Worldwatch Institute estimate that wind energy would be the least costly way to produce electricity if we were to include the harmful environmental and health costs of various energy resources in comparative cost estimates. This would be an important application of the full-cost pricing **principle of sustainability**.

Like any energy source, wind power has some drawbacks. Land areas with the greatest wind power potential are often far from cities. Roads must be built to deliver massive turbine blades and other parts to such areas. Wind turbines can also kill birds and bats—a problem that scientists and wind power developers have been working on (Science Focus 13.3).

To take advantage of its huge potential for using wind energy, the United States will have to invest in replacing its outdated electrical grid with a smart grid system that would connect cities to the country's growing number of wind farms (and solar power plants). A large number of wind farms in different areas connected to a smart grid could take up the slack when winds die down in any one area by sending power from windy areas.

Figure 13.44 lists the major advantages and disadvantages of using wind to produce electricity. According to many energy analysts, wind power has more benefits and fewer serious drawbacks than any other energy resource, except for increasing energy efficiency and reducing energy waste (Figure 13.23). **GREEN CAREER: Wind-energy engineering**

Producing Energy by Burning Solid Biomass

Energy can be produced by burning **biomass**, the organic matter found in plants, plant and animal wastes, and plant products such as scrap lumber. Biomass can be burned as a solid fuel or converted into liquid biofuels. Examples of biomass fuels include wood, wood wastes, charcoal made

Making Wind Turbines Safer for Birds and Bats

Wildlife ecologists and bird experts estimate that collisions with wind turbines kill as many as 234,000 birds and 600,000 bats each year in the United States. Such deaths are a legitimate concern.

However, according to studies by the Defenders of Wildlife, the U.S. Fish and Wildlife Service, and the Smithsonian Conservation Biology Institute, wind turbines are a minor source of bird and bat deaths compared to other human-related sources and account for only about 0.003% of such deaths. Each year, domestic and feral cats kill 1.4 billion to 3.7 billion birds; collisions with windows

1 billion; cars and trucks 89 million to 340 million; high-tension wires 174 million; and pesticides 72 million. Most of the wind turbines involved in bird and bat deaths were built years ago using outdated designs. Some were also built in bird migration corridors and in areas near large bat populations.

Developers of new wind farms avoid bird migration corridors, as well as areas with large bat colonies. Newer turbine designs reduce bird and bat deaths considerably by using slower blade rotation speeds and by not providing places for birds to perch or nest. Researchers are also

evaluating the use of ultraviolet light to deter birds and bats from turbines. Ultrasonic devices attached to turbine blades can scare bats away by emitting high-frequency sounds that we cannot hear. Another approach is to use radar to track large incoming flocks of migrating birds and to shut down the turbines until they pass.

CRITICAL THINKING

What would you say to someone who tells you that we should not depend on wind power because wind turbines can kill birds and bats?

Trade-Offs

Wind Power

Advantages	Disadvantages
High net energy	Needs backup or storage system when winds die down unless connected in a national electrical grid
Widely available	
Low electricity cost	Visual pollution for some people
Little or no direct emissions of CO_2 and other air pollutants	Low-level noise bothers some people
Easy to build and expand	Can kill birds and bats if not properly designed and located

FIGURE 13.44 Using wind to produce electricity has advantages and disadvantages (**Concept 13.5**). *Critical thinking:* Which single advantage and which single disadvantage do you think are the most important? Why? Do you think the advantages outweigh the disadvantages? Why?

Top: TebNad/Shutterstock.com. Bottom: T.W. van Urk/Shutterstock.com

energy used in less-developed countries, and 95% of the energy used in the poorest countries.

Wood is a renewable fuel only if it is not harvested faster than it is replenished. The problem is that about 2.7 billion people in 77 less-developed countries face a *fuelwood crisis*. To survive, they often meet their fuel needs by cutting trees faster than new ones can replace them.

One solution is to plant fast-growing trees, shrubs, or perennial grasses in *biomass plantations*. However, repeated cycles of growing and harvesting these plantations can deplete the soil of key nutrients. It can also allow the spread of some nonnative tree species that can become invasive species. Clearing forests and grasslands to provide the fuel also reduces biodiversity and the amount of vegetation that would otherwise capture climate-changing CO_2.

Burning wood and other forms of biomass produces CO_2 and other air pollutants such as fine particles in smoke. Figure 13.45 lists the major advantages and disadvantages of burning solid biomass as a fuel.

Using Liquid Biofuels to Power Vehicles

Biomass can be converted into liquid biofuels for use in motor vehicles. The two most common liquid biofuels are *ethanol* (ethyl alcohol produced from plants and plant wastes) and *biodiesel* (produced from vegetable oils). The three biggest biofuel producers are, in order, the United States (producing ethanol from corn), Brazil (producing ethanol from sugarcane residues), and the European Union (producing biodiesel from vegetable oils).

Biofuels have three major advantages over gasoline and diesel fuel produced from oil. *First*, biofuel crops can

from wood, and agricultural wastes such as sugarcane stalks, rice husks, and corncobs.

Most solid biomass is burned for heating and cooking. It can also be used to provide heat for industrial processes and to generate electricity. Biomass used for heating and cooking supplies 10% of the world's energy, 35% of the

FIGURE 13.45 Burning solid biomass as a fuel has advantages and disadvantages (**Concept 13.5**). *Critical thinking:* Which single advantage and which single disadvantage do you think are the most important? Why? Do you think the advantages outweigh the disadvantages? Why?

Top: Fir4ik/Shutterstock.com. Bottom: Eppic/Dreamstime.com.

be grown throughout much of the world, which can help more countries reduce their dependence on imported oil. *Second,* if growing biofuel crops keeps pace with harvesting them, there is no net increase in CO_2 emissions, unless existing grasslands or forests are cleared to plant biofuel crops. *Third,* biofuels are easy to store and transport through existing fuel networks and can be used in modified motor vehicles at little additional cost.

Since 1975, global ethanol production has increased sharply, especially in the United States and Brazil. Brazil makes ethanol from *bagasse,* a residue produced when sugarcane is crushed. This sugarcane ethanol has a medium net energy that is about 8 times higher than that of ethanol produced from corn. About 70% of Brazil's cars and light trucks run on ethanol or ethanol–gasoline mixtures produced from sugarcane grown on only 1% of the country's arable land. This has greatly reduced Brazil's dependence on imported oil.

In 2016, about 29% of the corn produced in the United States was used to produce ethanol, which is mixed with gasoline to fuel cars. In the United States, gasoline can contain no more than 10% by volume ethanol (known as E10).

Studies indicate that corn-based ethanol has a low net energy because of the large-scale use of fossil fuels to produce fertilizers, grow the corn, and convert it to ethanol. There is controversy over greenhouse gas emissions to the atmosphere from the production and use of corn-based ethanol. Some research indicates that burning fuel with

ethanol may add less greenhouse gases than burning pure gasoline. However, other research indicates an increase in greenhouse gas emissions when the greenhouse emissions from producing the corn are included.

According to a study by the Environmental Working Group, the heavily government-subsidized corn-based ethanol program in the United States has taken more than 2 million hectares (5 million acres) of land out of the soil conservation reserve (p. 247), an important topsoil conservation program. Growing corn also requires substantial amounts of water and land—resources that are in short supply in some areas.

Furthermore, scientists warn that large-scale biofuel farming could reduce biodiversity, degrade soil quality, and increase erosion. As a result, a number of scientists and economists call for withdrawing government subsidies for corn-based ethanol production and reducing the current limit of no more than 10% ethanol in U.S. gasoline. In contrast, corn growers and ethanol distillers have proposed allowing up to 30% ethanol in gasoline. They claim that the harmful environmental effects of corn-based ethanol are overblown and that this biofuel has many environmental and economic benefits.

An alternative to corn-based ethanol is *cellulosic ethanol,* which is produced from the inedible cellulose that makes up most of the biomass of plants in the form of leaves, stalks, and wood chips. Cellulosic ethanol can be produced from tall and rapidly growing grasses such as switchgrass and miscanthus that do not require nitrogen fertilizers and pesticides. They also do not have to be replanted because they are perennial plants and can be grown on degraded and abandoned farmlands. Ecologist David Tilman estimates that the net energy yield for cellulosic ethanol is about 5 times that of corn-based ethanol. However, producing cellulosic ethanol is not yet affordable and more research is needed to determine its possible environmental effects.

CONSIDER THIS ...

CONNECTIONS Biofuels and Climate Change

Nobel Prize–winning chemist Paul Crutzen has warned that intensive farming of biofuel crops could speed up atmospheric warming and climate change by producing more greenhouse gases than would be produced by burning fossil fuels instead of biofuels.

Another possible alternative to corn-based ethanol involves using *algae* to produce biofuels. As a crop, algae can grow rapidly in various aquatic environments. Algae store energy as natural oils in their cells. This oil can be extracted and refined to make a product very much like gasoline. Currently, extracting and refining the oil from algae is too costly. More research is needed to evaluate the potential for this possible biofuel option.

Figure 13.46 lists the major advantages and disadvantages of using ethanol and biodiesel biofuels.

Liquid Biofuels

Advantages	Disadvantages

Advantages

Reduced CO_2 emissions for some crops

Medium net energy for biodiesel from oil palms

Medium net energy for ethanol from sugarcane

Disadvantages

Fuel crops can compete with food crops for land and raise food prices

Fuel crops can be invasive species

Low net energy for corn ethanol and for biodiesel from soybeans

Higher CO_2 emissions from corn ethanol

FIGURE 13.46 Ethanol and biodiesel biofuels have advantages and disadvantages (**Concept 13.5**). ***Critical thinking:*** Which single advantage and which single disadvantage do you think are the most important? Why? Do you think the advantages outweigh the disadvantages? Why?

Top: tristan tan/Shutterstock.com

Tapping into the Earth's Internal Heat

Geothermal energy is heat stored in soil, underground rocks, and fluids in the earth's mantle. It is used to heat and cool buildings, heat water, and produce electricity.

Geothermal energy is available around the clock but is practical only at sites with high enough concentrations of underground heat.

A *geothermal heat pump* system (Figure 13.47) can heat and cool a house almost anywhere in the world. This system makes use of the temperature difference between the earth's surface and underground at a depth of 3–6 meters (10–20 feet), where the temperature typically is 10–20°C (50–60°F) year round. In winter, a closed loop of buried pipes circulates a fluid that extracts heat from the ground and carries it to a heat pump, which transfers the heat to a home's heat distribution system. In summer, this system works in reverse, removing heat from a home's interior and storing it underground.

According to the EPA, a well-designed geothermal heat pump system is the most energy-efficient, reliable, environmentally clean, and cost-effective way to heat or cool a space. Installation costs can be high but are generally recouped within 3 to 5 years, after which these systems save money for their owners. Initial costs can be added to a home mortgage to spread the financial burden over two or more decades.

Engineers have also learned how to tap into deeper, more concentrated *hydrothermal reservoirs* of geothermal energy. Wells are drilled into the reservoirs to extract dry steam (with a low water content), wet steam (with a high water content), or hot water. The steam or hot water is then used to heat buildings, provide hot water, grow vegetables in greenhouses, raise fish in aquaculture ponds, and spin turbines to produce electricity.

In 2105, geothermal energy was used to generate electricity in 24 countries and 70 countries used geothermal

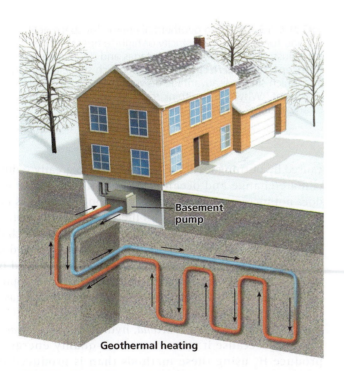

Geothermal heating

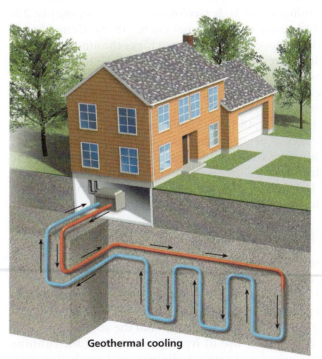

Geothermal cooling

FIGURE 13.47 Natural Capital: A geothermal heat pump system can heat or cool a house almost anywhere.

Andrés Ruzo—Geothermal Energy Sleuth and National Geographic Explorer

Andrés Ruzo is a geophysicist with a driving passion to learn about geothermal energy and to show how this renewable and clean energy resource can help us solve some of the world's energy problems. As a child growing up in Peru, he heard talk about a "boiling river" and did not believe it. However, in 2011 when he started working on a PhD in geology to determine Peru's geothermal potential, he learned that such a river exists. His research indicates that it is the result of magma rising in the earth's crust and heating the river's water.

Ruzio and his wife and field assistant, Sofia, have been gathering data to develop the first detailed map of the geothermal energy resources of northern Peru. He believes that geothermal energy is a hidden sleeping giant that can be tapped as an important renewable source of heat and electricity. He says that his goal in life is "to be a force of positive change in the world."

heating. The United States is the world's largest producer of geothermal energy from hydrothermal reservoirs. Iceland gets almost all of its electricity from renewable hydroelectric (69%) and geothermal power (29%) plants and about 90% of its space and water heating in homes from geothermal energy. In Peru, a National Geographic Explorer is carrying out research to develop that country's geothermal resources (Individuals Matter 13.1).

Another source of geothermal energy is *hot, dry rock* found 5 or more kilometers (3 or more miles) underground almost everywhere. Water is injected through wells drilled into this rock. Some of the water absorbs the heat and becomes steam that is brought to the surface and used to spin turbines to generate electricity. According to the U. S. Geological Survey, tapping just 2% of this source of geothermal energy in the United States could produce more than 2,000 times the amount of electricity currently used in the country. The limiting factor is its high cost, which could be brought down by more research and improved technology. **GREEN CAREER: Geothermal engineer**

Figure 13.48 lists the major advantages and disadvantages of using geothermal energy. The biggest factors limiting the widespread use of geothermal energy are the high cost of drilling the wells and building the plants and the lack of hydrothermal sites with concentrations of heat high enough to make it affordable.

Will Hydrogen Save Us?

Some scientists say that the fuel of the future is hydrogen gas (H_2). Most research has focused on using hydrogen fuel cells that combine H_2 and oxygen gas (O_2) to produce electricity and nonpolluting water vapor ($2H_2 + O_2 \rightarrow 2H_2O$ + energy) that is emitted into the atmosphere.

Widespread use of hydrogen as a fuel for running motor vehicles, heating buildings, and producing electricity would eliminate most of the outdoor air pollution

Trade-Offs

Geothermal Energy

Advantages	Disadvantages
Medium net energy and high efficiency at accessible sites	High cost except at concentrated and accessible sources
Lower CO_2 emissions than fossil fuels	Scarcity of suitable sites
Low operating costs at favorable sites	Noise and some CO_2 emissions

FIGURE 13.48 Using geothermal energy for space heating and for producing electricity or high-temperature heat for industrial processes has advantages and disadvantages (**Concept 13.5**). *Critical thinking:* Which single advantage and which single disadvantage do you think are the most important? Why? Do you think the advantages outweigh the disadvantages? Why?

caused by burning fossil fuels. It would also greatly reduce climate change and ocean acidification because its use does not increase atmospheric CO_2 as long as the H_2 is not produced with the help of fossil fuels or nuclear power.

Turning hydrogen into a major fuel source is a challenge for several reasons. *First*, there is hardly any hydrogen gas (H_2) in the atmosphere. H_2 can be produced by heating water or passing electricity through it; by stripping it from the methane (CH_4) found in natural gas and gasoline molecules; and through a chemical reaction involving coal, oxygen, and steam. *Second*, hydrogen has a *negative net energy* because it takes more high-quality energy to produce H_2 using these methods than is produced by burning it.

Third, fuel cells are the best way to use H_2 but current versions of fuel cells are expensive. However, progress in the development of nanotechnology (see Science Focus 12.1, p. 315), coupled with mass production, could lead to less expensive fuel cells.

Fourth, whether or not a hydrogen-based energy system produces less CO_2 and other outdoor air pollutants than a fossil fuel system depends on how the H_2 fuel is produced. Electricity from coal-burning and nuclear power plants can be used to decompose water into H_2 and O_2 but this approach does not avoid the harmful environmental effects associated with using coal and nuclear power. Research indicates that making H_2 from coal or stripping it from methane or gasoline adds much more CO_2 to the atmosphere per unit of heat generated than does burning the coal or methane directly.

Hydrogen's negative net energy is a serious limitation. It means that this fuel will have to be heavily subsidized for it to compete in the open market. However, this could change. Chemist Daniel Nocera has been learning from nature by studying how a leaf uses photosynthesis to produce the chemical energy used by plants, and he has developed an "artificial leaf." This credit-card-sized silicon wafer produces H_2 and O_2 when placed in a glass of tap water and exposed to sunlight. The hydrogen can be extracted and used to power fuel cells. Scaling up this or similar processes to produce large amounts of H_2 at an affordable price with an acceptable net energy could represent a tipping point for the use of solar energy and hydrogen fuel. Doing so would help implement the solar energy **principle of sustainability** on a global scale.

Figure 13.49 lists the major advantages and disadvantages of using hydrogen as an energy resource. **GREEN CAREER: Hydrogen energy development**

13.6 HOW CAN WE MAKE THE TRANSITION TO A MORE SUSTAINABLE ENERGY FUTURE?

CONCEPT 13.6 We could make the transition to a more sustainable energy future by improving energy efficiency, reducing energy waste, using a mix of renewable energy resources, and including the harmful environmental and health costs of energy resources in their market prices.

Establishing New Energy Priorities

Shifting from one major energy resource to another is not new. The world has shifted from wood to coal, then from coal to oil, and then to our current dependence on oil, natural gas, and coal as new technologies made these resources more available and affordable. Each of these

Trade-Offs

Hydrogen

Advantages	Disadvantages
Can be produced from plentiful water at some sites	Negative net energy
No CO_2 emissions if produced with the use of renewables	CO_2 emissions if produced from carbon-containing compounds
Good substitute for oil	High costs create need for subsidies
High efficiency in fuel cells	Needs H_2 storage and distribution system

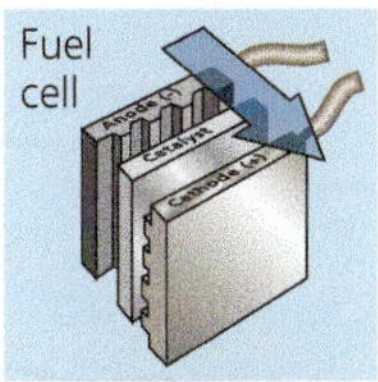

Fuel cell

Photo: LovelaceMedia/Shutterstock.com

FIGURE 13.49 Using hydrogen as a fuel for vehicles and for providing heat and electricity has advantages and disadvantages (**Concept 13.5**). *Critical thinking:* Which single advantage and which single disadvantage do you think are the most important? Why? Do you think that the advantages outweigh the disadvantages? Why?

shifts in key energy resources took about 50 to 60 years. As in the past, it takes an enormous investment in scientific research, engineering, technology, and infrastructure to develop and spread the use of new energy resources.

Currently, the world gets 86% and the United States gets 81% of the commercial energy they use from three carbon-containing fossil fuels—oil, coal, and natural gas (Figure 13.2). These fuels have supported tremendous economic growth and improved the lives of many people.

However, society is awakening to the fact that burning fossil fuels, especially coal, is largely responsible for three of the world's most serious environmental problems: air pollution, climate change, and ocean acidification. Fossil fuels are affordable because their market prices do not include these and other harmful health and environmental effects.

According to many scientists, energy experts, and economists, over the next 50 to 60 years and beyond, we need to make a new energy transition by (**1**) improving energy efficiency and reducing energy waste, (**2**) decreasing our dependence on nonrenewable fossil fuels, and (**3**) relying more on a mix of renewable energy from the sun, flowing water (hydropower), wind, biomass, biofuels, geothermal energy, and perhaps hydrogen. Countries also need to develop modern smart electrical grids to distribute electricity produced from these resources.

The use of fossil fuels can be reduced but they are not going to disappear, as shown for the United States in Figure 13.50. According to the IEA and British Petroleum (BP), between 2015 and 2035, the worldwide rates of energy

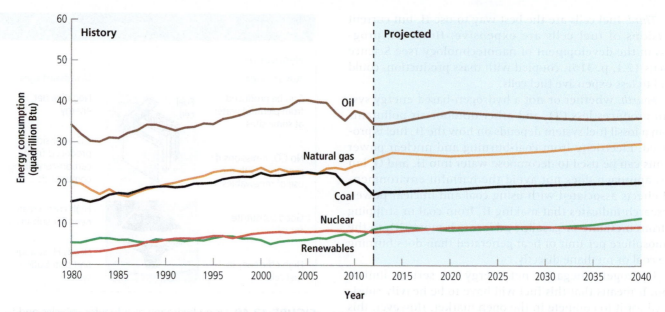

FIGURE 13.50 Energy consumption by source in the United States, 1980–2014, with projections to 2040.

(Compiled by the authors using data from U. S. Energy Information Administration.)

consumption are projected to decline for oil and coal, stay about the same for nuclear energy and hydropower, and increase for solar, wind, and natural gas. The IEA projects that in 2065, fossil fuels will provide at least 50% of the world's energy, compared to 86% today. The use of coal is likely to decline the most because of its harmful environmental effects and its key role in accelerating climate change and ocean acidification. Also, in many locations, it is cheaper to produce electricity by burning natural gas or by using wind farms. As a result, scientists and economists call for greatly increasing research and government subsidies to reduce the harmful environmental and health effects of using fossil fuels.

This restructuring of the global energy system and economy over the next 50 to 60 years and beyond will save money, create profitable business and investment opportunities, and provide jobs. It will also save lives by sharply reducing air pollution, help slow the increase in ocean acidity, and keep climate change from spiraling out of control and creating ecological and economic chaos. Finally, it will increase our positive environmental impact, and pass the world on to future generations in better shape than we found it, in keeping with the ethical **principle of sustainability**.

This energy shift is being driven by the availability of perpetual supplies of clean and increasingly cheaper solar and wind energy throughout the world. Advances in solar cell and wind turbine technology have been steadily reducing the cost of using wind and solar energy to produce electricity. This is in contrast to fossil fuels, which are dependent on finite supplies of nonrenewable oil, coal, and natural gas that are not widely distributed, are controlled by a few countries, and are subject to fluctuating prices based on supply and demand.

In this new technology-driven energy economy, an increasing percentage the world's electricity will be produced locally from available sun and wind and regionally from solar cell power plants and wind farms. It will be transmitted to consumers by modern, interactive smart electrical grids. Homeowners and businesses with solar panels on their land or their roofs (or traditional roof coverings that contain solar cells) can become independent electricity producers. They will be able to heat and cool their homes and businesses, run electrical devices, charge hybrid or electric cars, and sell excess electricity they produce. The United States could benefit economically, because making such a shift could set off an explosion of innovations in energy efficiency, renewable energy, and battery technologies by tapping into the country's ability to innovate.

Like any major societal change, making this shift will not be easy. However, to many analysts, the environmental, health, and economic benefits of making the shift far outweigh the harmful environmental, health, and economic effects of not making the shift.

This shift is underway and gaining momentum as the cost of electricity produced from the sun and wind continues its rapid fall and investors see a way to make money on these two of the world's fastest growing businesses. Germany (see the following Case Study), Sweden, Denmark, and Costa Rica have made the most progress in this transition. For example, Costa Rica gets more than 90% of its electricity from renewable resources, mostly hydropower, geothermal energy, and wind, and aims to generate all of its electricity from renewable sources by 2021. In 2017, while the U.S. government began to back away from a commitment to energy efficiency and renewable energy, the Chinese government planned a $360 billion investment in these technologies, projected to create 13 million jobs in China.

According to many scientists and economists, the shift to a new energy economy could be further accelerated if citizens, the leaders of emerging energy companies, and energy investors demanded the following from their elected officials:

- Use full-cost pricing to include the harmful health and environmental costs of using fossil fuels and all other energy resources in their market prices.

- Tax carbon emissions. This is supported by most economists and many business leaders, and is now done in 40 countries. Use the revenue to reduce taxes on income and wealth and to promote investments and research in new energy-efficient and renewable energy technologies.

- Sharply decrease and eventually eliminate government subsidies for fossil fuel industries, which are well established and profitable.

- Establish national feed-in-tariff systems that guarantee owners of wind farms, solar power plants, and home solar systems a long-term price for energy that they feed into their electrical grids (as is being done in more than 50 countries, many of them in Europe).

- Mandate that a certain percentage (typically 20–40%) of the electricity generated by utility companies must come from renewable resources (as is done in 24 countries and in 29 U.S. states).

- Increase government fuel efficiency (CAFE) standards for new vehicles to 43 kilometers per liter (100 miles per gallon) by 2040.

We have the creativity, wealth, and most of the technology needed to make the transition to a safer, more energy-efficient, and cleaner energy economy within your lifetime. The critical question is whether enough people have the political and ethical will to insist that elected officials and business leaders work together to make this transition.

Figure 13.51 lists ways to make this transition to a more environmentally and economically sustainable energy future. Figure 13.52 lists ways in which you can take part in making such an exciting transition to a more sustainable world.

Germany Is a Renewable Energy Superpower

Germany is phasing out nuclear power and, between 2006 and 2015, increased the percentage of its electricity produced by renewable energy, mostly from the sun and wind, from 6% to 31%. Its goals for 2050 are to: get more than 80% of its electricity from renewable energy; increase electricity efficiency by 50%; reduce its CO_2 emissions by 80% to get to 95% of 1990 levels; and sharply reduce its use of coal, which in 2015 produced 44% of its electricity.

Solutions

Making the Transition to a More Sustainable Energy Future

Improve Energy Efficiency	**More Renewable Energy**
Increase fuel-efficiency standards for vehicles, buildings, and appliances	Greatly increase the use of renewable energy
Provide large tax credits for buying efficient cars, houses, and appliances	Provide large subsidies and tax credits for use of renewable energy
	Greatly increase renewable energy research and development
Reward utilities for reducing demand for electricity	**Reduce Pollution and Health Risk**
	Phase out coal subsidies and tax breaks
	Levy taxes on coal and oil use
Greatly increase energy efficiency research and development	Phase out nuclear power subsidies, tax breaks, and loan guarantees

FIGURE 13.51 Energy analysts have made a number of suggestions for helping us make the transition to a more sustainable energy future (**Concept 13.6**). *Critical thinking:* Which five of these solutions do you think are the best ones? Why?

Top: andrea lehmkuhl/Shutterstock.com. Bottom: pedrosala/Shutterstock.com.

What Can You Do?

Shifting to More Sustainable Energy Use

- Walk, bike, or use mass transit or a car pool to get to work or school
- Drive only vehicles that get at least 17 kpl (40 mpg)
- Have an energy audit done in the place where you live
- Superinsulate the place where you live and plug all air leaks
- Use passive solar heating
- For cooling, open windows and use fans
- Use a programmable thermostat and energy-efficient heating and cooling systems, lights, and appliances
- Turn down your water heater's thermostat to 43–49°C (110–120°F) and insulate hot water pipes
- Turn off lights, TVs, computers, and other electronics when they are not in use
- Wash laundry in cold water and air dry it on racks

FIGURE 13.52 Individuals Matter: Ways you can use energy more sustainably. ***Critical thinking:*** Which three of these measures do you think are the most important ones to take? Why? Which of these steps have you already taken and which do you plan to take?

The biggest factor in this shift is Germany's use of a **feed-in-tariff** (also used by Great Britain and 47 other countries). Under a long-term contract, the government requires utilities to buy electricity produced by homeowners and businesses from renewable energy resources at a price that guarantees a good return and to feed it into the electrical grid. In 2016, the upper house of Germany's parliament passed a resolution calling for eliminating vehicles powered by gasoline or diesel by 2030.

German households own and make money from roughly 80% of the country's solar cell installations. The program is financed by all electricity users and costs less than $5 a month per household. It has created more than 370,000 new renewable energy jobs. In addition, the German government and private investors have subsidized research on technological improvements in solar and wind power and backup energy storage systems that could further lower costs. However, political opposition by Germany's coal industry and some electrical utility companies and the need for costly backup power could reduce the level of feed-in-tariffs and slow the country's shift to greater dependence on solar and wind power.

BIG IDEAS

- In evaluating energy resources, we should consider their net energy, the environmental and health impacts of using them, and their potential supplies.
- By phasing in a mix of renewable energy sources—especially solar, wind, flowing water, biomass, and geothermal energy—we could drastically reduce pollution, greenhouse gas emissions, ocean acidification, and biodiversity losses.
- Making the transition to a more sustainable energy future will require sharply increasing energy efficiency, reducing energy waste, using a mix of renewable energy resources, and including the harmful environmental and health costs of energy resources in their market prices.

Tying It All Together

Energy Resources and Sustainability

In the Core Case Study that opens this chapter, we looked at how horizontal drilling and hydrofracking have greatly increased U.S. production of oil and natural gas. Later we learned about the pros and cons of fracking. We looked at the advantages and disadvantages of an array of nonrenewable and renewable energy resources.

By relying mostly on nonrenewable fossil fuels, we violate the three **scientific principles of sustainability** (see inside back cover of this book). The technologies we use to obtain energy from these resources disrupt the earth's chemical cycles by diverting huge amounts of water, degrading or destroying terrestrial and aquatic ecosystems, and emitting large quantities of greenhouse gases and other air pollutants. Using these technologies also destroys and degrades biodiversity and ecosystem services.

By relying more on a diversity of direct and indirect forms of renewable solar energy and by greatly improving energy efficiency and reducing energy waste, we could implement the three **scientific principles of sustainability** during this century and greatly increase our beneficial environmental impact. Making this shift would also be in keeping with the three economic, political, and ethical **principles of sustainability**. By including the harmful environmental and health costs of fossil fuels, nuclear energy, and other energy resources in their market prices, society would be applying the *full-cost pricing principle*. This could be

Alfred Estes, Colorado School of Mines

accomplished with the aid of compromise and trade-offs in the political arena by applying the *win-win solutions principle*. Many analysts also argue that such solutions would greatly benefit life on earth now and in the long run—an application of the ethical principle of *responsibility to current and future generations*.

Chapter Review

Core Case Study

1. Describe the process of removing tightly held oil and natural gas through **horizontal drilling** and **hydraulic fracturing**, or **fracking** and the role of these technologies in greatly increasing the production of oil and natural gas in the United States.

Section 13.1

2. What are the two key concepts for this section? What is the earth's main source of energy? What is commercial energy and where does most of it come from in the world and in the United States? What percentage of the commercial energy used in the world and in the United States comes from nonrenewable fossil fuels. What is **net energy** and why is it important for evaluating energy resources? Use the net energy concept to explain why some energy resources need to be subsidized, and give an example of such a resource.

Section 13.2

3. What is the key concept for this section? What is **crude oil (petroleum)** and how is it extracted from the earth? Define **peak production** for an oil well or field. What is **refining**? What are **petrochemicals** and why are they important? What are **proven oil reserves**? What three countries have the world's largest proven crude oil reserves? What three countries are the world's largest producers of crude oil and what three countries are the largest consumers of crude oil? Summarize oil production and consumption in the United States and discuss the controversy over future oil production. What are the major advantages and disadvantages of using crude oil as an energy resource? What is **shale oil** and how is this heavy oil produced? What are **oil sands,** or **tar sands**? What is bitumen, and how is it extracted and converted to heavy oil? What are the

major advantages and disadvantages of using shale oil and heavy oils produced from oil sands as energy resources?

4. Define **natural gas**, **liquefied petroleum gas (LPG)**, and **liquefied natural gas (LNG)**. What three countries have the world's largest proven natural gas reserves? List the three counties that produce and the three countries that consume most of the world's natural gas? Describe the potential for greatly increasing natural gas production in the United States and list some pros and cons of fracking. What are four ways to reduce the environmental impact from the production of natural gas in the United States? What are the major advantages and disadvantages of using natural gas as an energy resource? What is **coal**, how is it formed, and how do the various types of coal differ? How does a coal-burning power plant work? What three countries have the world's largest proven coal reserves? What three countries are the world's largest producers and what three countries are the largest consumers of coal? Summarize the major environmental and health problems caused by the use of coal. How could society use full-cost pricing to reduce the harmful environmental impacts of coal? What are the major advantages and disadvantages of using coal as an energy resource?

Section 13.3

5. What is the key concept for this section? What is **nuclear fission**? How does a nuclear fission reactor work and what are its major safety features? Describe the **nuclear fuel cycle**. Explain how highly radioactive spent fuel rods are stored and what risks this presents. How has the United States dealt with the high-level radioactive waste problem from spent fuel rods? What can be done with worn-out nuclear power plants? Summarize the arguments over whether or not the widespread use of nuclear power could help slow projected climate change during this century. Summarize the arguments of experts who disagree over the future of nuclear power. What is the relationship between nuclear power plants and the spread of nuclear weapons? What is **nuclear fusion** and what is its potential as an energy resource?

Section 13.4

6. What are the two key concepts for this section? What is **energy efficiency**? What is **energy conservation**? What percentage of the energy used in the United States is unnecessarily wasted? Describe two widely used energy-inefficient technologies. What are the major benefits of improving energy efficiency? Define and give an example of **cogeneration**. List three other ways to save energy and money in industry. What is an energy-efficient smart electrical grid and how can it save energy and money? List three ways to save energy and money in transportation. Explain why the true cost of gasoline is much higher than what consumers pay at the pump. Distinguish among hybrid, plug-in hybrid, all electric, and fuel-cell motor vehicles. What is a **hydrogen fuel cell**? List four ways to save energy and money **(a)** in new buildings and **(b)** in existing buildings. List three ways in which you can save energy and money. Give three reasons why we are still wasting so much energy and money in producing and using energy.

Section 13.5

7. What is the key concept for this section? List four reasons why renewable energy is not more widely used. Distinguish between a **passive solar heating system** and an **active solar heating system** and discuss the major advantages and disadvantages of using such systems for heating buildings. What are three ways to cool buildings naturally? Define and give two examples of a **solar thermal system.** List the major advantages and disadvantages of using the centralized systems. What is a **solar cell** (**photovoltaic** or **PV cell**) and what are the major advantages and disadvantages of using such devices to produce electricity?

8. Define **hydropower** and summarize the potential for expanding it. What are the major advantages and disadvantages of using hydropower to produce electricity? What is the potential for using tides and waves to produce electricity? Summarize the potential for using wind power **(a)** globally and **(b)** in the United States. What are the major advantages and disadvantages of using wind to produce electricity?

What is **biomass** and what are the major advantages and disadvantages of using solid biomass to provide heat and electricity? What is the fuelwood crisis? What are the major advantages and disadvantages of using liquid biofuels such as ethanol and biodiesel to power motor vehicles? Describe the controversy over the use of ethanol as a biofuel in the United States? What is **geothermal energy** and what are three major sources of such energy? What are the major advantages and disadvantages of using geothermal energy as a source of heat and to produce electricity? List the major advantages and disadvantages of using hydrogen as a fuel.

Critical Thinking

1. Are you for or against the increased use of horizontal drilling and fracking (**Core Case Study**) to produce oil and natural gas in the United States? Explain. What are the alternatives?

2. Do you think that the estimated hidden costs of gasoline should be included in its price at the pump? Explain. Would you favor much higher gasoline taxes to accomplish this if payroll taxes and income taxes were reduced to balance gasoline tax increases, with no net additional cost to consumers? Explain.

3. List three things you could do to reduce your dependence on oil and gasoline. Which of these things do you already do or plan to do?

4. List five ways in which you unnecessarily waste energy during a typical day, and explain how these actions violate each of the three **scientific principles of sustainability**.

5. What do you think should be the top three energy resources? Explain. What do you think should be the three least-used energy resources? Explain.

6. Explain why you would support or oppose each of the following proposals made by various energy analysts:
   a. Government subsidies for all energy alternatives should be eliminated so that all energy choices

Section 13.6

9. What is the key concept for this section? Describe the need to make a shift in to a new and more sustainable energy future and list three components of such a shift proposed by energy experts.

10. What are this chapter's *three big ideas*? Explain how we can apply each of the six **principles of sustainability** to make a transition to a more sustainable energy future.

Note: Key terms are in bold type.

can compete on a level playing field in the marketplace.
   b. All government tax breaks and other subsidies for conventional fossil fuels, synthetic natural gas and oil, and nuclear power (fission and fusion) should be eliminated. They should be replaced with subsidies and tax breaks for improving energy efficiency and developing renewable energy resources.
   c. Development of renewable energy resources should be left to private enterprise and should receive little or no help from the federal government, but the nuclear power and fossil fuels industries should continue to receive large federal government subsidies and tax breaks.

7. How important is it to make the transition to a new energy future? Do you think it can be done? Explain. How would making such a transition might affect your life and the lives of any children or grandchildren you might have? How would not making such a transition affect your life and the lives of any children or grandchildren you might have?

8. Congratulations! You are in charge of the world. List the five most important features of your energy policy and explain why each of them is important and how they relate to each other.

Doing Environmental Science

Do a survey of energy use at your school, based on the following questions: How is the electricity generated? How are most of the buildings heated? How is water heated? How are most of the vehicles powered? How is the computer network powered? How could energy efficiency be improved, if at all, in each of these areas? If it does not already do so, how could your school make use of solar, wind, biomass, and other forms of renewable energy? Write up a proposal for using energy more efficiently and sustainably at your school and submit it to school officials.

Global Environment Watch Exercise

Go to your MindTap course to access the GREENR database. At the top of the page, do a "Basic Search" on *fracking*. Use the results to find information on the latest developments in public opposition to, and regulation of, fracking in the United States. Write a report on your findings.

Ecological Footprint Analysis

Study the table to the right and then answer these questions by filling in the blank columns in the table.

1. Convert the miles per gallon figures in the table to kilometers per liter (kpl).

2. How many liters (and how many gallons) of gasoline would each type of car use annually if it were driven 19,300 kilometers (12,000 miles) per year?

3. How many kilograms (and how many pounds) of carbon dioxide would be released into the atmosphere annually by each car, based on the fuel consumption calculated in question 2? Assume that the combustion of gasoline releases 2.3 kilograms of CO_2 per liter (19 pounds per gallon).

COMBINED CITY/HIGHWAY FUEL EFFICIENCY FOR 2017 MODELS

Model	Miles per Gallon (mpg)	Kilometers per Liter (kpl)	Annual Liters (Gallons) of Gasoline	Annual CO_2 Emissions
Chevrolet All-Electric Volt	106			
Nissan All-Electric Leaf	112			
Toyota Prius Prime Plug-in Hybrid	54			
Toyota Prius—Hybrid	52			
Chevrolet Cruze	34			
Honda Accord	29			
Jeep Patriot 4WD	24			
Ford F150 Pickup	22			
Chevrolet Camaro 8 cyl	20			
Ferrari F12	12			

Compiled by the authors using data from the U.S. Environmental Protection Agency Fuel Economy Report.

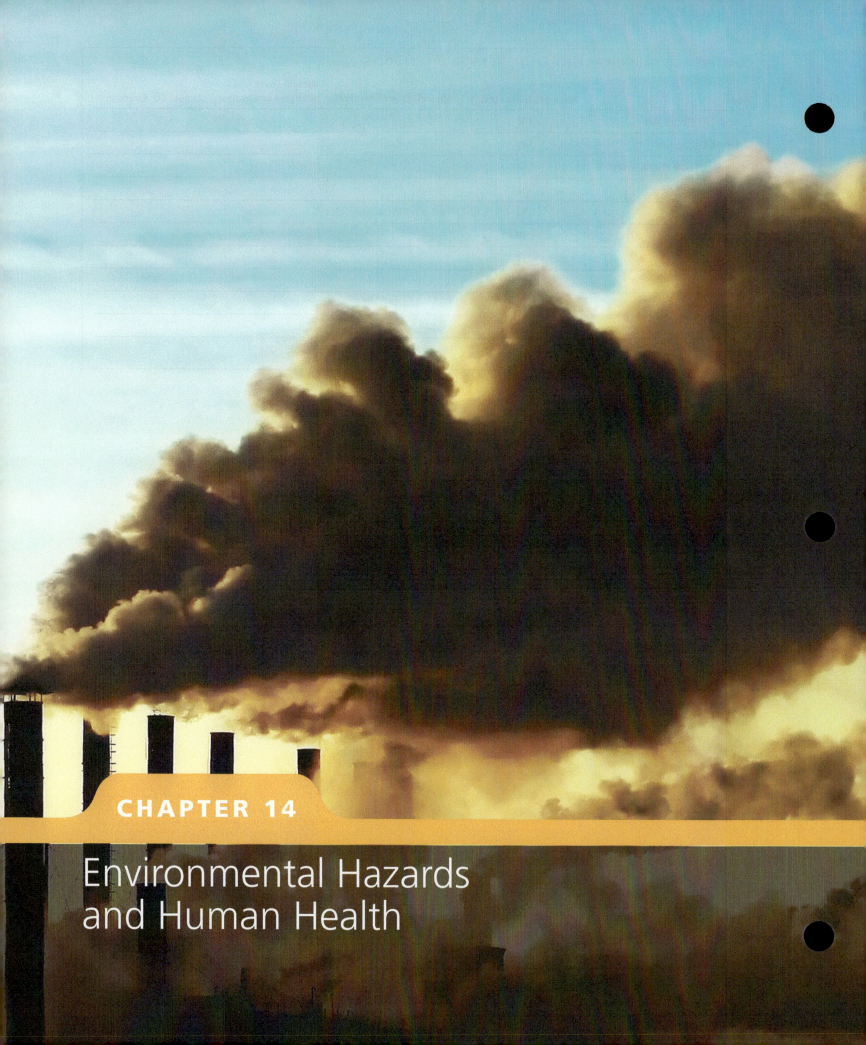

CHAPTER 14

Environmental Hazards and Human Health

Key Questions

14.1 What major health hazards do we face?

14.2 How do biological hazards threaten human health?

14.3 How do chemical hazards threaten human health?

14.4 How can we evaluate risks from chemical hazards?

14.5 How do we perceive and avoid risks?

Without effective air pollution control, coal-burning factories and power plants release toxic mercury and other air pollutants into the atmosphere.

Dudarev Mikhail/Shutterstock.com

Mercury's Toxic Effects

The metal mercury (Hg) and its compounds are toxic to humans. Research indicates that long-term exposure to high levels of mercury can permanently damage the human nervous system, brain, kidneys, and lungs. Exposure to low levels of mercury can cause birth defects and brain damages in fetuses and young children. Pregnant women, nursing mothers and their babies, women of childbearing age, and young children are especially vulnerable to the mercury's harmful effects.

Mercury is naturally released into the air from rocks, soil, and volcanoes and through vaporization from oceans. Such natural sources account for about one-third of the mercury that enters the atmosphere each year. The remaining two-thirds come from human activities.

Thousands of small-scale illegal mines in Asia, Latin America, and Africa (see Figure 12.12, p. 314) are the largest source of mercury air pollution. Miners use mercury to separate gold from its ore. Then they heat the mixture of gold and mercury to release the gold.

Other large sources of mercury in the atmosphere include emissions from coal-burning power and industrial plants (chapter-opening photo), cement kilns, smelters, and solid-waste incinerators.

Elemental mercury cannot be broken down or degraded. As a result, it builds up in soil, water, and tissues of humans and other animals. In the atmosphere, some elemental mercury is converted to more toxic inorganic and organic mercury compounds that can be deposited in lakes, in other aquatic environments, and on land.

Under certain conditions in aquatic systems, bacteria can convert inorganic mercury compounds to highly toxic *methylmercury* (CH_3Hg^+). These inorganic mercury compounds are typically deposited from the air or dumped into streams by small-scale gold miners. Like DDT (see Figure 8.12, p. 178), methylmercury can be biologically magnified in food chains and food webs. High levels of methylmercury are often found in the tissues of large fishes, such as tuna, swordfish, shark, and marlin, which feed at high trophic levels.

However, shrimp and salmon generally have low levels of mercury.

People are exposed to mercury in two major ways. They may eat fish and shellfish contaminated by methylmercury (Figure 14.1). This accounts for 75% of all human exposures to mercury. People may also inhale mercury vapors or inorganic particles present in the air, especially downwind from many coal-burning power and industrial plants (see chapter-opening photo) and solid-waste incinerators.

The greatest risk from exposure to low levels of methylmercury is brain damage in fetuses and young children. Studies estimate that 30,000–60,000 of the children born each year in the United States are likely to have reduced IQs and possible nervous system damage due to such exposure. Other health effects include poor balance and coordination, tremors, memory loss, insomnia, hearing loss, hair loss, and loss of peripheral vision.

This problem raises two important questions: How do scientists determine the potential harm from exposure to mercury and other chemicals? And how serious is the risk of harm from a particular chemical compared to other risks?

In this chapter, you will learn how scientists try to answer these and other questions about human exposure to chemicals. You will also learn about health threats from disease-causing bacteria, viruses, and protozoa, and from other environmental hazards that kill millions of people every year. Finally, we will consider ways to evaluate and avoid some risks. ●

FIGURE 14.1 Fish are contaminated with mercury in many lakes, including this one in Wisconsin.

14.1 WHAT MAJOR HEALTH HAZARDS DO WE FACE?

CONCEPT 14.1 We face health hazards from biological, chemical, physical, and cultural factors, and from the lifestyle choices we make.

We Face Many Types of Hazards

A **risk** is the probability of suffering harm from a hazard that can cause injury, disease, death, economic loss, or damage. Scientists often describe a risk in terms of its probability of causing harm in terms such as, "The lifetime probability of developing lung cancer from smoking one pack of cigarettes per day is 1 in 250." This means that 1 of every 250 people who smoke a pack of cigarettes every day will likely develop lung cancer over a typical lifetime (usually considered to be 70 years). Probability can also be expressed as a percentage, as in a 30% chance of developing a certain type of cancer.

Risk assessment is the process of using statistical methods to estimate how much harm a particular hazard can cause to human health or the environment. **Risk management** involves deciding whether and how to reduce a particular risk to a certain level and at what cost. Figure 14.2 summarizes how risks are assessed and managed.

Most people take avoidable risks every day. For example, they may choose to drive or ride in a car without a seatbelt or to text while driving. They may choose to eat foods that are high in cholesterol or that have too much sugar. They may drink too much alcohol or smoke.

Five major types of hazards pose risks to human health (**Concept 14.1**):

- *Biological hazards* from more than 1,400 **pathogens**, or microorganisms that can cause disease in other organisms. Examples are bacteria, viruses, parasites, protozoa, and fungi.

Risk Assessment	Risk Management
Hazard identification What is the hazard?	**Comparative risk analysis** How does it compare with other risks?
Probability of risk How likely is the event?	**Risk reduction** How much should it be reduced?
	Risk reduction strategy How will the risk be reduced?
Consequences of risk What is the likely damage?	**Financial commitment** How much money should be spent?

FIGURE 14.2 Risk assessment and risk management are used to estimate the seriousness of various risks and to help reduce such risks. *Critical thinking:* What is an example of how you have applied this process in your daily living?

- *Chemical hazards* from certain harmful chemicals in the air, water, soil, food, and human-made products (**Core Case Study**).

- *Natural hazards* such as fire, earthquakes, volcanic eruptions, floods, tornadoes, and hurricanes.

- *Cultural hazards* such as unsafe working conditions, criminal assault, and poverty.

- *Lifestyle choices* such as smoking, making poor food choices, and not getting enough exercise.

14.2 HOW DO BIOLOGICAL HAZARDS THREATEN HUMAN HEALTH?

CONCEPT 14.2 The most serious biological hazards we face are infectious diseases such as flu, acquired immunodeficiency syndrome (AIDS), tuberculosis, diarrheal diseases, and malaria.

Some Diseases Can Spread from One Person to Another

An **infectious disease** is a disease caused by a pathogen such as a bacterium, virus, or parasite invading the body and multiplying in its cells and tissues. **Bacteria** are single-cell organisms that are found everywhere and that can multiply rapidly on their own. Most bacteria are harmless and some are beneficial. However, those that cause diseases such as strep throat or tuberculosis are harmful.

A **virus** is a pathogen that invades a cell and takes over its genetic machinery to copy itself and spread throughout the body. Viruses can cause infectious diseases such as flu and AIDS. A **parasite** is an organism that lives on or inside another organism and feeds on it. Parasites range in size from one-celled organisms called protozoa to worms that are visible to the naked eye. They can cause an infectious disease such as malaria.

A **transmissible disease** is an infectious disease that can be transmitted from one person to another. Some transmissible diseases are bacterial diseases such as tuberculosis, many ear infections, and gonorrhea. Others are viral diseases such as the common cold, flu, and AIDS. Transmissible diseases can be spread through air, water, and food. They can also be transmitted by insects such as mosquitoes and by body fluids such as feces, urine, blood, semen, and droplets sprayed by sneezing and coughing.

A **nontransmissible disease** is caused by something other than a living organism and does not spread from person to person. Nontransmissible diseases include cardiovascular (heart and blood vessel) diseases, most cancers, asthma, and diabetes.

In 1900, infectious disease was the leading cause of death in the world. Since then, and especially since 1950, infectious disease rates and related death rates

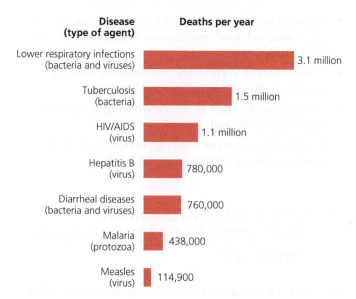

Disease (type of agent)	Deaths per year
Lower respiratory infections (bacteria and viruses)	3.1 million
Tuberculosis (bacteria)	1.5 million
HIV/AIDS (virus)	1.1 million
Hepatitis B (virus)	780,000
Diarrheal diseases (bacteria and viruses)	760,000
Malaria (protozoa)	438,000
Measles (virus)	114,900

FIGURE 14.3 Leading causes of death by infectious diseases in the world. *Data analysis:* How many people die from all seven of the infectious diseases every year? Every day?

(Compiled by the authors using data from the World Health Organization (WHO) and the U.S. Centers for Disease Control and Prevention)

have dropped significantly. This has been achieved mostly by a combination of improved sanitation, better health care, the use of antibiotics to treat bacterial diseases, and the development of vaccines to prevent the spread of some viral diseases.

Despite the declining risk of harm from infectious diseases, they remain serious a health threat, especially in less-developed countries. A large-scale outbreak of an infectious disease in an area or a country is called an *epidemic*. A global epidemic, like tuberculosis (see Case Study that follows) or AIDS, is called a *pandemic*. Figure 14.3 shows the annual death tolls from the world's seven deadliest infectious diseases (**Concept 14.2**).

CONSIDER THIS …

Learning from Nature

A shark's skin is covered with tiny bumps that somehow help it avoid bacterial infections. Scientists hope to use this information to create anti-bacterial films with a bumpy structure that could reduce human skin infections.

One reason why infectious disease is still a serious threat is that many disease-carrying bacteria have developed genetic immunity to widely used antibiotics (Science Focus 14.1). In addition, many disease-transmitting species of insects such as mosquitoes have become resistant to widely used pesticides such as DDT that once helped control their populations. However, pesticide producers are not developing new mosquito-killing insecticides. This is because

it costs insecticide producers about $250 million to develop and market each insecticide and annual sales of insecticides used to kill mosquitoes in the United States is $100 million.

CASE STUDY

The Global Threat from Tuberculosis

Tuberculosis (TB) is an ancient and highly contagious bacterial infection that destroys lung tissue and, without treatment, can lead to death. Many TB-infected people do not appear to be sick, and most of them do not know they are infected. Left untreated, each person with active TB typically infects a number of other people. Without treatment, about half of the people with active TB die from bacterial destruction of their lung tissue (Figure 14.4). According to the WHO, TB strikes about 10.4 million people a year and kills 1.5 million. In 2015, India had more new infections (2.8 million) and deaths (480,000) from TB than any other country, according to the WHO.

Several factors account for the spread of TB since 1990. One is a lack of TB screening and control programs, especially in less-developed countries where more than 90% of the new cases occur. However, researchers are developing new and easier ways to detect TB and to monitor its effects (Individuals Matter 14.1).

A second problem is that most strains of the TB bacterium have developed genetic resistance to the majority of the effective antibiotics (Science Focus 14.1). In addition, population growth, urbanization, and air travel have aided the spread of TB by greatly increasing person-to-person contact, especially in poor, crowded areas. A person with active TB might infect several people during a single bus or plane ride.

Slowing the spread of TB requires early identification and treatment of people with active TB, especially those with a chronic cough, which is the primary way in which the disease is spread from person to person. However, because many people do not show any symptoms of TB, they do not know they are infected and can infect other people. Treatment with a combination of four inexpensive drugs can cure 90% of individuals with active TB. To be effective, however, the drugs must be taken every day for 6–9 months. Symptoms often disappear after a few weeks, so many patients think they are cured and stop taking the drugs. This can allow TB to recur, in a more drug-resistant form, and to spread to others.

A deadly form of tuberculosis, known as *multidrug-resistant TB*, is on the rise. About 480,000 new cases occur every year, according to the WHO. Fewer than half of those cases are cured each year, and only with the best available medical care costing more than $500,000 per person on average. This form of TB kills about 150,000 people every year. Because this disease cannot be treated effectively with antibiotics, victims must be isolated from the rest of society, some permanently, and they pose a threat to health workers.

Genetic Resistance to Antibiotics

Antibiotics are chemicals that can kill bacteria. They have played an important role in the increase in life expectancy since 1950 in the United States and many other countries.

In 2014, the World Health Organization (WHO) issued a report warning that the age of antibiotics may be ending because many disease-causing bacteria are becoming genetically resistant to the antibiotics that have long been used to kill them. The reason for this is the astounding reproductive rate of bacteria. Some types of bacteria can grow from a population of 1 to well over 16 million in 24 hours. As a result, they can quickly become genetically resistant to an increasing number of antibiotics through natural selection (see Figure 4.9, p. 79). They pass such genetic resistance to their offspring, and research indicates that some bacteria can transfer such resistance to others of the same strain as well as to different strains of bacteria.

A major factor in the promotion of such genetic resistance, also called *antibiotic resistance*, is the widespread use of antibiotics on livestock raised in feedlots (see Figure 10.7, p. 234) and confined animal feeding operations, or CAFOs (see Figure 10.8, p. 234). Antibiotics are used to control disease and to promote growth among dairy and beef cattle, poultry, and hogs that are raised in large numbers in crowded conditions. The U.S. Food and Drug Administration (FDA) has estimated that about 80% of all antibiotics used in the United States and 50% of those used worldwide are added to the feed of healthy livestock. According to the Centers for Disease Control and Prevention (CDC), 22% of antibiotic-resistant illness in humans is linked to food, especially meat from livestock treated with antibiotics.

Another factor that can also promote genetic resistance to antibiotics is the overuse of antibiotics for colds, flu, and sore throats, most of which are caused by viruses that do not respond to treatment with antibiotics. In many countries, antibiotics are available without a prescription, which promotes excessive and unnecessary use. Another factor is the spread of bacteria around the globe by human travel and international trade. The growing use of antibacterial hand soaps and other antibacterial cleansers could also be promoting antibiotic resistance in bacteria. Such cleansers do not necessarily work any better than thorough hand washing with regular soap, according to the FDA.

Every major disease-causing bacterium has developed strains that resist at least 1 of the roughly 200 antibiotics in use. Furthermore, bacteria called superbugs, which resist all but a few antibiotics, are emerging. The CDC has estimated that each year, at least 2 million Americans get infectious diseases from superbugs, and at least 23,000 die. One of every 25 U.S. hospital patients picks up such an infection while in the hospital. A two-year British government study led by economist Jim O'Neil estimated that globally drug-resistant superbugs kill at least 700,000 people per year and, by 2050, could kill as many as 10 million people a year.

For example, a bacterium known as *methicillin-resistant Staphylococcus aureus,* commonly known as MRSA (or "mersa"), has become resistant to most common antibiotics. MRSA can cause severe pneumonia, a vicious rash, and a quick death if it gets into the bloodstream.

MRSA can be found in hospitals, nursing homes, schools, gyms, and college dormitories. It can be spread through skin contact, unsanitary use of tattoo needles, and contact with poorly laundered clothing and shared items such as towels, bed linens, athletic equipment, and razors. Another superbug found in hospitals is *Clostridium difficile*, or *C. diff*. It causes severe diarrhea and can live on surfaces such as bed rails and medical equipment. It causes about 250,000 infections and 14,000 deaths per year in the United States, according to the CDC.

Health officials warn that we could be moving into a post-antibiotic era of higher death rates. No major new antibiotics have been developed in recent years, mostly because drug companies have to spend millions of dollars to develop new antibiotics that are used for only a short time to treat infections. They make more money on drugs that users take daily for many years for diseases such as diabetes and high blood pressure. As a result, in 2016, just 5 of the world's 50 largest drug companies were developing new antibiotics. However, some governments and private groups are undertaking efforts to develop antibiotics and vaccines that are more effective.

CRITICAL THINKING

What are three steps that we could take to slow the rate at which disease-causing bacteria are developing resistance to antibiotics?

2 million Annual number of U.S. citizens who get infections that cannot be treated with any known antibiotics.

Hayat Sindi: Health Science Entrepreneur

Growing up in a home of humble means in Saudi Arabia, Hayat Sindi was determined to get an education, become a scientist, and do something for humanity. She was the first Saudi woman to be accepted at Cambridge University. She also was the first woman to earn a PhD in biotechnology at Cambridge, and she taught in the Cambridge's international medical program. She was named a United Nations Educational, Scientific, and Cultural Organization (UNESCO) Goodwill Ambassador for science education and a National Geographic Explorer.

As a visiting scholar, Sindi worked with a team of scientists at Harvard University that co-founded a nonprofit company called *Diagnostics for All* to bring low-cost health monitoring to remote, poor areas of the world. The Harvard team sought to develop simple and inexpensive diagnostic tools that could be used to detect certain illnesses and medical problems in remote areas.

One such tool is a piece of paper the size of a postage stamp, with tiny channels and wells etched into it. A technician loads the channels with diagnostic chemicals and then puts a drop of a patient's blood, urine, or saliva onto this paper. The fluid travels through the channels where the chemicals react with the fluid to change its color. Results show up in less than a minute and can easily be read to diagnose different medical infections and conditions such as declining liver function, which can result from taking drugs to combat TB, hepatitis, and HIV/AIDS. The test can be conducted by a technician with minimal training and requires no electricity, clean water, or special equipment. After the paper is used, it can be burned on the spot to prevent the spread of any infectious agents.

Dr. Sindi has a passion for inspiring women and girls, particularly those in the Middle East, to pursue science. As she explains, "I want all women to believe in themselves and know they can transform society."

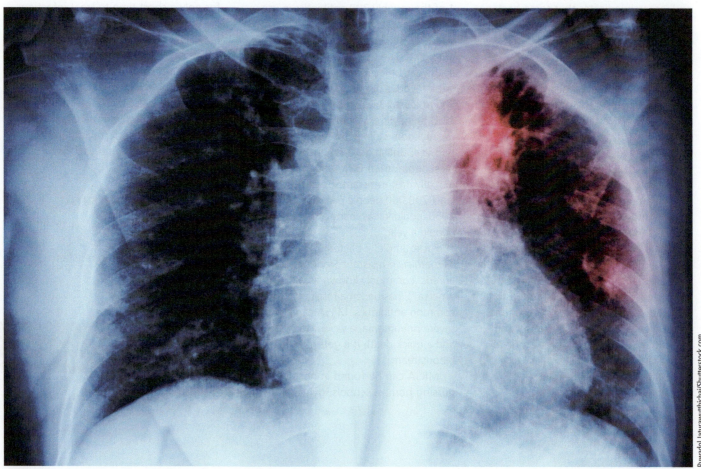

FIGURE 14.4 Colorized red areas in this chest X-ray show where TB bacteria have destroyed lung tissue in the upper left lung.

Some Viruses and Parasites Are Killers

Antibiotics do not affect viruses and some can be deadly. The biggest viral killer is the *influenza* or *flu virus* (**Concept 14.2**), because it often leads to fatal pneumonia. The flu virus is transmitted by the body fluids or by airborne droplets released when an infected person coughs or sneezes. Flu viruses are so easily transmitted that an especially potent flu virus could spread around the world in only a few months. This could cause a pandemic that could kill millions of people.

The second largest viral killer is the *human immunodeficiency virus*, or *HIV* (see Case Study that follows). According to the Joint United Nations Programme on HIV, in 2015, HIV infected about 2.1 million people and 1.1 million people died from AIDS-related illnesses (down from 2 million in 2005). HIV is transmitted by unsafe sex, the sharing of needles by drug users, infected mothers who pass the virus to their offspring before or during birth, and exposure to infected blood.

The third largest viral killer is the *hepatitis B virus (HBV)*, which damages the liver. According to the WHO, it kills more than 780,000 people each year. It is spread in the same ways that HIV is spread.

Another deadly viral disease is Ebola. In 2014, an outbreak of the disease in several West African countries infected 28,500 people and killed 11,300 of them. According to the WHO, the Ebola virus kills an average of 50% of those it infects within eight days. In 2016, an experimental Ebola vaccine was developed that gives 100% protection against the disease and is being evaluated by regulatory agencies. The chances of Ebola spreading in the United States and other more-developed countries are slim because hospitals, infection control, and safe burial procedures are more readily available than in many less-developed countries.

Another deadly virus is the *West Nile* virus, which is transmitted to humans by the bite of a common mosquito that is infected when it feeds on birds that carry the virus. Between 1999 and 2015, the virus caused severe illness in nearly 43,900 people and killed about 1,750 people in the United States, according to the CDC.

Between 2010 and 2016, the *Zika virus* spread rapidly in 42 countries, most in Latin America. In 2015, it affected more than a million people in Brazil. It is spread by the bite of a mosquito species that also spreads yellow fever and dengue fever. It can also be transmitted sexually through the semen of an infected male. The species is widespread in Latin America and by 2016, it had been found in 30 U.S. states, most of them warmer southern states. The most susceptible people live in parts of Texas, Louisiana, and Florida where the mosquito species that transmits the virus is found. The disease can spread rapidly in less-developed countries with warm climates where many houses have no window or door screens. The mosquitoes breed in standing water found near such homes.

The Zika virus has little effect on most adults. The main concern is a link between pregnant women carrying the virus and premature births or birth defects in some of their babies, including a shrunken head and brain and blindness. Scientists and U.S. health officials say there is little risk of a major outbreak of the disease in the United States because of the widespread use of window and door screens, air conditioning, and mosquito control programs. Pregnant women or women trying to get pregnant are advised not to travel to countries where the virus exists and is spreading.

Scientists estimate that throughout history, more than half of all infectious diseases were originally transmitted to humans from wild or domesticated animals. The development of such diseases has spurred the growth of the relatively new field of *ecological medicine*—devoted to tracking down disease connections between animals and humans. Scientists in this field have identified several human practices that encourage the spread of diseases among animals and people:

- The clearing or fragmenting of forests to make way for settlements, farms, and expanding cities.

- The hunting of wild game for food. In parts of Africa and Asia, local people who kill monkeys and other animals for bushmeat (see Figure 8.16, p. 183) regularly come in contact with primate blood and can be exposed to a simian (ape or monkey) strain of HIV, which causes AIDS.

- The illegal international trade in wild species.

- Industrialized meat production. For example, a deadly form of *E. coli* bacteria sometimes spreads from livestock to humans when people eat meat contaminated by animal manure. Salmonella bacteria found on animal hides and in poorly processed, contaminated meat also can cause food-borne disease.

Each of us can greatly reduce our chances of getting infectious diseases by washing our hands with plain soap frequently and thoroughly (for at least 20 seconds). We can greatly slow the spread of such diseases by not sharing personal items such as razors or towels, and by keeping cuts and scrapes covered with bandages until healed. It also helps to avoid contact with people who have flu or other viral diseases and to try not to touch your eyes, nose, and mouth.

Another growing health hazard is infectious diseases caused by parasites, especially malaria (see the second Case Study that follows).

CASE STUDY

The Global HIV/AIDS Epidemic

The spread of AIDS, caused by HIV infection, is a major global health threat. This virus cripples the immune system and leaves the body vulnerable to infections such as TB and rare forms of cancer such as *Kaposi's sarcoma*.

A person infected with HIV can live a normal life, especially with proper but costly treatment. In time, however, HIV can develop into AIDS, which can be fatal. An estimated 20% of all people infected with HIV are not aware of the infection and can spread the virus for years before being diagnosed.

Since HIV was identified in 1981, this viral infection has spread around the globe. According to the WHO, in 2015, about 36.7 million people worldwide (about 1.1 million in the United States) were living with HIV. In 2015, there were about 2.1 million new cases of AIDS (about 39,500 in the United States)—half of them in people ages 15 to 24.

Between 1981 and 2015, about 36 million people died of AIDS-related diseases, according to UNAIDS. According to the CDC, the U.S. death toll from AIDS for the same period was more than 698,000. In 2015, AIDS killed about 1.1 million people (about 7,000 in the United States)—down from a peak of 2.3 million in 2005. AIDS has reduced the life expectancy of the 750 million people living in sub-Saharan Africa from 62 to 47 years, on average, and to 40 years in the seven countries most severely affected by AIDS.

Worldwide, AIDS is the leading cause of death for people of ages 15 to 49. This affects the population age structures in several African countries, including Botswana (Figure 14.5), where 25% of all people between ages 15

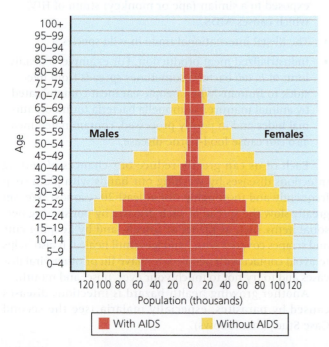

FIGURE 14.5 In Botswana, more than 25% of people ages 15–49 were infected with HIV in 2014. This figure shows two projected age structures for Botswana's population in 2020—one including the possible effects of the AIDS epidemic (red bars), and the other not including those effects (yellow bars). *Critical thinking:* How might this affect Botswana's economic development?

(Compiled by the authors using data from the U.S. Census Bureau, UN Population Division, and World Health Organization)

and 49 were infected with HIV in 2014, the third highest infection rate in the world. The premature deaths of many young, productive teachers, health-care workers, farmers, and other adults in these countries has contributed to declines in education, health care, food production, economic development, and political stability. They have also led to the disintegration of many families and to large numbers of orphaned children.

The treatment for an HIV infection includes a combination of antiviral drugs that can slow the progress of the virus. However, such drugs cost too much to be used widely in the less-developed countries where AIDS infections are widespread.

CASE STUDY

Malaria—The Spread of a Deadly Parasite

About 3.2 billion people—44% of the world's population—are at risk of getting malaria (Figure 14.6). Most of them live in poor African countries. People traveling to malaria-prone areas are also at risk because there is no vaccine that can prevent this disease.

Malaria is caused by a parasite that is spread by certain species of mosquitoes. When such a mosquito bites a person, it picks up the parasite in the person's blood and passes it on to the next person it bites. The parasite infects and destroys many of its victim's red blood cells, which causes intense fever, chills, drenching sweats, severe abdominal pain, vomiting, headaches, and greater susceptibility to other diseases.

According to the WHO and UNICEF, malaria killed about 438,000 people in 2015 and infected 214 million people. Some experts contend this total could be much higher, because public health records are incomplete in many areas. More than 90% of all malaria cases occur in sub-Saharan Africa, the area south of the Sahara Desert. Most cases involve children younger than 5 years old. Roughly every minute, a child under age 5 dies from malaria. Many children who survive malaria suffer brain damage or impaired learning ability.

Over the course of human history, malarial parasites probably have killed more people than all the wars ever fought. The spread of malaria slowed during the 1950s and 1960s, a time when widespread draining of swamps and marshes sharply reduced mosquito breeding areas. These areas were also sprayed with insecticides, and drugs were developed to kill the parasites in people's bloodstreams.

Since 1970, however, malaria has come roaring back. Most of the mosquito species that transmit malaria have become genetically resistant to most insecticides and the parasites have become genetically resistant to common antimalarial drugs. Climate change is expected to aid the spread of malaria by allowing malaria-carrying mosquitoes to spread from tropical areas to warming temperate areas.

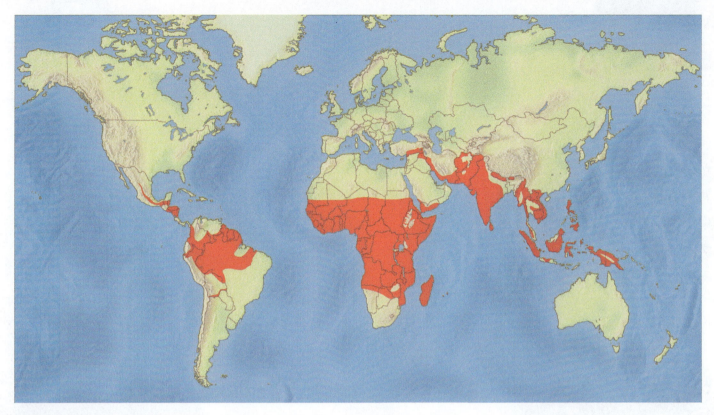

FIGURE 14.6 About 44% of the world's population live in areas in which malaria is prevalent. As the earth warms, malaria may spread to some temperate areas such as the southern half of the United States.

(Compiled by the authors using data from the World Health Organization and U.S. Centers for Disease Control and Prevention)

Scientists have made progress in developing a malaria vaccine, but currently no effective vaccine is available. Another approach is to provide poor people in malarial regions with free or inexpensive insecticide-treated bed nets (Figure 14.7) and window screens. Between 2000 and 2014, the percentage of Africa's population sleeping under mosquito nets increased from 2% to more than 50% and saved 6.2 million lives, according to the WHO.

Reducing the Incidence of Infectious Diseases

According to the WHO, the percentage of all deaths world-wide resulting from infectious diseases dropped from 35% to 16% between 1970 and 2015, primarily because a growing number of children were immunized against the major infectious diseases. Between 1990 and 2015, the estimated annual number of children younger than age 5 who died from infectious diseases dropped from nearly 12 million to 4.9 million. **GOOD NEWS**

Figure 14.8 lists measures that could help prevent or reduce the incidence of infectious diseases—especially in less-developed countries. The WHO has estimated that implementing the solutions listed in Figure 14.8 could save the lives of as many as 4 million children younger than age 5 a year. **GREEN CAREER: Infectious disease prevention**

CONNECTIONS: Drinking Water, Latrines, and Infectious Diseases

More than a third of the world's people—2.6 billion—do not have sanitary bathroom facilities. Nearly 1 billion people get their water for drinking, washing, and cooking from sources polluted by animal or human feces. A key to reducing sickness and premature death due to infectious disease is to focus on providing these people with simple latrines and access to safe drinking water.

14.3 HOW DO CHEMICAL HAZARDS THREATEN HUMAN HEALTH?

CONCEPT 14.3 Certain chemicals in the environment can cause cancers and birth defects and disrupt the human immune, nervous, and endocrine systems.

Some Chemicals May Cause Cancers, Mutations, and Birth Defects

There is growing concern about the effects of toxic chemicals on human health. A **toxic chemical** is an element or compound that can cause temporary or permanent harm

FIGURE 14.7 This baby in Senegal, Africa, is sleeping under an insecticide-treated mosquito net to reduce the risk of being bitten by malaria-carrying mosquitoes.

Olivier Asselin/Alamy Stock Photo

Solutions

Infectious Diseases

- Increase research on tropical diseases and vaccines
- Reduce poverty and malnutrition
- Improve drinking water quality
- Reduce unnecessary use of antibiotics
- Sharply reduce use of antibiotics on livestock
- Immunize children against major viral diseases
- Provide oral rehydration for diarrhea victims
- Conduct global campaign to reduce HIV/AIDS

FIGURE 14.8 Ways to prevent or reduce the incidence of infectious diseases, especially in less-developed countries. *Critical thinking:* Which three of these approaches do you think are the most important? Why?

Top: Omer N Raja/Shutterstock.com. Bottom: Rob Byron/Shutterstock.com.

or death to humans. The U.S. Environmental Protection Agency (EPA) has listed arsenic, lead, mercury (**Core Case Study**), vinyl chloride (used to make PVC plastics), and polychlorinated biphenyls (PCBs) as the top five toxic chemicals that are the most harmful to human health.

There are three major types of potentially toxic agents. **Carcinogens** are chemicals, some types of radiation, and certain viruses that can cause or promote *cancer*. Cancer is a disease in which malignant cells multiply uncontrollably and create tumors, or masses of abnormal cells. Tumors can damage the body and often lead to premature death. Examples of carcinogens are arsenic, benzene, formaldehyde, gamma radiation, PCBs, radon, ultraviolet (UV) radiation, vinyl chloride, and certain chemicals in tobacco smoke.

Typically, 10 to 40 years can pass between the initial exposure to a carcinogen and the appearance of detectable cancer symptoms. This time lag helps explain why many healthy teenagers and young adults have trouble believing that their own habits such as smoking and poor diet could lead to some form of cancer before they reach age 50.

Mutagens are the second major type of toxic substance. **Mutagens** include chemicals or forms of radiation that cause or increase the frequency of *mutations*, or changes, in the DNA molecules found in cells. Most mutations cause no harm, but some can lead to cancers and other disorders. For example, nitrous acid (HNO_2), formed by the digestion of nitrite (NO_2^-) preservatives in foods, can cause mutations linked to increases in stomach cancer in people who consume large amounts of processed foods and wine containing such preservatives. Harmful mutations occurring in reproductive cells can be passed on to offspring and to future generations.

Teratogens, a third type of toxic agent, are chemicals that harm a fetus or embryo or cause birth defects. Ethyl alcohol, an ingredient in alcoholic beverages, is a teratogen. Women who drink alcoholic beverages during pregnancy increase their risk of having babies with low birth weight and a number of physical, developmental, behavioral, and mental problems. Other teratogens are mercury (**Core Case Study**), lead, PCBs, formaldehyde, benzene, phthalates, and PCP (angel dust).

Some Chemicals Affect Our Immune and Nervous Systems

Since the 1970s, research on wildlife and laboratory animals along with some studies of humans has suggested that long-term exposure to some chemicals in the environment can disrupt important body systems, including our immune and nervous systems (**Concept 14.3**).

The *immune system* consists of specialized cells and tissues that protect the body against disease and harmful substances. For example, it forms *antibodies*, or specialized proteins, that detect and destroy invading agents. Some chemicals such as arsenic and methylmercury (**Core Case Study**) can weaken the human immune system. This leaves the body vulnerable to attacks by allergens and by infectious bacteria, viruses, and protozoa.

Neurotoxins are natural and synthetic chemicals that can harm the human *nervous system*, which includes the brain, spinal cord, and peripheral nerves. Neurotoxins can cause behavioral changes, learning disabilities, attention-deficit disorder, paralysis, and death. Examples of neurotoxins are PCBs, arsenic, lead, and certain pesticides.

Methylmercury (**Core Case Study**) is an especially dangerous neurotoxin because it persists in the environment and, like DDT, can be biologically magnified in food chains and food webs (Figure 14.9). According to the Natural Resources Defense Council, predatory fish such as tuna, marlin, orange roughy, swordfish, mackerel, grouper, and sharks can have mercury concentrations in their bodies that are 10,000 times higher than the levels in the water around them.

In one study, the EPA study found that almost half of the fish tested in 500 lakes and reservoirs across the United States had levels of mercury that exceeded safe levels (Figure 14.1). Similarly, a study by the U.S. Geological Survey of nearly 300 streams across the United States found mercury-contaminated fish in all of the streams surveyed, with one-fourth of the fish exceeding the safe levels determined by the EPA.

The EPA estimates that about 1 of every 12 women of childbearing age in the United States has enough mercury in her blood to harm a developing fetus. Figure 14.10 lists ways to prevent or reduce human inputs of mercury (**Core Case Study**) into the environment.

Some Chemicals Affect the Endocrine System

The *endocrine system* is a complex network of glands that release tiny amounts of *hormones* into the bloodstreams of humans and other vertebrate animals. Very low levels of these chemical messengers (often measured in parts per billion or parts per trillion) regulate bodily systems that control sexual reproduction, growth, development, learning ability, and behavior. Each type of hormone has a unique molecular shape that allows it to attach to certain parts of cells called *receptors*, and to transmit a chemical message.

Molecules of certain pesticides and other synthetic chemicals have shapes similar to those of natural hormones. This allows them to attach to the molecules of natural hormones and to disrupt the endocrine systems in humans and in some other animals (**Concept 14.3**). These molecules are called *hormonally active agents (HAAs)* or *endocrine disruptors*.

Examples of HAAs include some herbicides, organophosphate pesticides, dioxins, lead, phthalates, various fire retardants, and mercury (**Core Case Study**). Some HAAs, including BPA (Science Focus 14.2), act as hormone imposters, or *hormone mimics*. They are chemically similar to estrogens (female sex hormones) and can disrupt the endocrine system by attaching to estrogen receptor molecules. Other HAAs, called *hormone blockers*, disrupt the endocrine system by preventing natural hormones such as androgens (male sex hormones) from attaching to their receptors.

Estrogen mimics and hormone blockers can have a number of effects on sexual development and reproduction. Numerous studies involving wild animals, laboratory animals, and humans suggest that males of species

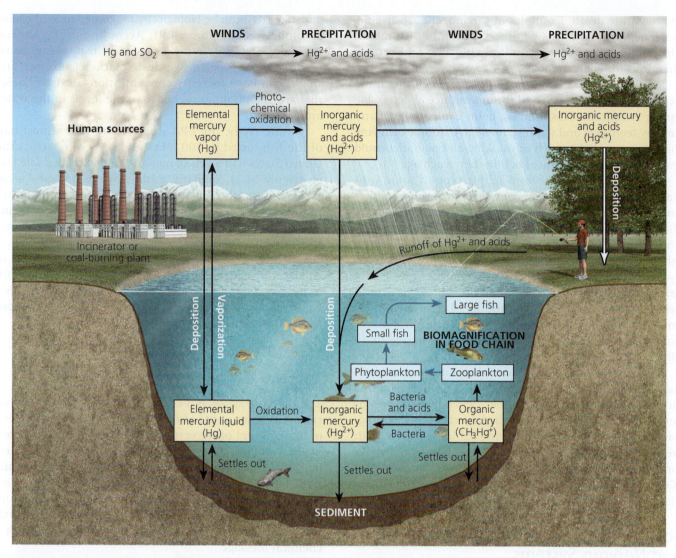

WINDS PRECIPITATION WINDS PRECIPITATION

Hg and SO₂ → Hg²⁺ and acids → Hg²⁺ and acids

Human sources

Elemental mercury vapor (Hg) — Photo-chemical oxidation → Inorganic mercury and acids (Hg²⁺) → Inorganic mercury and acids (Hg²⁺)

Incinerator or coal-burning plant

Deposition

Vaporization

Deposition

Runoff of Hg²⁺ and acids

Deposition

Large fish

Small fish **BIOMAGNIFICATION IN FOOD CHAIN**

Phytoplankton Zooplankton

Elemental mercury liquid (Hg) — Oxidation → Inorganic mercury (Hg²⁺) — Bacteria and acids → Organic mercury (CH₃Hg⁺)

Bacteria

Settles out Settles out Settles out

SEDIMENT

FIGURE 14.9 Movement of different forms of toxic mercury from the atmosphere into an aquatic ecosystem where it is biologically magnified in a food chain. *Critical thinking:* What is your most likely exposure to mercury?

Solutions

Mercury Pollution

Prevention	Control
Phase out waste incineration	Sharply reduce mercury emissions from coal-burning plants and incinerators
Remove mercury from coal before it is burned	
	Label all products containing mercury
Switch from coal to natural gas and renewable energy resources	Collect and recycle batteries and other products containing mercury

FIGURE 14.10 Ways to prevent or control inputs of mercury (**Core Case Study**) into the environment from human sources—mostly coal-burning power plants and incinerators. *Critical thinking:* Which four of these solutions do you think are the most important? Why?

Top: Mark Smith/Shutterstock.com. Bottom: tuulijumala/Shutterstock.com.

exposed to hormonal disrupters are generally becoming more feminized.

There is also growing concern about another group of HAAs that affect hormones generated by the thyroid gland. These pollutants, called *thyroid disrupters*, can cause growth, weight, brain, and behavioral disorders. Some of these chemicals are found in the nonstick surfaces of cookware and are used as flame retardants added to certain fabrics, furniture, plastics, and mattresses.

Scientists are also increasingly concerned about certain HAAs called *phthalates* (pronounced THALL-eights). These

The Controversy over BPA

The estrogen mimic *bisphenol A (BPA)* serves as a hardening agent in certain plastics that are used in a variety of products. They include some baby bottles, sipping cups, and pacifiers, as well as some reusable water bottles, sports drink and juice bottles, microwave dishes, and food storage containers. BPA is also used to make some dental sealants, as well as the plastic resins that line nearly all food and soft drink cans and cans holding baby formulas and foods. This type of liner allows containers to withstand extreme temperatures, keeps canned food from interacting with the metal in the cans, prevents rust in the cans, and helps preserve the canned food. People can also be exposed to BPA by touching thermal paper used to produce some cash register receipts.

A CDC study indicated that 93% of Americans age 6 and older had trace levels of BPA in their urine. Children and adolescents generally had higher urinary BPA levels than adults. These levels were well below the acceptable level set by the EPA. However, that level was established in the late 1980s, when little was known about the potential effects of BPA on human health.

Research indicates that the BPA in plastics can leach into water or food when the plastic is heated to high temperatures, microwaved, or exposed to acidic liquids. Harvard University Medical School researchers found a 66% increase in BPA levels in the urine of participants who drank from polycarbonate bottles regularly for 1 week.

By 2013, more than 90 published studies by independent laboratories had reported a number of significant adverse effects on test animals from exposure to very low levels of BPA. These effects include brain damage, early puberty, decreased sperm quality, certain cancers, heart disease, obesity, liver damage, impaired immune function, type 2 diabetes, hyperactivity, impaired learning, impotency in males, and obesity in test animals.

On the other hand, 12 studies funded by the chemical industry found no evidence or only weak evidence of adverse effects from low-level exposure to BPA in test animals. In 2008, the FDA concluded that BPA in food and drink containers was not a health hazard. In 2015, the European Food Safety Authority agreed, concluding that BPA is not appearing in people's body systems at high enough levels to cause harm. However, France has banned BPA from the lining of all food cans,

Most manufacturers offer BPA-free alternatives for products such as baby bottles, sipping cups, and sports water bottles. Many consumers are avoiding plastic containers with a #7 recycling code (which indicates that BPA can be present). People are also using powdered infant formula instead of liquid formula from lined metal cans. Some also choose glass bottles and food containers instead of those made of plastic or lined with plastic resins. In addition, some use glass, ceramic, or stainless steel coffee mugs instead of plastic cups.

Many manufacturers have replaced BPA with bisphenol S (BPS). However, studies indicate that BPS can have health effects similar to those of BPA, and BPS is now showing up in human urine at levels similar to those of BPA.

There are substitutes for the plastic resins containing BPA or BPS that line most of the food cans used in the United States. However, these replacements are more expensive, and the potential health effects of some chemicals they contain also need to be evaluated.

CRITICAL THINKING

Should plastics that contain BPA or BPS be banned from use in all children's products? Explain. Should such plastics be banned from use in the liners of canned food containers? Explain. What are the alternatives?

chemicals are used to make plastics more flexible and to make cosmetics easier to apply to the skin. They are found in a variety of products, including many detergents, perfumes, cosmetics, baby powders, body lotions for adults and babies, sunscreens, hair sprays, deodorants, soaps, nail polishes, shampoos for adults and babies, and the coatings on many timed-release drugs. They are also found in polyvinyl chloride (PVC) plastic products such as soft vinyl toys, teething rings, blood storage bags, intravenous (IV) drip bags, shower curtains, and some plastic food and drink containers.

Exposure of laboratory animals to high doses of various phthalates has caused birth defects, kidney and liver diseases, immune system suppression, and abnormal sexual development in these animals. Studies have linked exposure of human babies to phthalates with early puberty in girls and sperm damage in men. The European Union (EU) and at least 14 other countries have banned several phthalates. However, scientists, government regulators, and manufacturers in the United States are divided on the risks of phthalates to human health and reproductive systems.

Concerns about BPA, phthalates, and other HAAs show how difficult it can be to assess the potential harmful health effects from exposure to very low levels of various chemicals. Resolving these uncertainties will take decades of research. Some scientists argue that as a precaution during this period of research, people should sharply reduce their exposure to products that contain potentially harmful hormone disrupters, especially in products frequently used by pregnant women, infants, young children, and teenagers (Figure 14.11).

What Can You Do?

Exposure to Hormone Disrupters

- Eat certified organic produce and meats
- Avoid processed, prepackaged, and canned foods
- Use glass and ceramic cookware
- Store food and drinks in glass containers
- Use only natural cleaning and personal care products
- Use natural fabric shower curtains, not vinyl
- Avoid artificial air fresheners, fabric softeners, and dryer sheets
- Use only glass baby bottles and BPA-free, phthalate-free sipping cups, pacifiers, and toys

FIGURE 14.11 Individuals Matter: Ways to reduce your exposure to hormone disrupters. ***Critical thinking:*** Which three of these steps do you think are the most important ones to take? Why?

14.4 HOW CAN WE EVALUATE RISKS FROM CHEMICAL HAZARDS?

CONCEPT 14.4A Scientists use live laboratory animals, case reports of poisonings, and epidemiological studies to estimate the toxicity of chemicals, but these methods have limitations.

CONCEPT 14.4B Many health scientists call for much greater emphasis on pollution prevention to reduce our exposure to potentially harmful chemicals.

Many Factors Determine the Toxicity of Chemicals

Toxicology is the study of the harmful effects of chemicals on humans and other organisms. **Toxicity** is a measure of the ability of a substance to cause injury, illness, or death to a living organism. A basic principle of toxicology is that any synthetic or natural chemical can be harmful if ingested or inhaled in a large enough quantity. However, the critical question is: "What level of exposure to a particular toxic chemical will cause harm?"

This is a difficult question to answer because many variables must be considered in estimating the effects of human exposure to chemicals. A key factor is the **dose**, the amount of a harmful chemical that a person has ingested, inhaled, or absorbed through the skin at any one time.

Age is another variable that impacts how a person is affected by exposure to a particular chemical. For example, toxic chemicals usually have a greater effect on elderly adults. Fetuses, infants, and children are also more vulnerable to exposure to toxic chemicals than adults. Current research suggests that exposure to chemical pollutants in the womb may be related to increasing rates of autism, childhood asthma, and learning disorders.

Infants and young children are more susceptible to the effects of toxic substances than are adults, for three major reasons. *First,* they generally breathe more air, drink more water, and eat more food per unit of body weight than do adults. *Second,* they are exposed to toxins in dust and soil when they put their fingers, toys, and other objects in their mouths. *Third,* children usually have less well-developed immune systems and body detoxification processes than adults have.

The EPA proposes that in determining any risk, regulators should assume that the risk factor children is 10 times higher than the risk for adults. Some health scientists say that to be on the safe side, regulators should assume that this risk for children is 100 times higher than the risk for adults.

Toxicity also depends on *genetic makeup,* which determines an individual's sensitivity to a particular toxin. Some people are sensitive to a number of toxins—a condition known as *multiple chemical sensitivity* (MCS). Another factor is how well the body's detoxification systems, including the liver, lungs, and kidneys, are working.

Several other variables can affect the level of harm caused by a chemical. One is its *solubility.* Water-soluble toxins can move throughout the environment and get into water supplies, as well as the aqueous solutions that surround cells inside the body. Oil- or fat-soluble toxins can penetrate the membranes that surround our cells. Thus, oil- or fat-soluble toxins can accumulate in body tissues and cells.

Another factor is a substance's *persistence,* or resistance to breaking down. Many chemicals, including DDT and PCBs, were used widely because they are not easily broken down in the environment. This means that they are more likely to remain in the body and have long-lasting harmful health effects.

Biological accumulation and magnification (see Figure 8.12, p. 178) can also play a role in toxicity. Animals that eat higher on the food chain are more susceptible to the effects of fat-soluble toxic chemicals because of the magnified concentrations of the toxins in their bodies. Examples of chemicals that can be biomagnified include DDT, PCBs (Figure 14.12), and methylmercury (**Core Case Study**).

The health damage resulting from exposure to a chemical is called the **response**. An *acute effect* is an immediate or rapid harmful reaction ranging from dizziness to death. A *chronic effect* is a permanent or long-lasting consequence of exposure to a single dose or to repeated lower doses of a harmful substance. Kidney and liver damage are examples of chronic effects.

Natural and synthetic chemicals can be safe or toxic. In fact, many synthetic chemicals, including many of the medicines we take, are quite safe if used as intended, while many natural chemicals such as lead and mercury (**Core Case Study**) are deadly.

Water	Phytoplankton	Zooplankton	Rainbow smelt	Lake trout	Herring gull	Herring gull eggs
0.000002 ppm	0.0025 ppm	0.123 ppm	1.04 ppm	4.83 ppm	124 ppm	124 ppm

FIGURE 14.12 Biological magnification of polychlorinated biphenyls (PCBs) in an aquatic food chain in the Great Lakes.

Scientists Use Various Methods to Estimate Toxicity

The most widely used method for determining toxicity involves tests with live animals. Scientists expose a population of such animals to measured doses of a specific substance under controlled conditions. Laboratory-bred mice and rats are widely used because, as mammals, their systems function similarly to human systems. They are also small and can reproduce rapidly under controlled laboratory conditions.

Scientists estimate the toxicity of a chemical by determining the effects of various doses of the chemical on test organisms and plotting the results in a **dose-response curve** (Figure 14.13). One approach is to determine the *lethal dose*—the dose that will kill an animal. A chemical's *median lethal dose (LD50)* is the dose that can kill 50% of the animals (usually rats and mice) in a test population within a given time period, usually expressed in milligrams of the chemical per kilogram of body weight (mg/kg). Then scientists use mathematical models to *extrapolate,* or estimate, the effects of the chemical on humans, based on the results from testing the chemical on lab animals.

Chemicals vary widely in their toxicity (Table 14.1). Some can cause serious harm or death after a single very low dose. For example, swallowing a few drops of pure nicotine (found in e-cigarettes) would make you very sick and a teaspoon of it could kill you. Other chemicals such as water or table sugar cause such harm only at dosages so huge that it is nearly impossible to get enough into the body to cause injury or death. Most chemicals fall between these two extremes.

Animal tests have drawbacks. They typically take two to five years to complete and involve hundreds to thousands of test animals. They can cost as much as $2 million per substance tested. Some tests can be painful to the test animals and can harm or kill them. Animal welfare groups want to limit or ban the use of test animals and ensure that they are treated humanely.

Some scientists challenge the validity of extrapolating data from laboratory animals to humans. They argue that important differences exist between humans and the test animals. Other scientists say that such tests and models can work fairly well (especially for revealing cancer risks) when the correct experimental animal is chosen or when a chemical is toxic to several different test-animal species.

More humane methods for toxicity testing are being used in place of live animals. They include making computer simulations and using individual animal cells, instead of whole, live animals. High-speed robot testing

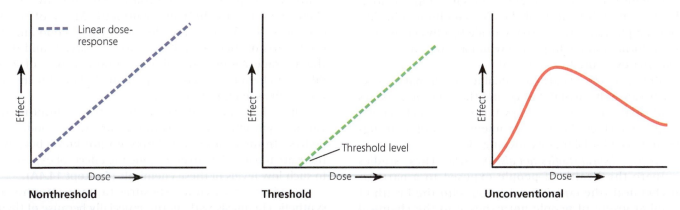

FIGURE 14.13 *Dose-response curves.* Scientists estimate the toxicity of various chemicals by determining how a chemical's harmful effects change as the dose increases. Some chemicals behave according to the *nonthreshold model* (left curve). Others behave according to the *threshold model* (center curve). Still others are unconventional in how they behave (right curve). For all of these graphs, the curves usually vary from being exactly linear, or straight. ***Critical Thinking:*** Can you think of commonly used chemicals that fit each of these models? What are they?

TABLE 14.1 Toxicity Ratings and Average Lethal Doses for Humans

Toxicity Rating	LD50 (milligrams per kilogram of body weight)*	Average Lethal Dose†	Examples
Supertoxic	Less than 5	Less than 7 drops	Nerve gases, botulism toxin, mushroom toxin, dioxin (TCDD)
Extremely toxic	5–50	7 drops to 1 teaspoon	Potassium cyanide, heroin, atropine, parathion, nicotine
Very toxic	50–500	1 teaspoon to 1 ounce	Mercury salts, morphine, codeine
Moderately toxic	500–5,000	1 ounce to 1 pint	Lead salts, DDT, sodium hydroxide, sodium fluoride, sulfuric acid, caffeine, carbon tetrachloride
Slightly toxic	5,000–15,000	1 pint to 1 quart	Ethyl alcohol, household cleansers, soaps
Essentially nontoxic	15,000 or greater	More than 1 quart	Water, glycerin, table sugar

*Dosage that kills 50% of individuals exposed.

†Amounts of substances in liquid form at room temperature that are lethal when given to a 70-kilogram (150-pound) human.

devices can now screen the biological activity of more than 1 million compounds a day to help determine their possible toxic effects.

The problems with estimating toxicities by using laboratory experiments get even more complicated (**Concept 14.4A**). In real life, each of us is exposed to a variety of chemicals, some of which can interact in ways that decrease or enhance their individual effects. Toxicologists already have great difficulty in estimating the toxicity of a single substance. Evaluating mixtures of potentially toxic substances, isolating the culprits, and determining how they interact, is overwhelming from a scientific and economic standpoint. For example, just studying the interactions among three of the 500 most widely used industrial chemicals would take 20.7 million experiments—a physical and financial impossibility.

Scientists use several other methods to get information about the harmful effects of chemicals on human health. For example, *case reports,* usually made by physicians, provide information about people who have become sick or died after exposure to a chemical.

Most case reports are not reliable for estimating toxicity because the actual dosage and the exposed person's health status are often unknown. However, such reports can provide clues about environmental hazards and suggest the need for laboratory investigations.

Epidemiological studies can also be useful. These studies compare the health of people exposed to a particular chemical (the *experimental group*) with the health of a similar group of people not exposed to the chemical (the *control group*). The goal is to determine whether the statistical association between exposure to a toxic chemical and a health problem is strong, moderate, weak, or undetectable.

Four factors can limit the usefulness of epidemiological studies. *First,* in many cases, too few people have been exposed to high enough levels of a toxic agent to detect statistically significant differences. *Second,* the studies usually take a long time. *Third,* closely linking an observed effect with exposure to a particular chemical is difficult because people are exposed to many different toxic agents throughout their lives and can vary in their sensitivity to such chemicals. *Fourth,* epidemiological studies cannot evaluate hazards from new technologies or chemicals to which people have not yet been exposed.

Are Trace Levels of Toxic Chemicals Harmful?

Almost everyone who lives in a more-developed country is exposed to potentially harmful chemicals (Figure 14.14) in their environment. Many of these chemicals build up to trace levels in their blood and in other parts of their bodies. CDC studies have found that the average American's blood contains traces of 212 different chemicals, including potentially harmful chemicals such as arsenic and BPA.

Should we be concerned about trace amounts of various synthetic chemicals in our air, water, food, and bodies? In most cases, we simply do not know because there are too few data to determine the effects of exposures to such low levels of these chemicals (**Concept 14.4A**).

Some scientists view exposures to trace amounts of synthetic chemicals with alarm, especially because of their potential long-term effects on the human body. Others scientists view the threats from such exposures as minor. These scientists contend that the concentrations of such chemicals are so low that they are harmless.

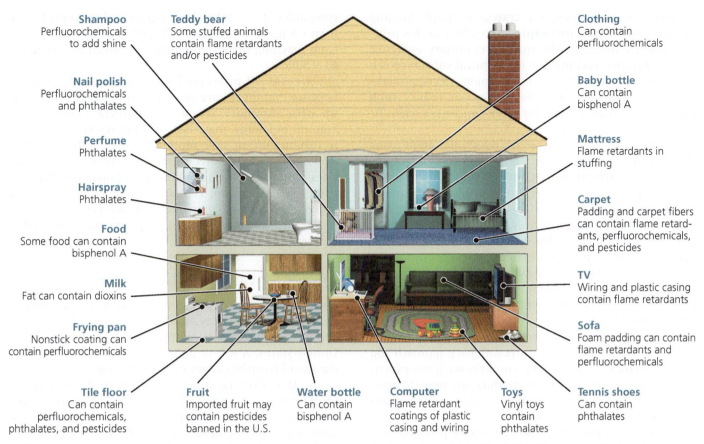

Shampoo
Perfluorochemicals to add shine

Teddy bear
Some stuffed animals contain flame retardants and/or pesticides

Clothing
Can contain perfluorochemicals

Nail polish
Perfluorochemicals and phthalates

Baby bottle
Can contain bisphenol A

Perfume
Phthalates

Mattress
Flame retardants in stuffing

Hairspray
Phthalates

Carpet
Padding and carpet fibers can contain flame retardants, perfluorochemicals, and pesticides

Food
Some food can contain bisphenol A

TV
Wiring and plastic casing contain flame retardants

Milk
Fat can contain dioxins

Frying pan
Nonstick coating can contain perfluorochemicals

Sofa
Foam padding can contain flame retardants and perfluorochemicals

Tile floor
Can contain perfluorochemicals, phthalates, and pesticides

Fruit
Imported fruit may contain pesticides banned in the U.S.

Water bottle
Can contain bisphenol A

Computer
Flame retardant coatings of plastic casing and wiring

Toys
Vinyl toys contain phthalates

Tennis shoes
Can contain phthalates

FIGURE 14.14 A number of potentially harmful chemicals are found in many homes. **Critical thinking:** Does the fact that we do not know much about the long-term harmful effects of these chemicals make you more likely or less likely to minimize your exposure to them? Why?

(Compiled by the authors using data from U.S. Environmental Protection Agency, Centers for Disease Control and Prevention, and New York State Department of Health)

Why Do We Know So Little About the Harmful Effects of Chemicals?

All methods for estimating toxicity levels and risks have serious limitations (**Concept 14.4A**), but they are all that we have. According to risk assessment expert Joseph V. Rodricks, "Toxicologists know a great deal about a few chemicals, a little about many, and next to nothing about most."

The U.S. National Academy of Sciences estimates that only 10% of the more than 85,000 registered synthetic chemicals in commercial use have been thoroughly screened for toxicity. Only 2% have been adequately tested to determine whether they are carcinogens, mutagens, or teratogens. Hardly any of the chemicals in commercial use have been screened for possible damage to the human nervous, endocrine, and immune systems.

Lack of data and high costs make regulation difficult. In fact, federal and state governments do not supervise the use of nearly 99.5% of the commercially available chemicals in the United States. The problem is much worse in less-developed countries.

Most scientists call for more research on the health effects of trace levels of synthetic chemicals. To minimize harm and take into account the uncertainty about health effects, scientists and regulators typically set allowed levels of exposure to toxic substances at 1/100th or even 1/1,000th of the estimated harmful levels.

Pollution Prevention and the Precautionary Principle

We know little about the potentially toxic chemicals around us and inside of us, and estimating their effects is very difficult, time-consuming, and expensive. So where does this leave us?

Some scientists and health officials, especially those in European countries, are pushing for greater emphasis on *pollution prevention* to protect human health (**Concept 14.4B**). They say chemicals that or known are suspected to cause significant harm should not be released into the environment at pollutant levels. Preventing such pollution requires finding harmless or less harmful substitutes for toxic and hazardous chemicals. It also requires recycling toxic chemicals within production processes to keep them from reaching the environment, as companies such as DuPont and 3M have done (see Case Study that follows).

Pollution prevention is a strategy for implementing the **precautionary principle**. According to this principle, when there is substantial preliminary evidence that an activity, technology, or chemical substance can harm living things or the environment, decision makers should take measures to prevent or reduce such harm, rather than waiting for more conclusive scientific evidence.

There is controversy over far we should go in using the precautionary principle. Those who favor a precautionary approach argue that anyone proposing to introduce a new chemical or technology should bear the burden of establishing its safety. This would require two major changes in the way we evaluate and manage risks. *First*, we would assume that new chemicals and technologies could be harmful until scientific studies show otherwise. *Second*, the existing chemicals and technologies that appear to have a strong chance of causing significant harm would be removed from the market until their safety is established.

Many manufacturers and businesses contend that widespread application of the precautionary approach and requiring pollution prevention would make it too expensive and almost impossible to introduce any new chemical or technology. They note that there is always some uncertainty in any scientific assessment of risk.

However, applying the precautionary principle can be good for business. It reduces health risks for employees and society, frees businesses from having to deal with pollution regulations, and reduces the threat of lawsuits from injured parties. It also focuses companies on finding solutions to pollution problems that are based on prevention rather than on cleanup. Companies might increase profits from sales of safer products and innovative technologies. They could also improve their images by operating in this manner.

Finally, proponents argue that society has an ethical responsibility to reduce known or potentially serious risks to human health, to the environment, and to future generations. This is in keeping with the ethical **principle of sustainability**.

Pollution Prevention Pays: The 3M Company

The U.S. based 3M Company makes 60,000 different products in 100 manufacturing plants around the world. In 1975, 3M began a Pollution Prevention Pays (3P) program. Since then, it has reformulated some of its products, redesigned equipment and processes, and reduced its use of hazardous raw materials. It has also recycled and reused more waste materials and sold some of its potentially hazardous but still useful wastes as raw materials to other companies. As of 2015, this program had prevented more than 1.8 million metric tons (2 million tons) of pollutants from reaching the environment and saved the company $1.9 billion.

The 3M 3P program has been successful largely because employees are rewarded if projects they come up with eliminate or reduce a pollutant; reduce the amount of energy, materials, or other resources required in production; or save money through reduced pollution control costs, lower operating costs, or increased sales of new or existing products. Employees at 3M have completed more than 10,000 3P projects. Since 1990, a growing number of companies have adopted similar pollution and waste prevention programs that have led to cleaner production.

Implementing Pollution Prevention

Pollution prevention programs by 3M and other companies are leading the way but there are major challenges in applying the precautionary principle more widely in the United States. A key to pollution prevention is banning the use of harmful chemicals or regulating their use.

At U.S. Congressional hearings in 2009, experts testified that the current regulatory system in the United States makes it virtually impossible for the government to limit or ban the use of toxic chemicals. Under this system, the EPA has required testing for only 200 of the more than 85,000 chemicals registered for use in the United States. In addition, it has issued regulations to control fewer than 12 of those chemicals.

However, there has been some progress. In 2011, after a 35-year delay promoted by politically powerful coal companies and utilities that burn coal to produce electricity, the EPA took a step in this direction by issuing a rule to control emissions of mercury (**Core Case Study**) and harmful fine-particle pollution from older coal-burning plants in 28 states. Many eastern states see high levels of deposition of mercury and harmful particles produced by coal-burning power and electric plants in the Midwest and blown eastward by prevailing winds (Figure 14.15). These new air pollution standards could prevent as many as 11,000 premature deaths, 200,000 non-fatal heart attacks, and 2.5 million asthma attacks. In 2014, the U.S. Supreme Court upheld these new EPA regulations but there have been efforts in Congress to delay implementation or exempt coal-burning power and industrial plants from the regulation.

Pollution prevention is also happening on an international scale. The Stockholm Convention of 2000 is an international agreement to ban or phase out the use of 12 of the most notorious *persistent organic pollutants (POPs)*, also called the *dirty dozen*. These highly toxic chemicals have been shown to produce numerous harmful effects, including cancers, birth defects, compromised immune systems,

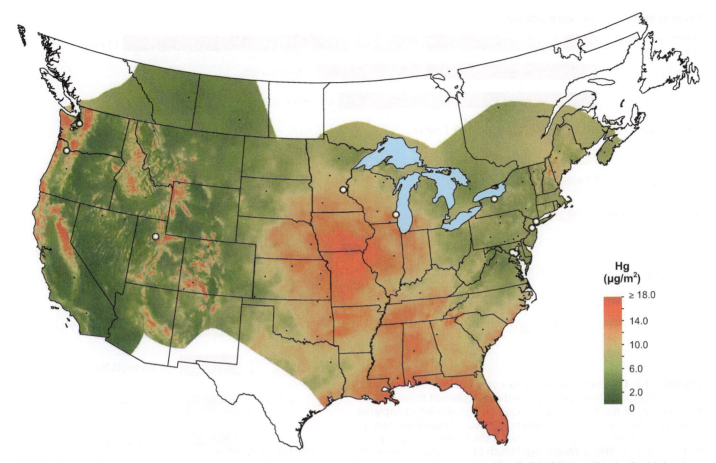

FIGURE 14.15 Atmospheric wet deposition of mercury in the lower 48 states in 2010. ***Critical thinking:*** Why do the highest levels occur mainly in the eastern half of the United States?

(Compiled by the authors using data from the Environmental Protection Agency and the National Atmospheric Deposition Program)

and declining sperm counts and sperm quality in men in a number of countries. The list includes DDT and eight other pesticides, PCBs, and dioxins. In 2009, nine more POPs were added, some of which are widely used in pesticides and in flame-retardants added to clothing, furniture, and other consumer goods. The treaty went into effect in 2004 but has not been formally approved or implemented by the United States.

A United Nations treaty known as the Minamata Convention seeks to curb most human-related inputs of mercury into the environment (**Core Case Study**). The overall goal is to reduce global mercury emissions by 15% to 35% in the next several decades. By January 2016, 128 countries had signed and 22 countries had formally approved the treaty, including the United States. It will go into effect after 50 countries have formally approved it. Once it does, participating countries must implement the best-available mercury emission-control technologies within five years. The treaty also restricts the use of mercury in common household products and measuring devices such as thermometers.

14.5 HOW DO WE PERCEIVE AND AVOID RISKS?

CONCEPT 14.5 We can reduce the major risks we face by becoming informed, thinking critically about risks, and making careful choices.

The Greatest Health Risks Come from Poverty, Gender, and Lifestyle Choices

Risk analysis involves identifying hazards and evaluating their associated risks (*risk assessment;* Figure 14.2, left), ranking risks (*comparative risk analysis*), determining options and making decisions about reducing or eliminating risks (*risk management;* Figure 14.2, right), and informing decision makers and the public about risks (*risk communication*).

Statistical probabilities based on experience, animal testing, and other assessments are used to estimate risks from older technologies and chemicals. To evaluate

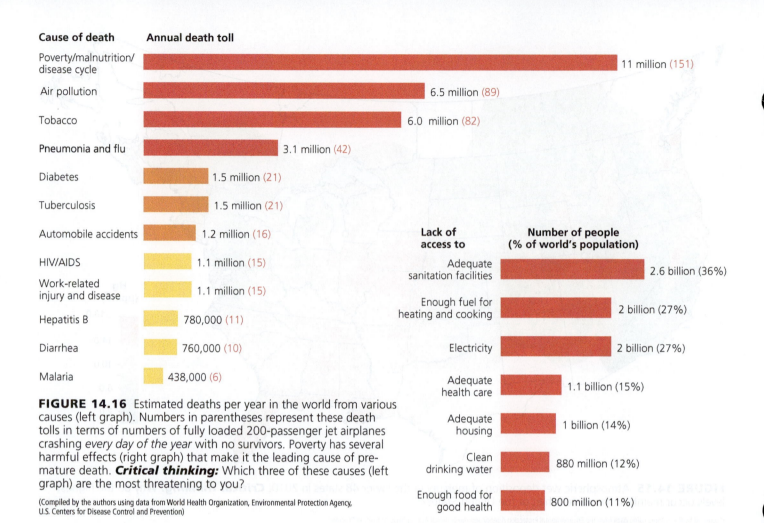

Cause of death	Annual death toll
Poverty/malnutrition/disease cycle	11 million (151)
Air pollution	6.5 million (89)
Tobacco	6.0 million (82)
Pneumonia and flu	3.1 million (42)
Diabetes	1.5 million (21)
Tuberculosis	1.5 million (21)
Automobile accidents	1.2 million (16)
HIV/AIDS	1.1 million (15)
Work-related injury and disease	1.1 million (15)
Hepatitis B	780,000 (11)
Diarrhea	760,000 (10)
Malaria	438,000 (6)

Lack of access to	Number of people (% of world's population)
Adequate sanitation facilities	2.6 billion (36%)
Enough fuel for heating and cooking	2 billion (27%)
Electricity	2 billion (27%)
Adequate health care	1.1 billion (15%)
Adequate housing	1 billion (14%)
Clean drinking water	880 million (12%)
Enough food for good health	800 million (11%)

FIGURE 14.16 Estimated deaths per year in the world from various causes (left graph). Numbers in parentheses represent these death tolls in terms of numbers of fully loaded 200-passenger jet airplanes crashing *every day of the year* with no survivors. Poverty has several harmful effects (right graph) that make it the leading cause of premature death. **Critical thinking:** Which three of these causes (left graph) are the most threatening to you?

(Compiled by the authors using data from World Health Organization, Environmental Protection Agency, U.S. Centers for Disease Control and Prevention)

new technologies and products, risk evaluators use more uncertain statistical probabilities, based on models rather than on actual experience and testing.

In terms of the number of deaths per year (Figure 14.16), *the greatest risk by far is poverty*. Many deaths due to poverty are caused by malnutrition, increased susceptibility to normally nonfatal infectious diseases, and often-fatal infectious diseases transmitted by unsafe drinking water.

Studies show the four greatest risks that shorten people's lives are living in poverty, being born male, smoking (see the Case Study that follows), and being obese. Some of the risks most likely to cause premature death stem from lifestyle choices that people make (Figure 14.17) (**Concept 14.1**). For example, overeating and lack of exercise can lead to obesity and type 2 diabetes, and long-term smoking can increase the risk of getting lung cancer.

CASE STUDY

Cigarettes and E-Cigarettes

Cigarette smoking is the world's most preventable and largest cause of suffering and premature death among adults.

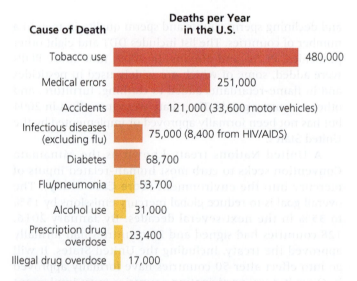

Cause of Death	Deaths per Year in the U.S.
Tobacco use	480,000
Medical errors	251,000
Accidents	121,000 (33,600 motor vehicles)
Infectious diseases (excluding flu)	75,000 (8,400 from HIV/AIDS)
Diabetes	68,700
Flu/pneumonia	53,700
Alcohol use	31,000
Prescription drug overdose	23,400
Illegal drug overdose	17,000

FIGURE 14.17 Leading causes of death in the United States. Some result from lifestyle choices and are preventable. **Critical thinking:** The number of deaths from tobacco use is how many times the number of deaths from flu/pneumonia?

(Compiled by the authors using data from U.S. Centers for Disease Control and Prevention)

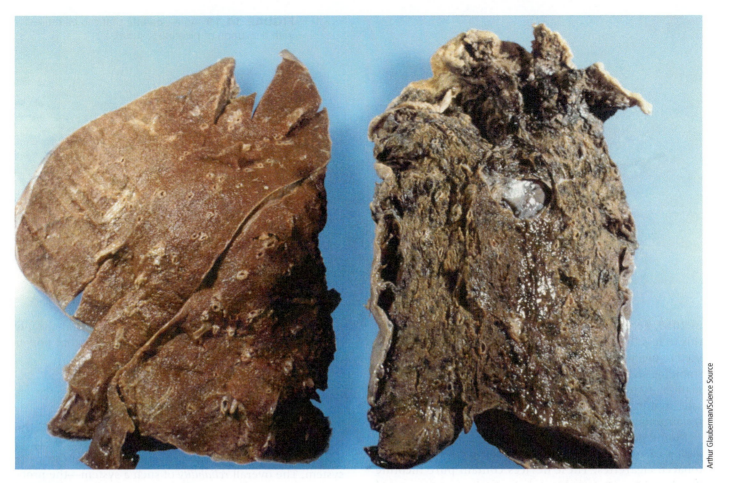

FIGURE 14.18 The startling difference between normal human lungs (left) and the lungs of a person who died of emphysema (right). The major causes of emphysema are prolonged smoking and exposure to air pollutants.

The WHO estimates that smoking contributed to the deaths of 100 million people during the 20th century and could kill 1 billion people during this century unless governments and individuals act to dramatically reduce smoking.

The WHO and the U.S. Surgeon General estimated that each year, tobacco contributes to the premature deaths of about 6 million people resulting from 25 illnesses, including heart disease, stroke, type 2 diabetes, lung and other cancers, memory impairment, bronchitis, and emphysema (Figure 14.18). This amounts to an average of more than 16,400 deaths every day, or about one every 5 seconds.

By 2030, the annual death toll from smoking-related diseases is projected to reach more than 8 million—an average of 21,900 preventable deaths per day—according to the CDC and WHO. About 80% of these deaths are expected to occur in less-developed countries, especially China, with 350 million smokers. The annual death toll from smoking in China is about 1.2 million, an average of about 137 deaths every hour. By 2050, the annual death toll from smoking in China could reach 3 million. There is little effort to reduce smoking in China, partly because cigarette taxes provide up to 10% of the government's annual revenues.

According to the CDC, smoking kills more than 480,000 Americans per year—an average of 1,315 deaths per day, or nearly one every minute (Figure 14.17). This death toll is roughly equivalent to more than six fully loaded 200-passenger jet planes crashing *every day of the year* with no survivors. Smoking also causes about 8.6 million illnesses every year in the United States.

The overwhelming scientific consensus is that the nicotine inhaled in tobacco smoke is highly addictive. A British government study showed that adolescents who smoke more than one cigarette have an 85% chance of becoming long-term smokers.

Studies indicate that cigarette smokers die, on average, 10 years earlier than nonsmokers. If people quit smoking by the age of 30, they can avoid nearly all the risk of dying prematurely.

A study by British researchers found that, globally, exposure to secondhand smoke contributes to about 600,000 deaths per year. In 2015, the CDC estimated that daily exposure to secondhand smoke is responsible for nearly 42,000 deaths per year in the United States.

In the United States, the percentage of adults who smoke dropped from more than 50% in the 1950s to 15%

FIGURE 14.19 An e-cigarette that can be refilled with a solution of nicotine (e-juice).

in 2015, according to the CDC, and the goal is to reduce this to less than 10% by 2025. This decline can be attributed to media coverage about the harmful health effects of smoking, sharp increases in cigarette taxes in many states, mandatory health warnings on cigarette packs, the ban on sales to minors, and bans on smoking in workplaces, bars, restaurants, and public buildings.

Some people are using various forms of *electronic cigarettes* or *e-cigarettes* (Figure 14.19) as a substitute for tobacco cigarettes. These devices contain pure nicotine dissolved in a syrupy solvent that contains one or more of over 7,000 chemicals to enhance taste and smell. A lithium-ion battery in the device heats the nicotine solution and converts it to a vapor containing liquid particles that the user inhales. Smoking e-cigarettes is called *vaping*. E-cigarettes can be refilled with solutions that vary from 2% to 10% in their concentrations of nicotine, which is a poison (Table 14.1).

Are e-cigarettes safe? No one knows, because they have not been around long enough to be thoroughly evaluated. E-cigarettes reduce or eliminate the inhalation of tar and numerous other harmful chemicals found in regular cigarette smoke. However, they still expose users to highly addictive nicotine, sometimes at levels of up to 5 times as high (10% nicotine) as that found in regular cigarettes (2% nicotine).

Preliminary research indicates that some e-cigarette vapors contain trace amounts of toxic cadmium, nickel, lead, and several substances that can cause cancer in test animals. Some of these toxins, not found in regular cigarette smoke, are toxic nanoparticles that are small enough to get past the body's defense systems and travel deep into the lungs. Like cigarette smoke, they can be inhaled as second-hand vapors by non-users. They probably come from flavorings and other additives. However, it will take much more research to establish any direct link between e-cigarettes and cancer. Another potential hazard to the mouth, face, and hands of e-cigarette users is that cheaply made and unregulated lithium-ion batteries that power e-cigarettes sometimes explode or catch fire.

Estimating Risks from Technologies

The more complex a technological system is, and the higher the number of people required to design and run it, the more difficult it is to estimate the risks of using the system. The overall *reliability* of such a system—the probability (expressed as a percentage) that the system will complete a task without failing—is the product of two factors:

System reliability (%) = Technology reliability (%) × Human reliability (%)

With careful design, quality control, maintenance, and monitoring, a highly complex system such as a nuclear power plant or a deep-sea oil-drilling rig can achieve a high degree of technological reliability. However, human reliability usually is much lower than technological reliability and is almost impossible to predict.

Suppose the estimated technological reliability of a nuclear power plant is 95% (0.95) and human reliability is 75% (0.75). Then the overall system reliability is 71% (0.95 × 0.75 = 71%). Even if we could make the technology 100% reliable (1.0), the overall system reliability would still be only 75% (1.0 × 0.75 = 75%).

We can make a system more foolproof, or fail-safe, by moving more of the potentially fallible elements from the human side to the technological side. However, chance events such as a lightning strike can knock out an automatic control system, and no machine or computer program can completely replace human judgment. In addition, the parts in any automated control system (such as the blowout protectors on the BP oil well that ruptured in the Gulf of Mexico in 2010, see Figure 11.29, p. 291) are manufactured, assembled, tested, certified, inspected, and maintained by fallible human beings. In addition, computer software programs used to monitor and control complex systems can be flawed because of human design error or can be deliberately sabotaged to cause their malfunction.

Most People Do a Poor Job of Evaluating Risks

Most people are not good at assessing the relative risks from the hazards that they encounter. Many people deny or shrug off the high-risk chances of death (or injury) from the activities they enjoy. These include *smoking* (1 in 250 by age 70 for a pack-a-day smoker), *hang gliding* (1 in 1,250), and *driving* (1 in 3,300 without a seatbelt and 1 in 6,070 with a seatbelt).

Indeed, the most dangerous thing that many people do each day is to drive or ride in a car. Yet, some of these same people may be terrified about their chances of being killed by getting pneumonia from *the flu* (a 1 in 130,000 chance), *a nuclear power plant accident* (1 in 200,000), *West Nile virus* (1 in 1 million), *lightning* (1 in 3 million), Ebola virus (1 in 4 million), *a commercial airplane crash* (1 in 9 million), *snakebite* (1 in 36 million), or *shark attack* (1 in 281 million).

Five factors can cause people to see a technology or a product as being more or less risky than experts judge it to be. The first factor is *fear*. Research shows that fear causes people to overestimate risks and to worry more about unusual risks than they do about common, everyday risks. Studies show that people tend to overestimate numbers of deaths caused by tornadoes, floods, fires, homicides, cancer, and terrorist attacks, and to underestimate death tolls from flu, diabetes, asthma, heart attack, stroke, and automobile accidents.

The second factor clouding risk estimation is the *degree of control* individuals have over a given situation. Many people have a greater fear of things over which they do not have personal control. For example, some individuals feel safer driving their own car for long distances than traveling the same distance on a plane. But look at the numbers. The risk of dying in a car accident in the United States while using a seatbelt is 1 in 6,070, whereas the risk of dying in a U.S. commercial airliner crash is about 1 in 9 million.

The third factor influencing risk evaluation is *whether a risk is catastrophic or chronic*. People usually are more frightened by news of catastrophic accidents such as a plane crash than a cause of death such as smoking, which has a much higher death toll spread out over time.

Fourth, some people have *optimism bias*, the belief that risks apply to other people but not to them. They may be upset when they see others driving erratically while talking on a cell phone or texting, for example, but believe they can do so without impairing their own driving ability.

A fifth factor affecting risk analysis is that many of the risky things people do are highly pleasurable and give *instant gratification*. The potential harm from such activities comes later. Examples are smoking cigarettes and eating too much food.

Guidelines for Evaluating and Reducing Risk

Here are four guidelines for evaluating and reducing risk and making better lifestyle choices (**Concept 14.5**):

- *Compare risks.* In evaluating a risk, the key question is not "Is it safe?" but rather "How risky is it compared to other risks?"

- *Determine how much risk you are willing to accept.* For most people, a 1 in 100,000 chance of dying or suffering serious harm from exposure to an environmental hazard is a threshold for changing their behavior. However, in establishing standards and reducing risk, the EPA generally assumes that a 1 in 1 million chance of dying from an environmental hazard is acceptable.

- *Evaluate the actual risk involved.* The news media usually exaggerate the daily risks we face in order to capture our interest and attract more readers, listeners, or television viewers. As a result, most people who are exposed to a daily diet of such exaggerated reports believe that the world is much more dangerous and risk-filled than it really is.

- *Concentrate on evaluating and carefully making important lifestyle choices.* When evaluating risk, it is important to ask, "Do I have any control over this?" There is no point worrying about risks over which we have little or no control. Factors over which individuals have at least some control include ways to reduce heart attack, stroke, and certain cancers, by deciding whether to smoke, what to eat, how much alcohol to drink, how much to exercise, and how safely to drive.

Mercury's Toxic Effects and Sustainability

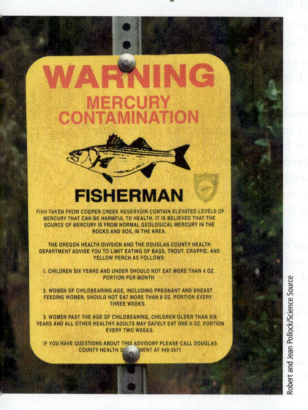

Robert and Jean Pollock/Science Source

In the Core Case Study that opens this chapter, we saw that mercury (Hg) and its compounds that occur regularly in the environment can permanently damage the human nervous system, kidneys, and lungs and harm fetuses and cause birth defects. In this chapter, we also learned of many other chemical hazards, as well as biological, physical, cultural, and lifestyle hazards in the environment. In addition, we saw how difficult it is to evaluate the nature and severity of threats from these various hazards.

One of the important facts discussed in this chapter is that on a global basis, the greatest threat to human health is poverty (often leading to malnutrition and disease), followed by the threats from smoking, air pollution, pneumonia and flu, and HIV/AIDS.

There are some threats that we can do little to avoid, but we can reduce other threats, partly by applying the three **scientific principles of sustainability** (see inside back cover of this book). For example, we can greatly reduce our exposure to mercury and other pollutants by shifting from the use of nonrenewable fossil fuels (especially coal) to wider use of a diversity of renewable energy resources, including solar energy and wind energy. We can reduce our exposure to harmful chemicals used in the manufacturing of various goods by cutting resource use and waste and by reusing and recycling material resources. We can also mimic biodiversity by using diverse strategies for solving environmental and health problems, and especially for reducing poverty and controlling population growth. In so doing, we also help preserve the earth's biodiversity and increase our beneficial environmental impact.

Chapter Review

Core Case Study

1. Describe the toxic effects of mercury and its compounds and explain how we are exposed to these toxins.

Section 14.1

2. What is the key concept for this section? Define and distinguish among **risk**, **risk assessment**, and **risk management**. Give an example of how scientists state probabilities. Give an example of a risk from each of the following: biological hazards, chemical hazards, natural hazards, cultural hazards, and lifestyle choices. What is a **pathogen**?

Section 14.2

3. What is the key concept for this section? Define **infectious disease**; define and distinguish among **bacteria**, **viruses**, and **parasites**; and give examples of diseases that each can cause. Define and distinguish between **transmissible disease** and **nontransmissible disease**, and give an example of each. In terms of death rates, what are the world's four most serious infectious diseases? List five factors that have contributed to genetic resistance in bacteria to commonly used antibiotics. What is MRSA and why is it so threatening?

4. Describe the global threat from tuberculosis and list three factors that have helped it spread. What is the biggest viral killer and how does it spread? Summarize the threats from the hepatitis B, Ebola, West Nile, and Zika viruses. What is the best way to reduce one's chances of getting an infectious disease? What is the focus of ecological medicine and what are some of its findings regarding the spread of diseases? Summarize the health threats from the global HIV/AIDS pandemic and its effects on the population age structure of Botswana.

5. What is malaria and how does it spread? How much of the human population is subject to this threat? List six major ways to reduce the global threat from infectious diseases.

Section 14.3

6. What is the key concept for this section? What is a **toxic chemical**? Define and distinguish among **carcinogens**, **mutagens**, and **teratogens**, and give an example of each. Describe the human immune, nervous, and endocrine systems, and for each of these systems, give an example of a chemical that can threaten it. What is a neurotoxin and why is methylmercury (**Core Case Study**) an especially dangerous one? Describe the process of biological magnification of chemicals in food chains and food webs. What are six ways to prevent or control environmental inputs of mercury? What are hormonally active agents (HAAs), what risks do they pose, and how can we reduce those risks? Summarize health scientists' concerns about exposure to bisphenol A (BPA) and the controversy over what to do about exposure to this chemical. Summarize the concerns over exposure to phthalates. List six ways to reduce your exposure to HAAs.

Section 14.4

7. What are the two key concepts for this section? Define **toxicology**, **toxicity**, **dose**, and **response**. What are three factors that affect the level of harm caused by a chemical? Give three reasons why children are especially vulnerable to harm from toxic chemicals. Describe how the toxicity of a substance can be estimated by testing laboratory animals, and explain the limitations of this approach. What is a **dose-response curve**? Explain how toxicities are estimated through use of case reports and epidemiological studies, and discuss the limitations of these approaches.

Critical Thinking

1. Assume that you are a national official with the power to set policy for controlling environmental mercury pollution from human sources (**Core Case Study**). List the goals of your policy and outline a plan for accomplishing those goals. List three or more possible problems that could result from implementing your policy.

2. What are three actions you would take to reduce the global threats to human health and life from each of the following: **(a)** tuberculosis, **(b)** HIV/AIDS, and **(c)** malaria?

3. Explain why you agree or disagree with each of the following statements:
   a. We should not worry much about exposure to toxic chemicals because almost any chemical, at a large enough dosage, can cause some harm.

8. Summarize the controversy over the effects of trace levels of chemicals. Why do we know so little about the harmful effects of chemicals? What is the **precautionary principle**? Explain why the use of pollution prevention based on the precautionary principle to deal with health threats from chemicals is controversial. Describe how pollution prevention paid off for the 3M Company. Describe some efforts to apply the precautionary principle on national and international levels. What is the Stockholm Convention? What is the Minamata Convention?

Section 14.5

9. What is the key concept for this section? What is **risk analysis**? In terms of premature deaths, what are the three greatest threats that people face? What are six ways in which poverty can threaten one's health? Describe the health threats from smoking and how we can reduce these threats. Summarize our knowledge of the health effects of using e-cigarettes. How can we reduce the threats resulting from the use of various technologies? What are five factors that can cause people to misjudge risks? List four guidelines for evaluating and reducing risk.

10. What are this chapter's *three big ideas*? Explain how we can lessen the threats of harm from mercury in the environment by applying the three **scientific principles of sustainability**.

Note: Key terms are in bold type.

b. We should not worry much about exposure to toxic chemicals because, through genetic adaptation, we can develop immunities to such chemicals.

c. We should not worry much about exposure to toxic chemicals because we can use genetic engineering to reduce our susceptibility to their effects.

d. We should not worry about exposure to a chemical such as bisphenol A (BPA) because it has not been absolutely proven scientifically that BPA has killed anyone.

4. Should we ban the use of hormone mimics such as BPA in making products to be used by children younger than age 5? Should such a ban be extended to all products? Explain.

5. Workers in a number of industries are exposed to higher levels of various toxic substances than are the public. Should we reduce the workplace levels allowed for such chemicals? What economic effects might this have?

6. Do you think that electronic cigarettes should be taxed and regulated like conventional cigarettes? Explain.

7. What are the three major risks you face from each of the following: (a) your lifestyle, (b) where you live, and (c) what you do for a living? Which of these risks are voluntary and which are involuntary? List three steps you could take to reduce each of these risks. Which of these steps do you already take or plan to take?

8. In deciding what to do about risks from chemicals in the area where you live, would you support legislation that requires the use of pollution prevention based on the precautionary principle and on the assumption that chemicals are potentially harmful until shown otherwise? Explain.

Doing Environmental Science

Pick a commonly used and potentially harmful chemical and use the library or Internet to learn about (a) what it is used for and how widely it is used, (b) its potential harm, (c) the scientific evidence for such claims, and (d) proposed solutions for dealing with this threat. Pick a study area, such as your dorm or apartment building, your block, or your city. In this area, try to determine the level of presence of the chemical you are studying. You could do this by finding four or five examples of items or locations containing the chemical and then estimating the total amount based on your sample. Write a report summarizing your findings.

Global Environment Watch Exercise

Go to your MindTap course to access the GREENR database. At the top of the page, do a "Basic Search" on *mercury pollution*. Research the latest developments in studies of the harmful health effects of mercury (**Core Case Study**). Find an example of an effort to prevent or control mercury pollution and write a short report summarizing your findings. Try to find reports of two studies that reach different conclusions about how mercury should be regulated. Summarize the arguments for these conclusions on both sides. Based on what you have found, do you think that mercury pollution should be regulated more strictly in the state or country where you live? Explain your reasoning.

Data Analysis

The graph below shows the effects of AIDS on life expectancy at birth in Botswana, 1950–2000, and projects these effects to 2050. Study the graph and answer the questions that follow.

1. By what percentage did life expectancy in Botswana increase between 1950 and 1995?

2. By what percentage did life expectancy in Botswana drop between 1995 and 2015?

3. By what percentage was life expectancy in Botswana projected to increase between 2015 and 2050?

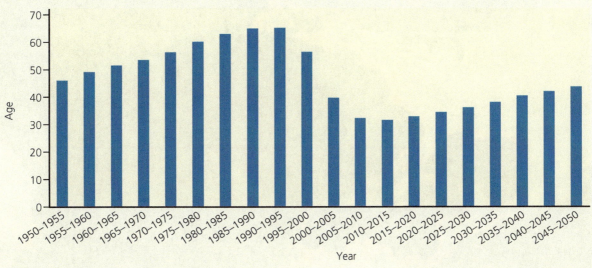

(Compiled by the authors using data from United Nations and U.S. Census Bureau.)

CHAPTER 15

Air Pollution, Climate Change,
and Ozone Depletion

Civilization has evolved during a period of remarkable climate stability, but this era is ending. We are entering a new era, a period of rapid and often-unpredictable climate change.

LESTER R. BROWN

Key Questions

15.1 What is the nature of the atmosphere?

15.2 What are the major air pollution problems?

15.3 How should we deal with air pollution?

15.4 How and why is the earth's climate changing?

15.5 What are the likely effects of climate change?

15.6 How can we slow climate change?

15.7 How have we depleted ozone in the stratosphere and what can we do about it?

Areas that could be flooded by the end of this century (shown in red) by a 1-meter (3-foot) rise in sea level due to projected climate change.

NASA

Melting Ice in Greenland

Greenland is the world's largest island with a population of 59,000 people. The ice that covers most of this mountainous island lies in glaciers that are as much as 3.2 kilometers (2 miles) thick.

Areas of the island's ice have been melting at an accelerating rate during Greenland's summers (Figure 15.1). Some of this ice is replaced by snow during winter months, but the annual net loss of Greenland's ice has increased during recent years.

Why does it matter that ice in Greenland is melting? It matters because considerable scientific evidence indicates that atmospheric warming is a key factor behind this melting. **Atmospheric warming** is the gradual rise in the average temperature of the atmosphere near the earth's surface over 30 years or more. During this century, atmospheric warming is projected to continue and to lead to dramatic **climate change**—measurable changes in global weather patterns based primarily on changes in the earth's atmospheric temperature *averaged over at least 30 years*. Climate scientists warn that, if no action is taken, the earth's climate system could reach tipping points that could change the earth's climate for hundreds to thousands of years.

Greenland's glaciers contain enough water to raise the global sea level by as much as 7 meters (23 feet) if all of it were to melt and drain into the sea. It is highly unlikely that this will happen. However, even a moderate loss of this ice over a century or more would raise sea levels considerably (see chapter-opening photo). Research indicates that Greenland's ice loss has already been responsible for nearly one-sixth of the global sea-level rise over the past 20 years. Climate scientists view Greenland's melting ice as an early warning that human activities are very likely to disrupt the earth's climate and our economies in ways that could threaten life as we know it, especially during the latter half of this century.

1982   20

FIGURE 15.1 The total area of Greenland's glacial ice that melted during the summer of 2012 (red area in right image) was much greater than the amount that melted during the summer of 1982 (left). This trend has continued since 2012.

Compiled by authors using data from Konrad Steffen and Russell Huff, University of Colorado, Boulder

In 1988, the World Meteorological Organization and the United Nations Environment Programme (UNEP) established the Intergovernmental Panel on Climate Change (IPCC) to document past climate changes and project future climate changes. The IPCC network includes more than 2,500 scientists working in climate studies and related disciplines from more than 130 countries.

After reviewing tens of thousands of research studies for more than 25 years, the IPCC and most of the world's major scientific bodies, including the U.S. National Academy of Sciences (NAS) and the British Royal Society, have come to three major conclusions about climate change: **(1)** it is real and is happening now, **(2)** human activities such as the burning of fossil fuels and the clearing of forests play an important role in current climate change, and **(3)** it is projected to accelerate and have harmful effects, such as rising seas, ocean acidification, species extinction, and more extreme weather, including more intense and longer lasting heat waves, unless we act now to slow it down.

In this chapter, we examine the nature of the atmosphere, air pollution, the likely causes and effects of projected climate change, and depletion of ozone in the stratosphere. We also look at some possible ways to deal with these serious environmental, economic, and political challenges.

15.1 WHAT IS THE NATURE OF THE ATMOSPHERE?

CONCEPT 15.1 The two innermost layers of the atmosphere are the *troposphere,* which supports life, and the *stratosphere,* which contains the protective ozone layer.

The Atmosphere Consists of Several Layers

Life exists under a thin blanket of gases surrounding the earth, called the **atmosphere**. It is divided into several spherical layers defined mostly by temperature differences (Figure 15.2).

About 75–80% of the earth's air mass is found in the **troposphere**, the atmospheric layer closest to the earth's surface. This layer extends about 17 kilometers (11 miles) above sea level at the equator and 6 kilometers (4 miles) above sea level over the poles. If the earth were the size of an apple, this lower layer containing the air you breathe would be no thicker than the apple's skin.

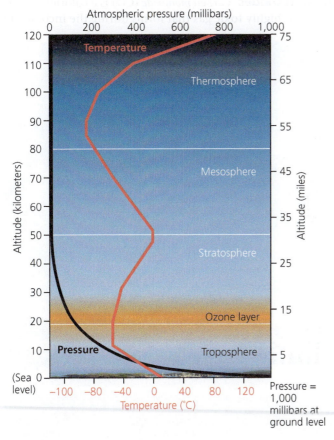

FIGURE 15.2 Natural Capital: The earth's atmosphere is a dynamic system that includes four layers. The average temperature of the atmosphere varies with altitude (red line) and with differences in the absorption of incoming solar energy. **Critical thinking:** Why do you think most of the planet's air is in the troposphere?

Take a deep breath. About 99% of the volume of air you inhaled consists of two gases: nitrogen (78%) and oxygen (21%). The remainder is 0.93% argon (Ar), 0.040% carbon dioxide (CO_2), and small amounts of water vapor (H_2O, varying from the equator to the poles), dust and soot particles, and other gases, including methane (CH_4), ozone (O_3), and nitrous oxide (N_2O). Several gases in the troposphere, including H_2O, CO_2, CH_4, and N_2O, are called **greenhouse gases** because they absorb and release energy that warms the troposphere and the earth's surface. Without this natural greenhouse effect, the earth would be too cold for life as we know it to exist.

The atmosphere's second layer is the **stratosphere**, which extends from about 17 to about 48 kilometers (from 11 to 30 miles) above the earth's surface (Figure 15.2). The stratosphere contains less matter than the troposphere but its composition is similar, with two notable exceptions. The stratosphere has a much lower volume of water vapor and a much higher concentration of ozone (O_3).

Most of the atmosphere's small amount of ozone (O_3) is concentrated in a portion of the stratosphere called the **ozone layer**, found roughly 17–26 kilometers (11–16 miles) above sea level. This UV filtering effect of ozone in the lower stratosphere acts as a "global sunscreen" that keeps about 95% of the sun's harmful UV radiation from reaching the earth's surface. The ozone layer allows life to exist on land and helps protect us from sunburn, skin and eye cancers, cataracts, and damage to our immune systems. It also prevents much of the oxygen in the troposphere from being converted to ground-level ozone, a harmful air pollutant. In other words, preserving the stratospheric ozone layer should be one of humanity's top priorities.

15.2 WHAT ARE THE MAJOR AIR POLLUTION PROBLEMS?

CONCEPT 15.2A Three major outdoor air pollution problems are *industrial smog* caused mostly by the burning of coal, *photochemical smog* caused by motor vehicle and industrial emissions, and *acid deposition* caused mainly by coal-burning power and industrial plants and motor vehicle emissions.

CONCEPT 15.2B The most threatening indoor air pollutants are smoke and soot from wood and coal fires (mostly in less-developed countries), cigarette smoke, and chemicals used in building materials and cleaning products.

Air Pollution Comes from Natural and Human Sources

Air pollution is the presence of chemicals in the atmosphere in concentrations high enough to harm organisms, ecosystems, or human-made materials, or to alter climate. Almost any chemical in the atmosphere can become a pollutant if it occurs in a high enough concentration. The effects of air pollution range from annoying to lethal.

Air pollutants come from natural and human sources. Natural sources include wind-blown dust, solid and gaseous pollutants from wildfires and volcanic eruptions, and volatile organic chemicals released by some plants. Most natural air pollutants spread out over the globe and become diluted or are removed by chemical cycles, precipitation, and gravity. However, pollutants emitted by large volcanic eruptions or forest fires can temporarily reach harmful levels.

Most human inputs of outdoor air pollutants occur in industrialized and urban areas where people, cars, and factories are concentrated. These pollutants are generated mostly by the burning of fossil fuels in power plants and industrial facilities *(stationary sources)* and in motor vehicles *(mobile sources)*.

Scientists classify outdoor air pollutants into two categories (Figure 15.3). **Primary pollutants** are chemicals or substances emitted directly into the air from natural processes and human activities at concentrations high enough to cause harm. While in the atmosphere, some primary pollutants react with one another and with other natural components of air to form new harmful chemicals, called **secondary pollutants**.

With their high concentrations of cars and factories, urban areas normally have higher outdoor air pollution levels than rural areas. However, prevailing winds can spread long-lived primary and secondary air pollutants from urban and industrial areas to the countryside and to other urban areas.

Over the past 40 years, the quality of outdoor air in most of the more-developed countries has improved, thanks mostly to grassroots pressure from citizens, in the 1960s and 1970s. This led governments in the United States and in most European countries to pass and enforce air-pollution-control laws.

Despite such efforts, air pollution is one of the world's most serious environmental problems. The WHO estimated that air pollution contributed to the deaths of 6.5 million people in 2015—3 million from outdoor air pollution and 3.5 million from indoor air pollution. Most people exposed to dangerous levels of air pollutants live in densely populated cities in less-developed countries where air pollution control laws do not exist or are poorly enforced.

Major Outdoor Air Pollutants

Carbon Oxides *Carbon monoxide* (CO) is a colorless, odorless, and highly toxic gas that forms during the incomplete

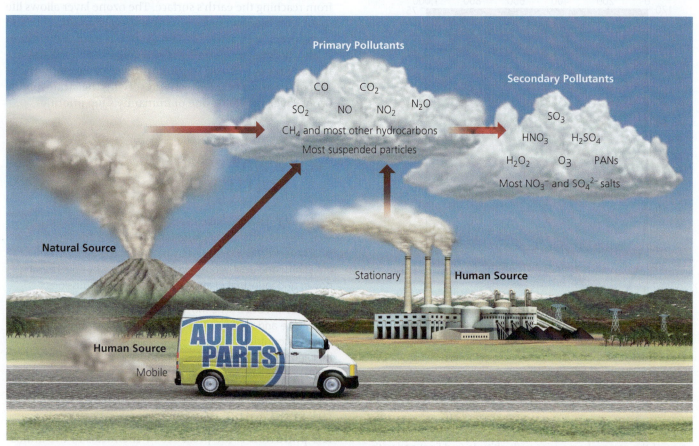

FIGURE 15.3 Human inputs of air pollutants come from *mobile sources* (such as cars) and *stationary sources* (such as industrial, power, and cement plants). Some *primary air pollutants* react with one another and with other chemicals in the air to form *secondary air pollutants*.

combustion of carbon-containing materials. Major sources of CO are motor vehicle exhaust, the burning of forests and grasslands, smokestacks of fossil fuel–burning power plants and industries, tobacco smoke, open fires, and inefficient stoves used for cooking or heating.

In the body, CO can combine with hemoglobin in red blood cells, which reduces the ability of blood to transport oxygen to body cells and tissues. Long-term exposure can trigger heart attacks and aggravate lung diseases such as asthma and emphysema. At high levels, CO can cause headache, nausea, drowsiness, confusion, collapse, coma, and death, which is why it is important to have CO detectors in your home.

Carbon dioxide (CO_2) is a colorless, odorless gas. About 93% of the CO_2 in the atmosphere is the result of the natural carbon cycle (see Figure 3.16, p. 59), and the rest comes from human activities such as the burning of fossil fuels, which adds CO_2 to the atmosphere, and the removal of forests and grasslands that help remove excess CO_2 from the atmosphere. Carbon dioxide is classified as an air pollutant because it has reached high enough levels to warm the atmosphere and bring about climate change, affecting human health. However, there is political pressure from the U.S. fossil fuel industry to reverse the EPA ruling that CO_2 is an air pollutant, despite overwhelming scientific evidence that it is.

Nitrogen Oxides and Nitric Acid *Nitric oxide* (NO) is a colorless gas that forms when nitrogen and oxygen gases react under high temperatures in automobile engines and coal-burning power and industrial plants. Lightning and certain bacteria in soil and water also produce NO as part of the nitrogen cycle (see Figure 3.17, p. 60).

In the air, NO reacts with oxygen to form *nitrogen dioxide* (NO_2), a reddish-brown gas. Collectively, NO and NO_2 are called *nitrogen oxides* (NO_x). Some of the NO_2 reacts with water vapor in the air to form *nitric acid* (HNO_3) and nitrate salts (NO_3^-), components of harmful *acid deposition*, discussed later in this chapter. Both NO and NO_2 play a role in the formation of *photochemical smog*—a mixture of chemicals formed under the influence of sunlight in cities with heavy traffic. *Nitrous oxide* (N_2O), a greenhouse gas, is emitted from fertilizers and animal wastes, and is produced by the burning of fossil fuels.

At high enough levels, nitrogen oxides can irritate the eyes, nose, and throat, aggravate lung ailments such as asthma and bronchitis, suppress plant growth, and reduce visibility when they are converted to nitric acid and nitrate salts.

Sulfur Dioxide and Sulfuric Acid *Sulfur dioxide* (SO_2) is a colorless gas with an irritating odor. About one-third of the SO_2 in the atmosphere comes from natural sources such as volcanoes. The other two-thirds (and as much as 90% in highly industrialized urban areas) come from human sources, mostly combustion of sulfur-containing coal in power and industrial plants, oil refining, and the smelting of sulfide ores.

In the atmosphere, SO_2 can be converted to *aerosols,* which consist of microscopic suspended droplets of *sulfuric acid* (H_2SO_4) and suspended particles of sulfate (SO_4^{2-}) salts that return to the earth as a component of acid deposition. Sulfur dioxide, sulfuric acid droplets, and sulfate particles reduce visibility and aggravate breathing problems. These chemicals can damage crops, trees, soils, and aquatic life in lakes. They also corrode metals and damage paint, paper, leather, and the stone used to build walls, statues, and monuments.

Particulates *Suspended particulate matter* (SPM) consists of a variety of solid particles and liquid droplets that are small and light enough to remain suspended in the air for long periods. The U.S. Environmental Protection Agency (EPA) classifies particles as fine, or PM-10 (with diameters less than 10 micrometers), and ultrafine, or PM-2.5 (with diameters less than 2.5 micrometers). About 62% of the SPM in outdoor air comes from natural sources such as dust, wildfires, and sea salt. The other 38% comes from human sources such as coal-burning power and industrial plants, motor vehicles, wind erosion from exposed topsoil, and road construction.

Particulate matter can irritate the nose and throat, damage the lungs, aggravate asthma and bronchitis, and shorten life spans. Particulates also reduce visibility, corrode metals, and discolor clothing and paints.

Ozone A major ingredient of photochemical smog is *ozone* (O_3), a colorless and highly reactive gas. Ozone can cause coughing and breathing problems, aggravate lung and heart diseases, reduce resistance to colds and pneumonia, and irritate the eyes, nose, and throat. Ozone also damages plants, rubber in tires, fabrics, and paints.

Scientific measurements show that human activities have decreased the amount of beneficial O_3 in the stratosphere and have increased the amount of harmful O_3 near ground-level—especially in some urban areas. Harmful ozone in the troposphere is a greenhouse gas that contributes to atmospheric warming and climate change. It also reduces photosynthesis by trees and other plants, which contributes to atmospheric warming by reducing the amount of excess CO_2 that they remove from the troposphere. We examine the serious issue of decreased stratospheric ozone in the final section of this chapter.

Volatile Organic Compounds (VOCs) Organic compounds that exist as gases in the atmosphere or that evaporate from sources on earth into the atmosphere are called *volatile organic compounds* (VOCs). Examples are hydrocarbons, emitted by the leaves of many plants, and *methane* (CH_4). As a greenhouse gas, CH_4 is about 25 times more effective per molecule than CO_2 is at warming the atmosphere. About a third of global methane emissions come from natural sources such as plants, wetlands, and termites. The rest come from human sources such as rice paddies, landfills, leaking natural gas wells and pipelines, and from cows (mostly from their belching).

Other VOCs are liquids that evaporate quickly into the atmosphere. Examples are benzene and other liquids used as industrial solvents, dry-cleaning fluids, and various components of gasoline, plastics, and other products.

Burning Coal Produces Industrial Smog

Seventy-five years ago, cities such as London, England, and the U.S. cities of Chicago, Illinois, and Pittsburgh, Pennsylvania, burned large amounts of coal in power plants and factories. They also burned coal to heat homes and often for cooking food. People in such cities, especially during winter, were exposed to **industrial smog**, consisting mostly of an unhealthy mix of sulfur dioxide, suspended droplets of sulfuric acid, and a variety of suspended solid particles in outside air. People who burned coal inside their homes were often exposed to dangerous levels of particulates and other indoor air pollutants.

Today, urban industrial smog is rarely a problem in most more-developed countries where coal is burned only in large power and industrial plants with reasonably good air pollution control. However, industrial smog remains a problem in industrialized urban areas of China, India,

Ukraine, Czechoslovakia (Figure 15.4), and other countries, where large quantities of coal are still burned in power plants, factories, and houses with inadequate pollution controls. Because of its heavy reliance on coal, China has some of the world's highest levels of industrial smog and 16 of the world's 20 most polluted cities.

Sunlight plus Cars Equals Photochemical Smog

Another type of smog is **photochemical smog**. It is a mixture of primary and secondary pollutants formed under the influence of UV radiation from the sun. In greatly simplified terms,

$$VOCs + NO_x + heat + sunlight \rightarrow \begin{array}{l} \text{ground-level ozone } (O_3) \\ + \text{ other photochemical oxidants} \\ + \text{ aldehydes} \\ + \text{ other secondary pollutants} \end{array}$$

The formation of photochemical smog begins when exhaust from morning commuter traffic releases large amounts of NO and VOCs into the air over a city. The NO is converted to reddish-brown NO_2, which is why photochemical smog is sometimes called *brown-air smog*. When

FIGURE 15.4 Severe industrial smog from an iron and steel factory in Czechoslovakia.

JAMES P. BLAIR/National Geographic Creative

FIGURE 15.5 Photochemical smog is a serious problem in Los Angeles, California, although air pollution laws have helped reduce the average number of severe smog days per year. **Question:** How serious is photochemical smog where you live?

iStock.com/Lee Pettet

exposed to ultraviolet radiation from the sun, some of the NO_2 reacts in complex ways with VOCs released by certain trees, motor vehicles, and businesses (especially bakeries and dry cleaners). The resulting mixture of pollutants, dominated by ground-level O_3, usually builds up to peak levels by late morning, irritating people's eyes and respiratory tracts. Some of these pollutants, known as *photochemical oxidants,* can damage lung tissue.

All modern cities have some photochemical smog, but it is more common in cities with sunny and warm climates and a large number of motor vehicles. Examples are Los Angeles, California (Figure 15.5), and Salt Lake City, Utah, in the United States; Sydney, Australia; São Paulo, Brazil; Bangkok, Thailand; and Mexico City, Mexico.

Several Factors Affect Air Pollution

Five natural factors help *reduce* outdoor air pollution. First, particles heavier than air settle out of the atmosphere. Second, *rain and snow* partially cleanse the air of pollutants.

Third, *salty sea spray* from the oceans washes out many pollutants from air that flows from land over the oceans. Fourth, *winds* sweep pollutants away and dilute them by mixing them with cleaner air. Fifth, some pollutants are removed by *chemical reactions.* For example, SO_2 can react with O_2 in the atmosphere to form SO_3, which reacts with water vapor to form droplets of H_2SO_4 that fall out of the atmosphere as acidic precipitation.

Six other factors can *increase* outdoor air pollution. First, *urban buildings* slow wind speed and reduce the dilution and removal of pollutants. Second, *hills and mountains* reduce the flow of air in valleys below them and allow pollutant levels to build up at ground level. Third, *high temperatures* promote the chemical reactions leading to the formation of photochemical smog. Fourth, *emissions of volatile organic compounds (VOCs)* from certain trees and plants in urban areas can promote the formation of photochemical smog.

The fifth factor that increases air pollution has to do with the *vertical movement of air.* During the day, the sun warms the air near the earth's surface. Normally, this warm air and most of the pollutants it contains rise to mix with the cooler air above it and are dispersed. Under certain atmospheric conditions, however, a layer of warm air can temporarily lie atop a layer of cooler air nearer the ground. This is called a **temperature inversion**. Because the cooler air near the surface is denser than the warmer air above, it does not rise and mix with the air above. If this condition persists, pollutants can build up to harmful and even lethal concentrations in the trapped layer of cool air near the ground.

The sixth factor is that *air pollution can move from one country to another*. Since 1992, levels of major air pollutants in the United States have decreased thanks to air pollution laws. However, a 2017 study by Lin Melyun and other atmospheric scientists found that since 1980, and especially since 1992, levels of photochemical smog in the western United States have increased because long-lived air pollutants released in China, India, and other Asian countries have moved across the Pacific Ocean. This has increased ozone levels from photochemical smog in parts of the western United States by as much as 65%.

Acid Deposition

Most coal-burning power plants, metal ore smelters, oil refineries, and other industrial facilities emit sulfur dioxide (SO_2), suspended particles, and nitrogen oxides (NO_x) into the atmosphere. In more-developed countries, these facilities usually use tall smokestacks to vent their exhausts high into the atmosphere where wind can dilute and disperse these pollutants. This reduces *local* air pollution, but it can increase *regional* air pollution in downwind areas because prevailing winds can transport SO_2 and NO_x pollutants as far as 1,000 kilometers (600 miles). During their trip, these compounds form secondary pollutants such as droplets of sulfuric acid (H_2SO_4), nitric acid vapor (HNO_3), and particles of acid-forming sulfate (SO_4^{2-}) and nitrate (NO_3^-) salts (Figure 15.3).

These acidic substances remain in the atmosphere for 2 to 14 days. During this period, they descend to the earth's surface in two forms: *wet deposition*, consisting of acidic rain, snow, fog, and cloud vapor, and *dry deposition*, consisting of acidic particles. The resulting mixture is called **acid deposition** (Figure 15.6)—often called *acid rain*.

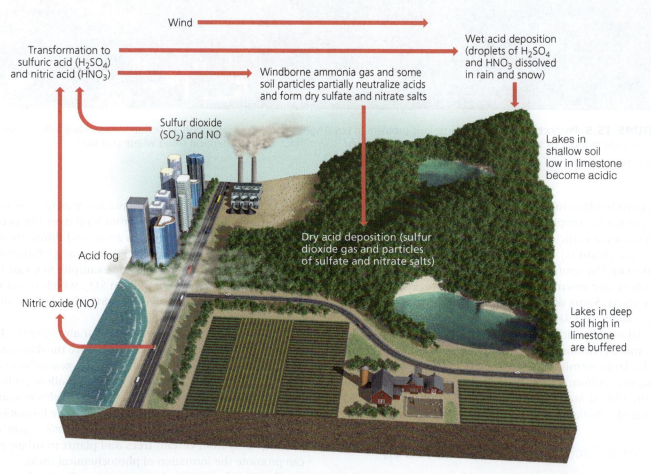

FIGURE 15.6 Natural Capital Degradation: *Acid deposition*, which consists of rain, snow, dust, and other particles with a pH lower than 5.6, is commonly called acid rain. ***Critical thinking:*** What are three ways in which your daily activities contribute to acid deposition?

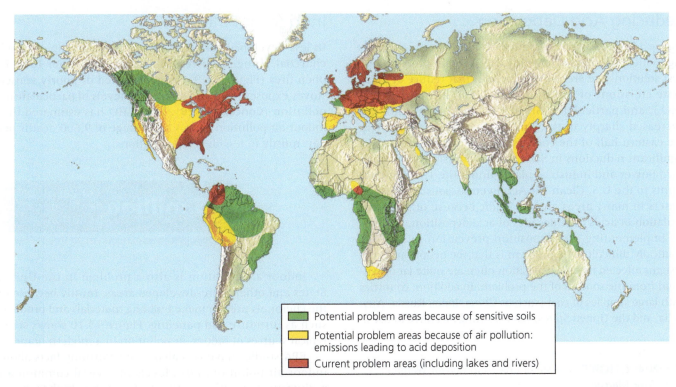

Potential problem areas because of sensitive soils

Potential problem areas because of air pollution: emissions leading to acid deposition

Current problem areas (including lakes and rivers)

FIGURE 15.7 This map shows regions where acid deposition is now a problem and regions with the potential to develop this problem. Such regions have large inputs of air pollution (mostly from power plants, industrial facilities, and ore smelters) or are sensitive areas with naturally acidic soils and bedrock that cannot neutralize (buffer) additional inputs of acidic compounds. *Question:* Do you live in or near an area that is affected by acid deposition or an area that is likely to be affected by acid deposition in the future?

(Compiled by the authors using data from World Resources Institute and U.S. Environmental Protection Agency)

Acid deposition is a regional air problem in areas that lie downwind from coal-burning facilities and from urban areas with large numbers of cars (**Concept 15.2A**). The map in Figure 15.7 shows areas of the world where acid deposition is, or is likely to be, a problem. In some areas, soils contain compounds that can react with and help neutralize, or *buffer*, some inputs of acids. The areas most sensitive to acid deposition are those with thin, acidic soils that provide no such natural buffering (Figure 15.7, all green and most red areas) and those where the buffering capacity of soils has been depleted by decades of acid deposition.

A combination of acid deposition and other air pollutants can harm crops and reduce plant productivity by removing essential plant nutrients such as calcium and magnesium from forest soils. They also cause soils to release ions of aluminum, lead, cadmium, and mercury, which are toxic to trees. These effects rarely kill trees directly, but can weaken trees and leave them vulnerable to stresses such as severe cold, diseases, insect attacks, and drought.

Acid deposition also damages statues and buildings, contributes to human respiratory diseases, and can leach toxic metals such as lead and mercury from soils and rocks into lakes used as sources of drinking water. These toxic metals can accumulate in the tissues of fish eaten by people (see Chapter 14, Core Case Study, p. 378). Because of excess acidity due to acid deposition, several thousand lakes in Norway and Sweden, and 1,200 lakes in Ontario, Canada, contain few if any fish. In the United States, several hundred lakes (most in the Northeast) are similarly threatened.

In the United States, older coal-burning power and industrial plants without adequate pollution controls, especially in the Midwest, emit the largest quantities of SO_2, particulates, and other pollutants that cause acid deposition. Prevailing winds in the United States carry these pollutants eastward. As a result, precipitation in parts of the eastern United States can be 10 times more acidic on average than precipitation in the western United States. Some mountaintop forests in the eastern United States and in areas to the east of large western U.S. cities are bathed in fog and dews that are about 1,000 times as acidic as normal precipitation.

Acid deposition has also become an international problem when acidic emissions from one country are transported to other countries by prevailing winds. The worst acid deposition occurs in Asia, especially in China, which gets about two-thirds of its energy and two-thirds of its electricity from burning coal, according to the U.S. Energy Information Administration.

Reducing Acid Deposition

Figure 15.8 lists ways to reduce acid deposition. According to most scientists who study the acid rain problem, the best solutions are *preventive approaches* that reduce or eliminate emissions of sulfur dioxide (SO_2), nitrogen oxides (NO_x), and particulates. Since 1994, acid deposition has decreased sharply in the United States and especially in the eastern half of the country. This is partly the result of significant reductions in SO_2 and NO_x emissions from coal-fired power and industrial plants under the 1990 amendments to the U.S. Clean Air Act. Even so, soils and surface waters in many areas are still acidic because of the accumulation of acids over decades of acid deposition.

Implementing acid deposition prevention solutions is politically difficult. One problem is that the people and ecosystems affected by acid deposition often are quite far downwind from the sources of the problem. In addition, countries with large supplies of coal (such as China, India, Russia, Australia, and the United States) have a strong incentive to use it.

CONSIDER THIS ...

CONNECTIONS Low-Sulfur Coal, Atmospheric Warming, and Toxic Mercury

Some U.S. power plants have lowered SO_2 emissions by switching from high-sulfur to low-sulfur coals such as lignite (see Figure 13.11, p. 337). However, because low-sulfur coal has a lower heat value, more coal must be burned to generate a given amount of electricity. This has led to increased CO_2 emissions, which contribute to atmospheric warming and climate change. Low-sulfur coal also has higher levels of toxic mercury and other trace metals, so when it is burned, more of these hazardous chemicals end up in the atmosphere.

Solutions		
Acid Deposition		
Prevention		**Cleanup**
Reduce coal use		Add lime to neutralize acidied lakes
Use natural gas and renewable energy resources in place of coal		Add phosphate fertilizer to neutralize acidied lakes
Remove SO_2 and NO_x from smokestack gases and remove NO_x from motor vehicular exhaust		
Tax SO_2 emissions		Add lime to neutralize acidied soils

FIGURE 15.8 Ways to reduce acid deposition and its damage. *Critical thinking:* Which two of these solutions do you think are the best ones? Why?

Top: Brittany Courville/Shutterstock.com. Bottom: racorn/Shutterstock.com.

Indoor Air Pollution

In less-developed countries, the indoor burning of wood, charcoal, dung, crop residues, coal, and other fuels in open fires (Figure 15.9) or in unvented or poorly vented stoves exposes people to dangerous levels of particulate air pollution (**Concept 15.2B**). The WHO has estimated that indoor air pollution causes an average of 9,600 deaths per day, mostly in less-developed nations.

3.5 million Annual global number of deaths due to indoor air pollution.

Indoor air pollution is also a problem in the United States and other more-developed areas, mostly because of the chemicals used to make building materials and products such as furniture and paneling. Figure 15.10 shows some typical sources of indoor air pollution in a modern home.

EPA studies have revealed some alarming facts about indoor air pollution. *First,* levels of several common air pollutants generally are two to five times higher inside U.S. homes and commercial buildings than they are outdoors. In some cases, they are as much as 100 times higher. *Second,* pollution levels inside cars in traffic-clogged urban areas can be up to 18 times higher than outside levels. *Third,* the health risks from exposure to such chemicals are

FIGURE 15.9 Burning wood to cook food inside this dwelling in Nepal exposes this woman and other occupants to dangerous levels of indoor air pollution.

Alain Lauga/Shutterstock.com

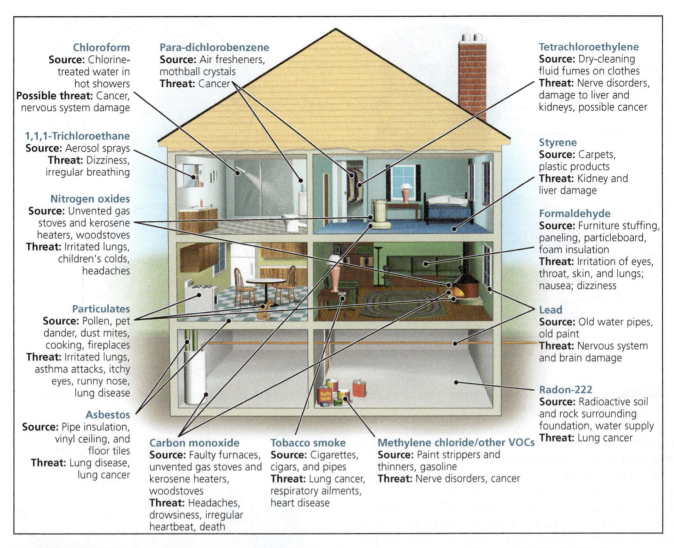

Chloroform
Source: Chlorine-treated water in hot showers
Possible threat: Cancer, nervous system damage

Para-dichlorobenzene
Source: Air fresheners, mothball crystals
Threat: Cancer

Tetrachloroethylene
Source: Dry-cleaning fluid fumes on clothes
Threat: Nerve disorders, damage to liver and kidneys, possible cancer

1,1,1-Trichloroethane
Source: Aerosol sprays
Threat: Dizziness, irregular breathing

Styrene
Source: Carpets, plastic products
Threat: Kidney and liver damage

Nitrogen oxides
Source: Unvented gas stoves and kerosene heaters, woodstoves
Threat: Irritated lungs, children's colds, headaches

Formaldehyde
Source: Furniture stuffing, paneling, particleboard, foam insulation
Threat: Irritation of eyes, throat, skin, and lungs; nausea; dizziness

Particulates
Source: Pollen, pet dander, dust mites, cooking, fireplaces
Threat: Irritated lungs, asthma attacks, itchy eyes, runny nose, lung disease

Lead
Source: Old water pipes, old paint
Threat: Nervous system and brain damage

Asbestos
Source: Pipe insulation, vinyl ceiling, and floor tiles
Threat: Lung disease, lung cancer

Carbon monoxide
Source: Faulty furnaces, unvented gas stoves and kerosene heaters, woodstoves
Threat: Headaches, drowsiness, irregular heartbeat, death

Tobacco smoke
Source: Cigarettes, cigars, and pipes
Threat: Lung cancer, respiratory ailments, heart disease

Methylene chloride/other VOCs
Source: Paint strippers and thinners, gasoline
Threat: Nerve disorders, cancer

Radon-222
Source: Radioactive soil and rock surrounding foundation, water supply
Threat: Lung cancer

FIGURE 15.10 Numerous indoor air pollutants can be found in many modern homes (**Concept 15.2B**). *Critical thinking:* To which of these pollutants are you exposed?

(Compiled by the authors using data from U.S. Environmental Protection Agency)

growing because most people in more-developed urban areas spend 70–98% of their time indoors or inside vehicles. Smokers, children younger than age 5, the elderly, the sick, pregnant women, people with respiratory or heart problems, and factory workers are especially at risk from indoor air pollution. **GREEN CAREER: Indoor air pollution specialist**

According to the EPA and public health officials, the four most dangerous indoor air pollutants in more-developed countries are *tobacco smoke* (see Chapter 14, pp. 396–397); *formaldehyde* emitted from many building materials and various household products (Figure 15.10); *radioactive radon-222 gas,* which can seep into houses from underground rock deposits; and *very small (ultrafine) particles* of various substances in emissions from motor vehicles, coal-burning facilities, wood fires, and forest and grass fires.

Air Pollution Is a Big Killer

Your respiratory system (Figure 15.11) helps protect you from air pollution. Hairs in your nose filter out large particles. Sticky mucus in the lining of your upper respiratory tract captures smaller (but not the smallest) particles and dissolves some gaseous pollutants. Hundreds of thousands of tiny, mucus-coated, hair-like structures, called *cilia,* line your upper respiratory tract. They continually move back and forth and transport mucus and the pollutants it traps to your throat where they are swallowed or expelled through sneezing and coughing.

Prolonged or acute exposure to air pollutants can overload or break down these natural defenses. Fine and ultrafine particulates can lodge deep in the lungs and contribute to lung cancer, asthma, heart attack, and stroke. Years of smoking or breathing polluted air can cause other

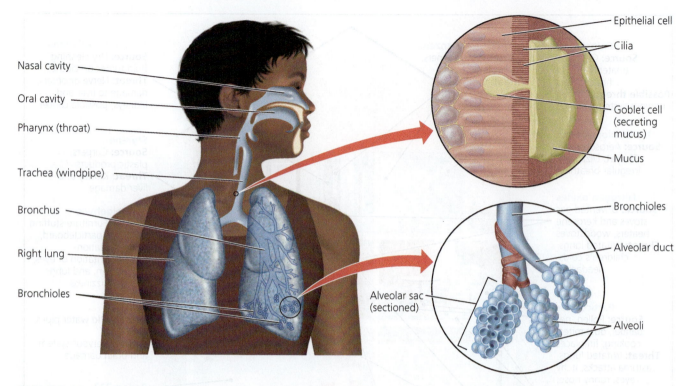

FIGURE 15.11 Certain components of the human respiratory system helps protect us from air pollution, but these defenses can be overwhelmed or breached.

lung ailments such as chronic bronchitis and emphysema, which lead to acute shortness of breath.

The WHO estimates that each year, indoor and outdoor air pollution kills about 6.5 million people, an average of 17,808 deaths per day. This illustrates why air pollution caused mostly by human activities is one of the world's most serious environmental problems. In China, air pollution kills 1.5 million people a year—an average of 4,110 deaths per day. This is largely because China burns almost as much coal, with inadequate pollution control, as all the world's other countries combined and has millions of motor vehicles—5 million in Beijing alone. China's leading direct causes of death related to air pollution are heart attacks, stroke, chronic obstructive pulmonary disease (COPD), and lung cancer. However, growing industrialization in India has increased its air pollution to the point where it rivals China's air pollution.

6.5 million Annual global number of deaths due to outdoor and indoor air pollution.

Steven Barrett and other researchers at the Massachusetts Institute of Technology (MIT) estimate that outdoor air pollution, mostly fine-particle pollution, contributes to the deaths of roughly 200,000 Americans every year. About half of these deaths are blamed on car and truck

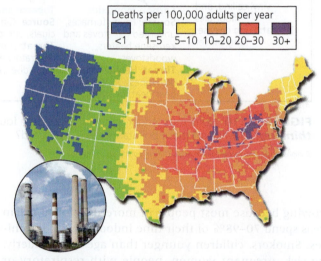

FIGURE 15.12 Distribution of premature deaths from air pollution in the United States, mostly from very small, fine, and ultrafine particles emitted into the atmosphere by coal-burning power plants. **Critical thinking:** Why do the highest death rates occur in the eastern half of the United States? If you live in the United States, what is the risk at your home or where you go to school?

Brittany Courville/Shutterstock.com

exhaust and the other half on coal-burning power and industrial plants (Figure 15.12). This death toll is roughly equivalent to that of two fully loaded, 275-passenger airliners crashing *every day* with no survivors. Millions more suffer from asthma attacks and other respiratory disorders brought on by indoor and outdoor air pollution.

According to EPA studies, each year, more than 125,000 Americans get cancer from breathing soot-laden diesel fumes emitted by buses, trucks, tractors, bulldozers and other construction equipment, trains, and ships. A large diesel truck emits as much particulate matter as 150 cars. A study led by Daniel Lack found that the world's 100,000 or more diesel-powered oceangoing ships emit almost half as much particulate pollution as do the world's 1 billion motor vehicles.

15.3 HOW SHOULD WE DEAL WITH AIR POLLUTION?

CONCEPT 15.3 Legal, economic, and technological tools can help us clean up air pollution, but the best solution is to prevent it.

Laws and Regulations Can Reduce Outdoor Air Pollution

The United States provides an example of how governments can reduce air pollution (**Concept 15.3**). The U.S. Congress passed the Clean Air Acts of 1970, 1977, and 1990. With these laws, the federal government established air pollution regulations for key outdoor air pollutants that are enforced by states and major cities.

Congress directed the EPA to establish air quality standards for six major outdoor pollutants—carbon monoxide (CO), nitrogen dioxide (NO_2), sulfur dioxide (SO_2), suspended particulate matter (SPM, smaller than PM-10), ozone (O_3), and lead (Pb). Each standard specifies the maximum allowable level for a pollutant, averaged over a specific period. The EPA has also established national emission standards for more than 188 *hazardous air pollutants (HAPs)* that can cause or contribute to serious health effects.

According to a 2016 EPA report, the combined emissions of the six major outdoor air pollutants decreased by about 65% between 1980 and 2015, even with significant increases during the same period in gross domestic product, vehicle miles traveled, population, and energy consumption (Figure 15.13).

The significant reduction of outdoor air pollution in the United States since 1970 has been successful mostly because of two factors. *First,* during the 1960s and early 1970s, U.S. citizens insisted that laws be passed and enforced to improve air quality. Prior to 1970, when Congress passed the Clean Air Act, air pollution control equipment for factories, power plants, and motor vehicles did not exist. *Second,* the country was affluent enough to afford such controls and improvements. For example, because of these factors, a new car today in the United States emits 75% less pollution than did a pre-1970 car.

Environmental scientists applaud this important success, but they call for strengthening U.S. air pollution laws by:

- Putting much greater emphasis on air pollution prevention. The power of prevention (**Concept 15.3**) was

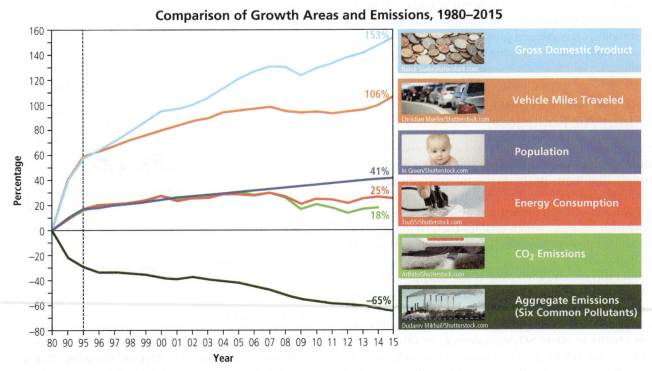

FIGURE 15.13 Levels of key air pollutants dropped sharply between 1980 and 2015, despite increases in other factors.

(Data and figure from U.S. Environmental Protection Agency)

made clear by the 99% drop in atmospheric lead emissions after lead in gasoline was banned in 1976.

- Sharply reducing emissions from approximately 20,000 older coal-burning power and industrial plants, cement plants, oil refineries, and waste incinerators that have not been required to meet the air pollution standards for new facilities under the Clean Air Acts.

- Reducing atmospheric emissions of toxic pollutants such as mercury.

- Continuing to improve fuel efficiency standards for motor vehicles, one of the most important steps needed to slow climate change.

- Stricter regulation of emissions from motorcycles and two-cycle gasoline engines used in devices such as chainsaws, lawnmowers, generators, scooters, and snowmobiles. The EPA estimates that running a typical gas-powered riding lawn mower for an hour creates as much air pollution as driving 34 cars for an hour.

- Setting much stricter air pollution regulations for airports and oceangoing ships.

- Sharply reducing indoor air pollution.

However, there is political pressure to weaken—not strengthen—U.S. air pollution control laws. Executives of companies that would be affected by stronger air pollution regulations claim that they would cost too much and would hinder economic growth. Proponents of stronger regulations contend that history has shown that most industry cost estimates for implementing U.S. air pollution control standards have been much higher than the costs actually proved to be. In addition, implementing such standards has created jobs and income by motivating many companies to develop new air pollution control technologies.

Using the Marketplace to Reduce Outdoor Air Pollution

One approach to reducing pollutant emissions is to allow producers of air pollutants to buy and sell government air pollution allotments in the marketplace (**Concept 15.3**). For example, with the goal of reducing SO_2 emissions, the Clean Air Act of 1990 authorized an *emissions trading*, or *cap-and-trade program,* which enables the 110 most polluting coal-fired power plants in 21 states to buy and sell SO_2 air pollution rights.

Under this system, each plant is annually given a number of pollution credits, which allow it to emit a certain amount of SO_2. A utility that emits less than its allotted amount has a surplus of pollution credits. That utility can use its credits to offset SO_2 emissions at its other plants, keep them for future plant expansions, or sell them to other utilities or private citizens or groups. Between 1990 and 2012, this emissions trading program helped reduce

SO_2 emissions from power plants in the United States by 76% at a cost of less than one-tenth of the cost projected by the utility industry, according to the EPA.

Proponents of this market-based approach say it is cheaper and more efficient than government regulation of air pollution. Critics contend that it allows utilities with older, dirtier power plants to buy their way out of their environmental responsibilities and to continue to pollute. The ultimate success of any emissions trading approach depends on two factors: how low the initial cap is set and how often it is lowered to promote continuing innovation in air pollution prevention and control.

Other Ways to Reduce Outdoor Air Pollution

Figure 15.14 summarizes several ways to reduce emissions of sulfur oxides, nitrogen oxides, and particulate matter from stationary sources such as coal-burning power plants and industrial facilities—the primary contributors to industrial smog.

Figure 15.15 lists several ways to prevent and reduce emissions from motor vehicles—the primary contributors to photochemical smog. In more-developed countries, many of these solutions have been successful. However, the already poor air quality in urban areas of many less-developed countries is worsening, because

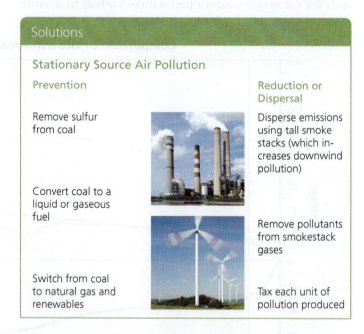

Solutions

Stationary Source Air Pollution

Prevention		Reduction or Dispersal
Remove sulfur from coal		Disperse emissions using tall smoke stacks (which increases downwind pollution)
Convert coal to a liquid or gaseous fuel		Remove pollutants from smokestack gases
Switch from coal to natural gas and renewables		Tax each unit of pollution produced

FIGURE 15.14 Ways to prevent, reduce, or disperse emissions of sulfur oxides, nitrogen oxides, and particulate matter from stationary sources, especially coal-burning power plants and industrial facilities (**Concept 15.3**). *Critical thinking:* Which two of these solutions do you think are the best ones? Why?

Top: Brittany Courville/Shutterstock.com. Bottom: racorn/Shutterstock.com.

Solutions

Motor Vehicle Air Pollution

Prevention

- Walk or bike or use mass transit
- Improve fuel efciency
- Get older, polluting cars off the road

Reduction

- Require emission control devices
- Inspect car exhaust systems twice a year
- Set strict emission standards

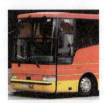

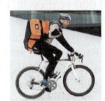

FIGURE 15.15 Ways to prevent or reduce emissions from motor vehicles (**Concept 15.3**). ***Critical thinking:*** Which two of these solutions do you think are the best ones? Why?

Top: egd/Shutterstock.com. Bottom: Tyler Olson/Shutterstock.com.

Solutions

Indoor Air Pollution

Prevention

- Ban indoor smoking
- Set stricter formaldehyde emissions standards for carpet, furniture, and building materials
- Prevent radon infiltration
- Use naturally based cleaning agents, paints, and other products

Reduction and Dilution

- Use adjustable fresh air vents for work spaces
- Circulate air more frequently
- Circulate a building's air through rooftop greenhouses
- Use solar cookers and efficient, vented woodburning stoves

FIGURE 15.16 Ways to prevent or reduce indoor air pollution (**Concept 15.3**). ***Critical thinking:*** Which two of these solutions do you think are the best ones? Why?

Top: Tribalium/Shutterstock.com. Bottom: PATSTOCK/AGE Fotostock.

of the sharp increase in the number of motor vehicles without adequate pollution control technology. Over the next 10 to 20 years, technology could help all countries clean up the air through improved engine and emission systems and all-electric, hybrid-electric, and plug-in hybrid vehicles (see Figure 13.24, p. 348) (**Concept 15.3**).

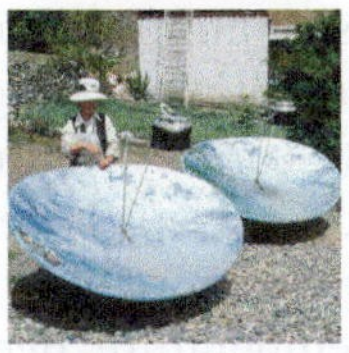

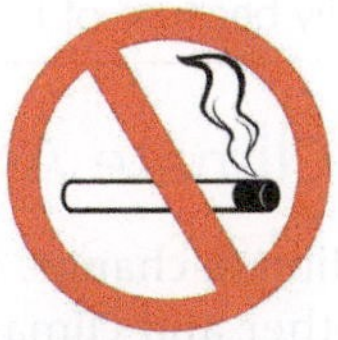

Reducing Indoor Air Pollution

Little effort has been devoted to reducing indoor air pollution even though it poses a much greater threat to human health than does outdoor air pollution (**Concept 15.2B**). Air pollution experts suggest several ways to prevent or reduce indoor air pollution, as shown in Figure 15.16.

In less-developed countries, indoor air pollution from open fires (Figure 15.9) and inefficient stoves could be reduced. More people could use inexpensive clay or metal ovens that burn fuels more efficiently and vent their exhausts to the outside, or they could use solar ovens and cookers (see Figure 13.35, p. 357) in sunny areas. Figure 15.17 lists some ways in which you can reduce your exposure to indoor air pollution.

What Can You Do?

Indoor Air Pollution

- Test for radon and formaldehyde inside your home and take corrective measures as needed
- Do not buy furniture and other products containing formaldehyde
- Test your home or workplace for asbestos fiber levels and check for any crumbling asbestos materials
- If you smoke, do it outside or in a closed room vented to the outside
- Make sure that wood-burning stoves, fireplaces, and kerosene- and gas-burning heaters are properly installed, vented, and maintained
- Install carbon monoxide detectors in all sleeping areas
- Use fans to circulate indoor air
- Grow house plants, the more, the better
- Do not store gasoline, solvents, or other volatile hazardous chemicals inside a home or attached garage
- Remove your shoes before entering your house to reduce inputs of dust, lead, and pesticides

FIGURE 15.17 Individuals Matter: You can reduce your exposure to indoor air pollution. ***Critical thinking:*** Which three of these actions do you think are the most important ones to take? Why?

15.4 HOW AND WHY IS THE EARTH'S CLIMATE CHANGING?

CONCEPT 15.4 Considerable scientific evidence strongly indicates that the earth's atmosphere is warming and changing the earth's climate primarily because of human activities.

Weather and Climate Are Not the Same

In thinking about climate change, it is important to distinguish between weather and climate. **Weather** consists of short-term changes in atmospheric variables such as the temperature and precipitation in a given area over a period of hours or days. By contrast, **climate**, as defined by the World Meteorological Society, is determined by the *average* weather conditions of the earth or a particular area, especially atmospheric temperature, over at least three decades. Scientists have used such long-term measurements to divide the earth into various climate zones (see Figure 7.2, p. 137).

Atmospheric warming (**Core Case Study**) does not mean that all areas of the earth are getting warmer. Instead, as the earth's average atmospheric temperature rises, some areas get warmer and some get cooler because of interactions in the planet's complex climate system. However, when the *global average* atmospheric temperature rises or falls over a period of at least three decades, the earth's climate is changing.

Climate Change Is Not New but Recently Has Accelerated

Climate change is neither new nor unusual. Over the past 3.5 billion years, many natural factors have played a role in past climate change. These natural factors include **(1)** massive volcanic eruptions and impacts by meteors and asteroids that cool the earth by injecting large amounts of debris into the atmosphere, **(2)** changes in solar input that can warm or cool the earth, **(3)** slight changes in the shape of the earth's orbit around the sun from mostly round to more elliptical over a 100,000 year cycle, **(4)** slight changes in the tilt of the earth's axis over a 41,000-year cycle; and **(5)** slight changes in the earth's wobbly orbit around the sun over a 20,000-year cycle. Factors 3, 4, and 5 together are known as the *Milankovitch cycles*. The earth's climate is also affected by **(6)** global air circulation patterns (see Figure 7.4, p. 139), **(7)** changes in the sizes of areas of ice (**Core Case Study**) that reflect incoming solar energy and cool the atmosphere, **(8)** changes in concentrations of greenhouse gases, and **(9)** occasional changes in ocean currents.

Scientific research reveals that the earth's climate has fluctuated over the past 900,000 years, slowly swinging back and forth between long periods of atmospheric warming and atmospheric cooling that led to ice ages (Figure 15.18, top left). These alternating freezing and thawing periods are known as *glacial* and *interglacial periods*.

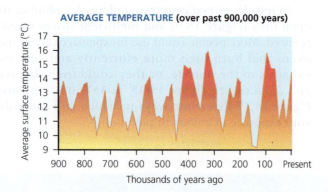

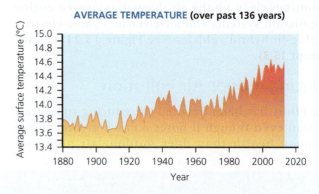

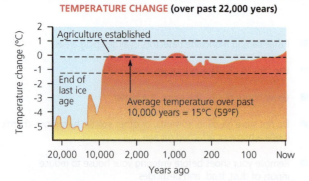

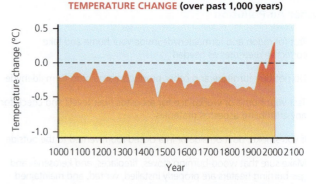

FIGURE 15.18 The global average temperature of the atmosphere near the earth's surface has changed significantly over different periods. The two graphs in the top half of this figure are estimates of global average temperatures, and the two graphs on the bottom are estimates of changes in the average temperature over different periods. **Critical thinking**: What are two conclusions that you can draw from these diagrams?

(Compiled by the authors using data from Goddard Institute for Space Studies, Intergovernmental Panel on Climate Change, National Academy of Sciences, National Aeronautics and Space Administration, National Center for Atmospheric Research, and National Oceanic and Atmospheric Administration)

For roughly 10,000 years, the earth has experienced an interglacial period with a generally steady global average surface temperature (Figure 15.18, bottom left). The resulting mostly stable climate allowed the human population to grow as agriculture developed and later as cities grew. For the past 1,000 years, the average temperature of the atmosphere near the earth's surface has remained fairly stable (Figure 15.18, bottom right). However, since 1975 the earth's average atmospheric temperatures have been rising (Figure 15.18, top right). In other words, we have been enhancing the natural greenhouse effect (Figure 3.3, p. 48) that supports the earth's life and human economies. As atmospheric levels of CO_2 rise, the gas becomes a pollutant that plays a role in the climate change and its harmful environmental, health, and economic effects.

Considerable scientific evidence indicates that increases in the CO_2 emissions from human activities such as burning fossil fuels has increased the earth's average atmospheric temperature and played an important role in climate change taking place since 1975 (Figure 15.18, top right). This is many times faster than past climate changes caused by natural factors that took place over hundreds to thousands of years (Figure 15.18, top left).

Scientists estimate past temperature changes such as those depicted in Figure 15.18 by analyzing evidence from many sources. They include radioisotopes in rocks and fossils; plankton and radioisotopes in ocean sediments; tiny bubbles, layers of soot, and other materials trapped in different layers of ancient air found in ice cores from glaciers (Figure 15.19); pollen from the bottoms of lakes and bogs; tree rings; and atmospheric temperature measurements taken regularly since 1861. These temperature measurements now include data from more than 40,000 measuring stations around the world, as well as from satellites.

Between 2007 and 2015, the world's leading scientific organizations—including the IPCC, NAS, British Royal Society, U.S. National Atmospheric and Oceanic Administration (NOAA), U.S. National Aeronautic and Space Administration (NASA), and the American Association for the Advancement of Science (AAAS), and at least 90% of the world's climate scientists—all reached the following four major conclusions:

1. The earth's climate is changing and very likely will accelerate unless we act now to slow it.

2. Human activities, such as burning fossil fuels (which adds CO_2 to the atmosphere) and the clearing of forests (which remove CO_2 from the atmosphere) play an important role in current climate change.

3. Average atmospheric temperatures are likely to increase and lead to more climate change, unless we act now to slow it.

4. Immediate and sustained action to curb climate change is possible and affordable and would bring major benefits for human health, economies, and the environment.

U.S. Geological Survey

FIGURE 15.19 *Ice cores* are extracted from deep holes drilled into ancient glaciers at various sites such as this one near the South Pole in Antarctica. Analysis of ice cores yields information about the past composition of the lower atmosphere, temperature trends such as those shown in Figure 15.2, upper right, solar activity, snowfall, and forest fire frequency.

Data from thousands of peer-reviewed scientific studies support the conclusion that human-influenced climate change is happening now. Below are a few pieces of such evidence.

- Between 1906 and 2016, the earth's average global surface temperature rose by 0.94°C (1.7°F), with much of this increase taking place since 1975 (Figure 15.18, upper right). By definition, this more than 40-year general increase in the earth's atmospheric temperature means that climate change has taken place.

- The ten warmest years on record since 1861 have taken place since 2005.

- In the Arctic, floating summer sea ice has been melting and shrinking in most years since 1979.

- In some parts of the world, glaciers that have existed for thousands of years (Figure 15.20), including Greenland's ice sheets (**Core Case Study**), are melting.

- In Alaska, glaciers are melting, frozen ground (permafrost) is thawing, loss of sea ice and rising sea levels are eating away at coastlines, and some communities will have to be relocated inland.

- The world's average sea level has been rising at an accelerated rate, especially since 1975. This rise is mostly due to the expansion of ocean water as it absorbed heat from the atmosphere and increasing runoff from melting land-based ice.

FIGURE 15.20 Between 1913 (top) and 2008 (bottom) much of the ice that covered Sperry Glacier in Montana's Glacier National Park melted.

Top: W. C. Alden/GNP Archives/US Geological Survey. Bottom: Lisa McKeon/US Geological Survey.

- Atmospheric levels of CO_2, CH_4, and other greenhouse gases that warm the troposphere have been rising sharply.

- As temperatures have risen, many terrestrial, marine, and freshwater species have migrated toward the poles and, on land, to cooler higher elevations. Species that cannot migrate face extinction.

You may have heard about a *climate change debate* among scientists as reported in the news media. This is misleading because after several decades of peer-reviewed research and discussion, there is no significant debate or disagreement among at least 90% of the world's more than 2,500 climate scientists on current climate change and its causes. Because scientists are highly skeptical and question the results of all research, this level of agreement among scientists in any field is rare.

However, there is intense debate and disagreement about climate change in the political arena among citizens, elected officials, and officials of companies that produce or burn fossil fuels that add CO_2 to the atmosphere. For over 40 years, the fossil fuel industry has financed efforts to cast doubt on whether the climate is changing and whether human activities (especially emissions of CO_2 from the burning of fossils fuels) play a significant role in climate change. Such disagreements focus on which, if any, political and economic actions should be taken to deal with climate change. This is a very difficult and important political, economic, and ethical issue because modern economies and lifestyles are built around burning fossil fuels that supply

90% of the world's commercial energy and that add massive amounts of climate-changing CO_2 to the atmosphere.

Climate and the Greenhouse Effect

The **greenhouse effect** (see Figure 3.3, p. 48) is a natural process that plays a major role in determining the earth's average atmospheric temperature and thus its climate. It occurs when some of the solar energy absorbed by the earth radiates into the atmosphere as infrared radiation (heat). As this radiation interacts with molecules of several *greenhouse gases* in the air, including water vapor (H_2O), carbon dioxide (CO_2), methane (CH_4), and nitrous oxide (N_2O), it increases their kinetic energy and warms the lower atmosphere and the earth's surface.

Numerous laboratory experiments and measurements of temperatures at different altitudes have confirmed the greenhouse effect—now one of the most widely accepted scientific theories in the atmospheric sciences. Life on the earth and our economic systems are dependent on the natural greenhouse effect because it keeps the planet at an average temperature of around 15°C (58°F). Without it, the planet would be a frozen, mostly uninhabitable place.

The atmospheric concentration of CO_2, as part of the carbon cycle (Figure 3.16, p. 59), plays a key role in determining the average temperature of the atmosphere. Measurements of CO_2 in bubbles in ice cores (Figure 15.19) at various depths in ancient glacial ice indicate that changes in the levels of this gas in the lower atmosphere have correlated closely with changes in the global average temperature near the earth's surface during the past 400,000 years (Figure 15.21). Scientists have noted a similar correlation between atmospheric temperatures and methane (CH_4) emissions.

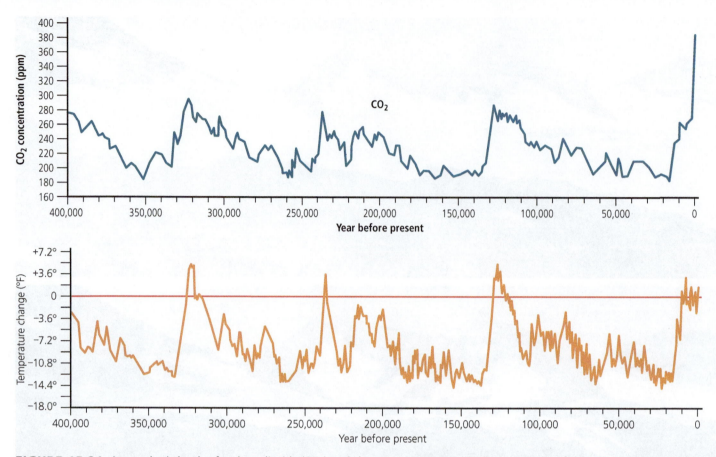

FIGURE 15.21 Atmospheric levels of carbon dioxide (CO_2) and changes in average global temperature of the atmosphere near the earth's surface over the past 400,000 years. These data were obtained by analysis of ice cores removed at Russia's Vostok Research Station in Antarctica.

(Compiled by the authors using data from Intergovernmental Panel on Climate Change, National Center for Atmospheric Research, and F. Vimeux, et al. 2002. *Earth and Planetary Science Letters*, vol. 203: 829–843)

Human Activities Play an Important Role in Current Atmospheric Warming

Research reveals that past changes in the earth's climate were due to the natural factors. However, since the beginning of the Industrial Revolution in the mid-1700s, human actions—mainly the burning of fossil fuels, deforestation, and agriculture—have led to significant increases in the concentrations of several greenhouse gases, especially CO_2, in the lower atmosphere (Figure 15.22). This is a long-lasting increase because about 80% of the CO_2 that human activities emit typically remains in the atmosphere for 100 years or more and about 20% of it remains for up to 1,000 years. After oscillating between 180 and 280 parts per million (ppm) for 400,000 years, average atmospheric levels of CO_2 reached 404 ppm in 2016—higher than at any time in the last 4.5 million years according to NOAA scientists.

Atmospheric levels of methane (CH_4), another greenhouse gas, have also increased greatly since the mid-1970s. Ice core analysis reveals that about 70% of global emissions of methane during the last 275 years were likely caused by human activities, including livestock production, rice production, natural gas production, use of landfills, and the flooding of land behind large dams. The other 30% came from natural sources. Methane remains in the atmosphere for about 12 years compared to at least 100 years for CO_2. However, each molecule of methane warms the air 25 times more than a molecule of CO_2.

In 2016, the three largest emitters of energy-related CO_2 were, in order, China, the United States, and India, according to the U.N. Statistics Division. In comparing CO_2 emissions sources, scientists use the concept of a **carbon footprint**. It refers to the amount of CO_2 generated by an individual, a country, a city, or any other entity over a given period. A **per capita carbon footprint** is the average footprint per person in a population. China has the largest national carbon footprint, followed by the United States and India. The United States has the largest per capita carbon footprint and since 1850, has emitted far more CO_2 than any other country.

110 million Average number of metric tons of CO_2 that human activities pump into the atmosphere every day.

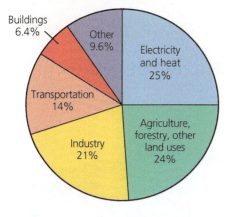

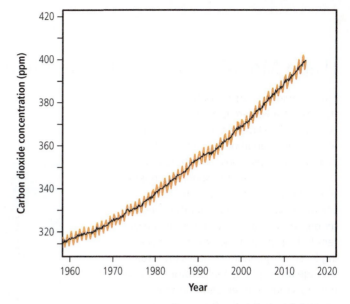

FIGURE 15.22 Annual emissions of carbon dioxide (CO_2) from various sources (left) have contributed to rising global atmospheric levels of CO_2 (right), 1880–2015.

(Compiled by the authors using data from International Energy Agency and U.S. Department of Energy Carbon Dioxide Information Analysis Center (left); Earth Policy Institute BP, *Statistical Review of World Energy*, 2016, Intergovernmental Panel on Climate Change, and National Center for Atmospheric Research (right)).

Most of the evidence that scientists use in projecting future atmospheric temperatures and thus climate change involves the use of climate models (Science Focus 15.1).

Role of the Sun in Current Atmospheric Warming

The energy output of the sun plays the key role in the earth's temperature, and this output has varied over millions of years. However, climate researchers Claus Froehlich, Mike Lockwood, and Ben Santer all concluded in separate studies that most of the rise in global average atmospheric temperatures since 1975 (Figure 15.18, top right) could not be the result of increased solar output. Instead, they determined that the energy output of the sun has dropped slightly during the past several decades. A detailed data analysis by physicist Richard Muller and his colleagues confirmed this conclusion.

Froehlich noted that, according to satellite and weather balloon measurements since 1975, the troposphere has warmed while the stratosphere has cooled. This is the opposite of what a hotter sun would do, which would be to heat the atmosphere from the top down. Instead, the data show that the atmosphere is now heating from the bottom up, which indicates that inputs at the earth's surface (most likely from human activities) play the more important role in atmospheric warming.

Effects of the Oceans on Current Atmospheric Warming

The world's oceans have played a key role in slowing the rate of recent atmospheric warming and climate change.

Research indicates that the oceans remove roughly 25% of the CO_2 pumped into the lower atmosphere by human activities. Most of it is stored as carbon compounds in marine algae and vegetation and in coral reefs and eventually is stored for several hundred million years in carbon compounds in ocean bottom sediments.

The oceans also absorb heat from the lower atmosphere. According to a 2016 study by researchers at Lawrence Livermore National Laboratory, more than 90% of the heat held in the atmosphere by greenhouse gas pollution since the 1970s has ended up in the oceans. Thus, the average temperature of the oceans has also risen since 1970.

The uptake of both CO_2 and heat by the world's oceans has slowed the rate of atmospheric warming and the rate of climate change. However, this has resulted in the increasingly serious problem of ocean acidification, which could have harmful effects on marine ecosystems (see Science Focus 9.3, p. 214).

Effects of Cloud Cover on Atmospheric Warming

Warmer temperatures increase evaporation of surface water, which raises the relative humidity of the atmosphere in various parts of the world. This creates more clouds that can either cool or warm the atmosphere. An increase in thick and continuous *cumulus clouds* at low altitudes would have a cooling effect by reflecting more sunlight back into space. An increase in thin, wispy *cirrus clouds* at high altitudes would warm the lower atmosphere by allowing more sunlight to reach the earth's surface and preventing some heat from escaping into space.

Using Models to Project Future Changes in Atmospheric Temperatures

There is widespread scientific evidence that the atmosphere is warming and changing the climate and that human activities play an important role in these changes. The key question is how much the earth's average atmospheric temperatures are likely to change in the future. Scientists have developed mathematical models of the earth's complex climate system to help project the effects of increasing levels of greenhouse gases on future average atmospheric temperatures. These models simulate interactions among incoming sunlight, the three natural Milankovitch cycles, clouds, landmasses, glaciers, floating sea ice, ocean currents, clouds, greenhouse gases, air pollutants, and other factors within the earth's complex climate system.

Scientists run these continually improving models on supercomputers and compare the results to known past average atmospheric temperatures. They use these data to project future changes in the earth's average atmospheric temperature. Figure 15.A gives a greatly simplified summary of some of the key interactions in the global climate system used in climate models.

Climate models provide *projections* of what is likely to happen to the average temperature of the lower atmosphere, based on available data and different assumptions about future changes such as CO_2 and CH_4 levels in the atmosphere. How well the projections match what actually happens depends on the validity of the assumptions, the variables built into the models (Figure 15.A), and the accuracy of the data used.

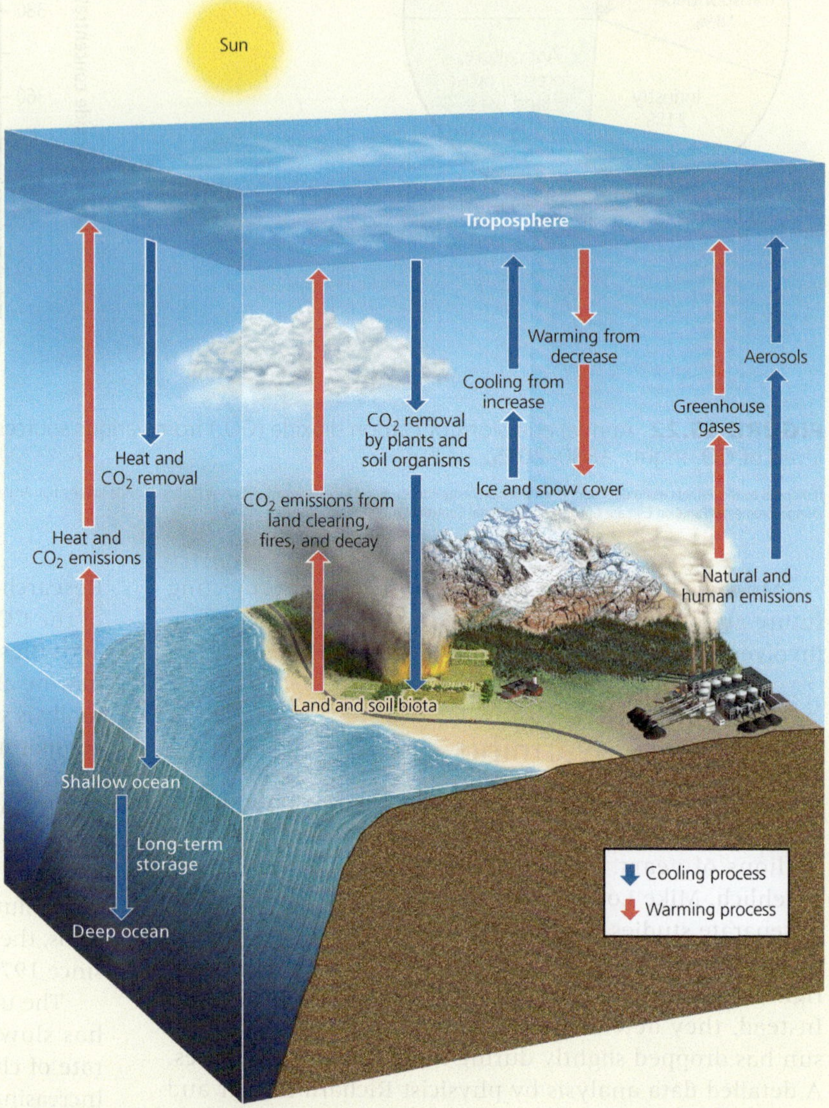

Recall that scientific research does not provide absolute proof or certainty. Instead, science provides us with varying levels of certainty. According to the 2014 IPCC report, based on analysis of past climate data and the use of more than two dozen climate models, it is *extremely likely* that human activities,

FIGURE 15.A A simplified model of some major processes that interact to affect the earth's climate by determining the average temperature and greenhouse gas content of the lower atmosphere. Red arrows show processes that warm the atmosphere and blue arrows show those that cool it. ***Critical thinking:*** Why do you think a decrease in snow and ice cover is adding to the warming of the atmosphere?

especially the burning of fossil fuels, have played an important role in the observed atmospheric warming since 1975 (Figure 15.18, top right). Researchers based this conclusion on thousands of peer reviewed research studies and on the fact that,

after thousands of times running the models, the only way to get the model results to match actual past temperature measurements is to include the human activities factor (Figure 15.B).

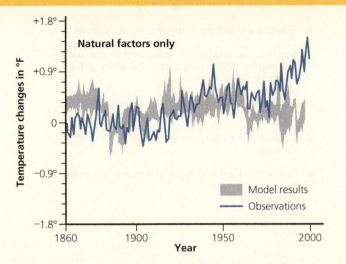
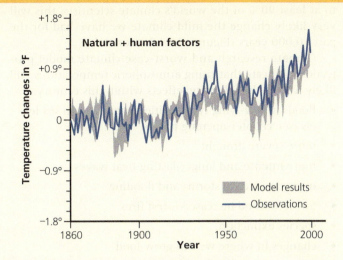

FIGURE 15.B Comparison of actual climate data with modeled projections for the period between 1860 and 2000 using natural factors only (left) and a combination of natural and human factors (right). Scientists have found that actual data match projections far more closely when human factors are included in the models.

(Compiled by the authors using data from Intergovernmental Panel on Climate Change.)

However, there is a high degree of scientific uncertainty about projected changes in the earth's average atmospheric temperature. According to current climate models, by the end of the century, the temperature of the earth's atmosphere is projected to rise from 1.5°C (2.7°F) to 4.5°C (8.1°F). Despite their limitations, these models are the best and only tools that we have for projecting likely average atmospheric temperatures in coming decades. This also shows the urgent need to increase data and research on climate change and on improving climate models.

CRITICAL THINKING

If the earth's highest projected temperature increase of (4.5°C or 8.1°F) is reached, what are three major ways in which this will likely affect your lifestyle and those of any children or grandchildren you might have?

According to a 2014 NAS report on climate, the latest scientific research indicates that the net global effect of cloud cover changes is likely to increase atmospheric warming. More research is needed to evaluate this effect and project how it will affect future atmospheric temperatures.

Effects of Outdoor Air Pollution on Atmospheric Warming

According to the 2014 IPCC report, air pollution in the form of *aerosols* (suspended microscopic droplets and solid particles) from human activities can hinder or enhance both the greenhouse effect and cloud formation, depending on factors such as their size and reflectivity.

Most aerosols, such as light-colored sulfate particles produced by fossil fuel combustion, tend to reflect incoming sunlight and cool the lower atmosphere. However, black carbon particles, or *soot*—emitted into the air by combustion of fossil fuels, diesel exhaust, cooking fires, and forest fires, absorb solar energy—warm the lower atmosphere.

Climate scientists do not expect aerosols and soot particles to affect the earth's average atmospheric temperature very much over the next 50 years for two reasons. *First,* aerosols and soot fall back to the earth or are washed out of the lower atmosphere within weeks or months, whereas CO_2 typically remains in the lower atmosphere for 100 years or longer. *Second,* aerosol and soot emissions are being reduced by air pollution control regulations, especially in more-developed countries, because of their harmful impacts on plants and humans.

15.5 WHAT ARE THE LIKELY EFFECTS OF CLIMATE CHANGE?

CONCEPT 15.5 The current and projected change in the atmosphere's temperature and the resulting climate change could have severe and long-lasting consequences, including rising sea levels, shifts in the locations of croplands and wildlife habitats, and more extreme weather.

Rapid Atmospheric Warming Could Have Serious Consequences

Most past changes in the temperature of the lower atmosphere took place over thousands of years (Figure 15.18, top left). What makes the current problem urgent is that humanity faces *a rapid projected increase in the average temperature of the lower atmosphere during this century*. According to at least 90% of the world's climate scientists, this will very likely change the mild climate we have had for the past 10,000 years (Figure 15.18, bottom left).

Climate research and worst-case climate model projections indicate that rising atmospheric temperatures will likely lead to the following effects within this century:

- floods in low-lying coastal cities from a rise in sea levels (see chapter-opening photo)
- more severe drought
- more intense and longer-lasting heat waves
- more destructive storms and flooding
- forest loss and increased forest fires
- species extinction
- changes in where we can grow food

These effects will likely reduce food security and increase poverty and social conflict in many poorer nations that are the least responsible for atmospheric warming and the least able to deal with its harmful consequences. The models indicate that we will have to deal simultaneously with many of the disruptive effects of climate change within this century—an incredibly short time to bring about a major shift in the way we live and interact with our life-support system.

Scientists have identified several components of the earth's climate system that could exceed **climate change tipping points**—thresholds beyond which climate change could last for hundreds to thousands of years. Figure 15.23 lists several climate change tipping points.

Let us look more closely at some projected consequences of not acting now to slow climate change, according to the latest climate models.

Increased Melting of Ice and Snow

Models project that climate change will be the most severe in the world's polar regions. During the past 50 years, temperature measurements indicate that the atmosphere above the poles has warmed much more than the atmosphere in the rest of the world.

As warming increases in the polar regions, more snow and ice will melt, exposing darker and less reflective land and ocean water. This will further increase atmospheric warming above the poles, melt more ice and snow, and increase the temperature of the polar oceans in a runaway positive feedback loop (see Figure 2.12, p. 38).

Mostly because of the increased temperature of the arctic atmosphere and Arctic Ocean waters, the area covered by floating sea ice and the volume of summer sea ice measured every October have decreased (see graphs in Figure 15.24). In October 2016, the area of Arctic sea ice was the lowest since scientists began monitoring the ice coverage via satellite in 1981. Because of annual

- Atmospheric carbon level of 450 ppm
- Melting of all arctic summer sea ice
- Collapse and melting of the Greenland ice sheet
- Severe ocean acidification, collapse of phytoplankton populations, and a sharp drop in the ability of the oceans to absorb CO_2
- Massive release of methane from thawing arctic permafrost and from the arctic seafloor
- Collapse and melting of most of the western Antarctic ice sheet
- Severe shrinkage or collapse of Amazon rain forest

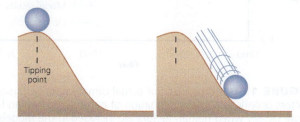

FIGURE 15.23 Climate scientists have come up with this list of possible climate change tipping points.

changes in short-term weather conditions, summer arctic sea ice coverage is likely to fluctuate. However, the overall projected long-term trends are for the Arctic to continue warming, the average summer sea ice coverage to decrease, and the ice to become thinner.

One of the climate-change tipping points that concern scientists is the complete melting of floating summer Arctic sea ice. If the current trend continues, summer Arctic sea ice may be gone by 2050, according to the 2014 IPCC report. This would lead to dramatic and long-lasting changes in weather and climate that could affect the entire planet.

Another effect of arctic warming is faster melting of polar land-based ice, including that in Greenland (**Core Case Study**). This melting is adding freshwater to the northern seas, and is likely to contribute to a projected rise in sea level during this century. Nature photographer James Balog (Individuals Matter, 15.1) has created a compelling visual record of dramatic melting of glaciers around the world.

Another great storehouse of ice is the earth's mountain glaciers. During the past 25 years, many of these glaciers have been shrinking wherever summer melting exceeds the addition of ice from precipitation in winter. Mountain glaciers play a vital role in the water cycle (see Figure 3.15, p. 58) by storing water as ice during cold seasons and releasing it slowly to streams during warmer seasons.

About 80% of the mountain glaciers in South America's Andes range are slowly shrinking. If this continues, 59 million people in Bolivia, Peru, and Ecuador who rely on melt water from the glaciers for irrigation and hydropower could face severe water, power, and food shortages.

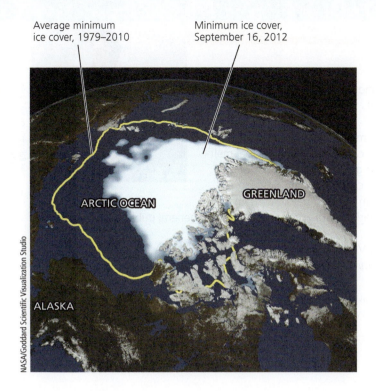

Average minimum ice cover, 1979–2010

Minimum ice cover, September 16, 2012

ARCTIC OCEAN

GREENLAND

ALASKA

NASA/Goddard Scientific Visualization Studio

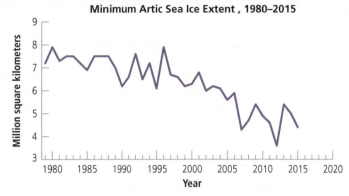

Minimum Artic Sea Ice Extent, 1980–2015

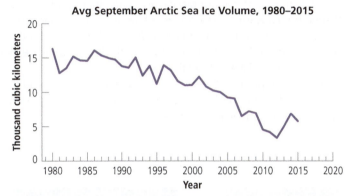

Avg September Arctic Sea Ice Volume, 1980–2015

FIGURE 15.24 *The big melt:* Rising average atmospheric and ocean temperatures have caused more and more arctic sea ice to melt during the summer months. The yellow line added to this satellite image (left) shows the average summer minimum area of ice during the period 1979–2010, in contrast to the white, ice-covered summer minimum in 2012. The graphs (right) show that the sea ice melt generally has been increasing. **Critical thinking:** Which is declining faster, the summer sea ice area, or the summer sea ice volume?

(Compiled by the authors using data from U.S. Goddard Space Flight Center, National Aeronautics and Space Administration, National Snow and Ice Data Center)

In the United States, according to climate models, people living in the Columbia, Sacramento, and Colorado River basins could face similar threats as the winter snow packs that feed these rivers are projected to shrink by as much as 70% by 2050. Glacier National Park in the U.S. state of Montana once had 150 glaciers, but by 2016, only 25 remained (Figure 15.20).

Methane Emissions from Thawing Permafrost

Permafrost exists in soils found beneath about 25% of the exposed land in Alaska, Canada, and Siberia in the Northern Hemisphere. Huge amounts of carbon are locked up in organic material below these permafrost soils. When exposed to warmer temperatures permafrost thaws and softens but does not melt.

This thaw is already happening in parts of Alaska and Siberia. If this trend continues, a significant amount of organic material found below the permafrost will be decomposed by microorganisms and release huge amounts of CH_4 and CO_2 into the atmosphere. Research by Ted Schuur and other scientists indicates that between 5% and 15% of the carbon stored mostly as CO_2 in the world's permafrost is

likely to be released during this century. This would accelerate atmospheric warming, thaw more permafrost, and lead to increased atmospheric warming and climate change in another runaway positive feedback loop that could lead to a climate change tipping point (Figure 15.23). Some scientists are also concerned about release of the greenhouse gas CH_4 from a thawing layer of permafrost on the Arctic Sea floor and on arctic lake bottoms (Figure 15.25).

Rising Sea Levels

Good news: The world's oceans have helped slow climate change by absorbing about 90% of the extra heat added to the atmosphere as a result of increased emissions of greenhouse gases from the burning of fossil fuels and other human activities. **Bad news:** As the oceans store more heat they expand. This, along with water flowing into the ocean from melting land-based glaciers, raises the world's average sea level. Scientific research indicates that over the last decade thermal expansion of ocean water has caused about one-third of the observed rise in the planet's sea level.

In 2014, the IPCC projected that, according to climate models, the average global sea level is likely to rise by

James Balog: Watching Glaciers Melt

James Balog has used his creativity and skills as a world-renowned photographer to alert humanity to major environmental problems such as losses in biodiversity, threats to North America's old-growth forests, and climate change as revealed by melting glaciers.

In 2007, he began working on his Extreme Ice Survey, the world's most wide-ranging photographic study of the rapid melting of many of the world's glaciers. To capture images of glacial melting, Balog developed camera systems that take daylight photos of 22 glaciers in locations around the world every half hour or hour. Balog used the resulting images to create stunning time-lapse videos of melting glaciers, where huge volumes of ice can disappear in hours or days. These images provided important data for glaciologists and other scientists and clearly and dramatically showed the general public that climate change from a warmer atmosphere is having a major impact now.

Balog's work has been featured in *National Geographic*, in the NOVA documentary *Extreme Ice*, and in *Chasing Ice*, an internationally acclaimed and award-winning full-length film. His book *Ice: Portraits of the World's Vanishing Ice* describes the astounding loss of ice from various glaciers through 2012.

Balog's non-profit *Earth Vision Trust* has a mission to spread the visual message of climate change, to finance studies of melting glaciers, and to spur people to action. He says, "Seeing is believing. I was a climate change skeptic until I saw the evidence in the ice. Climate change is real and the time to act is now."

James Balog/Extreme Ice Survey/ Earth Vision Institute

FIGURE 15.25 Scientists ignite a large bubble of methane gas released from an arctic lake bottom in Alaska.

MARK THIESSEN/National Geographic Creative

0.4–0.6 meters (1.3–2 feet) by the end of this century—about 10 times the rise that occurred in the 20th century. A 2016 study by a team of scientists led by Benjamin Strauss at Climate Central, using a larger set of data, estimated that sea levels could rise by 1.1–1.2 meters (3–4 feet) by 2100, with 50% to 66% of this rise coming from the melting of Greenland's ice (**Core Case Study**). However, accelerated melting could lead to seas rising by 2.4 meters (8 feet) according to a 2017 worst-case projection by the National Oceanic and Atmospheric

Administration, depending on how much of the land-based ice in Greenland and perhaps West Antarctica melt as the global atmospheric temperature continues to rise during this century.

According to climate models, sea-level rise will not be uniform throughout the world because of factors such as ocean currents and winds. For example, coastal scientist John Pethick projects that by 2100, Bangladesh's sea level could rise by as much as 4 meters (13 feet), several times higher than the projected global average sea-level rise. In addition, the sea level is rising as much as 4 times faster than the global average along parts of the U.S. Atlantic coast, according to the U.S. Geological Survey.

According to the 2014 IPCC and NAS reports on climate change, a 1-meter (3-foot) rise in sea level during this century (excluding the additional effects of the resulting higher storm surges) could cause the following serious effects:

- Degradation or destruction of at least one-third of the world's coastal estuaries, wetlands, coral reefs, and deltas where much of the world's rice is grown.

- Disruption of many of the world's coastal fisheries.

- Saltwater contamination of freshwater coastal aquifers resulting in degraded supplies of groundwater used for drinking and irrigation.

- Flooding in large areas of low-lying countries such as Bangladesh, one of the world's poorest and most densely populated nations.

- Flooding and erosion of low-lying barrier islands and gently sloping coastlines, especially in U.S. coastal states such as Florida (Figure 15.26), Texas, Louisiana, New Jersey, South Carolina, and North Carolina.

FIGURE 15.26 If the average sea level rises by 1 meter (3 feet), the areas shown here in red in the U.S. state of Florida will be flooded.

(Compiled by the authors using data from Jonathan Overpeck, Jeremy Weiss, and the U.S. Geological Survey)

- Flooding of some of the world's largest coastal cities (see red-shaded areas in chapter-opening photo), such as Venice, London, and New Orleans, and displacement of at least 150 million people—a number almost equal to half of the current U.S. population.

- Submersion of low-lying island nations such as the Maldives (Figure 15.27) and Fiji.

More Severe Drought

Drought occurs when evaporation due to higher atmospheric temperatures exceeds precipitation for a prolonged period. A study by National Center for Atmospheric Research scientist Aiguo Dai and his colleagues found that severe and prolonged drought has affected at least 30% of the earth's land (excluding Antarctica)—an area the size of Asia. According to a study by climate researchers at NASA, up to 45% of the world's land area could be experiencing severe drought by 2059.

It is not possible to tie a specific severe drought to atmospheric warming because natural cyclical processes can also cause severe droughts. However, the extra heat in the atmosphere evaporates water from soils. According to a study by climate scientists Richard Seager and Martin Hoerling, this depletion of soil moisture in drier areas prolongs droughts and makes them more severe, regardless of their causes.

Prolonged drought can decrease the growth of trees and other plants, which decreases their removal of CO_2 from the atmosphere. It also dries out forests and grasslands, which can increase the frequency of wildfires that add CO_2 to the atmosphere. Climate scientists project that these effects from more prolonged and more intense drought could speed up atmospheric warming and lead to more prolonged droughts, drier conditions, and more forest fires. This is another example of a runaway positive feedback loop that could exceed one or more climate change tipping points (Figure 15.23).

More Extreme Weather

There is not enough evidence to link any specific extreme weather event such as a heat wave, severe flood, or hurricane to climate change. However, with more heat in the atmosphere, climate scientists project that climate change will likely increase the overall chances and the intensity of extreme weather events. For example, storms will get stronger and heat waves will be hotter. In addition, climate models project that a more intensified water cycle with more water vapor in the atmosphere will bring higher levels of flooding to some areas due to heavier snowfall or rainfall.

Since 1950, heat waves have become longer, more frequent, and in some cases more intense. Because atmospheric warming increases the kinetic energy of the molecules of gases in the atmosphere, this trend is likely to continue in some areas. This could raise the number of heat-related deaths, reduce crop production, and expand deserts. For example, a 2013 heat wave in Europe caused 70,000 premature deaths. At the same time, because a warmer atmosphere can hold more water vapor, other areas, including the eastern half of the United States, will likely experience higher flood levels from heavy and prolonged rainfall or snowfall. In 2010, a World Meteorological Organization panel of experts concluded that warmer ocean temperatures are likely to lead to fewer but stronger hurricanes and typhoons that could cause more damage in many heavily populated coastal areas.

Threats to Biodiversity

According to the latest IPCC reports, projected climate change is likely to alter ecosystems and take a toll on biodiversity on every continent. Up to 85% of the Amazon rain forest—one of the world's major centers of biodiversity—could be lost and converted to savanna if the global atmospheric temperature rises by the highest projected amount (Science Focus 15.1), according to a scientific study led by climate scientist Chris Jones.

Research indicates that the most vulnerable ecosystems are coral reefs, polar seas, coastal wetlands, high-elevation forests, and alpine and arctic tundra. As the atmosphere warms, at least 17% of the world's species could face extinction by 2100. The hardest hit species will be

- Cold-climate plant and animal species, including the polar bear and walruses in the Arctic and penguins in Antarctica.

- Species that live at higher elevations.

- Species with limited ranges.

- Species with limited tolerance for temperature change, such as corals.

FIGURE 15.27 For low-lying island nations like the Maldives (population 295,000) in the Indian Ocean, rising sea levels could spell disaster.

Malbert/Dreamstime.com

The primary cause of these extinctions will be loss of habitat. On the other hand, the populations of plant and animal species that thrive in warmer climates could grow.

A warmer climate is also increasing populations of insects and fungi that damage trees, especially in areas where winters are no longer cold enough to control their populations. According to scientists, this helps explain the recent severe insect damage to large areas of pine forests in Canada and the American West and spruce forests in Alaska (see Case Study that follows). Climate change also threatens the biodiversity found in many existing state and national parks, wildlife reserves, wilderness areas, and wetlands.

Climate change also threatens marine ecosystems. Coral reefs are especially vulnerable because of their limited tolerance for water temperature increases, which bleach colorful coral reefs when excess heat expels the algae that help provide food for the tiny polyps that build the reefs. Many coral reefs are dying for this reason, including a growing area of Australia's Great Barrier Reef, according to a 2017 study by coral reef expert Terry Hughes.

Alaska: A Preview of the Effects of Climate Change

Alaska is a leading indicator for most of the effects of climate change. It is warming at about twice the average rate of the rest of the United States, and its warming is accelerating.

As a result, all but five of Alaska's glaciers were shrinking in 2015. Coastal sea ice is melting earlier in the summer and refreezing later in the fall over the past several years. As the sea ice shrinks away, coastal villages suffer from erosion and more flooding and damage from storms.

The government has identified at least 31 Alaskan coastal villages and towns, some of them home for Inuit and other aboriginal groups in danger of being flooded from rising sea levels. The government plans to relocate some of them further inland on higher ground if they can find a suitable site and the money needed to do so. However, relocation can take years. Meanwhile, one big winter storm can flood such a village.

Increased summer thawing of the permafrost soil that covers 85% of Alaska is releasing the greenhouse gases CH_4 and CO_2 into the atmosphere. The softer soil makes it easier for trees to become uprooted and for roads and buildings to shift and sink. Lakes and wetlands have drained in some areas where permafrost has thawed. These changes are forcing some species of birds, fish, mammals, and trees to shift to new habitat or face extinction.

Alaska's warmer climate has led to explosions of beetle populations that have wiped out large areas of white spruce forests. It has also dried out forests, which along with insect damage, is likely contributing to greatly increased fire damage.

Threats to Food Production

Farmers face dramatic changes due to shifting climates and an intensified hydrologic cycle, if the atmosphere keeps warming as projected. According to the IPCC, crop productivity is projected to increase slightly with moderate warming at middle to high latitudes in areas like midwestern Canada, Russia, and Ukraine. However, the projected rise in crop productivity might be limited because the soils in these northern regions generally lack sufficient plant nutrients. Crop production will decrease if the warming goes too far.

Higher temperatures, more intense droughts and heavier downpours could reduce crop yields in some farming areas. Shrinking irrigation water supplies and growing pest populations in a warmer world could also cut crop yields in some areas.

Climate change models project a decline in agricultural productivity and food security in tropical and subtropical regions, especially in Southeast Asia and Central America, largely because excessive heat. Flooding of river deltas due to rising sea levels could reduce crop production, partly because some aquifers that supply irrigation water will be infiltrated by salt water. This flooding would also affect fish production in coastal aquaculture ponds. Food production in farm regions dependent on rivers fed by melting glaciers will drop. In arid and semiarid areas, food production will be lowered by more intense and longer droughts.

According to the IPCC, food is likely to be plentiful for a while in a warmer world, because of the longer growing season in northern regions. However, scientists warn that during the latter half of this century, food production will likely decrease as temperatures continue to rise. As a result, several hundred million of the world's poorest people could face starvation and malnutrition due to a drop in food production caused by climate change.

Threats to Human Health, National Security, and Economies

According to IPCC and other reports, more frequent and prolonged heat waves in some areas will increase deaths and illnesses—especially among older people, people in poor health, and the urban poor who cannot afford air conditioning. On the other hand, fewer people will die from cold weather. However, research at the Harvard University School of Public Health suggests that during the latter half of this century, the projected rise in the number of heat-related deaths will likely exceed the projected drop in the number of cold-related deaths.

A warmer and more CO_2-rich atmosphere will likely favor rapidly multiplying insects, including mosquitoes and ticks that transmit diseases such as West Nile virus, Lyme disease, and dengue fever (which has increased 10-fold since 1973). Warming will also favor microbes, toxic molds, and fungi, as well as pollen-producing plants that cause allergies and asthma attacks. In addition, insect pests and weeds are likely to multiply, spread, and reduce crop yields.

Higher atmospheric temperatures and levels of water vapor will contribute to increased photochemical smog in many urban areas. This could increase pollution-related deaths and illnesses due to heart ailments and respiratory problems.

Recent studies by the U.S. Department of Defense and the NAS warn that the effects of climate change could affect U.S. national security. These effects include increased food and water scarcity, poverty, environmental degradation, unemployment, social unrest, mass migration of tens of millions of environmental refugees, political instability, and the weakening of fragile governments. All of these factors could increase terrorism, according to the U.S. Department of Defense, which includes climate change as a major factor in its planning.

Climate change will also take a toll on human economies. According to a 2015 survey of 750 risk experts conducted by the World Economic Forum, failure to slow and adapt to climate change tops the list of threats to the global economy, and the risks are increasing. For example, the area likely to be flooded along the Gulf of Mexico by a 1-meter (39-inch) rise in gulf waters is home to 90% of U.S. offshore energy production, 30% of the country's oil and natural gas supply, and a port that serves 31 states, and 2 million people. Thus, flooding in this area is a threat to the country's energy supply, economic security, and national security.

A growing number of economists and major multinational companies are recognizing that more intense drought, flooding, and extreme weather events will contribute to lower economic productivity, disruption of supply chains, higher costs for food and other commodities, and increased financial risk for companies and investors. These harmful effects of climate change will also disrupt lives and economies by forcing tens of millions of people to migrate to other areas.

15.6 HOW CAN WE SLOW CLIMATE CHANGE?

CONCEPT 15.6 We can reduce greenhouse gas emissions and the threat of climate change while saving money and improving human health if we cut energy waste and rely more on cleaner renewable energy resources.

Dealing with Climate Change Is Difficult

According to at least 90% of the world's climate scientists and many and other analysts, reducing the threat of projected climate change is one of the most urgent scientific, political, economic, and ethical issues that humanity faces. However, the following characteristics of this complex problem make it difficult to tackle:

- *The problem is global.* Dealing with this threat will require unprecedented and prolonged international cooperation.

- *The problem is a long-term political issue.* Climate change is happening now and is already having harmful impacts, but most voters and elected officials do not view it as an urgent problem. Current elected officials tend to focus on short-term problems and will not be in office when the most damaging effects of climate change kick in during the last half of this century. In addition, most of the people who will suffer the most serious harm from projected climate change during the latter half of this century have not been born yet.

- *The harmful and beneficial impacts of climate change are not spread evenly.* Higher-latitude nations such as Canada, Russia, and New Zealand may temporarily have higher crop yields, fewer deaths in winter, and lower heating bills. But other mostly poor nations such as Bangladesh could see more flooding and higher death tolls.

- *Proposed solutions, such as sharply reducing the use of fossil fuels, are controversial.* They could disrupt economies and lifestyles and threaten the profits of the economically and politically powerful fossil fuel and utility companies.

- *The projected temperature changes and effects are uncertain.* Current climate models lead to a wide range in the projected temperature increase (Science Focus 15.1) and sea-level rise. Thus, there is uncertainty over whether the harmful effects of climate change will be moderate or catastrophic. This makes it difficult to plan for avoiding or managing the risks. It also highlights the urgent need for more scientific research to reduce the uncertainty in climate models. However, in the United States, there is political pressure to cut back on government funding for climate change research and atmospheric measurements.

What Are Our Options?

There are two ways to deal with global climate change. One, called *mitigation*, is to slow down climate change to reduce or avoid its most harmful effects. The other approach, called *adaptation*, recognizes that some climate change is unavoidable because we have waited too long to act and that people will have to adapt to some of its harmful effects. Most analysts call for a combination of both approaches.

Regardless of the approach taken, most climate scientists argue that the most urgent priority is to avoid any and all irreversible climate change tipping points (Figure 15.23). For example, if we continue adding CO_2 to the atmosphere at the current rate, we will very likely exceed the estimated tipping point of 450 ppm of atmospheric CO_2 within a few decades and could reach 900 ppm by 2050 according to a 2017 study by Gavin Foster and other climate scientists. Climate models project that this would likely lock in significant climate change for hundreds or perhaps thousands of years.

Reducing Greenhouse Gas Emissions

Climate scientists generally agree that to avoid some of the severest harmful effects of climate change, we need to limit the global average temperature increase to 2.0°C (3.6°F) over the preindustrial global average temperature. Staying under this temperature limit requires significantly reducing CO_2 emissions by reducing the use of fossil fuels (especially coal), greatly increasing energy efficiency, and shifting to much greater dependence on cleaner and low-carbon energy resources such as solar power, wind power, and geothermal energy (**Concept 15.6**).

The problem is that world's nations and energy companies together hold reserves of fossil fuels that, if burned, would emit nearly 5 times the amount of CO_2 that climate scientists estimate that we can safely emit. Reducing fossil fuel use to avoid exceeding this temperature climate tipping point means leaving 82% of the world's coal reserves and 50% of all natural gas and all arctic oil reserves in the ground. Currently, the economic well-being of the politically powerful fossil-fuel companies and most of the world's national economies depend on using all or most of these reserves of fossil fuels. This is what drives the intense political, economic, and ethical debate over whether to, and how to, slow climate change.

A growing number of scientists and other analysts recognize that shifting away from our human dependence on fossil fuels will be difficult but contend that it can be done over the next 50 years. They point out that humans have shifted energy resources before—first from wood to coal, then to oil, and now to natural gas—with each shift taking about 50 years. Humans have the knowledge and ability to shift to a reliance on energy efficiency and renewable energy over the next 50 years.

Figure 15.28 lists ways to slow climate change caused by human activities over the next 50 years (**Concept 15.6**). The items in the left column are prevention approaches for reducing CO_2 emissions and those in the right column are cleanup solutions for removing excess CO_2 from the atmosphere.

According to the 2014 IPCC report, there is good news related to dealing with the threat of climate change by reducing our CO_2 emissions:

- We could shift to hybrid, plug-in hybrid, and electric cars over the next 20–30 years, and charge their batteries with electricity produced from low-carbon renewable energy sources such as the sun and wind.

- The shift to renewable energy is accelerating as the prices for electricity from low-carbon wind turbines and solar cells are falling rapidly, and investments in these technologies are growing. Between 2008 and 2015, the cost of produce electricity from solar cells dropped more than 80%, wind electricity production costs fell more than 50%, and battery costs dropped more than 70%. In the United States, Texas produces

Slowing Climate Change

Prevention	Cleanup
Cut fossil fuel use (especially coal)	Sequester CO_2 by planting trees and preserving forests and wetlands
Shift from coal to natural gas	
Repair leaky natural gas pipelines and facilities	Sequester carbon in soil using biochar
Improve energy efficiency	Sequester CO_2 deep underground (with no leaks allowed)
Shift to renewable energy resources	
Reduce deforestation	Sequester CO_2 in the deep ocean (with no leaks allowed)
Use more sustainable agriculture and forestry	
Put a price on greenhouse gas emissions	Remove CO_2 from smokestack and vehicle emissions

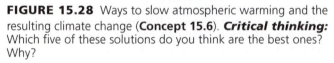

FIGURE 15.28 Ways to slow atmospheric warming and the resulting climate change (**Concept 15.6**). *Critical thinking:* Which five of these solutions do you think are the best ones? Why?

Top: Mark Smith/Shutterstock.com. Center: racorn/Shutterstock.com. Bottom: pedrosala/Shutterstock.com.

more electricity from wind and solar power than all but five of the world's countries.

- Engineers have designed affordable net-zero carbon emission buildings and know how to reduce the carbon footprints of existing buildings.

- Dealing with climate change will create jobs and profitable businesses.

- Many business leaders see slowing climate change as a global investment opportunity.

Removing CO_2 from the Atmosphere

Some scientists and engineers are designing cleanup strategies for removing some of the CO_2 from the atmosphere or from smokestack emissions, and *sequestering*, or storing, it in other parts of the environment (Figure 15.28, right). One strategy, known as **carbon capture and storage (CCS)**, would remove some of the CO_2 gas from smokestack emissions of coal-burning power and industrial plants and convert it to a liquid to be pumped under pressure into underground storage sites (Figure 15.29).

Another approach being tested in Iceland involves dissolving extracted CO_2 in water and pumping it into porous rocks such as basalt where the CO_2 reacts with calcium, magnesium, or iron in the rock to form a mineral called calcite. This locks away the CO_2 permanently. However, the process takes a lot of water and requires transporting the water-CO_2 mixture to the right kind of rock deposits.

Four major problems with CCS schemes are that, with current technology,

- they can remove and store only part of the CO_2 from smokestack emissions, at great cost,

- they do not address the massive emissions of CO_2 from motor vehicle exhausts, food production, and the deliberate burning of forests to provide land for growing food,

- they require a lot of energy, which could lead to greater use of fossil fuels and higher emissions of CO_2 and other air pollutants, and

- the CO_2 that is removed would have to remain sequestered from the atmosphere forever. Large-scale leaks and smaller continuous leaks from CO_2 storage sites could dramatically increase atmospheric warming and climate change in a short time.

So far, the experimental projects for capturing and storing CO_2 have not been very effective and have been quite costly. The high costs are making U.S. utilities reluctant to build CCS plants since it would greatly increase the cost of electricity to consumers.

Other approaches for removing CO_2 include planting large areas of trees and fertilizing the ocean with iron pellets to boost populations of phytoplankton, which remove CO_2 from the atmosphere. Preliminary experiments indicate the iron pellet scheme may not work and could disrupt marine ecosystems.

According to some environmental scientists, most forms of CSS are costly, risky, and ineffective *cleanup solutions* (Figure 15.28, right) to a serious problem that can be dealt with more effectively by using cheaper, quicker, and safer, *prevention* approaches (Figure 15.28, left).

Geoengineering Solutions

Other proposed solutions fall under the umbrella of **geoengineering**, or trying to manipulate certain natural conditions to help counter the human enhancement of the earth's greenhouse effect. Some of these proposals are shown in Figure 15.30. One proposal calls for injecting large amounts of sulfate particles into the stratosphere to reflect some of the incoming sunlight into space in order to slow the warming of the troposphere. Other scientists have called for placing a series of giant mirrors in orbit above the earth to reflect incoming sunlight for the same purpose. Another scheme is to deploy a large fleet of computer-controlled ocean-going ships to inject saltwater spray high into the sky to make clouds whiter and more reflective.

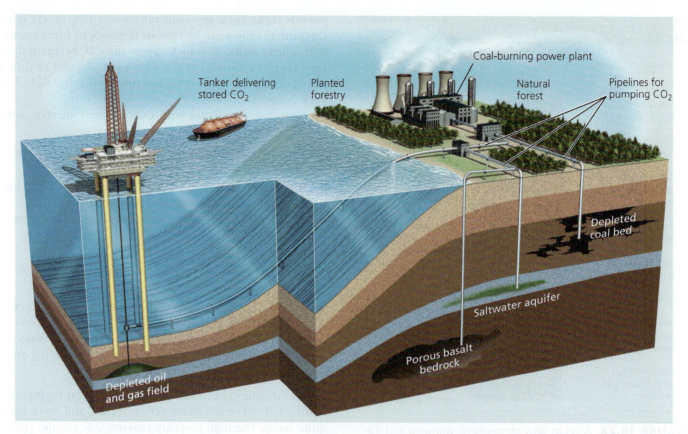

FIGURE 15.29 Some proposed carbon capture and storage (CCS) schemes for removing some of the carbon dioxide from smokestack emissions and from the atmosphere and storing (sequestering) it in soil, plants, deep underground reservoirs, and sediments beneath the ocean floor. **Critical thinking:** Which of these proposed strategies do you think would work best and which would be the least effective? Why?

Some climate scientists reject the idea of launching sulfates into the stratosphere as being too risky because of possible unknown effects. For example, if the sulfates reflected too much sunlight, they could reduce evaporation enough to alter global rainfall patterns and worsen the already dangerous droughts in certain areas. In addition, a study by atmospheric scientist Simone Tilmes indicated that chlorine released by reactions involved in this scheme could speed up the thinning of the earth's vital stratospheric ozone layer, a problem discussed in Section 15.7.

According to some scientists, a major problem with most of these technological fixes is that, if they succeed, they could be used to justify the continued rampant use of fossil fuels. This would allow CO_2 levels in the lower atmosphere to continue building and it would add to the serious problem of ocean acidification.

In addition, thinking that we can use geoengineering schemes to slow or prevent climate change could seriously delay a shift from relying on fossil fuels to greatly improving energy efficiency and relying more on a mix of low-carbon renewable energy resources over the next 50 years. Most climate scientists and economists say we cannot afford such a delay.

Government Actions to Reduce Greenhouse Gas Emissions

Governments can use seven major strategies to promote the solutions listed in Figure 15.28 (**Concept 15.6**). They can:

- Regulate carbon dioxide (CO_2) and methane (CH_4) as climate changing air pollutants that can harm public health and welfare based on a 2014 Supreme Court decision affirming the right of the EPA to regulate greenhouse gas emissions. Opponents are trying to delay or weaken such regulations or have Congress change the Clean Air Act to forbid the regulation of greenhouse gases and reverse the EPA ruling that CO_2 is an air pollutant.

- Phase out the most polluting coal-burning power plants over the next 50 years and replace them with cleaner natural gas and renewable energy alternatives such as wind power and solar power.

- Put a price on carbon emissions by phasing in taxes on each unit of CO_2 or CH_4 emitted or by phasing in energy taxes on each unit of fossil fuel burned (Figure 15.31). These tax increases could be offset by reductions in taxes on income, wages, and profits or by

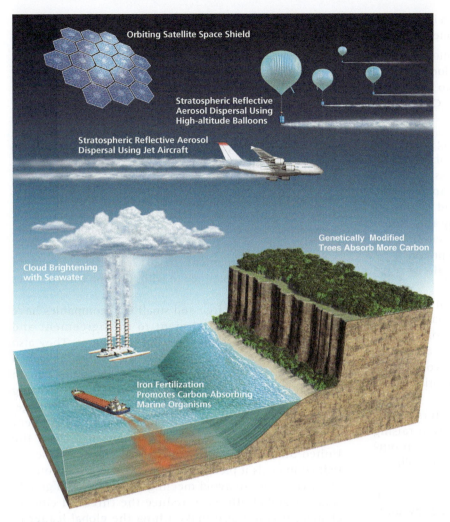

FIGURE 15.30 Geoengineering schemes include ways to reflect more sunlight into space. **Critical thinking:** Of the approaches shown here, which three do you think are the most workable? Why?

Image labels: Orbiting Satellite Space Shield; Stratospheric Reflective Aerosol Dispersal Using High-altitude Balloons; Stratospheric Reflective Aerosol Dispersal Using Jet Aircraft; Genetically Modified Trees Absorb More Carbon; Cloud Brightening with Seawater; Iron Fertilization Promotes Carbon-Absorbing Marine Organisms

direct quarterly tax rebates to American consumers. In other words, *tax pollution and energy waste, not payrolls and profits.* Ireland, Sweden, and British Columbia already have carbon taxes. In 2014, China and 72 other nations, the World Bank, and more than 1,000 corporations called for putting a price on carbon emissions. However, carbon taxes will not be effective if the tax level is not high enough and the level is not raised as needed to meet CO_2 emission reduction goals.

- Use a cap-and-trade system (see p. 418 and Figure 15.32), which would use the marketplace to help reduce emissions of CO_2 and CH_4. The government would place a cap on emissions in a country or region, issue permits to emit certain levels of these pollutants, and let polluters trade permits in the marketplace. This approach works only if the original caps are set low enough to encourage a serious reduction in emissions and are lowered on a regular basis to promote the development of more effective technologies for reducing carbon dioxide emissions.

- Over 10-20 years, phase out government subsidies and tax breaks for the fossil-fuel industry and unsustainable industrialized food production. Phase in subsidies and tax breaks for energy-efficient technologies, low-carbon renewable energy, and more sustainable food production (Figure 10.28, p. 251).

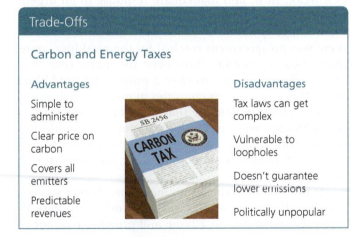

Trade-Offs

Carbon and Energy Taxes

Advantages	Disadvantages
Simple to administer	Tax laws can get complex
Clear price on carbon	Vulnerable to loopholes
Covers all emitters	Doesn't guarantee lower emissions
Predictable revenues	Politically unpopular

FIGURE 15.31 Using carbon and energy taxes or fees to help reduce greenhouse gas emissions has advantages and disadvantages. **Critical thinking:** Which two advantages and which two disadvantages do you think are the most important and why?

Trade-Offs

Cap-and-Trade Policies

Advantages	Disadvantages
Clear legal limit on emissions	Revenues not predictable
Rewards cuts in emissions	Vulnerable to cheating
Record of success	Rich polluters can keep polluting
Low expense for consumers	Puts variable price on carbon

FIGURE 15.32 Using a cap-and-trade policy to help reduce greenhouse gas emissions has advantages and disadvantages. **Critical thinking:** Which two advantages and which two disadvantages do you think are the most important and why?

- Focus research and development efforts on innovations that lower the cost of clean energy alternatives.

- Work out agreements to finance and monitor efforts to reduce deforestation—which accounts for 12–17% of global greenhouse gas emissions—and to promote global tree-planting efforts (see Chapter 9, Core Case Study, p. 194).

Environmental economists and a growing number of business leaders, along with the President of the World Bank, call for putting a price on carbon emissions as the best way to curb them before it is too late to avoid the projected catastrophic environmental and economic effects of significant climate change. Doing this would promote including the estimated harmful environmental and health costs of using fossil fuels in fuel prices, in keeping with the full-cost pricing **principle of sustainability**.

However, establishing laws and regulations that raise the price of fossil fuels is politically controversial and difficult because of the immense political and economic power of the fossil fuel and electric utility industries and society's dependence on fossil fuels for 90% of its commercial energy. Opponents also argue that raising the price of fossil fuels would hurt economies and consumers.

However, economists and other supporters of this approach contend the economic benefits from raising fossil fuel prices enough to slow climate change far outweigh the economic drawbacks for two reasons. First is the cost savings to consumers and governments from the resulting environmental and health benefits. Second, higher fossil fuel prices would help economies by spurring innovation in finding ways to reduce carbon emissions, improve energy efficiency, and phase in a mix of affordable low-carbon renewable energy resources.

International Climate Change Treaties

Governments have entered into international climate negotiations. In December 1997, delegates from 161 nations met in Kyoto, Japan, to negotiate a treaty to slow atmospheric warming and climate change. The first phase of the resulting *Kyoto Protocol* went into effect in February 2005 with 187 of the world's 194 countries (not including the United States) ratifying the agreement by late 2009.

The 37 participating more-developed countries agreed to cut their emissions of CO_2, CH_4, and N_2O to certain levels by 2012, when the treaty was to expire. However, 16 of the nations failed to do so. Less-developed countries, including China, were excused from this requirement, because such reductions would curb their economic growth.

In 2015, delegates from 195 countries met in Paris, France, in another attempt to achieve a global climate change agreement. In a historic agreement, the governments agreed to the following:

- Accepting a goal to keep the increase in global average temperatures below 2°C (3.6°F).

- Pledging to reduce their greenhouse gas emissions by a set amount.

- Meeting every 5 years to evaluate progress and raise their goals.

The agreement went into effect in November 2016 after more than 55 countries (not including the United States) ratified it.

Germany, China, and India are on target to exceed their emissions reduction goals by reducing their use of coal and by investing heavily in wind and solar energy. Some fossil fuel companies and climate change deniers are exerting pressure on the U.S. government to withdraw from the agreement. In 2017, the U. S. President announced that he planned to withdraw the United States from the treaty.

More than 365 large and small U.S. companies have reaffirmed their commitment to the agreement and oppose U.S. withdrawal from the agreement. They also urge the U.S. government to meet the country's pledged greenhouse gas emissions reduction goals and to remain a global leader in dealing with the serious and growing environmental and economic threats of climate change. However, because of efforts to reverse U.S. government measures to deal with climate change since 2017, the United States likely will not reach the greenhouse gas reduction goals it pledged to meet. This could encourage other countries to avoid meeting their goals and greatly weaken global efforts to reduce the threat of climate change. It could also make China the global leader in dealing with climate change.

Climate scientists applaud the Paris climate agreement, reached after 18 years of international negotiations. However, some see this international agreement as a weak, slow, and inadequate response to an urgent global environmental and economic problem. Countries are not legally bound to reach their goals. In addition, there was no agreement reached for the wealthier countries whose economies have been the major contributors to climate change to raise a proposed $500 billion by 2020 to assist poorer countries in meeting their goals. Climate scientists estimate that even if all of the treaty's commitments were to be fully honored, it would not reduce atmospheric temperatures enough to prevent serious environmental and economic problems during this century.

As mentioned at the beginning of this chapter section, slowing climate change is difficult and controversial. However, according to a large and growing number of climate and environmental scientists, economists, and business leaders, the benefits of slowing climate change (Figure 15.33) far outweigh the long-term economic and environmental risks of not doing so.

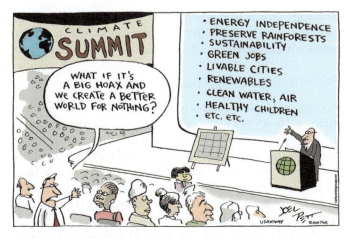

FIGURE 15.33 One take on the benefits of slowing climate change.

Some Countries, States, Cities, and Companies Are Reducing Their Carbon Footprints

Some nations are leading others in facing the challenges of projected climate change. Costa Rica aims to be the first country to become *carbon neutral* by cutting its net carbon emissions to zero by 2030. The country generates 78% of its electricity with renewable hydroelectric power and another 18% from renewable wind and geothermal energy.

China emits more greenhouse gases than any other country. However, it is also rapidly becoming the world leader in developing and selling low-carbon solar cells, solar water heaters, wind turbines, advanced batteries, and plug-in and all-electric cars. China sees this as a way to help reduce its greenhouse gas emissions and its dependence on coal, and to boost its economy by becoming the leader in some of this century's most rapidly growing businesses.

Some U.S. state and local governments are moving ahead in dealing with climate change. By 2016, at least 32 U.S. states had set goals for reducing greenhouse gas emissions or were involved in multi-state programs. California plans to get 33% of its electricity from low-carbon renewable energy sources by 2030. That state is showing that it is possible to implement policies that cut carbon emissions and create jobs. Since 1990, local governments in more than 650 cities around the world (including more than 450 U.S. cities) have established programs to reduce their greenhouse gas emissions.

Leaders of some big U.S. companies, including Alcoa, DuPont, Ford Motor Company, General Electric, and Shell Oil, have joined with leading environmental organizations to form the U.S. Climate Action Partnership. The partnership calls for the U.S. government to enact strong national climate change legislation, saying, "In our view, the climate change challenge will create more economic opportunities than risks for the U.S. economy." Each company is working on reducing its carbon footprint as a way to help slow climate change and to save money. Many companies now recognize that "there is gold in going green."

Colleges and Universities Are Reducing Their Carbon Footprints

Many colleges and universities are also taking action. Arizona State University boasts the largest collection of solar panels of any U.S. university. Maine's College of the Atlantic gets 100% of its electricity from renewable sources and has been carbon-neutral since 2007. Students there built a wind turbine that powers a nearby organic farm, which offers organic produce to the campus, to local schools, and to food banks.

The University of California, Irvine, reduced its energy consumption by 24% between 2008 and 2015. Students at the University of Washington in Seattle agreed to an increase in their fees to help the school buy electricity from renewable energy sources. And a growing number of campus groups are urging the administrators at their schools to help slow climate change by ending their endowment fund investments in fossil fuel companies.

CONSIDER THIS ...

THINKING ABOUT What Your School Can Do

What are three steps that you think your school should take to help reduce its carbon footprint? What steps, if any, is it now taking?

Every Individual Can Make a Difference

Each of us will play a part in the atmospheric warming and climate change projected to occur during this century. Whenever we use energy generated by fossil fuels, for example, we add CO_2 to the atmosphere. However, nearly two-thirds of the average American's carbon footprint comes from carbon released during the manufacture and delivery of food, shelter, clothing, cars, computers, and every other consumer products and services.

An important aspect of your carbon footprint is your diet. Foods vary in the greenhouse gases that result from their production and delivery. For example, processed foods require much more energy to produce and thus have higher greenhouse gas emissions than do foods such as fresh fruit and vegetables. Meat production, especially if it is done on factory farms, involves far higher greenhouse gas emissions than production of grains and vegetables. In addition, greenhouse gas emissions from beef production and consumption are 12 times higher than those associated with producing and eating the same amount of chicken. By choosing foods carefully, you can reduce your carbon footprint.

You can learn about your carbon footprint by using a footprint calculator, several of which are available online. Figure 15.34 lists some ways in which you can cut your CO_2

What Can You Do?

Reducing CO₂ Emissions

- Calculate your carbon footprint (there are several helpful websites)
- Drive a fuel-efficient car, walk, bike, carpool, and use mass transit
- Reduce garbage by reducing consumption, recycling, and reusing more items
- Use energy-efficient appliances and LED lightbulbs
- Wash clothes in warm or cold water and hang them up to dry
- Close window curtains to keep heat in or out
- Use a low-flow showerhead
- Eat less meat or no meat
- Heavily insulate your house and seal all air leaks
- Use energy-efficient windows
- Set your hot-water heater to 49°C (120°F)
- Plant trees
- Buy from businesses working to reduce their emissions

FIGURE 15.34 You can reduce your annual emissions of CO₂. **Critical thinking:** Which of these steps, if any, do you take now or plan to take in the future?

emissions. One person taking each of the steps makes a small contribution to reducing greenhouse gas emissions, but when millions of people take such steps, global change can happen.

Preparing for Climate Change

According to global climate models, the world needs to make a 50–85% cut in emissions of greenhouse gases by 2050 to stabilize concentrations of these gases in the atmosphere. This would help prevent the atmosphere from warming by more than 2°C (3.6°F) and would head off projected rapid and long-lasting changes in the world's climate along with the projected harmful environmental, economic, and health effects.

However, because of the political difficulty of making such large reductions, many analysts believe that while we work to slash greenhouse gas emissions, we should also prepare for the likely harmful effects of projected climate change. Figure 15.35 shows some ways to do this.

For example, organizations are carrying out projects such as expanding mangrove forests as buffers against storm surges. They are building shelters on high ground and planting trees on slopes to help prevent landslides in the face of projected higher levels of precipitation and rising sea levels. Low-lying countries such as Bangladesh are planning for what to do with millions of environmental refugees who will be flooded out by rising sea levels and more intense storms.

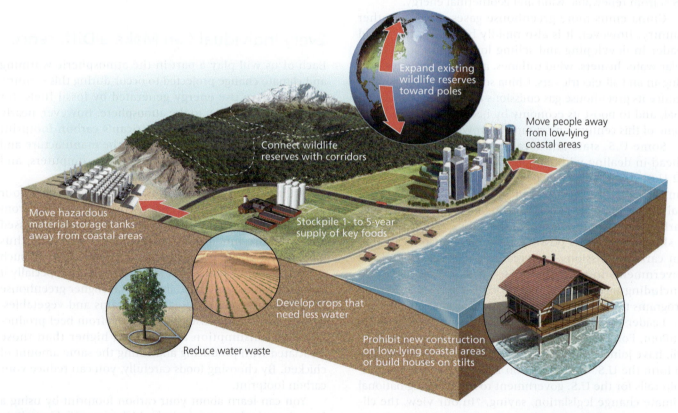

FIGURE 15.35 Solutions: Ways for us to prepare for the possible long-term harmful effects of climate change. **Critical thinking:** Which three of these adaptive steps do you think are the most important ones to take? Why?

Some coastal communities in the United States now require that new houses and other new construction be built high enough off the ground or further back from the current shoreline to survive such hazards. In the aftermath of flooding caused by Hurricane Sandy, in 2012, New York City is planning new floodwalls and floodgates for its subway system. In anticipation of rising sea levels, Boston has elevated one of its sewage treatment plants. Some cities plan to establish cooling centers to shelter residents during increasingly intense heat waves.

The Netherlands, with most of its population living below sea level, is famous for building dikes to hold back the rising North Sea for more than 800 years. The country's government and people have been formulating a 200-year plan for dealing with climate change.

15.7 HOW HAVE WE DEPLETED OZONE IN THE STRATOSPHERE AND WHAT CAN WE DO ABOUT IT?

CONCEPT 15.7A Widespread use of certain chemicals has reduced ozone levels in the stratosphere and allowed more harmful ultraviolet (UV) radiation to reach the earth's surface.

CONCEPT 15.7B To reverse ozone depletion, we must stop producing ozone-depleting chemicals and adhere to the international treaties that ban such chemicals.

The Use of Certain Chemicals Threatens the Ozone Layer

The ozone in the stratosphere (Figure 15.2) keeps about 95% of the sun's harmful ultraviolet (UV-A and UV-B) radiation from reaching the earth's surface and harming us and many other species.

However, measurements taken by researchers revealed a considerable seasonal depletion, or thinning, of ozone concentrations in the stratosphere above Antarctica (Figure 15.36) and above the Arctic since the 1970s. Similar measurements revealed a slight overall ozone thinning everywhere except over the tropics. The loss of ozone over Antarctica has been called an *ozone hole*. A more accurate term is *ozone thinning* because the ozone depletion varies with altitude and location.

Based on these measurements and on mathematical and chemical models, the overwhelming consensus of researchers in this field is that ozone depletion in the stratosphere poses a serious threat to humans, other animals, and some primary producers (mostly plants) that use sunlight to support the earth's food webs (**Concept 15.7A**).

The origin of this dangerous environmental threat began in 1930 with the accidental discovery of the first chlorofluorocarbon (CFC), a compound that contains

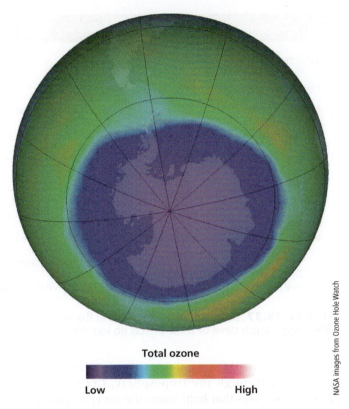

Total ozone

Low　　　　　　　　　　　High

NASA images from Ozone Hole Watch

FIGURE 15.36 Natural Capital Degradation: This colorized satellite image shows ozone thinning over Antarctica during September of 2016. Ozone depletion of 50% or more occurred in the darker blue area.

carbon, chlorine, and fluorine. Chemists soon developed similar compounds to create a family of highly useful CFCs, known by their trade name Freons.

These chemically unreactive, odorless, nonflammable, nontoxic, and noncorrosive compounds were thought to be dream chemicals. Inexpensive to manufacture, they became popular as coolants in air conditioners and refrigerators, propellants in aerosol spray cans, cleansers for electronic parts such as computer chips, fumigants for granaries and ships' cargo holds, and gases used to make insulation and packaging.

It turned out that CFCs were too good to be true. Starting with the 1974 research of chemists Sherwood Rowland and Mario Molina (Individuals Matter 15.1), scientists showed that CFCs are persistent chemicals that can reach the stratosphere and destroy some of its protective ozone. Satellite data and other measurements and models indicate that 75–85% of the observed ozone losses in the stratosphere since 1976 resulted from people releasing CFCs and other ozone-depleting chemicals into the troposphere beginning in the 1950s.

After entering the troposphere, these long-lived chemicals eventually reached the stratosphere. There they began destroying ozone faster than it was being formed. Such ozone depletion is a disruption of one of the earth's most important forms of natural capital that helps sustain

Effects of Ozone Depletion

Human Health and Structures

- Worse sunburns

- More eye cataracts and skin cancers

- Immune system suppression

Food and Forests

- Reduced yields for some crops

- Reduced seafood supplies due to smaller phytoplankton populations

- Decreased forest productivity for UV-sensitive tree species

Wildlife

- More eye cataracts in some species

- Shrinking populations of aquatic species sensitive to UV radiation

- Disruption of aquatic food webs due to shrinking phytoplankton populations

Air Pollution and Climate Change

- Increased acid deposition

- Increased photochemical smog

- Degradation of outdoor painted surfaces, plastics, and building materials

- While in troposphere, CFCs act as greenhouse gases

FIGURE 15.37 Decreased levels of ozone in the stratosphere can have a number of harmful effects (**Concept 15.7A**). *Critical thinking:* Which three of these effects do you think are the most threatening? Why?

life and the world's economies. During their upward movement through the troposphere, CFCs also act as greenhouse gases that help warm the lower troposphere.

Why Does Ozone Depletion Matter?

Why should we care about ozone depletion? Figure 15.37 lists some of the harmful effects of stratospheric ozone thinning. One effect is that more biologically damaging UV-A and UV-B radiation will reach the earth's surface (**Concept 15.7A**). This increased radiation is a likely contributor to rising numbers of eye cataracts, damaging sunburns, and skin cancers. Figure 15.38 lists ways you can protect yourself from harmful UV radiation.

Another serious threat from ozone depletion and the resulting increase in UV radiation reaching the planet's surface is the possible impairment or destruction of phytoplankton, especially in Antarctic waters. These tiny marine plants play a key role in removing CO_2 from the atmosphere and they form the base of many ocean food webs. Greatly decreasing their populations would eliminate the vital ecological services they provide. This loss of plankton could accelerate atmospheric warming and climate change by reducing the capacity of the oceans to remove large amounts of the CO_2 that humanity is adding to the atmosphere.

Reversing Stratospheric Ozone Depletion

According to researchers in this field, we should immediately stop producing all ozone-depleting chemicals (**Concept 15.7B**). However, models and measurements indicate that even with immediate and sustained action, it will take at least 60 years for the earth's ozone layer to recover the levels of ozone it had in 1960s, and it could take about 100 years for it to recover to pre-1950 levels.

In 1987, representatives of 36 nations met in Montreal, Canada, and developed the *Montreal Protocol*. This treaty's goal was to cut emissions of CFCs (but not other ozone-depleting chemicals) by 35% between 1989 and 2000. After hearing more bad news about seasonal ozone thinning above Antarctica in 1989, representatives of 93 countries had more meetings and in 1992 adopted the *Copenhagen Amendment*, which accelerated the phase-out of CFCs and added some other key ozone-depleting chemicals to the agreement.

The Montreal Protocol is viewed as the world's most successful global environmental agreement. It set an important precedent because nations and companies

What Can You Do?

Reducing Exposure to UV Radiation

- Stay out of the sun, especially between 10 A.M. and 3 P.M.

- Do not use tanning parlors or sunlamps

- When in the sun, wear clothing and sunglasses that protect against UV-A and UV-B radiation

- Be aware that overcast skies do not protect you

- Do not expose yourself to the sun if you are taking antibiotics or birth control pills

- When in the sun, use a sunscreen with a protection factor of at least 15

- Buy from businesses working to reduce their CFC emissions

FIGURE 15.38 Ways you can reduce your exposure to harmful UV radiation. Which of these precautions do you already take?

Sherwood Rowland and Mario Molina—A Scientific Story of Expertise, Courage, and Persistence

In 1974, calculations by the late Sherwood Rowland (left) and Mario Molina (right), chemists at the University of California–Irvine, indicated that chlorofluorocarbons (CFCs) were lowering the average concentration of ozone in the stratosphere. These scientists decided they had an ethical obligation to go public with the results of their research. They shocked both the scientific community and the $28-billion-per-year CFC industry by calling for an immediate ban of CFCs in spray cans, for which substitutes were available.

The research of these two scientists led them to four major conclusions. *First,* once CFCs are put into the troposphere, these persistent chemicals remain there for a long time.

Second, over 11–20 years, these compounds rise into the stratosphere through convection, random drift, and the turbulent mixing of air in the troposphere.

Third, once they reach the stratosphere, the CFC molecules break down under the influence of high-energy UV radiation. This releases highly reactive chlorine atoms (Cl), as well as atoms of fluorine (F) and bromine (Br), all of which accelerate the breakdown of ozone (O_3) into O_2 and O in a cyclic chain of chemical reactions. This destroys ozone faster than it forms in the stratospheric ozone layer.

Fourth, each CFC molecule can last in the stratosphere for 65–385 years, depending on its type. During that time, each chlorine atom released during the breakdown of CFCs can break down hundreds of O_3 molecules.

The CFC industry (led by DuPont) was a powerful, well-funded adversary with a lot of profits and jobs at stake. It attacked Rowland's and Molina's calculations and conclusions, but the two researchers held their ground, expanded their research, and explained their results to other scientists, elected officials, and the media. After 14 years of delaying tactics, DuPont officials acknowledged in 1988 that CFCs were depleting the ozone layer. They also agreed to stop producing CFCs and to sell the higher-priced alternatives that their chemists had developed.

In 1995, Rowland and Molina received the Nobel Prize in chemistry for their work on CFCs.

worked together and used a *prevention approach* to solve a serious environmental problem.

This approach worked for three reasons. *First,* there was convincing and dramatic scientific evidence of a serious problem. *Second,* CFCs were produced by a small number of international companies and this meant there was less corporate resistance to finding a solution. *Third,* the certainty that CFC sales would decline over a period of years because of government bans unleashed the economic and creative resources of the private sector to find even more profitable substitute chemicals.

The most widely used substitutes are hydrofluorocarbons (HFCs), which also act as greenhouse gases during their trip to the stratosphere. An HFC molecule can be up to 10,000 times more potent in warming the atmosphere than a molecule of CO_2. The IPCC has warned that global use of HFCs is growing rapidly and that they need to be quickly replaced with substitutes that do not deplete ozone in the stratosphere or act as greenhouse gases while they are in the troposphere. Several companies have developed HFC substitutes that have to be evaluated.

The international agreements on protecting stratospheric ozone are working. According to 2016 study by NASA scientists, between 2000 and 2015, the area of ozone thinning in the stratosphere above Antarctica (Figure 15.36), which peaks in September and October, had shrunk by an area equal to about one-third the area of the continental United States. If this continues, the ozone layer could return to 1980 levels by 2050.

The landmark international agreements on stratospheric ozone, now signed by all 196 of the world's countries, are important examples of successful global cooperation in response to a serious global environmental problem (**Concept 15.7B**). This is also an example of the win-win **principle of sustainability** in action.

GOOD NEWS

SUSTAINABILITY

BIG IDEAS

- We need to give top priority to preventing and reducing outdoor and indoor air pollution and reducing stratospheric ozone depletion.

- Reducing the projected harmful effects of rapid climate change during this century requires emergency action to sharply reduce greenhouse gas emissions, increase energy efficiency, and rely more on low-carbon renewable energy resources.

- We can prepare for some climate change but we could realize important economic, ecological, and health benefits by drastically reducing greenhouse gas emissions with the goal of slowing climate change.

Melting Ice in Greenland and Sustainability

Konrad Steffen University of Colorado/CIRES

In this chapter, you learned that human activities such as the burning of fossil fuels and the widespread clearing and burning of forests for planting crops and grazing cattle have contributed to higher levels of greenhouse gases in the atmosphere. This has contributed to atmospheric warming and climate change, which are projected to increase rapidly during this century.

The effects of climate change could include rapid melting of land-based ice (**Core Case Study**) and arctic sea ice, longer and more intense drought, rising sea levels, declining biodiversity, and severe threats to food supplies, human health, economies, and national security. Many of these effects could further accelerate climate change in worsening spirals of change.

You also learned how the widespread use of certain chemicals has thinned the stratospheric ozone layer, which has led to plant and animal health problems. It is encouraging that the world's nations implemented an international effort to reverse this effect.

We can apply the six **principles of sustainability** (see inside back cover of this book) to help reduce the harmful effects of air pollution, climate change, and stratospheric ozone depletion. We can reduce emissions of pollutants, greenhouse gases, and ozone-depleting chemicals by relying more on direct and indirect forms of solar energy than on fossil fuels; recycling and reusing matter resources much more widely than we do now; and mimicking biodiversity by using a variety of often locally available low-carbon renewable energy resources in place of fossil fuels and nuclear power. We can make progress in achieving these goals by including the harmful environmental and health costs of fossil fuel use in market prices; seeking win-win solutions to these threats that will benefit both the economy and the environment; and passing along to future generations an environment and life-support system in a condition as good as or better than what we now enjoy.

Chapter Review

Core Case Study

1. Define **atmospheric warming** and **climate change.** Summarize the story of Greenland's melting glaciers, how this process is related to climate change, and its possible effects on the world during this century.

Section 15.1

2. What is the key concept for this section? Define and distinguish among **atmosphere, troposphere, stratosphere,** and **ozone layer.** Define and give two examples of **greenhouse gases.** Why is the ozone layer important?

Section 15.2

3. What are the two key concepts for this section? What is **air pollution**? Distinguish between **primary** **pollutants** and **secondary pollutants** and give an example of each. List the major outdoor air pollutants and their harmful effects. Distinguish between **industrial smog** and **photochemical smog.** List and briefly explain five natural factors that help reduce outdoor air pollution and six natural factors that make it worse. What is a **temperature inversion** and how can it affect outdoor air pollution levels? What is **acid deposition**, how does it form, and what are its major environmental impacts on vegetation, lakes, human-built structures, and human health? List three major ways to reduce acid deposition.

4. What is the most threatening indoor air pollutant in many less-developed countries? What are the four most dangerous indoor air pollutants in the United States? Briefly describe the human body's defenses

against air pollution, how they can be overwhelmed, and illnesses that can result. In the world and in the United States, about how many people die prematurely from outdoor and indoor air pollution each year?

Section 15.3

5. What is the key concept for this section? Summarize the use of air pollution control laws in the United States and how they could be improved. List the advantages and disadvantages of using an emissions trading program to reduce outdoor air pollution. List the major ways to reduce emissions from power plants and motor vehicles. What are four ways to reduce indoor air pollution?

Section 15.4

6. What is the key concept for this section? Define and distinguish between **weather** and **climate**. Summarize the trends in atmospheric warming and cooling during the past 900,000 years, the past 10,000 years, 1,000 years, and since 1975. How do scientists get information about past temperatures and climates? List three major conclusions of the IPCC and other scientific bodies regarding changes in the temperature of the earth's atmosphere. List eight pieces of scientific evidence that support the conclusion that human-influenced climate change is happening now. Why is the phrase "climate change debate by scientists" misleading?

7. What is the **greenhouse effect** and why is it so important to life on the earth? What role does CO_2 emissions play in atmospheric warming and what are two major sources of these emissions? Define **carbon footprint**. Define **per capita carbon footprint**. Explain how scientists use models to project future changes in atmospheric temperatures. Describe how each of the following might affect average atmospheric temperatures and projected climate change during this century: **(a)** the oceans, **(b)** cloud cover, and **(c)** outdoor air pollution.

Section 15.5

8. What is the key concept for this section? Define **climate change tipping point**, and list five possible

examples. Summarize the projections of scientists on how climate change is likely to affect each of the following: ice and snow cover, permafrost, sea levels, severe drought, extreme weather events, biodiversity, food production, human health, economies, and national security.

Section 15.6

9. What is the key concept for this section? What are five reasons why dealing with projected climate change is a difficult problem? What are two basic approaches to dealing with climate change? List five major prevention strategies and four cleanup approaches for slowing projected climate change. List five pieces of good news related to dealing with climate change. What is **carbon capture and storage (CCS)**? List four major problems associated with capturing and storing carbon dioxide emissions. Define **geoengineering** and describe two proposed geoengineering strategies for dealing with the threat of climate change. What are the main potential problems with relying on geoengineering strategies? List seven things that governments can do to help slow projected climate change. What are the major advantages and disadvantages of using **(a)** carbon or energy taxes and **(b)** a cap-and-trade system to help reduce the threat of climate change? Summarize the 2015 Paris international agreement on climate change and its limitations. List five ways in which you can reduce your carbon footprint. List five ways in which we can prepare for the possible long-term harmful effects of projected climate change.

Section 15.7

10. What are the two key concepts for this section? How have human activities depleted ozone in the stratosphere? List five harmful effects of such depletion. Explain how scientists Sherwood Roland and Mario Molina alerted the world to this threat. What has the world done to reduce the threat of ozone depletion in the stratosphere? What are the *three big ideas* for this chapter? Explain how we can apply the six **principles of sustainability** to the problems of air pollution, climate change, and ozone depletion.

Note: Key terms are in bold type.

Critical Thinking

1. If you had convincing evidence that at least half of Greenland's glaciers (**Core Case Study**) were sure to melt during this century, would you argue for taking serious actions now to slow projected climate change? Summarize your arguments for or against such actions.

2. Suppose someone tells you that carbon dioxide (CO_2) should not be classified as an air pollutant because it is a natural chemical added to the atmosphere every time we exhale. Would you consider this faulty reasoning? Explain.

3. China's burning of coal has caused major and growing air pollution problems for the country and for its neighboring nations, and it has contributed to projected climate change. In addition, air pollution generated in China now sometimes spreads across the Pacific Ocean to the west coast of North America. Do you think China is justified in developing its coal resource aggressively as other countries—including the United States—have done with their coal resources? Explain. What are China's alternatives? If you think that China should sharply reduce its dependence on coal, would you also call for the United States to sharply reduce its use of coal? Explain.

4. Explain why you agree or disagree with at least 90% the world's climate scientists that **(a)** climate change is happening now, **(b)** human activities play an important role in this climate change, and **(c)** human actions can slow down the rate of climate change and avert or delay its projected harmful environmental, health, and economic effects.

5. Explain why you would support or oppose each of the strategies listed in Figure 15.28 for slowing projected climate change caused by atmospheric warming.

6. Some scientists have suggested that we could help cool the warming atmosphere by annually injecting huge quantities of sulfate particles into the stratosphere. This might have the effect of reflecting some incoming sunlight back into space. Explain why you would support or oppose this geoengineering scheme.

7. What are three consumption patterns or other aspects of your lifestyle that directly add greenhouse gases to the atmosphere? Which, if any, of these habits would you be willing to give up in order to help slow projected climate change?

8. Congratulations! You are in charge of the world. Explain your strategy for dealing with each of the following problems: **(a)** outdoor air pollution, **(b)** indoor air pollution, **(c)** climate change, and **(d)** stratospheric ozone depletion.

Doing Environmental Science

Gather data on the trends in average annual temperatures and average annual precipitation over the past 30 years in the area where you live. (Possible sources include weather sites on the Internet, your school library, local TV and radio meteorologists, and local or regional weather bureaus.) Try to find data for as many of these years as possible and plot these data to determine whether the average temperature and precipitation during this period has increased, decreased, or stayed about the same. Write a report summarizing your search for data, your results, and your conclusions.

Global Environment Watch Exercise

Go to your Mind Tap course to access the GREENR database. Starting on the home page under "Browse Issues and Topics" select *Climate Change* and click on the pin for Greenland. Find the latest information on the melting of glaciers in Greenland (Core Case Study). What is the rate of melting of the glaciers and how might this affect sea levels during this century? Summarize the evidence used by scientists to support their statements about melting ice in Greenland. Summarize any information you find about ongoing studies of Greenland's ice.

Data Analysis

Coal often contains sulfur (S) as an impurity that is released as gaseous SO_2 during combustion, and SO_2 is one of six primary air pollutants monitored by the EPA. The U.S. Clean Air Act limits sulfur emissions from large coal-fired boilers to 0.54 kilograms (1.2 pounds) of sulfur per million Btus (British thermal units) of heat generated. (1 metric ton = 1,000 kilograms = 2,200 pounds = 1.1 ton; 1 kilogram = 2.2 pounds.)

1. Given that coal burned in power plants has a heating value of 27.5 million Btus per metric ton (25 million Btus per ton), determine the number of kilograms (and pounds) of coal needed to produce 1 million Btus of heat.

2. If all of the sulfur in the coal is released to the atmosphere during combustion, what is the maximum percentage of sulfur that the coal can contain and still allow the utility to meet the standards of the Clean Air Act?

CHAPTER 16

SOLID AND HAZARDOUS WASTE

Follow nature's example; realize waste's potential.
GUNTER PAULI

Key Questions

16.1 What environmental problems are related to solid and hazardous wastes?

16.2 How should we deal with solid waste?

16.3 Why are refusing, reducing, reusing, and recycling so important?

16.4 What are the advantages and disadvantages of burning or burying solid waste?

16.5 How should we deal with hazardous waste?

16.6 How can society shift to a low-waste economy?

Cradle-to-Cradle Design

The life cycle of a product begins when it is manufactured (its cradle) and ends when it is discarded as solid waste, typically in a landfill or as litter (its grave).

Designer William McDonough wants us to abandon this *cradle-to-grave* view of the life cycle of products. He argues for a *cradle-to-cradle* approach, in which we think of products as parts of a continuing cycle instead of as materials that become solid waste that is burned or buried in landfills or that ends up as litter. This approach, first explored in the 1970s by business analyst Walter Stahel, is the basis for much of McDonough's work. He envisions an economy where all products or their parts will be reused repeatedly in other products. Parts that are no longer useful would be degradable so that natural nutrient cycles could recycle their materials and chemicals. The degradable parts would be thought of as *biological nutrients* (Figure 16.1, left) and those parts that are reused would be *technical nutrients* (Figure 16.1, right).

In their books, *Cradle to Cradle* and *The Upcycle,* McDonough and chemist Michael Braungart lay out this vision as a way not just to lessen our harmful environmental impact but also to have a beneficial environmental impact. They call for us to think

of solid wastes and pollution as potentially useful and economically valuable materials and chemicals. Instead of asking, "How do I get rid of these wastes?" they say we need to ask, "How much money can I get for these resources?" and "How can I design products that don't end up as wastes or pollutants?"

This way of thinking means designing products so they can be recycled or reused, much like nutrients in the biosphere. With this approach, people might think of trash cans and garbage trucks as resource containers, and of landfills as urban mines filled with stuff we can recycle, as the earth does. They might think of garbage not as something to throw away but as economically valuable materials to reuse for other purposes.

Cradle-to-cradle design is a form of biomimicry (see Chapter 1, Core Case Study, p. 4) because it helps implement the earth's chemical cycling **principle of sustainability** (see Inside Back Cover). For example, a chair manufacturer applying this approach designs and builds its chairs such that when one part breaks, most of the other parts can be reused in the manufacture of a new chair. As much as possible, only

biodegradable materials are used so that worn-out, discarded parts will break down in the environment and become part of nature's nutrient cycles. As McDonough likes to say, in nature, waste equals food.

There are many ways to apply this approach. One important way is to *design toxic substances out* of products and processes. If a product requires the use of a toxic heavy metal, for example, it should be redesigned to make use of a nontoxic substitute for that ingredient. Another strategy is to *sell services instead of products.* For example, think of carpeting not as a product to be used and discarded, but as a floor covering service. The carpet company owns the carpeting and leases it to the user. The company then replaces worn carpeting on a regular basis as part of the service and recycles the materials in the worn-out pieces to make new carpeting.

In this chapter, we consider the problems of solid and hazardous wastes resulting from human activities. We also consider ways to make the transition to a more sustainable low-waste economy by preventing and reducing the production of such wastes as a way to apply the cradle-to-cradle approach. ●

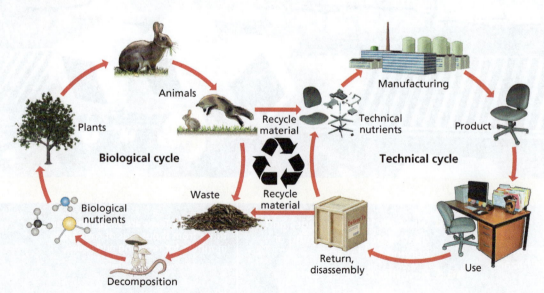

FIGURE 16.1 Cradle-to-cradle design and manufacturing aims to make all products reusable and all components that must be discarded biodegradable. By connecting technical and biological nutrient cycles, it mimics nature and essentially eliminates waste by converting it to nutrients.

16.1 WHAT ENVIRONMENTAL PROBLEMS ARE RELATED TO SOLID AND HAZARDOUS WASTES?

CONCEPT 16.1A Solid waste contributes to pollution and includes valuable resources that could be reused or recycled.

CONCEPT 16.1B Hazardous waste contributes to pollution, as well as to natural capital degradation, health problems, and premature deaths.

Solid Waste Is Piling Up

In the natural world, there is essentially no waste because the wastes of one organism become nutrients or raw materials for others in food chains and food webs. This natural cycling of nutrients is the basis of the chemical cycling **principle of sustainability**.

Humans violate this principle by producing huge amounts of solid wastes that are burned, buried in landfills, or end up as litter. Studies and experience indicate that by mimicking nature, we could reduce this waste of potential resources, money, and the resulting environmental harm by up to 80%.

One major category of waste is **solid waste**—any unwanted or discarded material that is not a liquid or a gas. There are two major types of solid waste. The first is **industrial solid waste** produced by mines (see Figure 12.8, p. 310), farms, and industries that supply people with goods and services. It also includes construction and demolition waste. The second is **municipal solid waste (MSW)**, often called *garbage* or *trash. It* consists of the combined solid wastes produced by homes and workplaces other than factories. Examples of MSW include paper, cardboard, food wastes, cans, bottles, yard wastes, furniture, plastics, metals, glass, wood, and electronic, or e-waste. Much of the world's MSW ends up as litter in rivers, lakes, the ocean (see the second Case Study that follows), and natural landscapes (Figure 16.2). Some resource experts suggest that we change the name of the trash we produce from MSW to MWR—mostly wasted resources.

In more-developed countries, most MSW is collected and buried in landfills or burned in incinerators. In many less-developed countries, much of it ends up in open dumps, where poor people often eke out a living finding items they can use or sell. The United States is the world's largest producer of MSW (see the Case Study that follows).

FIGURE 16.2 *Municipal solid waste:* Various types of solid waste have been dumped in this isolated mountain area.

Mikadun/Shutterstock.com

CASE STUDY

Solid Waste in the United States

The United States leads the world in total production of industrial and municipal solid waste, as well as industrial and MSW per person production. With only 4.3% of the world's people, the United States produces about 25% of the world's industrial and municipal solid waste. According to the U.S. Environmental Protection Agency (EPA), about 98.5% of all solid waste produced in the United States is industrial solid waste from mining (76%), agriculture (13%), and industry (9.5%). The remaining 1.5% of U.S. solid waste is MSW.

Every year, Americans generate enough MSW to fill a bumper-to-bumper convoy of garbage trucks long enough to encircle the earth's equator almost six times. Consider some of the solid wastes that consumers throw away each year, on average, in the high-waste economy of the United States:

- Enough tires to encircle the earth's equator almost three times.

- An amount of disposable diapers that, if linked end to end, would reach to the moon and back seven times.

- Enough carpet to cover the state of Delaware.
- Enough nonreturnable plastic bottles to form a stack that would reach from the earth to the moon and back about six times.
- About 100 billion plastic shopping bags, or 274 million per day, an average of nearly 3,200 every second.
- Enough office paper to build a wall 3.5 meters (11 feet) high across the country from New York City to San Francisco, California.
- 25 billion plastic foam coffee cups—enough, if lined up end to end, to circle the earth's equator 436 times.
- $165 billion worth of food

Most of these wastes break down very slowly, if at all. Lead, mercury, glass, and plastic foam do not break down. An aluminum can take 500 years to disintegrate. Disposable diapers can take 550 years to break down, and a plastic shopping bag may stick around for up to 1,000 years.

CASE STUDY

Ocean Garbage Patches: There Is No Away

In 1997, ocean researcher Charles Moore discovered two gigantic, slowly rotating masses of plastic and other solid wastes in the middle of the North Pacific Ocean near the Hawaiian Islands. It is known as the *North Pacific Garbage Patch* (Figure 16.3). Its wastes are mostly small particles floating on or just beneath the ocean's surface. They are trapped there by a vortex where rotating ocean currents called *gyres* meet.

About 80% of this waste is washed or blown off beaches, pours out of storm drains, and floats down streams and rivers that empty into the sea. Most of the rest comes from wastes dumped into the ocean from cargo and cruise ships.

The North Pacific Garbage Patch—viewed as the planet's largest human trash dump—occupies an area estimated to be at least the size of Texas. Such estimates are difficult to verify because this continuously swirling plastic-laden soup consists mostly of small particles of plastic suspended just beneath the surface and difficult to see and measure.

Research shows that the tiny plastic particles ultimately degrade into microscopic particles that can contain potentially hazardous chemicals. Some long-lived toxins in these microscopic plastic particles can build up to high concentrations in food chains and webs and can end up in fish sandwiches and other forms of seafood.

Research shows that the tiny plastic particles can be harmful to marine mammals, albatrosses and other seabirds, and fish that mistake the particles for food and swallow them. Because they cannot digest the plastic, it can fill up their stomachs and cause them to die from starvation or poisoning.

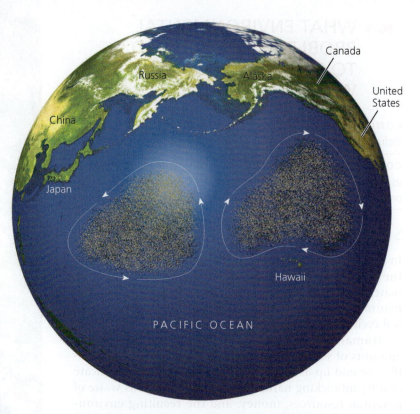

FIGURE 16.3 The North Pacific Garbage Patch consists of two vast slowly swirling masses of small plastic particles floating just under the water. Five other huge garbage patches have been discovered in the world's other major oceans.

Since the Great Pacific Garbage Patch was discovered, five other huge swirling garbage patches have been found in gyres in the world's other oceans. In total, these garbage patches cover an area of ocean greater than all of the earth's land area—the massive pollution legacy of a throwaway human culture. Scientists estimate that the amount of plastics entering the ocean will double by 2050.

Each of us risks contributing to such patches every time we use and discard a plastic item. We may think we have thrown it away but there is no away.

Unfortunately, there is no practical or affordable way to clean up this gigantic amount of marine litter. The only useful approach is to prevent the garbage patches from growing by reducing the production of solid waste.

Hazardous Waste Is a Serious and Growing Problem

Another major category of waste is **hazardous**, or **toxic, waste**. It is any discarded material or substance that threatens human health or the environment because it is toxic, corrosive, or flammable, can undergo violent or explosive chemical reactions, or can cause disease. Examples include industrial solvents, hospital medical waste,

What Harmful Chemicals Are in Your Home?

Cleaning
Disinfectants
Drain, toilet, and window cleaners
Spot removers
Septic tank cleaners

Paint Products
Paints, stains, varnishes, and lacquers
Paint thinners, solvents, and strippers
Wood preservatives
Artist paints and inks

General
Dry-cell batteries (mercury and cadmium)
Glues and cements

Gardening
Pesticides
Weed killers
Ant and rodent killers
Flea powders

Automotive
Gasoline
Used motor oil
Antifreeze
Battery acid
Brake and transmission fluid

FIGURE 16.4 Harmful chemicals are found in many homes. The U.S. Congress has exempted the disposal of many of these household chemicals and other items from government regulation. **Question:** Which of these chemicals could you find in the place where you live?

car batteries (containing lead and acids), household pesticide products, dry-cell batteries (containing mercury and cadmium), and ash and sludge from incinerators and coal-burning power and industrial plants. You might be surprised to learn that there are hazardous chemicals in many household products (Figure 16.4).

Another form of extremely hazardous waste is the highly radioactive waste produced by nuclear power plants and nuclear weapons facilities (see Chapter 13, pp. 342–343). Such waste must be stored safely for at least 10,000 years. After 60 years of research, scientists and governments have not found a scientifically and politically acceptable way to safely isolate these dangerous wastes for any such period of time.

According to the U.N. Environment Programme (UNEP), more-developed countries produce 80–90% of the world's hazardous wastes. The United States produces the most hazardous waste. China is closing in on the number one spot as it continues to industrialize rapidly without adequate pollution controls.

E-Waste—A Serious Hazardous Waste Problem

What happens to your cell phone, computer, television set, and other electronic devices when they are no longer useful (see chapter-opening photo) or when new models come out? They become *electronic waste*, or *e-waste*—the fastest-growing solid waste problem in the United States and in the world. The two leading producers of e-waste are the United States and China.

Between 2000 and 2014, the recycling of U.S. e-waste increased from 10% to 29%. Much of the remaining e-waste ends up in landfills and incinerators and much of it contains gold, rare earth metals, and other valuable materials that could be recycled or reused.

Much of the e-waste in the United States that is not recycled, buried, or incinerated is shipped to China, India, and other Asian and African countries for processing. Labor is cheap and environmental regulations are weak in those countries. Workers there—many of them children—dismantle, burn, and treat e-waste with acids to recover valuable metals and reusable parts. This exposes them to toxic metals such as lead and mercury and other harmful chemicals. The remaining scrap is dumped into waterways and fields or burned in open fires that also expose people to toxic chemicals.

The transfer of such hazardous waste from more-developed to less-developed countries is banned under the International Basel Convention. Despite this ban, much of the world's e-waste is not officially classified as hazardous waste, or it is illegally smuggled out of some countries. The United States can export its e-waste legally because it has not ratified the Basel Convention.

16.2 HOW SHOULD WE DEAL WITH SOLID WASTE?

CONCEPT 16.2 A sustainable approach to solid waste is first to produce less of it, then to reuse or recycle it, and finally to safely dispose of what is left.

Waste Management

Society can deal with the solid wastes it creates in two ways. One is **waste management**, which focuses on reducing their environmental harm. This approach begins with the question, "What do we do with solid waste?" It typically involves mixing wastes together and then burying them, burning them, or shipping them to another location.

The other approach is **waste reduction,** focused on producing much less solid waste and reusing, recycling, or composting what is produced as much as possible (**Concept 16.2** and **Core Case Study**). This approach begins with questions such as "How can we avoid producing so much solid waste?" and "How can we use the solid waste we produce as a resource, as nature does?"

Many waste experts prefer using **integrated waste management**—a variety of coordinated strategies for waste management and waste reduction (Figure 16.5). Figure 16.6 compares the science-based waste management goals of the EPA and National Academy of Sciences with waste management trends based on actual waste data.

Let us look more closely at these options in the order of priorities suggested by scientists.

The Four Rs of Waste Reduction

A more sustainable approach to handling solid waste is to produce less of it, reuse or recycle it, and safely dispose of what is left (Figure 21.7, left). This waste reduction approach (**Concept 16.2**) is based on four Rs, listed below in an order of priority suggested by scientists:

- **Refuse:** Don't use it.
- **Reduce:** Use less of it.
- **Reuse:** Use it over and over.

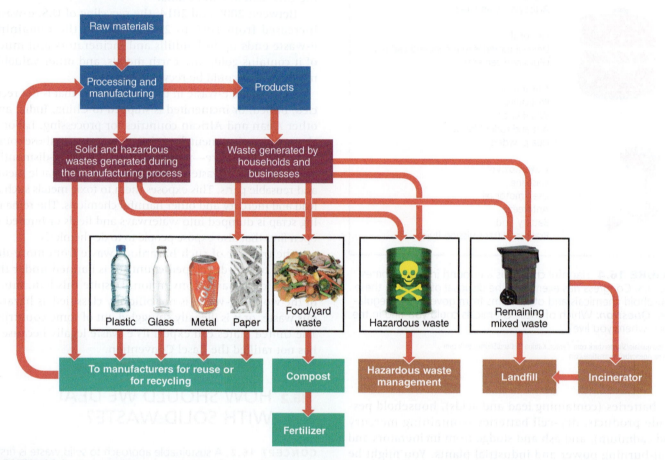

FIGURE 16.5 We can *reduce* wastes by refusing or reducing resource use and by reusing, recycling, and composting what we discard, or we can *manage* them by burying them in landfills or incinerating them. Most countries rely primarily on burial and incineration. *Critical thinking:* What happens to the solid waste you produce?

Mariyana M/Shutterstock.com.; Sopotnicki/Shutterstock.com.; Scanrail1/Shutterstock.com.; chris kolaczan/Shutterstock.com.; vilax/Shutterstock.com.; MrGarry/Shutterstock.com.; Le Do/Shutterstock.com.

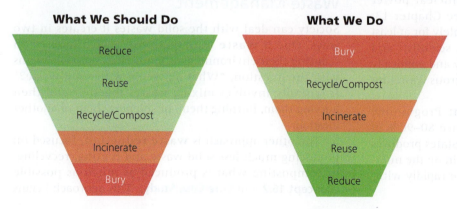

FIGURE 16.6 Priorities recommended by the U.S. National Academy of Sciences for dealing with municipal solid waste (left) compared with actual waste-handling practices in the United States (right). *Critical thinking:* Why do you think most countries do not follow most of the scientific-based priorities listed on the left?

(Compiled by the authors using data from U.S. Environmental Protection Agency, U.S. National Academy of Sciences, Columbia University, and *BioCycle*)

- **Recycle:** Convert used resources to useful items and buy products made from recycled materials. An important form of recycling is **composting**, which mimics nature by using bacteria and other decomposers to break down yard trimmings, vegetable food scraps, and other biodegradable organic wastes into materials that can be used to improve soil fertility.

The first three Rs are preferred because they are *waste prevention* approaches that tackle the problem of waste production before it occurs. Recycling is important, but it deals with waste after it has been produced. Some scientists and economists estimate that society could eliminate up to 80% of the solid waste it produces by following the four Rs strategy. This would mimic the earth's chemical cycling **principle of sustainability**. Figure 16.7 lists ways in which you can use the four Rs of waste reduction to reduce your output of solid waste.

Here are six strategies that some industries and communities use to reduce resource use, waste, and pollution and promote the cradle-to-cradle-approach to design, manufacturing, and marketing (**Core Case Study**).

First, *change industrial processes to eliminate or reduce the use of harmful chemicals*. Since 1975, the 3M Company has taken this approach and, in the process, saved $1.9 billion (see Chapter 14, Case Study, p. 394).

Second, *redesign manufacturing processes and products to use less material and energy*. For example, the weight of a typical car has been reduced by about one-fourth since the 1960s through the use of lighter steel, aluminum, magnesium, plastics, and composite materials.

Third, *develop products that are easy to repair, reuse, remanufacture, compost, or recycle*. For example, some Xerox photocopiers that are leased by businesses are made of reusable or recyclable parts that allow for easy remanufacturing. They are projected to save the company $1 billion in manufacturing costs.

Fourth, *establish cradle-to-cradle responsibility laws* that require companies to take back various consumer products such as electronic equipment, appliances, and motor vehicles for recycling or remanufacturing, as Japan and many European countries do.

Fifth, *eliminate or reduce unnecessary packaging*. Use the following hierarchy for product packaging: no packaging, reusable packaging, and recyclable packaging.

Sixth, *use fee-per-bag waste collection systems* that charge consumers for the solid waste they throw away but provide free pickup of recyclable and reusable items.

What Can You Do?

Solid Waste

- Follow the four Rs of resource use: Refuse, Reduce, Reuse, Recycle

- Ask yourself whether you really need what you're buying and refuse packaging wherever possible

- Rent, borrow, or barter goods and services when you can, buy secondhand, and donate or sell unused items

- Buy things that are reusable, recyclable, or compostable, and be sure to reuse, recycle, and compost them

- Buy products with little or no packaging and recycle any packaging as much as possible

- Avoid disposables such as paper and plastic bags, plates, cups, and utensils, disposable diapers, and disposable razors whenever reusable versions are available

- Cook with whole, fresh foods, avoid heavily packaged processed foods, and buy products in bulk whenever possible

- Discontinue junk mail as much as possible and read online newspapers and magazines and e-books

FIGURE 16.7 Individuals Matter: You can save resources and money by reducing your output of solid waste and pollution. *Critical thinking:* Which three of these steps do you think are the most important ones to take? Why? Which of these things do you already do?

16.3 WHY ARE REFUSING, REDUCING, REUSING, AND RECYCLING SO IMPORTANT?

CONCEPT 16.3 By refusing and reducing resource use and by reusing and recycling what we use, we decrease our consumption of matter and energy resources, reduce pollution and natural capital degradation, and save money.

Alternatives to the Throwaway Economy

People in today's industrialized societies have increasingly substituted throwaway items for reusable ones, which has resulted in growing masses of solid waste. By applying the four Rs, society can slow or stop this trend. Individuals can guide and reduce their consumption of resources along with pollution and solid wastes by asking questions such as:

- Do I really need this? (refusing)

- How many of these do I actually need? (reducing)

- Is this something I can use more than once? (reusing)

- Can this be converted into something else when I am done with it? (recycling)

Cradle-to-cradle design (**Core Case Study**) elevates reuse to a new level. According to William McDonough (Individuals Matter 16.1), the key to shifting to a reuse economy is to design for it. Some manufacturers of computers, photo copiers, motor vehicles, and other products have designed their products so that when they are no longer useful, they can be retrieved from consumers for repair or remanufacture.

William McDonough

PHOTOCOME-US/SIPA/Sipa Press/Beijing China/Newscom

William McDonough is an architect, designer, and visionary thinker, devoted to the earth-friendly design of buildings, products, and cities.

McDonough views wastes as resources out of place as a result of poor design. He also notes that humans have been releasing a growing number of chemicals to the environment faster than the earth's natural chemical cycles can remove them.

McDonough's approach to design has been applied in numerous projects, including the Adam Joseph Lewis Center for Environmental Studies at Oberlin College. Architects and designers view it as one of the most important and inspiring examples of environmentally friendly design. It uses recycled and nontoxic materials that can be further recycled. It gets heat from the sun and the earth's interior and electricity from solar cells, and it produces 13% more energy than it consumes. The building's greenhouse contains an ecosystem of plants and animals that purify the building's sewage and wastewater. Rainwater is collected and used to irrigate the surrounding green space, which includes a restored wetland, a fruit orchard, and a vegetable garden.

McDonough has been recognized by *Time* magazine as a "Hero for the Planet." He has also received numerous design awards and three presidential awards. He believes we can use cradle-to-cradle design to leave the world better off than we found it.

CONSIDER THIS …

Learning from Nature

McDonough and his business partner, chemist Michael Braungart, call for using environmentally and economically sustainable design to mimic nature by reusing and recycling the chemicals and products we make with the goal of zero waste.

One way to implement cradle-to-cradle design is for governments to ban or severely restrict the disposal of certain items. For example, the European Union (EU) has led the way by banning e-waste from landfills and incinerators and requiring electronics manufacturers to take back their products at the end of their useful lives. To cover the costs of these programs, consumers pay a recycling tax on electronic products, an example of implementing the full-cost pricing **principle of sustainability**. Japan and China also use the take-back approach. In the United States, there is no federal take-back law, but more than 20 states have such laws and several more are considering them.

Governments have also banned the use of certain throwaway items. For example, Finland banned all beverage containers that cannot be reused, and consequently, 95% of that country's soft drink, beer, wine, and spirits containers are refillable. The use of rechargeable batteries is cutting toxic waste by reducing the amount of conventional batteries that are thrown away. The newest rechargeable batteries come fully charged, can hold a charge for up to two years when not in use, and can be recharged in about 15 minutes.

In many countries, the landscape is littered with plastic bags. They can take 400 to 1,000 years to break down and can kill animals that try to eat them or become ensnared in them. Huge quantities of plastic bags and other plastic products and solid wastes end up in the ocean (Figure 16.3). Many people are using reusable cloth bags instead of throwaway paper or plastic bags to carry groceries and other items they buy.

To encourage the use of reusable bags, the governments of Denmark, Ireland, Taiwan, Great Britain, and the Netherlands tax plastic shopping bags. In Ireland, a tax of about 25¢ per bag cut plastic bag litter by 90% as people switched to reusable bags. In 2014, the European Union passed a directive aimed at cutting the use of single-use plastic bags by 80%.

The U.S. state of California joined Hawaii in banning single-use plastic bags in grocery stores and other selected retail outlets. They have also been banned in 133 U.S. cities or counties, despite intense lobbying against such bans by the plastics industry. Figure 16.8 lists some other ways to reuse various items.

What Can You Do?

Reuse

- Buy beverages in refillable glass containers
- Use reusable lunch containers
- Store refrigerated food in reusable containers
- Use rechargeable batteries and recycle them when their useful life is over
- When eating out, bring your own reusable container for leftovers
- Carry groceries and other items in a reusable basket or cloth bag
- Buy used furniture, cars, and other items, whenever possible

FIGURE 16.8 Individuals Matter: Ways to reuse the items we purchase. *Questions:* Which of these suggestions have you tried and how did they work for you?

FIGURE 16.9 Large-scale municipal composting site.

imging/Shutterstock.com

Recycling

The cradle-to-cradle approach (**Core Case Study**) gives the highest priority to reuse but also relies on recycling. Worn-out items from the technical cycle of cradle-to-cradle manufacturing are recycled or sent into the biological cycle where ideally they degrade and become biological nutrients (Figure 16.1).

McDonough and Braungart break recycling down into two categories: *upcycling* and *downcycling*. Ideally, all discarded items would be upcycled—recycled into a form that is more useful than the recycled item was. In downcycling, the recycled product is still useful, but not as useful or long-lived as the original item.

Households and workplaces produce five major types of materials that can be recycled: paper products, glass, aluminum, steel, and some plastics. These materials can be reprocessed into new, useful products in two ways. **Primary recycling** involves using materials again for the same purpose. An example is recycling used aluminum cans into new aluminum cans. **Secondary recycling** involves downcycling or upcycling waste materials to make different products such as downcycling used tires to make sandals.

Recycling involves three steps: collection of materials for recycling, conversion of recycled materials to new products, and selling and buying products that contain recycled material. Recycling is successful environmentally and economically only when all three of these steps are carried out.

Composting is another form of recycling that mimics nature's recycling of nutrients. It involves using bacteria to decompose yard trimmings, vegetable food scraps, and other biodegradable organic wastes into humus. This organic material can be added to soil to supply plant nutrients, slow soil erosion, retain water, and improve crop yields.

People can compost organic wastes in composting piles that must be turned over occasionally or in small composting drums that can be rotated to mix the wastes and speed up the decomposition process. In the United States, about 3,000 municipal composting programs recycle about 60% of the yard wastes in the country's MSW (Figure 16.9). However, such programs must exclude toxic materials that can contaminate the compost and make it unsafe for fertilizing crops and lawns.

Bioplastics

Because plastics are made to last, they do not completely break down once they are discarded. In addition, most of today's plastics are made from organic polymers that are produced from petroleum-based chemicals. Processing these chemicals creates hazardous waste and causes water and air pollution. However, some products are now being made from *bioplastics*. This type of plastic is usually more environmentally friendly because it is made from biologically based chemicals.

Henry Ford, who developed the first Ford car and founded Ford Motor Company, supported research on the development of a bioplastic made from soybeans and another made from hemp. A 1914 photograph shows him using an ax to strike the body of a Ford car made from a soy bioplastic to demonstrate its strength and resistance to denting. However, as oil became cheaper and widely available, petrochemical plastics took over the market.

Now, confronted with climate change and other environmental problems associated with the use of petroleum, chemists are stepping up efforts to make biodegradable and more environmentally sustainable plastics. These *bioplastics* can be made from plants such as corn, soy, sugarcane, switchgrass, chicken feathers, and some components of garbage.

Some bioplastics are more environmentally friendly than others. For example, some are made from corn raised through industrialized agriculture, which requires great amounts of energy, water, and petrochemical fertilizers and thus has a very large ecological footprint. In evaluating and choosing bioplastics, scientists urge consumers to learn how they were made, how long they take to degrade, and whether they degrade into harmful chemicals.

CRITICAL THINKING

How could bioplastics pollute the environment?

Recent research based on actual data instead of models indicates that the United States recycles or composts about 24% of its MSW, which is significantly lower than the EPA estimate of 34%. Experts say that with education and proper economic incentives, Americans could recycle and compost at least 80% of their MSW, in keeping with the chemical cycling **principle of sustainability**.

According to the Organization for Economic Cooperation and Development (OECD), Germany leads the world in recycling. It recycles 65% of its MSW, with consumers separating recyclable items into different categories and depositing them in color-coded bins found throughout the country. South Korea comes in second and recycles 59% of its MSW. Austria, Switzerland, Sweden, Belgium, and the Netherlands all recycle at least 50% of their MSW. Turkey, which recycles only 1% of its waste, is in last place.

Currently, only 7% by weight of all plastic wastes in the United States is recycled. This percentage is low because there are many different types of plastic resins, which are difficult to separate from products that contain several kinds of plastic. However, progress is being made in the recycling of plastics and in the development of more degradable bioplastics (Science Focus 16.1).

Engineer Mike Biddle has developed a 16-step automated commercial process for recycling high-value plastics. It separates plastic items from nonplastic items in mixed solid waste, separates plastic types from one another, and converts them to pellets that can be sold and used to make new plastics products. For his work,

Biddle has been named a Technology Pioneer by the World Economic Forum and has received some of the world's most important environmental rewards.

Figure 16.10 lists the advantages and disadvantages of recycling (**Concept 16.3**). Whether recycling makes economic sense depends on its economic and environmental benefits and costs. Critics of recycling programs argue that recycling is costly and adds to the taxpayer burden in communities where recycling is funded through taxation.

Trade-Offs

Recycling

Advantages	Disadvantages
Reduces energy and mineral use and air and water pollution	Can cost more than burying in areas with ample landfill space
Reduces greenhouse gas emissions	Reduces profits for landfill and incinerator owners
Reduces solid waste	Inconvenient for some

Jacqui Martin/Shutterstock.com

FIGURE 16.10 Recycling solid waste has advantages and disadvantages (**Concept 16.3**). *Critical thinking:* Which single advantage and which single disadvantage do you think are the most important? Why?

Proponents of recycling point to studies showing that the net economic, health, and environmental benefits of recycling (Figure 16.10, left) far outweigh the financial costs. The EPA estimates that each year, recycling and composting in the United States reduce emissions of climate-changing carbon dioxide by an amount roughly equal to that emitted by 36 million passenger vehicles. In addition, the U.S. recycling industry employs about 1.1 million people, and doubling the U.S. recycling rate would create about 1 million new jobs.

Cities that make money by recycling and that have higher recycling rates tend to use a *single-pickup system* for both recyclable and nonrecyclable materials, instead of a more expensive two-truck system. Successful systems also use a pay-as-you-throw approach. They charge by weight for picking up trash but not for picking up recyclable or reusable materials, and they require citizens and businesses to sort their trash and recyclables by type, as Germany does. San Francisco, California, uses such a system and recycles, composts, or reuses 80% of its MSW.

16.4 WHAT ARE THE ADVANTAGES AND DISADVANTAGES OF BURNING OR BURYING SOLID WASTE?

CONCEPT 16.4 Technologies for burning and burying solid wastes are well developed, but burning can contribute to air and water pollution and greenhouse gas emissions, and buried wastes can contribute to water pollution.

Burning Solid Waste Has Advantages and Disadvantages

Globally, MSW is burned in more than 800 large *waste-to-energy incinerators* (Figure 16.11), 86 of them in the United States. The United States incinerates about 9% of its MSW. One reason for this low percentage is that in the past, incineration earned a bad reputation because of highly polluting and poorly regulated incinerators. By contrast,

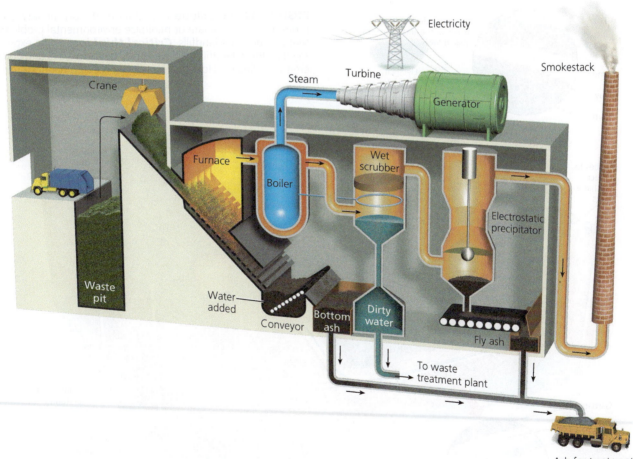

FIGURE 16.11 Solutions: A waste-to-energy incinerator with pollution controls burns mixed solid wastes and recovers some of the energy to produce steam for use in heating buildings or producing electricity. *Critical thinking:* Would you invest in such a project? Why or why not?

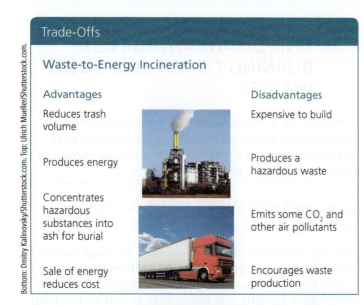

Trade-Offs

Waste-to-Energy Incineration

Advantages	Disadvantages
Reduces trash volume	Expensive to build
Produces energy	Produces a hazardous waste
Concentrates hazardous substances into ash for burial	Emits some CO_2 and other air pollutants
Sale of energy reduces cost	Encourages waste production

FIGURE 16.12 Incinerating solid waste has advantages and disadvantages (**Concept 16.4**). These trade-offs also apply to the incineration of hazardous waste. ***Critical thinking:*** Which single advantage and which single disadvantage do you think are the most important? Why?

Denmark incinerates over half of its MSW in state-of-the-art waste-to-energy incinerators, that far exceed European air pollution standards. However, all incinerators produce an ash that contains toxic chemicals and must be stored safely somewhere, essentially forever.

Figure 16.12 lists the advantages and disadvantages of using incinerators to burn solid waste. Despite the availability of state-of-the-art incinerators, many U.S. citizens, local governments, and environmental scientists remain opposed to waste incineration because incinerators require a large, steady stream of waste to be profitable. This high demand for burnable wastes undermines efforts to reduce waste and increase reuse and recycling.

Burying Solid Waste Has Advantages and Disadvantages

In the United States, about 53% of all MSW, by weight, is buried in sanitary landfills, compared to 80% in Canada, 15% in Japan, and 4% in Denmark. There are two types of landfills. In newer landfills, called **sanitary landfills** (Figure 16.13), solid waste is spread out in thin layers, compacted, and covered daily with a fresh layer of clay

FIGURE 16.13 Solutions: A state-of-the-art sanitary landfill is designed to eliminate or minimize environmental problems that plague older landfills. ***Critical thinking:*** Some experts say that these landfills will eventually develop leaks and could emit toxic liquids. How do you think this could happen?

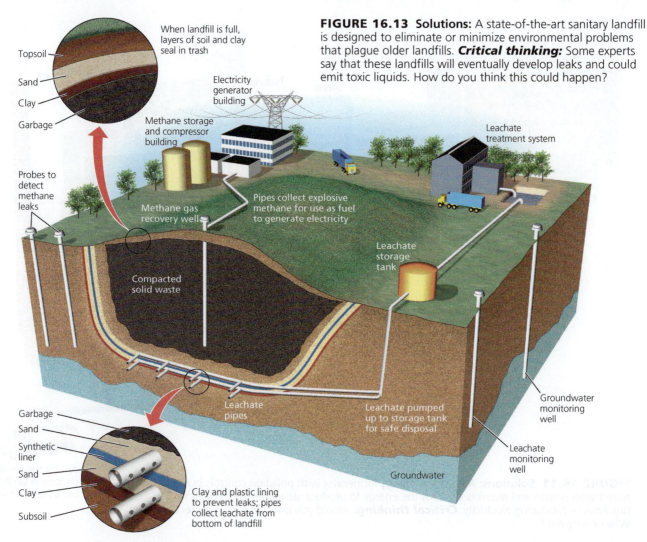

When landfill is full, layers of soil and clay seal in trash

Topsoil
Sand
Clay
Garbage

Probes to detect methane leaks

Electricity generator building

Methane storage and compressor building

Methane gas recovery well

Pipes collect explosive methane for use as fuel to generate electricity

Leachate treatment system

Leachate storage tank

Compacted solid waste

Leachate pipes

Leachate pumped up to storage tank for safe disposal

Groundwater monitoring well

Leachate monitoring well

Groundwater

Garbage
Sand
Synthetic liner
Sand
Clay
Subsoil

Clay and plastic lining to prevent leaks; pipes collect leachate from bottom of landfill

Trade-Offs

Sanitary Landfills

Advantages	Disadvantages
Low operating costs	Noise, traffic, and dust
Can handle large amounts of waste	Releases greenhouse gases (methane and CO_2) unless they are collected
Filled land can be used for other purposes	Output approach that encourages waste production
No shortage of landfill space in many areas	Eventually leaks and can contaminate groundwater

FIGURE 16.14 Using sanitary landfills to dispose of solid waste has advantages and disadvantages (**Concept 16.4**). *Critical thinking:* Which single advantage and which single disadvantage do you think are the most important? Why?

or plastic foam. This process keeps the material dry, cuts down on odors, reduces the risk of fire, and helps keep rats and other pests away from the wastes.

The bottoms and sides of well-designed sanitary landfills have strong double liners and containment systems that collect the liquids leaching from them. Some landfills also have systems for collecting methane, a potent greenhouse gas that is produced when the wastes decompose in the absence of oxygen. The collected methane can be burned as a fuel to generate electricity although this adds climate-changing CO_2 to the atmosphere.

Paper products represent the largest percentage of landfill materials. Other common materials include yard waste, plastics, metals, wood, glass, and food waste. Some types of solid waste are not accepted at U.S. landfills. Examples are tires, waste oil, oil filters, items containing mercury such as compact fluorescent light bulbs and thermometers, electronics, and medical waste. Some retired landfill sites are being used as sites for solar cell power plants. Figure 16.14 lists the advantages and disadvantages of using sanitary landfills to dispose of solid waste.

The second type of landfill is an **open dump**, essentially a field or large pit where garbage is deposited and sometimes burned. Open dumps are rare in more-developed countries, but are widely used near major cities in many less-developed countries. China disposes of much of its rapidly growing mountains of solid waste mostly in rural open dumps or in poorly designed and poorly regulated landfills.

16.5 HOW SHOULD WE DEAL WITH HAZARDOUS WASTE?

CONCEPT 16.5 A more sustainable approach to hazardous waste is first to produce less of it, then to reuse or recycle it, then to convert it to less-hazardous materials, and finally to safely store what is left.

Hazardous Waste Requires Special Handling

Figure 16.15 shows an integrated management approach suggested by the U.S. National Academy of Sciences. It establishes three priority levels for dealing with hazardous waste: produce less; convert as much of it as possible to less-hazardous substances; and put the rest in long-term, safe storage (**Concept 16.5**). Denmark follows these priorities, but most countries do not.

As with solid waste, the top priority for hazardous waste management should be pollution prevention and waste reduction. Using this approach, industries try to find substitutes for toxic or hazardous materials. Then they reuse or recycle the hazardous materials they use within industrial processes, whenever possible. They may also sell their hazardous wastes as raw materials for making other products, in keeping with the cradle-to-cradle approach (**Core Case Study**).

At least 33% of industrial hazardous wastes produced in the European Union are exchanged through clearing houses where they are sold as raw materials for use by other industries. The producers of these wastes do not have to pay for their disposal and recipients get low-cost

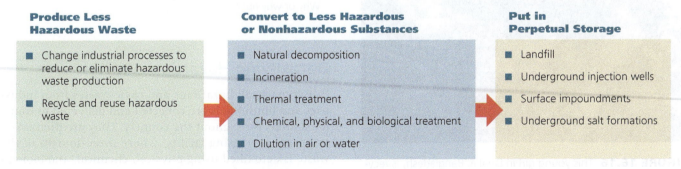

Produce Less Hazardous Waste	Convert to Less Hazardous or Nonhazardous Substances	Put in Perpetual Storage
■ Change industrial processes to reduce or eliminate hazardous waste production	■ Natural decomposition	■ Landfill
■ Recycle and reuse hazardous waste	■ Incineration	■ Underground injection wells
	■ Thermal treatment	■ Surface impoundments
	■ Chemical, physical, and biological treatment	■ Underground salt formations
	■ Dilution in air or water	

FIGURE 16.15 *Integrated hazardous waste management:* The U.S. National Academy of Sciences has suggested these priorities for dealing with hazardous waste (**Concept 16.5**).

raw materials. About 10% of the hazardous waste in the United States is exchanged through such clearing houses, an amount that could be increased significantly. E-waste can also be recycled because it contains valuable materials (see Case Study that follows).

Recycling E-Waste

In some countries, workers in e-waste recycling operations—many of them children (Figure 16.16)—are often exposed to toxic chemicals as they dismantle the electronic trash to extract its valuable metals or other parts that can be sold for reuse or recycling.

According to the United Nations, much of the world's e-waste is shipped to China. One popular destination is the small port city of Guiyu, where the air reeks of burning plastic and acid fumes. There, more than 5,500 small-scale e-waste businesses employ more than 30,000 people, including children. They work for very low wages in dangerous conditions to extract valuable metals like gold, silver, copper, and various rare earth metals (see Chapter 12,

James P. Blair/National Geographic Creative

FIGURE 16.16 This young girl in Dhaka, Bangladesh, is recycling batteries by hammering them apart to extract tin and lead. The workers at this shop are mostly women and children.

Case Study, p. 306) from millions of discarded computers, television sets, and cell phones.

These workers usually wear no masks or gloves, work in rooms with no ventilation, and are exposed to toxic chemicals. They carry out dangerous activities such as smashing TV picture tubes with large hammers to recover certain components—a method that releases large amounts of toxic lead dust into the air. They burn computer wires to expose copper. They also melt circuit boards in metal pots over coal fires to extract lead and other metals and douse the boards with strong acid to extract gold. After the valuable metals are removed, leftover parts are burned or dumped into rivers or onto the land. An estimated 82% of the Guiyu area's children younger than age six suffer from lead poisoning.

The United States is the world's largest producer of e-waste and in 2014 recycled about 29% of it (up from 19% in 2009), according to the Consumer Electronics Association. By 2015, a total of 28 states and the District of Columbia had banned the disposal of computers and TV sets in landfills and incinerators. These measures set the stage for an emerging, highly profitable *e-cycling* industry. By 2015, 28 states along with New York City had laws requiring electronics manufacturers to take back their products for recycling. In 2016, Apple introduced Liam—a 29-armed robot capable of taking apart 1.2 million iPhones a year, saving components that can be recycled.

Some call for a U.S. federal law to institute a cradle-to-cradle approach (**Core Case Study**) that would require manufacturers to take back all electronic devices they produce and recycle them domestically. It could be similar to laws in the European Union, where a recycling fee typically covers the costs of such programs. Without such a law, there is little incentive for recycling e-waste.

Take-back programs are important. But the only real long-term solution is a cradle-to-cradle approach (**Core Case Study**) through which electrical and electronic products would be designed to be produced and easily repaired, remanufactured, or recycled, without the use of toxic materials.

CONSIDER THIS ...

THINKING ABOUT E-Waste Recycling

Would you support a recycling fee on all electronic devices? Why or why not?

Detoxifying Hazardous Wastes

In Denmark, all hazardous and toxic wastes from industries and households are collected and delivered to transfer stations throughout the country. They are then taken to a large processing facility, where three-fourths of the waste is detoxified using physical, chemical, and biological methods. The rest is buried in a carefully designed and monitored landfill.

Physical methods for detoxifying hazardous wastes may involve using charcoal or resins to filter out harmful solids, distilling liquid wastes to separate out harmful chemicals, and precipitating such chemicals from solution. Especially deadly wastes, such as those contaminated by mercury, can be encapsulated in glass, cement, or ceramics and put in secure storage sites. *Chemical methods* are used to convert hazardous chemicals to harmless or less harmful chemicals through chemical reactions.

Some scientists and engineers consider *biological methods* for treatment of hazardous waste to be the wave of the future. One such approach is *bioremediation,* in which bacteria and enzymes are used to destroy toxic or hazardous substances or to convert them to harmless compounds. Bioremediation is often used on contaminated soil. It usually takes longer to work than most physical and chemical methods, but it costs much less.

Phytoremediation is another biological method for treating hazardous wastes. It involves using natural or genetically engineered plants to absorb, filter, and remove contaminants from polluted soil and water. This method is still being evaluated and is slow compared to other alternatives.

Hazardous wastes can also be incinerated to break them down and convert them to harmless or less harmful chemicals. This has the same advantages and disadvantages as incinerating solid wastes (Figure 16.12).

Plasma gasification is another thermal method that uses arcs of electrical energy to produce very high temperatures to vaporize hazardous wastes in the absence of oxygen. The process reduces the volume of a given amount of waste by 99%, produces a synthetic gaseous fuel, and encases toxic metals and other materials in glassy lumps of rock. It is not widely used because of its high cost.

Storing Hazardous Waste

After reduction, reuse, and recycling options have been pursued (Figure 16.15 and **Concept 16.5**), remaining hazardous waste can be buried on land or stored long-term in secure vaults. In reality, burial is the most widely used method in the United States and in most countries, largely because of its lower cost.

The most common form of burial is *deep-well disposal* in which liquid hazardous wastes are pumped under very high pressure through a pipe into dry, porous rock formations deep underground. The rock formations lie far beneath aquifers that are tapped for drinking and irrigation water (see Figure 11.25, p. 285). Theoretically, the hazardous liquids soak into the porous rock material and are isolated from overlying groundwater by essentially impermeable layers of clay and rock. The cost is low.

However, there are a limited number of such sites and limited space within them. Sometimes the wastes leak into groundwater from the well shaft used to transfer the wastes to disposal sites or migrate into groundwater in unexpected ways. Deep-well disposal also encourages the production of hazardous waste rather than the reduction of such waste.

In the United States, almost two-thirds of all liquid hazardous wastes are injected into deep disposal wells. This amount is increasing sharply with the use of hydraulic fracturing, or fracking (Figure 13.1, p. 328) to produce natural gas and oil trapped in shale rock. Fracking produces large volumes of liquid hazardous waste, which has the potential to contaminate groundwater and increase the risk of earthquakes (see Science Focus 13.1, p. 336).

Many scientists view current regulations for deep-well disposal in the United States as inadequate (see the Case Study that follows). Figure 16.17 lists the advantages and disadvantages of using deep-well disposal of liquid hazardous wastes.

Some liquid hazardous wastes are stored in lined ponds, pits, or lagoons, called *surface impoundments*. Sometimes impoundments include liners to help contain the waste. Where liners are not used or when they leak, the concentrated wastes can seep into groundwater. Studies conducted by the EPA found that 70% of all U.S. hazardous waste storage ponds lack liners and could threaten groundwater supplies. The EPA also warns that eventually all impoundments are likely to leak.

Because surface impoundments are not covered, harmful chemicals can evaporate and pollute the air. In addition, flooding from heavy rains and storms can cause such ponds to overflow. Figure 16.18 lists the advantages and disadvantages of using this method.

Sometimes highly toxic materials (such as mercury; see Chapter 14, Core Case Study, p. 378) that cannot be destroyed, detoxified, or safely burned are buried in carefully designed and monitored *secure hazardous waste landfills* (Figure 16.19). This is the least-used method because of the expense involved. Figure 16.20 lists some ways in which you can reduce your output of hazardous waste.

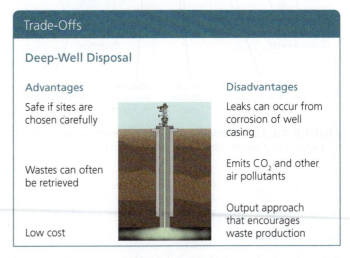

Trade-Offs

Deep-Well Disposal

Advantages	Disadvantages
Safe if sites are chosen carefully	Leaks can occur from corrosion of well casing
Wastes can often be retrieved	Emits CO_2 and other air pollutants
Low cost	Output approach that encourages waste production

FIGURE 16.17 Injecting liquid hazardous wastes into deep underground wells has advantages and disadvantages. **Critical thinking:** Which single advantage and which single disadvantage do you think are the most important? Why?

Trade-Offs

Surface Impoundments

Advantages	Disadvantages
Low cost	Water pollution from leaking liners and overflows
Wastes can often be retrieved	Air pollution from volatile organic compounds
Can store wastes indenitely with secure double liners	Output approach that encourages waste production

Jim West/Alamy Stock Photo

FIGURE 16.18 Storing liquid hazardous wastes in surface impoundments has advantages and disadvantages. ***Critical thinking:*** Which single advantage and which single disadvantage do you think are the most important? Why?

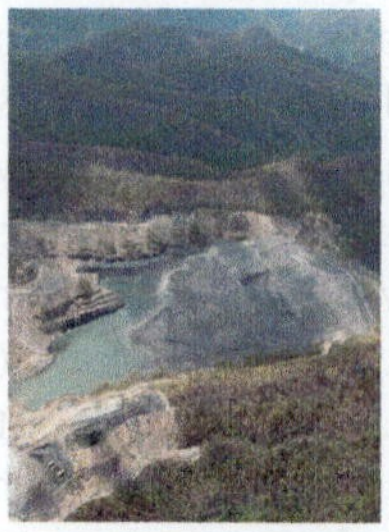

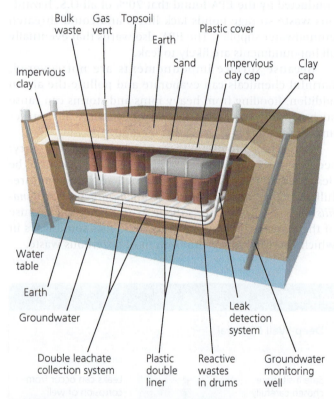

FIGURE 16.19 Solutions: Hazardous wastes can be isolated and stored in a secure hazardous waste landfill.

CASE STUDY

Hazardous Waste Regulation in the United States

Several U.S. laws help regulate the management and storage of hazardous waste. About 5% of all hazardous waste produced in the United States is regulated under the Resource Conservation and Recovery Act (RCRA, pronounced "RICK-ra"), passed by the U.S. Congress in 1976 and amended in 1984.

Under RCRA, the EPA sets standards for the management of several types of hazardous waste. It also issues permits to companies that allow them to produce and dispose of a certain amount of that waste by approved methods. Permit holders must use a *cradle-to-grave* system. This means tracking its transfer from a point of generation (cradle) to an approved off-site disposal facility (grave), and submitting proof of this disposal to the EPA. RCRA is a good start, but 95% of the hazardous and toxic wastes, produced in the United States, including e-waste, are not regulated.

What Can You Do?

Hazardous Waste

- Avoid using pesticides and other hazardous chemicals, or use them in the smallest amounts possible
- Use less harmful substances instead of commercial household cleaners. For example, use vinegar to polish metals, clean surfaces, and remove stains and mildew, and baking soda to clean utensils and to deodorize and remove stains.
- Do not dump pesticides, paints, solvents, oil, antifreeze, or other hazardous chemicals down the toilet, down the drain, into the ground, into the garbage, or down storm drains. Free hazardous waste disposal services are available in many cities.
- Do not throw old fluorescent lightbulbs (which contain mercury) into regular trash. Many communities and home product retailers offer free recycling of these bulbs.

FIGURE 16.20 Individuals Matter: You can reduce your output of hazardous wastes (**Concept 16.5**). ***Critical thinking:*** Which two of these measures do you think are the most important ones to take? Why?

CONSIDER THIS ...

THINKING ABOUT Hazardous Waste

Why do you think that 95% of the hazardous waste produced in the United States is not regulated? Do you favor regulating more of such wastes? What do you think would be the economic consequences of doing so? Would this promote the cradle-to-cradle approach to reducing solid and hazardous wastes?

The Toxic Substances Control Act (TSCA) has also been in place since 1976. Its purpose is to regulate and ensure the safety of the thousands of chemicals used in the manufacture of, or as ingredients in, many products. Under this law, companies must notify the EPA before introducing a new chemical into the marketplace, but they are not required to provide any data about its safety. In other words, any new chemical is viewed as safe unless the EPA can show it is harmful.

Under intense pressure from manufacturers, when the TSCA was passed in 1976, it allowed the approximately 62,000 chemicals then on the market to continue being used without being tested for safety. The law also made it very difficult for the EPA to demonstrate that a new chemical is hazardous enough to ban. In addition, the EPA has had limited funding for evaluating the safety of old or new chemicals.

As a result, since 1976, the EPA has used this act to ban only 5 of the at least 85,000 chemicals now in use. In 2016, Congress revised the TSCA, requiring the EPA to

- review all existing and new chemicals, identify those that pose unreasonable risks, and regulate or eliminate those risks,

- quickly review chemicals known to persist in the environment and to build up in humans,

- determine whether a new chemical meets certain safety standards before it enters the market,

- consider a chemical's effects on vulnerable populations such as infants, children, pregnant women, chemical workers, and the elderly, and

- require companies to make data on their chemicals available.

These are helpful improvements but critics point out that the law does not require enough funding from the chemical industry for the EPA to evaluate thousands of chemicals. An analysis by the Environmental Working Group estimated that even with adequate funding, the EPA needs 28 years to evaluate the risks of 90 priority chemicals, 20 additional years to finalize regulations for those chemicals, and 35 more years to implement the resulting rules. The rules will then be subject to lengthy lawsuits from manufacturers. In other words, evaluating and regulating just 90 of the 85,000 chemicals in use under the 2016 update of this law will take an estimated 83 years.

In 1980, Congress passed the Comprehensive Environmental Response, Compensation, and Liability Act (CERCLA). It is commonly known as the Superfund Act and is administered by the EPA. The goals of the act are to identify sites, called Superfund sites, where hazardous waste has contaminated the environment and to clean them up, using EPA-approved methods. Superfund sites that represent an immediate and severe threat to human health are put on a *National Priorities List* and scheduled for the earliest possible cleanup.

As of July of 2017, there were 1,336 sites on the Superfund list, along with 53 proposed new sites, and 393 sites had been cleaned up and removed from the list. The Waste Management Research Institute estimates that at least 10,000 sites should be on the priority list and that cleanup of these sites would cost about $1.7 trillion, not including legal fees. These environmental and economic costs show why it is important to emphasize waste reduction and pollution prevention over the "end-of-pipe" regulation and cleanup approach that the United States and most countries rely on.

In 1984, Congress amended the Superfund Act to give citizens the right to know what toxic chemicals are being stored or released in their communities. This required large manufacturing facilities to report their annual releases of any of nearly 650 toxic chemicals. If you live in the United States, you can find out what toxic chemicals are being stored and released in your neighborhood by going to the EPA's *Toxic Release Inventory* website.

The Superfund Act, designed to make polluters pay for cleaning up abandoned hazardous waste sites, has greatly reduced the number of illegal dumpsites around the country. It also forced waste producers who were fearful of liability claims to reduce their production of such waste and to recycle or reuse much more of it. However, in 1995, under pressure from polluters, the U.S. Congress did not renew the tax on oil and chemical companies, which financed the Superfund legislation. Since then taxpayers, not polluters, have been paying for cleanups (with an average cost of $26 million per site) when responsible parties cannot be found. As a result, the pace of cleanup has slowed.

16.6 HOW CAN WE SHIFT TO A LOW-WASTE ECONOMY?

CONCEPT 16.6 Shifting to a low-waste economy will require individuals and businesses to reduce resource use and to reuse and recycle most solid and hazardous wastes at local, national, and global levels.

Citizens Can Take Action

In the United States, individuals and groups have organized grassroots (bottom-up) campaigns to prevent the construction of hundreds of incinerators, landfills, treatment plants for hazardous and radioactive wastes, and chemical plants in or near their communities. These campaigns have organized sit-ins, concerts, and protest rallies. They have also gathered signatures on petitions and presented them to lawmakers.

Health risks from incinerators and landfills, when averaged over the entire country, are quite low. However, the risks for people living near such facilities are higher. Manufacturers and waste industry officials point out that something must be done with the toxic and hazardous wastes created in the production of certain goods and services. They contend that even if local citizens adopt a "not in my back yard" (NIMBY) approach, the waste will always end up in someone's back yard.

Many citizens do not accept this argument. Their view is that the best way to deal with most toxic and hazardous waste is to produce much less of it by focusing on pollution and waste prevention as suggested by the U.S. National

Academy of Sciences (Figure 16.15). They argue that the goal should be "not in anyone's back yard" (NIABY) or "not on planet Earth" (NOPE).

International Treaties

For decades, countries have regularly shipped hazardous wastes to other countries for disposal or processing. Since 1992, an international treaty known as the Basel Convention has banned participating countries from shipping hazardous waste (including e-waste) to or through other countries without their permission.

By 2016, this agreement had been ratified (formally approved and implemented) by 183 countries. The United States has signed but has not ratified the convention. In 1995, the treaty was amended to outlaw all transfers of hazardous wastes from industrial countries to less-developed countries. This ban will help, but it will not wipe out the highly profitable illegal shipping of hazardous wastes. Hazardous waste smugglers evade the laws by using an array of tactics, including bribes, false permits, and mislabeling hazardous of wastes as recyclable materials.

In 2001, delegates from 122 countries developed a global treaty known as the Stockholm Convention on Persistent Organic Pollutants (POPs). POPS are organic chemicals produced by manufacturers that persist in the environment. The treaty started by regulating the use of 12 widely used persistent organic pollutants that can accumulate in the fatty tissues of humans and other animals that occupy high trophic levels in food webs. Because they persist in the environment, POPs can also be transported long distances by wind and water.

The original list of 12 hazardous chemicals, called the *dirty dozen,* included DDT and eight other chlorine containing persistent pesticides, PCBs, dioxins, and furans. Since then, other chemicals have been added. Based on blood tests and statistical sampling, medical researchers at New York City's Mount Sinai School of Medicine concluded that nearly every person on earth likely has detectable levels of some POPs in their bodies. By 2016, 180 countries (not including the United States) had ratified a strengthened version of the treaty. The list of regulated POPs is expected to grow. However, the long-term health effects of this involuntary global chemical experiment are largely unknown.

In 2000, the Swedish Parliament enacted a law that, by 2020, will ban all potentially hazardous chemicals that are persistent in the environment and can accumulate in living tissue. This law also requires industries to perform risk assessments on the chemicals they use and to show that these chemicals are safe to use, as opposed to requiring the government to show that they are dangerous. In other words, chemicals are assumed guilty until proven innocent—the reverse of the current policy in the United States and most other countries. There is strong opposition to this approach in the United States, especially from most of the industries that produce and use potentially hazardous chemicals.

Encouraging Reuse and Recycling

Why aren't reuse and recycling more common? *First,* these strategies must compete with the use of cheap, disposable products that do not include their hidden harmful environmental and health costs in their market prices. This is a violation of the full-cost pricing **principle of sustainability**.

Second, the economic playing field is uneven, because in most countries, resource extraction industries receive more government tax breaks and subsidies than do reuse and recycling industries.

Third, the demand and thus the price paid for recycled materials fluctuates, mostly because it is not a high priority for most governments, businesses, and individuals to buy goods made of recycled materials.

How can we encourage reuse and recycling? Governments can *increase* subsidies and tax breaks for reusing and recycling materials and *decrease* subsidies and tax breaks for making items from virgin resources. Another strategy is to ramp up use of the fee-per-bag waste collection system that charges households for the trash they throw away by weight but not for their recyclable and reusable wastes. When Fort Worth, Texas, instituted such a program, the proportion of households recycling their trash went from 21% to 85%. The city went from losing $600,000 in its recycling program to making $1 million a year because of increased sales of recycled materials to industries.

Governments can also pass laws requiring companies to take back and recycle or reuse packaging and electronic waste discarded by consumers. Japan and some European Union countries have such laws. Another strategy is to encourage or require government purchases of recycled products to help increase demand for, and help lower prices of, these products.

With or without government intervention, some industries are finding ways to save money by reusing and recycling their materials. Some are learning from nature and becoming part of a resource exchange web (see the Case Study that follows).

CASE STUDY

Biomimicry and Industrial Ecosystems: Copying Nature

An important goal for a more sustainable society is to make its industrial manufacturing processes cleaner and more sustainable by redesigning them to mimic the way nature deals with wastes—an approach called *biomimicry* (see Chapter 1 Core Case Study, p. 4). According to the chemical cycling **principle of sustainability,** the waste outputs of one organism become the nutrient inputs of another organism. This explains why there is essentially no waste in nature.

Biomimicry involves two major steps. The first is to study how natural systems have responded to changes in environmental conditions over many millions of years. The second step is to try to copy or adapt these responses within human systems in order to deal with various environmental challenges.

One way for industries to mimic nature is to reuse or recycle most of the materials and chemicals they use, instead of burying or burning them or shipping them somewhere. Industries can set up *resource exchange webs,* in which the wastes of one manufacturer become the raw materials for another. This approach is similar to food webs in natural ecosystems and is a direct application of the cradle-to-cradle concept (**Core Case Study**).

This is happening in Kalundborg, Denmark, where an electric power plant and nearby industries, farms, and homes are collaborating to save money and to reduce their outputs of waste and pollution within what is called an *ecoindustrial park,* or *industrial ecosystem.* They exchange waste outputs and convert them into resources, as shown in Figure 16.21. This cuts pollution and waste and reduces the flow of nonrenewable mineral and energy resources through the local economy.

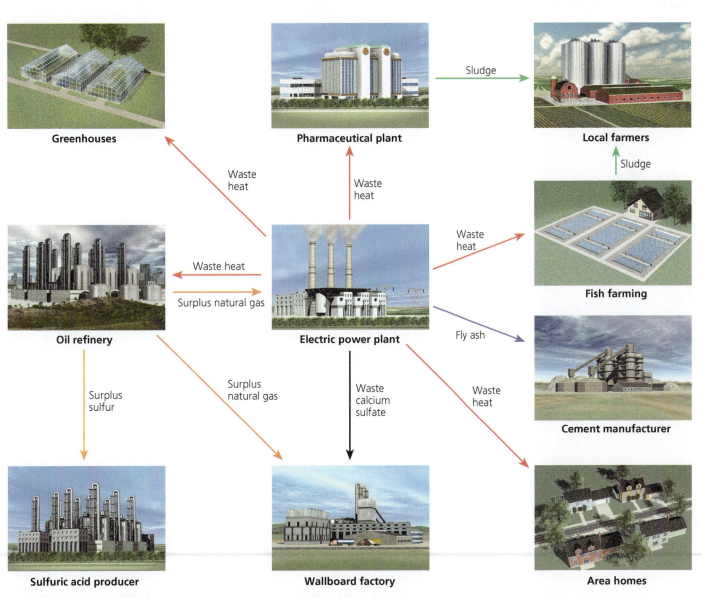

FIGURE 16.21 Solutions: This *industrial ecosystem* in Kalundborg, Denmark, reduces waste production by mimicking a natural ecosystem's food web. The wastes of one business become the raw materials for another, thus mimicking the way nature recycles chemicals. ***Critical thinking:*** Is there an industrial ecosystem near where you live or go to school? If not, think about where and how such a system could be set up.

Ecoindustrial parks provide many economic benefits for businesses. By encouraging recycling and waste reduction prevention, they cut the costs of managing solid wastes, controlling pollution, and complying with pollution regulations. They also reduce a company's chances of being sued because of damages, to people or to the environment, caused by their actions. In addition, companies improve the health and safety of workers by reducing their exposure to toxic and hazardous materials, thereby reducing company health insurance costs. Biomimicry also encourages companies to come up with new, environmentally beneficial, and less resource-intensive chemicals, processes, and products that they can sell worldwide. Today, more than 100 such parks operate in various places around the world, including the United States and China, and more are being built or planned.

- The order of priorities for dealing with solid waste should be first to minimize production of it, then to reuse and recycle as much of it as possible, and finally to safely burn or bury what is left.

- The order of priorities for dealing with hazardous waste should be first to minimize production of it, then to reuse or recycle it, to convert it to less-hazardous material, and to safely store what is left.

- We can view solid wastes as wasted resources, and hazardous wastes as materials that we want to avoid producing in the first place.

Tying It All Together

The Cradle-to-Cradle Approach and Sustainability

The cradle-to-cradle approach to design, manufacture, and use of materials is an important strategy for reducing the amount of solid and hazardous wastes we produce. By mimicking nature, this approach views all discarded materials and substances as nutrients that circulate within industrial and natural cycles. It also allows us the opportunity to convert the harmful environmental impacts of human activities to beneficial impacts. The challenge is to make the transition from an unsustainable high-waste, throwaway economy to a more sustainable low-waste, reducing–reusing–recycling economy as soon as possible.

Such a transition will require applying the six **principles of sustainability**. We can reduce our outputs of solid and hazardous waste by relying much less on fossil fuels and nuclear power (which produces long-lived, hazardous radioactive wastes) while relying much more on renewable energy from the sun, wind, and flowing water. We can mimic nature's chemical cycling processes by reusing and recycling materials as much as possible. Integrated waste management,

which uses a diversity of approaches and emphasizes waste reduction and pollution prevention, is a way to mimic nature's use of biodiversity and implement the cradle-to-cradle approach.

Including more of the harmful environmental and health costs of the consumer economy in market prices helps apply the full-cost pricing **principle of sustainability** while encouraging people to refuse, reduce, reuse, and recycle. Doing this benefits the environment, creates new jobs and businesses capitalizing on the four Rs, and provides health and environmental benefits for us, thus finding win-win solutions. This could also lead to lower levels of resource use per person, and

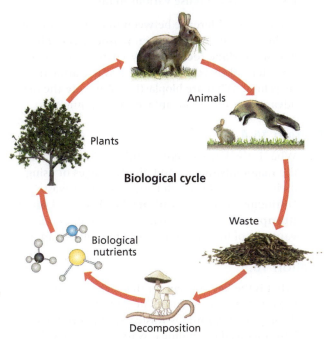

Biological cycle

Plants — Animals — Waste — Decomposition — Biological nutrients

thus lower levels of solid and hazardous waste production. All these measures together would help us pass along to future generations an environment that is as livable or more so than the one we have enjoyed.

Chapter Review

Core Case Study

1. Explain the concept of cradle-to-cradle design. Why is it a true form of biomimicry? List and briefly describe two strategies for employing it.

Section 16.1

2. What are the two key concepts for this section? Distinguish among **solid waste, industrial solid waste, municipal solid waste (MSW)**, and **hazardous (toxic) waste**, and give an example of each. Summarize the types and sources of solid waste generated in the United States and explain what happens to it. Why is e-waste a growing problem? What is the Great Pacific Garbage Patch and how did it come to be? How does it harm marine life and how can the growth of such patches be prevented?

Section 16.2

3. What is the key concept for this section? Define and distinguish among **waste management, waste reduction**, and **integrated waste management**. Summarize the priorities that prominent scientists suggest we should use for dealing with solid waste and compare them to actual practices in the United States. Distinguish among **refusing, reducing, reusing, recycling, and composting** in dealing with the solid wastes we produce. Why are refusing, reducing, and reusing the preferred actions from an environmental standpoint? List six ways in which industries and communities can reduce resource use, waste, and pollution.

Section 16.3

4. What is the key concept for this section? Explain why refusing, reducing, reusing, and recycling are so important and give examples of each. How has cradle-to-cradle design elevated reuse to a new level? List five ways to reuse various items.

5. What is the difference between upcycling and down-cycling? Distinguish between **primary recycling** and **secondary recycling**. What are three important steps of recycling? Why is e-waste attractive for recycling? What are bioplastics? What are the major advantages and disadvantages of recycling?

Section 16.4

6. What is the key concept for this section? What are the major advantages and disadvantages of using incinerators to burn solid and hazardous waste? Distinguish between **sanitary landfills** and **open dumps**. What are the major advantages and disadvantages of burying solid waste in sanitary landfills?

Section 16.5

7. What is the key concept for this section? What are the priorities that scientists suggest we should use in dealing with hazardous waste? Summarize the problems involved in sending e-wastes to less-developed countries for recycling. Describe three ways to detoxify hazardous wastes. What is bioremediation? What is phytoremediation? What are the major advantages and disadvantages of incinerating hazardous wastes?

8. What are the major advantages and disadvantages of storing liquid hazardous wastes in deep underground wells and in surface impoundments? What is a secure hazardous waste landfill? List four ways to reduce your output of hazardous waste. Summarize the regulation of hazardous wastes in the United States.

Section 16.6

9. What is the key concept for this section? How has grass-roots action led to improved solid and hazardous waste management in the United States? What are three factors that discourage recycling? What are three ways to encourage recycling and reuse? Give three examples of how people are saving or making money through reuse, recycling, and composting. Describe regulation of hazardous wastes at the global level through the Basel Convention and the treaty to control persistent organic pollutants (POPs). How is biomimicry being applied to the waste problem? What is an industrial ecosystem and what are its benefits?

10. What are this chapter's three big ideas? Explain how cradle-to-cradle design and manufacturing (**Core Case Study**) can help us apply the six **principles of sustainability**.

Note: Key terms are in bold type.

Critical Thinking

1. Find three products that you regularly use that could be made using cradle-to-cradle design and manufacturing (**Core Case Study**). For each of these products, sketch out a rough plan for how you would design and build it so that its parts could be reused many times or recycled in such a way that they would not harm the environment.

2. Do you think that manufacturers of computers, television sets, cell phones, and other electronic products should be required to take their products back at the end of their useful lives for repair, remanufacture, or recycling in a manner that is environmentally responsible and that does not threaten the health of recycling workers? Explain. Would you be willing to pay more for these products to cover the costs of such a take-back program? If so, what percentage more per purchase would you be willing to pay for these products?

3. Think of three items that you regularly use once and then throw away. Are there reusable items that you could use in place of these disposable items?

4. Do you think that you could consume less by refusing to buy some of the things you regularly buy? If so,

what are three of those things? Do you think that this is something you ought to do? Explain.

5. A company called Changing World Technologies has built a pilot plant to test a process it has developed for converting a mixture of discarded computers, old tires, turkey bones and feathers, and other wastes into oil by mimicking and speeding up natural processes for converting biomass into oil. Explain how this recycling process, if successful, could lead to increased waste production.

6. Would you oppose having **(a)** a sanitary landfill, **(b)** a hazardous waste surface impoundment, **(c)** a hazardous waste deep-injection well, or **(d)** a solid waste incinerator in your community? For each of these facilities, explain your answer. If you oppose having such facilities in your community, how do you think the solid and hazardous wastes generated in your community should be managed?

7. How does your school dispose of its solid and hazardous wastes? Does it have a recycling program? How well does it work? Does your school encourage

reuse? If so, how? Does it encourage waste prevention? If so, how? Does it have a hazardous waste collection system? If so, describe it. List three ways in which you would improve your school's waste reduction and management systems.

8. Congratulations! You are in charge of the world. List the three most important components of your strategy for dealing with **(a)** solid waste and **(b)** hazardous waste.

Doing Environmental Science

Collect the trash (excluding food waste) that you generate in a typical week. Measure its total weight. Sort it into major categories such as paper, plastic, metal, and glass. Then weigh each category and calculate its percentage by weight of the total amount of trash that you have measured. What percentage by weight of this waste consists of materials that could be recycled? What percentage consists of materials for which you could have used a reusable substitute, such as a coffee mug instead of a disposable cup? What percentage by weight of the items could you have done without? Compare your answers to these questions with those of your classmates. Together with your classmates, combine all the results and do the same analysis for the entire class. Use these results to estimate the same values for the entire student population at your school.

Global Environment Watch Exercise

Go to your Mind Tap course and access the GREENR database. Starting on the homepage, under "Browse Issues and Topics", click on *Pollution*, then select *E-Waste*. Use this portal to research and find statistics on how rapidly the world's production of e-waste is growing and how rapidly e-waste production is growing in the United States. Write a brief report on what the United States and one other country of your choice are doing to deal with this growing waste problem. Include statistics on how much e-waste is generated in each country, on how much of it is recycled, and on how much of it goes to landfills. Compare the two approaches in terms of how successful they are.

Ecological Footprint Analysis

Researchers estimate that the average daily municipal solid waste production per person in the United States is 2 kilograms (4.4 pounds). Use the data in the pie chart below (showing the latest available date from the EPA) to get an idea of a typical annual MSW ecological footprint for each American by calculating the total weight in kilograms (and pounds) for each category generated during 1 year (1 kilogram = 2.20 pounds). Use the table (below, right) to enter your answers.

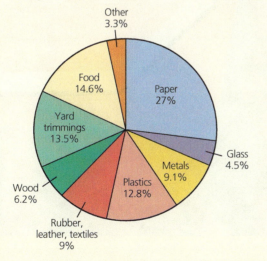

Composition of a typical sample of U.S. municipal solid waste, 2013 (latest available data).

(Compiled by the authors using data from the U.S. Environmental Protection Agency)

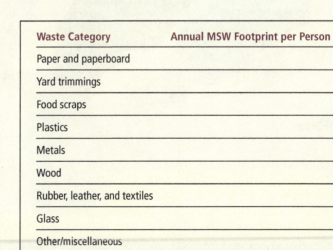

Waste Category	Annual MSW Footprint per Person
Paper and paperboard	
Yard trimmings	
Food scraps	
Plastics	
Metals	
Wood	
Rubber, leather, and textiles	
Glass	
Other/miscellaneous	

CENGAGE brain.com To access course materials, including Aplia homework, please visit www.cengagebrain.com

CHAPTER 17

Environmental Economics,
Politics, and Worldviews

When it is asked how much it will cost to protect the environment, one more question should be asked: How much will it cost our civilization if we do not?
GAYLORD NELSON

Key Questions

17.1 How are economic systems related to the biosphere?

17.2 How can we use economic tools to deal with environmental problems?

17.3 How can we implement more sustainable and just environmental policies?

17.4 What are some major environmental worldviews?

17.5 How can we live more sustainably?

Marchers in New York City protesting natural gas fracking.
Richard Levine/Alamy Stock Photo

The Greening of American Campuses

Since the mid-1980s, there has been a boom in environmental awareness on college campuses and in public and private schools around the world. In the United States, hundreds of colleges and universities have taken the lead in a quest to become more sustainable and to educate their students about **sustainability**—the capacity of the earth's natural systems and human cultural systems to survive, flourish, and adapt to changing environmental conditions into the very long-term future.

For example, at Oberlin College in Ohio, a group of students worked with faculty members and architects to design a more sustainable environmental studies building (Figure 17.1) powered by solar panels, which produce 30% more electricity than the building uses. Closed-loop underground geothermal wells provide heating and cooling. In its solar greenhouse, a series of open tanks populated by plants and other organisms purifies the building's wastewater. The building collects rainwater for irrigating the surrounding grasses, gardens, and meadow, which contain a diversity of plant and animal species.

In 2016, the Sierra Club rated Maine's College of the Atlantic as the country's greenest university. More than 35% of its classes are related to environmental issues and 75% of its faculty members carry out sustainability research. The school gets all of its electricity from renewable resources (93% from wind and 7% from solar). Since 2013, the college's endowment has had no investments in fossil fuels.

At the University of Washington in Seattle, more than half of the food served on campus comes from the campus farm and other small local producers. All eggs served are organic from cage-free hens. This saves the school money and cuts it energy use and greenhouse gas emissions.

The University of California, San Diego (UCSD), uses only drought-tolerant native plants for all of its new landscaping, which saves the campus a great deal of water that has historically been used to water grass in this drought-stricken area of the country. More than a third of UCSD's vehicle fleet is all-electric and the school runs 55 of its vehicles on biofuel.

The University of Wisconsin–Oshkosh uses a biodigester to convert manure from nearby farms to fuel that supplies 20% of the energy used for heating the campus buildings. The school also gets 20% of its electricity from wind power.

The University of Connecticut offers almost 600 sustainability classes and 40% of its faculty members carry out research related to environmental sustainability. Students run an organic farm that provides some of the food for its dining halls.

In addition to making their campuses greener, colleges are increasingly offering environmental sustainability courses and programs. At Pfeiffer University, many students have accompanied Professor Luke Dollar, a National Geographic Explorer, on trips to Madagascar to take part in his research on that country's endangered species and ecosystems.

These are just a few examples of the hundreds of institutions educating students who will provide leadership in working to make our societies and economies more sustainable during the next few decades. This chapter is about the economic, political, and ethical aspects of environmental problems and solutions. ●

Robb Williamson/NREL

FIGURE 17.1 The Adam Joseph Lewis Center for Environmental Studies at Oberlin College in Oberlin, Ohio.

17.1 HOW ARE ECONOMIC SYSTEMS RELATED TO THE BIOSPHERE?

CONCEPT 17.1 Ecological economists and most sustainability experts regard human economic systems as subsystems of the biosphere.

Economic Systems Depend on Natural Capital

Economics is the social science that deals with the production, distribution, and consumption of goods and services to satisfy people's needs and wants. In a market-based economic system, buyers and sellers interact to make economic decisions about how goods and services are produced, distributed, and consumed. In a true *free-market* economic system, all economic decisions are governed solely by the competitive interactions of *supply* and demand (Figure 17.2). *Supply* is the amount of a good or service that producers offer for sale at a given price. *Demand* is the amount of a good or service that people are willing and able to buy at a given price. If the demand for a good or service is greater than the supply, the price rises. If the supply is greater than demand, the price falls.

Most economic systems use three types of *capital,* or resources, to produce goods and services (Figure 17.3). **Natural capital** (see Figure 1.3, p. 7) includes resources and ecosystem services produced by the earth's natural processes that support all life and all economies. **Human capital** includes the physical and mental talents of the people who provide labor, organizational and management skills, and innovation. **Manufactured capital** includes the machinery, materials, and factories that people create using natural resources.

Models of Economies

Economic growth is an increase in the capacity of a nation, state, city, or company to provide goods and services to people. Today, a typical industrialized country depends on a **high-throughput economy**, which attempts to boost economic growth by increasing the flow of natural matter and energy resources through the economic system to produce more goods and services (Figure 17.4). Such an economy produces valuable goods and services. However, it also converts large quantities of high-quality matter and energy resources into waste, pollution, and low-quality heat, which tend to flow into planetary sinks (air, water, soil, and organisms).

Economic development focuses on creating economies that serve to improve human well-being by meeting basic human needs for items such as food, shelter, physical and economic security, and good health. The world's countries vary greatly in their levels of economic growth and development.

For more than 200 years, economists have debated whether there are limits to economic growth. *Neoclassical economists* assume that the potential for economic growth

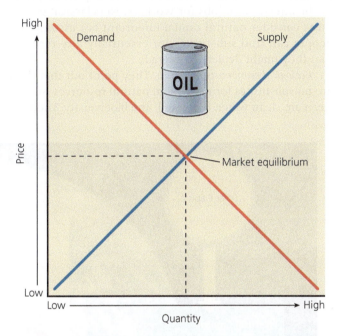

FIGURE 17.2 Supply and demand curves for a saleable product in a free market economic system. If all factors except supply, demand, and price are held fixed, *market equilibrium* occurs at the point where the supply and demand curves intersect. *Data analysis:* How would an increase in the available supply of oil shift the market equilibrium point on this diagram?

| Natural Capital | Manufactured Capital | Human Capital | Goods and Services |

FIGURE 17.3 Three types of capital are used to produce goods and services in most economic systems.

Center: Elena Elisseeva/Shutterstock.com. Right center: Michael Shake/Shutterstock.com. Right: iStock.com/Yuri.

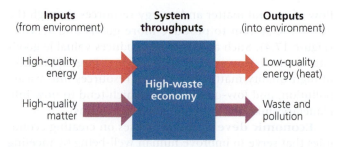

Inputs
(from environment)

System throughputs

Outputs
(into environment)

High-quality energy

High-quality matter

High-waste economy

Low-quality energy (heat)

Waste and pollution

FIGURE 17.4 The *high-throughput economies* of most of the world's more-developed countries rely on a continually increasing the flow of energy and matter resources to promote economic growth. **Critical thinking:** What are three ways in which you regularly add to this throughput of matter and energy through your daily activities?

is essentially unlimited and is necessary to provide profits for businesses and jobs for workers. Neoclassical economists consider natural capital important but assume that people can find substitutes for essentially any resource that they might deplete or degrade.

Ecological economists disagree. They point out that there are no substitutes for many vital natural resources, such as clean air, clean water, fertile soil, and biodiversity. They also see no substitutes for crucial ecosystem services such as climate control, air and water purification, pollination, topsoil renewal, and nutrient cycling. In contrast to neoclassical economists, they view human economic systems as subsystems of the biosphere that depend heavily on the natural resources and ecosystem services that make up earth's irreplaceable natural capital (Figure 17.5). (**Concept 17.1**)

According to ecological economists, economic growth becomes unsustainable when it depletes or degrades various irreplaceable forms of natural capital, on which all human economic systems depend. Research indicates that human actions have probably exceeded four major planetary boundaries or ecological tipping points (see Science Focus 3.3, p. 63). According to some estimates, humanity is currently using the renewable resources of 1.5 planet Earths and could be using that of 2 planet Earths by 2030. This will not be sustainable for very long beyond that time. In other words, we are borrowing renewable resources from future generations by using more than we have been allotted.

2 Number of Earths that could be needed to sustain the world's projected population and total renewable resource use in 2030.

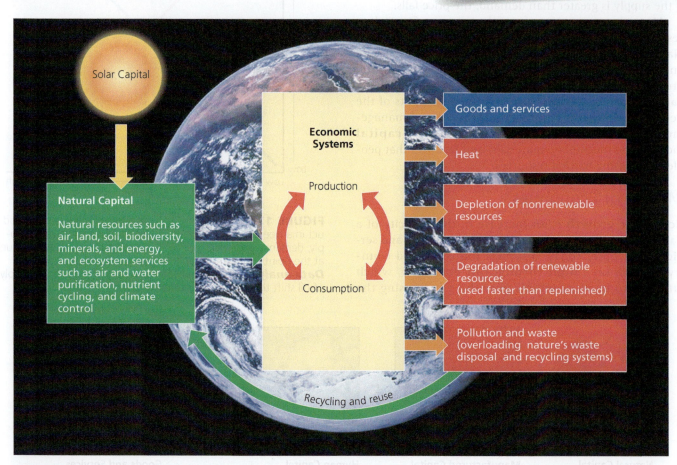

FIGURE 17.5 *Ecological economists* see all human economies as subsystems of the biosphere that depend on natural resources and services provided by the sun and earth. **Critical thinking:** Do you agree or disagree with this model? Explain.

Photo: NASA.

Ecological economist Robert Costanza and his colleagues estimated the value of 17 ecosystem services provided by the earth's major biomes to be at least $125 trillion per year—1.7 times more than the $75.2 billion that entire world spent on goods and services in 2016. Figure 23.6 shows conservative estimates of the monetary values of some of the world's ecosystem services.

Most ecological and environmental economists call for **environmentally sustainable economic development**. It uses political and economic systems to encourage environmentally beneficial and more sustainable forms of economic growth and to discourage environmentally harmful forms of economic growth that degrade natural capital.

CONSIDER THIS ...

THINKING ABOUT Economic Growth

Do you think that the economy of the country where you live is sustainable or unsustainable? Explain.

Estimating the Future Value of a Resource

One tool used by economists, businesses, and investors to determine the value of a resource is the *discount rate*—an estimate of a resource's future economic value compared to its present value. It is based on the idea that today's value of a resource may be higher than its value in the future. Thus, its future value should be discounted. The size of the discount rate (usually given as a percentage) is a key factor affecting how a resource such as a forest is used or managed.

At a zero discount rate, for example, the timber from a stand of redwood trees worth $1 million today will still be worth $1 million 50 years from now. However, the U.S. Office of Management and Budget, the World Bank, and most businesses typically use a 10% annual discount rate to estimate the future value of a resource. At this rate as the years go by, the timber in a stand of redwood trees will be worth increasingly less, and within 45 years, it will be worth less than $10,000. Using this discount rate, it makes sense from an economic standpoint for the owner of this resource to cut these trees down as quickly as possible.

However, this economic analysis does not take into account the immense economic value of the ecosystem services provided by forests (see Figure 9.2, left, p. 195, and Figure 17.6). Such services include the absorption of precipitation and gradual release of water and other nutrients, natural flood control, water and air purification, prevention of soil erosion, removal and storage of excess atmospheric carbon dioxide, and protection of biodiversity within a variety of forest habitats.

A high discount rate makes it difficult to sustain these important ecological and economic ecosystem services. If these economic values were included, it would make more sense now and in the future to preserve large areas of redwoods for the ecosystem services they provide and to find substitutes for redwood products. However, while these ecosystem services are vital for the earth as a whole and for future generations, they do not provide the current owner of the redwoods with any monetary return.

Setting discount rates is difficult and controversial. Proponents cite several reasons for using high (5–10%) discount rates. One argument is that inflation can reduce the value of future earnings on a resource. Another is that innovation or changes in consumer preferences can make a product or resource obsolete. For example, the plastic composites made to look like redwood may reduce the future use and market value of this timber.

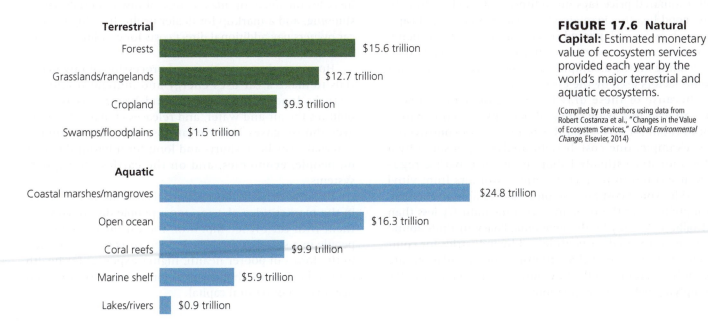

FIGURE 17.6 Natural Capital: Estimated monetary value of ecosystem services provided each year by the world's major terrestrial and aquatic ecosystems.

(Compiled by the authors using data from Robert Costanza et al., "Changes in the Value of Ecosystem Services," *Global Environmental Change*, Elsevier, 2014)

Terrestrial

- Forests — $15.6 trillion
- Grasslands/rangelands — $12.7 trillion
- Cropland — $9.3 trillion
- Swamps/floodplains — $1.5 trillion

Aquatic

- Coastal marshes/mangroves — $24.8 trillion
- Open ocean — $16.3 trillion
- Coral reefs — $9.9 trillion
- Marine shelf — $5.9 trillion
- Lakes/rivers — $0.9 trillion

Critics point out that high discount rates encourage rapid exploitation of resources for immediate payoffs, thus making long-term sustainable use of most renewable natural resources virtually impossible. They argue that a 0% or even a negative discount rate should be used to protect unique, scarce, and irreplaceable resources such as old-growth forests. A negative discount rate would result in the economic value of a forest or other resource *increasing* over time. Some economists argue that as ecosystem services continue to be degraded, they will become more valuable, so a negative discount rate is the only type that makes sense. They point out that zero or negative discount rates of −1 to −3% would make it profitable to use nonrenewable and renewable resources more slowly and in more sustainable ways.

CONSIDER THIS ...

THINKING ABOUT Discount Rates

If you owned a forested area, would you want the discount rate for resources such as trees from the forest to be high, moderate, zero, or negative? Explain.

Cost–Benefit Analysis

Another widely used tool for making economic decisions about how to control pollution and manage resources is *cost–benefit analysis*. In this process, analysts compare estimated costs and benefits of actions such as implementing a pollution control regulation, building a dam on a river, and preserving an area of forest.

Making a cost–benefit analysis involves determining who benefits and who is harmed by a particular regulation or project and estimating the current and future monetary values (costs) of those benefits and harms. Direct costs involving land, labor, materials, and pollution-control technologies are often easy to estimate. Analysts can put estimated price tags on indirect costs such as human life, good health, clean air, clean water, and natural capital such as an endangered species, a forest, or a wetland. However, such monetary value estimates vary widely depending on the assumptions, value judgments, and discount factors used by the estimators.

Because of these drawbacks, a cost–benefit analysis can lead to a wide range of benefits and costs with a lot of room for error, and this is a source of controversy. For example, one cost–benefit analysis sponsored by a U.S. industry estimated that compliance with a regulation written to protect American workers from vinyl chloride would cost $65 billion to $90 billion. In the end, complying with the regulation cost the industry less than $1 billion. A study by the Economic Policy Institute found that the estimated costs projected by industries for complying with proposed U.S. environmental regulations are often inflated in an effort by industries to avoid or delay complying with such regulations.

If conducted fairly and accurately, cost–benefit analysis can be a helpful tool for making economic decisions, but it always includes uncertainties. Environmental economists advocate using the following guidelines to minimize possible abuses and errors in any cost–benefit analysis involving some part of the environment:

- State all assumptions used.
- Include estimates of the ecosystem services provided by the ecosystems involved.
- Estimate short- and long-term benefits and costs for all affected population groups.
- Compare the costs and benefits of alternative courses of action.

17.2 HOW CAN WE USE ECONOMIC TOOLS TO DEAL WITH ENVIRONMENTAL PROBLEMS?

CONCEPT 17.2 We can use resources more sustainably by including the harmful environmental and health costs of producing goods and services in their market prices *(full-cost pricing)*, by subsidizing environmentally beneficial goods and services, and by taxing pollution and waste instead of wages and profits.

Full-Cost Pricing

The *market price,* or *direct price,* that people pay for products or services usually does not include all of the *indirect,* or *external, costs* of harm to the environment and human health associated with providing and using them. Such costs are often called *hidden costs.*

For example, when someone buys a new car, the price includes the *direct,* or *internal, costs* of raw materials, labor, shipping, and a markup for dealer profit. In using the car, car owners pay additional direct costs for gasoline, maintenance, repairs, and insurance.

However, the extraction and processing of raw materials to make a car uses energy and mineral resources, disturbs land, produces solid and hazardous wastes, pollutes the air and water, and releases climate-changing greenhouse gases into the atmosphere. These hidden external costs have short- and long-term harmful effects on people, economies, and on the earth's life-support systems.

Because these harmful external costs are not included in the market price of a car, most people do not connect them with car ownership. Still, the car buyer and other people in a society pay these hidden costs sooner or later, in the forms of poorer health, higher expenses for health care and insurance, higher taxes for pollution control, and degradation of natural capital.

FIGURE 17.7 Most of the harmful environmental effects of strip-mining coal and burning it to produce electricity are not included in the cost of electricity.

Andreas Reinhold/Shutterstock.com

Ecological economists and environmental experts call for including estimated costs of harm to the environment and human health in the market prices of goods and services. This practice is called **full-cost pricing**, and is one of the six **principles of sustainability**. Failure to include the harmful health and environmental costs in the market prices of goods and services is viewed one of the major causes of the environmental problems we face (see Chapter 1, pp. 16–17).

According to its proponents, full-cost pricing would reduce resource waste, pollution, and environmental degradation and improve human health. It would also encourage producers to invent more resource-efficient and less-polluting methods of production, and it would inform consumers about the environmental and health effects of goods and services they buy. For example, if the harmful environmental and health costs of mining and burning coal to produce electricity (Figure 17.7) were included in the market prices of coal-fired electricity, coal would be much more expensive and likely would be replaced by improved energy efficiency and less environmentally harmful resources such as natural gas and solar and wind power.

Putting full-cost pricing into practice would result in some industries and businesses disappearing or remaking themselves. New businesses would also appear. This is a normal and revitalizing process in a dynamic and creative capitalist economy. Shifting to full-cost pricing over a decade or two would give some environmentally harmful businesses enough time to transform themselves into profitable, environmentally beneficial businesses.

There are three reasons why full-cost pricing is not used more widely. *First*, most producers of harmful products and services would have to charge more for them, and some would go out of business. Naturally, these producers oppose such pricing. *Second,* many environmental and health costs are difficult to estimate. *Third,* many environmentally harmful businesses use their political and economic power to obtain government subsidies and tax breaks that help them increase their profits and in some cases stay in business

Shifting to Environmentally Beneficial Subsidies

Some subsides, called *perverse subsidies*, lead to environmental damage and harmful health effects. Examples include depletion subsidies and tax breaks for extracting minerals and fossil fuels, cutting timber on public lands, and irrigating with low-cost water.

According to environmental scientist Norman Myers, such *perverse subsidies and tax breaks* cost the world's governments (taxpayers) at least $2 trillion a year—an average $3.8 million a minute! Myers also estimates that perverse government subsidies and tax breaks cost the average American taxpayer $2,000 per year.

Environmental scientists and ecological economists call for phasing out environmentally harmful subsidies and tax breaks and phasing in environmentally beneficial subsidies and tax breaks. More subsidies and tax breaks would go businesses involved in pollution prevention, waste prevention, sustainable forestry and agriculture, conservation of water supplies, energy-efficiency improvements, renewable energy use, and measures to slow projected climate change.

Making such a *subsidy shift* over two to three decades would not be easy because the powerful interests that receive harmful environmental subsidies and tax breaks spend a lot of time and money *lobbying,* or trying to influence governments to continue and even increase their subsidies. They also lobby governments to deny subsidies to their more environmentally beneficial competitors such as solar and wind energy. Countries including Japan, France, and Belgium have started making some small subsidy shifts.

CONSIDER THIS ...

THINKING ABOUT Subsidies

Do you favor phasing out environmentally harmful government subsidies and tax breaks, and phasing in environmentally beneficial ones over the next two to three decades? Explain. What are three things you could do to help bring this about? How might such subsidy shifting affect your lifestyle?

Taxing Pollution and Wastes Instead of Wages and Profits

To many analysts, the tax systems in most countries are backward. They *discourage* what people want more of— jobs, income, and profit-driven innovation—and *encourage* what people want less of—pollution, resource waste, and environmental degradation. A more environmentally sustainable economic and political system would *lower* taxes on labor, income, and wealth, and *raise* taxes on environmental activities that produce pollution, wastes, and environmental degradation. Some 2,500 economists, including eight Nobel Prize winners in economics, have endorsed this *tax-shifting* concept.

Proponents list three requirements for the successful shift to more environmentally sustainable, or green, taxes:

- Phase in green taxes over 10 to 20 years to allow business to plan for change.

- Reduce income, payroll, or other taxes by an amount equal to that of the green taxes so that there would be no net increase in taxes.

- Design a safety net for the poor and lower-middle class individuals who would suffer financially from any new taxes on essentials such as fuel, water, electricity, and food.

Figure 17.8 lists some of the advantages and disadvantages of using green taxes.

In Europe and the United States, polls indicate that when tax shifting is explained to voters, 70% of them support the idea. Costa Rica, Sweden, Germany, Denmark, Spain, and the Netherlands have raised taxes on several environmentally harmful activities while cutting taxes on income, wages, or both. **GOOD NEWS**

The U.S. Congress has not enacted green taxes, mostly because of opposition by the automobile, fossil

Trade-Offs

Environmental Taxes and Fees

Advantages	Disadvantages
Help bring about full-cost pricing	Low-income groups are penalized unless safety nets are provided
Encourage businesses to develop environmentally beneficial technologies and goods	Hard to determine optimal level for taxes and fees
Easily administered by existing tax agencies	If set too low, wealthy polluters can absorb taxes as costs

FIGURE 17.8 Using green taxes to help reduce pollution and resource waste has advantages and disadvantages. **Critical thinking:** Which single advantage and which single disadvantage do you think are the most important? Why?

Top: Chuong Vu/Shutterstock.com. Bottom: EduardSV/Shutterstock.com.

fuel, mining, chemical, and other politically powerful industries. These opponents claim that green taxes will harm the economy and consumers by forcing producers to raise the prices of their goods and services. In addition, most voters have been conditioned to oppose any new taxes and have not been educated about the economic and environmental benefits of a tax-shifting approach that would improve environmental quality with no net increase in their taxes.

Environmental Indicators

Economic growth is usually measured by the percentage of change per year in a country's **gross domestic product (GDP)**: the annual market value of all goods and services produced by all firms and organizations, foreign and domestic, operating within a country. A country's economic growth per person is measured by changes in the **per capita GDP**: the GDP divided by the country's total population at midyear.

GDP and per capita GDP indicators provide a standardized, useful method for measuring and comparing the economic outputs of nations. However, the GDP was deliberately designed to measure such outputs without taking into account their beneficial or harmful environmental and health impacts. Many ecological economists and environmental scientists call for the development and widespread use of new indicators—called *environmental indicators*—to help monitor environmental quality and human well-being.

One such indicator is the *genuine progress indicator (GPI)*—the GDP plus the estimated value of beneficial transactions that meet basic needs, minus the estimated harmful environmental, health, and social costs of all transactions. Examples of beneficial transactions included in the GPI are unpaid volunteer work, health care provided by family members, childcare, and housework. Harmful costs that are subtracted to arrive at the GPI include the costs of pollution, resource depletion and degradation, and crime.

Another environmental indicator is Global Green Economy Index (GGEI). It measures the performances of 60 nations in areas of leadership on climate change, energy efficiency, markets and investments, and natural capital, based on analysis by a panel of experts. In 2016, the top-five-ranked countries on the GGEI were Sweden, Norway, Finland, Switzerland, and Germany. The United States ranked 30th.

These and other environmental indicators being developed are far from perfect. However, without such indicators, it will be difficult to monitor the overall effects of human activities on human health, on the environment, and on the planet's natural capital and to evaluate the effectiveness of solutions to the environmental problems humanity faces.

Using the Marketplace to Reduce Pollution and Resource Waste

Governments can act to reduce pollution and waste. In one incentive-based regulation system, the government decides on acceptable levels of total pollution or resource use; sets limits, or *caps*, to maintain these levels; and gives or sells companies a certain number of *tradable pollution* or *resource-use permits* governed by the caps.

With this *cap-and-trade* approach, a permit holder that does not use its entire allocation can save credits for future expansion, use them in other parts of its operation, or sell them to other companies. The United States has used this approach to reduce the emissions of sulfur dioxide (see Chapter 15, p. 418) and several other air pollutants. Tradable rights could also be established among countries to help preserve biodiversity and reduce emissions of greenhouse gases and other regional and global pollutants.

Figure 17.9 lists the advantages and disadvantages of using tradable pollution and resource-use permits. The effectiveness of such programs depends on how high or low the initial cap is set and on the rate at which the cap is regularly reduced to encourage further innovation.

Selling Services Instead of Products

One approach to working toward more environmentally beneficial economies is to sell certain services in place of certain products that provide those services. With this approach, a manufacturer or service provider makes more money if the production of its product involves minimal material use and pollution and if the product lasts, is energy

Trade-Offs

Tradable Environmental Permits

Advantages	Disadvantages
Flexible and easy to administer	Wealthy polluters and resource users can buy their way out
Encourage pollution prevention and waste reduction	Caps can be too high and not regularly reduced to promote progress
Permit prices to be determined by market transactions	Self-monitoring of emissions can allow cheating

FIGURE 17.9 Using tradable pollution and resource-use permits to reduce pollution and resource waste has advantages and disadvantages. ***Critical thinking:*** Which two advantages and which two disadvantages do you think are the most important? Why?

M. Shcherbyna/Shutterstock.com

efficient, produces as little pollution as possible while in use, and is easy to maintain, repair, reuse, or recycle.

Such an economic shift is under way in some businesses. Since 1992, Xerox has been leasing most of its copy machines as part of its mission to provide *document services* instead of selling photocopiers. When a customer's service contract expires, Xerox takes the machine back for reuse or remanufacture. It has a goal of sending no material to landfills or incinerators. To save money, Xerox designs machines to have the fewest possible parts, be energy efficient, and emit as little noise, heat, ozone, and chemical waste as possible.

CONSIDER THIS ...

Learning from Nature

At the flooring service company Interface, engineers studied the floors of tropical forests to design a best-selling, nature-based carpet pattern that reduces carpet waste and installation time. It became a best seller.

In Europe, Carrier has been shifting from selling heating and air conditioning equipment to providing indoor heating and cooling services. The company makes higher profits by leasing and installing energy-efficient equipment that is durable and easy to rebuild or recycle. Carrier also makes money through helping clients save energy by adding insulation, eliminating heat losses, and boosting energy efficiency in their offices and homes.

Reducing Poverty Helps the Environment and Human Health

Poverty occurs when people do not have enough money to meet their basic needs for food, water, shelter, health care, and education. According to the World Bank, poverty is the way of life for 2.5 billion, or nearly one of every three of the world's people. They live on incomes of less than $3.10 per day. One fifth of the world's people live in extreme poverty (Figure 17.10), struggling to survive on incomes of less than $1.90 a day.

Some analysts are alarmed at the widening gap between rich and poor countries and between superrich individuals and the rest of the world. According to a 2016 study by Oxfam, the world's richest eight people had wealth equal to that of the poorest half of the world's population. Some economists say that part of this wealth will trickle down to the poor and middle class. Others point out that for almost three decades, instead of trickling down, most wealth has been flowing up at an increasing rate to rich individuals, corporations, and countries.

Poverty causes a number of harmful health effects such as hunger, malnutrition, infectious disease, and a shorter life span (see Figure 14.16, right, p. 396). Poverty has also been identified as one of the five major causes of the environmental problems we face (p. 16). To reduce poverty and its harmful effects, governments, businesses,

JAMES P. BLAIR/National Geographic Creative

FIGURE 17.10 This 3-year-old girl was sleeping in her family's shack in a slum in Port-Au-Prince, Haiti.

international lending agencies, and wealthy individuals could undertake the following:

- Mount a massive global effort to combat malnutrition and the infectious diseases that annually kill millions of people.

- Provide universal primary school education for all children and for the world's nearly 800 million illiterate adults. Illiteracy can foster terrorism and strife within countries by contributing to the creation of large numbers of unemployed individuals who have little hope of improving their lives or those of their children.

- Help less-developed countries reduce their population growth, mostly by elevating the social and economic status of women, reducing poverty, and providing access to family planning.

- Focus on sharply reducing the high total and per capita ecological footprints of countries such as the United States and China.

- Make large investments in small-scale infrastructure such as solar-cell power facilities for rural villages and sustainable agriculture projects to help less-developed nations work toward more energy-efficient and environmentally beneficial economies.

- Encourage lending agencies to make small, low-interest loans to poor people who want to increase their incomes (see Case Study that follows).

Ecologist and National Geographic Explorer Sasha Kramer has been working in the impoverished and ecologically degraded nation of Haiti to attack the problems of hunger, soil depletion, and water pollution all at once. Her nonprofit organization has distributed waterless composting toilets throughout the country to collect human wastes and transform them into compost, which Haitian farmers can use to rebuild depleted soil and boost food production. This process also keeps human wastes out of Haiti's water supply and reduces the dangerous threat of waterborne infectious diseases.

GOOD NEWS

CASE STUDY

Microlending

Most of the world's able-bodied poor people want to work and earn enough to climb out of poverty and make a better life for themselves and their families. With small loans, they could buy what they need to start farms or small businesses. However, few of them have credit records or assets that they could use as collateral to secure the loans.

For over three decades, an innovation called *microlending,* or *microfinance,* has helped a number of people living in poverty to deal with this problem. In 1983, economist Muhammad Yunus started the Grameen Bank in Bangladesh, a country with a high poverty rate and a rapidly growing population. Unlike commercial banks, the Grameen Bank is essentially owned and run by borrowers and the Bangladeshi government. Since it was founded, the bank has provided more than $8 billion in microloans of $50 to $500 at low interest rates to 7.6 million impoverished people in Bangladesh who do not qualify for loans at traditional banks.

Most of these loans have been used by women to start small businesses, plant crops, buy small irrigation pumps, buy cows and chickens for producing and selling milk and eggs, and buy bicycles for transportation. Microloans are also used to develop day-care centers, health-care clinics, reforestation projects, drinking water supply projects, literacy programs, and small-scale solar- and wind-power systems in rural villages (Figure 17.11).

The Grameen Bank's average repayment rate on its microloans has been 95% or higher. That is nearly twice the average repayment rate for loans by conventional commercial banks—and the Grameen Bank has consistently made a profit. Typically, about half of Grameen's borrowers move above the poverty line within 5 years of receiving their loans.

GOOD NEWS

In 2006, Yunus and his colleagues at the bank jointly won the Nobel Peace Prize for their pioneering use of microcredit loans that change people's lives. He has stated "Unleashing the energy and creativity in each human being is the answer to poverty." Banks based on the Grameen microcredit model have spread to 58 countries (including the United States) with an estimated 500 million participants.

National Renewable Energy Laboratory

FIGURE 17.11 A microloan helped these women in a rural village in India to buy a small solar-cell panel (installed on the roof behind them) that provides electricity to help them make a living, thus applying the solar energy **principle of sustainability**.

Millennium Development Goals

In 2000, the world's nations set goals—called *Millennium Development Goals*—for sharply reducing hunger and poverty, improving health care, achieving universal primary education, empowering women, and moving toward environmental sustainability by 2015. That year, the United Nations published its Progress Chart showing mixed results in reaching the goals. Most countries did well in expanding primary education, while women's representation in national parliaments did not improve in most places. Many countries did well in bringing clean drinking water to all of their citizens, while some countries did very poorly.

More-developed countries pledged to donate 0.7%—or $7 of every $1,000—of their annual national income to less-developed countries to help them in achieving these goals. So far, Denmark, Luxembourg, Sweden, Norway, and the Netherlands have donated what they had promised. The United States—the world's richest country—gives only 0.16% of its national income and Japan, another wealthy country, gives only 0.18% compared with the 0.9% given by Sweden.

CONSIDER THIS ...

THINKING ABOUT The Millennium Development Goals

Do you think the country where you live should devote at least 0.7% of its annual national income toward achieving the Millennium Development Goals? Explain.

Shifting to More Environmentally Sustainable Economies

The three scientific laws governing matter and energy changes (see pp. 35 and 37–38) and the six **principles of sustainability** (see inside back cover) suggest that the best long-term solution to our environmental and resource problems is to shift away from high-throughput (high-waste) economies based on ever-increasing matter and energy flow (Figure 17.3) over the next several decades. The goal would be to develop more sustainable **low-throughput (low-waste) economies** based on improving energy efficiency and recycling and reusing matter (Figure 17.12). Figure 17.13 shows some components of this form of more environmentally sustainable economic development.

The drive to improve environmental quality and to work toward environmental sustainability has created new major growth industries along with profits and large numbers of new *green jobs* (Figure 17.14). According to the Ecotech Institute, there were 3.8 million green jobs in the United States in 2014, mostly in the rapidly growing wind and solar energy sector.

Making the shift to more sustainable economies will require governments and industries to greatly increase their spending on research and development—especially in the areas of energy efficiency and renewable energy—as Germany has done in recent years.

Ray Anderson (1934–2011), founder of the American company Interface, the world's largest commercial manufacturer of carpet tiles, led the way in making businesses more sustainable. In 1994, he announced plans to develop the nation's first truly sustainable corporation. Within 16 years, Interface had cut water usage by 74%, net greenhouse gas emissions by 32%, solid waste by 63%, fossil fuel use by 60%, and energy use by 44%. These efforts have saved Interface more than $433 million, and the company's profits tripled. Anderson also created a consulting group as part of Interface to help other businesses start on the path toward becoming more sustainable.

17.3 HOW CAN WE IMPLEMENT MORE SUSTAINABLE AND JUST ENVIRONMENTAL POLICIES?

CONCEPT 17.3 Individuals can work together to take part in political processes that influence how environmental policies are made and implemented.

Environmental Laws and Regulations

Governments play a key role in dealing with environmental problems. They do this by developing **environmental policy,** which consists of environmental laws, regulations, and programs that are designed, implemented, and enforced by one or more government agencies (see Case Study that follows). It involves enacting and enforcing laws that set pollution standards, regulate the release of toxic chemicals into the environment, and protect certain slowly renewed resources such as public forests, parks, and wilderness areas from unsustainable use. A typical *policy life cycle* consists of four stages: **(1)** problem recognition, **(2)** policy formation, **(3)** policy implementation, and **(4)** policy adjustment based on experience.

Most environmental regulation in the United States and in many other countries involves passing laws that

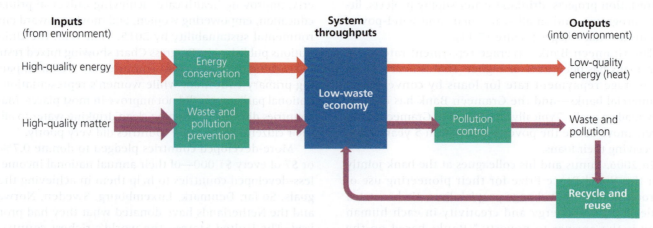

FIGURE 17.12 Solutions: Learning and applying lessons from nature can help us design and manage more sustainable low-throughput economies. ***Critical thinking:*** What are three ways in which your school could decrease any unsustainable economic and environmental practices, and three ways in which it could promote more sustainable economic and environmental practices?

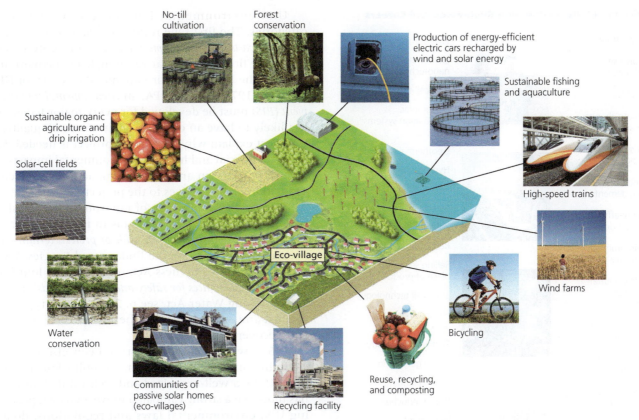

FIGURE 17.13 Solutions: Some of the components of more environmentally sustainable economic development. *Critical thinking:* What are three new types of jobs that could be generated by such an economy?

No-till cultivation

Forest conservation

Production of energy-efficient electric cars recharged by wind and solar energy

Sustainable fishing and aquaculture

Sustainable organic agriculture and drip irrigation

Solar-cell fields

High-speed trains

Eco-village

Wind farms

Water conservation

Bicycling

Communities of passive solar homes (eco-villages)

Recycling facility

Reuse, recycling, and composting

Photos going clockwise starting at "No-till cultivation": Jeff Vanuga/National Resource Conservation Service. Natalia Bratslavsky/Shutterstock.com. Pi-Lens/Shutterstock.com. Vladislav Gajic/Shutterstock.com. hxdbzxy/Shutterstock.com. Varina C/Shutterstock.com. Kalmatsuy/Shutterstock.com. Brenda Carson/Shutterstock.com. Alexander Chaikin/Shutterstock.com. National Renewable Energy Laboratory/U.S. Department of Energy. iStock.com/anhong. pedrosala/Shutterstock.com. Robert Kneschke/Shutterstock.com

are typically enforced through a *command-and-control* approach. Critics say that this strategy can unnecessarily increase costs and discourage innovation, because many of these government regulations concentrate on cleanup instead of prevention. Some regulations also set compliance deadlines that are too short to allow companies to find innovative ways to reduce pollution and waste.

A different approach favored by many economists and environmental and business leaders, is to use *incentive-based environmental regulations*. Rather than to require all companies in a particular market to follow the same fixed procedures or use the same technologies, governments can establish long-term goals and heavy penalties for not achieving the goals. This approach uses the economic forces of the marketplace to encourage businesses to be innovative in reducing pollution and resource waste.

In passing laws, developing budgets, and formulating regulations, elected and appointed government officials must deal with pressure from competing *special interest groups.* Each of these groups pushes for laws, subsidies or tax breaks, and regulations favorable to its cause. Such groups also seek to weaken or repeal laws, subsidies, tax breaks, and regulations that are unfavorable to their positions. Examples of special-interest groups include *profit-making organizations,* such as corporations; *nongovernmental organizations (NGOs),* most of which are nonprofit organizations such as environmental groups; *labor unions,* representing the interests of workers; and *trade associations,* representing various industries.

The design for stability and gradual change in democracies is highly desirable but several features of democratic governments hinder their ability to deal with environmental problems. Problems such as biodiversity loss and climate change are complex, invisible, and difficult to understand. They develop over decades, have long-lasting effects, and require integrated, long-term solutions. However, because local and national elections are held as often as every two years in most democracies, most politicians spend much of their time seeking reelection. They tend to focus on short-term, isolated issues rather than on long-term, complex problems.

Some analysts argue that environmental science should play a major role in the formulation of environmental policy. However, this is difficult when many political leaders have little or no understanding of how the earth's natural systems work and how those systems support all life, economies, and societies.

Environmentally Sustainable Businesses and Careers

Aquaculture	Environmental law
Biodiversity protection	Environmental nanotechnology
Biofuels	Fuel-cell technology
Climate change research	Geographic information systems (GIS)
Conservation biology	Geothermal geologist
Ecotourism management	Hydrogen energy
	Hydrologist
Energy-efficient product design	Marine science
	Pollution prevention
Environmental chemistry	Reuse and recycling
	Selling services in place of products
Environmental design and architecture	Solar-cell technology
	Sustainable agriculture
Environmental economics	Sustainable forestry
Environmental education	Urban gardening
	Urban planning
Environmental engineering	Waste reduction
Environmental entrepreneur	Watershed hydrologist
	Water conservation
Environmental health	Wind energy

FIGURE 17.14 *Green careers:* Some key environmental businesses and careers are expected to flourish during this century, while environmentally harmful, or sunset, businesses are expected to decline. See the website for this book for more information on various environmental careers. ***Critical thinking:*** How could some of these careers help you apply any of the **principles of sustainability**?

CASE STUDY

U.S. Environmental Laws

During the 1950s and 1960s, the United States experienced severe pollution and environmental degradation as its economy grew rapidly without pollution control laws and regulations. This changed in the late 1960s and in the 1970s when massive protests by citizens led the U.S. Congress to pass a number of major environmental laws (Figure 17.15). Most of them were enacted in the 1970s, known as the *decade of the environment* in the United States. Implementing these laws has provided millions of jobs and profits from many new technologies for reducing pollution and environmental degradation. This has also improved the health of U.S. citizens.

U.S. environmental laws generally fit into five categories. The first type *requires evaluation of the environmental impacts of certain human activities.* It is represented by one of the first and most far-reaching federal environmental laws, the National Environmental Policy Act, or NEPA, passed in 1970. Under NEPA, an *environmental impact statement (EIS)* must be developed for every major federal project likely to have an effect on environmental quality. The EIS must explain why a proposed project is needed, identify its beneficial and harmful environmental impacts, suggest ways to lessen any harmful impacts, and present an evaluation of alternatives to the project.

The second major type of environmental legislation *sets standards for pollution levels* (as in the Clean Air Acts, see p. 417). A third type *sets aside or protects certain species, resources, and ecosystems* (the Endangered Species Act, see p. 185, and the Wilderness Act, see p. 207). A fourth type *screens new substances for safety and sets standards* (as in the Safe Drinking Water Act, see p. 288). A fifth type *encourages resource conservation* (see the Resource Conservation and Recovery Act, see p. 464).

U.S. environmental laws have been effective, especially in controlling some forms of pollution. However, since 1980, a well-organized and well-funded movement has mounted a strong campaign to weaken or repeal existing U.S. environmental laws and regulations, do away with the EPA, and change the ways in which public lands (see Case Study that follows) are used.

Three major groups strongly opposed to environmental laws and regulations are:

- Corporate leaders and other powerful people who see laws and regulations as threats to their profits, wealth, and power.

- Citizens who view environmental laws as threatening to their private property rights and jobs.

- State and local government officials who resent having to implement state and federal laws and regulations with little or no federal funding, or who disagree with specific regulations.

Another problem working against additional environmental laws and regulations is that the focus of environmental issues has shifted away from easy-to-see dirty smokestacks and filthy rivers to complex, long-term, and less visible environmental problems. These include biodiversity loss, groundwater pollution, and climate change.

Since 2000, efforts to weaken U.S. environmental laws and regulations have escalated. However, independent polls show that more than 80% of the U.S. public strongly supports the country's environmental laws and regulations. But the polls also show that less than 10% of the U.S. public (and in hard economic times only about 2–3%) considers the environment to be one of the nation's most pressing problems. As a result, environmental concerns often are not transferred to the ballot box or to personal spending decisions.

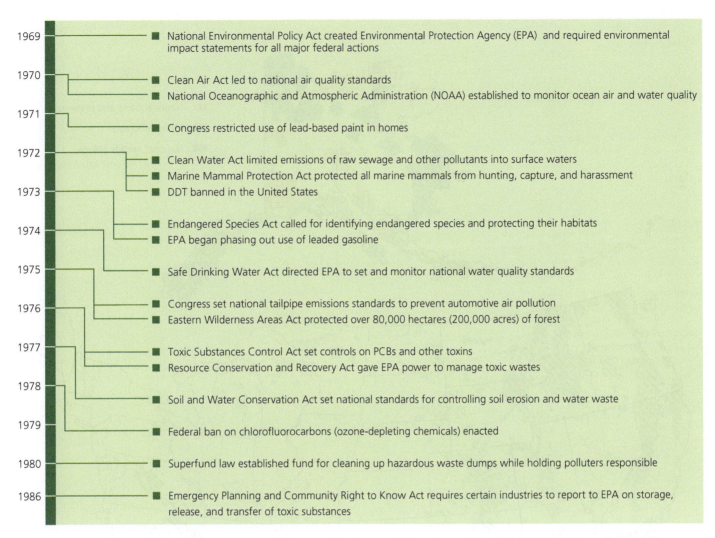

FIGURE 17.15 Some of the major environmental laws and their amended versions enacted in the United States from 1969 to 1986.

To make a transition to a more environmentally sustainable society, U.S. citizens (and citizens in other democratic countries) will have to elect and support ecologically literate and environmentally concerned leaders. A rapidly growing number of citizens are also insisting that elected leaders work across party lines to end the political deadlock that has virtually immobilized the U.S. Congress since 1980, with respect to environmental issues and other important societal concerns.

Managing Public Lands in the United States—Politics in Action

No nation has set aside as much of its land for public use, resource extraction, enjoyment, or wildlife habitat as has the United States. About 28% of the country's land is jointly owned by all U.S. citizens and managed for them by the federal government. About three-fourths of this federal public land is in Alaska and another fifth is in the western states (Figure 17.16).

Some federal public lands are used for many purposes. For example, the *National Forest System* consists of 155 national forests and 22 national grasslands. These lands, managed by the U.S. Forest Service (USFS), are used for logging, mining, livestock grazing, farming, oil and gas extraction, recreation, and conservation of watershed, soil, and wildlife resources.

The Bureau of Land Management (BLM) manages large areas of land—40% of all land managed by the federal government and 13% of the total U.S. land surface—mostly in the western states and Alaska. These lands are used primarily for mining, oil and gas extraction, and livestock grazing.

The U.S. Fish and Wildlife Service (USFWS) manages 562 *National Wildlife Refuges*. Most refuges protect habitats and breeding areas for waterfowl and big game to provide a harvestable supply for hunters. Permitted activities in most refuges include hunting, trapping, fishing, oil and

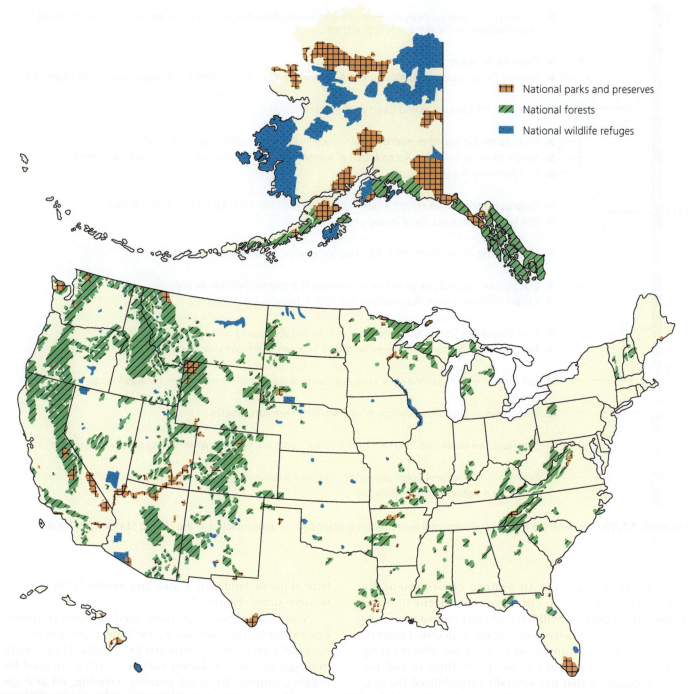

FIGURE 17.16 Natural Capital: National forests, parks, and wildlife refuges managed by the U.S. federal government.
Critical thinking: Do you think U.S. citizens should jointly own more or less of the nation's land? Explain.

(Compiled by the authors using data from U.S. Geological Survey and U.S. National Park Service)

Legend:
- National parks and preserves
- National forests
- National wildlife refuges

gas development, mining, logging, grazing, farming, and some military activities.

The uses of some other public lands are more restricted. The *National Park System,* managed by the National Park Service (NPS), includes 59 major parks (Figure 17.17) and 358 national recreation areas, monuments, memorials, battlefields, historic sites, parkways, trails, rivers, seashores, and lakeshores. Only camping, hiking, sport fishing, and boating can take place in the national parks, whereas sport hunting, mining, and oil and gas drilling are allowed in national recreation areas.

The most restricted public lands are 762 roadless areas that make up the *National Wilderness Preservation System.* These areas lie within the other public lands and are managed by the agencies in charge of those surrounding lands. Most of these areas are open only for recreational activities

FIGURE 17.17 Yosemite National Park, California, with El Capitan rising above the Merced River.

RICHARD NOWITZ/National Geographic Creative

such as hiking, sport fishing, camping, and non-motorized boating.

Many federal public lands contain valuable oil, natural gas, coal, geothermal, timber, and mineral resources. Since the 1800s, there has been intense controversy over how to use and manage these resources on the country's public lands.

Most conservation biologists and ecological economists believe that four principles should govern use of public lands:

1. Protect biodiversity, wildlife habitats, and ecosystems as the top priority.

2. Do not provide government subsidies or tax breaks for using or extracting resources on public lands.

3. Require users of public lands to reimburse the American people for use of their property and the resources it contains.

4. Hold all users or extractors of resources on public lands fully responsible for any environmental damage they cause.

There is strong and effective opposition to these principles. Developers, resource extractors, many economists, and many citizens tend to view public lands in terms of their usefulness in providing mineral resources, timber, grazing land, and other resources and increasing short-term economic growth. They have succeeded in blocking implementation of the four principles listed above. For example, in recent years, analyses of budgets reveal that the government has spent an average of $1 billion a year—an average of $2.7 million a day—on subsidies and tax breaks for privately owned interests that use U.S. public lands for mining, fossil fuel extraction, logging, and livestock grazing.

Some developers and resource extractors have sought to go further in opening up federal lands for more economic development and resource extraction and reducing or eliminating federal regulation of these lands. Here are five proposals that they have presented to Congress over the past several years:

1. Sell public lands or their resources to corporations or private individuals or turn them over to state and local governments.

2. Slash federal funding for the administration of regulations related to public lands.

3. Cut diverse old-growth national forests for timber and for making biofuels and replace them with tree plantations.

4. Open national parks, national wildlife refuges, and wilderness areas to oil drilling, mining, off-road vehicles, and commercial development.

5. Eliminate or take regulatory control away from the National Park Service. Launch a 20-year construction program in the national parks to build new concessions and theme parks that would be run by private firms.

CONSIDER THIS ...

THINKING ABOUT U.S. Public Lands

Explain why you would support or oppose each of the five proposals for changing the use and management of U.S. public lands, listed above.

Environmental Justice

Environmental justice is an ideal whereby every person is entitled to protection from environmental hazards regardless of race, gender, age, national origin, income, social class, or any political factor.

Studies show that a large share of polluting factories, hazardous waste dumps, incinerators, and landfills in the United States are located in communities populated mostly by minorities. Other research shows that, in general, toxic waste sites in white communities are cleaned up faster and more completely than similar sites in communities populated by African Americans, Latinos, and Native Americans.

Such environmental discrimination in the United States and many other parts of the world has led to a growing effort known as the *environmental justice movement*. Supporters of this movement pressure governments, businesses, and environmental organizations to become aware of environmental injustice and act to prevent it. This movement has made some progress toward their goals.

Some politicians and business representatives suggest that economics should be the main factor in decisions about where to locate new power plants, freeways, landfills, incinerators, and other such potentially disruptive facilities. Often, however, these areas are home to low-income residents who have much less political power than developers and corporations have. Many analysts argue that an ethical principle of environmental justice should carry as much weight as economic factors do in such decisions. This political struggle remains unresolved in most areas of the world.

CONSIDER THIS ...

THINKING ABOUT Environmental Justice

Do you think that the principles of environmental justice should get equal weight, more weight, or less weight in political decisions about where to locate potentially environmentally harmful facilities such as incinerators? Explain.

Environmental Policy Principles

Environmental scientists, economists, and political scientists have proposed several principles designed to reduce environmental harm and help legislators and individuals evaluate existing or proposed environmental policies:

- *Reversibility principle:* Avoid making decisions that cannot be reversed if they turn out to be harmful. Two essentially irreversible actions are the production of toxic coal ash from coal-burning power plants and the production of deadly radioactive wastes in nuclear power plants. In both cases, the resulting hazardous wastes must be stored safely for thousands of years.

- *Precautionary principle:* When substantial evidence indicates that an activity threatens human health or the environment, take measures to prevent or reduce such harm, even if some of the evidence is not conclusive.

- *Prevention principle:* Make decisions that prevent a problem from occurring or becoming worse.

- *Net energy principle:* Prohibit or limit widespread use of energy resources and technologies with low or negative net energy (see Figure 13.3, p. 330) that need government subsidies and tax breaks to compete in the marketplace. Examples include nuclear power (considering the whole fuel cycle), tar sands, shale oil, ethanol made from corn, and hydrogen fuel.

- *Polluter-pays principle:* Develop regulations and use economic tools, such as green taxes, to ensure that polluters bear the costs of dealing with the pollutants and wastes they produce. This stimulates the development of innovative ways to reduce and prevent pollution and wastes.

- *Environmental justice principle:* Do not allow any group of people to bear an unfair share of the burden created by pollution, environmental degradation, or the execution of environmental laws.

- *Holistic principle:* Focus on long-term solutions that address root causes of interconnected problems instead of on short-term and often ineffective fixes that treat each problem separately.

- *Triple bottom line principle:* Balance economic, environmental, and social needs when making policy decisions.

Implementing such principles is not easy. It requires policy makers throughout the world to become more environmentally literate. It also requires robust debate among politicians and citizens, mutual respect for diverse beliefs, and a dedication to dealing with environmental problems by implementing the win-win **principle of sustainability.** This would replace the more polarized I-win-you-lose approach, which fails to recognize that the solutions to urgent environmental problems will require openness, innovation, and compromise among political players and other people with divergent views.

CONSIDER THIS ...

THINKING ABOUT Environmental Policy Principles

Which three of the principles do you think are the most important? Why? Which ones do you think could influence legislators in your city, state, or country? Why?

You Can Make a Difference

A major theme of this book is that *individuals matter.* History shows that significant change usually comes from the *bottom up* when individuals work together to bring about change. Without this grassroots pressure, individual citizens and organized citizen groups (see chapter-opening photo) pollution and environmental degradation would be much worse today (**Concept 17.3**).

With the growth of the Internet, digital technology, and social media, individuals have become more empowered. Partly because of this social networking, the number of citizens' groups, national and global action networks, and NGOs focused on environmental and other problems has grown rapidly. Figure 17.18 lists ways in which individuals living in democracies can influence and change government policies.

You can provide environmental leadership in several different ways. First, you can *lead by example,* using your own lifestyle and values to show others that beneficial environmental change is possible (Individuals Matter 17.1). You can buy only what you need, use fewer disposable products, eat sustainably produced food, practice the 4Rs of resource use (refuse, reduce, reuse, recycle), adjust your lifestyle to reduce your carbon footprint, and walk, bike, or take mass transit to work or school.

Second, you can *work within existing economic and political systems to bring about environmental improvement* by campaigning and voting for informed and ecologically literate candidates, and by communicating with elected officials. As environmental writer and activist Bill McKibben says, "First change your politicians, then worry about your light bulbs." You can also send a message to

What Can You Do?

Influencing Environmental Policy

- Become informed on issues

- Make your views known at public hearings

- Make your views known to elected representatives and understand their positions on environmental issues

- Contribute money and time to candidates who support your views

- Vote

- Run for office

- Form or join nongovernment organizations (NGOs) seeking change

- Support reform of election campaign financing that reduces undue influence by corporations and wealthy individuals

FIGURE 17.18 Individuals Matter: Ways in which you can influence environmental policy. *Critical thinking:* Which three of these actions do you think are the most important? Which ones, if any, have you taken?

companies that you think are harming the environment through their production processes or products. You can do this by *voting with your wallet.* Do not buy their products or services and let them know why. You can also work to improve environmental quality by choosing one of the many rapidly growing environmental careers highlighted throughout this book and listed in Figure 17.14 and on this book's companion website.

Third, you can *run for some sort of local office.* Look in the mirror. Maybe you are someone who can make a difference as an officeholder.

Fourth, you can *propose and work for better solutions to environmental problems.* Leadership is much more than just taking a stand for or against something. It also involves coming up with solutions to problems and persuading people to work together to achieve them.

Some environmentally active citizens and leaders are motivated by two important findings. *First,* research by social scientists indicates that social change requires active support by only 5–10% of the population, which often is enough to lead to a political tipping point. *Second,* experience has shown that reaching such a critical mass can bring about social change much faster than most people think.

GOOD NEWS

Roles of Environmental Groups

The spearheads of the global conservation, environmental, and environmental justice movements are the tens of thousands of nonprofit NGOs working at the international, national, state, and local levels. The growing influence of

Xiuhtezcatl Roske-Martinez

Mark Sagliocco/Getty Images Entertainment/Getty Images

Xiuhtezcatl Roske-Martinez, born in 2000, learned about environmentalism from his parents while spending much of his early childhood enjoying the beautiful forests and streams near his Boulder, Colorado, home. His father is an Aztec who believes that all life is sacred and should be respected and cared for. His mother co-founded the nonprofit Earth Guardians as part of her commitment to protecting the earth's water, air, and atmosphere.

As a young child, Xiuhtezcatl (pronounced "shu-TEZ-cot") heard a lot about the harmful effects of human activities and noticed that the forest around him was changing. Trees were dying—killed by beetles whose populations were exploding because winter temperatures did not get low enough to kill them off. Dead trees fueled large fires that destroyed more trees.

To Xiuhtezcatl these effects of climate change were real and scary, and they motivated him. At age 6, he gave his first speech at a climate change rally. Since then, he has used his natural leadership ability to become a dynamic and highly effective environmental activist.

He helped persuade the Boulder City Council to stop using pesticides in parks, impose a fee on plastic bag use, and require a power company to depend more on renewable energy. He organized and spoke at press conferences, created a multimedia presentation about the harmful environmental effects of plastic bags, spoke at city council meetings and before an EPA hearing, and went door-to-door to organize dozens of rallies and marches. At age 12, he was invited to speak about climate change at the 2012 Rio+20 United Nations Summit on sustainable development in Brazil. As youth leader for Earth Guardians, Xiuhtezcatl has set up Earth Guardian Crews in many parts of the world to promote environmental education and awareness and to encourage other people to act.

these organizations is one of the most important changes affecting environmental decisions and policies.

NGOs range in size from grassroots groups with just a few members to *mainline* organizations. The World Wildlife Fund (WWF) is a 5-million member global conservation organization that operates in 100 countries, with 5 million members globally, and with 1.2 million members in the United States. Other international groups with large memberships include Greenpeace, the Nature Conservancy, Conservation International, and the Natural Resources Defense Council (NRDC).

In the United States, more than 8 million citizens belong to more than 30,000 NGOs that deal with environmental issues. They range from small grassroots groups to large, heavily funded *mainstream* groups, the latter usually staffed by expert lawyers, scientists, economists, lobbyists, and fund-raisers.

The largest environmental groups have become powerful and important forces within the U.S. political system. They have helped persuade Congress to pass and strengthen environmental laws (Figure 17.15), and they fight attempts to weaken or repeal these laws. According to political analyst Konrad von Moltke, "There isn't a government in the world that would have done anything for the environment if it weren't for the citizen groups." A worldwide network of grassroots NGOs is creating bottom-up political, social, economic, and environmental change and can be viewed as an emerging citizen-based *global sustainability movement*.

Student Environmental Groups

Hundreds of campus environmental groups and many high school environmental groups are leading the way to make their schools and local communities more sustainable (**Core Case Study**). Most of these groups work with members of their school's faculty and administration to bring about environmental improvements in their schools.

At Northland College in Ashland, Wisconsin, students helped design the green Environmental Living and Learning Center (Figure 17.19), which houses 150 students and features a wind turbine, solar panels, furniture made of recycled materials, and waterless (composting) toilets. Northland students voted to impose a *green fee* of $40 per semester on themselves to help finance the college's sustainability programs.

Dickinson College in Carlisle, Pennsylvania, integrates sustainability throughout its curriculum and uses wind power to offset all of its electricity use. Since 1990, De Anza Community College in Cupertino, California, has been integrating sustainability concepts into its curriculum. In addition, a team of students, faculty, administrators, and members of the local community worked together to build the LEED-platinum-certified Kirsch Center for Environmental Studies.

Many student groups conduct *environmental audits* of their campuses or schools. They use the resulting data to propose changes to make their campuses or schools more environmentally sustainable while usually saving money in

FIGURE 17.19 The Environmental Living and Learning Center is a residence hall and meeting space at Northland College in Ashland, Wisconsin. Northland students had a major role in designing the building to be more sustainable than conventional buildings.

Courtesy of Northland College

the process. Such audits have focused on implementing or improving recycling programs, convincing university food services to buy more food from local organic farms, improving the energy efficiency of buildings, shifting from fossil fuels to renewable energy, and implementing concepts of environmental sustainability throughout the curriculum.

Other students have focused on institutional investments. In 2015, more than 400 student-led campaigns were pressuring colleges and universities to stop investing their endowment funds in environmentally harmful industries, such as coal-fired electricity production. They also work toward getting their schools to increase their investments in renewable energy and other environmentally beneficial businesses in their endowment funds.

CONSIDER THIS ...

THINKING ABOUT The Greening of Your Campus

What major steps is your school taking to increase its environmental sustainability (**Core Case Study**) and to educate its students about environmental sustainability?

Global Environmental Security

Countries are legitimately concerned with *military security* and *economic security*. However, ecologists and many economists point out that all economies are supported by the

earth's natural capital (Figure 1.3, p. 7). Thus, environmental security, economic security, and national security are interrelated.

According to environmental scientist Norman Myers: "National security is no longer about fighting forces and weaponry alone. It relates increasingly to watersheds, croplands, forests, genetic resources, climate, and other factors that, taken together, are as crucial to a nation's security as are military factors." Myers and other analysts call for all countries to make environmental security a major focus of diplomacy and government policy at all levels.

International Environmental Organizations

A number of international environmental organizations help shape global environmental policy and improve environmental security and sustainability. Perhaps the most influential is the United Nations, which houses a large family of organizations including the UN Environment Programme (UNEP), the World Health Organization (WHO), the UN Development Programme (UNDP), and the Food and Agriculture Organization (FAO).

Other organizations that make or influence environmental decisions are the World Bank, the Global Environment Facility (GEF), and the International Union for the Conservation of Nature (IUCN). Despite their limited funding, these and other international organizations have played important roles in:

- expanding the understanding of environmental issues,
- gathering and evaluating environmental data,
- developing and monitoring international environmental treaties,
- providing grants and loans for sustainable economic development and reduction of poverty, and
- helping more than 100 nations develop environmental laws and institutions.

Some analysts have called for an environmental security council, on the order of the UN Security Council, to raise the priority level for dealing with environmental problems such as climate change that threaten long-term international security.

17.4 WHAT ARE SOME MAJOR ENVIRONMENTAL WORLDVIEWS?

CONCEPT 17.4 Major environmental worldviews differ on which is more important—human needs and wants, or the overall health of ecosystems and the biosphere, which support life and human economies.

Differing Environmental Worldviews

People disagree on how serious our environmental problems are, as well as what we should do about them. These conflicts arise partly from differences among environmental worldviews. As we noted in Chapter 1, your **environmental worldview** is the assumptions and beliefs that you have about how the natural world works and how you think you should interact with the environment.

Your environmental worldview is determined partly by your **environmental ethics**—what you believe about what is right and what is wrong in your behavior toward the environment. According to environmental ethicist Robert Cahn:

The main ingredients of an environmental ethic are caring about the planet and all of its inhabitants, allowing unselfishness to control the immediate self-interest that harms others, and living each day so as to leave the lightest possible footprints on the planet.

People with differing environmental worldviews can study the same data, be logically consistent in their analysis of those data, and arrive at quite different conclusions. This happens because they begin with different assumptions and values.

Human-Centered Environmental Worldviews

Human-centered environmental worldviews focus primarily on the needs and wants of people. One such worldview, called the *planetary management worldview*, sees humans as the planet's most important, intelligent, and dominant species. It believes that the human species should and can manage and dominate the earth. This view holds that other species and parts of nature should be evaluated according how useful they are to humans.

Another human-centered environmental worldview is the *stewardship worldview*. This view assumes that people have an ethical responsibility to be responsible managers, or *stewards,* of the earth. It also calls for us to encourage environmentally beneficial forms of economic growth and development and discourage environmentally harmful forms. According to the stewardship view, when people deplete or degrade the earth's natural capital, they are borrowing from the earth and future generations. This means that they have an ethical responsibility to pay this debt by leaving the earth's life-support system in as good a condition as what they enjoy, or better.

Some people believe that any human-centered worldview will eventually fail because it wrongly assumes humans now have or can gain enough knowledge to become effective managers or stewards of the earth. Critics of human-centered worldviews point out that we are living unsustainably by taking over most of the earth's land and water, changing the earth's climate, acidifying the global ocean, greatly increasing species extinction, and likely exceeding four of the earth's planetary boundaries, or ecological tipping points (see Science Focus 3.3, p. 63).

We are doing this even though we have much to learn about how the earth works, how it supports all life and our economies (Figure 17.4), and what goes on in a handful of soil, a patch of forest, the bottom of the ocean, and most other parts of the planet. As biologist David Ehrenfeld puts it, "In no important instance have we been able to demonstrate comprehensive successful management of the world, nor do we understand it well enough to manage it even in theory." The failure of the Biosphere 2 project (Science Focus 17.1) supports this view.

Life-Centered and Earth-Centered Environmental Worldviews

Life-centered worldviews hold that all forms of life have value as participating members of the biosphere, regardless of their potential or actual use to humans. **Earth-centered worldviews** also hold this view and expand it to include the entire biosphere, especially its ecosystems and the ecological services they provide.

Eventually, all species become extinct. However, most people with a life-centered worldview believe we have an ethical responsibility to avoid hastening the extinction of

Biosphere 2: A Lesson in Humility

In 1991, eight scientists (four men and four women) were sealed inside Biosphere 2, a $200 million glass and steel enclosure designed to be a self-sustaining life-support system (Figure 17.A). The goal of the project was to increase our understanding of Biosphere 1: the *earth's* life-support system.

The sealed system of interconnected domes was built in the desert near Tucson, Arizona. It contained artificial ecosystems, including a tropical rain forest, a savanna, a desert, a lake, streams, freshwater and saltwater wetlands, and a mini-ocean with a coral reef.

Biosphere 2 was designed to mimic the earth's natural chemical cycling systems. Water evaporated from its ocean and other aquatic systems and condensed to provide rainfall over the tropical rain forest. The precipitation trickled through soil into the marshes and back into the ocean before beginning the cycle again.

The facility was stocked with more than 4,000 species of plants and animals, including small primates, chickens, cats, and insects, selected to help maintain life-support functions. Human and animal excrement and other wastes were treated and recycled as fertilizer to help support plant growth. Sunlight and external natural gas–powered generators provided energy. The Biospherians were to be isolated for 2 years and raise their own food, using intensive organic agriculture. They were to breathe air that was purified by plants and to drink water cleansed by natural chemical cycling processes.

From the beginning, many unexpected problems cropped up and the life-support system began to unravel. The level of oxygen in the air declined with soil organisms converting it to carbon dioxide. Additional oxygen had to be pumped in from the outside to keep the Biospherians from suffocating.

Tropical birds died after the first freeze. An ant species invaded the enclosure, proliferated, and killed off most of the system's original insect species. In total, 19 of the Biosphere's 25 small animal species (76%) became extinct. Before the 2-year period was over, all plant-pollinating insects went extinct, thereby dooming to extinction most of the plant species.

Despite many problems, the facility's waste and wastewater were recycled. With much hard work, the Biospherians were also able to produce 80% of their food supply, despite rampant weed growths, spurred by higher CO_2 levels that crowded out food crops. However, they suffered from persistent hunger and weight loss.

FIGURE 17.A Biosphere 2, constructed near Tucson, Arizona, was designed to be a self-sustaining life-support system.

In the end, an expenditure of $200 million failed to maintain a life-support system for eight people for 2 years. Ecologists Joel E. Cohen and David Tilman, who evaluated the project, concluded, "No one yet knows how to engineer systems that provide humans with life-supporting services that natural ecosystems provide for free."

CRITICAL THINKING

Do you think that science and engineering ever will be able to provide humans with the life support systems that nature now provides at no cost? Explain.

---

any species, for two reasons. One is that each species is a unique part of the diverse genetic information that helps the earth's life to continue by changing in response to changes in environmental conditions. Another reason is that every species has the potential for providing economic benefits through its participation in providing ecosystem services.

People with an earth-centered worldview believe that we have an ethical responsibility to take a wider view and preserve the earth's biodiversity, ecosystem services, and the functioning of its life-support systems for the benefit of the earth's life, now and in the future. They argue that humans are not in charge of the world and that human economies are subsystems of the biosphere (Figure 17.5) and are utterly dependent on the earth's natural capital.

Some people with the stewardship worldview and the earth-centered worldview believe that we have an ethical obligation to save the earth. American farmer, philosopher, and poet Wendell Berry calls this "arrogant ignorance." He and others point out that the earth does not need saving. It has sustained an incredible variety of life for 3.8 billion years despite major changes in environmental conditions through natural factors and increasing environmental degradation caused by human activities.

According to Berry and other analysts, what we need to do is change our civilization to avoid degrading our life-support system and threatening up to half of the world's species with extinction. To these analysts, it is the human civilization that needs saving, and perhaps the human species, if we go too far in degrading the biosphere that keeps us alive and supports our economies.

One earth-centered worldview is called the *environmental wisdom worldview*, which in many ways is the opposite of planetary management worldview. According to this view:

- We are part of—not apart from—the community of life and the ecological processes that sustain all life.

- We are not in charge of the world. Instead, we are subject to nature's scientific laws that cannot be broken.

- Human economies and other systems are subsystems of the earth's life-support systems (Figure 17.5)

- We need to learn how to work with nature instead of trying to conquer it. We can do this by learning how life has sustained itself on the earth for 3.8 billion years and use these lessons from nature (environmental wisdom) to guide us in living more simply and sustainably.

- By not degrading the earth's life-support system, we act in our own self-interest. Earth care is self-care.

- We have an ethical responsibility to leave the earth in a condition as good as or better than what we inherited—in keeping with the ethical **principle of sustainability**.

17.5 HOW CAN WE LIVE MORE SUSTAINABLY?

CONCEPT 17.5 We can live more sustainably by becoming environmentally literate, learning from nature, living more simply and lightly on the earth, and becoming active environmental citizens.

Becoming More Environmentally Literate

There is widespread scientific evidence and agreement that we are a species that is degrading the earth's life-support system on which we and other species depend. During this century, this behavior will very likely threaten human civilization and wipe out up to half of the world's other species. Part of the problem stems from our incomplete understanding of how the earth's life-support system works, how our actions affect its life-sustaining systems, and how we can change our behavior toward the earth and thus toward ourselves. Improving this understanding begins by grasping three important ideas that form the foundation of *environmental literacy*:

1. Natural capital matters because it supports the earth's life and human economies.

2. Human ecological footprints are immense and are expanding rapidly.

3. Once we exceed planetary boundaries, or ecological tipping points (Science Focus 3.3), the resulting harmful consequences could last for hundreds to thousands of years.

Learning how to live more sustainably requires a foundation of environmental education with the goal of producing environmentally literate citizens. Acquiring environmental literacy involves being able to answer certain key questions and having a basic understanding of certain key topics, as summarized in Figure 17.20. We hope that, with this course and textbook, you have begun to build your foundation of environmental literacy and that you use it to live more sustainably.

Questions to answer

- How does life on earth sustain itself?

- How am I connected to the earth and other living things?

- Where do the things I consume come from and where do they go after I use them?

- What is environmental wisdom?

- What is my environmental worldview?

- What is my environmental responsibility as a human being?

Components

- Basic concepts: sustainability, natural capital, exponential growth, carrying capacity

- Principles of sustainability

- Environmental history

- The two laws of thermodynamics and the law of conservation of matter

- Basic principles of ecology: food webs, nutrient cycling, biodiversity, ecological succession

- Population dynamics

- Sustainable agriculture and forestry

- Soil conservation

- Sustainable water use

- Nonrenewable mineral resources

- Nonrenewable and renewable energy resources

- Climate disruption and ozone depletion

- Pollution prevention and waste reduction

- Environmentally sustainable economic and political systems

- Environmental worldviews and ethics

FIGURE 17.20 Achieving environmental literacy involves being able to answer certain questions and having an understanding of certain key topics (**Concept 17.5**). **Question:** After taking this course, do you feel that you can answer the questions asked here and have a basic understanding of each of the key topics listed in this figure?

FIGURE 17.21 Experiencing nature can help us understand the need to protect the earth's natural capital and to live more sustainably.

djgis/Shutterstock.com

Learning from the Earth

Formal environmental education is important, but is it enough? Many analysts say *no*. They call for people to appreciate not only the economic value of nature but also its ecological, aesthetic, and spiritual values. To these analysts, the problem is not just a lack of environmental literacy but also a lack of intimate contact with nature. This can reduce people's ability to act more responsibly toward the earth and thus toward themselves and other people.

A growing chorus of analyst suggests that people have much to learn from nature. They call for us to acquire a sense of awe, wonder, mystery, excitement, and humility by exploring a forest, taking in the majesty of the sea, or enjoying a beautiful scene in nature (Figure 17.21). You might pick up a handful of topsoil and try to sense the teeming microscopic life within it that helps keep you alive by supporting food production. You might look at a tree, a mountain, a rock, or a bee, or listen to the sound of a bird and sense how each of them is connected to you and you to them, through the earth's life-sustaining processes.

Direct experiences with nature can reveal parts of the complex web of life that cannot be bought, recreated with technology, or reproduced with genetic engineering. Understanding and directly experiencing the precious and free gifts we receive from nature can help people make an ethical commitment to live more sustainably on the earth and thus to preserve our own species and cultures.

According to some psychologists and other analysts, experiencing and understanding nature is necessary

Juan Martinez—Reconnecting People with Nature

National Geographic Explorer Juan Martinez learned firsthand about the value of connecting with nature. Now he is instilling that value in others, particularly disadvantaged youths.

Martinez grew up in a poor area of Los Angeles, California, where as a boy he was in danger of becoming absorbed by a gang culture. One of his teachers recognized Martinez's potential and gave him a chance to pass a class that he was failing by joining the school's Eco Club.

Martinez took that opportunity and when the club planned a field trip to see the Grand Teton Mountains of Wyoming, he jumped at the chance. As a result, he says, "I still can't find words to describe the first moment I saw those mountains rising up from the valley. Watching bison, seeing a sky full of stars, and hiking through that scenery was overwhelming."

The experience transformed Martinez's life. Today, he is the Director of Leadership Development in the Children and Nature Network, an organization creating links between environmental organizations, corporations, government, education, and individuals to reconnect children with nature. His work as an environmental leader has inspired many others to do similar work.

Martinez has received a great deal of recognition for his efforts, including invitations to White House forums on environmental education. His greatest reward, however, is in seeing how his efforts help others.

REBECCA HALE/National Geographic Creative

for healthy living. Journalist Richard Louv coined the term **nature-deficit disorder** to describe a wide range of problems, including anxiety, depression, and attention-deficit disorders that can result from a lack of contact with nature. Louv argues that the problem is especially apparent among children who play mostly indoors and at best view the natural world digitally—something new in the history of humankind.

Urban living, along with extensive use of the Internet, cell phones, and other electronic devices, contribute to nature-deficit disorder, which is viewed as one of the five major causes of environmental problems (pp. 17–18). Many environmental leaders are helping people connect directly with nature (Individuals Matter 17.2).

Earth-focused philosophers say that to be rooted, each of us needs to find a sense of place—a stream, a mountain, a patch of forest, a yard, a neighborhood lot—any piece of the earth that we know, experience emotionally, and love. According to biologist Stephen Jay Gould, "We will not fight to save what we do not love." When we become part of a place, it becomes a part of us. Then we are driven to defend it from harm and to help heal its wounds (Figure 17.22). We might discover and tap into what conservationist Aldo Leopold called "the green fire that burns in our hearts" and use this as a force for respecting and working with the earth and with one another.

Living More Simply

On a timescale of hundreds of thousands to millions of years, the earth is resilient and can survive and heal many wounds. Mostly because of human actions, we are living on a planet with a warmer and sometimes harsher climate, less dependable supplies of water, more acidic oceans, extensive soil degradation, increased mass extinction of species, degradation of key ecosystem services, and widespread ecological disruption. Unless we change our course, scientists warn that these and other harmful environmental changes will intensify. Figure 17.23 lists 12 guidelines—the *sustainability dozen*—developed by environmental scientists and ethicists for living

Joel W. Rogers/Corbis Documentary/Getty Images

FIGURE 17.22 This woman and others in Vancouver, Canada, are protesting the clear-cutting of old-growth forests in the Canadian province of British Columbia.

Guidelines for living more sustainably

- Mimic the ways nature sustains itself by using the earth as a model and teacher.

- Protect the earth's natural capital and repair ecological damage caused by human activities.

- Focus on preventing pollution and resource waste.

- Reduce resource consumption, waste, and pollution by reducing demand and using matter and energy resources more efficiently.

- Recycle, reuse, and repair everything and thus copy nature by having our wastes become resources.

- Rely more on clean, renewable energy resources such as solar and wind energy.

- Slow climate change.

- Reduce population growth and gradually reduce population size.

- Celebrate and protect biodiversity and cultural diversity.

- Promote social justice for humans and ecological justice for other species that keep us alive.

- End poverty.

- Leave the earth in a condition that is as good as or better than what we inherited.

FIGURE 17.23 *Sustainability dozen:* Guidelines for living more sustainably.

more sustainably by converting environmental concerns, literacy, and lessons from the earth into environmentally responsible action for current and future generations.

Some analysts urge people who have a habit of consuming excessively to learn how to live more simply and sustainably. Seeking happiness through the pursuit of material things is considered folly by almost every major religion and philosophy. Yet, today's avalanche of advertising messages encourages people to buy more and more things to fill a growing list of wants as a way to achieve happiness. As American humorist and writer Mark Twain (1835–1910) observed: "Civilization is the limitless multiplication of unnecessary necessities."

Some people are adopting a lifestyle of *voluntary simplicity.* It should not be confused with poverty, which is *involuntary simplicity.* Voluntary simplicity involves people learning to live with less stuff, using products and services that have smaller harmful environmental impacts, and creating beneficial environmental impacts (**Concept 17.5**). Their goals are to consume less, share more, live simply, make friends, treasure family, and enjoy life. Their motto is: "Shop less, live more."

Practicing voluntary simplicity is a way to apply the Indian philosopher and leader Mahatma Gandhi's *principle of enoughness:* "The earth provides enough to satisfy every person's need but not every person's greed.... When we take more than we need, we are simply taking from each other, borrowing from the future, or destroying the environment and other species." Most of the world's major religions have similar teachings.

Living more simply and sustainably starts with asking the question: What do I really need? This is not an easy question to answer, because people in affluent societies are conditioned to view excessive material possessions as needs instead of wants. Many people have become addicted to buying more and more stuff as a way to find meaning in their lives and they often run up large debts to feed their stuff habit.

Throughout this text, you have encountered lists of ways we can live more lightly on the earth by reducing the size and impact of our ecological footprints on the earth. Figure 17.24 lists eight key ways in which some people are choosing to live more simply and sustainably.

The seriousness of the environmental problems we face can be overwhelming and lead many people to feel guilty, fearful, apathetic, and powerless. We can move beyond these immobilizing feelings by recognizing and avoiding the following two common mental traps that lead to denial, indifference, and inaction:

- *Gloom-and-doom pessimism (it is hopeless)*

- *Blind technological optimism (science and technological fixes will save us)*

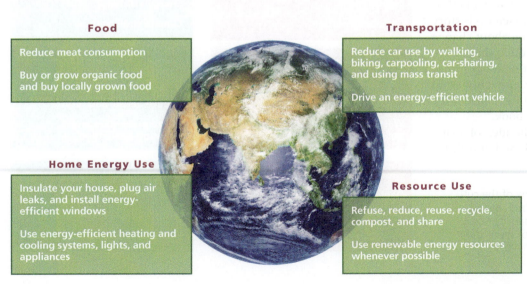

Food

Reduce meat consumption

Buy or grow organic food and buy locally grown food

Home Energy Use

Insulate your house, plug air leaks, and install energy-efficient windows

Use energy-efficient heating and cooling systems, lights, and appliances

Transportation

Reduce car use by walking, biking, carpooling, car-sharing, and using mass transit

Drive an energy-efficient vehicle

Resource Use

Refuse, reduce, reuse, recycle, compost, and share

Use renewable energy resources whenever possible

FIGURE 17.24 *Living more lightly:* Eight ways to shrink our ecological footprints and expand our beneficial environmental impacts (**Concept 17.5**). *Critical thinking:* Which of these steps do you already take? Which, if any, do you hope to take in the future?

Avoiding these traps helps us hold on to, and be inspired by, empowering feelings of realistic hope, rather than to be immobilized by feelings of despair and fear.

CONSIDER THIS ...

THINKING ABOUT Mental Traps

Have you fallen into either of these traps? If so, are you aware that you have, and how do you think you could free yourself from either of them?

Here is what business entrepreneur and environmental writer Paul Hawken told a graduating class at the University of Portland:

> *When asked if I am pessimistic or optimistic about the future, my answer is always the same: If you look at the science about what is happening on the earth and aren't pessimistic, you don't understand the data. But if you meet the people who are working to restore this earth and the lives of the poor, and you aren't optimistic, you haven't got a pulse.... You join a multitude of caring people.*

Bringing About a Sustainability Revolution in Your Lifetime

Figure 17.25 lists some of the major cultural shifts in emphasis that could help bring about a sustainability revolution. One of the leaders in the movement to develop and promote detailed plans for making the shift to more sustainable ways of living is Lester R. Brown. The *Washington Post* has called him "one of the world's most influential thinkers," and in 2011, *Foreign Policy* named him one of the Top Global Thinkers. Brown's Plan B for shifting to a more environmentally and economically sustainable future has four main goals: **(1)** stabilize population growth, **(2)** stabilize climate change, **(3)** eradicate poverty, and **(4)** restore the earth's natural support systems.

We know what needs to be done and we can change the way we treat the earth and thus our life-support system and ourselves. History also shows that people can bring about change faster than we might think, once they leave behind ideas and practices that no longer work and nurture solutions to the environmental problems that we all face. We have presented such solutions throughout this book.

While some skeptics say the idea of a sustainability revolution is idealistic and unrealistic, entrepreneur Paul Hawken has argued that "the most unrealistic person in the world is the cynic, not the dreamer." In addition, according to the late Steve Jobs, cofounder of Apple Inc., "The people who are crazy enough to think they can change the world are the ones who do." If these and other individuals had not had the courage to forge ahead with ideas that others called idealistic and unrealistic, very few of the environmental and other achievements that we now celebrate would have happened.

The key to a sustainability revolution is that individuals matter. Each of our choices and actions makes a difference, we are all in this together, and the situation is not hopeless. We can work together to become the generation that averts environmental chaos and leaves the earth—our only home—in better shape than it is now. What an incredibly exciting time to be alive.

BIG IDEAS

- A more sustainable economic system would include in market prices the harmful environmental and health costs of producing and using goods and services, subsidize environmentally beneficial goods and services, tax pollution and waste instead of wages and profits, and reduce poverty.
- Individuals can work together to become part of the political processes that influence how environmental policies are made and implemented.
- Living more sustainably means becoming environmentally literate, learning from nature, living more simply, and becoming an active environmental citizen.

Unsustainable Path
Sustainable Path
Energy and Climate
Fossil fuels
Direct and indirect solar energy
Energy waste
Energy efficiency
Climate disruption
Climate stabilization
Matter
High resource use and waste
Less resource use
Consume and throw away
Reduce, reuse, and recycle
Waste disposal and pollution control
Waste prevention and pollution prevention
Life
Deplete and degrade natural capital
Protect natural capital
Reduce biodiversity
Protect biodiversity
Population growth
Population stabilization

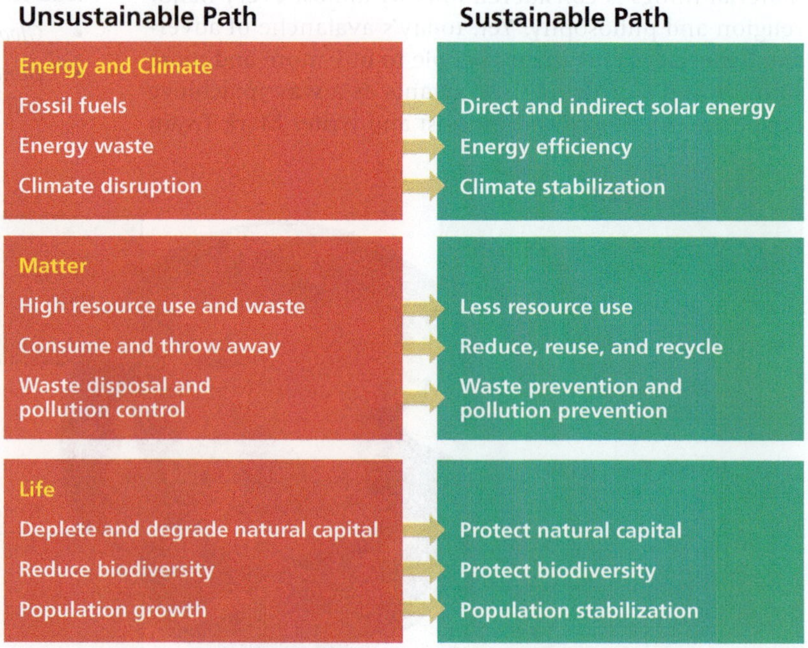

FIGURE 17.25 Solutions: Some of the cultural shifts in emphasis that scientists say will be necessary to bring about a sustainability revolution. ***Critical thinking:*** Which of these shifts do you think are most important? Why?

Tying It All Together

Greening College Campuses and Sustainability

Sailorr/Shutterstock.com

College students around the world have shown that it is possible to create sustainable environmental policies, at least in the communities in and around many college campuses (**Core Case Study**). The human population has the abilities and resources to implement policies that would help eradicate poverty and malnutrition, eliminate illiteracy, sharply reduce infectious diseases, stabilize human populations, and protect the earth's natural capital. We can do this by applying the three **scientific principles of sustainability** (see Inside Back Cover)—relying much more on solar energy and other renewable energy sources, reusing and recycling much more of what we produce, and respecting, restoring, and protecting as much as possible of the biodiversity that supports our lives and economies.

National and international policy makers could also be guided by the three economic, political, and ethical **principles of sustainability** (see Inside Back Cover). In the political arena, they will have to try harder to find win-win solutions that benefit the largest numbers of people while also benefiting the environment. Such solutions will likely have to include internalizing the harmful environmental and health costs of producing and using goods and services (full-cost pricing). If we are truly interested in long-term sustainability, we also have the ethical responsibility to leave the world in a condition at least as good as or better than what we inherited.

Making a shift to more sustainable societies and economies can occur much more rapidly than we think. We can choose to take part in this change by becoming politically aware, informed, and active with regard to issues that affect our environmental and political futures.

Chapter Review

Core Case Study

1. Define **sustainability** and give four examples of how colleges and universities are playing an important role in promoting it.

Section 17.1

2. What is the key concept for this section? What is **economics**? What are supply and demand and how are they related to price? Distinguish among **natural capital**, **human capital**, and **manufactured capital**. Define **economic growth**. What is a **high-throughput economy**? Distinguish between **economic development** and **environmentally sustainable economic development**. Summarize the controversy over the use of discount rates to estimate the future value of a resource. Describe the use and limitations of cost-benefit analysis.

Section 17.2

3. What is the key concept for this section? Why do products and services actually cost more than most people think? What is **full-cost pricing** and what are some benefits of using it to determine the market values of goods and services? Give three reasons why it is not widely used. Define **subsidies**. What are perverse subsidies and tax breaks and how do they contribute to environmental problems? Describe the proposed shift from environmentally harmful to environmentally beneficial subsidies and tax breaks over the next two to three decades. Why do some analysts find the typical tax system backwards? How could a tax shift improve environmental quality? List three requirements for making such a tax shift. What are the major advantages and disadvantages of green taxes?

4. Define and distinguish between **gross domestic product (GDP)** and **per capita GDP**. What is the **genuine progress indicator (GPI)** and how does it differ from the GDP? Why are environmental indicators important? What are the major advantages and disadvantages of using the cap-and-trade approach to implementing environmental regulations for controlling pollution and resource use? What are some environmental benefits of selling services instead of goods? Give two examples of this approach.

5. What is **poverty** and how is it related to population growth, environmental degradation, and human health? List six ways in which governments, businesses, lenders, and individuals can help reduce poverty. What is *microlending* and how can it benefit the poor and the environment? What are the Millennium Development Goals and what progress has been made in meeting them? What is a **low-throughput (low-waste) economy**? List six components of more environmentally sustainable economic development. Name five green businesses or careers that would be important in more environmentally sustainable economies.

Section 17.3

6. What is the key concept for this section? Define **environmental policy.** List the four components of the policy life cycle. Distinguish between command-and-control and incentive-based, or innovation-friendly, government regulations and list the advantages of the second approach. What are special interest groups? Give four examples. Why is it difficult to develop and implement environmental policy in countries such as the United States with a democratic form of government? What role did U.S. citizens play in the passage of most U.S. environmental laws in the 1970s? List five categories of U.S. environmental laws. Describe the effectiveness of the anti-environmental movement in the United States since the 1980s. What three groups are strongly opposed to U.S. environmental laws and regulations? What are four major types of public lands in the United States? Summarize the political controversy over managing these lands.

7. What is **environmental justice** and why is it important? List eight principles that decision makers can use in making environmental policy. What are four ways in which individuals in democracies can influence environmental policy? What are three ways to provide environmental leadership? Describe the roles of grassroots and mainstream environmental organizations. Give two examples of how students

and educational institutions have led the way in shifting to more sustainable ways of living. Explain the importance of environmental security, relative to economic and military security. What are four important organizations that help shape global environmental policy? List five accomplishments of international environmental organizations.

Section 17.4

8. What is the key concept for this section? What is an **environmental worldview**? What are **environmental ethics**? What is a **human-centered worldview**? Distinguish between the planetary management and stewardship environmental worldviews. Summarize the debate over whether we can effectively manage the earth. Summarize the ecological lessons learned from the failure of the Biosphere 2 project. Define **life-centered worldview** and **earth-centered worldview**. Why is it a fallacy to think that we need to save the earth? What do we need to save? What are the key principles of the environmental wisdom worldview?

Section 17.5

9. What is the key concept for this section? What are three basic principles of environmental literacy? Give two examples of how we can learn from the earth. What is **nature-deficit disorder** and why is it important? What does it mean to adopt a lifestyle of voluntary simplicity? What is the first step to living more lightly? List six of the proposed 12 guidelines for living more sustainably. List eight important steps that individuals can take to help make the transition to living more sustainably. List six important cultural shifts that could be part of a sustainability revolution.

10. What are this chapter's three big ideas? Explain how college students have changed the policies of their institutions by applying some of the six **principles of sustainability**.

Note: Key terms are in bold type.

Critical Thinking

1. In making a shift to a more sustainable future, what do you think are the three most important things that **(a)** your college or university should do, and **(b)** you should do?

2. Suppose that over the next 20 years, the environmental and health costs of goods and services are gradually added to their market prices until those prices more closely reflect their total costs. What harmful effects and what beneficial effects might such a full-cost pricing process have on your lifestyle and on the lives of any children, grandchildren, and great grandchildren you eventually might have?

3. Which three of the components of more environmentally sustainable economic development shown in Figure 17.13 do you think are the most important ones to promote? Explain your choices. Which three do you think are the least important? Explain your choices.

4. Explain why you agree or disagree with each of the eight principles listed on pp. 490–491, which some analysts recommend for making environmental policy decisions. Which three of these principles do you think are the most important? Why?

5. Explain why you agree or disagree with **(a)** each of the four principles that biologists and some economists have suggested for using public lands in the United States (p. 489) and **(b)** each of the five suggestions made by developers and resource extractors for managing and using U.S. public lands (p. 490).

6. This chapter summarized several different environmental worldviews. Which, if any, of these environmental worldviews do you favor? If you disagree with all of them, what is your environmental worldview? Has your environmental worldview changed because of your taking this course? If so, how? Compare your answers to these questions with those of your classmates.

7. Explain why you agree or disagree with the following statements: **(a)** everyone has the right to have as many children as they want; **(b)** all people have a right to use as many resources as they want; **(c)** individuals should have the right to do whatever they want with land they own, regardless of whether such actions harm the environment, their neighbors, or the local community; **(d)** other species exist to be used by humans; **(e)** all forms of life have a right to exist regardless of their usefulness to humans. Are your answers consistent with the beliefs making up your environmental worldview, which you described in question 6? If not, explain.

8. Do you think we have a reasonable chance of bringing about a sustainability revolution within your lifetime? Explain. If you are nearing the end of this course, is your view of the future more hopeful or less hopeful than it was when you began the course? Compare your answers with those of your classmates.

Doing Environmental Science

Choose an environmental issue, such as water pollution, climate change, population growth, biodiversity loss, or any other issue that you have studied in this course. Conduct a poll of students, faculty, staff, and local residents by asking them the questions that follow relating to your particular environmental issue. Poll as many people as you can in order to get a large sample. Create categories. For example, note whether each respondent is male or female. By creating such categories, you are placing each person into a *respondent pool*. You can add other questions about age, political leaning, and other factors to refine your pools as you see fit.

Poll Questions

Question 1   On a scale of 1 to 10, how knowledgeable are you about environmental issue X?

Question 2   On a scale of 1 to 10, how aware are you of ways in which you, as an individual, impact environmental issue X?

Question 3   On a scale of 1 to 10, how important is it for you to learn more about environmental issue X?

Question 4   On a scale of 1 to 10, how sure are you that an individual can have a positive influence on environmental issue X?

Question 5   On a scale of 1 to 10, how sure are you that the government is providing the appropriate level of leadership with regard to environmental issue X?

1. Collect your data and analyze your findings to measure any differences among the respondent pools.

2. List any major conclusions you would draw from the data.

3. Publicize your findings on your school's website or in the local newspaper.

Global Environment Watch Exercise

Go to your MindTap course to access the GREENR database. Use the "World Map" link at the top of the page to navigate to information about specific countries. Click on the pins for the United States, China, India, and one other country of your choice and research what each of them is doing to try to become more sustainable. Which of these sustainability programs are working well? Which ones are not working well? Write a report comparing these programs.

Ecological Footprint Analysis

Working with classmates, conduct an ecological footprint analysis of your campus. Work with a partner, or in small groups, to research and investigate one or more aspects of your school, such as recycling or composting; water use; food service practices; energy use; building management and energy conservation; transportation for both on- and off-campus trips; or grounds maintenance. Depending on your school and its location, you may want to add more areas to the investigation. You can also decide to study the campus as a whole, or to break it down into smaller research areas, such as dorms, administrative buildings, classroom buildings, grounds, and other areas.

1. After deciding on your group's research area, conduct your analysis. As part of your analysis, develop a list of questions that will help determine the ecological impact related to your chosen topic. For example, with regard to water use, you might ask how much water is used, what is the estimated amount that is lost through leaking pipes and faucets, and what is the average monthly water bill for the school, among other questions. Use such questions as a basis for your research.

2. Analyze your results and share them with the class to determine what can be done to shrink the ecological footprint of your school within the area you have chosen.

3. Arrange a meeting with school officials to share your action plan with them.

Glossary

acid deposition The falling of acids and acid-forming compounds from the atmosphere to the earth's surface. Commonly known as *acid rain*, a term that refers to the wet deposition of droplets of acids and acid-forming compounds.

acidic solution Any water solution that has more hydrogen ions (H^+) than hydroxide ions (OH^-); any water solution with a pH less than 7. Compare *basic solution, neutral solution*.

acidity Chemical characteristic that helps determine how a substance dissolved in water (a solution) will interact with and affect its environment; based on the comparative amounts of hydrogen ions (H^+) and hydroxide ions (OH^-) contained in a particular volume of the solution. See *pH*.

acid rain See *acid deposition*.

active solar heating system System that uses solar collectors to capture energy from the sun and store it as heat for space heating and water heating. Liquid or air pumped through the collectors transfers the captured heat to a storage system such as an insulated water tank or rock bed. Pumps or fans then distribute the stored heat or hot water throughout a dwelling as needed. Compare *passive solar heating system*.

adaptation Any genetically controlled structural, physiological, or behavioral characteristic that helps an organism survive and reproduce under a given set of environmental conditions. See *biological evolution, mutation, natural selection*.

adaptive trait See *adaptation*.

aerobic respiration Complex process that occurs in the cells of most living organisms, in which nutrient organic molecules such as glucose ($C_6H_{12}O_6$) combine with oxygen (O_2) to produce carbon dioxide (CO_2), water (H_2O), and energy. Compare *photosynthesis*.

age structure Percentage of the population (or number of people of each gender) at each age level in a population.

agrobiodiversity The genetic variety of animal and plant species used on farms to produce food.

air pollution One or more chemicals in high enough concentrations in the air to harm humans, other animals, vegetation, or materials. Such chemicals or physical conditions are called air pollutants. See *primary pollutant, secondary pollutant*.

alley cropping Planting of crops in strips with rows of trees or shrubs on each side.

amphibians Class of animals that includes frogs, toads, and salamanders.

anaerobic respiration Form of cellular respiration in which some decomposers get the energy they need through the breakdown of glucose (or other nutrients) in the absence of oxygen. Compare *aerobic respiration*.

animal manure Dung and urine of animals used as a form of organic fertilizer. Compare *green manure*.

Anthropocene New era in which humans have become major agents of change in the functioning of the earth's life support system as their ecological footprints have spread over the earth. Compare *Holocene*.

anthropocentric Human-centered. Compare *biocentric*.

aquaculture Growing and harvesting of fish and shellfish for human use in freshwater ponds, irrigation ditches, and lakes, or in cages or fenced in areas of coastal lagoons and estuaries, or in the open ocean.

aquatic life zone Marine and freshwater portions of the biosphere. Examples include freshwater life zones (such as lakes and streams) and ocean or marine life zones (such as estuaries, coastlines, coral reefs, and the open ocean).

aquifer Porous, water-saturated layers of sand, gravel, or bedrock that can yield an economically significant amount of water.

area strip mining Type of surface mining used where the terrain is flat. An earthmover strips away the overburden and a power shovel digs a trench to remove the mineral deposit. The trench is then filled with overburden, and a new trench is made parallel to the previous one. The process is repeated over the entire site. Compare *mountaintop removal mining, open-pit mining, subsurface mining*.

artificial selection Process by which humans select one or more desirable genetic traits in the population of a plant or animal species and then use *selective breeding* to produce populations containing many individuals with the desired traits. Compare *genetic engineering, natural selection*.

asthenosphere Zone within the earth's mantle made up of hot, partly melted rock that flows and can be deformed like soft plastic.

atmosphere The whole mass of air surrounding the earth. See *stratosphere, troposphere*. Compare *biosphere, geosphere, hydrosphere*.

atmospheric warming An increase in the average temperature of the atmosphere near the earth's surface over 30 years or more.

atom The smallest unit of an element that can exist and still have the unique characteristics of that element; made of subatomic particles, it is the basic building block of all chemical elements and thus all matter. Compare *ion, molecule*.

atomic number Number of protons in the nucleus of an atom. Compare *mass number*.

atomic theory Idea that all elements are made up of atoms; the most widely accepted scientific theory in chemistry.

autotroph See *producer*.

background extinction Normal extinction rate of various species as a result of changes in environmental conditions. Compare *mass extinction*.

bacteria Prokaryotic, one-celled organisms, some of which transmit diseases. Most act as decomposers and get the nutrients they need by breaking down complex organic compounds in the tissues of living or dead organisms into simpler inorganic nutrient compounds.

basic solution Water solution with more hydroxide ions (OH^-) than hydrogen ions (H^+); water solution with a pH greater than 7. Compare *acid solution, neutral solution*.

bioaccumulation An increase in the concentration of a chemical in specific organs or tissues at a level higher than would normally be expected. Compare *biomagnification*.

biocentric Life-centered. Compare *anthropocentric*.

biodegradable pollutant Material that can be broken down into simpler substances (elements and compounds) by bacteria or other decomposers. Paper and most organic wastes such as animal manure are biodegradable but can take decades to biodegrade in modern landfills. Compare *nondegradable pollutant*.

biodiversity Variety of different species (*species diversity*), genetic variability among individuals within each species (*genetic* *diversity*), variety of ecosystems (*ecological diversity*), and functions such as energy flow and matter cycling needed for the survival of species and biological communities (*functional diversity*).

biodiversity hotspot An area especially rich in plant species that are found nowhere else and are in great danger of extinction.

biofuel Gas (such as methane) or liquid fuel (such as ethyl alcohol or biodiesel) made from plant material (biomass).

biogeochemical cycle Natural processes that recycle nutrients in various chemical forms from the nonliving environment to living organisms and then back to the nonliving environment. Examples include the carbon, nitrogen, phosphorus, sulfur, and hydrologic cycles.

biological community See *community*.

biological diversity See *biodiversity*.

biological evolution Change in the genetic makeup of a population of a species in successive generations. If continued long enough, it can lead to the formation of a new species. Note that populations, not individuals, evolve. See also *adaptation, natural selection, theory of evolution*.

biological extinction Complete disappearance of a species from the earth. It happens when a species cannot adapt and successfully reproduce under new environmental conditions or when a species evolves into one or more new species. Compare *speciation*. See also *endangered species, mass extinction, threatened species*.

biomagnification Increase in concentration of DDT, PCBs, and other slowly degradable, fat-soluble chemicals in organisms at successively higher trophic levels of a food chain or web. Compare *bioaccumulation*.

biomass Organic matter produced by plants and other photosynthetic producers; total dry weight of all living organisms that can be supported at each trophic level in a food chain or web; dry weight of all organic matter in plants and animals in an ecosystem; plant materials and animal wastes used as fuel.

biome A terrestrial region distinguished by the predominance of certain types of vegetation and other forms of life. Examples include various types of deserts, grasslands, and forests.

biomimicry Process of observing certain changes in nature, studying how natural systems have responded to such changing conditions over many millions of years, and applying what is learned to dealing with some environmental challenge.

biosphere Zone of the earth where life is found. It consists of parts of the atmosphere (the troposphere), hydrosphere (mostly surface water and groundwater), and lithosphere (mostly soil and surface rocks and sediments on the bottoms of oceans and other bodies of water). Compare *atmosphere, geosphere, hydrosphere*.

birth rate See *crude birth rate*.

buffer Substance that can react with hydrogen ions in a solution and thus hold the acidity or pH of a solution fairly constant. See *pH*.

carbon capture and storage (CCS) Process of removing carbon dioxide gas from coal-burning power and industrial plants and storing it somewhere (usually underground or under the seabed). To be effective, it must be stored so that it cannot be released into the atmosphere, essentially forever.

carbon cycle Cyclic movement of carbon in different chemical forms from the environment to organisms and then back to the environment.

carbon footprint The amount of carbon dioxide generated by an individual, an organization, or a geographically or politically defined area (such as a city, state, or country) in a given period of time.

carcinogen Chemicals, ionizing radiation, and viruses that cause or promote the development of cancer. Compare *mutagen, teratogen*.

carnivore Animal that feeds on other animals. Compare *herbivore, omnivore*.

carrying capacity (K) Maximum population of a particular species that a given habitat can support over a given period. Compare *cultural carrying capacity*.

CCD See *colony collapse disorder*.

CCS See *carbon capture and storage*.

cell Smallest living unit of an organism. Each cell is encased in an outer membrane or wall and contains genetic material (DNA) and other substances that enable it to perform its life function. Organisms such as bacteria consist of only one cell, but most organisms contain many cells.

cell theory The idea that all living things are composed of cells; the most widely accepted scientific theory in biology.

CFCs See *chlorofluorocarbons*.

chemical change Interaction between chemicals in which the chemical composition of the elements or compounds involves changes. Compare *nuclear change, physical change*.

chemical cycling The circulation of chemicals from the environment (mostly from soil and water) through organisms and back to the environment.

chemical formula Shorthand way to show the number of atoms (or ions) in the basic structural unit of a compound. Examples include H_2O, $NaCl$, and $C_6H_{12}O_6$.

chemical reaction See *chemical change*.

chemosynthesis Process in which certain organisms (mostly specialized bacteria) extract inorganic compounds from their environment and convert them into organic nutrient compounds without the presence of sunlight. Compare *photosynthesis*.

chlorofluorocarbons (CFCs) Organic compounds, made up of atoms of carbon, chlorine, and fluorine, that can deplete the ozone layer when they slowly rise into the stratosphere and react with ozone molecules.

CHP (combined heat and power) See *cogeneration*.

chromosome A grouping of genes and associated proteins in plant and animal cells that carry certain types of genetic information. See *genes*.

chronic malnutrition Faulty nutrition, caused by a diet that does not supply an individual with enough protein, essential fats, vitamins, minerals, and other nutrients needed for good health. Compare *overnutrition, chronic undernutrition*.

chronic undernutrition Condition suffered by people who cannot grow or buy enough food to meet their basic energy needs. Most chronically undernourished children live in less-developed countries and are likely to suffer from mental retardation and stunted growth, and to die from infectious diseases. Compare *chronic malnutrition, overnutrition*.

clear-cutting Method of timber harvesting in which all trees in a forested area are removed in a single cutting. Compare *selective cutting, strip cutting*.

climate General pattern of atmospheric conditions in a given area over periods ranging from at least 30 years to thousands of years. The two main factors determining an area's climate are its average temperature, with its seasonal variations, and the average amount and distribution of precipitation. Compare *weather*.

climate change Measurable changes in global weather patterns based primarily on changes in the earth's atmospheric temperature averaged over at least 30 years.

climate change tipping point Point related to climate change at which an environmental problem reaches a threshold level where scientists fear it could cause irreversible climate disruption.

coal Solid, combustible mixture of organic compounds with 30–98% carbon by weight, mixed with various amounts of water and small amounts of sulfur and nitrogen compounds. It forms in several stages as the remains of plants are subjected to heat and pressure over millions of years.

coal gasification Conversion of solid coal to synthetic natural gas (SNG).

coal liquefaction Conversion of solid coal to a liquid hydrocarbon fuel such as synthetic gasoline or methanol.

coastal wetland Land along a coastline, extending inland from an estuary that is covered with salt water all or part of the year. Examples include marshes, tidal flats, and mangrove swamps. Compare *inland wetland*.

coastal zone Warm, nutrient-rich, shallow part of the ocean that extends from the high-tide mark on land to the edge of a shelf-like extension of continental land masses known as the continental shelf. Compare *open sea*.

coevolution Evolution in which two or more species interact and exert selective pressures on each other that can lead each species to undergo adaptations. See *evolution, natural selection*.

cogeneration Production of two useful forms of energy, such as high-temperature heat or steam and electricity, from the same fuel source.

colony collapse disorder (CCD) The occurrence of very large losses of European honeybees colonies in the United States and in parts of Europe.

combined heat and power (CHP) production See *cogeneration*.

commensalism An interaction between organisms of different species in which one type of organism benefits and the other type is neither helped nor harmed to any great degree. Compare *mutualism*.

commercial forest See *tree plantation*.

community Populations of all species living and interacting in an area at a particular time.

complex carbohydrate Two or more monomers of simple sugars (such as glucose) linked together. Two examples are starches and cellulose.

compost Partially decomposed organic plant and animal matter used as a soil conditioner or fertilizer.

composting A form of recycling that mimics nature by using bacteria to decompose yard trimmings, vegetable food scraps, and other biodegradable organic wastes into materials than can be used to increase soil fertility.

compound Combination of atoms, or oppositely charged ions, of two or more elements held together by attractive forces

called chemical bonds. Examples are NaCl, CO_2, and $C_6H_{12}O_6$. Compare *element*.

coniferous evergreen plants Cone-bearing plants (such as spruces, pines, and firs) that keep most of their narrow, pointed leaves (needles) all year. Compare *broadleaf deciduous plants, broad-leaf evergreen plants*.

coniferous trees Cone-bearing trees, mostly evergreens, that have needle-shaped or scale-like leaves. They produce wood known commercially as softwood. Compare *deciduous plants*.

conservation biology Multidisciplinary science created to deal with the crisis of accelerating losses of the genes, species, communities, and ecosystems that make up earth's biological diversity.

conservation-tillage farming Crop cultivation in which the soil is disturbed little (minimum-tillage farming) or not at all (no-till farming) in an effort to reduce soil erosion, lower labor costs, and save energy. Compare *conventional-tillage farming*.

consumer Organism that cannot synthesize the organic nutrients it needs and gets its organic nutrients by feeding on the tissues of producers or of other consumers; generally divided into *primary consumers* (herbivores), *secondary consumers* (carnivores), *tertiary (higher-level) consumers*, *omnivores*, and *detritivores* (decomposers and detritus feeders). In economics, one who uses economic goods. Compare *producer*.

continental drift The slow movement of the continents atop the earth's mantle.

contour farming Plowing and planting across the changing slope of land, rather than in straight lines, to help retain water and reduce soil erosion.

contour strip mining Form of surface mining used on hilly or mountainous terrain. A power shovel cuts a series of terraces into the side of a hill. An earthmover removes the overburden, and a power shovel extracts the coal. The overburden from each new terrace is dumped onto the one below. Compare *area strip mining, mountaintop removal, open-pit mining, subsurface mining*.

convergent plate boundary Area where the earth's tectonic plates are pushed together. Compare *divergent plate boundary, transform fault*.

coral reef Formation produced by massive colonies containing billions of tiny coral animals, called polyps, that secrete a stony substance (calcium carbonate) around themselves for protection. When the corals die, their empty outer skeletons form layers; reefs grow with the accumulation of such layers. They are found in the coastal zones of warm tropical and subtropical oceans.

core Inner zone of the earth that consists of a solid inner core and a liquid outer core. Compare *crust, mantle*.

corrective feedback loop See *negative feedback loop*.

crop rotation Planting a field, or an area of a field, with different crops from year to year to reduce soil nutrient depletion. A plant such as corn, tobacco, or cotton, which removes large amounts of nitrogen from the soil, is planted one year. The next year a legume such as soybeans, which adds nitrogen to the soil, is planted.

crown fire Extremely hot forest fire that burns ground vegetation and treetops. Compare *ground fire, surface fire*.

crude birth rate Annual number of live births per 1,000 people in the population of a geographic area at the midpoint of a given year. Compare *crude death rate*.

crude death rate Annual number of deaths per 1,000 people in the population of a geographic area at the midpoint of a given year. Compare *crude birth rate*.

crude oil Gooey liquid consisting mostly of hydrocarbon compounds and small amounts of compounds containing oxygen, sulfur, and nitrogen. Extracted from underground accumulations, it is sent to oil refineries, where it is converted to heating oil, diesel fuel, gasoline, tar, and other materials.

crust Solid outer zone of the earth. It consists of oceanic crust and continental crust. Compare *core, mantle*.

cultural carrying capacity The limit on population growth that would allow most people in an area or in the world to live in reasonable comfort and freedom without impairing the ability of the planet to sustain future generations.

cultural eutrophication Process in which human inputs of plant nutrients (mostly nitrates and phosphates) through the atmosphere and in runoff from urban and agricultural areas within a lake's watershed accelerate the eutrophication of the lake. See *eutrophication*.

dam A structure built across a river to control the river's flow or to create a reservoir. See *reservoir*.

data Factual information collected by scientists.

death rate See *crude death rate*.

deciduous plants Trees, such as oaks and maples, and other plants that survive during dry or cold seasons by shedding their leaves. Compare *coniferous trees, succulent plants*.

decomposer Organism that digests parts of dead organisms, as well as cast-off fragments and wastes of living organisms, by breaking down the complex organic molecules in those materials into simpler inorganic compounds and then absorbing the soluble nutrients. Producers return most of these chemicals to the soil and water for reuse. Decomposers consist of various bacteria and fungi. Compare *consumer, detritivore, producer*.

deforestation Removal of trees from a forested area.

demographic transition Hypothesis that countries, as they become industrialized, have declines in death rates followed by declines in birth rates.

depletion time The time it takes to use a certain fraction (usually 80%) of the known or estimated supply of a nonrenewable resource at an assumed rate of use. Finding and extracting the remaining 20% usually costs more than it is worth.

desalination Purification of salt water or brackish (slightly salty) water by removal of dissolved salts.

desert Biome in which evaporation exceeds precipitation and the average amount of precipitation is less than 25 centimeters (10 inches) per year. Such areas have little vegetation or have widely spaced, mostly low vegetation. Compare *forest, grassland*.

desertification Conversion of rangeland, rain-fed cropland, or irrigated cropland to desert-like land, with a drop in agricultural productivity of 10% or more. It usually is caused by a combination of overgrazing, soil erosion, prolonged drought, and climate change.

detritivore Consumer organism that feeds on detritus, parts of dead organisms, and cast-off fragments and wastes of living organisms. Examples include earthworms, termites, and crabs. Compare *decomposer*.

detritus Parts of dead organisms and cast-off fragments and wastes of living organisms.

detritus feeder See *detritivore*.

dissolved oxygen (DO) content Amount of oxygen gas (O_2) dissolved in a given volume of water at a particular temperature and pressure, often expressed as a concentration in parts of oxygen per million parts of water.

divergent plate boundary Area where the earth's tectonic plates move apart in opposite directions. Compare *convergent plate boundary, transform fault*.

dose Amount of a potentially harmful substance an individual ingests, inhales, or absorbs through the skin. Compare *response*. See *dose-response curve*.

dose-response curve Plot of data showing the effects of various doses of a toxic agent on a group of test organisms. See *dose, response*.

drainage basin See *watershed*.

drought Condition in which an area does not get enough water because of lower-than-normal precipitation or higher-than-normal temperatures that increase evaporation.

earth-centered environmental worldview Worldview holding that we are part of, and dependent on, nature; that the earth's life-support system exists for all species, not just for us; that our economic success and the long-term survival of our cultures and our species depend on learning how the earth has sustained itself for billions of years and integrating such lessons from nature into the ways we think and act. Compare *human-centered environmental worldview* and *life-centered environmental worldview*.

earthquake Shaking of the ground resulting from the fracturing and displacement of subsurface rock, which produces a fault, or from subsequent movement along the fault.

eco-city City in which long-term environmental sustainability is the main focus. Residents are able to walk, bike, or use low-polluting mass transit for most of their travel; buildings, vehicles, and appliances meet high energy-efficiency standards; trees and plants adapted to the local climate and soils are planted throughout the city to provide shade, beauty, and wildlife habitats, and to reduce air pollution, noise, and soil erosion.

ecological diversity The variety of forests, deserts, grasslands, oceans, streams, lakes, and other biological communities. See *biodiversity*. Compare *functional diversity, genetic diversity, species diversity*.

ecological footprint Amount of biologically productive land and water needed to supply a population with the renewable resources it uses and to absorb or dispose of the wastes from such resource use. It is a measure of the average environmental impact of populations in different countries and areas. See *per capita ecological footprint*.

ecological niche Total way of life, or role of a species in an ecosystem. It includes all physical, chemical, and biological conditions that a species needs in order to live and reproduce in an ecosystem.

ecological restoration Deliberate alteration of a degraded habitat or ecosystem to restore as much of its ecological structure and function as possible.

ecological succession Process in which communities of plant and animal species in a particular area are replaced over time by a series of different and often more complex communities. See *primary ecological succession, secondary ecological succession*.

ecological tipping point Point at which an environmental problem reaches a threshold level, which causes an often irreversible shift in the behavior of a natural system.

ecologist Biological scientist who studies relationships between living organisms and their environment.

ecology Biological science that studies the relationships between living organisms and their environment.

economic depletion Exhaustion of 80% of the estimated supply of a nonrenewable resource. Finding, extracting, and processing the remaining 20% usually costs more than it is worth. May also apply to the depletion of a renewable resource, such as a fish or tree species.

economic development Improvement of human living standards by economic growth. Compare *economic growth*.

economic growth Increase in the capacity to provide people with goods and services; an increase in gross domestic product (GDP). Compare *economic development*. See *gross domestic product*.

economics Social science that deals with the production, distribution, and consumption of goods and services to satisfy people's needs and wants.

ecosystem One or more communities of different species interacting with one another and with the chemical and physical factors making up their nonliving environment.

ecosystem services Natural services that support life on the earth and are essential to the quality of human life and the functioning of the world's economies. Examples are the chemical cycles, natural pest control, and natural purification of air and water. See *natural resources*.

electromagnetic radiation Forms of kinetic energy traveling as electromagnetic waves. Examples include radio waves, TV waves, microwaves, infrared radiation, visible light, ultraviolet radiation, X-rays, and gamma rays.

electron (e) Tiny particle moving around outside the nucleus of an atom. Each electron has one unit of negative charge and almost no mass. Compare *neutron, proton*.

element Chemical, such as hydrogen (H), iron (Fe), sodium (Na), carbon (C), nitrogen (N), or oxygen (O), whose distinctly different atoms serve as the basic building blocks of all matter.

endangered species Wild species with so few individual survivors that the species could soon become extinct in all or most of its natural range. Compare *threatened species*.

Endangered Species Act (ESA) U.S. law passed in 1973, designed to identify and protect endangered species in the United States and abroad. It creates recovery programs for the species it lists with the goal of helping the populations of protected species recover to levels where legal protection is no longer needed.

endemic species Species that is found in only one area. Such species are especially vulnerable to extinction.

energy Capacity to do work or to transfer heat.

energy conservation Reducing or eliminating unnecessary waste of energy.

energy efficiency Percentage of the total energy input that does useful work and is not converted into low-quality, generally useless heat in an energy conversion system or process. See *energy quality*, *net energy*.

energy quality Ability of a form of energy to do useful work. High-temperature heat and the chemical energy in fossil fuels are examples of concentrated high-quality energy. Low-quality energy such as low-temperature heat is dispersed or diluted and cannot do much useful work. See *high-quality energy*, *low-quality energy*.

environment All external conditions, factors, matter, and energy, living and nonliving, that affect any living organism or other specified system.

environmental degradation Depletion or destruction of a potentially renewable resource such as soil, grassland, forest, or wildlife that is used faster than it is naturally replenished. If such use continues, the resource becomes nonrenewable (on a human time scale) or nonexistent (extinct). See also *natural capital degradation*, *sustainable yield*.

environmental ethics Human beliefs about what is right or wrong with how we treat the environment.

environmentalism Social movement dedicated to protecting the earth's life support systems for us and other species.

environmental justice Fair treatment and meaningful involvement of all people, regardless of race, color, sex, national origin, or income, with respect to the development, implementation, and enforcement of environmental laws, regulations, and policies.

environmentally sustainable economic development An approach that uses political and economic systems to encourage environmentally beneficial and more sustainable forms of economic growth and to discourage environmentally harmful forms of economic growth that degrade natural capital.

environmental movement Citizens organized to demand that political leaders enact laws and develop policies to curtail pollution, clean up polluted environments, and protect unspoiled areas from environmental degradation.

environmental policy Laws, rules, and regulations related to an environmental problem that are developed, implemented, and enforced by a particular government body or agency.

environmental resistance All of the limiting factors that act together to limit the growth of a population. See *limiting factor*.

environmental science Interdisciplinary study that uses information and ideas from the physical sciences (such as biology, chemistry, and geology) as well as those from the social sciences and humanities (such as economics, politics, and ethics) to learn how nature works, how we interact with the environment, and how we can to help deal with environmental problems.

environmental scientist Scientist who uses information from the physical sciences and social sciences to understand how the earth works, learn how humans interact with the earth, and develop solutions to environmental problems. See *environmental science*.

environmental wisdom worldview Worldview holding that humans are part of and totally dependent on nature and that nature exists for all species, not just for us. Our success depends on learning how the earth sustains itself and integrating such environmental wisdom into the ways we think and act. See *earth-centered environmental worldview*. Compare *frontier worldview*, *planetary management worldview*, *stewardship worldview*.

environmental worldview Set of assumptions and beliefs about how people think the world works, what they think their role in the world should be, and what they believe is right and wrong environmental behavior (environmental ethics). See *earth-centered environmental worldview*, *human-centered environmental worldview*, and *life-centered environmental worldview*.

EPA U.S. Environmental Protection Agency; responsible for managing federal efforts to control air and water pollution, radiation and pesticide hazards, environmental research, hazardous waste, and solid waste disposal.

erosion Process or group of processes by which loose or consolidated earth materials, especially topsoil, are dissolved, loosened, or worn away and removed from one place and deposited in another. See *weathering*.

estuary Partially enclosed coastal area at the mouth of a river from which freshwater, carrying fertile silt and runoff from the land, mixes with salty seawater.

eutrophication Physical, chemical, and biological changes that take place after a lake, estuary, or slow-flowing stream receives inputs of plant nutrients—mostly nitrates and phosphates—from natural erosion and runoff from the surrounding land basin. See *cultural eutrophication*.

eutrophic lake Lake with a large or excessive supply of plant nutrients, mostly nitrates and phosphates. Compare *oligotrophic lake*.

evaporation Conversion of a liquid into a gas.

evergreen plants Plants that keep some of their leaves or needles throughout the year. Examples include cone-bearing trees (conifers) such as firs, spruces, pines, redwoods, and sequoias. Compare *deciduous plants*, *succulent plants*.

evolution See *biological evolution*.

experiment Procedure a scientist uses to study some phenomenon under known conditions. Scientists conduct some experiments in the laboratory and others in nature. The resulting scientific data or facts must be verified or confirmed by repeated observations and measurements, ideally by several different investigators.

exponential growth Growth in which some quantity, such as population size or economic output, increases at a constant rate per unit of time. An example is the growth sequence 2, 4, 8, 16, 32, 64, and so on, which increases by 100% at each interval. When the increase in quantity over time is plotted, this type of growth yields a curve shaped like the letter J.

external cost Harmful environmental, economic, or social effect of producing and using an economic good that is not included in the market price of the good. Compare *internal cost*. See *full cost pricing*.

extinction See *biological extinction*.

extinction rate Percentage or number of species that go extinct within a certain period of time such as a year.

family planning Providing information, clinical services, and contraceptives to help people choose the number and spacing of children they want to have.

feedback Any system output of matter, energy, or information that, when fed back into the system, increases or decreases a change to the system. See *feedback loop*, *negative feedback loop*, and *positive feedback loop*.

feedback loop The process that occurs when an output of matter, energy, or information is fed back into the system as an input and leads to changes in that system. See *feedback*. Compare *negative feedback loop* and *positive feedback loop*.

feed-in-tariff Long-term contract that requires utilities to buy electricity produced by homeowners and businesses from renewable energy resources at a price that guarantees a good return and to feed it into the electrical grid.

feedlot Confined outdoor or indoor space used to raise hundreds to thousands of domesticated livestock.

first law of thermodynamics Whenever energy is converted from one form to another in a physical or chemical change, no energy is created or destroyed, but energy can be changed from one form to another; you cannot get more energy out of something than you put in; in terms of energy quantity, you cannot get something for nothing. This law does not apply to nuclear changes, in which large amounts of energy can be produced from small amounts of matter. See *second law of thermodynamics*.

fishery Concentration of particular aquatic species suitable for commercial harvesting in a given ocean area or inland body of water.

fish farming See *aquaculture*.

fishprint Area of ocean needed to sustain the consumption of an average person, a nation, or the world. Compare *ecological footprint*.

floodplain Flat valley floor next to a stream channel. For legal purposes, the term often applies to any low area that has the potential for flooding, including certain coastal areas.

flows See *throughputs*.

food chain Series of organisms in which each eats or decomposes the preceding one. Compare *food web*.

food insecurity Condition under which people live with chronic hunger and malnutrition that threatens their ability to lead healthy and productive lives. Compare *food security*.

food security Condition under which every person in a given area has daily access to enough nutritious food to have an active and healthy life. Compare *food insecurity*.

food web Complex network of many interconnected food chains and feeding relationships. Compare *food chain*.

forest Biome with enough average annual precipitation to support the growth of tree species and smaller forms of vegetation. Compare *desert, grassland*.

fossil fuel Product of partial or complete decomposition of plants and animals; occurs as crude oil, coal, natural gas, or

heavy oil as a result of exposure to heat and pressure in the earth's crust over millions of years. See *coal, natural gas*.

fossils Skeletons, bones, shells, body parts, leaves, seeds, or impressions of such items that provide recognizable evidence of organisms that lived long ago.

fracking See *hydraulic fracturing*.

freshwater Relatively pure water containing few dissolved salts.

freshwater life zones Aquatic systems where water with a dissolved salt concentration of less than 1% by volume accumulates on or flows through the surfaces of terrestrial biomes. Examples include *standing* (lentic) bodies of freshwater such as lakes, ponds, and inland wetlands and *flowing* (lotic) systems such as streams and rivers. Compare *biome*.

full-cost pricing Setting market prices of goods and services to include hidden harmful environmental and health costs of producing and using them. See *external cost, internal cost*.

functional diversity Biological and chemical processes or functions such as energy flow and matter cycling needed for the survival of species and biological communities. See *biodiversity, ecological diversity, genetic diversity, species diversity*.

GDP See *gross domestic product*.

generalist species Species with a broad ecological niche. They can live in many different places, eat a variety of foods, and tolerate a wide range of environmental conditions. Examples include flies, cockroaches, mice, rats, and humans. Compare *specialist species*.

genes Coded units of information about specific traits that are passed from parents to offspring during reproduction. They consist of segments of DNA molecules found in chromosomes.

genetically modified organism (GMO) Organism whose genetic makeup has been altered by genetic engineering.

genetic diversity Variability in the genetic makeup among individuals within a single species. See *biodiversity*. Compare *ecological diversity, functional diversity, species diversity*.

genetic engineering Insertion of an alien gene into an organism to give it a beneficial genetic trait. Compare *artificial selection, natural selection*.

genetic variability Variety in the genetic makeup of individuals in a population.

genuine progress indicator (GPI) GDP plus the estimated value of beneficial transactions that meet basic needs, but in which no money changes hands, minus the estimated harmful environmental,

health, and social costs of all transactions. Compare *gross domestic product*.

geoengineering Any use of technology to try to manipulate certain natural conditions in order to reduce atmospheric warming and climate change.

geographic isolation Separation of populations of a species into different areas for long periods of time. Compare *reproductive isolation*.

geology Study of the earth's dynamic history. Geologists study and analyze rocks and the features and processes of the earth's interior and surface.

geosphere Earth's intensely hot core, thick mantle composed mostly of rock, and thin outer crust that contains most of the earth's rock, soil, and sediment. Compare *atmosphere, biosphere, hydrosphere*.

geothermal energy Heat transferred from the earth's underground concentrations of dry steam (steam with no water droplets), wet steam (a mixture of steam and water droplets), or hot water trapped in fractured or porous rock.

global warming Warming of the earth's lower atmosphere (troposphere) because of increases in the concentrations of one or more greenhouse gases. It can result in climate change that can last for decades to thousands of years. See *greenhouse effect, greenhouse gases, natural greenhouse effect*.

GMO See *genetically modified organism*.

GPI See *genuine progress indicator*.

GPP See *gross primary productivity*.

graph A tool for conveying information that we can summarize numerically by illustrating that information in a visual format.

grassland Biome found in regions where enough annual average precipitation to support the growth of grass and small plants but not enough to support large stands of trees. Compare *desert, forest*.

gray water Used water from bathtubs, showers, sinks, dishwashers, and clothes washers that can be stored in a holding tank and reused to irrigate lawns and non-edible plants, flush toilets, and wash cars.

greenhouse effect Natural effect that releases heat in the atmosphere near the earth's surface. Water vapor, carbon dioxide, ozone, and other gases in the lower atmosphere (troposphere) absorb some of the infrared radiation (heat) radiated by the earth's surface. Their molecules vibrate and transform the absorbed energy into longer-wavelength infrared radiation in the troposphere. If the atmospheric concentrations of these greenhouse gases increase and other natural processes do not remove them, the average temperature of the lower atmosphere will increase.

greenhouse gases Gases in the earth's lower atmosphere (troposphere) that cause the greenhouse effect. Examples include carbon dioxide, chlorofluorocarbons, ozone, methane, water vapor, and nitrous oxide.

green manure Freshly cut or still-growing green vegetation that is plowed into the soil to increase the organic matter and humus available to support crop growth. Compare *animal manure*.

green revolution Popular term for the introduction of scientifically bred or selected varieties of grain (rice, wheat, corn) that, with adequate inputs of fertilizer and water, can greatly increase crop yields.

gross domestic product (GDP) Annual market value of all goods and services produced by all firms and organizations, foreign and domestic, operating within a country. See *per capita GDP*. Compare *genuine progress indicator (GPI)*.

gross primary productivity (GPP) Rate at which an ecosystem's producers capture and store a given amount of chemical energy as biomass in a given length of time. Compare *net primary productivity*.

ground fire Fire that burns decayed leaves or peat deep below the ground's surface. Compare *crown fire, surface fire*.

groundwater Water that sinks into the soil and is stored in slowly flowing and slowly renewed underground reservoirs called *aquifers*; underground water in the zone of saturation, below the water table. Compare *runoff, surface water*.

habitat Place or type of place where an organism or population of organisms lives. Compare *ecological niche*.

habitat fragmentation Breakup of a habitat into smaller pieces, usually as a result of human activities.

hazardous chemical Chemical that can cause harm because it is flammable or explosive, can irritate or damage the skin or lungs (such as strong acidic or alkaline substances), or can cause allergic reactions of the immune system (allergens). See also *toxic chemical*.

hazardous waste Any solid, liquid, or containerized gas that can catch fire easily, is corrosive to skin tissue or metals, is unstable and can explode or release toxic fumes, or has harmful concentrations of one or more toxic materials that can leach out. Sometimes called *toxic waste*.

heat Total kinetic energy of all randomly moving atoms, ions, or molecules within a given substance, excluding the overall motion of the whole object. Heat always flows spontaneously from a warmer sample of matter to a colder sample of matter. This is one way to state the *second law of thermodynamics*.

herbivore Plant-eating organism. Examples include deer, sheep, grasshoppers, and zooplankton. Compare *carnivore, omnivore*.

heterotroph See *consumer*.

high-grade ore Ore containing a large amount of a desired mineral. Compare *low-grade ore*.

high-input agriculture See *industrialized agriculture*.

high-quality energy Energy that is concentrated and has great ability to perform useful work. Examples include high-temperature heat and the energy in electricity, coal, oil, gasoline, sunlight, and nuclei of uranium-235. Compare *low-quality energy*.

high-quality matter Matter that is concentrated and contains a high concentration of a useful resource. Compare *low-quality matter*.

high-throughput economy Economic system in most advanced industrialized countries, in which ever-increasing economic growth is sustained by maximizing the rate at which matter and energy resources are used, with little emphasis on pollution prevention, recycling, reuse, reduction of unnecessary waste, and other forms of resource conservation. Compare *low-throughput economy*.

high-waste economy See *high-throughput economy*.

HIPPCO Acronym used by conservation biologists for the six most important secondary causes of premature extinction: **H**abitat destruction, degradation, and fragmentation; **I**nvasive (nonnative) species; **P**opulation growth (too many people consuming too many resources); **P**ollution; **C**limate change; and **O**verexploitation.

Holocene Period of relatively stable climate and other environmental conditions that followed the long period of glaciation; it has allowed the human population to grow, develop agriculture, and take over a large and growing share of the earth's land and other resources. Compare *Anthropocene*.

horizontal drilling A method of drilling first vertically to a certain point, then bending the flexible well bore and drilling horizontally to gain access to oil and natural gas deposits trapped within layers of shale or other rock deposits.

human capital People's physical and mental talents that provide labor, innovation, culture, and organization. Compare *manufactured capital, natural capital*.

human-centered (anthropocentric) environmental worldview Worldview that sees the natural world primarily as a support system for human life. Includes the *planetary management worldview*, which holds that humans are separate from and in charge of nature and that we can

manage the earth mostly for our benefit, into the distant future; and the *stewardship worldview*, which holds that we can and should manage the earth for our benefit, but we have an ethical responsibility to be caring and responsible managers, or *stewards*, of the earth. Compare *earth-centered environmental worldview* and *life-centered environmental worldview*.

human resources See *human capital*.

hunger See <u>chronic</u> undernutrition.

hydraulic fracturing Process of injecting water mixed with sand and some toxic chemicals underground through horizontal natural gas wells and then using explosives and high pressure to fracture the deep rock and free up the natural gas stored there. The gas flows out of the well along with much of the water and a mix of compounds pulled from the rocks, including salts, toxic heavy metals, and naturally occurring radioactive materials. Commonly referred to as *fracking*.

hydroelectric power plant Structure in which the energy of falling or flowing water spins a turbine generator to produce electricity.

hydrogen fuel cell Device that uses hydrogen gas (H_2) as a fuel to produce electricity when it reacts with oxygen gas (O_2) in the atmosphere; it emits only water vapor.

hydrologic cycle Biogeochemical cycle that collects, purifies, and distributes the earth's fixed supply of water from the environment to living organisms and then back to the environment.

hydropower Electrical energy produced by falling or flowing water. See *hydroelectric power plant*.

hydrosphere Earth's *liquid water* (oceans, lakes, other bodies of surface water, and underground water), *frozen water* (polar ice caps, floating ice caps, and ice in soil, known as permafrost), and *water vapor* in the atmosphere. See also *hydrologic cycle*. Compare *atmosphere, biosphere, geosphere*.

igneous rock Rock formed when molten rock material (magma) wells up from the earth's interior, cools, and solidifies into rock masses. Compare *metamorphic rock, sedimentary rock*. See *rock cycle*.

immigration Migration of people into a country or area to take up permanent residence.

indicator species Species whose decline serves as an early warning that a community or ecosystem is being degraded. Compare *keystone species, native species, nonnative species*.

industrialized agriculture (high-input agriculture) Production of large quantities of crops and livestock for domestic and foreign sale; involves use of large inputs of energy from fossil fuels (especially oil and natural gas), water, fertilizer, and pesticides.

industrial smog Type of air pollution consisting mostly of a mixture of sulfur dioxide, suspended droplets of sulfuric acid formed from some of the sulfur dioxide, and suspended solid particles. Compare *photochemical smog*.

industrial solid waste Solid waste produced by mines, factories, refineries, food growers, and businesses that supply people with goods and services. Compare *municipal solid waste*.

inertia (persistence) The ability of a living system such as a grassland or a forest to survive moderate disturbances.

infant mortality rate Number of babies out of every 1,000 born each year who die before their first birthday.

infectious disease Disease caused when a pathogen such as a bacterium, virus, or parasite invades the body and multiplies in its cells and tissues. Examples are flu, HIV, malaria, tuberculosis, and measles. See *transmissible disease*. Compare *nontransmissible disease*.

inland wetland Land away from the coast, such as a swamp, marsh, or bog, that is covered all or part of the time with freshwater. Compare *coastal wetland*.

inorganic compounds All compounds not classified as organic compounds. See *organic compounds*.

inorganic fertilizer Commercially prepared mixture of inorganic plant nutrients such as nitrates, phosphates, and potassium applied to the soil to restore fertility and increase crop yields. Compare *organic fertilizer*.

inputs Matter and energy from the environment that is put into a system. See *system*. Compare *outputs* and *throughputs*.

instrumental value Value of an organism, species, ecosystem, or the earth's biodiversity based on its usefulness to humans. Compare *intrinsic value*.

integrated pest management (IPM) Combined use of biological, chemical, and cultivation methods in proper sequence and timing to keep the size of a pest population below the level that causes economically unacceptable loss of a crop or livestock animal.

integrated waste management Variety of strategies for both waste reduction and waste management designed to deal with the solid wastes we produce.

internal cost Direct cost paid by the producer and the buyer of an economic good. Compare *external cost*. See *full cost pricing*.

interspecific competition Attempts by members of two or more species to use the same limited resources in an ecosystem.

intrinsic value Value of an organism, species, ecosystem, or the earth's biodiversity based on its existence, regardless of whether it has any usefulness to humans. Compare *instrumental value*.

invasive species See *nonnative species*.

ion Atom or group of atoms with one or more positive (+) or negative (–) electrical charges. Examples are Na$^+$ and Cl$^-$. Compare *atom, molecule*.

IPM See *integrated pest management*.

irrigation Supplying water to crops by artificial means rather than by relying on natural rainfall.

isotopes Two or more forms of a chemical element that have the same number of protons but different mass numbers because they have different numbers of neutrons in their nuclei.

keystone species Species that play roles affecting many other organisms in an ecosystem. Compare *indicator species, native species, nonnative species*.

kinetic energy Energy that matter has because of its mass and speed, or velocity. Compare *potential energy*.

K-selected species Species that tend to do well in competitive conditions when their population size is near the carrying capacity (*K*) of their environment. They tend to reproduce later in life and have a small number of offspring with fairly long life spans. Examples include elephants, whales, humans, birds of prey, saguaro cactus, and most tropical rain forest trees. Compare *r-selected species*.

lake Large natural body of standing freshwater formed when water from precipitation, land runoff, or groundwater flow fills a depression in the earth created by glaciation, earth movement, volcanic activity, or a giant meteorite. See *eutrophic lake, oligotrophic lake*.

landfill See *sanitary landfill*.

land-use planning Planning to determine the best present and future uses of each parcel of land.

law of conservation of energy See *first law of thermodynamics*.

law of conservation of matter In any physical or chemical change, matter is neither created nor destroyed but merely changed from one form to another; in physical and chemical changes, existing atoms are rearranged into different spatial patterns (physical changes) or different combinations (chemical changes).

law of nature See *scientific law*.

LD50 See *median lethal dose*.

less-developed country Country that has low to moderate industrialization and low to moderate per capita GDP. Most are located in Africa, Asia, and Latin America. Compare *more-developed country*.

life-centered environmental worldview Worldview holding that all species have value as participating members of the biosphere, regardless of their potential or actual use to humans; includes the belief that we have an ethical responsibility to avoid hastening the extinction of species through our activities. Compare *earth-centered environmental worldview* and *human-centered environmental worldview*.

life expectancy Average number of years a newborn infant can be expected to live.

limiting factor Single factor that limits the growth, abundance, or distribution of the population of a species in an ecosystem. See *limiting factor principle*.

limiting factor principle Too much or too little of any abiotic factor can limit or prevent growth of a population of a species in an ecosystem, even if all other factors are at or near the optimal range of tolerance for the species.

lipids A chemically diverse group of large organic compounds that do not dissolve in water. Examples are fats and oils for storing energy, waxes for structure, and steroids for producing hormones.

liquefied natural gas (LNG) Natural gas converted to liquid form by cooling it to a very low temperature.

liquefied petroleum gas (LPG) Mixture of liquefied propane (C$_3$H$_8$) and butane (C$_4$H$_{10}$) gas removed from natural gas and used as a fuel.

lithosphere Outer shell of the earth, composed of the crust and the rigid, outermost part of the mantle outside the asthenosphere. Compare *crust, geosphere, mantle*.

LNG See *liquefied natural gas*.

logistic growth Pattern in which exponential population growth occurs when the population is small, and population growth decreases steadily with time as the population approaches the carrying capacity.

low-grade ore Ore containing a small amount of a desired mineral. Compare *high-grade ore*.

low-quality energy Energy that is dispersed and has little ability to do useful work. An example is low-temperature heat. Compare *high-quality energy*.

low-quality matter Matter that is dilute or dispersed or contains a low concentration of a useful resource. Compare *high-quality matter*.

low-throughput (low-waste) economy Economy based on working with nature by recycling and reusing discarded matter; preventing pollution; conserving matter and energy resources by reducing unnecessary waste and use; and building things that are easy to recycle, reuse, and repair. Compare *high-throughput economy*.

low-waste economy See *low-throughput economy*.

LPG See *liquefied petroleum gas*.

malnutrition See *chronic malnutrition*.

mangrove forest Ecosystem, found on some coastlines in warm tropical climates, that may contain any of 69 species that can live partly submerged in the salty environment of coastal swamps.

mantle Zone of the earth's interior between its core and its crust. Compare core, crust. See *geosphere, lithosphere*.

manufactured capital Manufactured items made from natural resources and used to produce and distribute economic goods and services; includes tools, machinery, equipment, factories, and transportation and distribution facilities. Compare *human capital, natural capital*.

marine life zone See *saltwater life zone*.

mass extinction Catastrophic, widespread, often global event in which major groups of species are wiped out over a short time compared with normal (background) extinctions. Compare *background extinction*.

mass number Sum of the number of neutrons and the number of protons in the nucleus of an atom. It gives the approximate mass of that atom. Compare *atomic number*.

matter Anything that has mass (the amount of material in an object) and takes up space. On the earth, where gravity is present, we weigh an object to determine its mass.

median lethal dose (LD50) Amount of a toxic material per unit of body weight of test animals that kills half the test population in a certain time.

megacity City with 10 million or more people.

metamorphic rock Rock produced when a preexisting rock is subjected to high temperatures (which may cause it to melt partially), high pressures, chemically active fluids, or a combination of these agents. Compare *igneous rock, sedimentary rock*. See *rock cycle*.

microorganisms Organisms such as bacteria that are so small that it takes a microscope to see them.

migration Movement of people into and out of specific geographic areas.

mineral Any naturally occurring inorganic substance found in the earth's crust as a crystalline solid. See *mineral resource*.

mineral resource Concentration of mineral material in or on the earth's crust in a form and amount such that extracting and converting it into useful materials or

items is currently or potentially profitable. Mineral resources are classified as *metallic* (such as iron and tin ores) or *nonmetallic* (such as sand and salt).

minimum-tillage farming See *conservation-tillage farming*.

mining Any of a variety of processes by which mineral resources are extracted from the earth's crust. See *mountaintop removal mining, open-pit mining, strip-mining, subsurface mining, surface mining*.

model Approximate representation or simulation of a system being studied.

molecule Combination of two or more atoms of the same chemical element (such as O_2) or different chemical elements (such as H_2O) held together by chemical bonds. Compare *atom, ion*.

monoculture Cultivation of a single crop, usually on a large area of land. Compare *polyculture*.

more-developed country Country that is highly industrialized and has a high per capita GDP. Compare *less-developed country*.

mountaintop removal mining Type of surface mining that uses explosives, massive power shovels, and large machines called draglines to remove the top of a mountain and expose seams of coal underneath a mountain. Compare *area strip mining, contour strip mining*.

MSW See *municipal solid waste*.

municipal solid waste (MSW) Solid materials discarded by homes and businesses in or near urban areas. See *solid waste*. Compare *industrial solid waste*.

mutagen Agent such as a chemical or form of radiation that increases the frequency of mutations in the DNA molecules found in cells. See *carcinogen, mutation, teratogen*.

mutation Random change in DNA molecules making up genes that can alter anatomy, physiology, or behavior in offspring. See *mutagen*.

mutualism Type of species interaction in which both participating species generally benefit. Compare *commensalism*.

nanotechnology Uses science and engineering to manipulate and create materials out of atoms and molecules at the ultrasmall scale of less than 100 nanometers.

native species Species that normally live and thrive in a particular ecosystem. Compare *indicator species, keystone species, nonnative species*.

natural capital Natural resources and natural services that keep us and other species alive and support our economies. See *natural resources, natural services*.

natural capital degradation The waste, depletion, or destruction of any of the

earth's natural capital. See *environmental degradation*.

nature deficit disorder A wide range of problems, including anxiety, depression, and attention deficit disorders that might be resulting at least partially from a lack of contact with nature.

natural gas Underground deposits of gases consisting of 50–90% by weight methane gas (CH_4) and small amounts of heavier gaseous hydrocarbon compounds such as propane (C_3H_8) and butane (C_4H_{10}).

natural greenhouse effect See *greenhouse effect*.

natural income Renewable resources such as plants, animals, and soil provided by natural capital.

natural resources Materials, such as air, water, and soil, and forms of energy in nature that are essential or useful to humans. See *natural capital*.

natural selection Process by which a particular beneficial gene (or set of genes) is reproduced in succeeding generations more than other genes. The result of natural selection is a population that contains a greater proportion of organisms better adapted to certain environmental conditions. See *adaptation, biological evolution, mutation*.

negative feedback loop The process that occurs when an output of matter, energy, or information is fed back into the system as an input and slows or stops a change occurring to the system or causes the system to change in the opposite direction. See *system*. Compare *positive feedback loop*.

net energy The amount of high-quality energy available from an energy resource minus the high-quality energy needed to make the energy available.

net primary productivity (NPP) Rate at which all the plants in an ecosystem produce net useful chemical energy; it is equal to the difference between the rate at which the plants in an ecosystem produce useful chemical energy (gross primary productivity) and the rate at which they use some of that energy through cellular respiration. Compare gross *primary productivity*.

neutral solution Water solution containing an equal number of hydrogen ions (H^+) and hydroxide ions (OH^-); water solution with a pH of 7. Compare *acid solution, basic solution*.

neutron (n) Elementary particle in the nuclei of all atoms (except hydrogen-1). It has a relative mass of 1 and no electric charge. Compare *electron, proton*.

niche See *ecological niche*.

nitric oxide (NO) Colorless gas that forms when nitrogen and oxygen gas in air react at the high-combustion temperatures in automobile engines and coal-burning plants. Lightning and certain bacteria in soil and water also produce NO as part of the *nitrogen cycle*.

nitrogen cycle Cyclic movement of nitrogen in different chemical forms from the environment to organisms and then back to the environment.

nitrogen dioxide (NO_2) Reddish-brown gas formed when nitrogen oxide reacts with oxygen in the air.

nitrogen oxides (NO*x*) The collective term for nitric oxide and nitrogen dioxide. See *nitric oxide* and *nitrogen dioxide*.

noise pollution Any unwanted, disturbing, or harmful sound that impairs or interferes with hearing.

nondegradable pollutant Material that is not broken down by natural processes. Examples include the toxic elements lead and mercury. Compare *biodegradable pollutant*.

nonnative species Species that migrate into an ecosystem or are deliberately or accidentally introduced into an ecosystem by humans. Compare *native species*.

nonpoint source Broad and diffuse area, rather than a specific point, from which pollutants enter bodies of surface water or air. Examples include runoff of chemicals and sediments from cropland, livestock feedlots, logged forests, urban streets, parking lots, lawns, and golf courses. Compare *point source*.

nonrenewable resource Resource that exists in a fixed amount (stock) in the earth's crust and has the potential for renewal by geological, physical, and chemical processes taking place over hundreds of millions to billions of years. Examples include copper, aluminum, coal, and oil. We classify these resources as exhaustible because we are extracting and using them at a much faster rate than they are formed. Compare *renewable resource*.

nontransmissible disease Disease that is not caused by living organisms and does not spread from one person to another. Examples include most cancers, diabetes, cardiovascular disease, and malnutrition. Compare *transmissible disease*.

no-till farming See *conservation-tillage farming*.

NPP See *net primary productivity*.

nuclear change Process in which nuclei of certain isotopes spontaneously change, or are forced to change, into one or more different isotopes. The three principal types of nuclear change are radioactive decay, nuclear fission, and nuclear fusion. Compare *chemical change*, *physical change*.

nuclear fission Nuclear change in which the nuclei of certain isotopes with large mass numbers (such as uranium-235 and plutonium-239) are split apart into lighter nuclei when struck by a neutron. This process releases more neutrons and a large amount of energy. Compare *nuclear fusion*.

nuclear fuel cycle Includes the mining of uranium, processing and enriching the uranium to make fuel, using it in the reactor, safely storing the resulting highly radioactive wastes for thousands of years until their radioactivity falls to safe levels, and retiring the highly radioactive plant by taking it apart and storing its high- and moderate-level radioactive material safely for thousands of years.

nuclear fusion Nuclear change in which two nuclei of isotopes of elements with a low mass number (such as hydrogen-2 and hydrogen-3) are forced together at extremely high temperatures until they fuse to form a heavier nucleus (such as helium-4). This process releases a large amount of energy. Compare *nuclear fission*.

nucleic acid Large polymer molecule made by linking hundreds to thousands of four types of monomers called nucleotides. Two nucleic acids—DNA (deoxyribonucleic acid) and RNA (ribonucleic acid)—participate in the building of proteins and carry hereditary information used to pass traits from parent to offspring.

nucleus Extremely tiny center of an atom, making up most of the atom's mass. It contains one or more positively charged protons and one or more neutrons with no electrical charge (except for a hydrogen-1 atom, which has one proton and no neutrons in its nucleus).

nutrient Any chemical an organism must take in to live, grow, or reproduce.

nutrient cycle See *biogeochemical cycle*.

nutrient cycling See *chemical cycling*.

ocean currents Mass movements of surface water produced by prevailing winds blowing over the oceans.

oil sands See *tar sands*.

old-growth (primary) forest Virgin and old, second-growth forests containing trees that are often hundreds—sometimes thousands—of years old. Examples include forests of Douglas fir, western hemlock, giant sequoia, and coastal redwoods in the western United States.

oligotrophic lake Lake with a low supply of plant nutrients. Compare *eutrophic lake*.

omnivore Animal that can use both plants and other animals as food sources. Examples include pigs, rats, cockroaches, and humans. Compare *carnivore, herbivore*.

open access renewable resource Renewable resource owned by no one and available for use by anyone at little or no charge. Examples include clean air, underground water supplies, the open ocean and its fish, and the ozone layer. Compare *common property resource*.

open dump Fields or holes in the ground where garbage is deposited and sometimes covered with soil. They are rare in more-developed countries, but are widely used in many less-developed countries. Compare *sanitary landfill*.

open-pit mining Removing minerals such as gravel, sand, and metal ores by digging them out of the earth's surface and leaving an open pit behind. Compare *area strip mining, contour strip mining, mountaintop removal, subsurface mining*.

open sea Part of any ocean that lies beyond the continental shelf. Compare *coastal zone*.

ore Part of a metal-yielding material that can be economically extracted from a mineral; typically contains two parts: the ore mineral, which contains the desired metal, and waste mineral material (gangue). See *high-grade ore, low-grade ore*.

organic agriculture Growing crops without the use of synthetic pesticides, synthetic inorganic fertilizers, and genetically modified crops; raising livestock without use of antibiotics or growth hormones; and using organic fertilizer (manure, legumes, compost) and natural pest controls (bugs that eat harmful bugs, plants that repel bugs and environmental controls such as crop rotation).

organic compounds Compounds containing carbon atoms combined with each other and with atoms of one or more other elements such as hydrogen, oxygen, nitrogen, sulfur, phosphorus, chlorine, and fluorine. All other compounds are called *inorganic compounds*.

organic farming See *organic agriculture* and *sustainable agriculture*.

organic fertilizer Organic material such as animal manure, green manure, and compost applied to cropland as a source of plant nutrients. Compare *inorganic fertilizer*.

organism Any form of life.

outputs Matter and energy that leaves a living system and enters the environment. See *system*. Compare *inputs* and *throughputs*.

overburden Layer of soil and rock overlying a mineral deposit. Surface mining removes this layer.

overgrazing Destruction of vegetation when too many grazing animals feed too long on a specific area of pasture or rangeland and exceed the carrying capacity of a rangeland or pasture area.

overnutrition Diet so high in calories, saturated (animal) fats, salt, sugar, and processed foods, and so low in vegetables and fruits that the consumer runs a high risk of developing diabetes, hypertension, heart disease, and other health hazards. Compare *chronic malnutrition* and *chronic undernutrition*.

ozone (O₃) Colorless and highly reactive gas and a major component of photochemical smog. Also found in the ozone layer in the stratosphere. See *photochemical smog*.

ozone depletion Decrease in concentration of ozone (O₃) in the stratosphere. See *ozone layer*.

ozone layer Layer of gaseous ozone (O₃) in the stratosphere that protects life on earth by filtering out most harmful ultraviolet radiation from the sun.

parasite Consumer organism that lives on or in, and feeds on, a living plant or animal, known as the host, over an extended period. The parasite draws nourishment from and gradually weakens its host; it may or may not kill the host. See *parasitism*.

parasitism Interaction between species in which one organism, called the parasite, preys on another organism, called the host, by living on or in the host.

particulates Also known as suspended particulate matter (SPM); variety of solid particles and liquid droplets small and light enough to remain suspended in the air for long periods. About 62% of the SPM in outdoor air comes from natural sources such as dust, wild fires, and sea salt. The remaining 38% comes from human sources such as coal-burning electric power and industrial plants, motor vehicles, plowed fields, road construction, unpaved roads, and tobacco smoke.

passive solar heating system System that, without the use of mechanical devices, captures sunlight directly within a structure and converts it into low-temperature heat for space heating or for heating water for domestic use. Compare *active solar heating system*.

pasture Managed grassland or enclosed meadow that usually is planted with domesticated grasses or other forage to be grazed by livestock.

pathogen Living organism that can cause disease in another organism. Examples include bacteria, viruses, and parasites.

PCBs See *polychlorinated biphenyls*.

peak production Point in time when the pressure in an oil well drops and its rate of conventional crude oil production starts declining, usually a decade or so; for a group of wells or for a nation, the point at which all wells on average have passed peak production.

peer review Process of scientists reporting details of the methods and models they used, the results of their experiments, and the reasoning behind their hypotheses for other scientists working in the same field (their peers) to examine and criticize.

per capita carbon footprint Average carbon footprint per person in a population. Compare *carbon footprint*.

per capita ecological footprint Amount of biologically productive land and water needed to supply each person in a population with the renewable resources he or she uses and to absorb or dispose of the wastes from such resource use. It measures the average environmental impact of individuals in populations in different countries and areas. Compare *ecological footprint*.

per capita GDP Annual gross domestic product (GDP) of a country divided by its total population at midyear. See *gross domestic product*. Compare *genuine progress indicator (GPI)*.

perennial Plant that can live for more than 2 years. Compare *annual*.

permafrost Perennially frozen layer of the soil that forms when the water there freezes. It is found in arctic tundra.

perpetual resource Resource that is essentially inexhaustible on a human time scale because it is renewed continuously. Solar energy is an example. Compare *nonrenewable resource*, *renewable resource*.

persistence See *inertia*.

pest Unwanted organism that directly or indirectly interferes with human activities.

pesticide Any chemical designed to kill or inhibit the growth of an organism that people consider undesirable.

petrochemicals Chemicals obtained by refining (distilling) crude oil. They are used as raw materials in manufacturing most industrial chemicals, fertilizers, pesticides, plastics, synthetic fibers, paints, medicines, and many other products.

petroleum See *crude oil*.

pH Numeric value that indicates the relative acidity or alkalinity of a substance on a scale of 0 to 14, with the neutral point at 7. Acid solutions have pH values lower than 7; basic or alkaline solutions have pH values greater than 7.

phosphorus cycle Cyclic movement of phosphorus in different chemical forms from the environment to organisms and then back to the environment.

photochemical smog Complex mixture of air pollutants produced in the lower atmosphere by the reaction of hydrocarbons and nitrogen oxides under the influence of sunlight. Especially harmful components include ozone, peroxyacyl nitrates (PANs), and various aldehydes. Compare *industrial smog*.

photosynthesis Process taking place in cells of green plants in which radiant energy from the sun is used to combine carbon dioxide (CO₂) and water (H₂O) to produce oxygen (O₂), carbohydrates (such as glucose, C₆H₁₂O₆), and other nutrient molecules. Compare *aerobic respiration*.

photovoltaic (PV) cell Device that converts radiant (solar) energy directly into electrical energy. Also called a solar cell.

physical change Process that alters one or more physical properties of an element or a compound without changing its chemical composition. Examples include changing the size and shape of a sample of matter (crushing ice and cutting aluminum foil) and changing a sample of matter from one physical state to another (boiling and freezing water). Compare *chemical change, nuclear change*.

phytoplankton Small, drifting plants, mostly algae and bacteria, found in aquatic ecosystems. Compare *plankton, zooplankton*.

pioneer species First hardy species—often microbes, mosses, and lichens—that begin colonizing a site as the first stage of ecological succession. See *ecological succession*.

planetary management worldview Worldview holding that humans are separate from nature, that nature exists mainly to meet our needs and increasing wants, and that we can use our ingenuity and technology to manage the earth's life-support systems, mostly for our benefit. It assumes that economic growth is unlimited. Compare *environmental wisdom worldview, stewardship worldview*.

plankton Small plant organisms (phytoplankton) and animal organisms (zooplankton) that float in aquatic ecosystems.

plantation agriculture Growing specialized crops such as bananas, coffee, and cacao in tropical less-developed countries, primarily for sale to more-developed countries.

plate tectonics Theory of geophysical processes that explains the movements of lithospheric plates and the processes that occur at their boundaries. See *lithosphere, tectonic plates*.

point source Single identifiable source that discharges pollutants into the environment. Examples include the smokestack of a power plant or an industrial plant, drainpipe of a meatpacking plant, chimney of a house, or exhaust pipe of an automobile. Compare *nonpoint source*.

policies Programs, and the laws and regulations through which they are enacted, that a government enforces and funds.

politics Process through which individuals and groups try to influence or control government policies and actions that affect the local, state, national, and international communities.

pollutant Particular chemical or form of energy that can adversely affect the health, survival, or activities of humans or other living organisms. See *pollution*.

pollution Undesirable change in the physical, chemical, or biological characteristics of air, water, soil, or food that can adversely affect the health, survival, or activities of humans or other living organisms.

pollution cleanup Device or process that removes or reduces the level of a pollutant after it has been produced or has entered the environment. Examples include automobile emission control devices and sewage treatment plants. Compare *pollution prevention*.

pollution prevention Device, process, or strategy used to prevent a potential pollutant from forming or entering the environment or to sharply reduce the amount entering the environment. Compare *pollution cleanup*.

polychlorinated biphenyls (PCBs) Group of 209 toxic, oily, synthetic chlorinated hydrocarbon compounds that can be biologically amplified in food chains and webs.

polyculture Complex form of intercropping in which a large number of different plants maturing at different times are planted together. Compare *monoculture*.

population Group of individual organisms of the same species living in a particular area.

population change Increase or decrease in the size of a population. It is equal to (Births + Immigration) − (Deaths + Emigration).

population crash Dieback of a population that has used up its supply of resources, exceeding the carrying capacity of its environment. See *carrying capacity*.

population density Number of organisms in a particular population found in a specified area or volume.

population size Number of individual organisms in a population at a given time.

positive feedback loop The process that occurs when an output of matter, energy, or information is fed back into the system as an input and causes the system to change further in the same direction. See *system*. Compare *negative feedback loop*.

potential energy Energy stored in an object because of its position or the position of its parts. Compare *kinetic energy*.

poverty Inability of people to meet their basic needs for food, clothing, and shelter.

prairie See *grassland*.

precautionary principle When there is substantial preliminary evidence that an activity, technology, or chemical substance can can cause harm, decision makers should act to prevent harm to humans and to the environment. See *pollution prevention*.

precipitation Water in the form of rain, sleet, hail, and snow that falls from the atmosphere onto land and bodies of water.

predation Interaction in which an organism of one species (the predator) captures and feeds on some or all parts of an organism of another species (the prey).

predator Organism that captures and feeds on some or all parts of an organism of another species (the prey).

predator–prey relationship Relationship that has evolved between two organisms, in which one organism has become the prey for the other, the latter called the predator. See *predator*, *prey*.

prey Organism that is killed by an organism of another species (the predator) and serves as its source of food.

primary consumer Organism that feeds on some or all parts of plants (herbivore) or on other producers. Compare *detritivore*, *omnivore*, *secondary consumer*.

primary ecological succession Ecological succession in an area without soil or bottom sediments. See *ecological succession*. Compare *secondary ecological succession*.

primary pollutant Chemical that has been added directly to the air by natural events or human activities and occurs in a harmful concentration. Compare *secondary pollutant*.

primary productivity See *gross primary productivity*, *net primary productivity*.

primary recycling Process in which materials are recycled into new products of the same type—turning used aluminum cans into new aluminum cans, for example. Compare *secondary recycling*.

primary sewage treatment Mechanical sewage treatment in which large solids are filtered out by screens and suspended solids settle out as sludge in a sedimentation tank. Compare *secondary sewage treatment*.

principles of sustainability Principles by which nature has sustained itself for billions of years by relying on solar energy, biodiversity, and nutrient recycling.

probability Mathematical statement about how likely it is that something will happen.

producer Organism that uses solar energy (green plants) or chemical energy (some bacteria) to manufacture the organic compounds it needs as nutrients from simple inorganic compounds obtained from its environment. Compare *consumer*, *decomposer*.

protein Large polymer molecules formed by linking together long chains of monomers called amino acids.

proton (p) Positively charged particle in the nuclei of all atoms. Each proton has a relative mass of 1 and a single positive charge. Compare *electron*, *neutron*.

proven oil reserves Identified deposits from which conventional crude oil can be extracted profitably at current prices with current technology.

PV cell See *photovoltaic cell*.

pyramid of energy flow Diagram representing the flow of energy through each trophic level in a food chain or food web. With each energy transfer, only a small part (typically 10%) of the usable energy entering one trophic level is transferred to the organisms at the next trophic level.

radioactive decay Change of a radioisotope to a different isotope by the emission of radioactivity.

radioactive waste Waste products of nuclear power plants, research, medicine, weapons production, and other processes involving nuclear reactions. See *radioactivity*.

radioactivity Nuclear change in which unstable nuclei of atoms spontaneously shoot out "chunks" of mass, energy, or both at a fixed rate. The three principal types of radioactivity are gamma rays and fast-moving alpha particles and beta particles.

rain shadow effect Low precipitation on the leeward side of a high mountain range when prevailing winds flow up and over these mountains, dropping their moisture on the windward side and creating semiarid and arid conditions on their leeward side.

rangeland Land that supplies forage or vegetation (grasses, grass-like plants, and shrubs) for grazing and browsing animals. Compare *pasture*.

reconciliation ecology Science of inventing, establishing, and maintaining habitats to conserve species diversity in places where people live, work, or play.

recycle To collect and reprocess discarded materials so that they can be made into new products. An example is collecting aluminum cans, melting them down, and using the aluminum to make new cans or other aluminum products. See *primary* recycling, secondary recycling. Compare reduce and reuse.

reduce To consume less and live a simpler lifestyle. Compare recycle and reuse.

refining A complex process of heating crude oil to separate it into various fuels and other components with different boiling points.

reliable science Concepts and ideas that are widely accepted by experts in a particular field of the natural or social sciences. Compare tentative science, unreliable science.

reliable surface runoff Surface runoff of water that generally can be counted on as a stable source of water from year to year. See runoff.

renewable resource Resource that can be replenished rapidly (in hours to several decades) through natural processes as long as it is not used up faster than it is replaced. Examples include trees in forests, grasses in grasslands, wild animals, fresh surface water in lakes and streams, most groundwater, fresh air, and fertile soil. If such a resource is used faster than it is replenished, it can be depleted. Compare nonrenewable resource and perpetual resource. See also environmental degradation.

replacement-level fertility Average number of children a couple must bear to replace themselves. The average for a country or the world usually is slightly higher than two children per couple (2.1 in the United States and 2.5 in some less-developed countries) mostly because some children die before reaching their reproductive years. See also total fertility rate.

representative democracy A government by the people through elected officials and representatives.

reproductive isolation Condition under which mutation and change by natural selection operate independently in the gene pools of geographically isolated populations. Compare geographic isolation. See speciation.

reserves Resources that have been identified and from which a usable mineral can be extracted profitably at present prices with current mining or extraction technology.

reservoir Artificial lake created when a stream is dammed. See dam.

resilience The ability of a living system to be restored through secondary succession after a more severe disturbance.

resource Anything obtained from the environment to meet human needs and wants. It can also be applied to other species.

resource partitioning Process of dividing up resources in an ecosystem so that species with similar needs (overlapping ecological niches) use the same scarce resources at different times, in different ways, or in different places. See ecological niche.

response Amount of health damage caused by exposure to a certain dose of a harmful substance or form of radiation. See dose, dose-response curve.

restoration ecology Research and scientific study devoted to restoring, repairing, and reconstructing damaged ecosystems.

reuse To use a product over and over again in the same form. An example is collecting, washing, and refilling glass beverage bottles. Compare reduce and recycle.

riparian zone A thin strip or patch of vegetation that surrounds a streams. These zones are very important habitats and resources for wildlife.

risk Probability that something undesirable will result from deliberate or accidental exposure to a hazard. See risk analysis, risk assessment, risk management.

risk analysis Identifying hazards, evaluating the nature and severity of risks associated with the hazards (risk assessment), ranking risks (comparative risk analysis), and using this and other information to determine options and make decisions about reducing or eliminating risks (risk management).

risk assessment Process of gathering data and making assumptions to estimate short- and long-term harmful effects on human health or the environment from exposure to hazards associated with the use of a particular product or technology. See risk, risk analysis.

risk management Use of risk assessment and other information to determine options and make decisions about reducing or eliminating risks. See risk, risk analysis.

rock Any solid material that makes up a large, natural, continuous part of the earth's crust. See mineral.

rock cycle Slowest of the earth's cycles, consisting of geologic, physical, and chemical processes that form and modify rocks in the earth's crust over millions of years.

r-selected species Species that have a capacity for a high rate of population growth (r). They tend to have many, usually small, offspring and to give them little or no parental care or protection. Examples include algae, bacteria, and most insects with short life spans. Compare K-selected species.

runoff Freshwater from precipitation and melting ice that flows on the earth's surface into streams, lakes, wetlands, and reservoirs. See reliable surface runoff, surface runoff, surface water. Compare groundwater.

salinity Concentration of various salts dissolved in a given volume of water.

salinization Accumulation of salts in soil that can eventually make the soil unable to support plant growth.

saltwater life zones Aquatic life zones associated with oceans: oceans and their accompanying bays, estuaries, coastal wetlands, shorelines, coral reefs, and mangrove forests.

sanitary landfill Waste disposal site on land in which waste is spread in thin layers, compacted, and covered with a fresh layer of clay or plastic foam each day. Compare open dump.

science The pursuit of discovering order in nature and using that knowledge to make predictions about what is likely to happen in nature. See data, reliable science, scientific hypothesis, scientific law, scientific model, scientific theory, tentative science, unreliable science.

scientific hypothesis A proposed explanation of a scientific law or certain scientific observations. Compare scientific law, scientific model, scientific theory.

scientific law Description of what scientists find happening in nature repeatedly in the same way, without known exception. See first law of thermodynamics, law of conservation of matter, second law of thermodynamics. Compare scientific hypothesis, scientific model, scientific theory.

scientific model A simulation of complex processes and systems. Many are mathematical models that are run and tested using computers. Compare scientific hypothesis, scientific law, scientific theory.

scientific theory A well-tested and widely accepted scientific hypothesis or group of related hypotheses. Compare scientific hypothesis, scientific law, scientific model.

secondary consumer Organism that feeds only on primary consumers. Compare detritivore, omnivore, primary consumer.

secondary ecological succession Ecological succession in an area in which natural vegetation has been removed or destroyed but the soil or bottom sediment has not been destroyed. See ecological succession. Compare primary ecological succession.

secondary pollutant Harmful chemical formed in the atmosphere when a primary air pollutant reacts with normal air components or other air pollutants. Compare primary pollutant.

secondary recycling A process in which waste materials are converted into different products; for example, used tires can be shredded and turned into rubberized road surfacing. Compare *primary recycling*.

secondary sewage treatment Second step in most waste treatment systems in which aerobic bacteria decompose as much as 90% of degradable, oxygen-demanding organic wastes in wastewater. Compare *primary sewage treatment*.

second-growth forest A stand of trees resulting from secondary ecological succession; these forests develop after the trees in an area have been removed by human activities, such as clear-cutting for timber or cropland, or by natural forces such as fire, hurricanes, or volcanic eruptions.

second law of thermodynamics Whenever energy is converted from one form to another in a physical or chemical change, we end up with lower-quality or less usable energy than we started with. In any conversion of heat energy to useful work, some of the initial energy input is always degraded to lower-quality, more dispersed, less useful energy—usually low-temperature heat that flows into the environment. See *first law of thermodynamics*.

sedimentary rock Rock that forms from the accumulated products of erosion and in some cases from the compacted shells, skeletons, and other remains of dead organisms. Compare *igneous rock*, *metamorphic rock*. See *rock cycle*.

selective cutting Cutting of intermediate-aged, mature, or diseased trees in an uneven-aged forest stand, either singly or in small groups. This encourages the growth of younger trees and maintains an uneven-aged stand. Compare *clear-cutting, strip cutting*.

septic tank Underground tank for treating wastewater from a home, used in rural and suburban areas. Bacteria in the tank decompose organic wastes, and the sludge settles to the bottom of the tank. The effluent flows out of the tank into the ground through a field of drainpipes.

shale oil Slow-flowing, dark brown, heavy oil obtained when kerogen in oil shale is vaporized at high temperatures and then condensed. Shale oil can be refined to yield gasoline, heating oil, and other petroleum products.

smart growth Form of urban planning that recognizes that urban growth will occur but uses zoning laws and other tools to prevent sprawl, direct growth to certain areas, protect ecologically sensitive and important lands and waterways, and develop urban areas that are more environmentally sustainable and more enjoyable places to live.

smelting Process in which a desired metal is separated from the other elements in an ore mineral.

SNG See *synthetic natural gas*.

soil Complex mixture of inorganic minerals (clay, silt, pebbles, and sand), decaying organic matter, water, air, and living organisms.

soil conservation Methods used to reduce soil erosion, prevent depletion of soil nutrients, and restore nutrients previously lost by erosion, leaching, and excessive crop harvesting.

soil erosion Movement of soil components, especially topsoil, from one place to another, usually by wind, flowing water, or both. This natural process can be greatly accelerated by human activities that remove vegetation from soil. Compare *soil conservation*.

soil horizons Horizontal zones, or layers, that make up a particular mature soil. Each horizon has a distinct texture and composition that vary with different types of soils. See *soil profile*.

soil profile Cross-sectional view of the horizons in a soil. See *soil horizon*.

soil salinization Soil degradation process in which repeated applications of irrigation water in dry climates lead to the gradual accumulation of salts in the upper soil layers; can stunt crop growth, lower crop yields, and eventually kill plants and ruin the land.

solar cell See *photovoltaic cell*.

solar energy Direct radiant energy from the sun; produces indirect forms of solar energy, including wind (resulting from differences in temperature between air masses), falling and flowing water (hydropower), and biomass (solar energy converted into chemical energy stored in the chemical bonds of organic compounds in trees and other plants)—none of which would exist without direct solar energy.

solar thermal system System that collects and concentrates solar energy and uses it to boil water and produce steam for generating electricity; also known as *concentrating solar power (CSP)* system.

solid waste Any unwanted or discarded material that is not a liquid or a gas. See *industrial solid waste, municipal solid waste*.

specialist species Species with a narrow ecological niche. They may be able to live in only one type of habitat, tolerate only a narrow range of climatic and other environmental conditions, or use only one type or a few types of food. Compare *generalist species*.

speciation Formation of two species from one species because of divergent natural selection in response to changes in environmental conditions; usually takes thousands of years. Compare *extinction*.

species Group of similar organisms, and for sexually reproducing organisms, a set of individuals that can mate and produce fertile offspring. Every organism is a member of a certain species.

species diversity Number of different species (species richness) combined with the relative abundance of individuals within each of those species (species evenness) in a given area. See *biodiversity, species evenness, species richness*. Compare *ecological diversity, functional diversity, genetic diversity*.

species evenness Degree to which comparative numbers of individuals of each of the species present in a community are similar. See *species diversity*. Compare *species richness*.

species richness Variety of species, measured by the number of different species contained in a community. See *species diversity*. Compare *species evenness*.

spoils Unwanted rock and other waste materials produced when a material is removed from the earth's surface or subsurface by mining, dredging, quarrying, or excavation.

stewardship worldview Worldview holding that we can manage the earth for our benefit but that we have an ethical responsibility to be caring and responsible managers, or stewards, of the earth. It calls for encouraging environmentally beneficial forms of economic growth and discouraging environmentally harmful forms. Compare *environmental wisdom worldview, planetary management worldview*.

stratosphere Second layer of the atmosphere, extending about 17–48 kilometers (11–30 miles) above the earth's surface. It contains a layer of gaseous ozone (O_3), which filters out about 95% of the incoming harmful ultraviolet radiation emitted by the sun. Compare *troposphere*. See *ozone layer*.

strip-cropping Planting regular crops and close-growing plants, such as hay or nitrogen-fixing legumes, in alternating rows or bands to help reduce depletion of soil nutrients.

strip-cutting Variation of clear-cutting in which a strip of trees is clear-cut along the contour of the land, with the corridor being narrow enough to allow natural regeneration within a few years. After regeneration, another strip is cut above the first, and so on. Compare *clear-cutting, selective cutting*.

strip mining Form of surface mining in which bulldozers, power shovels, or stripping wheels remove large chunks of the earth's surface in strips. See *area strip mining, contour strip mining, surface mining*. Compare *subsurface mining*.

subsidence Slow or rapid sinking of part of the earth's crust that is not slope-related.

subsidies Payments and protections of various forms that help businesses and industries to survive and thrive.

subsurface mining Extraction of a metal ore or fuel resource such as coal from a deep underground deposit. Compare *surface mining*.

sulfur cycle Cyclic movement of sulfur in various chemical forms from the environment to organisms and then back to the environment.

sulfur dioxide (SO_2) Colorless gas with an irritating odor. About one-third of the SO_2 in the atmosphere comes from natural sources as part of the sulfur cycle. The other two-thirds come from human sources, mostly combustion of sulfur-containing coal in electric power and industrial plants and from oil refining and smelting of sulfide ores.

sulfuric acid Compound containing hydrogen, sulfur, and oxygen; a hazardous chemical that is often a component of acid precipitation. See *acid deposition*.

surface fire Forest fire that burns only undergrowth and leaf litter on the forest floor. Compare *crown fire, ground fire*.

surface mining Removing soil, subsoil, and other strata and then extracting a mineral deposit found fairly close to the earth's surface. See *area strip-mining, contour strip-mining, mountaintop removal, open-pit mining, strip-mining*. Compare *subsurface mining*.

surface runoff Water flowing off the land into bodies of surface water. See *reliable surface runoff*.

surface water Precipitation that does not infiltrate the ground or return to the atmosphere by evaporation or transpiration. Found in streams, rivers, lakes, and wetlands. See *runoff*. Compare *groundwater*.

survivorship curve Line graph that shows the percentages of the members of a population surviving at different ages. There are three generalized types of survivorship curves: *late loss, early loss*, and *constant loss*.

suspended particulate matter See *particulates*.

sustainability Ability of earth's various systems, including human cultural systems and economies, to survive and adapt to changing environmental conditions indefinitely.

sustainable agriculture Method of growing crops and raising livestock by relying on organic fertilizers, soil conservation, water conservation, biological pest control, and minimal use of nonrenewable fossil-fuel energy.

sustainable yield Highest rate at which a potentially renewable resource can be used indefinitely without reducing its available supply. See also *environmental degradation*.

synfuels Synthetic gaseous and liquid fuels produced from solid coal or sources other than natural gas or crude oil.

synthetic biology Technology that enables scientists to make new sequences of DNA and to use such genetic information to design and create artificial cells, tissues, body parts, and organisms not found in nature.

synthetic fertilizers Manufactured chemicals that contain nutrients such as nitrogen, phosphorus, potassium, calcium, and several others.

synthetic natural gas (SNG) Gaseous fuel containing mostly methane produced from solid coal.

synthetic pesticides Chemicals manufactured to kill or control populations of organisms that interfere with crop production.

system A set of components that function and interact in some regular way. Most living systems have inputs, throughputs, and outputs of matter and energy from the environment. See *inputs, outputs, and throughputs*.

tailings Rock and other waste materials removed as impurities when waste mineral material is separated from the metal in an ore.

tar sands Deposit of a mixture of clay, sand, water, and varying amounts of a tar-like heavy oil known as bitumen. Bitumen can be extracted from tar sands by heating. It is then purified and upgraded to synthetic crude oil. Also called oil sands.

tectonic plates Various-sized pieces of the earth's lithosphere that move slowly around atop the mantle's flowing asthenosphere. Most earthquakes and volcanoes occur around the boundaries of these plates. See *lithosphere*.

temperature A measure of the average heat or thermal energy of the atoms, ions, or molecules in a sample of matter.

temperature inversion Layer of dense, cool air trapped under a layer of less dense, warm air. It prevents upward-flowing air currents from developing. In a prolonged inversion, air pollution in the trapped layer may build up to harmful levels.

tentative science Preliminary scientific data, hypotheses, and models that have not been widely tested and accepted. Compare *reliable science, unreliable science*.

teratogen Chemical, ionizing agent, or virus that causes birth defects. Compare *carcinogen, mutagen*.

terracing Planting crops on a long, steep slope that has been converted into a series of broad, nearly level terraces with short vertical drops from one to another that run along the contour of the land to retain water and reduce soil erosion.

tertiary (higher-level) consumers Animals that feed on animal-eating animals. They feed at high trophic levels in food chains and webs. Examples include hawks, lions, bass, and sharks. Compare *detritivore, primary consumer, secondary consumer*.

theory of evolution Widely accepted scientific idea that all life forms developed from earlier life forms. It is the way most biologists explain how life has changed over the past 3.8 billion years and why it is so diverse today.

theory of island biogeography Widely accepted scientific theory holding that the number of different species (species richness) found on an island is determined by the interactions of two factors: the rate at which new species immigrate to the island and the rate at which species become extinct, or cease to exist, on the island. See *species richness*.

thermal energy See *heat*.

thermal inversion See *temperature inversion*.

threatened species Wild species that is still abundant in its natural range but is likely to become endangered because of a decline in numbers. Compare *endangered species*.

throughputs Matter and energy that flowing through a system. See *system*. Compare *inputs* and *outputs*.

tipping point Threshold level at which an environmental problem causes a fundamental and irreversible shift in the behavior of a system. See *climate change tipping point, ecological tipping point*.

total fertility rate (TFR) Estimate of the average number of children who will be born alive to a woman during her lifetime if she passes through all her childbearing years (ages 15–44). More simply, it is an estimate of the average number of children that women in a given population will have during their childbearing years.

toxic chemical Chemical that can cause harm to an organism. See *carcinogen, mutagen, teratogen*.

toxicity Measure of the harmfulness of a substance.

toxicology Study of the adverse effects of chemicals on health.

toxic waste See *hazardous waste*.

traditional intensive agriculture Production of enough food for a farm family's survival and a surplus that can be sold. This type of agriculture uses higher inputs of labor, fertilizer, and water than traditional subsistence agriculture. See *traditional subsistence agriculture*. Compare *industrialized agriculture*.

traditional subsistence agriculture Production of enough crops or livestock for a farm family's survival and, in good years, a surplus to sell or put aside for hard times. Compare *industrialized agriculture, traditional intensive agriculture*.

tragedy of the commons Depletion or degradation of a potentially renewable resource to which people have free and unmanaged access. An example is the depletion of commercially desirable fish species in the open ocean beyond areas controlled by coastal countries. See *common-property resource, open access renewable resource*.

trait Characteristic passed on from parents to offspring during reproduction in an animal or plant.

transform fault Area where the earth's lithospheric plates move in opposite but parallel directions along a fracture (fault) in the lithosphere. Compare *convergent plate boundary, divergent plate boundary*.

transmissible disease Disease that is caused by living organisms (such as bacteria, viruses, and parasitic worms) and can spread from one person to another by air, water, food, or body fluids, or in some cases by insects or other organisms. Compare *nontransmissible disease*.

transpiration Process in which water is absorbed by the root systems of plants, moves up through the plants, passes through pores (stomata) in their leaves or other parts, and evaporates into the atmosphere as water vapor.

tree farm See *tree plantation*.

tree plantation Site planted with one or only a few tree species in an even-aged stand. When the stand matures, it is usually harvested by clear-cutting and then replanted. These farms normally raise rapidly growing tree species for fuelwood, timber, or pulpwood. Compare *old-growth forest, second-growth forest*.

trophic level All organisms that are the same number of energy transfers away from the original source of energy (for example, sunlight) that enters an ecosystem. For example, all producers belong to the first trophic level and all herbivores belong to the second trophic level in a food chain or a food web.

troposphere Innermost layer of the atmosphere. It contains about 75% of the mass of earth's air and extends about 17 kilometers (11 miles) above sea level. Compare *stratosphere*.

tsunami Series of large waves generated when part of the ocean floor suddenly rises or drops, typically due to an earthquake.

undernutrition See *chronic undernutrition*.

unreliable science Scientific results or hypotheses presented as reliable science without having undergone the rigors of the peer review process. Compare *reliable science, tentative science*.

upwelling Movement of nutrient-rich bottom water to the ocean's surface. It can occur far from shore but usually takes place along certain steep coastal areas where the warm surface layer of ocean water is pushed away from shore and replaced by cold, nutrient-rich bottom water.

urban area Geographic area containing a community with a population of 2,500 or more. The minimum number of people used in this definition varies among countries, from 2,500 to 50,000.

urban growth Rate of growth of an urban population.

urban heat island Condition in cities in which the heat generated by cars, factories, furnaces, lights, air conditioners, and heat-absorbing dark roofs and streets creates an island of heat surrounded by cooler suburban and rural areas.

urbanization Creation or growth of urban areas, or cities, and their surrounding developed land. See *degree of urbanization, urban area*.

urban sprawl Growth of low-density development on the edges of cities and towns. See *smart growth*.

virtual water Water that is not directly consumed but is used to produce food and other products.

virus Infectious agent that is smaller than a bacterium; it works by invading a cell and taking over its genetic machinery to copy itself. It then multiplies and spreads throughout one's body, causing a viral disease such as flu or AIDS.

volatile organic compounds (VOCs) Organic compounds that exist as gases in the atmosphere and act as pollutants, some of which are hazardous.

volcano Vent or fissure in the earth's surface through which magma, liquid lava, ash, and gases are released into the environment.

waste management Managing wastes to reduce their environmental harm without seriously trying to reduce the amount of waste produced. See *integrated waste management*. Compare *waste reduction*.

waste reduction Reducing the amount of waste produced; wastes that are produced are viewed as potential resources that can be reused, recycled, or composted. See *integrated waste management*. Compare *waste management*.

wastewater water that contains sewage and other wastes from homes, farms, and industries.

water cycle See *hydrologic cycle*.

water footprint A rough measure of the volume of water used directly and indirectly to keep a person or group alive and to support their lifestyles.

waterlogging Saturation of soil with irrigation water or excessive precipitation so that the water table rises close to the surface.

water pollution Any physical or chemical change in surface water or groundwater that can harm living organisms or make water unfit for certain uses.

watershed (drainage basin) Land area that delivers water, sediment, and dissolved substances via small streams to a major stream (river).

water table Upper surface of the zone of saturation, in which all available pores in the soil and rock in the earth's crust are filled with water. See *zone of aeration, zone of saturation*.

weather Short-term changes in the temperature, barometric pressure, humidity, precipitation, sunshine, cloud cover, wind direction and speed, and other conditions in the troposphere at a given place and time. Compare *climate*.

wetland Land that is covered all or part of the time with salt water or freshwater, excluding streams, lakes, and the open ocean. See *coastal wetland, inland wetland*.

wilderness Area where the natural ecosystems have not been seriously disturbed by human activities.

windbreak Row of trees or hedges planted to partially block wind flow and reduce soil erosion on cultivated land.

wind farm Cluster of wind turbines in a windy area on land or at sea, built to capture wind energy and convert it into electrical energy.

worldview How people think the world works and what they think their role in the world should be. See *environmental wisdom worldview, planetary management worldview, stewardship worldview*.

yield The amount of food produced per unit of land.

zone of aeration Zone in soil that is not saturated with water and that lies above the water table. See *water table, zone of saturation*.

zone of saturation Zone below the water table where all available pores in soil and rock in the earth's crust are filled by water. See *water table, zone of aeration*.

zoning Designating parcels of land for particular types of use.

zooplankton Animal plankton; small floating herbivores that feed on plant plankton (phytoplankton). Compare *phytoplankton*.

Index

Note: Page numbers in **boldface** type
indicate key terms. Page numbers
followed by italicized *f*, *t*, or *b*
indicate figures, tables, and boxes.

A

AAAS. *See* American Association for the
   Advancement of Science
Abiotic environment, 48, 50*f*
Abyssal zone, 153, 153*f*
Accidents, motor vehicle, 125
Acid deposition (acid rain), **412**–413,
      412*f*–413*f*
   buffering by soils, 413, 413*f*
   formation of, 412, 412*f*
   global regions affected by, 413*f*
   harmful effects of, 413
   nitrogen oxides, 60, 409
   preventing, 414, 414*f*
   reducing, 414, 414*f*
   wet and dry forms of, 412
Acidic solution, 33
Acidification, ocean, 156*b*
Acidity, **32**
Acid mine drainage, 310
Acquired immune deficiency syndrome
      (AIDS), 383, 383–384, 384*f*
Active solar heating systems, 353*f*, **354**
Acute effect, of chemical exposure, 390
Adaptation
   to climate change, 434
   defined, **79**
   limits of adaptation through natural
      selection, 80
Advanced light-water reactor, 344
Aerobic respiration
   in carbon cycle, 58
   defined, **49**
Aerosols, 409, 427
Affluence
   environmental impact and, 12, 15–16
   harmful and beneficial effects of, 15–16
   overconsumption and, 110*b*
Africa
   irrigation systems in, 254*b*
   Lifestraw™ use in, 287, 287*f*
African elephants, 146*b*, 146*f*, 174*f*, 179
African honeybees, 74
Age structure, **97**, **113**
   affecting growth/decline, 113–115
   aging populations and, 114–115
   diagrams, 113*f*, 114*f*
   U.S. baby boom generation, 114, 114*f*
Agribusiness, 232. *See also* Industrialized
      agriculture
Agriculture. *See also* Food production
   agrobiodiversity, 239
   cash crops, 229
   climate change related decline in
      production, 433
   community-supported agriculture
      (CSA), 254
   conservation-tillage, 247
   crossbreeding and artificial selection, 232

energy use in U. S., 234
fish and shellfish production, 233
genetically engineered foods, 232–233
genetic engineering and, 232
green revolution, 230–231, 240
hidden costs of, 232
high-input, 229
increase in production, 228–229
industrialized (*see* Industrialized
      agriculture)
meat production, 233, 240
monocultures, 229, 251
multiple cropping, 230
organic, 229–230, 231*f*
pest control, 242–247
plantation, 229
polycultures, 229, 252*b*
subsidies, 232, 253
subsistence, 229
sustainability improvements, 250–252,
      251*f*, 255*b*
traditional intensive, 229
traditional subsistence, 229
urban food production, 226*b*, 226*f*,
      255, 256*b*
water footprint of, 265
water pollution by, 282, 291
yield, 229
Agrobiodiversity, **239**
Agroforestry, 247
A horizon, 52*f*
AIDS. *See* Acquired immune deficiency
      syndrome
Air, vertical movement of, 411
Air circulation
   air pollution and, 411
   climate affected by, 420
Air pollutants
   from coal-burning factories, 410*f*
   hazardous air pollutants (HAPs), 417
   mercury, 395*f*
   natural and human sources of,
      407–408, 408*f*
   outdoor, 408–410, 408*f*
   primary, 408, 408*f*
   secondary, 408, 408*f*
Air pollution
   acid deposition, 412–413, 412*f*–413
   agriculture contribution to, 238–239
   allotments in the marketplace, 418
   concentration in cities, 122
   dealing with, 417–419, 417*f*–419*f*
   health effects of, 16
   human health and, 409
   industrial smog, 410, 410*f*
   from meat production, 241
   in Mexico City, 124
   from mining, 310
   mobile sources of, 408, 408*f*
   motor vehicles and, 124, 408, 419*f*
   natural and human sources of,
      407–408, 408*f*
   photochemical smog, 410–411, 411*f*
   reducing indoor, 419, 419*f*

reducing outdoor, 417–419, 417*f*–419*f*
regional, 412
stationary sources of, 408, 408*f*
Alaska
   Denali National Park, 192*f*–193*f*
   wilderness areas, 207
Algae
   bloom, 289
   cultural eutrophication of lakes and, 284
   ocean pollution and, 289
Allen, Will, 226*b*, 226*f*, 230, 247, 248
Alley cropping, 247, 248*f*
Alloy, 305
Aluminum, 34, 305
   recycling, 39, 314
   time for disintegration of can, 452
Amazon region, deforestation in, 8*f*,
      203, 312
American alligator, 75–76, 76*f*
American Association for the Advancement
      of Science (AAAS), 421
American baby boom (case study),
      114, 114*f*
American Water Works Association, 278
Ammonia, 60, 60*f*
Ammonium ions, 60
Amphibians
   decline in, 70*f*, 78*b*
   as indicator species, 74, 78*b*
   sustainability and, 85*b*
Amplitude, earthquake, 319
Anderson, Ray, 484
Andes Mountains, glaciers of, 428
Anemia, iron deficiency, 228
Animal manure, **248**
Animals. *See also specific animals*
   arctic tundra/cold grasslands, 144
   in cold/northern coniferous forests, 148
   desert, 144*b*
   savanna, 144
   in temperate deciduous forest, 147
Anopheles mosquitoes, 243
Antarctica
   ice cores, 424*f*
   ozone thinning, 441, 441*f*
Anthropocene, 63*b*
Antibiotics, 79, 79*f*
   resistance to, 381*b*
   used in aquaculture, 241
   used in meat production, 241
Antibodies, 387
Antinoise technologies, 123
Appliances
   energy-efficient, 349, 351
   water-saving, 278
Aquaculture, 228, **233**, 235*f*
   advantages and disadvantages of, 242*f*
   aquatic ecosystem, 241–242
   open-ocean, 250
   polyaquaculture, 250
   recirculating systems, 250
   sustainable, 250, 250*f*
Aquaculture Stewardship Council
      (ASC), 250

Aquariums, 186–187
Aquatic biodiversity, 213–220
    extinctions, 215
    human activities, impact of,
        invasive species introduction215, 215f
        ocean acidification, 214b
        overfishing, 215
    marine biodiversity, protecting and
        sustaining, 216–217
Aquatic biomes. See Aquatic ecosystems
Aquatic ecosystems
    aquaculture as threat to, 241–242
    biodiversity (see Aquatic biodiversity)
    ecosystem and economic services
        provided by, 152, 152f
    freshwater, 157–159, 157f–159f
    human impacts on
        freshwater systems, 161–162
        marine systems, 155, 157, 157f
    limiting factors in, 98
    marine, 152–157, 152f–157f
Aquatic life zones, 152. See also Aquatic
        ecosystems; Freshwater life zones;
        Marine life zones
    net primary productivity, 55, 56f
Aqueducts, 273, 274f
Aquifers, 57, 263–264
    contamination of, 286
    depletion of, 269
    non-renewable, 264
    recharge of, 264
Aral Sea, 275, 275f
Architecture, green, 349
Arctic fox, 82f
Arctic tundra, 144, 145f,
Area strip mining, 309, 311f
Arenal Volcano National Park,
        Costa Rica, 194f
Arid conditions, 139
Arizona, San Pedro River in, 206f
Arizona State University, 439
Artificial selection, 81–82, 232
ASC. See Adventurers and Scientists for
        Conservation
ASC. See Aquaculture Stewardship
        Council
Asian/Indian elephant, 174f
Asthenosphere, 303, 317f
Atmosphere, 47, 47f, 407
    air circulation, 138–140, 420
    climate and (see Climate; Climate
        change)
    energy transfer by convection in, 138f
    greenhouse gases, effects of, 137 (see
        also Greenhouse gases)
    kinetic energy in, 423
    layers of, 407, 407f
    ozone, 47, 407, 409, 441–443
    temperature inversion, 412
Atmospheric temperature, 420–421, 423,
        426b–427b
Atmospheric warming, 406b. See also
        Climate change
    climate change, 420–422, 427–433
    cloud cover and, 425, 427
    of lower by greenhouse gases, 137
        (see also Greenhouse gases)
    low-sulfur coal, atmospheric warming,
        and toxic mercury, 414
Atomic number, 32
Atomic theory, 32
Atoms, 32, 32f, 35, 49f
Attenborough, David, 107

Audits, environmental, 492
Audubon Society, 295b
Australia
    coal use by, 335
    Great Barrier Reef, 98f
Automobiles. See also Motor vehicles
    battery, lithium-ion, 306, 316b, 349b
    energy efficient, 347–348, 348f
    fuel cells in, 348
    fuel economy, 347
    hybrid-electric and electric cars, 307f,
        347–348, 348f
    ultralight, ultrastrong composite
        materials in, 348

B

Baby boom, in United States, 114, 114f
Background extinction rate, 82, 169
Back-to-the-tap movement, 288
Bacteria, 379
    actions of, 51b
    genetic resistance to antibiotics, 381b
    methicillin-resistant Staphylococcus aureus
        (MRSA), 381b
    in nitrogen cycle, 60, 60f
    tuberculosis, 380
Bagasse, 364
Balanced chemical equation, 35
Ballard, Robert, 318b
Balog, James, 430b
Bangladesh, 10, 17f, 430, 440, 462f
Barrett, Steven, 416
Basel Convention, 453, 466
Basic solution, 33
Bathyal zone, 153, 153f
Bats, 93, 94f
Battery
    for electric vehicles, 348, 348f
    lithium-ion, 306, 316b, 349b
    rechargeable, 456
    search for better, 349b
    ultracapacitors, 349b
Battery engineer, 349
Bedrock, 52b
Bees, 72f
Behavioral strategies of prey species, 93
Belize, 212
Benyus, Janine, 9b
B horizon, 52f
Bias, 31, 399
Bicycling, 126, 127f
Biddle, Mike, 458
Big bluestem, root system of, 253f
Big Ideas
    biodiversity and evolution, 84b
    climate, terrestrial and aquatic systems, 162b
    economics, politics, and environmental
        policy, 500b
    ecosystems, 63b
    energy resources, 370b
    environmental hazards and human
        health, 399b
    environmental problems and
        sustainability, 21b
    extinction, 188b
    food production, 255b
    geology and mineral resources, 321b
    human population growth and urban
        areas, 129b
    laws of thermodynamics and matter
        conservation, 39b
    pollution, climate change, and ozone,
        443b

    saving ecosystems and ecosystem
        services, 219b
    solid and hazardous waste, 467b
    species interactions and populations
        sizes, 101b
    water resources and water pollution,
        295b
Bioaccumulation, 178
Biodiesel, 363
Biodiversity, 5, 68–85
    agrobiodiversity, 239
    aquatic (see Aquatic biodiversity)
    biological evolution, 77–80
    climate and, 134–163
    climate change threats to, 431–433
    components of, 71–73
        ecosystem diversity, 71, 71f
        functional diversity, 71f, 72
        genetic diversity, 71, 71f
        species diversity, 71, 71f
    Convention on Biological Diversity
        (CBD), 185
    of coral reefs, 152, 156b
    in Costa Rica, , 208, 220b
    defined, 71
    E. O. Wilson and, 73b
    extinction, 82–83
    geological processes and, 81b
    hotspots, 210, 211f
    illegal killing, capturing, and selling of
        wild species as threat to, 179–182,
        179f, 181f
    mountains as islands of, 150
    number of species, 169–170
    origin of term, 62
    parks as islands of, 207
    principle of sustainability, 5, 6f
    saving ecosystems and ecosystem
        services, 192–219
    saving species, 166–189 (see also
        Species approach to sustaining
        biodiversity)
    speciation, 80–81, 82f
    species roles in ecosystems, 74–77
        American alligator case study,
            75–76, 76f
        generalist species, 74, 74f
        indicator species, 74–75
        invasive species, 74
        keystone species, 75, 76
        native species, 74
        nonnative species, 74
        sharks case study, 76–77, 77f
        specialist species, 74, 74f
    sustaining, 192–219
Biodiversity hotspots, 210, 211f
Biofuels
    climate change and, 364
    ethanol, 363, 365f
    liquid, 363–364
Biogas digesters, 251
Biogeochemical cycles, 56. See also
        Nutrient cycles
Biological capacity, 12
Biological diversity. See Biodiversity
Biological evolution, 77. See also Evolution
Biological extinction, 82, 169. See also
        Extinction
Biological hazards, 379–385. See also
        Infectious disease
Biological pest control, 242, 246f
Biomagnification
    of DDT, 178f

in food chain, 388f
Biomass, **362**
  burning solid, 362–363, 363f
  as fuelwood crisis, 203, 363
Biomass plantations, 363
Biomes, **72**, **141**. *See also* Aquatic
    ecosystems; Terrestrial biomes/
    ecosystems; *specific biomes*
  climate and, 141–152, 142f, 151f
  as mosaic of patchwork of areas, 141
  transition zone (ecotone), 141
  variety along the 39th parallel across
      the United States, 73f
Biomimicry, 466–467
Biomining, 308
Bioplastics, 458b
Bioprospectors, 172
Bioremediation, 463
Biosphere, **47**, 47f, **48**, 49f
  economic systems related to, 475f–477f,
      475–478
Biosphere 2, 495b, 495f
Biosphere reserves, 208
Biotic environment, 48, 50f
Birds
  bluebirds, nesting boxes for, 212
  DDT effects on, 178, 178f
  economic and ecosystem services
      provided by, 195
  as indicator species, 184
  population decline, 183–184
  shorebirds, specialized niches of, 74
Birth control methods, availability of, 112
Birth rate
  demographic momentum and, 113–114
  demographic transition, 115, 115f
  factors affecting, 111–112
Bisphenol A (BPA), 389b, 393f
Bisphenol S (BPS), 389b
Bitumen, 333
Bituminous coal, 304
Black howler monkey, 212
Black smokers, 308, 308f
Bleaching
  of coral, 156b, 156f
  in sewage treatment, 293
Blindness, vitamin A deficiency, 228
BLM. *See* Bureau of Land Management
Bluebirds, nesting boxes for, 212
Blue Legacy International, 162b
Boar, European wild, 175f, 208
Bolivia, lithium reserves in, 306
Boomerang effect, 245
Boreal forests, 148
Bormann, F. Herbert, 28b, 29, 32
Borneo, 170, 172f, 244
Botanical gardens, 186
Botswana, HIV/AIDS in, 384, 384f
Bottled water, 287
Boulder, Colorado, 279
BPA. *See* Bisphenol A
BPS. *See* Bisphenol S
Braungart, Michael, 450
Brazil
  African honeybees, 74
  cerrado, 239
  Curitiba, 128–129, 129f
  deforestation in, 203
  farm subsidies in, 253
  hydropower use in, 359
  integrated pest management (IPM)
      use in, 246
  smart cards for water credits, 279

Broadleaf deciduous trees, 147
Broadleaf evergreen plants, 147
Broad-spectrum agents, 243
Brown, Lester R., 500
Brown-air smog, 410
Browsing animals, 144
Buffering of acid deposition, 413
Buffer zone concept, 208
Buildings. *See also* Homes
  acid deposition damage, 413
  cooling naturally, 354
  energy efficiency in building design,
      348–349
  existing buildings, 349–351
  geothermal heat pumps for, 350
  heating with solar energy, 353–354, 353f
Bureau of Land Management (BLM), 487
Burney, Jennifer, 254, 254b
Burning solid waste, 459–461, 459f–461f
Buses, 127f, 129
Bushmeat, 182, 183f
Butane, in natural gas, 333, 335
Butterfly, mimicry in, 93, 93f

C
CAFOs. *See* Concentrated animal feeding
      operations
Cahn, Robert, 494
Calcium carbonate, 156b, 213, 214b,
      214f, 304
California
  aquifer depletion in, 271drip irrigation
      use, 277
  San Andreas fault, 317f
  Sequoia National Park, 3f, 27f
  solar thermal systems, 356, 356f
  southern sea otter, 90b, 90f, 100, 101b,
      101f
  waste treatment, 295b
California condor, 171f, 186
California State Water Project, 273, 274f
Camels, 144b
Camouflage, 92, 93f
Canada
  landfills, 460
  monarch butterfly, 83
  Montreal Protocol, 442
  more-developed countries, 10
  oil reserves, 332
  water supply of, 263
Cancer, , 386, 417
Canopy, 147, 150f
Cap-and-trade, 418, 437, 437f, 481
Capital, 475
  human, 475, 475f
  manufactured, 475, 475f
  natural, 475, 475f (*see also* Natural capital)
Caps, 481
Captive breeding, 186–187
Car(s). *See* Automobiles; Motor vehicles
Carbohydrates
  complex, 33
  formed in photosynthesis, 48
  simple, 33
Carbon
  atomic structure, 32f
  in coal, 34
  graphene, 315b, 315f
  isotopes, 32
Carbonate ions, 213, 214b
Carbon capture and storage (CCS), **435**, 436f
Carbon cycle, 57–59, 59f, 199
Carbon dioxide

from agricultural activities, 239
air pollutant, 409
atmospheric, 424–425, 435
from burning fossil fuels, 409
carbon capture and storage (CCS),
    435, 436f
in carbon cycle, 58–59, 59f
from coal burning, 335, 336f
as greenhouse gas, 423
ocean acidification and, 214b
ocean uptake of, 425
top emitters of, 424, 425f
Carbon footprint, **424**
  reduction by
    colleges and universities, 439
    countries, states, and cities, 439
    individuals, 439–440
Carbonic acid, 214b
Carbon monoxide, 408–409
Carbon taxes, 437
Car-centered cities, 124
Carcinogens, **386**
Carnivores, **49**, 53f, 91, 146b
Carnivorous plants, 95f
Carrier, 482
Carrying capacity, **99**, 99f–100f
  for human population, 110b
  of Yellowstone National Park for gray
      wolves, 209b
Car-sharing networks, 125–126
Carson, Rachel, 20, 20f
Cascade of extinctions, 184
Case reports, 392
Case study. *See also* Core case study
  Alaska, climate change effects, 432
  American alligator, 75–76, 76f
  American baby boom, 114, 114f
  Aral Sea disaster, 275–276, 275f
  biodiversity in Costa Rica, identifying
      and protecting, 208
  bird population declines, 183–184
  bottled water, 287–288
  cigarettes and e-cigarettes, 396–398, 398f
  Colorado River, 268, 268f
  drinking water, 288–289
  eco-city concept in Curitiba, Brazil,
      128–129, 129f
  ecological restoration of tropical dry
      forest in Costa Rica, 212
  Endangered Species Act of 1973,
      185–186
  forest cover in United States, 198–199, 199f
  freshwater resources in United States,
      265–266, 266f
  Germany, renewable energy, 369–370
  global HIV/AIDS epidemic, 383–384,
      384f
  hazardous waste problem (e-waste), 453
  hazardous waste regulation in United
      States, 464–465
  how dams can kill delta, 272
  India, population growth, 116–117
  industrial ecosystems and biomimicry,
      466–467
  industrialized food production in United
      States, 232
  law of unintended consequences of law,
      244–245
  malaria, 384–385, 385f
  managing public lands in United States,
      487–490, 488f
  marine ecosystems, jellyfish invasions,
      215–216, 216f

Case study. *(continued)*
   Mexico City, 124
   microlending, 483, 483*f*
   monarch butterfly, 83–84, 84*f*
   North Pacific Garbage Patch, 452
   Ogallala aquifer, 269, 271, 270*f*
   Oil production and consumption in
      United States, 332
   pollution prevention, 394
   rare earth metals, 306–307, 307*f*
   recycling e-waste, 462
   sharks as keystone species, 76–77, 77*f*
   slowing population growth in China,
      117–118
   solid waste in United States, 451–452
   stresses on U.S. public parks, 207–208
   tuberculosis, 380
   urbanization in United States, 119–120,
      120*f*
   U.S. environmental laws, 486–487, 487*f*
   U.S. population growth, 110–111
Cash crops, 229
Cattle, overgrazing by, 200
CBD. *See* Convention on Biological
   Diversity
CCD. *See* Colony collapse disorder
CCS. *See* Carbon capture and storage
CDC. *See* Centers for Disease Control and
   Prevention
Cells (biological), **33**, 34*f*
Cell theory, 33
Cellulosic ethanol, 364
Center for Biological Diversity, 244
Center pivot irrigation method, 277*f*
Centers for Disease Control and
   Prevention (CDC)
   obesity estimates by, 228
   resistance to antibiotics, 381*b*
   on tobacco use, 397
CERCLA. *See* Comprehensive
   Environmental Response,
   Compensation, and Liability Act
Certainty, in science, 31
Change the Course campaign, 280*b*
Chemical bonds, 32
Chemical change, **34**
Chemical composition, 34
Chemical cycling, **5**, 6*f*
   principle of sustainability, 5, 6*f*
Chemical equation, 34
   balancing, 35
Chemical formula, **33**
Chemical hazards, 379, 385–395
   bisphenol A (BPA), 389*b*
   cigarettes and e-cigarettes (case study),
      396–398, 398*f*
   dirty dozen pollutants, 394, 466
   endocrine system, effects on, 387–389
   evaluating, 390–395
   case reports, 392
      epidemiological studies, 392
      estimating toxicity, 391, 391*f*, 392*t*
      trace levels of chemicals, 392
   knowledge of toxic chemicals, 393
   pollution prevention (case study),
      394–395
   trace levels, 392
Chemical reaction, **34**
   pollutant removal by, 411
Chemical warfare, 92, 93*f*
Chernobyl, Ukraine, 343
Chickens
   industrialized agriculture and, 234*f*
   urban areas, 255

Children
   cost of raising and educating, 112
   educating, 116
   food security and, 253
   in labor force, 111
   number of children per couple, 112
Chimpanzees, Jane Goodall and, 30*b*
China
   acid deposition and, 413–414
   air pollution, 416
   carbon dioxide emissions, 424
   carbon footprint, 424
   coal reserves, 335
   coal use, 335
   cost of gold in, 302*b*
   cultural eutrophication of lakes in, 284
   ecological footprint, 118, 482
   electronic waste (e-waste), 453, 462
   e-waste, 453, 462
   grain production in, 231
   hazardous waste, 453
   hydropower use in, 359
   industrial smog, 410
   meat consumption in, 233
   natural gas reserves, 335
   nuclear power reactors in, 343
   one-child policy, 118
   paper production in, 202
   pollution from mining, 310
   pollution of coastal areas, 289
   poverty in, 118
   rare earth metals, supplies of, 307
   renewable energy use, 353
   slowing population growth in
      (case study), 118
   solar cell use in, 359*f*
   tree plantations, 202
   urban areas, 119
   water pollution in, 286
   water supply of, 263
   wind power use in, 353
Chlorination, in sewage treatment, 293
Chlorofluorocarbon (CFC), 441
C horizon, 52*f*
CHP. *See* Combined heat and power system
Chromium, 305
Chromosomes, **34**, 34*f*
Chronic effect, of chemical exposure, 390
Chronic malnutrition, **227**
Chronic undernutrition, **227**
Chytrid fungus, 78*b*
Cilia, 415, 416*f*
Circle of poison, 245
Cirrus clouds, 425
CITES. *See* Convention on International
   Trade in Endangered Species
Cities. *See also* Urban areas
   car-centered, 124
   compact cities, 124
   dispersed cities, 124
   heat islands, 140
Citizen environmental groups, role of, 492
Clean Air Act, 414, 417, 418, 436, 486
Clean Air Task Force, 335
Clean Water Act, 291–292, 313*b*
Clear-cutting, 197, 198*f*
Climate, **137**, **420**
   biodiversity and, 134–163
   biomes and, 141, 142*f*, 146*b*, 151*f*
   cities affect on local, 123
   factors influencing, 137–140
   air circulation, 138
   greenhouse effect, 138
   ocean currents, 137, 137*f*

   surface features, 139–140, 140*f*
   greenhouse gas effect on, 407
   latitude and elevation effects on, 141
   local, surface features affecting,
      139–140, 140*f*
   in tropical forests, 199–201
   weather distinguished from, 137, 420
Climate change
   acceleration in, 420–423
   agriculture contribution to, 238–239
   atmospheric warming, 406*b*, 427–433
   biofuels and, 364
   cloud cover and, 425, 427
   consequences of, 427–433
      biodiversity, threats to, 431–432
      extreme weather, 431
      food production, threats to, 432–433
      human health, national security, and
         economies, 433
      melting ice, 428–429
      permafrost melting, 429
      sea level rise, 429–431
   contribution to species extinction,
      177–179
   Earth's history of, 420–427
   greenhouse gases and, 138*b*,
      423–424*f*
   human activities and, 424–425
   melting ice, 406*b*, 406*f*
   models to predict future changes,
      426*b*–427*b*, 426*f*–427*f*
   nuclear power and, 343
   ocean acidification and, 214*b*
   oceans and, 425
   outdoor air pollution and, 427
   preparing for, 440–441
   sea level rise, 406*b*, 429–431
   slowing, 435, 439*f*
      carbon capture and storage (CCS),
         435, 436*f*
      by colleges and universities, 439
      by countries, states, and cities, 439
      geoengineering schemes, 435–436
      government strategies for, 436–438
      greenhouse gas emissions,
         prevention and control of, 434,
         436–438
      by individual choices, 439–440
      solutions for, 435*f*
   as threat to coral reefs, 136*b*
   tipping points, 428, 428*f*
   water cycle and, 263
   water shortages, 266
   worst-case scenario, 428, 430
Climate change tipping points, **428**, 428*f*
Climate graphs
   deserts, 143*f*
   forests, 149*f*
   grasslands, 145*f*
Climate zones,137*f*
Cloud cover and climate change, 425
Clownfish, 89*f*, 94–95
Coal, **335**, 337–339, 337*f*
   acid deposition and, 412
   advantages and disadvantages of use,
      337*f*
   bituminous, 304
   electricity from, 335, 338*f*, 479*f*
   formation of, 337*f*
   industrial smog from, 410
   low-sulfur, 414
   mountaintop removal mining, 310, 312*f*
   pollution from, 335, 337, 339*f*
   reserves, 335

Coal ash, 337
Coal-burning power plants, 337–393
Coastal areas
    dead zones, 289, 296b
    marshes, 152, 154f
    pollution of, 289, 292f
    sea levels (see Sea level rise)
    wetlands, 152
Coastal coniferous forests, 148
Coastal marshes, 152, 154f
Coastal wetlands, **152**
Coastal zone, **152**, 153f
Cobalt, 305
Cod fishery collapse, 217f
Coevolution, **93**
Cogeneration, **346**
Cohen, Joel E., 495b
Cold deserts, 142, 143f
Cold forests, 148
Cold grasslands, 144, 145f
College of the Atlantic, 439
Colleges and universities
    carbon footprint reduction by, 439
    environmental policy, role in, 493–493
    greening of American campuses, 474b
Colony collapse disorder (CCD),
    168b, 180b
Colorado, water meter use in Boulder, 279
Colorado River, 268, 268f, 272
Columbia University, 228, 266
Command-and-control approach, 485
Commensalism, 91, **95**
Commercial energy, 36, 329, 345f
Commercial forest, 195
Communities, **48**. See also Ecosystem(s)
    response to environmental changes,
    95–97
Community-supported agriculture (CSA),
    254
Compact cities, 124
Compact fluorescent lighting, 351
Comparative risk analysis, 395
Competition, 91
    interspecific, 91
    intraspecific, 91
Compost, **248**
Composting, **455**, 457, 457f, 485f, 492
    toilet systems, 293
Compounds
    chemical formula, 33
    defined, **31**
    organic, 33
    used in book, 33t
Comprehensive Environmental Response,
    Compensation, and Liability Act
    (CERCLA), 465
Computers, energy-efficient, 351
Concentrated animal feeding operations
    (CAFOs), 233, 241f
Concentrating solar power (CSP), 355
Congressional Research Service, 344
Coniferous evergreen trees (conifers), 148
Connections
    atmospheric warming and water
        pollution, 282
    coloration and poison, 92
    corn, ethanol, and soil conservation, 247
    drinking water, latrines, and infectious
        diseases, 385
    drones, elephants and poachers, 180
    energy flow and feeding people, 53
    exponential growth and doubling time, 15
    lionfish, coral reef destruction, and
        economics, 215

low-sulfur coal, atmospheric warming,
    and toxic mercury, 414
meat production and ocean dead zones,
    239
metal prices and thievery, 307
mountains and climate, 151
nutrient cycles and life, 56
pesticides and food choices, 244
poverty and population growth, 16
urban living and biodiversity awareness,
    122
water leaks and water bills, 278
Conservation, in Costa Rica, 194b
Conservation concessions, 203
Conservation International, as citizen
    environmental group, 492
Conservationist school, 19
Conservation of energy, law of, **37**
Conservation of matter, law of, **35**
Conservation-tillage farming, 247
Constant loss population, 100
Consumer Electronics Association, 462
Consumers, **49**, 50f
    predation (see Predation)
    primary, 49, 50f
    secondary, 49, 50f
    tertiary, 49
Container Recycling Institute, 287
Containment shell, 340, 340f
Containment structure, for nuclear plants,
    343
Continental crust, 303f, 304
Continental shelf, 153, 153f
Contour planting, 247, 248f
Contour strip mining, **309**–310, 311f
Control group, 28b, 392
Controlled experiment, 28b, 28f
Control rods, 339, 340f
Control site, 28b, 28f
Convection, 138f
Convection cells/currents, 303
Convention on Biological Diversity (CBD),
    185
Convention on International Trade in
    Endangered Species (CITES), 184–185
Cooling
    buildings naturally, 354
    geothermal, 365f
Copenhagen Amendment, 442
Copper
    mine, open-pit, 301f,
    mining low-grade ore, 308
Coral bleaching, 156b
Coral reef
    biodiversity of, 156b
    bleaching, 156b, 156f
    degradation of, 157f
    human impact on, 157f
    ocean acidification and, 156b
    Red Sea, 135f
Core, Earth's, 47, 47f, **303**, 303f
Core case study
    amphibians, 70b
    conservation in Costa Rica, 194b, 194f
    cost of gold, 302b
    cradle-to-cradle design, 450b, 450f
    greening of American campuses, 474b,
        474f
    Growing Power, urban food oasis, 226b
    Gulf of Mexico, dead zone, 262b, 262f
    honeybee decline, 168b
    human population growth, 108b
    hydrofracking, oil and natural gas
        production, 328b, 328f

learning from Earth, 4b
    melting ice in Greenland, 406b, 406f
    mercury's toxic effects, 378b, 378f
    southern sea otters, 90b, 90f
    tropical rain forests, 46b
Corn, 229
    ethanol from, 363–364
Corrective feedback loop, **39**
Costanza, Robert, 196b, 477
Costa Rica
    biodiversity in, 194b, 208, 220b
    carbon neutral, 439
    conservation in, 194b
    ecological restoration of tropical dry
        forest in, 212
    golden toad, 78f
    Monteverde Cloud Forest Reserve, 78f
    nature reserves, 194b, 210f
    rain forest, 194f
Costs
    direct or internal, 478
    full-cost pricing, 478
    hidden, 478
    indirect or external, 478
Cousteau, Alexandra, 162b
Cover crops, 247
Coyotes, in Yellowstone National Park, 209b
Cradle-to-grave system, 450b, 464
Creosote, 144b
Critical thinking, 30
Crop(s). See also Crop production
    cover crops, 247
    perennial, 251, 253b
Croplands, 228
Crop production, 228–233. See also
    Agriculture
    cash crops, 229
    crossbreeding and, 232
    genetic engineering and, 232
    green revolution and, 230
    industrialized, 228–229
    monocultures, 229
    multiple cropping, 230
    organic, 229–230
    polycultures, 229, 251, 252b
    traditional, 229
Crop rotation, 249
Crossbreeding, 81, 232
Crown fire, 197, 199f
Crude oil, 290, **330**–333, 331f. See also Oil
    advantages and disadvantages of use,
        332, 333f
    conventional (light), 331
    dependence on, 332
    extraction of, 332
    peak production, 331
    refining, 331, 331f
    reserves of, 332
Crude petroleum, 289–290
Cruise ships, pollution by, 289
Crust, Earth's, 47, 47f, 47f, **302b**, **304**
    tectonic plates, 316, 317f
Crystalline solid, 304
CSA. See Community-supported
    agriculture
CSP. See Concentrated solar power
Cuba, integrated pest management (IPM)
    use in, 246
Cultural eutrophication, **159**, 161, **284**
Cultural hazards, 379
Cultural norms, birth rate and, 112
Cumulus clouds, 425
Curitiba, Brazil, 128–129, 129f
Cuyahoga River, 20, 283

Cyanide, 302b
Cyprus, drip irrigation use in, 277
Czechoslovakia, industrial smog, 410, 410f

D

Dai, Aiguo, 431
Dairy products, sustainable production of, 249
Dams and reservoirs, 151f
  advantages and disadvantages of, 271–272
  Colorado River, 268, 268f
  definitions, **271**
  how dams can kill delta, 272
  hydroelectric production, 268
  hydropower and, 359
  tidal energy, 359
Darwin, Charles, 79, 89
Data, **29**
Data centers, energy waste of, 346
DDT, 20, 178, 178f, 243, 245, 378b, 380, 387, 390, 395
Dead zones, 239, 262b, 289, 296b
De Anza Community College, 492
Death rate, 108b
  demographic transition, 115, 115f
  factors affecting, 112–113
Death Valley, 140
Debt-for-nature swap, 203
Decade of the environment, 21
Decibels, 123, 123f
Decomposers, **49**, 51f–54f
  in tropical rain forest, 148
Decomposition, 150
Decomposition zone, 283f
Deepwater Horizon, 289, 291f
Deep-well disposal, 463, 463f
Deforestation, **189**
  in Costa Rica, 194b
  defined, **197**
  experimenting with a forest (case study), 28b, 28f
  in Haiti, 203f
  of hillsides, 280, 281f
  mining and, 312
  natural capital degradation, 199f
  tropical forests, 199–201
Degradation. See Natural capital degradation
Degree of control, 399
Delta, **159**, 272
Demand, 475
Demographers, 109
Demographic momentum, 113
Demographic transition, **115**, 115f
Denali National Park, Alaska, 193f
Denmark
  cogeneration, 346
  hazardous waste, 461–462
  integrated pest management (IPM) use in, 246
  Kalundbord industrial ecosystem, 467, 468f
  landfills, 460
  reusable bags, 456
  waste-to-energy incineration, 460
  wind power use in, 361
Density-dependent factors, 98
Density-independent factors, 98
Department of Energy, 333
Depletion allowances, 308
Depletion time, **306**, 306f
Desalination, 276
Desertification, 238
  reducing, 249

Deserts
  adaptations to survival in, 144b
  biomes, 141
  climate graphs, 143f
  cold, 142, 143f
  human impacts on, 151f
  temperature, 142, 143f
  tropical, 142, 143f
Detritivores, **49**, 51f, 72b
Detritus feeders, **49**, 53f–55f, 60
Development. See also Economic development
  biodiversity friendly, 210
Diablo Lake in North Cascades National Park, 207f
Dickinson College, 492
Dieback, 99
Dieldrin, 244–245
Diet, 180b
Dioxins, 241
Direct price, 478
Dirty dozen pollutants, 394, 466
Discharge trading policy, 291
Disease
  infectious, 379
    drinking water, latrines, and infectious diseases, 385
    genetic resistance to antibiotics, 381b
    global HIV/AIDS epidemic (case study), 383–384, 384f
    malaria (case study), 384–385, 385f
    reducing, 385
    solutions, 386f
    tuberculosis (case study), 380
    viral, 383
    water pollution and, 282
  nontransmissible, 379
  transmissible, 379
  water pollution and, 282
Disinfection, in sewage treatment, 293
Dispersed cities, 124
Dissolved oxygen content, 98, 152
Distillation, 276
The Diversity of Life (Wilson), 73b
DNA (deoxyribonucleic acid), 33, 34
  in cells, genes, and chromosomes, 33–34, 34f
  genetic variability, 79
Dolomite, 304
Doomsday Seed Vault, 240
Dose, **390**
Dose-response curve, **391**, 391f
Doubling time, 15
Drainage basin, **157**, **264**
Drinking water
  back-to-the-tap movement, 288
  bottled, 287–288
  drinking water, latrines, and infectious diseases, 385
  purifying, 286
  Safe Drinking Water Act, 288, 486
Drip irrigation, 254b, 277f, 277–278
Drones, 62, 180
Drought, 266, 268, 275, 279
Dry casks, at nuclear plants, 341f, 342
Dry deposition, 412
Drylands, degradation of, 238
Dumps, 451, **461**
Durant, Will, 301

E

Earle, Sylvia A., 217, 218b, 289
Early loss population, 100
Earth

biodiversity components, 71f
  global air circulation, 139f
  life-support system, 110b
biodiversity hotspots, 211f
biomes and climate, 142f
human ecological footprint, 11f
ocean currents and climate zones, 137f
tectonic plates, 317f
topsoil erosion, 237f
wildlands, 207f
rotation of, 137
tapping internal heat of, 365–366
tectonic plates of, 81b, 81f
Earth-centered environmental worldview, **18**, **494**–496
Earth Day, 20–21
Earth Policy Institute, 118, 199
Earthquakes, 81b, **319**–321, 319f–321f
  tsunamis, **319**–321, 319f–321f
Earth's life-support system, 47–48, 110b
Echolocation, 93
Eco-city concept, 126, 128–129
Ecoindustrial park, 467
Ecological deficit, 12
Ecological economists, 476, 476f
Ecological footprint
  China, 118, 482
  contribution to species extinction, 177–179
  defined, **12**
  fishprint, 223
  per capita, 12
  pollution and, 12–13
  sustainability, 5–6, 8–9
  United States, 112f, 471, 482
  unsustainability of human, 10–12
  urban areas, 108b
  water footprint, 265
Ecological medicine, 383
Ecological niches, **74**
  of American alligator, 75
  of bird species in coastal wetland, 75f
  of generalist and specialty species, 74, 74f
  grasslands, 142
  overlap, 91
  specialized in tropical rain forest, 147, 150f
  species roles in ecosystems, 74–77
  in temperate deciduous forest, 147
  in temperate rain forest, 148
  in tropical rain forest, 146–147
Ecological restoration, **211**
Ecological sewage treatment systems, 294
Ecological Society of America, 233
Ecological succession, **95**–96
  natural ecosystems, 211
  primary, 95, 96f
  secondary, 95, 97f
  in rangelands, 204–205, 205f
Ecological tipping point, **39**, 46b, 63b, 96
Ecologist, 62
Ecology, **5**, **48**
  reconciliation, 212
Economic development, **475**–457
  biodiversity friendly, 210
  demographic transition and, 115, 115f
  environmentally sustainable, **477**
  slowing human population growth by, 118
  urbanization advantages, 121
Economic growth, **475**
  sustainability of, 475–477
Economics, 472–501, **475**
  cost–benefit analysis, 478
  dependence on natural capital, 475

models of economies, 455–477
  using economic tools to deal with
    environmental problems, 478–485
  environmental economic indicators,
    481
  environmental laws and regulations,
    484–485
  full-cost pricing, 478–480
  marketplace use to reduce pollution
    and resource waste, 481, 481f
  microlending (case study), 483
  Millennium Development Goals, 483
  poverty reduction, 482–483
  selling services instead of products,
    481–482
  subsidy shift, 480
  taxing pollution and waste, 480–481,
    480f
  using lessons from nature, 482
Economic services
  of bird species, 183–184
  of coral reefs, 157f
  of forests, 195f, 196b
  of freshwater systems, 158f
  of marine ecosystems, 152f, 157f
  provided by species, 170–173
Economists
  ecological, 476, 476f
  environmental, 477
  neoclassical, 476
Economy
  alternatives to throwaway, 455–456
  free-market, 475
  high-throughput, 475, 476f
  low-throughput (low-waste),
    484, 484f
Ecosystem(s), **5**, 44–64, **48**
  artificial, creating, 211
  components of, 48–52
  Earth's life support system and, 47–48
  energy in, 53–56, 53f–54f
  industrial, 466–467
  matter in, 56–61
  net primary productivity in, 55, 56f
  nutrient (biogeochemical) cycles, **56**
  organization of matter in, 49f
  Red List of Ecosystems, IUCN, 213
  restoration of, 211–212
  species roles in, 74–77
    American alligator case study,
      75–76, 76f
    generalist species, 74, 74f
    indicator species, 74–75
    keystone species, 75
    native species, 74
    nonnative species, 74
    sharks case study, 76–77, 77f
    specialist species, 74, 74f
  study by scientists, 62–63
    direct study of nature, 62
    field research, 62
    laboratory research, 62–63
  sustainability through constant change,
    96–97
  trophic levels in, 53f–54f
Ecosystem approach to sustaining
    biodiversity, 195–220
  aquatic biodiversity, 213–220
  conservation in Costa Rica, 194b,
    194f, 208
  forests
    ecological restoration of tropical dry
      forest in Costa Rica (case study),
      212

    major threats to, 195–201
    managing and sustaining, 201–204
  grasslands, managing and sustaining,
    204–205, 205f
  parks and nature preserves, managing
    and sustaining, 207, 208
  pricing nature's ecosystem services,
    196b
  terrestrial biodiversity and ecosystem
    services, 208–211
    biodiversity hotspots, 210
    emergency action strategy, 210
    five point plan, 208, 210
    reconciliation ecology, 212
    restoration of damaged ecosystems,
      211
    what can you do?, 212f
  wilderness protection, 206, 207f
Ecosystem diversity, 71–72
Ecosystem modeler, 63
Ecosystem services, **6**–8, 7f
  biodiversity, 71
  of bird species, 183
  ecological succession, 95
  extinction and, 173
  of forests, 195, 196b
  of freshwater systems, 158f
  of honeybees, 168b
  of marine ecosystems, 152, 152f
  natural enemies of pests, 242
  pollination, 72b, 168b
  pricing, 196b
  provided by species, 169–170
  reconciliation ecology to protect, 212
  water cycle and, 56–57
  of wetlands, 160
Ecotone, 141
Ecotourism, 172, 208, 212
Ecuador, San Lucas marsupial frog in,
  69f
Edge effect, **141**
Education
  family planning, 116
  of women, 116
Egg pulling, 186
Egypt, Red Sea of, 135f
Ehrlich, Paul, 12
Einstein, Albert, 29, 30
EIS. *See* Environmental impact statement
El-Ashry, Mohamed, 276
Electrical grid, energy-efficient, 347
Electric cars, 307f
Electricity
  from electric motor, 346
  energy-efficient electrical grid, 347
  as kinetic energy, 35, 36f
  net energy yield, 330f
  produced by
    coal, 335, 337, 339, 339f
    hydropower, 358–360
    nuclear power, 330
    solar cells, 357–358, 358f–359f
    solar power, 353
    solar thermal systems, 355, 356f
    wind power, 360–362, 360f–362f
Electromagnetic radiation, **35**
Electromagnetic spectrum, 36f
Electronic cigarettes (e-cigarettes), 398,
  398f
Electronic waste (e-waste), 453
  recycling of, 454f, 455, 457–459, 458f
Electrons, **32**
Elements, **31**–33, 32t
Elephants, 146b–147b, 146f, 180

Elevation, effect on climate and
  vegetation, 141
El Niño-Southern Oscillation (ENSO),
  139
Emergency action strategy, 210
Emigration, 97, 110
Emissions trading, 418
Endangered natural capital, 171f, 211f
Endangered species, **170**, 171f
  American alligator, 75
  biodiversity hotspots, 210
  bird species, 75f, 184
  California condor, 171f
  captive breeding of, 186
  Convention on International Trade
    in Endangered Species (CITES),
    184–185
  egg pulling, 186
  giant panda, 179
  gorilla, 179, 182, 183f
  gray wolf, 209b, 209f
  hyacinth macaw, 182
  illegal trade in wildlife, 181, 185
  Mexican gray wolf, 171f
  natural capital and, 170–173
  orangutans, 171, 172f
  San Lucas marsupial frog, 69f
  scalloped hammerhead shark, 77f
  southern sea otters, 90b, 90f, 92b, 101b,
    101f, 102b
  Sumatran tiger, 171f, 181
  threats from invasive species, 176
  U.S. Endangered Species Act, 185
  White rhinoceros, 181f
  whooping crane, 171f
Endangered Species Act, 185–186, 486
Endemic species, **82**
  in mountains, 150
Endocrine system, chemical effects on,
  387–389
End-of-pipe cleanup approach, 465
Energy
  from aerobic respiration, 49
  cogeneration, 346
  commercial, 36, 329, 345f
  conservation of, 37
  defined, **35**
  efficient use of, 345–352
  industrialized agriculture use of, 229
  kinetic, 35–36, 36f–37f
  net, 329–330, 329f
  potential, 35, 36, 37f
  solar (*see* Solar energy)
  thermal, 35
  thermodynamic laws and, 37–38
  transfer by convection, 138f
  transition to sustainable future,
    367–369, 369f
  use for bottled water production, 288
Energy content, 35
Energy efficiency, **345**–352
  benefits of improving, 346f
  in building design, 348–349
  electrical grid, 347
  energy waste, 345–346, 351–352
  in existing buildings, 349–351
  improving in industries and utilities,
    346
  of solar cells, 357
  in transportation, 347
Energy flow
  one-way, 50
  through ecosystems in food chains and
    food webs, 53, 53f

Energy quality, **36**–37
  high-quality, 36*f*–37*f*
  low-quality, 37
Energy resources, 326–371. *See also specific resources*
  advantages and disadvantages of using
    coal, 339*f*
    conventional natural gas, 337*f*
    conventional nuclear fuel cycle, 342*f*
    conventional oil, 333*f*
    fossil fuels, 330–339
    geothermal energy, 366*f*
    heavy oils, 334*f*
    hydrogen, 367*f*
    hydropower, 360*f*
    liquid biofuels, 365*f*
    nuclear power, 339–345
    renewable energy resources, 352–367
    solar cells, 359*f*
    solar heating systems, 355*f*
    solar thermal systems, 356*f*
    solid biomass, 364*f*
    wind power, 363*f*
  coal, 335–339, 337*f*–339*f*
  efficient use (*see* Energy efficiency)
  fossil fuels, 330–339 (*see also* Fossil fuels)
  laws of thermodynamics and, 329
  marketplace competition, 329
  natural gas, 328*b*, 32333–335, 336*b*
  net energy, 329–330, 329–330*f*
  nonrenewable energy resources, 328*b*, 329
  nuclear power, 339–345
  oil, 328*b*, 328*f*, 330–333, 331*f*, 333*f*
  renewable energy resources, 329, 352–367
  transition to sustainable future, 367–370
Energy use by source, 329*f*
ENSO. *See* El Niño-Southern Oscillation
Environment, **5**
Environmental audits, 492
Environmental conservation and protection, rise in U.S., 18–21
Environmental degradation, **10**, 115–116. *See also* Natural capital degradation
Environmental economic indicators, 481
Environmental economics. *See* Economics
Environmental economists, 478
Environmental ethicist, career as, 494
Environmental ethics, **18, 494**
Environmental hazards and human health, 378–399
  biological hazards, 379–385
  chemical hazards, 379, 385–390
  risk assessment and risk management, 379, 379*f*
  risk perception and evaluation, 395–399
Environmental impact, 10, 12–13
Environmental impact statement (EIS), 486
Environmentalism, **5**
Environmental justice, **490**
Environmental laws, 184–185. *See also specific laws and regulations*
  air pollution, 408, 411*f*, 412
  innovation discouraged or encouraged by, 466
  to protect humans from harmful effects of pesticides, 245
  on recycling, 462, 464
  on water quality, 282, 288

Environmental laws, United States. *See also specific laws and regulations*
  air pollution and, 408
  Clean Air Act, 414, 417–418, 436
  Clean Water Act, 291, 313*b*
  Comprehensive Environmental Response, Compensation, and Liability Act (CERCLA), 465
  Endangered Species Act, 185–186, 486
  Federal Insecticide, Fungicide, and Rodenticide Act (FIFRA), 245
  Federal Water Pollution Control Act, 291
  hazardous waste regulation in, 464
  National Environmental Policy Act (NEPA), 486
  Resource Conservation and Recovery Act (RCRA), 464, 486
  Safe Drinking Water Act, 486
  Toxic Substances Control Act (TSCA), 464
  Water Quality Act, 291
  Wilderness Act, 207, 486
Environmental literacy, 496
Environmentally sustainable economic development, **477**
Environmentally sustainable society, **21**
Environmental nanotechnology, career in, 315*b*
Environmental policy, 484–494
  citizen environmental groups, role of, 492
  defined, **484**
  environmental justice, 490
  environmental security, 493
  influencing, 491*f*
  principles guiding, 490–491
  students and educational institutions, role of, 492–493, 493*f*
  U.S. environmental laws and regulations (case study), 486–487, 487*f*
Environmental problems, 5–21. *See also specific issues*
  causes of, 14–21
    industrialized food production, 233–242
    isolation from nature, 17–18
    population growth, 14–15, 15*f*
    poverty, 16
    pricing, 16–17
    unsustainable resource use, 15–16
  ecological footprint and, 12–14, 14*f*
  sustainability principles, 5–10, 6*f*, 9*f*
  urban, 119–124
  using economic tools to deal with, 478–484
    environmental indicators, 481
    environmental laws and regulations, 484–486, 487*f*
    full-cost pricing, 478–480
    marketplace use to reduce pollution and resource waste, 481
    microlending (case study), 483
    Millennium Development Goals, 483–484
    poverty reduction, 482–483
    selling services instead of products, 481–482
    subsidy shift, 480
    taxing pollution and waste, 480, 480*f*
    using lessons from nature, 484*f*
Environmental Protection Agency (EPA)
  establishment of, 21

  on geothermal heat pump, 365
  on groundwater contamination, 286
  hazardous waste regulation, 464
  on hazardous waste storage, 463
  on indoor air pollution, 414
  on mercury, 386
  on nonpoint sources of water pollution, 282
  on outdoor air pollution, 417
  particulate classification by, 409
  on pesticide use in United States, 243
  on risk evaluation, 399
  on solid waste sources in United States, 451
  on toxic chemicals, 386
  Toxic Release Inventory, 465
Environmental refugees, 113
Environmental regulations, 484–485
  incentive-based, 485
  innovation-friendly, 485
Environmental resistance, **99**
Environmental science, **5**
Environmental security, 493
Environmental Working Group, 244, 364, 465
Environmental worldview, **18, 494**
  earth-centered, 18, 494
  human-centered, 18, 494
  life-centered, 18, 494
  planetary management, 18, 494
  stewardship, 18, 494
Environmental writer, career as, 491
EPA. *See* Environmental Protection Agency
Epidemic, 380
Epidemiological studies, 392
Epiphytes, 95
Erosion
  rock formation, 304–305
  topsoil, 236–238, 238*f*, 247
  wind, 229, 237
Erosion hotspots, 247
Eruption, volcano, 81*b*, 318*f*
Estes, James, 92*b*
Estuary, 154*f*, 268
Ethanol
  as biofuel, 363–364
  cellulosic, 364
Ethics, 9, 9*f*, 18
Euphotic zone, 153, 153*f*
European honeybee. *See* Honeybees
European Union
  bisphenol A (BPA) ban, 389
  electronic waste (e-waste), 456
  hazardous waste, 461
  laws on recycling, 462
  plastic bag use, 456
European wild boar (feral pig), 175*f*, 208
Eutrophication, **159**
  cultural, 284
Eutrophic lake, **159**, 160*f*
Evaporation, 144*b*
  Hadley cell creation by, 138
  in water cycle, 56
Everglades National Park, invasive species in, 176, 177*f*
Evolution, 77–80
  adaptations and, 80
  biological, **77**
  coevolution, 93
  defined, **77**
  fossils and the fossil record, 78
  myths about, 80
  natural selection, **78**, 79–80, 79*f*
  wilderness areas as centers for, 207

Experiment, controlled, 28*b*, 28*f*
Experimental group, 28*b*, 392
Experimental site, 28*b*
Exponential growth, **14**, 15*f*, 99, 99*f*–100*f*
Extinction, 82–83
  amphibian species, 78*b*
  of aquatic species, 213
  background rate of, 169
  cascade of extinctions, 184
  characteristics of species vulnerable to, 173
  dams and reservoirs associated with, 272
  defined, 169
  human activities that hasten
    climate change, 177–179
    habitat loss and fragmentation, 173–174, 174*f*
    illegal killing, capturing, and selling of wild species, 179–182,
    invasive (non-native) species introduction, 174–177, 175*f*
    pollution, 177–179
    population growth and resource use, 177–179
  mass, 82, **169**
  rates
    background, 169
    estimating, 170*b*
Exxon Valdez, 289

F

Falling and flowing water, energy from, 358–360, 360*f*
Family planning, **116**
  in China, 118
  in India, 116
FAO. *See* Food and Agriculture Organization of the United Nations
Farm Act (Food Security Act), 247
Farming. *See* Agriculture
Farms
  tree, 195, 196*f*
  urban, 226*b*, 226*f*, 256*b*
  wildlife, 186
Farm-to-City Market Basket program, 226*b*
Fats, 227
Fault, 319
FDA. *See* Food and Drug Administration
Fear, risk evaluation and, 399
Fecal snow, 124
Federal Insecticide, Fungicide, and Rodenticide Act (FIFRA), 245
Federal Water Pollution Control Act, 291
Feedback, **38**
Feedback loop, **38**
  negative, **39**, 39*f*
  positive, **38**, 38*f*
Feeding level, 48. *See also* Trophic level
Feedlots, 228, 233, 240–241, 241*f*,
Ferreira, Juliana Machado, 182*b*
Fertility rate, 110–111, 111*f*
  factors affecting, 111–112
  total, **110**, 111*f*
  in United States, 110, 112*f*
Fertilizer
  compost, 248
  manure, 248
  nitrogen in, 60
  organic, 248
  phosphorus in, 61
  synthetic inorganic, 249
  water pollution from, 291

Field research, 62
FIFRA. *See* Federal Insecticide, Fungicide, and Rodenticide Act
Finland, refillable container use in, 456
Fire, 197, 199*f*
  crown, 197, 199*f*
  management of forest fires, 202
  prescribed burns, 202
  surface, 197, 199*f*
Firewood, gathering and hauling of, 116*f*
First law of thermodynamics, **37**
Fish
  illegal trade in, 182
  mercury in, 378*b*, 378*f*, 387
Fish and Wildlife Service. *See* U.S. Fish and Wildlife Service (USFWS)
Fishery, **215**, 217*f*, **233**
Fish farming, **233**, 235*f*, 468*f*
Fishing
  commercial methods, 215, 216*f*
  overfishing, 215
  trawlers, habitat damage from, 213, 213*f*
Fishprint, 223
Fish production. *See* Aquaculture
Fissure, 316
Fitness, 80
Flooding
  climate change and, 430–431
  reducing pollution from, 280, 281*f*
  sea level rise and, 429–431, 431*f*
  urban areas, 122
Flood irrigation, 277
Floodplains, 160, 280
Floodplain zone, 159, 161*f*
Florida
  Everglades, 76, 176, 177*f*
  sea level rise and, 431*f*
Flowing water, as indirect form of solar energy, 358–360
Flows within systems, **38**, 38*f*
Flu virus, 383
Focus, earthquake, 319
Food. *See also* Food production
  genetically engineered, 232–233
  hunger, 227
  labeling, 230
  specialization, 229
  waste, 248
Food and Agriculture Organization (FAO) of the United Nations, 200
  on aquaculture, 241
  on fisheries, 233
  on meat production, 233, 249
  on overgrazing, 204, 241
  on soil salinization, 238
  on undernutrition and malnutrition, 228
  on waterlogging, 238
  on water pollution, 239
  on water use for waste removal, 279
Food and Drug Administration (FDA)
  on antibiotic use in animals, 241
  on antibiotic use in feed, 381*b*
  bisphenol A (BPA), 389*b*
  pesticide regulation, 245
Food chain, **53**, 53*f*
Food desert, 226*b*, 227, 256*b*
Food insecurity, **227**
Food prices, control of, 252–253
Food production, 224–256. *See also* Agriculture
  agrobiodiversity, loss of, 239
  community-supported agriculture (CSA), 254

crop production229–230, 230*f*
  cash crops, 229
  industrialized, 229
  monocultures, 229
  multiple cropping, 230
  organic, 229–230, 291
  polycultures, 229, 241, 252*b*
  traditional, 229
crossbreeding and artificial selection, 232
environmental problems caused by, 233–242
fish and shellfish production, 233
genetic engineering and, 232–233
government policies to improve, 252–253
green revolution, 230–231, 240
hidden costs of, 232
increase in, 228–229
meat production, 233, 234*f*, 240–241, 241*f*
monocultures, 229, 251
organic, 229–230, 291
pest control, 242–247
polycultures, 229, 241, 252*b*
specialization, 229
subsidies, 232, 253
sustainability improvements, 247–252, 251*f*–252*f*
  in aquaculture, 250, 250*f*
  erosion reduction, 247
  in meat and dairy product production, 249–250
  reducing soil salinization and desertification, 249, 249*f*
  shifting to more sustainable production, 250–252, 251*f*–252*f*
  soil fertility restoration, 248–249
  what can you do?, 252*f*
traditional, 229
urban, 226*b*, 226*f*, 255, 256*b*
yield, 229
Food Quality Protection Act, 245
Food security, **227–228**
  improving, 252–255
  problems of human, 228
Food Security Act (Farm Act), 247
Food waste, 254
Food web, **53**, 54*f*
Forage, 204
Ford, Henry, 458*b*
Forests
  acid deposition and, 413
  age and makeup of, 195–196
  biomes, 141, 142*f*, 151*f*
  clear-cutting, 197, 198*f*, 198–199
  climate change threats to, 431–432
  coastal, 148
  cold, 148, 149*f*
  commercial, 195
  conservation in Costa Rica (core case study), 194*b*, 194*f*
  deforestation, 28*f*, 194*b*, 197–198, 199*f*
    in Costa Rica, 194*b*
    defined, 197
    experimenting with a forest (case study), 28*b*, 28*f*
    in Haiti, 203*f*
    of hillsides, 280, 281*f*
    mining and, 312
    natural capital degradation, 200*f*
    reducing tropical, 203–204, 204*f*
    tropical forests, 62*b*, 199–201, 204*f*

Forests (continued)
  ecological restoration of tropical dry
     forest in Costa Rica, 212
  economic and ecosystem services, 195,
     196b
  experimenting with (case study),28b
  fire, 197, 199f
  harvesting methods, 197, 198f
  human impacts on, 151f
  major threats to, 195–201
  managing and sustaining, 201–204, 202f
    certification of sustainably grown
      timber, 201
    by fire management, 202
    by logging practices, 203
    by reducing demand for harvested
      trees, 202–203
    tropical forests, 203–204, 204f
  mangrove, 152, 154f
  natural capital, 195f
  old-growth forest, 195, 195f
  regrowth in United States, 197–198
  second-growth forest, 195
  selective-cutting, 197, 198f
  strip cutting, 197, 198f
  temperate, 148, 149f
  tree plantation, 195, 196f
  tropical, 146–147, 149f, 196f, 199–201
Forest Stewardship Council (FSC), 201
Formaldehyde, 415, 415f
Formosan termite, 175f
Fossil fuels, 47. See also specific fuel type
  advantages and disadvantages of using,
    330–339
  burning
    air pollutants from, 408
    ocean acidification and, 214b
  in carbon cycle, 59, 59f
  coal, 335–337, 338f–339f
  commercial energy from, 36, 329
  industrialized agriculture and, 229
  natural gas, 329, 333, 335, 337f
  oil, 330, 331f, 332, 333, 333f, 334f, 335,
Fossil record, 78
Fossils, 78, 81b
Fox, 82f
Fracking, 285, 328b, 328f, 336b, 336f, 473f
Frandsen, Vestergaard, 287
Fraunhofer Institute for Solar Energy
    Systems, 357
Free-market economic system, 475
Freons, 441
Freshwater, 263. See also Freshwater
    systems; Groundwater; Water
    resources
  access to, 263
  available amount of, 263
  cost of, 276–277
  dams and reservoirs to manage,
    271–272, 272f, 273f
  desalinization, 276
  distribution of, 263, 266
  hydrologic cycle and, 263
  increasing use of, 264–265, 265f
  as irreplaceable resource, 263
  purifying drinking water, 286–287, 287f
  rainwater, capture and storage of, 278
  resources in United States (case study),
    265–266, 266f
  shortages, 266–267, 266f
  surface runoff, 264
  sustainable use of, 276–280
    benefits of, 276–277
    flood damage reduction, 280

    government subsidies, 276–277
    in industries and homes, 278–279,
      279f
    irrigation efficiency improvements,
      277–278, 278f
    reducing losses, 279f
    in waste removal, 279
    water transfers, 272–275, 274f
Freshwater life zones, 152
Freshwater scarcity stress, 266, 267f
Freshwater systems
  ecosystem and economic services of,
    158f, 160–161
  extinctions, 213
  human impacts on, 161–162
  lakes, 157, 158f–159f
  in land wetlands, 160, 161
  as natural capital, 158f
  rivers and streams, 159–160, 160f
  standing and flowing systems, 157–159
Frogs, coloration differences in, 93f
FSC. See Forest Stewardship Council
Fuel assemblies, 339, 341f
Fuel cell, 348, 366–367
Fuel-cell technology, career in, 348
Fuel rods, 339
  spent, 341–342, 342f
Fuelwood use in less-developed countries,
    202 363
Fukushima Daiichi nuclear power plant
    accident, 344
Full-cost pricing, 9, 9f, 478–480
  pricing nature's ecosystem services,
    196b
  reducing automobile use, 125
Functional diversity, 71f, 72
Fungi
  decomposers, 51b, 51f
  disease in amphibians, 78b
Fungicides, 243. See also Pesticides
Federal Insecticide, Fungicide, and
    Rodenticide Act (FIFRA), 245

G

Gandhi, Mahatma, 499
Garbage. See also Municipal solid waste
    (MSW)
  ocean garbage patches, 452, 452f
Gasoline-electric hybrid cars, 347, 348f
Gasoline tax, 125
GDP. See Gross domestic product
Generalist species, 74, 74f
Gene revolutions, 232
Genes, 34, 34f
Gene splicing, 82, 232
Genetically modified (GM) crops, 232–233
Genetically modified organisms (GMOs),
    82, 232
Genetic diversity, 71, 71f, 72f
Genetic engineering, 82
  biomining, 308
  controversy over GM food production,
    232–233
  implanting genetic resistance to pests,
    245
  to produce new varieties of crops and
    livestock, 232
Genetic information, 34
Genetic makeup, toxicity and, 390
Genetic resistance, 79
  to antibiotics, 381b
Genetic variability, 79
Genuine progress indicators (GPI), 481
Geoengineering, 435–436

Geographic information system (GIS), 62
Geographic isolation, 80, 82f
Geological processes, effect on biodiversity,
    81b
Geologic processes, 303–305
Geology, 300–322
  defined, 303
  dynamic nature of Earth's, 303–304,
    303f
  earthquakes, 319–321, 319f–320f
  Earth's structure, 303–304, 303f
  rock cycle, 304, 305f
  volcanoes, 316–319, 318f
Geosphere, 47, 47f
Geothermal energy, 365–366, 366b, 366f
Geothermal engineer, career as, 366
Geothermal heat pumps, 354, 365, 365f
Germany
  bottled water in, 288
  car-sharing networks, 126
  temperate deciduous forest, 149f
  wind power use in, 361
Giant panda, 74, 74f, 179
GIS. See Geographic information system
GIS analyst, 62
Glacial periods, 420
Glacier National Park, 422f, 429
Glaciers, 57, 96f, 157, 275
  in Greenland, 406b, 406f
  melting, 406b, 406f, 421, 430b, 432
Gleick, Peter, 287
Glen Canyon Dam, 268f
Global Coral Reef Monitoring Network,
    156b
Global Environment Facility (GEF), 493
Global Forest Watch, 199
Global ocean, 152
Global positioning system (GPS), 62, 252
Global satellite positioning equipment, use
    in industrial fishing, 215
Global sustainability movement, 492
Global Water Policy Project, 279, 280b,
    284
Glucose, 49
  in aerobic respiration, 49
  as simple carbohydrate, 33
GM. See Genetically modified crops
GMOs See Genetically modified organisms
Gobi Desert, Mongolia, 142, 143f
GOI. See Genuine progress indicators
Gold, 305
  Canada mine, 311f
  as element, 31f
  Illegal mining in Africa, 314f
  real cost, 302b, 322b
  top producing countries, 302b
Golden toad, 78f
Goodall, Jane, 30b
Goods and services, 16–17, 475, 475f
Gorilla, 183f
Government
  greenhouse gas emissions, 436–438
  laws (see Environmental laws)
  policies to improve food production,
    252–253
  taxation (see Tax; Tax breaks)
Governmental agencies. See specific
    departments or agencies
Governmental policy. See Environmental
    policy
Government subsidies. See Subsidies
GPP. See Gross primary productivity
GPS. See Global positioning system
Grade, metal ore, 309

Grains, 227, 227f
  countries producing, 269
  green revolutions and, 230–231, 231f
  for meat production, 249, 249f
  uses of, 228,
  water used for production of, 265
Grameen Bank, 483
Granite, 304
Graphene, 315b, 315f
Grasslands
  biomes, 136b, 141, 151f, 163b
  cerrado, 239
  climate graphs, 145f
  cold, 144, 145f
  human impacts on, 151f
  inertia and resilience in, 96
  managing and sustaining, 204–205, 205f
  natural capital degradation in, 148f, 151f
  savanna, 144, 145f
  temperate, 144, 145f, 148f
  tropical, 144, 145f
Grassroots action, for solid and hazardous waste management, 465
Gravel, 305
Gravity, 30, 48
Gray fox, 82f
Graying of America, 114
Gray water, 278
Gray wolf
  as endangered species, 171f, 209b, 209f
  in Yellowstone National Park, 208, 209b
Grazing
  overgrazing, 204–205, 205f, 238
  rotational, 204
Great Lakes, invasive species in, 176
Great Pacific Garbage Patch, 452, 452f
Great Smoky Mountains National Park, 208
Green architecture, 349
Green Belt Movement, 204
Green Building Council, 349
Green careers, 486f
  battery engineer, 349
  bioprospecting, 172
  ecological medicine, 383
  ecologist, 62
  ecosystem modeler, 63
  environmental ethicist, 494
  environmental nanotechnology, 315b
  fuel-cell technology, 348
  geothermal engineer, 366
  GIS analyst, 62
  hydrogen energy development, 367
  indoor air pollution specialist, 415
  infectious disease prevention, 385
  integrated pest management, 246
  remote sensing analyst, 62
  small-scale sustainable agriculture, 254
  solar-cell technology, 357
  sustainable aquaculture, 250
  sustainable environmental design and architecture, 349
  sustainable forestry, 201
  wastewater purification, 286
  water conservation specialist, 278
  wind-energy engineering, 362
Greenhouse effect, 47, 48f, 58, 138b, 423
  climate change and, 423
  human activities and, 137
Greenhouse gases, 48f, 407
  climate change and, 138b, 423
  from livestock production, 239

prevention and control of emissions, 436–438
Greening of American campuses, 474b, 474f
Greenland, melting ice in, 406b, 406f, 444b
Green manure, 248
Greenpeace, 492
Green revolution, 230–231
  limits on the expansion of, 240
Green roof, 349, 350f
Green taxes, 480–481, 480f
Gross domestic product (GDP), 481
Gross primary productivity (GPP), 53
Groundwater, 57, 263–264
  aquifers, 269, 271, 271f
  as critical resource, 263–264
  depleting/withdrawing, 269, 269f, 272f
  pollution, 284–286, 285f–286f
Growing Power, 226b, 226f, 229, 247, 250, 251, 254, 256b
Guanacaste National Park, Costa Rica, 212
Guatemala, age-structure diagram in, 113f
Gulf of Mexico
  dead zone in, 239, 262b, 262f
  oil spill in, 289, 267f
Gunnoe, Maria, 313b
Guo, Yu-Guo, 316b

H
HAAs. See Hormonally active agents
Habitat, 74
Habitat degradation/destruction
  aquatic biodiversity and, 213–215
  extinction and, 170b, 173
Habitat fragmentation, 173
  amphibians decline, 78b
Habitat islands, 173
Habitat loss
  amphibians decline, 78b
  bird population decline and, 183–184
  extinction and, 173
Hadley cells, 138, 139f
Haiti
  deforestation in, 203f
  poverty in, 482f, 483
Hammerhead sharks, 76, 77f
Hardin, Garrett, 12
Hawaiian Islands
  plastic bag ban, 456
  species extinction and threatened species, 173
Hawken, Paul, 500
Hazardous air pollutants (HAPs), 417
Hazardous waste, 452
  chemicals in your home, 453f
  dealing with, 453–455, 454f
  detoxifying, 462–463
  electronic waste (e-waste), 453
  grassroots action and, 465
  integrated management, 454, 454f
  international treaties on, 466
  radioactive waste, 453
  regulation in United States, 464–465
  storing, 463, 464f
  what can you do?, 455f
Hazards and human health, 376–400
  biological hazards, 379–385
  chemical hazards, 379, 385–390
  risk assessment and risk management, 379, 379f
  risk perception and evaluation, 390–399
Headwater streams, 159
Health. See Human health

Heat, 35
Heating
  buildings and water with solar energy, 353–354, 355f
  energy-efficient, 350–351
  net energy yield, 330f
  solar thermal systems, 355, 356f
Heat islands, 123
Heat pump, 365, 365f
Heavy oil, 333, 333f–334f
Hepatitis B virus (HBV), 383
Herbicides, 243. See also Pesticides
  pollution from, 244
  resistant crops, 243
Herbivores, 49, 53f, 91
Heritable trait, 79
Herpetologists, 78b
Hidden costs, 478
  of transportation, 347
High-grade ore, 304
High-input agriculture, 229. See also Industrialized agriculture
High-quality energy, 36, 47, 53
High-temperature industrial heat, net energy yield from, 330f
High-throughput economy, 475, 476f
Hillsides, deforestation of, 281f
HIPPCO (habitat destruction, degradation, and fragmentation: invasive [nonnative] species population growth and increasing use of resources, pollution, climate change, overexploitation), 173, 183–184
HIV. See Human immunodeficiency virus
Hoerling, Martin, 431
Holdren, John, 12, 344
Holistic principle, 490
Holocene, 63b
Homes
  energy efficiency in
    building design, 348–349
    existing buildings, 349–351, 350f
    saving energy where you live, 352f
  geothermal heat pumps for, 365, 365f
  harmful chemicals in, 453f
  indoor air pollution, 414, 415f
  for pest enemies, 245
  reducing freshwater losses by, 278, 279f
  solar heating systems, 353–354, 353f–354f
  toxic chemicals in, 393f
Honeybees, 168b, 168f, 180b, 180f
  African, 74, 175f
  colony collapse disorder, 168b
  core case study, 168b
  decline in, 72b, 168b, 180b
  introduction into North America, 174
  pesticides as threat to, 244
  pollination by, 72b, 72f, 168b, 189b
  sustainability and, 189b
Horizons, soil, 50
Horizontal drilling, 328b, 328f
Hormonally active agents (HAAs), 387
Hormone blockers, 387
Hormone disrupters, 389, 390f
Hormone mimics, 387
Hormones, 387
  chemical effects on, 387–389, 389b, 390f
  for pest control, 246
Horwich, Robert, 212
Host, for parasite, 93
Hubbard Brook Experimental Forest, New Hampshire, 28b, 33f, 38

Human(s)
    health (*see* Human health)
    major adaptations of, 79–80
    population (*see* Human population;
        Human population growth)
    respiratory system, 415–417, 416*f*
Human activities, impact of
    air pollution from, 407–408, 408*f*
    aquatic biodiversity, 213–215
    climate change and, 424–425, 430*f*, 433
    cultural eutrophication of lakes, 284
    environmental impact of, 12
    extinction rates and, 169–170, 173–184
        climate change, 177–179
        habitat loss and fragmentation, 173
        illegal killing, capturing, and selling of
            wild species, 179–182, 181*f*
        invasive (non-native) species
            introduction, 173–177, 174*f*, 175*f*
        pollution, 177–179, 177*f*
        population growth and resource use,
            177–179
    freshwater systems, 161–162
    marine ecosystems, 155, 157, 157*f*
    terrestrial ecosystems, 151–152, 151*f*
Human capital, **475**, 475*f*
Human-centered environmental
    worldview, **18**, **494**
Human health, 376–400
    air pollution and, 415–417
    biological hazards, 379–385
    case study
        cigarettes and e-cigarettes, 396–398,
            397*f*
        global HIV/AIDS epidemic, 383–384,
            384*f*
        malaria, 384–385, 385*f*
        pollution prevention, 394–395
        tuberculosis, 380, 382*f*
    chemical hazards, 379, 385–390
    climate change threat to, 433
    cultural hazards, 379
    indoor air pollution, 419, 419*f*
    infant mortality rate, 112–113
    infectious disease concentration in
        cities, 122
    lifestyle choices, 379
    mercury, toxic effects of, 378*b*, 378*f*
    nanotechnology and, 315*b*
    natural hazards, 379
    pesticides as threat to, 244
    poverty and, 396, 396*f*, 482–483
    risk assessment and risk management,
        379, 379*f*
    risk perception and evaluation, 390–399
        difficulty of evaluating risks, 399
        estimating risks from technology,
            398–399
        principle for, 399
    water access and, 263
    water pollution and, 284
Human immunodeficiency virus (HIV),
    383–384, 385*f*
Humanities, 15
Human population. *See also* Human
    population growth
    age structure
        aging populations and, 114–115
        defined, 113
        diagrams, 113*f*, 114*f*
        projections based on, 113–114
        U.S. baby boom generation, 114, 114*f*
    in biodiversity hotspots, 210, 211*f*
    carrying capacity of Earth for, 110*b*

declines in countries, 113*f*, 113–114
demographic transition, 115, 115*f*
factors influencing size, 109–113
    birth rate, 111–112
    death rate, 112–113
    fertility rate, 110, 111*f*
    migration, 113
food security and, 227
nature's controls on, 101
population change, calculating, 110
urban areas, 119–124
Human population growth
    baby boom in United States, 110, 111*f*,
        114
    contribution to species extinction,
        177–179
    demographic momentum, 113–114
    distribution of, 109
    environmental degradation and, 116
    exponential, **14**, 15*f*, 109
    in less-developed countries, 109, 110*b*
    limits on, 109, 110*b*
    natural capital degradation, 111*f*
    poverty and, 16, 116
    rate of, 109, 109*f*
    slowing, 115–119
        in China (case study), 118–119, 118*f*
        by economic development, 115–116,
            115*f*
        by empowerment of women, 116,
            116*f*
        in India (case study), 116–117
        by promoting family planning, 116
    threat to bird populations, 183–184
    United States (case study), 110–111,
        112*f*
Humus, 51
Hunger, **227**
Hyacinth macaw, 182, 182*f*
Hybrid-electric cars, 307, 307*f*
Hydraulic fracturing, 285, **328***b*, 328*f*, 463
Hydrocarbons, 33
Hydrogen
    as a fuel, 366–367, 367*f*
    fuel cell, 348 366–367
    nuclear fusion, 344
Hydrogen bonds, 34*f*
Hydrogen energy development, career in,
    367
Hydrogen ions, 32–33
Hydrogen sulfide, 49, 334
Hydrologic cycle, **56**, 58*f*, 151, 263, 432
Hydrologists, 276
Hydropower, 358–359, 360*f*
Hydrosphere, **47**, 47*f*
Hydrothermal ore deposits, 308, 308*f*
Hydrothermal reservoirs, 365
Hydroxide ions, 32–33
Hypercities, 119
Hypothesis, scientific, **29**

I

Ice, 47
    density of, 57*b*
    glaciers, 57
    melting, 406*b*, 406*f*, 428–429, 429*f*,
        430*b*
Ice cores, 421, 421*f*, 424*f*
Igneous rock, **304**, 305*f*
Illegal trade in wildlife, 181, 185
Immigration, 97
    population change and, 110
    into United States, 110–111
    into urban areas, 119

Immune system, 387
Impala, 94*f*
Incineration of wastes, 459–460, 460*f*
Income, natural, **21**
India
    carbon dioxide emissions, 424, 425*f*
    children in labor force, 111–112
    grain production in, 231
    homeless people, 117*f*
    population pressure, 107*f*
    poverty, 117
    rice production, 239
    slowing population growth in (case
        study), 116–117
    water carried by women, 264*f*
    water pollution in, 284
Indian (Bengal) tiger, 174*f*, 180
Indicator species, **74**, 184
    amphibians as, 74, 78*b*, 85*b*
Individuals Matter, 8, 491
    Ballard, Robert, 318*b*
    Balog, James, 430*b*
    Benyus, Janine, 9*b*
    Burney, Jennifer, 254*b*
    Cousteau, Alexandria, 162*b*
    Earle, Sylvia, 218*b*
    Ferreira, Juliana Machado, 182*b*
    Goodall, Jane, 30*b*
    Gunnoe, Maria, 313*b*
    Guo, Yu-Guo, 316*b*
    Lovejoy, Thomas E., 62*b*
    Martinez, Juan, 498*b*
    McDonough, William, 456*b*
    Molina, Mario, 443*b*
    Postel, Sandra, 280*b*
    Roske-Martinez, Xiuhtezcatl, 492*b*
    Rowland, Sherwood, 443*b*
    Ruzo, Andrés, 366*b*
    saving energy where you live, 352*f*
    Sekercioğlu, Çağan Hakki, 184*b*
    Sereivathana, Tuy, 147*b*
    Sindi, Hayat, 382*b*
    Tilman, David, 252*b*
    Wilson, E. O., 73*b*
Indonesia, 2004 tsunami and, 320–321,
    320*f*
Indoor air pollutants, 415, 415*f*
Indoor air pollution, 414–415
    health issues with, 414–415, 414*f*
    reducing, 419, 419*f*
    sources of, 414–415, 414*f*
Indoor air pollution specialist, career as,
    415
Industrial ecosystems (case study),
    466–467
Industrialized agriculture, **229**
    agribusiness, 232
    agrobiodiversity, loss of, 239
    energy inputs, 234–235
    environmental problems caused by,
        233–242, 234*f*
        air pollution and climate change,
            238–239
        aquaculture, 241, 242*f*
        biodiversity loss, 239–240
        desertification, 238
        genetically engineered foods,
            232–233
        irrigation excesses, 238, 239*f*
        meat production, 240–241, 241*f*
        topsoil erosion, 236–238, 237*f*–238*f*
    genetic engineering, 232, 232
    goal of increased yield, 229
    green revolution in, 230–231

hidden costs of, 232
meat production, 233, 236, 239f, 240–241, 241f
monocultures, 229
organic agriculture compared to, 231f
plantation agriculture, 229
subsidies, 232
in United States (case study), 232
Industrial Revolution, 63b, 424
Industrial smog, **410**, 410f
Industrial solid waste, **451**
Industry
air pollution source, 415–417, 416f
energy efficiency in, 346
net energy yield of high-temperature industrial heat, 330f
reducing freshwater losses by, 278–279, 279f
Inertia, **96**
Inexhaustible resources, **6**, 7, 7f
Infant mortality rate, **112**–113
Infectious disease, **379**–385
case study
global HIV/AIDS epidemic, 383–384, 384f
malaria, 384–385, 385f
tuberculosis, 380, 382f
deadliest, 380f
drinking water, latrines, and infectious diseases, 385
genetic resistance to antibiotics, 381b
reducing, 385
solutions, 386f
viral, 383
water pollution and, 283
Infectious disease prevention, career in, **385**
Influenza virus, 383
Infrastructure, 120
Inland wetlands, **160**–162
Inputs into systems, **38**, 38f
Insecticides, 243. *See also* Pesticides
Federal Insecticide, Fungicide, and Rodenticide Act (FIFRA), 245
pollution from, 244
Insects
desert, 144b
predator avoidance strategies, 93
roles of, 72b, 72f
Instant gratification, risk evaluation and, 399
Insulation, building, 349
Integrated pest management (IPM), **246**
Integrated waste management, **454**, 469b
Interface, 484
Interglacial periods, 420
Intergovernmental Panel on Climate Change (IPCC), 406b, 421, 426b, 427–434, 443
Internal combustion engine, 348, 348f
International Basel Convention, 453
International Center for Technology Assessment, 125, 347
International Energy Agency, 332
International treaties. *See* Treaties
International Union for the Conservation of Nature (IUCN)
amphibian declines and, 70b
on biodiversity hotspots, 210
bird trade, 181
endangered species and, 170
environmental security and, 494
extinctions of aquatic species, 213

marine protected areas, 217
on parks and nature preserves, 207
polar bears and, 179
Red List of Endangered Species, 183
Red List of Threatened Species, 213
sharks, 77
International Water Association, 284
Interspecific competition, **91**
Intraspecific competition, 91
Introduced species, 174, 175f. *See also* Invasive species
Invasive species, **74**, 176–177
aquatic biodiversity, threat to, 213
bird population declines from, 183–184
Burmese python, 175f, 176, 177f
controlling, 176–177, 178f
kudzu, 175f, 176, 178f
lionfish,215f
natural capital restoration of rangelands, 205
in U.S. national parks, 208
zebra mussel, 175f, 176
Iodine deficiency, 228, 228f
Ion(s), **32**, 33f, 33t
Iowa
topsoil loss in, 237f
wind power use in, 361
IPAT model, 12–13
IPCC. *See* Intergovernmental Panel on Climate Change
IPM. *See* Integrated pest management
Ireland, tax on plastic bags in, 456
Iron, 305
Iron deficiency, 228
Irrigation
Aral Sea disaster, 275–276, 275f
center-pivot, 277f
drip, 254b, 277f
efficiency improvements, 277–278, 277f
flood, 277
soil salinization and, 238, 239f
solar, 254b
waterlogging, 238
Islands of biodiversity, mountains as, 150
Island species, vulnerability of, 173
Isle Royal, Michigan, 96f
Isotopes, **32**
Israel
irrigation in, 277, 278
wastewater use, 278
Italy
energy-efficiency, 352
solar power providers, 357
IUCN. *See* International Union for the Conservation of Nature
Ivory, 179

J
Jackson, Wes, 253b
Janzen, Daniel, 212
Japan
energy efficiency in, 352
Fukushima Daiichi nuclear power plant accident, 344
landfills, 460
life expectancy in, 112
population decline and aging population, 113f, 114
rare earth metals, supplies of, 307
urban population, 119
wind farms, 361
Jobs, Steve, 500
Jones, Chris, 431

J-shaped curve of population growth, 15f, 99, 99f

K
Kalundborg, Denmark, 467, 468f
Kangaroo rats, 144b
Kauffman, Matthew, 209b
Kelp forests, 90b, 90f, 91, 92b, 102b
Kenaf, 202
Kenya, tree planting in, 204
Kerogen, 333
Keystone species, **75**
American alligator as, 75–76, 76f
gray wolf, 209b
sharks, 76–77, 77f
southern sea otters, 90b, 90f, 91, 100, 101b, 102b
Kinetic energy, **35**–36, 36f
in atmosphere, 431
of flowing or falling water, 358
in wind, 360f
Kramer, Sasha, 483
K-selected species, **100**
Kudzu, 175f, 176
KuzeyDoğa, 184b
Kyoto Protocol, 438

L
Laboratory research, 62–63
Labor force, children in, 111
Labor unions, 485
Lack, Daniel, 417
Lake Mead, 272
Lake Powell reservoir, 272
Lakes, **157**–159, 159f
acid deposition and, 412f, 413
classified by nutrient content, 157
cultural eutrophication, 284
eutrophic, 159, 160f
oligotrophic, 157, 284
water pollution of, 284
zones of life, 152, 153f
Lake trout, 94f
Landfills, **460**, 460f–461f
secure hazardous waste, 463
Land Institute, 253b
Landscaping, water-thrifty, 278
Land subsidence, 271, 310
Lanthanum, 307, 307f, 313
Las Vegas, Nevada, 121f, 279
Late loss population, 100
Lateral recharge of aquifers, 264
Latitude, effect on climate and vegetation, 141
Lava, 304, 318f
Lava rock, 304, 316
Law(s), 184–185. *See also* Environmental laws
Lawns, water use on, 278
Law of conservation of energy, **37**
Law of conservation of matter, **35**
Law of gravity, 30
Law of nature, **30**
Law of unintended consequences, 244–245
Laws of thermodynamics, 37–38, 329
Leadership in Energy and Environmental Design (LEED), 349
Leaks, water, 278
LED lighting, 346
Legumes, 249
Leopold, Aldo, 19, 20f, 167, 225, 498
Lerner, Jaime, 129

Less-developed countries, **10**
   children in labor force, 111
   death rate declines in, 112
   demographic transition, 115
   empowerment of women in, 116
   environmental impact and, 16
   fuelwood use in, 202, 363
   genetic engineering, 254
   hazardous waste transfer to, 453, 466
   human population growth in, 109, 110b
   infectious diseases, 385, 386f
   indoor air pollution, 414, 419
   laws establishing, 286
   malnutrition, 17f
   municipal solid waste (MSW) handling, 451
   number of children per couple in, 112
   parks and nature preserves, 207
   percentage of income spent on food, 232
   population growth in, 114, 482
   poverty in urban areas, 123
   second green revolution in, 231
   smoking in, 397
   traffic congestion in, 125
   water leakage in, 278
   water pollution in, 294
Lethal dose, 381
Levees, 280
Levin, Donald, 170
Levin, Philip, 170
Lianas, 147
Life-centered environmental worldview, **18**, **494**
Life cycle, of metal product, 309, 309f
Life expectancy, **112**
Lifestraw™, 287, 287f
Lifestyle choices, human health and, 379, 395–396
Life-support system, Earth's, 47, 47f, 63b, 110b
Lighting
   compact fluorescent, 351
   energy-efficient, 351
LED, 307, 346, 351
Light-water reactor, 339, 340f
   advanced, 344
Lignite, 304
Likens, Gene, 28b, 29, 32
Limestone, 156b, 304, 305
Limiting factors, **98**–99
Lion, 50f
Lionfish, 215f
Liquefied natural gas (LNG), **335**
Liquefied petroleum gas (LPG), **335**
Liquid biofuels, 363–364, 365f
Literacy, environmental, 496
Lithium, 306
Lithium-ion battery, 316b, 349b
Lithosphere, **304**
Litter, 450b, 451, 452
Livestock
   antibiotic use in feed, 381b
   environmental problems caused by, 236, 236f
   genetic engineering to produce new varieties of, 232, 234f
   meat production, 233
Living machine, 295b, 295f
Living roof, 349
Living components, 48, 96–97
LNG. *See* Liquefied natural gas
Lobbying, 480
Locavores, 254

Logging, 28b, 197, 201, 203, 204
   in parks and nature preserves, 208
Logical reasoning, 30
Logistic growth, 99, 99f
Los Angeles, California
   photochemical smog, 411f
   wastewater reuse by, 279
Louv, Richard, 498
Lovejoy, Thomas E., 62b
Low-grade ore, **304**
Low-quality energy, **37**, 53
Low-throughput (low-waste) economy, **484**
LPG. *See* Liquefied petroleum gas
Lu, Xi, 361

M

Maathai, Wangari, 203
Macronutrients, 227
Magma, 304, 305f
Magnitude, earthquake, 319
Malaria, 244–245, 384–385, 385f
Malnutrition, 16, 17f, 113
   chronic, **227**
Manganese, 305
Manganese nodules, 309
Mangrove forests, 152
Mantle, Earth's, 47, 47f, **303**–304, , 303f
Manufactured capital, **475**, 475f
Manure, 248
Mara, Peter, 183
Marble, 304
Marine ecosystems. *See also* Oceans
   biodiversity, protecting and sustaining, 210–211
   coral reefs
      biodiversity of, 156b
      bleaching, 156b, 156f
      degradation of, 156b
      human impact on, 155, 157f
      ocean acidification and, 156b
      Red Sea, 135f
      threats to, 156b, 157f
   ecosystem and economic services of, 152–153, 152f
   human impacts on, 155, 157, 157f
   meat production and ocean dead zones, 239
   as natural capital, 152f
   ocean pollution, 289, 291f
   protected areas, 217–218
   zones, 152, 153f
Marine life zones, **152**
Marine protected areas (MPAs), 217
Marine reserves, 217
Marine snow, 155
Marketplace, outdoor air pollution reduction by, 418
Market price, 478
Marriage, average age at, 112
Marsh, artificial, 295b
Marsh, coastal, 154f
Marsh, George Perkins, 18
Marshes, 160
Martinez, Juan, 498b
Massachusetts, 262b, 275
Mass extinction, **82**, **169**
Mass number, **32**
Mass transit, 126, 128f
Mass transit rail, 128f
Materials revolution, 313
Mathematical models, 31, 63, 170b.
   See also Models
   climate change, 426b, 426f

   models to project future changes in atmospheric temperatures, 426b, 426f
Matter, **31**–34
   atoms, 32–33, 32f
   in cells, genes and chromosomes, 33–34, 34f
   chemical forms of, 31
   compounds, **31**, 32–3333t
   conservation of, 35
   elements, 31, 31f, 32t
   ions, 32, 33f, 33t
   molecules, 32
   physical, chemical, and nuclear changes in, 34
   physical states of, 31
   in systems, 38, 38f
McDonough, William, 450b, 455, 456b, 457
McKibben, Bill, 198, 491
MCS. *See* Multiple chemical sensitivity
Meat production, 233
   environmental problems caused by, 240–241, 249–250
   grain required for, 249f
   ocean dead zones and, 239
   sustainability issues, 250–252, 251f
Median lethal dose (LD50), 391
Megacities, 119, 120f
Megaregions, 119
Megareserves, 208, 210f
Melting, rock formation by, 304
Mercury
   atmospheric deposition of, 395f
   from coal-burning factories, 377f
   as element, 31f
   in gold mining, 312, 378b
   Minamata Convention, 395
   pollution, 388f
   in sharks, 77
   toxic effects of, 378b, 378f, 388f, 400b, 414f
Mesquite, 144b
Metals
   life cycle of metal product, 309, 309f
   rare earth metals and oxides, 306–307, 307f
Metamorphic rock, **304**, 305f
Metamorphism, 304
Methane
   from decomposition of submerged vegetation, 359
   as greenhouse gas, 407, 409, 423
   as hydrocarbon, 33
   from livestock, 241
   in natural gas, 333–335
   as organic compound, 33
   release from arctic lake, 429, 430f
   release from permafrost melting, 429
   stripping hydrogen from, 366
   volatile organic compound, 409
Methicillin-resistant *Staphylococcus aureus* (MRSA), 381b
Methylmercury, 378b, 387
Mexican gray wolf, 171f
Mexico, Colorado River basin and, 268, 268f
Microclimates, of cities, 140
Microfiltration, 276
Microlending, 483
Micronutrients, 227
Microorganisms (microbes), 51b
Migration, population change and, 110, **113**

Millennium Development Goals, 483, 484*f*
Millennium Ecosystem Assessment (2005), 10, 151
Milwaukee, Wisconsin, 226*b*, 226*f*
Mimicry, 93, 93*f*
Minamata Convention, 395
Mineral(s), **304**
  mining (*see* Mining)
  nonmetallic, 305
  rare earth metals and oxides, 306–307, 307*f*
Mineral deficiency, 228, 228*f*
Mineral resources, **304**. *See also* Rock
  dependence on, 304–305
  depletion allowances, 308
  environmental effects of use, 309–312
  life cycle of metal product, 309, 309*f*
  mining (*see* Mining)
  nonrenewable, 300–322
  per capita mineral use, 306
  recycling, 306–307, 306*f*
  rock cycle, 304, 305*f*
  strategic metal resources, 306
  substitutes for, 313
  supplies of, 305–309
    depletion of, 306, 306*f*
    mining lower-grade ores, 307
    oceans as source of, 308–309
    price effect on supply, 307–308
    rare earth minerals, 307
  sustainable use of, 312–314
  types of, 304
Mining
  biomining, 308
  copper mine, open-pit, 301*f*
  deep-sea, 208
  lower-grade ores, 308
  natural capital degradation, 310*f*
  in parks and nature preserves, 208
  seafloor, 308
  subsidies, 308
  techniques
    area strip mining, 309, 311*f*
    contour strip mining, 309–310, 311*f*
    mountaintop removal, 310, 312*f*
    open-pit mining, 309
    surface mining, 310
  water pollution by, 282, 310–312
Mirrors, in solar thermal systems, 356, 356*f*
Mission Blue campaign, 218*b*
Mississippi River basin, Gulf of Mexico annual dead zone and, 262*b*, 262*f*, 296*b*
Miss Waldron's red colobus monkey, 182
Mitigation, of climate change effects, 434
Models
  climate change, 426*b*–427*b*, 426*f*–427*f*
  defined, **29**
  mathematical, 31, 63
  of a system, 38*f*
Mojave Desert, 140, 356
Molecules, **32**, 57*b*
Molina, Mario, 441, 443*b*
Monarch butterfly, 83–84, 84*f*
Monocultures, 63*b*, 148*f*, **229**, 251
Monomers, 33
Monterey Bay, California, 90*f*
Monteverde Cloud Forest Reserve, Costa Rica, 78*f*
Montreal Protocol, 442
Moore, Charles, 452
More-developed countries, **10**
  air pollution reduction in, 427

cost of raising and educating children, 112
  environmental impact and, 16
  fertilizer use, 249
  first green revolution in, 231
  hazardous waste transfer from, 453
  indoor air pollution, 414
  industrialized agriculture in Kyoto Protocol, 438
  municipal solid waste (MSW) handling, 451
  overconsumption in, 110*b*
  water pollution laws, 294
Mortality rate, infant, **112**–113
Mostly wasted resources (MWR), 451
Moths, predator avoidance by, 93, 94*f*
Motor, electric, 347, 348*f*
Motor vehicles
  accidents, 125
  air pollution and, 125, 415–417
  car-centered nation, 124
  car-sharing networks, 125–126
  fuel efficiency standards, 418
  pros and cons of, 125
  reducing use, 125–126
Mountain goats, 208
Mountains
  climate and, 151
  ecological roles of, 150–151, 151*f*
  endemic species in, 150
  glaciers, 151, 422*f*, 428–429
  human impacts on, 151*f*
  as islands of biodiversity, 150
  rain shadow effect, 139–140, 140*f*
Mountaintop removal, **310**, 312*f*
Mount Hood, Oregon, 159*f*
Moving energy. *See* Kinetic energy
MPAs. *See* Marine protected areas
MRSA. *See* Methicillin-resistant *Staphylococcus aureus*
Muir, John, 19, 19*f*, 135
Multiple chemical sensitivity (MCS), 390
Multiple cropping, 230
Mumford, Lewis, 128
Municipal solid waste (MSW), **451**
  burning, 453, 459–461
  composting, 459
  recycling, 457–459
  single-pickup system, 459
Mutagens, 79, **387**
Mutations, **79**, 387
Mutualism, **94**–95, 94*f*
MWR. *See* Mostly wasted resources
Myers, Norman, 480, 493

N

Namib Desert, 142
Nanotechnology, 315*b*–316*b*, 322*b*, 349*b*, 357, 367
Narrow-spectrum agents, 243
National Academy of Sciences, 186
  on integrated management of hazardous waste, 461, 461*f*
  on integrated pest management, 246
  synthetic pesticides, 244–245
  on toxicity of synthetic chemicals, 393
  on waste management goals, 454, 454*f*
National Center for Atmospheric Research, 431
National Environmental Policy Act (NEPA), 486, 487
National Forest System, 198, 487
National laws. *See* Environmental laws
National Marine Fisheries Service (NMFS), 185

National Oceanic and Atmospheric Administration (NOAA), 218*b*
National parks. *See also* Parks and nature preserves; *specific parks*
  environmental threats to, 208
  stresses on U.S. public parks (case study), 207–208
National Park Service, U.S., 208, 199*b*, 488, 490
National Park System, 207–208, 488, 489*f*
National Priorities List, 465
National Renewable Energy Laboratory (NREL), 353
National security, climate change treat to, 433
National Wilderness Preservation System, 207, 207*f*, 488
National Wildlife Refuge System, 186, 187*f*
Native species, **74**
Natural capital, **5**–6, 7*f*, **475**, 475*f*
  atmosphere, 407*f*
  average precipitation and average temperature, 141*f*
  biodegradable wastes, 283*f*
  biodiversity, 71–73, 71*f*
  biological pest control, 246*f*
  biomes and climate, 142*f*
  carbon cycle, 59*f*
  climate zones and ocean currents, 137*f*
  ecosystem components, 52*f*
  endangered, 171*f*, 211*f*
  forests, 195*f*
  freshwater systems, 158*f*
  geothermal heat pump, 365*f*
  groundwater, 265*f*
  human ecological footprint and, 11*f*, 13*f*
  hydrothermal deposits, 308*f*
  life-support system, 47*f*
  marine ecosystems, 152*f*
  natural resources and ecosystem services, 7*f*
  nitrogen cycle, 60*f*
  phosphorus cycle, 61*f*
  plant species, 173*f*
  rock cycle, 305*f*
  soil formation and profile, 52*f*
  terrestrial and aquatic ecosystems, 477*f*
  water cycle (hydrologic cycle), 58*f*
  wolf-spider, 242*f*
Natural capital degradation, **10**
  acid deposition, 412*f*
  Aral Sea shrinkage, 275*f*
  city inputs and outputs, 123*f*
  cod fishery collapse, 217*f*
  deforestation, 199*f*
  deforestation in Haiti, 203*f*
  deforestation of hillsides, 281*f*
  food production, 236*f*
  freshwater scarcity stress, 267*f*
  grasslands, 148*f*
  groundwater contamination, 285*f*
  harvesting timber, 197*f*
  human ecological footprint and, 11*f*
  marine ecosystems and coral reefs, 157*f*
  mining, 310*f*312*f*
  needs to meet, 111*f*
  orangutans, 172*f*
  overgrazing, 200*f*, 205*f*
  ozone depletion effects, 442*f*
  ozone thinning, 441*f*
  pollution of coastal waters, 290*f*
  of renewable natural resources, 11*f*
  sea-bottom habitats, 213*f*
  soil salinization, 239*f*

Natural capital degradation *(continued)*
　terrestrial ecosystems, 151*f*
　topsoil erosion, 237*f*, 238*f*
　tropical rain forests, 46*b*, 46*f*
　urban sprawl, 122*f*
　wildlife species reduction, 174*f*
Natural capital depletion of mineral
　　resources, 306*f*
Natural capital restoration of riparian
　　zones, 205, 206*f*
Natural ecological restoration, 96
Natural gas, **333**–335, 337*f*
　advantages and disadvantages of, 337*f*
　climate change and, 335
　extraction of, 335
　liquefied natural gas (LNG), 335
　liquefied petroleum gas (LPG), 335
　production and fracking in United
　　States, 328*b*, 328*f*
　reserves, 335
Natural hazards, 379
Natural income, **21**
Natural recharge of aquifers, 269
Natural resources, **6**, 7*f*. *See also* Resources
Natural Resources Defense Council
　　(NRDC)
　on bottled water, 287
　environmental group, 492
Natural selection, **78**
　genetic diversity and, 79–80, 79*f*
　limits of adaptation through, 80
　myths concerning, 80
Nature, isolation from, 17–18
Nature Conservancy, 185, 218, 492
Nature-deficit disorder, 18, **498**
Nature reserves
　biosphere reserves, 208
　buffer zone concept, 208
　in Costa Rica, 194*b*, 208, 210*f*
NCA. *See* National Climate Assessment
Negative feedback loop, 39, 39*f*
Nelson, Gaylord, 473
Neoclassical economists, 475
NEPA. *See* National Environmental
　　Policy Act
Nervous system, 387
Net energy, **329**–330
　crude oil, 332
　of hydrogen, 366
　negative, 366
　of solar thermal systems, 356
Net energy principle, 490
Net energy yield
　of cellulosic ethanol, 364
　energy resources and systems, 330*f*
Net primary productivity (NPP), **55**, 56*f*
　of lakes, 157, 159
　oceans, 155
　of tropical rain forests, 147
Neurotoxins, 387
Neutral solution, 32
Neutrons, **32**, 32*f*
New Zealand
　climate change, 434
　end of farm subsidies in, 253
NGOs. *See* Nongovernmental organizations
NIABY. *See* Not in anyone's backyard
Niche, **74**. *See also* Ecological niches
Nicotine, 391
NIMBY. *See* Not in my backyard
Nitrate ions, 32, 33*f*, 59–60
Nitrates, ocean pollution by, 282*t*, 284,
　　286
Nitric acid, 60, 409

Nitric oxide, 60, 409
Nitrogen
　atmospheric concentration, 407
　in fertilizers, 248
Nitrogen cycle, 59–60, 60*f*
Nitrogen dioxide, 60, 409
Nitrogen oxides
　acid deposition and, 412, 414
　air pollutant, 409, 415*f*, 418*f*
　photochemical smog, 409
Nitrous oxide
　air pollutant, 409
　as greenhouse gas, 407, 423
NMFS. *See* National Marine Fisheries
　　Service
NOAA. *See* National Oceanic and
　　Atmospheric Administration
Nocera, Daniel, 367
Noise pollution, **122**–123
Nongovernmental organizations (NGOs),
　　485, 491–492
Nonliving components, 5, 48, 50*f*
Nonmetallic minerals, 305
Nonnative species, **74**, 173–176
Nonpoint sources pollution, **282**–283, 291
Nonrenewable energy resources, 329, 329*f*
Nonrenewable (exhaustible)
　resources, **6**, 7*f*
　mineral resources, 300–322
　nonrenewable aquifers, 264
Nontransmissible disease, **379**
NOPE. *See* Not on planet Earth
North Carolina, natural restoration, 97*f*
North Cascades National Park,
　　Washington, 207*f*
Northern coniferous forests, 148
Northland College, 492, 493*f*
Not in anyone's backyard (NIABY), 466
Not in my backyard (NIMBY), 465
Not on planet Earth (NOPE), 466
NPP. *See* Net primary productivity
NRCS. *See* National Resources
　　Conservation Service
NRDC. *See* Natural Resources Defense
　　Council
NREL. *See* National Renewable Energy
　　Laboratory
Nuclear fission, **339**
Nuclear fission reactor
　decommissioning, 343
　Fukushima Daiichi nuclear power plant
　　accident, 344
　light-water, 344
　safety issues, 344
　structure and function of, 339–341, 340*f*
Nuclear fuel cycle, **341**, 341*f*, 342*f*
Nuclear fusion, **344**, 345*f*
Nuclear power
　advanced light-water reactor, 344
　advantages and disadvantages of using,
　　339–345
　climate change and, 343
　dealing with radioactive nuclear wastes,
　　342–343, 342*f*
　electricity from, 341, 341*f*
　fuel cycle, 330, 341, 341*f*, 342*f*
　Fukushima Daiichi nuclear power plant
　　accident, 344
　future of, 343–344
　net energy yield, 330*f*
　nuclear fission reactor structure and
　　function, 339–341, 340*f*
　power plants, 339–344, 340*f*
　safety issues, 344

Nuclear weapons, 344
Nucleic acids, 33
Nucleotides, 33
Nucleus (atomic), **32**, 32*f*
Nutria, 175*f*
Nutrient cycles, **56**, 450*b*, 450*f*
　carbon cycle, 57–59, 59*f*
　nitrogen cycle, 59–60, 60*f*
　phosphorus cycle, 60–61, 61*f*
　water (hydrologic) cycle, 56–57, 58*f*
Nutrient cycling, **5**–6, 8, 50, 52*f*
Nutrients, **5**, **47**, 284
Nutrition
　malnutrition, 227
　overnutrition, 228
　undernutrition, 227
　vitamin and mineral deficiencies, 228,
　　228*f*

O
Oberlin College, Ohio, 474*b*, 474*f*
Obese/overweight, 228
Ocean acidification, 156*b*, 214*b*
Ocean currents, **137**–138, 137*f*
　ocean garbage patches and, 452, 452*f*
Oceanic crust, 303*f*, 304
Ocean pollution, 289, 291*f*
　from oil, 289–291, 291*f*
　reducing and preventing, 291
Oceans
　climate change, role in, 425
　coral reefs, 152, 156*b*
　dead zones, 289, 296*b*
　ecosystem and economic services of,
　　152, 152*f*
　garbage patches, 452, 452*f*
　global ocean, 152
　human impacts on, 157*f*, 213–215
　hydropower from, 359
　marine reserves, 217
　meat production and ocean dead zones,
　　239
　minerals from, 308–309, 308*f*
　open sea, 153
　pollution, 289–291, 291*f*
　zones, 152–153, 153*f*
Ocean thermal-energy conversion
　　(OTEC), 360
Ogallala Aquifer, 269, 270*f*, 271
Oil, 328*b*, 330–332
　crude oil, 289, , **330**–332, 331*f*
　　advantages and disadvantages of use,
　　　332
　　conventional (light), 331, 333*f*
　　dependence on, 330–331
　　extraction of, 333*f*
　　oil production and consumption in
　　　United States, 332
　　peak production, 331
　　refining, 331, 331*f*
　　reserves of, 332
　dependence on, 330–331
　heavy, 330–331, 333, 334*f*
　ocean pollution from, 289–291, 291*f*
　shale oil, **333**, 334*f*
　tar sands, 333, 334*f*
　tight, 328*b*, 328*f*
Oil sands, **333**, 334*f*
Old-growth forest, **195**, 195*f*, 197–198
Oligotrophic lake, **157**, 284
Olympic National Park, 208
Omnivores, **49**
One-child policy, of China, 118–119
One-way energy flow, 50

On the Origin of Species by Means of Natural Selection (Darwin), 79
OPEC. *See* Organization of Petroleum Exporting Countries
Open-access renewable resources, 12
Open dump, **461**
Open-pit mining, **309**
Open sea, 153, 153*f*
Opportunists, 100
Optimism bias, risk evaluation and, 399
Optimum level/range, 98, 99*f*
Orangutans, 171, 172*f*
Orcas, 101*b*
Orchids, 182
Ore
    high-grade, **304**
    low-grade, **304**
Oregon, Trillium Lake in, 159*f*
Organic agriculture, **229**–230, 231*f*
    industrialized agriculture compared to, 231*f*
    labeling of products, 230
    farming, water pollution prevention and, 291
Organic compounds, **33**
Organic fertilizer, **248**
Organisms, 5, **48**, 49*f*
Organization of Petroleum Exporting Countries (OPEC), 332
Outdoor air pollutants, 408–410, 408*f*
    hazardous air pollutants (HAPs), 417
    photochemical oxidants, 411
Outdoor air pollution
    atmospheric warming, 427
    increase, factors responsible for, 411–412
    reducing, 417–419
Outputs from systems, **38**, 38*f*
Overburden, **309**, 311*f*
Overconsumption, 110*b*
Overexploitation, of bird populations, 184
Overfishing, 215
Overgrazing, **204**, 205*f*, 238
Overhunting, of amphibians, 78*b*
Overnutrition, 228
Overshoot, population, 99, 100*f*
Oxpeckers, 94*f*
Oxygen
    in aerobic respiration, 49
    atmospheric concentration, 407
Oxygen-depleted zones, 289
Oxygen sag curve, 283, 283*f*
Ozone, 47
    air pollutant, 409
    depletion, 441–444, 441*f*–442*f*, 444*b*
    photochemical, 409
    stratosphere, 407, 409, 442
    UV filtering effect of, 441
Ozone hole, 441
Ozone layer, 47, **407**, 407*f*

P

Pacific Institute, 288
Palm oil, 229
Pandemic, 380
Pangaea, 81*b*, 81*f*
Paper, trees for production of, 202
Paracelsus, 377
Parasites, 94, 101*b*, 180*b*, **379**
    of amphibians, 78*b*
    malaria, 384
Parasitism, **93**, 94*f*
Parks and nature preserves
    biosphere reserves, 208

buffer zone concept, 208
    in Costa Rica, 208, 210*f*
    designing and managing, 208
    ecological restoration of tropical dry forest in Costa Rica (case study), 212
    environmental threats to national parks, 208
    in less-developed countries, 207
    megareserves, 208, 210*f*
    stresses on U.S. public parks (case study), 207–208
    wilderness protection, 206–207, 207*f*
Particulates, as air pollutants, 409
Passive solar heating systems, **353**, 353*f*
Pastures, **204**, 228
Patches, in biomes, 141
Pathogens, **379**
Pauli, Gunter, 449
PCBs, 241
Peak production, **331**
Peer review, **30**, 31
Peers, 30
Pelican Island National Wildlife Refuge, Florida, 186, 187*f*
Pennsylvania, natural gas production by fracking in, 336*f*
Pension systems, 112
Per capita ecological footprint, **12**, 111
Per capita GDP, **481**
Per capita resource use, in United States, 111
Perennial crops, 251, 253*b*
Permafrost, 47, **144**, 421, 429
Persistence, **96**, 443*b*
    of pesticides, 243
    of toxins, 390
Persistent organic pollutants (POPs), 245, 394, 466
Pest(s), **242**–247
    alternatives to synthetic pesticide use, 245–246
    genetic resistance and, 245
    insects, 72*b*
    integrated pest management (IPM), 246–247
    natural enemies of, 242, 246, 232*f*, 235, 235*f*
Pesticides, **242**–244
    alternatives to synthetic pesticide use, 245–246
    benefits of synthetic, 243, 243*f*
    circle of poison (boomerang effect), 245
    food choices and, 244
    honeybee declines and, 189*b*
    inconsistent results from use of, 245
    persistence of, 243
    problems with synthetic, 243–244
    reducing your exposure to, 244*f*
    spectrum of, 243
    types, 242–243
    unintended consequences of use, 244–245
    use in integrated pest management (IPM), 246–247
    water pollution from, 291
Pest management, 246–247
Pethick, John, 430
Petrochemicals, **331**
Petroleum, **330**
Pfeiffer University, 474*b*
Pheromones, 246
Phosphate ions, 61
Phosphates, 284, 305

Phosphorus, in fertilizers, 248
Phosphorus cycle, 60–**61**, 61*f*
Photochemical oxidants, 411
Photochemical smog, **410**–411, 411*f*
Photosynthesis, **48**, 64*b*, 153
    in carbon cycle, 58, 59*f*
Photovoltaic (PV) cells, **357**
Phthalates, 388
Physical change, **34**, 35
Physical states of matter, 31
Phytoplankton, 49, 51*b*, 54*f*, 153, 157, 391*f*
Phytoremediation, 463
Pimentel, David, 173, 245
Pimm, Stuart, 169
Pinchot, Gifford, 19
Pitcher plant, 95*f*
Plague, bubonic, 101
Planetary boundaries, 63*b*, 64*b*, 476, 494, 496
Planetary management worldview, 18, 494
Plantation agriculture, **229**
Plants. *See also* Forests
    acid deposition and, 412
    carnivorous, 95*f*
    desert, 144*b*
    epiphytes, 95
    green manure, **248**
    illegal trade in, 182
    parasites, 94
    phytoremediation, 463
    seed banks, 186
    sewage treatment 292, 293*f*
    from tropical rain forests, 147
Plasma gasification, 463
Plastics
    bags, 456
    bioplastics, 458*b*
    bisphenol A (BPA), 389*b*
    bottled water, 287–288
    ocean garbage patches, 452, 452*f*
    recycling, 457, 458*b*
    as water pollutants, 282*t*
Plate tectonics, 61*f*
Plug-in hybrid-electric vehicle, 347, 348*f*, 349*b*
Plutonium, 342, 343
Poachers, 180
Poaching, 180, 181, 185, 186, 208
Poincare, Henri, 27
Point sources of pollution, **280**–281, 281*f*, 291–292
Poison dart frog, 93*f*
Polar bear, 179*f*
Policies. *See* Environmental policy
Policy life cycle, 484
Politics
    environmental groups, role of, 491–492
    environmental justice, 490
    environmental policy principles, 490–491, 491*f*
    managing public lands in United States, 487–490, 488*f*
    transition to sustainable energy future, role in, 369, 370
Pollination, 72*b*, 72*f*, 170*b*
    by honeybees, 168*b*
    as mutualism, 94
Pollutants
    air
        from coal-burning factories, 377*f*
        hazardous air pollutants (HAPs), 417
        mercury, 388*f*
        natural and human sources of, 407–408

Pollutants *(continued)*
    outdoor, 408–410
    primary, 408, 408*f*
    secondary, 408, 408*f*
    from coal-burning factories, 377*f*
        dirty dozen, 394, 466
        persistent organic pollutants (POPs),
            245, 466
        primary, 408, 408*f*
        secondary, 408, 408*f*
    water, 282*t*
        mercury, 378*b*, 378*f*
        plastics, 282
Polluter-pays principle, 490
Pollution. *See also* Air pollution; Water
            pollution
    amphibian declines and, 78*b*
    from coal, 335–337
    concentration in cities, 122, 123*f*
    contribution to species extinction, 178,
            178*f*
    marketplace use to reduce, 481, 481*f*
    from meat production, 241
    in Mexico City, 124
    by mining, 309–312
    noise, 122–123
    nonpoint sources, 282, 291
    from pesticides, 244
    point sources, 280–281, 281*f*, 291–292
    reducing by selling services instead of
            products, 481
    soil, 237
    taxing, 480, 480*f*
    threat to bird populations, 184
Pollution prevention
    implementing, 394–395
    with integrated pest management, 246
    precautionary principle, 393–394
Polyaquaculture, 250
Polycultures, 229, 253*b*
Polymers, 33
Polyps, 156*b*
Polyvinyl chloride (PVC), 389
POPs. *See* Persistent organic pollutants
Population(s), **48**, 49*f*
    age structure of, 97
    crash, 99, 100*f*
    defined, **97**
    evolution of, 79*f*
    human (*see* Human population)
    range of tolerance and, 98, 99*f*
    reproductive patterns of, 100
    survivorship curves for, 100, 100*f*
Population change, **110**, 113
Population crash, **99**, 100*f*
Population density, **98**
Population growth
    age structure and, 97
    exponential, 14, 15*f*
    human (*see* Human population growth)
    J-shaped and S-shaped curves, 15*f*,
            99–100, 99*f*
    limits on, 98
Population size, **97**
    age structure and, 97
    carrying capacity, 99
    environmental impact and, 12
    limiting factors, 98
    range of tolerance of, 98, 99*f*
    of southern sea otters, 101*b*, 101*f*
Portland, Oregon
    bicycle-friendly city, 126, 127*f* water
            treatment facilities, 286
Portugal, solar-cell power plants, 357

Positive feedback loop, **38**, 38*f*
Postel, Sandra, 279, 280*b*
Post-reproductive ages, in age structure
            diagrams, 113, 113*f*
Post-reproductive stage, 97
Potassium, in fertilizers, 248
Potential energy, 35, **36**, 37*f*
Poverty
    in China, 118
    defined, **16**, **482**
    demographic transition and, 115–116,
            115*f*
    dependence on forests, 195
    food insecurity and, 227
    harmful and beneficial effects of
            affluence, 15–16
    human health and, 395–396, 396*f*,
            482–483
    reducing, 482–483
    risks associated with, 395–396, 396*f*
    in urban living, 123
    in women, 116
Power plants
    coal-burning, 335, 337–339, 337*f*–338*f*
    nuclear, 339–341, 339*f*–340*f*
Prairie, 145*f*
Prairie potholes, 160
Praying mantis, 72*b*, 72*f*
Precautionary principle, **394**
    environmental policy, 490
    pollution prevention, 393–394
Precipitation
    aquifer recharge from, 264
    biomes and, 140–152, 141*f*–142*f*, 151*f*
    natural capital, 141*f*
    rain shadow effect, 139, 140*f*
    in United States, 266*f*
    in water cycle, 56–57
Predation, **91**–93
Predator, **91**
    behavioral strategies, 93
    keystone, gray wolf as, 209*b*
    nonnative of amphibians, 78*b*
    top, 75
Predator–prey relationship, **91**–93
    coevolution, 93
Pre-reproductive ages, in age structure
            diagrams, 113, 113*f*
Pre-reproductive stage, 97
Prescribed burns, 202
Preservationist view, 19
Prevailing winds, 138, 139*f*–140*f*, 413
Prevention principle, 490
Prey, **91**
    predator avoidance strategies, 93, 93*f*
Price
    direct, 478
    full-cost pricing, 478–480
    market, 478
    nonrenewable energy resources, 353
Price-Anderson Act, 344
Pricing
    environmental and health costs of,
            16–17
    full-cost pricing, 9, 9*f*
    nature's ecosystem services, 196*b*
    nonrenewable mineral resource supply
            and, 306
    of water, 272
Prickly-pear cactus, 55*f*
Primary consumers, **49**, 53*f*–54*f*
Primary ecological succession, **95**, 96*f*
Primary forest, 195
Primary pollutants, **408**, 408*f*

Primary recycling, **457**
Primary sewage treatment, **292**
Principle of enoughness, 499
Probability, in science, 31
Producers, **48**–49, 50*f*–54*f*
Profit-making organizations, 485
Projections, 426*b*
Proof, scientific, 31
Propane, in natural gas, 333, 335
Protected areas, 186
    marine ecosystems, 216–217
Proteins, 33,
    fish as source of, 215
Protons, **32**, 32*f*
Proven oil reserves, **332**
Public lands, managing in United States,
            487–491, 488*f*
Purple loosestrife, 175*f*
Purple sea urchin, 92*f*
PVC. *See* Polyvinyl chloride
Pyramid of energy flow, **53**, 54*f*

R

Raccoon, 74, 74*f*
Radioactive nuclear wastes, 342–343
Radioactive radon-222 gas, 415
Rail
    mass transit rail, 128*f*
    rapid rail, 128*f*
Rainfall in tropical forests, 199
Rain forest
    temperate, 148
    tropical (*see* Tropical forests, Rain
            forests)
Rainforest Alliance, 204
Rain shadow effect, **139**, 140*f*
Rainwater, in urban areas, 278
Rangelands, **204**, 228
    managing and sustaining, 204–205, 205*f*
Range of tolerance, **98**, 99*f*
Rapid rail, 128*f*
Rare earth metals and oxides, 306
    In China, 310
    importance of, 306–307
    substitutes for, 313
    supplies of, 307
RCRA. *See* Resource Conservation and
            Recovery Act
Reactor. *See* Nuclear fission reactor
Receptors, 387
Recirculating aquaculture systems, 250
Reconciliation ecology, **212**
    water-thrifty landscaping, 278
Recycling, **455**
    advantages and disadvantages, 458, 458*f*
    aluminum, 39
    in Curitiba, Brazil, 129
    downcycling, 457
    encouraging, 466
    energy efficiency with, 346
    e-waste, 453, 454*f*
    mineral depletion time and, 306, 306*f*
    mineral resources, 313–314
    as negative, or corrective, feedback
            loop, 38–39
    plastics, 458*b*
    primary (closed-loop), 457
    secondary, 457
    single-pickup system and, 459
    trade-offs, 458*f*
    in United States, 459
    upcycling, 457
Red Sea, 135*f*
Refillable containers, 456

Refined oil, 290
Refining, **331**, 331*f*
Refugees, environmental, 113
Refuge system, U. S., 186
Refusing to use/buy, **454**
Rehabilitation, ecosystem, 211
Reich, Peter, 229
Reilly, William K., 292
Reintroduction of gray wolf in
    Yellowstone National Park, 209*b*
Reliability, system, 398
Reliable science, **30**
Reliable surface runoff, **264**
Religious beliefs, birth rate and, 112
Remote sensing, 62
Remote sensing analyst, **62**
Renewable energy resources, 329. *See also*
        specific resources
    advantages and disadvantages of using,
        352–367
    cooling buildings naturally, 354
    falling and flowing water, 358–359
    geothermal energy, 365–366, 366*f*
    heating buildings and water with solar
        energy, 353–354, 353*f*
    hydrogen, 366–367, 367*f*
    hydropower, 359–360, 360*f*
    liquid biofuels, 363–364, 365*f*
    solar cells, 357–358, 358*f*–359*f*
    solar thermal systems, 355–357, 356*f*
    solid biomass, 362–363, 364*f*
    wind power, 360–362, 360*f*–363*f*
Renewable resources, 6, 11*f. See also specific*
        resources
    degradation of, 11–12
    open-access, 12
    shared, 12
    sustainable yield, 6
Replacement, of damaged ecosystem,
        211
Reproductive ages, in age structure
        diagrams, 113, 113*f*
Reproductive isolation, **80**, 82*f*
Reproductive patterns, 100
Reproductive stage, 97
Reptiles, desert, 144*b*
Reserves
    coal, 335
    crude oil, 332
    mineral, 308
    natural gas, 335
Reservoirs, 37*f*
    hydropower and, 359
    hydrothermal, 365
    silt buildup, 272
    water pollution of, 284
Resilience, **96**
Resource Conservation and Recovery Act
        (RCRA), 464, 486
Resource exchange webs, 467
Resource partitioning, **91**, 91*f*
Resources, 7*f. See also* Natural capital;
        specific resources
    degradation (*see* Natural capital
        degradation)
    inexhaustible, 6, 7*f*
    nonrenewable (exhaustible), 6, 7*f*
    renewable, 6, 7*f*
    sustainable yield, 6
    tragedy of the commons, 12
Resource-use permits, 481, 481*f*
Resource use per person, contribution to
        species extinction, 177
Respiratory system, 415, 416*f*

Response, to chemical exposure, 390
Responsibility to future generations,
    9, 9*f*
Restoration, ecological restoration,
    211–212
Reuse, **454**
    encouraging, 466
    of e-waste, 454*f*
    refillable containers, 456
    what can you do?, 456*f*
Reverse osmosis, 276
Reversibility principle, 490
Rhinoceros, 181*f*
Rice, 229, 231*f*
Richter scale, 319
Riparian zones, overgrazing damage to,
    205, 206*f*
Risk, **379**
Risk analysis, **395**
    death, leading causes of, 396*f*
    difficulty of evaluating risks, 399
    estimating risks from technology,
        398–399
    factors in, 399
Risk assessment, **379**, 379*f*
Risk communication, 395
Risk management, **379**, 379*f*
Risk perception and evaluation, 395–399
Rivers and streams, 157–162
    channelized, 281*f*
    dams and reservoirs, 161
        advantages and disadvantages of, 360*f*
        Colorado River, 268, 268*f*
        how dams can kill delta, 272
        hydropower and, 271
        tidal energy, 359
    deltas, 159
    overgrazing damage to riparian zones,
        205, 206*f*
    pollution of, 283–284
    zones, 159, 161*f*
RNA, 33
Robert Wood Johnson Foundation, 228
Rock, **304**. *See also* Mineral resources
    Earth's structure and, 303–304, 303*f*
    hot, dry for geothermal energy, 366
    igneous, 304, 305*f*
    metamorphic, 304, 305*f*
    molten, 303
    ore, 304
    sedimentary, 304, 305*f*
    shale, 328, 328*f*
Rock cycle, **304**, 305*f*
Rockstrom, Johan, 63*b*
Rodenticides, 243. *See also* Pesticides
Rodricks, Joseph V., 393
Roof, green, 349, 350*f*
Roosevelt, Theodore (Teddy), 19, 20*f*,
    186
Rosenzweig, Michael L., 212
Roske-Martinez, Xiuhtezcatl, 492*b*
Rotational grazing, 204
Rotation of the earth on its axis, 139*f*
Rowland, Sherwood, 441, 443*b*
r-selected species, **100**
Rule of 70, 15
Runoff, **157**
Russia
    coal reserves, 335
    crude oil, 332
    gold producer, 302*b*
    natural gas reserves, 335
    nuclear fuel, 343
Ruzo, Andrés, 366*b*

S
Safe Drinking Water Act, 288, 486
Saguaro cactus, 143*f*, 144*b*
Sahara Desert, 384
Salinity, 152
Salinization of soil, **238**
Salmonella, 181, 383
Salt, removing from seawater, 276
Saltwater life zones, **152**
San Andreas fault, 317*f*
San Clemente Island, California, 155*f*
Sand, 305
A Sand County Almanac (Leopold), 19
Sandstone, 304
Sanitary landfill, **460**, 461*f*
San Lucas marsupial frog, 69*f*
San Pedro River, Arizona, 206*f*
Santa Cruz, Bolivia, 46*f*
Saudi Arabia
    aquifer depletion, 270*f*
    oil reserves, 332
    wheat production, 269
Savanna, 136*b*, 136*f*, 146*b*
Scalloped hammerhead shark, 77*f*
Science, **29**–31. *See also* Science Focus
    limitations of, 31
    processes of, 29*f*
    reliable, 30
    tentative, 30
    unreliable, 30
Science Focus
    amphibian decline, 78*b*
    biomimicry principles, 19*b*
    bioplastics, 458*b*
    Biosphere 2, 495*b*, 495*f*
    bisphenol A (BPA), 389*b*
    California's southern sea otters,
        101*b*, 101*f*
    coral reefs, 156*b*, 156*f*
    environmental effects of natural gas
        production and fracking, United
        States, 336*b*
    estimating extinction rates, 170*b*
    genetic resistance to antibiotics, 381*b*
    geological processes affect biodiversity,
        81*b*
    gray wolf reintroduction in Yellowstone
        National Park, 209*b*
    greenhouse gases and climate, 138*b*
    honeybee losses, 180*b*, 180*f*
    human population growth, 110*b*
    insects, 72*b*, 72*f*
    models to project future changes
        in atmospheric temperatures,
        426*b*–427*b*, 426*f*–427*f*
    nanotechnology revolution, 315*b*
    ocean acidification, 214*b*
    organisms, invisible, 51*b*
    perennial polyculture and the Land
        Institute, 253*b*
    planetary boundaries, 63*b*
    pricing nature's ecosystem services,
        196*b*
    revisiting Savanna, 146*b*
    search for better batteries, 349*b*
    sewage treatment by working with
        nature, 295*b*, 295*f*
    staying alive in the desert, 144*b*
    threats to kelp forests, 92*b*
    water's unique properties, 57*b*
    wind turbines for birds and bats, 363*b*
Scientific hypothesis, **29**
Scientific laws, 28*b*
    governing energy changes, 37–38

Scientific principles of sustainability, **5**, 6f
Scientific process, 29–30, 29f
Scientific proof, 31
Scientific theory, **29**, 30
Scientists
　characteristics of, 29
　field research by, 62
　laboratory research by, 62–63
　use of the scientific process, 29, 29f
SCS. *See* Soil Conservation Service
Sea anemone, 89f, 94–95
Seafood production, 233, 235, 235f
Seager, Richard, 266, 431
Sea-grass beds, 152, 155f
Sea lamprey, 94, 94f, 175f
Sea level rise, 406b, 429–431, 431f, 434
Sea otters, southern, 90b, 90f, 91, 101b,
　　101f, 102b
Sea turtles, 176
Sea urchin, 92b, 92f
Seawater
　mineral extraction from, 308–309
　salt removal from, 276
Secondary consumers, **49**, 50f, 53f–54f
Secondary ecological succession, **95**, 97f
Secondary pollutants, **408**, 408f
Secondary recycling, **457**
Secondary sewage treatment, **292**, 293f
Second-growth forest, **195**
Secondhand smoke, 397
Second law of thermodynamics, **38**, 53
Security, environmental, 493
Sedimentary rock, **304**, 305f
Sediments, 282t, 304
Seed banks, 186, 240
Seed morgue, 240
Segura, Paola, 254
Seismic waves, 319
Sekercioğlu, Çağan Hakki, 184b
Selection, artificial, **81**–82, 232
Selective breeding, 81
Selective cutting, 197, 198f, 202
Selling services instead of products, 481
Semiarid/arid conditions, 139–140
Septic tank, **292**
Sequoia National Park, California, 3f, 27f
Sereivathana, Tuy, 147b
Services, selling, 481
Sewage
　ocean pollution by, 289
　water pollution from, 291–292
Sewage treatment
　ecological sewage treatment systems,
　　294
　improving conventional, 293–294
　primary, 292, 293f
　secondary, 292, 293f
　by working with nature, 295b, 295f
Sex attractants, 246
Shale, 304
Shale oil, **333**, 333f
Shantytowns, 123
Shared resources, 12
Sharks, 76–77, 77f, 101b
Shellfish production, 233
Shelterbelts, 247
Silicon, in solar cells, 357
Silicon dioxide, 304
Silt buildup in reservoirs, 272
Simones, Cid, 254
Simple carbohydrates, 33
Simple sugars, 33
Simplicity, voluntary, 499
Sindi, Hayat, 382b

Sinkhole, 271
Slate, 304
Slums, 117, 119
Smart cards for water credits, 279
Smart grid, 347
Smart growth, **126**, 128f
Smelting, **311**
Smog
　brown-air, 410
　industrial, **410**, 410f
　photochemical, 410–411, 411f
Smoking, 396–398, 396f
Snail, genetic diversity in, 72f
Snakes, as invasive species, 175f, 176
Snow
　marine, 155
　melting of, 428–429
Snowpacks, 275
Social sciences, 475
Society, environmentally sustainable, **21**
Soil(s), **50**
　of boreal forests, 148
　buffering of acid deposition, 413
　desertification, 238, 249
　erosion (*see* Soil erosion)
　fertility (*see* Soil fertility)
　formation, 52f
　as foundation of life on land (Science
　　Focus), 50–52
　mature, 50, 52f
　pollution, 237
　profile, **50**
　subsoil, 51, 52f
　topsoil (*see* Topsoil)
　tundra, 146
Soil conservation, **247**, 248f
Soil erosion, **236**
　hotspots, 247
　reducing, 249
　from wind, 237f
Soil fertility
　loss with topsoil erosion, 237
　restoring, 248–249
Soil salinization, **238**
　reducing, 249, 249f
Solar cells, **357**–358
　advantages and disadvantages of, 359f
Solar-Cell technology, career in, **357**
Solar cookers, 357
Solar energy, **5**–6, 6f, 36, 64b
　climate and, 137
　cooling with, 354
　for drinking water purification, 286–287
　electricity from, 355–357
　global air circulation and, 139f
　greenhouse effect, 48f
　heating buildings and water with,
　　353–354, 354f
　indirect forms of,
　　biomass, 362–363
　　flowing water, 358–360
　　wind, 360–362
　in photosynthesis, 48
　principle of sustainability, 5, 6f
　solar cells, 357–358, 358f
　solar thermal systems, 355, 356f
Solar heating systems
　active, 353f, 354
　passive, 353, 353f
Solar irrigation systems, 254b
Solar thermal systems, **355**, 356f
Solid biomass, 362–363, 364f
Solid waste, **451**
　burning, 459–460

　burying, 460–461
　dealing with, 453–455
　industrial, 451
　municipal solid waste (MSW), 451, 451f
　ocean garbage patches, 452, 452f
　refusing, reducing, reusing, and
　　recycling of, 455–459, 456f, 458
　in United States, 451–452
　what can you do?, 455f
Solubility, 390
Solution, pH of, 32–33
Solutions
　acid deposition, 414f
　coastal water pollution, 292f
　groundwater depletion, 272f
　groundwater pollution, 286f
　improving energy efficiency, 346f
　indoor air pollution, 419f
　infectious disease, 386f
　mercury pollution, 388f
　more sustainable food production, 251f
　motor vehicle air pollution, 419f
　reducing flood damage, 281f
　reducing irrigation water losses, 278f
　reducing water losses, 279f
　slowing climate change, 435f
　smart growth tools, 128f
　soil salinization, 249f
　stationary source air pollution, 418f
　sustainable aquaculture, 250f
　sustainable forestry, 202f
　sustainable use of nonrenewable
　　minerals, 316f
　sustaining tropical forests, 204f
　transition to sustainable energy future,
　　369f
　water pollution, 294f
Soot, 335, 427
Soulé, Michael, 172
Source zone, 159, 161f
South Africa, lifetime rates for water
　　pricing in, 276
South Dakota, gold mine in, 302f
Southern sea otters, 90b, 90f, 92b, 101b,
　　101f, 102b
South Korea, solar-cell power plants, 357
Soybeans, 232
Space heating, net energy yield for, 330f
Special interest groups, 485
Specialist species, **74**, 74f
Speciation, **80**
　rate, 170
Speciation crisis, 169
Species, **5**
　economic services provided by, 170–173
　ecosystem roles of, 74–77
　American alligator case study, 75–76,
　　76f
　generalist species, 74, 74f
　indicator species, 74–75
　keystone species, 75
　native species, 74
　nonnative species, 74
　sharks case study, 76–77, 77f
　specialist species, 74, 74f
　ecosystem services provided by, 170
Species approach to sustaining
　　biodiversity, 166–189
　case study
　　bird population declines, 183–184
　　U.S. Endangered Species Act,
　　　185–186
　extinction (*see* Extinction)
　honeybee decline core case study, 180b

international treaties and national laws and, 184–185
  protecting species
    protected areas, 186, 188, 188*f*
    questions raised by efforts to protect, 188
    seed banks, botanical gardens, and wildlife farms, 186
    treaties and laws, 184–185
    zoos and aquariums, 186–187
Species-area relationship, 170*b*
Species diversity, 71, 71*f*
Species interactions, 91–95
  coevolution, 93, 94*f*
  commensalism, 95, 95*f*
  interspecific competition, 91
  mutualism, 94–95, 94*f*
  parasitism, 93–94, 94*f*
  predation, 91–93
  resource partitioning, 91, 91*f*
Sperry Glacier, 422*f*
Spiders, pest control by, 242
SPM. *See* Suspended particulate matter
Spoils, **309**, 310*f*, 310
Squatter settlements, 123–124
S-shaped curve of population growth, 15*f*, 99, 99*f*
Stability, in living systems, 96
Standby mode, 351
State of the Birds (2011 study), 183
Statistical tools, 31
Steel, 305
Stewards, 18, 494
Stewardship worldview, 18, 494
Stockholm Convention on Persistent Organic Pollutants, 466
Stored energy. *See* Potential energy
Strategic metal resources, 306
Stratosphere, **47**, **407**, 407*f*
  ozone in, 407, 409
Streams. *See also* Rivers and streams
  buried by mining spoils, 310
  most serious pollutants threatening, 284
  pollution by mining, 310
Strip-cropping, 247, 248*f*
Strip cutting, 197, 198*f*, 201
Strip mining, **309**, 311*f*
Students, role in environmental policy, 492–493, 493*f*
Subatomic particles, 32
Subsidence, land, 271, 310
Subsidies, 17
  energy resources, 330
  for ethanol, 364
  farm, 232, 253
  freshwater, 276
  perverse, 480
  renewable energy, 353
  for reuse and recycling, 466
  shifting to environmentally beneficial, 480
  for solar thermal systems, 356
  subsurface mining, 310
Subsidy shifts, 17, 480
Subsistence agriculture, 229
Subsoil, 51, 52*f*
Suburbs, 119, 120*f*
Succession. *See* Ecological succession
Succulent (fleshy) plants, 144*b*
Sugarcane, 364
Sugars, simple, 33
Sulfat ion, 33*t*
Sulfur dioxide
  acid deposition, 412

aerosols, 409
air pollutant, 409
from coal burning, 335, 337*f*
Sulfuric acid, 409
Sulfur oxides, emissions trading for, 418, 418*f*
Sumatran tiger, 171*f*, 181
Superfund Act, 465
Superinsulation, 349
Supply, 475
Surface fire, 197, 199*f*
Surface impoundments, 463, 464*f*
Surface mining, **309**–310
Surface runoff, **57**, **264**
  captured and stored by dams and reservoirs, 271
  ocean pollution from, 289
  reliable, 264
Surface water, **157**, **264**
  groundwater and, 263–264
  reducing and preventing pollution, 294
Survivorship curve, **100**, 100*f*
Suspended particulate matter (SPM), 409
Sustainability, **4***b*
  amphibians and, 85*b*
  changes in living systems and, 96–97
  components of, 5–8
  cradle-to-cradle approach and, 469*b*
  dead zones and, 296*b*
  eco-city concept, 126–128
  energy resources and, 371*b*
  global sustainability movement, 492
  greening American campuses, 501*b*
  Growing Power and, 256*b*
  honeybee and, 189*b*
  Hubbard Brook forest experiment and, 40*b*
  learning from Earth and, 22*b*
  low-waste economy, 465–467
  melting ice in Greenland and, 444*b*
  mercury's toxic effects and, 400*b*
  population growth, urbanization and, 130*b*
  principles of, 5–10, 6*f*, 9*f*
  real cost of gold and, 322*b*
  smart growth, 126
  southern sea otters and, 102*b*
  tropical African Savanna and, 163*b*
  tropical rain forests and, 64*b*
  unsustainable living, 10, 12
  urban areas, 124–126
Sustainability Action Network, 204
Sustainability principles, 5–10, 6*f*, 9*f*
  biodiversity, 5, 6*f*
  chemical cycling, 5, 6*f*
  full-cost pricing, 9, 9*f*
  responsibility to future generations, 9, 9*f*
  solar energy, 6, 7*f*
  win-win, 9, 9*f*
Sustainability revolution, 500, 500*f*
Sustainable environmental design and architecture, career in, **349**
Sustainable food production, 247–252, 251*f*–252*f*
  in aquaculture, 250, 250*f*
  in meat and dairy product production, 249–250
  protecting topsoil, 247
  reducing soil salinization and desertification, 249, 249*f*
  shifting to more sustainable production, 250–252, 251*f*
  soil fertility restoration, 248–249

what can you do?, 252*f*
Sustainable living, 496–
  environmental literacy, 496
  learning from the earth, 497–498, 497*f*
  living more simply and lightly, 498–500
  sustainability revolution, 500, 500*f*
Sustainable yield, **6**
Svalbard Global Seed Vault, 186, 240*f*
Swamps, 160
Sweden
  integrated pest management (IPM) use in, 246
  pesticide use in, 245
  superinsulation use, 349
Switchgrass, 364
Synthetic biology, 81–**82**
Synthetic inorganic fertilizer, 248–249
System(s), **38**–39
  feedback loops in, 38–39, 38*f*–39*f*
  inputs, flows, and outputs of, 38
  model of, 38*f*
System reliability, 398

T

Taigas, 148
Tapeworms, 94
Tar sands, **333**
Tax
  carbon, 437*f*
  energy, 437*f*
  farm subsidies and, 232
  gasoline, 125, 347
  green, 480
  on plastic bags, 456
  taxing pollution and waste, 480–481, 480*f*
Tax breaks
  climate change and, 437
  energy waste, 352
  forests and, 201
  mining, 308
  renewable energy, 353
  for reuse and recycling, 466
Tax-shifting, 480
Tax shifts, 17, 125, 481
Technology, environmental impact and, 12
Tectonic plates, 81*b*, 81*f*, **316**, 317*f*
Temperate deciduous forest, 147, 149*f*
Temperate deserts, 142, 143*f*
Temperate grassland, 144, 145*f*
Temperate rain forests, 148
Temperature
  atmosphere (*see* Atmospheric warming)
  biomes and, 142*f*, 151*f*, 163*b*
  natural capital, 141*f*
Temperature inversion, **412**
Tentative science, **30**
Teratogens, **387**
Terracing, 247, 248*f*
Terrestrial biodiversity and ecosystem services, 205–212
Terrestrial biomes/ecosystems
  biodiversity (*see* Terrestrial biodiversity)
  climate and, 138*b*, 137–140
  deserts, 142, 143*f*, 144*b*
  ecosystem approach to sustaining biodiversity
    biodiversity hotspots, 210, 211*f*
    ecological restoration of tropical dry forest in Costa Rica (case study), 212
    emergency action strategy, 210
    five point plan, 208, 210
    forests, major threats to, 195–201

Terrestrial biomes *(continued)*
    forests, managing and sustaining,
       201–204
    grasslands, managing and sustaining,
       204
    parks and nature preserves, managing
       and sustaining, 207, 207*f*
    pricing nature's ecosystem services,
       196*b*
    reconciliation ecology, 212
    restoration of damaged ecosystems,
       211
    terrestrial biodiversity and ecosystem
       services, 205–212
    what can you do?, 212*f*
    wilderness protection, 206–207, 207*f*
  forests, 197–198, 197*f*–198*f*
  grasslands, 142, 144, 145*f*, 146
  human activities affecting, 151*f*, 151–
    152
  latitude and elevation in, 141
  mountains, 150–151, 151*f*
  net primary productivity, 55, 56*f*
Tertiary consumers, **49**, 53*f*–54*f*
Texas, wind power use in, 361
TFR. *See* Total fertility rate
Thailand
  family planning, 116
  mangrove forest, 154*f*
Thermal energy, **35**
Thermodynamics, 37–38
  first law of, **37**
  second law of, **38**
Thinking About
  cost of sustaining ecosystems, 219
  disadvantages of urbanization, 123
  discount rates, 478
  economic growth, 477
  environmental justice, 490
  environmental political principles, 491
  e-waste recycling, 462
  GM corps, 233
  greening your campus, 493
  hazardous waste, 464
  marine reserves, 217
  meat consumption, 250
  mental traps, 500
  Millennium Development Goals, 484
  poor, affluent and environmental
     harm, 16
  responsibilities, 18
  r-selected and *K*-selected species, 100
  scientific proof, 31
  subsidies, 480
  survivorship curves, 100
  tropical forests, 201
  tigers, 181
  urban sprawl, 121
  urban trends, 119
  U.S. population, 106
  U.S. public lands, 490
  what you eat, 49
  what your school can do, 439
Threatened species, **170**
  Red List of Threatened Species, 213
  whale shark, 77*f*
three dimensional seismic maps, 331
3M Company, 394
Throughputs of systems, **38**, 38*f*
Tidal waves, 320
Tides, energy from, 359
Tiger, 174*f*, 181
Tight oil, 328*b*
Tilman, David, 229, 252*b*, 364, 495*b*

Tilmes, Simone, 436
Tipping points
  climate change, 428, 428*f*
  ecological, **39**, 63*b*
Tobacco smoke, as pollutant, 415
Tobacco use, 396*f*
Todd, John, 295*b*
Toilets, composting toilet systems, 293
Tokyo, population of, 119, 120*f*
Topsoil
  defined, **236**
  erosion, 236–238, 237*f*–238*f*
  A horizon, 52*f*
  nutrient cycling, 6
  soil salinization, 238
Total fertility rate (TFR), **110**, 111*f*
  factors affecting, 111–112
  in United States, 110, 111*f*
Toxic chemicals, **385**. *See also* Chemical
     hazards; Toxins
  dirty dozen pollutants, 394, 466
  estimating toxicity of, 391–392, 392*t*
  in homes, 393*f*
  knowledge of, 393
  trace levels, 392
Toxicity, **390**
  dose-response curve, 391, 391*f*
  estimating, 391–392, 392*t*
  ratings and average lethal doses, 392*t*
Toxicology, **390**
Toxic Release Inventory, 465
Toxic Substances Control Act (TSCA), 464
Toxic waste, **452**. *See also* Hazardous waste
Toxins. *See also* Chemical hazards
  acid deposition, 413
  in coal ash, 337
  in electronic waste (e-waste), 453
  groundwater contamination, 286
  harmful chemicals in your homes, 453*f*
  mercury, 378*b*, 414*f*
  mountaintop removal mining and, 310
  in ocean tides, 289
  in sewage, 293
  in suspended particulate matter (SPM),
    409
Tradable environmental permits, 481*f*
Trade associations, 485
Trade-Offs, 8
  animal feedlots, 241*f*
  aquaculture, 242*f*
  bicycles, 127*f*
  buses, 127*f*
  CAFOs, 241*f*
  Cap-and-trade policies, 437
  carbon and energy taxes, 437*f*
  coal, 339*f*
  conventional natural gas, 337*f*
  conventional nuclear fuel cycle, 342*f*
  conventional oil, 333*f*
  deep-well disposal, 463*f*
  environmental taxes and fees, 480*f*
  geothermal energy, 366*f*
  heavy oils from oil sands, 334*f*
  hydrogen, 367*f*
  hydropower, 360*f*
  liquid biofuels, 365*f*
  mass transit rail, 128*f*
  organic farming, 251*f*
  passive/active solar heating, 355*f*
  rapid rail, 128*f*
  recycling, 458*f*
  sanitary landfills, 461*f*
  solar cells, 359*f*
  solar thermal systems, 356*f*

solid biomass, 364*f*
  surface impoundments, 464*f*
  synthetic pesticides, 243
  tradable environmental permits, 481*f*
  waste-to-energy incineration, 460*f*
  wind power, 363*f*
  withdrawing groundwater, 269*f*
Traditional intensive agriculture, **229**
Traditional subsistence agriculture, **229**
Traditions, birth rate and, 112
Traffic congestion, 125
Tragedy of the commons, 12
Trait, **34**
Transition zone, 141, 159, 161*f*
Transmissible disease, **379**
Transpiration, 144*b*
Transportation
  bicycling, 126, 127*f*
  energy efficiency, 347
  hidden costs, 347
  motor vehicles
    accidents, 125
    advantages and disadvantages of use,
      125
    air pollution and, 408417–419, 419*f*
    car-centered nation, 124
    car-sharing networks, 125
    fuel efficiency standards, 418
    reducing use, 125–126
  net energy yield, 330*f*
  priorities, 126*f*
  trade-offs, 127*f*
  in urban areas, 124–126
Trash. *See* Municipal solid waste (MSW)
Trawlers, 213, 213*f*
Treaties, 184–185
  Copenhagen Amendment, 442
  hazardous waste, 466
  Kyoto Protocol, 438
  Minamata Convention, 395
  Montreal Protocol, 442
  on ozone depletion, 442–443
Tree plantations, 195, 195*f*
Trees. *See* Forests
Trillium Lake, Oregon, 159*f*
Trophic level, **48**, 53*f*
Tropical deserts, 142, 143*f*
Tropical forests
  biome, 142*f*, 151*f*
  deforestation, 197
    reducing, 203–204
  ecological restoration of tropical dry
     forest in Costa Rica, 212
  rain forests
    canopy of, 147, 149*f*
    conservation in Costa Rica, 194*b*, 194*f*
    degradation of, 46*f*
    destruction of, 147
    inertia and resilience in, 96
    net primary productivity, 55, 56*f*, 147
    specialized niches in, 147, 150*f*
    sustainability and, 64*b*
    vegetation removal effects on
      weather and climate, 59
  sustaining, 204*ff*
Troposphere, **47**, **407**, 407*f*
Trout, range of tolerance of, 94*f*
Tsunamis, **319**–321, 319*f*–321*f*
  Fukushima Daiichi nuclear power plant
    accident, 344
Tundra, 146
Twain, Mark, 499
Tyedmers, Peter, 235
Tying It All Together

amphibians and sustainability, 85*b*
cradle-to-cradle approach and sustainability, 469*b*
dead zones and sustainability, 296*b*
energy sources and sustainability, 371*b*
greening college campuses and sustainability, 501*b*
Growing Power and sustainability, 256*b*
honeybees and sustainability, 189*b*
Hubbard Brook forest experiment and sustainability, 40*b*
learning from the Earth and sustainability, 22*b*
melting ice in Greenland and sustainability, 444*b*
mercury's toxic effects and sustainability, 400*b*
population growth, and sustainability, 130*b*
real cost of gold and sustainability, 322*b*
southern sea otters and sustainability, 102*b*
sustaining Costa Rica's biodiversity, 220*b*
tropical African Savanna and sustainability, 163*b*
tropical rain forests and sustainability, 64*b*

U

Ultracapacitors, 349*b*
Ultraviolet radiation
    amphibian decline and, 78*b*
    harmful effects of, 407
    ozone filtering of, 407, 442
    reducing exposure to, 442*f*
Undernutrition, 113, **227**
UNEP. *See* United Nations Environmental Programme
UN International Seabed Authority, 308
United Nations
    on access to clean water, 267
    biosphere reserves, 208
    environmental security and, 493
    on e-waste, 462
    on hydropower use, 359
United Nations Children's Fund (UNICEF), 253–254
United Nations Environmental Programme (UNEP), 198, 204
    on agriculture's use of resources, 236
    on hazardous waste production by, 453
    Intergovernmental Panel on Climate Change (IPCC), 406*b*
    on pesticides and human health, 244
    on pollution of coastal areas, 289
    on topsoil erosion, 237
United States
    aquifer depletion (case study), 269, 270*f*
    average life expectancy in, 112
    baby boom in, 114, 114*f*
    bottled water in, 287
    carbon dioxide emissions, 424
    carbon footprint, 424
    as car-centered nation, 124
    coal reserves, 339
    conservation-tillage farming in, 247
    death, leading causes of, 380*f*
    drip irrigation use, 277
    earthquake risk, 317*f*
    electronic waste (e-waste), 453, 453*f*
    energy use by, 329, 329*f*
    energy waste in, 345, 345*f*

environmental laws and regulations (case study), 486–487, 487*f*
    e-waste, 453, 453*f*
    fertility rate, 111*f*
    first green revolution in, 230–231
    forest cover in (case study), 198
    geothermal energy use in, 366
    grain production in, 231
    hazardous waste, 453
    hazardous waste regulation, 464–465
    hydropower use in, 359
    industrialized food production in, 232
    infant mortality rate, 113
    invasive species, 175*f*
    landfills, 460
    managing public lands (case study), 487–490, 488*f*
    mountaintop removal mining, 310, 312*f*
    natural gas reserves, 335
    obesity in, 228
    outdoor air pollution reduction, 408
    per capita mineral use, 306
    percentage of disposable income spent on food, 232
    pollution of coastal areas, 284
    population growth case study, 110–111
    precipitation in, 266*f*
    premature deaths from air pollution, 416*f*
    rare earth metals, supplies of, 307
    recycling in, 453
    rise of environmental conservation and protection, 18–21
    septic tanks in, 289
    share of atmospheric warming, 414
    soil salinization in, 238
    solid waste in, 451
    stresses on U.S. public parks (case study), 207–208
    urbanization in, 119–120
    wastewater purification, 286
    water scarcity hotspots, 266, 267*f*
    water use in, 265–266
    Wilderness Act, 207
    wind power, 361
University of California, Santa Cruz, 92*b*
University of Connecticut, 474
University of Washington, 439, 474
University of Wisconsin–Oshkosh, 474*b*
UN Population Division, 116
Unreliable science, **30**
Unsustainable living, 21
The Upcycle (McDonough and Braungart), 450
Upcycling, 457
Upwelling, 137*f*, 155
Uranium, 33, 327*f*, 328–329, 328*f*, 329*f*
Urban areas
    aging infrastructure in, 120
    air pollution, 408, 409, 411
    compact cities, 124
    dispersed cities, 124
    eco-city concept, 126–129, 129*f*
    food production, 226*b*, 226*f*, 255
    growth outward and upward, 124
    megacities, 119, 120*f*
    Mexico City (case study), 124
    poverty in, 119
    smart growth, 126, 128*f*
    smog, 410, 410*f*
    sustainable, 126*f*, 126–129
    transportation in, 125–126
    trends in global, 119
    in United States, 119–120, 120*f*

Urban heat island, 123
Urbanization, 112
    advantages of, 121
    disadvantages of, 122
        climate effects, 123
        concentration of pollution and health problems, 122
        noise, 122–123
        vegetation lack, 122
        water problems, 122
    number of children per couple and, 111
    in United States (case study), 119–120
Urban sprawl, 121, 121*f*–122*f*
U.S. Agency for International Development (USAID), 182, 202
U.S. Census Bureau, 111, 114
U.S. Department of Agriculture (USDA)
    on animal waste, 241
    on conservation-tillage farming, 247
    on genetically engineered food, 232
    integrated pest management, 246
    pesticides and, 244, 245
    soil conservation, 247
    on urban food production, 255
U.S. Department of the Interior
    on freshwater shortages, 266
    on surface mining, 310
U.S. Energy Information Administration, 336*b*, 413
U.S. Fish and Wildlife Service (USFWS), 176, 487
    on bioinvaders, 215
    Endangered Species Act and, 185
    gray wolf reintroduction in Yellowstone National Park, 209*b*
U.S. Forest Service (USFS), 19, 199, 487
U.S. Geological Survey (USGS), 265
    on geothermal energy, 366
    on mercury contamination, 387
    on nonrenewable minerals, 304
    on per capita mineral use, 306
    on rare earth metal recovery, 307
    on U.S. coal reserves, 339
U.S. National Ocean Service, 217
User-pays approach, 125, 276
Utilities, energy efficiency in, 346

V

Variables, 28*b*, 31, 97
Varroa mite, 180*b*
Vegetation, 150. *See also* Plants; Terrestrial biomes/ecosystems
    lack in urban areas, 122
Vines, 147
Viral disease, of amphibians, 78*b*
Virtual water, **265**, 266*f*
Viruses, **379**, 383–385
Vitamin A deficiency, blindness from, 228
Vitamin deficiency, 228
Volatile organic compounds (VOCs)
    air pollutant, 409–410
    emission, release from plants, 411
    photochemical smog, 410
Volatile organic hydrocarbons, 290
Volcanic eruptions, 81*b*
Volcanoes, **316**, 318*f*, 318–319
Voluntary simplicity, 499

W

Wallace, Alfred Russel, 79
Warblers, 91*f*
Warning coloration, 92, 93*f*

Washington
  Diablo Lake in North Cascades National
      Park, 207f
  Olympic National Park, 208
Waste, 428–449
  animal, 240–241
  from aquaculture, 241
  burning, 459–462, 460f–462f
  cradle-to-grave system, 464
  electronic waste (e-waste), 453, 462
  end-of-pipe cleanup approach, 465
  food, 250
  hazardous (see Hazardous waste)
  marketplace use to reduce, 481
  from mining, 309
  radioactive nuclear wastes, 342–343
  recycling of (see Recycling)
  reducing by selling services instead of
      products, 481–482
  reuse of (see Reuse)
  solid (see Solid waste)
  taxing, 480–481
  transition to low-waste economy,
      465–466
  wastewater (see Wastewater)
Waste management, 453
  grassroots action and, 465
  industrial ecosystems (case study),
      466–467
  integrated, 454
Waste Management Research Institute,
      465
Waste prevention, 455
Waste reduction, 453
  refusing, reducing, reusing, and
      recycling of, 454–459,
  by selling services instead of products,
      481–482
Waste removal, water use for, 279
Waste-to-energy incinerators, 459, 459f
Wastewater, 283
  from fracking, 336b
  mountaintop removal mining and, 310
  purification, 286
  treatment, 293
Wastewater purification, career in, 286
Water. See also Water resources
  bottled, 287–288
  chemical composition, 34
  energy from flowing (see Hydropower)
  for fracking, 336b
  fresh (see Freshwater)
  gray, 278
  groundwater (see Groundwater)
  heating with solar energy, 353–354,
      354f
  hydrologic cycle, 56–57
  hydropower, 348–360
  irrigation, 238
  ocean currents, 137–138, 139f
  pH of, 33
  pollution of (see Water pollution)
  problems of urban areas, 122
  properties, 57b
  purifying drinking water, 286–287, 287f
  use in United States, 265–266
  virtual, 265
Water conservation specialist, 278
Water cycle, 56–57, 58f
Water-filled pools, at nuclear plants, 342,
      342f
Water footprint, 265
Water heater
  solar, 354

tankless instant, 350
Water leaks, 278
Waterlogging, 238
Water meters, 279
Water pollutants, 282t
  mercury, 378b, 378f
  most serious threats, 284
  plastics, 282
Water pollution, 280–296
  from agriculture, 282, 284, 289, 291
  atmospheric warming and, 282
  concentration in cities, 122
  defined, 280
  from flooding, 280
  from fracking, 336b
  groundwater, 284–286, 285f–286f
  of lakes and reservoirs, 284, 284f
  laws, 288
  from meat production, 240–241
  by mining, 309–312
  nonpoint sources, 282, 291
  ocean pollution, 289–291
  point sources, 280–281, 281f, 291–292
  reducing and preventing
    nonpoint sources, 282
    ocean pollution, 289–291
    point sources, 280–281
    sewage treatment, 292–294, 293f
    solutions for, 294f
    sustainable ways for, 294
    what can you do?, 294f
  sea otters, effects on, 101b
  from sewage, 292, 295b
  sources of, 280–282, 281f, 282t
  of streams and rivers, 283–284, 283f
  threats to kelp forests from, 92b
  topsoil erosion, 247
Water quality
  most serious pollutants threatening, 284
  water cycle and, 56–57
Water resources, 260–296
  Colorado River, 268, 268f
  dams and reservoirs, 268, 273f
  increasing freshwater supplies, 263–268
  increasing use of, 264–265, 265f
  pollution of, 280–296
  reliable surface runoff, 264
  sustainable use of, 276–280
  United States (case study), 265–266,
      266f
  virtual water, 265
  water scarcity hotspots in United States,
      266–267267f
  water transfers, 272–275, 274f
Watershed, 157, 264
Water table, 263
Water transfers, 272–275
Water vapor, 47
  atmospheric concentration, 407
  as greenhouse gas, 407, 423
Wavelength, 35, 36f
Waves, energy from, 359
Weather, 137, 420
  climate distinguished from, 137, 420
  extreme events, 431
Weixin, Luan, 289
West Virginia, mountaintop removal
      mining in, 310, 312f
Wet deposition, 412
Wetlands
  coastal, 152
  draining of, 280
  inland, 160–161
Whale shark, 77f

What Can You Do?
  controlling invasive species, 178f
  hazardous waste, 464f
  hormone disruptors, 390f
  indoor air pollution, 419f
  influencing environmental policy, 491f
  protecting species, 188f
  reducing CO₂ emissions, 440f
  reducing exposure to pesticides, 244f
  reducing water pollution, 294f
  reuse, 456f
  shifting to sustainable energy use, 370f
  solid waste, 455f
  sustainable food production, 252f
  sustaining terrestrial biodiversity, 212f
  UV radiation, reducing exposure to,
      442f
  water use and waste, 279f
Wheat, 229
  root system of, 253f
  Saudi Arabia production of, 269
WHO. See World Health Organization
Whooping crane, 171f
WildAid, 77, 179
Wilderness, 206
  areas as centers for evolution, 206–207
  Diablo Lake in North Cascades National
      Park, 207f
  protecting, 206
Wilderness Act, 207, 486
Wilderness Society, 19
Wildlife farms, 186
Wildlife refuges, 186, 187f
Wilson, Edward O., 73b, 92
  biodiversity and, 169, 206
  ecosystem approach to sustaining
      aquatic biodiversity, 218–219
  species–area relationship and, 170b
Wind. See also Air circulation
  air pollution spread by, 411
  erosion from, 142, 247
  pollutant removal/dilution, 411
  prevailing, 138–139, 139f–140f
Wind-energy engineering, career in, 362
Wind erosion, 142, 247
Wind farms, , 360
Wind power, 360–362, 363f
  advantages and disadvantages of, 363f
  electricity from, 360–362, 360f–362f
  offshore, 362
  in United States, 361–362, 361f
Wind turbines, 36f, 360f, 360–362, 363b
Win-win solutions, 9, 371b
Wolf spider, 242f
Women
  educational and employment
      opportunities for, 112
  empowerment of, 116, 116f
  family planning, 116
  role in slowing population growth, 113f
Woods Hole National Fisheries Service,
      215
World Bank
  environmental security and, 493
  on poverty, 482
World Commission on Water, 284
World Glacier Monitoring Service, 272
World Health Organization (WHO)
  on access to clean water, 267
  access to water as global health issue,
      263
  on air pollution, 16, 408
  dieldrin use by, 244–245
  on drinking water purification, 286

on genetic resistance to antibiotics, 381*b*
on HIV/AIDS, 383, 384
indoor air pollution, 16, 414, 416
on infectious disease, 385
iron deficiency, 228
on malaria, 384–385, 385*f*
outdoor pollution, 416
overnutrition and obesity, 228
on pesticides, 243, 244
on tobacco use, 397
on tuberculosis, 380
vitamin A deficiency, blindness from, 228
World Meteorological Organization, 406*b*, 431
World Resources Institute (WRI), 197, 237, 276

Worldwatch Institute, 126, 202, 362
World Wildlife Fund (WWF), 170
    as citizen environmental group, 492
    dam-and-reservoir systems, 272
    Global Footprint Network, 12, 16
    Living Planet Report, 14, 25
WRI. *See* World Resources Institute

X

Xerox, 455, 482

Y

Yellowstone National Park, gray wolf in, 208, 209*b*
Yield
    in food production, 229, 232
    green revolution, 230–231

net energy, 329–330, 330*f*
    sustainable, **6**
Yucca Mountain, 343
Yunus, Muhammad, 483

Z

Zebra, 91, 145*f*
Zebra mussel, 175*f*, 176
Zone of saturation, **263**
Zoning laws, 126
Zooplankton, 49, 54*f*, 153, 295*b*, 391*f*
Zoos, 186–187
Zooxanthellae, 156*b*

Three scientific principles of sustainability can guide us in making a shift to a more sustainable society.

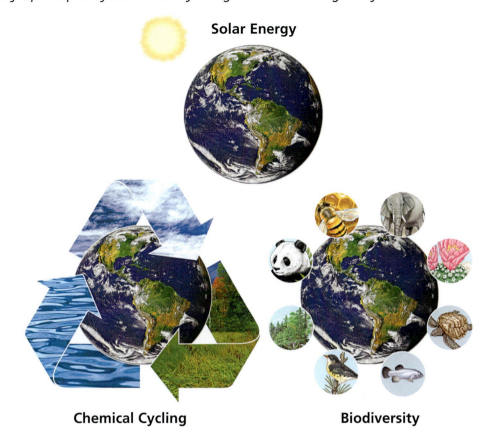

Solar Energy

Chemical Cycling

Biodiversity

Three other principles of sustainability can guide us in making a shift to a more sustainable society.

ECONOMICS
Full-cost pricing

POLITICS
Win-win results

ETHICS
Responsibility to future generations